Examples in

Analysis

SECOND EDITION

Examples in Structural Analysis

SECOND EDITION

William M. C. McKenzie

CRC Press
Taylor & Francis Group
Boca Raton London New York

CRC Press is an imprint of the
Taylor & Francis Group, an **informa** business

A SPON BOOK

CRC Press
Taylor & Francis Group
6000 Broken Sound Parkway NW, Suite 300
Boca Raton, FL 33487-2742

© 2014 by Taylor & Francis Group, LLC
CRC Press is an imprint of Taylor & Francis Group, an Informa business

No claim to original U.S. Government works

Printed on acid-free paper
Version Date: 20131114

International Standard Book Number-13: 978-1-4665-9526-2 (Paperback)

Visit the Taylor & Francis Web site at
http://www.taylorandfrancis.com

and the CRC Press Web site at
http://www.crcpress.com

Contents

Preface

Prior to the development of quantitative structural theories in the mid-18th century and since, builders relied on an intuitive and highly developed sense of structural behaviour. The advent of modern mathematical modelling and numerical methods has to a large extent replaced this skill with a reliance on computer generated solutions to structural problems. Professor Hardy Cross[1] aptly expressed his concern regarding this in the following quote:

> *'There is sometimes cause to fear that the scientific technique, the proud servant of the engineering arts, is trying to swallow its master.'*

It is inevitable and unavoidable that designers will utilize continually improving computer software for analyses. However, it is essential that the use of such software should only be undertaken by those with the appropriate knowledge and understanding of the mathematical modelling, assumptions and limitations inherent in the programs they use.

Students adopt a variety of strategies to develop their knowledge and understanding of structural behaviour, e.g. the use of:

- computers to carry out sensitivity analyses,
- physical models to demonstrate physical effects such as buckling, bending, the development of tension and compression and deformation characteristics,
- the study of worked examples and carrying out analyses using 'hand' methods.

This textbook focuses on the provision of numerous fully detailed and comprehensive worked examples for a wide variety of structural problems. In each chapter a résumé of the concepts and principles involved in the method being considered is given and illustrated by several examples. A selection of problems is then presented which students should undertake on their own prior to studying the given solutions.

Students are strongly encouraged to attempt to visualise/sketch the deflected shape of a loaded structure and predict the type of force in the members prior to carrying out the analysis; i.e.

(i) in the case of pin-jointed frames identify the location of the tension and compression members,

(ii) in the case of beams/rigid-jointed frames, sketch the shape of the bending moment diagram and locate points of contraflexure indicating areas of tension and compression.

A knowledge of the location of tension zones is vital when placing reinforcement in reinforced concrete design and similarly with compression zones when assessing the effective buckling lengths of steel members.

When developing their understanding and confirming their own answers by studying the solutions provided, students should also analyse the structures using a computer analysis, and identify any differences and the reasons for them.

The methods of analysis adopted in this text represent the most commonly used 'hand' techniques with the exception of the direct stiffness method in Chapter 7. This matrix based method is included to develop an understanding of the concepts and procedures adopted in most computer software analysis programs. A method for inverting matrices is given in Appendix 3 and used in the solutions for this chapter – it is *not* necessary for students to undertake this procedure. It is included to demonstrate the process involved when solving the simultaneous equations as generated in the direct stiffness method.

Whichever analysis method is adopted during design, it must always be controlled by the designer, i.e. not a computer! This can only be the case if a designer has a highly developed knowledge and understanding of the concepts and principles involved in structural behaviour. The use of worked examples is one of a number of strategies adopted by students to achieve this.

In this 2nd Edition the opportunity has been taken to modify the x-y-z co-ordinate system/ symbols and Chapter 6 on buckling instability, to reflect the conventions adopted in the structural Eurocode EN 1993-1-1 for steel structures, i.e.

x-x   along the member,
y-y   the major principal axis of the cross-section (e.g. parallel to the flange in a steel beam) and
z-z   the minor principal axes of the cross-section (e.g. perpendicular to the flange in a steel beam).

Local and flexural buckling equations as given in the EN 1993-1-1 are also considered.

Chapter 4 for the analysis of beams has been expanded to include moment redistribution and moment envelopes. Chapter 5 has been expanded to include the analysis of singly-redundant, rigid-jointed frames using the unit load method.

In addition, two new chapters have been added: Chapter 9 relating to the construction and use of influence lines for beams and Chapter 10, the use of approximate methods of analysis for pin-jointed frames, multi-span beams and rigid-jointed frames.

1   Cross, H. *Engineers and Ivory Towers*. New York: McGraw Hill, 1952

William M.C. McKenzie

To Karen, Gordon, Claire and Eilidh

Acknowledgements

I wish to thank Caroline for her endless support and encouragement.

William M. C. McKenzie is a lecturer in structural engineering at Edinburgh Napier University on undergraduate and postgraduate courses, including the MSc course in Advanced Structural Engineering. He graduated with a 1st Class Honours degree and a Ph.D. from Heriot-Watt University, Edinburgh and has been involved in consultancy, research and teaching for more than 35 years.

His publications include research papers relating to stress analysis using holographic interferometry. He is also the author of six design textbooks relating to both the British Standards and the Eurocodes for structural design and one structural analysis textbook.

As a member of the Institute of Physics he is both a Chartered Engineer and a Chartered Physicist. He has presented numerous CPD courses/seminars and guest lectures to industry throughout the UK, in China, Singapore and Malaysia, in relation to the use of the Structural Eurocodes.

1.   Structural Analysis and Design

1.1   *Introduction*

The design of structures, of which analysis is an integral part, is frequently undertaken using computer software. This can only be done safely and effectively if those undertaking the design fully understand the concepts, principles and assumptions on which the computer software is based. It is vitally important therefore that design engineers develop this knowledge and understanding by studying and using hand-methods of analysis based on the same concepts and principles, e.g. equilibrium, energy theorems, elastic, elasto-plastic and plastic behaviour and mathematical modelling.

In addition to providing a mechanism for developing knowledge and understanding, hand-methods also provide a useful tool for readily obtaining approximate solutions during preliminary design and an independent check on the answers obtained from computer analyses.

The methods explained and illustrated in this text, whilst not exhaustive, include those most widely used in typical design offices, e.g. method-of-sections/joint resolution/unit load/McCaulay's method/moment distribution/plastic analysis etc.

In Chapter 7 a résumé is given of the direct stiffness method; the technique used in developing most computer software analysis packages. The examples and problems in this case have been restricted and used to illustrate the processes undertaken when using matrix analysis; this is **not** regarded as a hand-method of analysis.

1.2   *Equilibrium*

All structural analyses are based on satisfying one of the fundamental laws of physics, i.e.

$$F = ma \qquad\qquad \text{Equation (1)}$$

where
F is the force system acting on a body
m is the mass of the body
a is the acceleration of the body

Structural analyses carried out on the basis of a force system inducing a **dynamic** response, for example structural vibration induced by wind loading, earthquake loading, moving machinery, vehicular traffic etc., have a non-zero value for 'a' the acceleration. In the case of analyses carried out on the basis of a **static** response, for example stresses/deflections induced by the self-weights of materials, imposed loads which do **not** induce vibration etc., the acceleration 'a' is equal to **zero**.

Static analysis can be regarded as a special case of the more general dynamic analysis in which:

$$F = ma = 0 \qquad\qquad \text{Equation (2)}$$

F can represent the applied force system in any direction; for convenience this is normally considered in either two or three mutually perpendicular directions as shown in Figure 1.1.

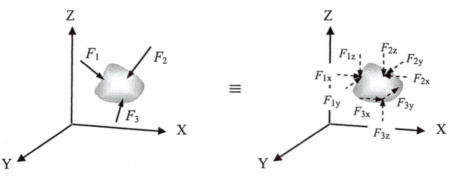

Figure 1.1

The application of Equation (2) to the force system indicated in Figure 1.1 is:

Sum of the forces in the direction of the X-axis	$\Sigma F_x = 0$	Equation (3)
Sum of the forces in the direction of the Y-axis	$\Sigma F_y = 0$	Equation (4)
Sum of the forces in the direction of the Z-axis	$\Sigma F_z = 0$	Equation (5)

Since the structure is neither moving in a linear direction, nor in a rotational direction a further three equations can be written down to satisfy Equation (2):

Sum of the moments of the forces about the X-axis	$\Sigma M_x = 0$	Equation (6)
Sum of the moments of the forces about the Y-axis	$\Sigma M_y = 0$	Equation (7)
Sum of the moments of the forces about the Z-axis	$\Sigma M_z = 0$	Equation (8)

Equations (3) to (8) represent the static equilibrium of a body (structure) subject to a three-dimensional force system. Many analyses are carried out for design purposes assuming two-dimensional force systems and hence only two linear equations (e.g. equation (3) and equation (5) representing the x and z axes) and one rotational equation (e.g. equation (7) representing the y-axis) are required. The x, y and z axes must be mutually perpendicular and can be in any orientation, however for convenience two of the axes are usually regarded as horizontal and vertical, (e.g. gravity loads are vertical and wind loads frequently regarded as horizontal). It is usual practice, when considering equilibrium, to assume that clockwise rotation is positive and anti-clockwise rotation is negative. The following conventions have been adopted in this text:

x-direction: horizontal direction   - positive is left-to right   ⟶ +ve

z- direction: vertical direction   - positive is upwards   ↑ +ve

y- direction: rotation about the y-axis - positive is clockwise   ↷ +ve

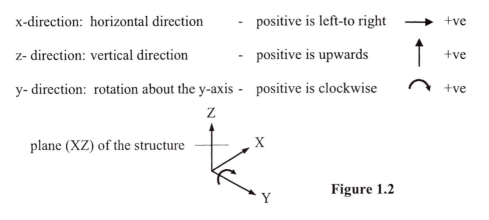

plane (XZ) of the structure

Figure 1.2

Structures in which all the member forces and external support reactions can be determined using only the equations of equilibrium are '*statically determinate*' otherwise they are '*indeterminate structures*'. The degree-of-indeterminacy is equal to the number of unknown variables (i.e. member forces/external reactions) which are in excess of the equations of equilibrium available to solve for them, see Section 1.5

The availability of current computer software enables full three-dimensional analyses of structures to be carried out for a wide variety of applied loads. An alternative, more traditional, and frequently used method of analysis when designing is to consider the stability and forces on a structure separately in two mutually perpendicular planes, i.e. a series of plane frames and ensure lateral and rotational stability and equilibrium in each plane. Consider a typical industrial frame comprising a series of parallel portal frames as shown in Figure 1.3. The frame can be designed considering the X-Z and the Y-Z planes as shown.

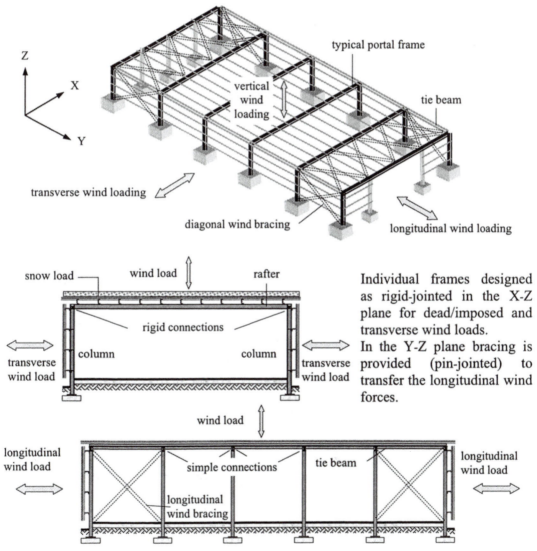

Individual frames designed as rigid-jointed in the X-Z plane for dead/imposed and transverse wind loads.

In the Y-Z plane bracing is provided (pin-jointed) to transfer the longitudinal wind forces.

Figure 1.3

1.3   *Mathematical Modelling*

The purpose of mathematical modelling is to predict structural behaviour in terms of loads, stresses and deformations under any specified, externally applied force system. Since actual structures are physical, three-dimensional entities it is necessary to create an idealized model which is representative of the materials used, the geometry of the structure and the physical constraints e.g. the support conditions and the externally applied force system.

The precise idealisation adopted in a particular case is dependent on the complexity of the structure and the level of the required accuracy of the final results. The idealization can range from simple two-dimensional '*beam-type*' and '*plate*' elements for pin-jointed or rigid jointed plane frames and space frames to more sophisticated three-dimensional elements such as those used in *grillages* or *finite element* analyses adopted when analysing for example bridge decks, floor-plates or shell roofs.

It is essential to recognise that irrespective of how advanced the analysis method is, it is always an approximate solution to the real behaviour of a structure.

In some cases the approximation reflects very closely the actual behaviour in terms of both stresses and deformations whilst in others, only one of these parameters may be accurately modelled or indeed the model may be inadequate in both respects resulting in the need for the physical testing of scaled models.

1.3.1 Line Diagrams

When modelling it is necessary to represent the form of an *actual structure* in terms of idealized structural members, e.g. in the case of plane frames as beam elements, in which the beams, columns, slabs etc. are indicated by line diagrams. The lines normally coincide with the centre-lines of the members. A number of such line diagrams for a variety of typical plane structures is shown in Figures 1.4 to 1.9. In some cases it is sufficient to consider a section of the structure and carry out an approximate analysis on a sub-frame as indicated in Figure 1.8.

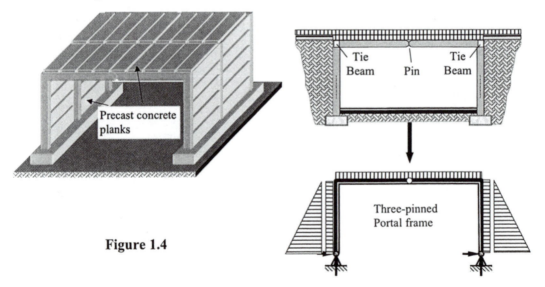

Figure 1.4

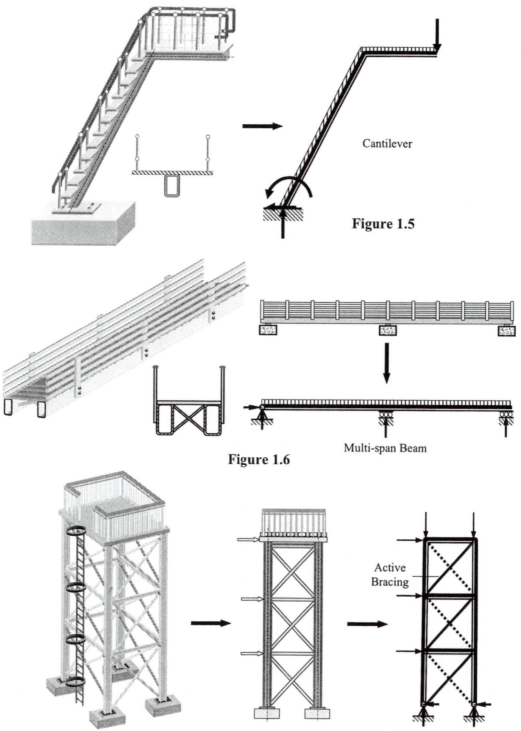

Cantilever

Figure 1.5

Multi-span Beam

Figure 1.6

Active Bracing

Figure 1.7

Braced pin-jointed frame

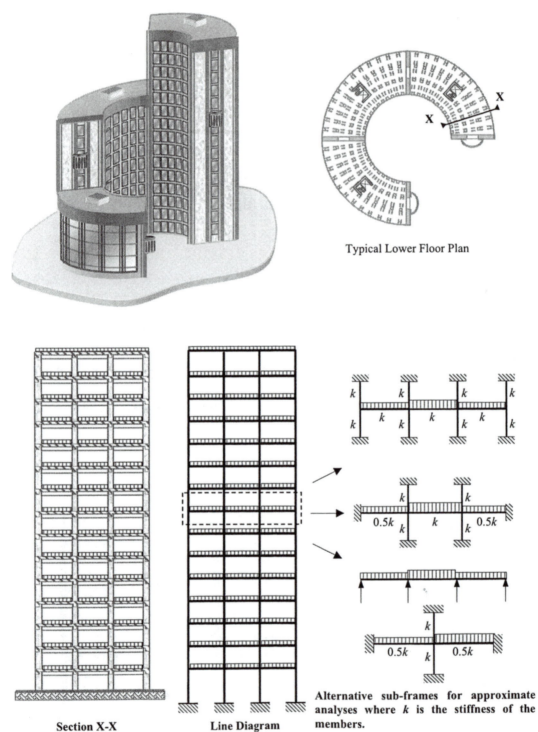

Typical Lower Floor Plan

Section X-X

Line Diagram

Alternative sub-frames for approximate analyses where *k* is the stiffness of the members.

Figure 1.8

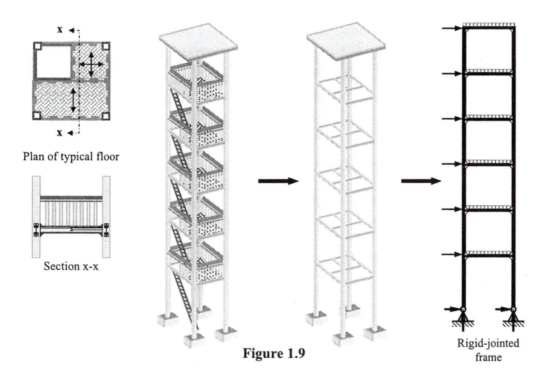

Figure 1.9

Plan of typical floor

Section x-x

Rigid-jointed frame

1.3.2 Load Path

The support reactions for structures relate to the restraint conditions against linear and rotational movement. Every structural element and structure must be supported in order to transfer the applied loading to the foundations where it is dissipated through the ground. For example beams and floor slabs may be supported by other beams, columns or walls which are supported on foundations which subsequently transfer the loads to the ground. It is important to trace the **load path** of any applied loading on a structure to ensure that there is no interruption in the flow as shown in Figure 1.10.

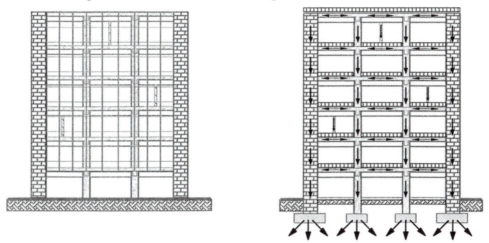

Load path for a typical frame

Figure 1.10

The loads are transferred between structural members at the joints using either simple or rigid connections (i.e. moment connections). In the case of simple connections axial and/or shear forces are transmitted whilst in the case of rigid connections in addition to axial and shear effects, moments are also transferred.

The type of connections used will influence the degree-of-indeterminacy and the method of analysis required (e.g. determinate, indeterminate, pin-jointed frame, rigid-jointed frame). Connection design, reflecting the assumptions made in the analysis, is an essential element in achieving an effective load path.

1.3.3 Foundations

The primary function of all structural members/frames is to transfer the applied dead and imposed loading, from whichever source, to the foundations and subsequently to the ground. The type of foundation required in any particular circumstance is dependent on a number of factors such as the magnitude and type of applied loading, the pressure which the ground can safely support, the acceptable levels of settlement and the location and proximity of adjacent structures.

In addition to purpose made pinned and roller supports the most common types of foundation currently used are indicated Figure 1.11. The support reactions in a structure depend on the types of foundation provided and the resistance to lateral and rotational movement.

square pad foundations

rectangular pad foundation

rectangular combined foundation

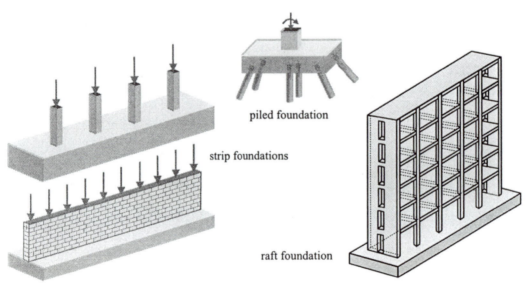

piled foundation

strip foundations

raft foundation

Figure 1.11

1.4   *Structural Loading*

All structures are subjected to loading from various sources. The main categories of loading are: permanent (e.g. self-weight) and variable loads (e.g. imposed and wind loads). In some circumstances there may be other loading types which should be considered, such as settlement, fatigue, temperature effects, dynamic loading, or impact effects (e.g. when designing bridge decks, crane-gantry girders or maritime structures). In the majority of cases design considering combinations of permanent, imposed and wind loads is the most appropriate.

Most floor systems are capable of lateral distribution of loading. In situations where lateral distribution is not possible, the effects of the concentrated loads should be considered with the load applied at locations which will induce the most adverse effect, e.g. maximum bending moment, shear and deflection. In addition, local effects such as crushing and punching should be considered where appropriate.

In multi-storey structures it is very unlikely that all floors will be required to carry the full imposed load at the same time. Statistically it is acceptable to reduce the total floor loads carried by a supporting member by varying amounts depending on the number of floors or floor area carried. Dynamic loading is often represented by a system of equivalent static forces which can be used in the analysis and design of a structure.

The primary objective of structural analysis is to determine the distribution of internal moments and forces throughout a structure such that they are in equilibrium with the applied design loads.

Mathematical models which can be used to idealise structural behaviour include: two- and three-dimensional elastic behaviour, elastic behaviour considering a redistribution of moments, elasto-plastic/plastic behaviour and non-linear behaviour. The following chapters illustrate most of the hand-based techniques commonly used to predict structural member forces and behaviour.

In braced structures (i.e. those in which structural elements have been provided specifically to transfer lateral loading) where floor slabs and beams are considered to be simply supported, vertical loads give rise to different types of beam loading. Floor slabs can be designed as either one-way spanning or two-way spanning as shown in Figures 1.12(a) and (b).

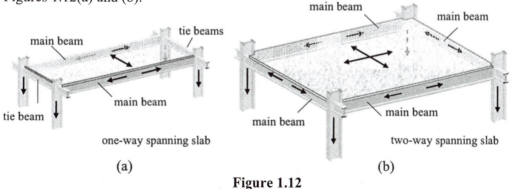

Figure 1.12

In the case of one-way spanning slabs the entire load is distributed to the two main beams. Two-way spanning slabs distribute load to main beams along all edges. These differences give rise to a number of typical beam loadings in floor slabs as shown in Figures 1.13.

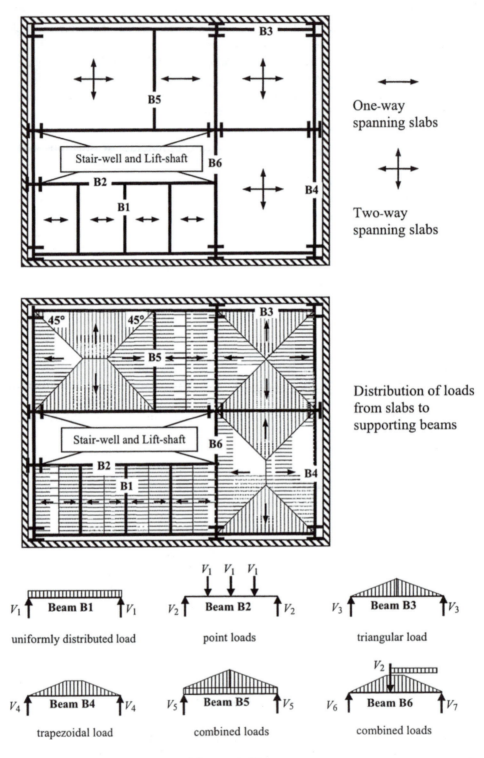

Figure 1.13

1.5   *Statical Indeterminacy*

Any plane-frame structure which is in a state of equilibrium under the action of an externally applied force system must satisfy the following three conditions:

- the sum of the horizontal components of all applied forces must equal zero,
- the sum of the vertical components of all applied forces must equal zero,
- the sum of the moments (about any point in the plane of the frame) of all applied forces must equal zero.

This is represented by the following '*three equations of static equilibrium*'

Sum of the horizontal forces equals zero   +ve $\Sigma F_x = 0$ ⟶

Sum of the vertical forces equals zero   +ve $\Sigma F_z = 0$ ↑

Sum of the moments about a point in the plane of the forces equals zero +ve $\Sigma M = 0$ ↷

In *statically determinate* structures, all internal member forces and external reactant forces can be evaluated using the three equations of static equilibrium. When there are more unknown member forces and external reactant forces than there are available equations of equilibrium a structure is *statically indeterminate* and it is necessary to consider the compatibility of structural deformations to fully analyse the structure.

A structure may be indeterminate due to redundant components of reaction and/or redundant members. i.e. a redundant reaction or member is one which is not essential to satisfy the minimum requirements of stability and static equilibrium, (**Note:** it is not necessarily a member with zero force).

The degree-of-indeterminacy (referred to as I_D in this text) is equal to the number of unknown variables (i.e. member forces/external reactions) which are in excess of the equations of equilibrium available to solve for them.

1.5.1 *Indeterminacy of Two-Dimensional Pin-Jointed Frames*

The external components of reaction (r) in pin-jointed frames are normally one of two types:

i)   a roller support providing one degree-of-restraint, i.e. perpendicular to the roller,
ii) a pinned support providing two degrees-of-restraint, e.g. in the horizontal and vertical directions.

as shown in Figure 1.14

roller supports: providing one restraint perpendicular to the roller.

pinned supports: providing two mutually perpendicular restraints

Figure 1.14

It is necessary to provide three non-parallel, non-concentric, components of reaction to satisfy the three equations of static equilibrium. Consider the frames indicated in Figures 1.15 and 1.16

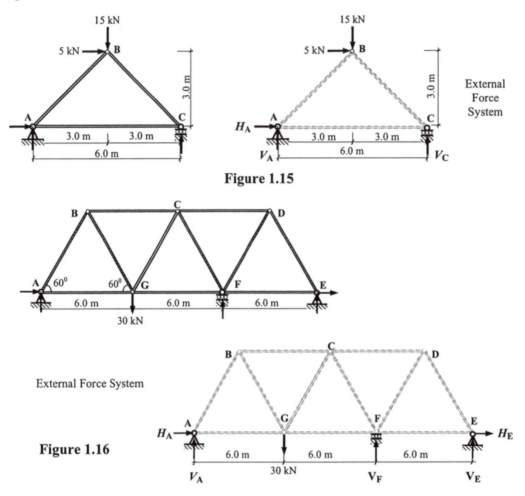

Figure 1.15

Figure 1.16

In Figures 1.15 and 1.16 the applied forces and the external components of reaction represent co-planar force systems which are in static equilibrium. In Figure 1.15 there are *three* unknowns, (H_A, V_A and V_C), and three equations of equilibrium which can be used to determine their values: there are no redundant components of reaction.

In Figure 1.16 there are *five* unknowns components of reaction, (H_A, V_A, V_F, H_E and V_E), and only three equations of equilibrium; there are two redundant reactions in this case.

The internal members of pin-jointed frames transfer either tensile or compressive axial loads through the nodes to the supports and hence reactions. A *simple* pin-jointed frame is one in which the minimum number of members is present to ensure stability and static equilibrium.

Consider the basic three member pinned-frame indicated in Figure 1.15. There are three nodes and three members. A triangle is the basis for the development of all pin-jointed frames since it is an inherently stable system, i.e. only one configuration is possible for any given three lengths of the members.

Consider the development of the frame shown in Figure 1.17:

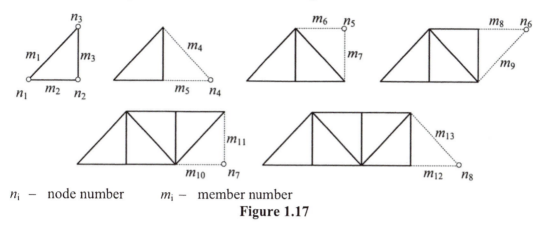

n_i − node number m_i − member number

Figure 1.17

Initially there are three nodes and three members. If the number of members in the frame is to be increased then for each node added, two members are required to maintain the triangulation. The minimum number of members required to create a simple frame can be determined as follows:

m = the initial three members + (2 × number of additional joints)
   $= 3 + 2(n - 3)$ ➤ $m = (2n - 3)$
e.g. in this case $n = 8$ and therefore the minimum number of members = $[(2 \times 8) - 3)]$
                                                    ∴ $m = 13$

Any members which are added to the frame in addition to this number are redundant members and make the frame statically indeterminate; e.g.

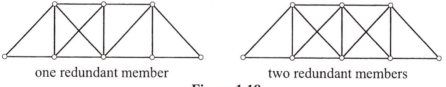

one redundant member   two redundant members

Figure 1.18

It is also essential to consider the configuration of the members in a frame to ensure that it is triangulated. The simple frames indicated in Figure 1.19 are unstable.

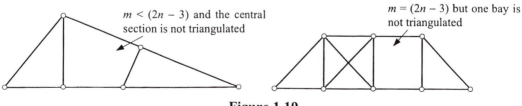

$m < (2n - 3)$ and the central section is not triangulated

$m = (2n - 3)$ but one bay is not triangulated

Figure 1.19

As indicated previously, the minimum number of reactant forces to maintain static equilibrium is three and consequently when considering a simple, pin-jointed plane-frame and its support reactions the combined total of members and components of reaction is equal to:

$$\Sigma \text{ (number of members + support reactions)} = (m + r) = (2n - 3) + 3 = \mathbf{2n}$$

Consider the frames shown in Figure 1.20 with pinned and roller supports as indicated.

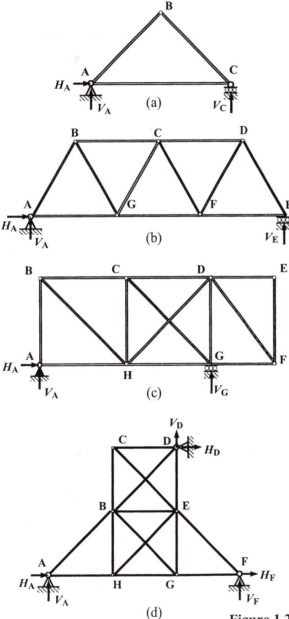

number of joints $n = 3$
number of members $m = 3$
       $(2n - 3) = 3$
number of support reactions $r = 3$
       $(m + r) = 6 = 2n$
The frame is statically determinate
$$I_D = 0$$

number of joints $n = 7$
number of members $m = 11$
       $(2n - 3) = 11$
number of support reactions $r = 3$
       $(m + r) = 14 = 2n$
The frame is statically determinate
$$I_D = 0$$

number of joints $n = 8$
number of members $m = 14$
       $(2n - 3) = 13$
number of support reactions $r = 3$
       $(m + r) = 17 > 2n$
The frame is statically indeterminate with one redundant internal member
$$I_D = 1$$

number of joints $n = 8$
number of members $m = 15$
       $(2n - 3) = 13$
number of support reactions $r = 6$
       $(m + r) = 21 > 2n$
The frame is statically indeterminate and has 5 redundancies:
(2 internal members + 3 external reactions)
$$I_D = 5$$

Figure 1.20

The degree of indeterminacy $I_D = (m + r) - 2n$

Compound trusses which are fabricated from two or more simple trusses by a structural system involving no more than three, non-parallel, non-concurrent, unknown forces can also be stable and determinate. Consider the truss shown in Figure 1.21(a) which is a simple truss and satisfies the relationships $m = (2n - 3)$ and $I_D = 0.$

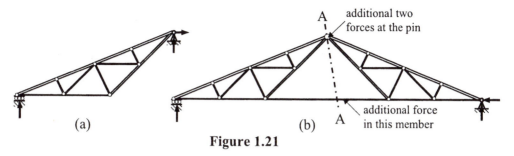

Figure 1.21

This truss can be connected to a similar one by a pin and an additional member as shown in Figure 1.21(b) to create a compound truss comprising two statically determinate trusses. Since only an additional three unknown forces have been generated the three equations of equilibrium can be used to solve these by considering a section A-A as shown (see Chapter 3 – Section 3.2. - Method of Sections for Pin-Jointed Frames: Problem 3.4).

1.5.2 Indeterminacy of Two-Dimensional Rigid-Jointed Frames

The external components of reaction (r) in rigid-jointed frames are normally one of three types:

   i)   a roller support providing one degree-of-restraint, i.e. perpendicular to the roller,

   ii)   a pinned support providing two degrees-of-restraint, e.g. in the horizontal and vertical directions,

   iii)   a fixed (encastre) support providing three degrees-of-restraint, i.e. in the horizontal and vertical directions and a moment restraint

as shown in Figure 1.22

| **roller supports:** providing one restraint perpendicular to the roller. | **pinned supports:** providing two mutually perpendicular restraints | **fixed supports:** providing two mutually perpendicular restraints and one moment restraint. |

Figure 1.22

In rigid-jointed frames, the applied load system is transferred to the supports by inducing axial loads, shear forces and bending moments in the members. Since three components of reaction are required for static equilibrium the total number of unknowns is equal to: $[(3 \times m) + r]$.   At each node there are three equations of equilibrium, i.e.

Σ the vertical forces $F_z = 0;$

Σ the horizontal forces $F_x = 0;$

Σ the moments $M = 0,$ providing $(3 \times n)$ equations.

The degree of indeterminacy $I_D = [(3m) + r] - 3n$

Consider the frames shown in Figure 1.23

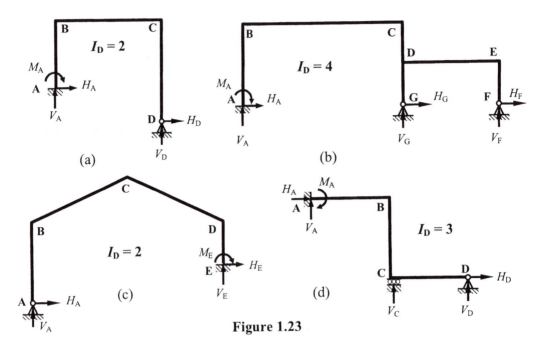

Figure 1.23

The existence of an internal pin in a *member* in a rigid-frame results in only shear and axial loads being transferred through the frame at its location. This reduces the number of unknowns and hence redundancies since an additional equation is available for solution due to the sum of the moments about the pin being equal to zero, i.e. $\Sigma M_{pin} = 0$
Consider the effect of introducing pins in the frames shown in Figure 1.24

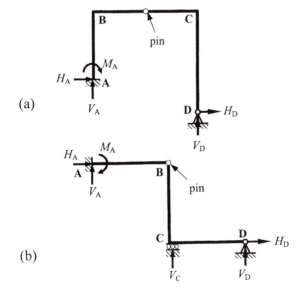

$I_D = \{[(3m) + r] - 3n\} - 1$ due to the release of the moment capacity at the position of the pin.
$I_D = \{[9 + 5] - 12\} - 1 = 1$

$I_D = \{[(3m) + r] - 3n\} - 1$ due to the release of the moment capacity at the position of the pin.
$I_D = \{[9 + 6] - 12\} - 1 = 2$

Figure 1.24

The existence of an internal pin at a *node* with two members in a rigid-frame results in the release of the moment capacity and hence one additional equation as shown in Figure 1.25(a). When there are three members meeting at the node then there are effectively *two* values of moment, i.e. M_1 and M_2 and in the third member $M_3 = (M_1 + M_2)$ The introduction of a pin in one of the members produces a single release and in two members (effectively all three members) produces two releases as shown in Figure 1.25(b).

In general terms the introduction of '*p*' pins at a joint introduces '*p*' additional equations. When pins are introduced to all members at the joint the number of additional equations produced equals (number of members at the joint − 1).

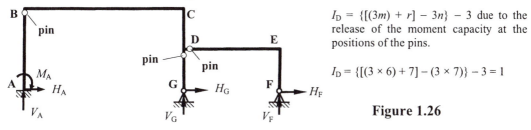

(a)   (b)

Figure 1.25

Consider the frame shown in Figure 1.26.

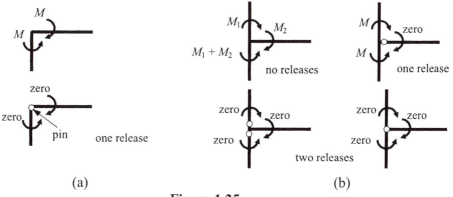

$I_D = \{[(3m) + r] - 3n\} - 3$ due to the release of the moment capacity at the positions of the pins.

$I_D = \{[(3 \times 6) + 7] - (3 \times 7)\} - 3 = 1$

Figure 1.26

The inclusion of an internal roller within a member results in the release of the moment capacity and in addition the force parallel to the roller and hence provides two additional equations. Consider the continuous beam ABC shown in Figure 1.27. in which a roller has been inserted in member AB

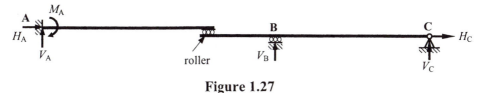

Figure 1.27

$I_D = \{[(3m) + r] - 3n\} - 2$ due to the release of the moment and axial load capacity at the roller   ∴ $I_D = \{[(3 \times 2) + 6] - (3 \times 3) - 2 = 1$

Consider the same beam AB with a pin added in addition to the roller.

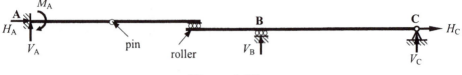

Figure 1.28

$I_D = \{[(3m) + r] - 3n\} - 3$ due to the release of the moment capacity at the position of the pin and the release of the moment and axial load capacity at the roller

$I_D = \{[(3 \times 2) + 6] - (3 \times 3) - 3 = 0$ The structure is statically determinate.

A similar approach can be taken for three-dimensional structures; this is not considered in this text.

1.6   *Structural Degrees-of-Freedom*

The degrees-of-freedom in a structure can be regarded as the possible components of displacements of the nodes including those at which some support conditions are provided. In pin-jointed, plane-frames each node unless restrained, can displace a small amount δ which can be resolved in to horizontal and vertical components δ_H and δ_V as shown in Figure 1.29.

Each component of displacement can be regarded as a separate degree-of-freedom and in this frame there is a total of three degrees-of-freedom i.e. the vertical and horizontal displacement of node B and the horizontal displacement of node C as indicated.

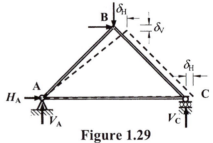

Figure 1.29

In a pin-jointed frame there are effectively two possible components of displacement for each node which does not constitute a support. At each roller support there is an additional degree-of-freedom due to the release of one restraint. In a *simple*, i.e. **statically determinate** frame, the number of degrees-of-freedom is equal to the number of members. Consider the two frames indicated in Figures 1.20(a) and (b):

In Figure 1.20 (a): the number of members $m = 3$
               possible components of displacements at node B $= 2$
               possible components of displacements at node support C $= 1$
               Total number of degrees-of-freedom $(= m) = 3$

In Figure 1.20 (b): the number of members $m = 11$
               possible components of displacements at nodes $= 10$
               possible components of displacements at support E $= 1$
               Total number of degrees-of-freedom $(= m) = 11$

In the case of indeterminate frames, the number of degrees-of-freedom is equal to the (number of members $- I_D$); consider the two frames indicated in Figures 1.20 (c) and (d):

In Figure 1.20 (c): the number of members $\qquad$ m **= 14**
$\qquad$ possible components of displacements at nodes $\qquad$ = 12
$\qquad$ possible components of displacements at support G $\qquad$ = 1
$\qquad$ degree-of-indeterminacy $\qquad$ I_D **= 1**
$\qquad$ **Total number of degrees-of-freedom** $\qquad$ $(m - I_D)$ **= 13**

In Figure 1.20 (d): the number of members $\qquad$ m **= 15**
$\qquad$ possible components of displacements at nodes $\qquad$ = 10
$\qquad$ degree-of-indeterminacy $\qquad$ I_D **= 5**
$\qquad$ **Total number of degrees-of-freedom** $\qquad$ $(m - I_D)$ **= 10**

In rigid-jointed frames there are effectively three possible components of displacement for each node which does not constitute a support; they are rotation and two components of translation e.g. θ, δ_H and δ_V. At each pinned support there is an additional degree-of-freedom due to the release of the rotational restraint and in the case of a roller, two additional degrees-of-freedom due to the release of the rotational restraint and a translational restraint. Consider the frames shown in Figure 1.23.

In Figure 1.23 (a): the number of nodes (excluding supports) $\qquad$ = 2
$\qquad$ possible components of displacements at nodes $\qquad$ = 6
$\qquad$ possible components of displacements at support D $\qquad$ = 1
$\qquad$ **Total number of degrees-of-freedom** $\qquad$ = 7

In Figure 1.23 (b): the number of nodes (excluding supports) $\qquad$ = 4
$\qquad$ possible components of displacements at nodes $\qquad$ = 12
$\qquad$ possible components of displacements at support G $\qquad$ = 1
$\qquad$ possible components of displacements at support F $\qquad$ = 1
$\qquad$ **Total number of degrees-of-freedom** $\qquad$ = 14

In Figure 1.23 (c): the number of nodes (excluding supports) $\qquad$ = 3
$\qquad$ possible components of displacements at nodes $\qquad$ = 9
$\qquad$ possible components of displacements at support A $\qquad$ = 1
$\qquad$ **Total number of degrees-of-freedom** $\qquad$ = 10

In Figure 1.23 (d): the number of nodes (excluding supports) $\qquad$ = 1
$\qquad$ possible components of displacements at nodes $\qquad$ = 3
$\qquad$ possible components of displacements at support C $\qquad$ = 2
$\qquad$ possible components of displacements at support D $\qquad$ = 1
$\qquad$ **Total number of degrees-of-freedom** $\qquad$ = 6

The introduction of a pin in a member at a node produces an additional degree-of-freedom. Consider the typical node with four members as shown in Figure 1.30. In (a) the node is a rigid connection with no pins in any of the members and has the three degrees-of-freedom indicated. In (b) a pin is present in one member, this produces an additional degrees-of-freedom since the rotation of this member can be different from the remaining three, similarly with the other members as shown in (c) and (d).

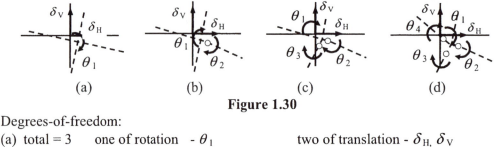

| (a) | (b) | (c) | (d) |

Figure 1.30

Degrees-of-freedom:
(a) total = 3   one of rotation - θ_1 two of translation - δ_H, δ_V
(b) total = 4   two of rotation - θ_1, θ_2 two of translation - δ_H, δ_V
(c) total = 5   three of rotation - $\theta_1, \theta_2, \theta_3$ two of translation - δ_H, δ_V
(d) total = 6   four of rotation - $\theta_1, \theta_2, \theta_3, \theta_4$ two of translation - δ_H, δ_V

In many cases the effects due to axial deformations is significantly smaller than those due to the bending effect and consequently an analysis assuming axial rigidity of members is acceptable. Assuming axial rigidity reduces the degrees-of-freedom which are considered; consider the frame shown in Figure 1.31.

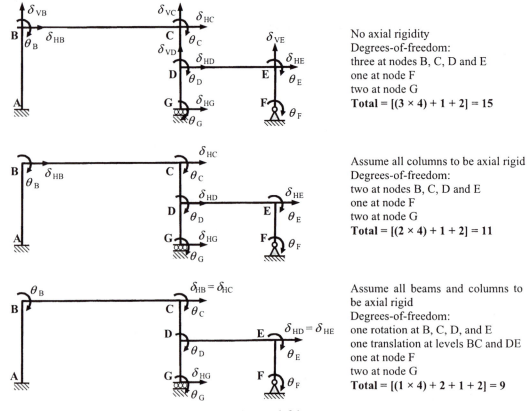

No axial rigidity
Degrees-of-freedom:
three at nodes B, C, D and E
one at node F
two at node G
Total = [(3 × 4) + 1 + 2] = 15

Assume all columns to be axial rigid
Degrees-of-freedom:
two at nodes B, C, D and E
one at node F
two at node G
Total = [(2 × 4) + 1 + 2] = 11

Assume all beams and columns to be axial rigid
Degrees-of-freedom:
one rotation at B, C, D, and E
one translation at levels BC and DE
one at node F
two at node G
Total = [(1 × 4) + 2 + 1 + 2] = 9

Figure 1.31

1.6.1 Problems: Indeterminacy and Degrees-of-Freedom

Determine the degree of indeterminacy and the number of degrees-of-freedom for the pin-jointed and rigid-jointed frames indicated in Problems 1.1 to 1.3 and Problems 1.4 to 1.6 respectively.

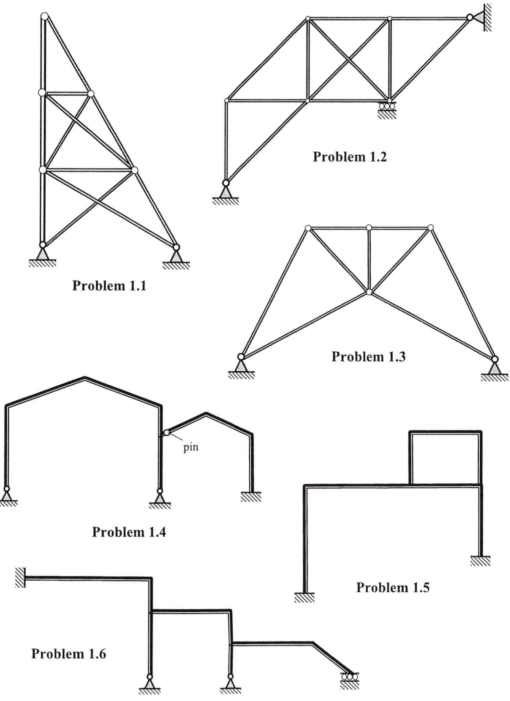

Problem 1.2

Problem 1.1

Problem 1.3

pin

Problem 1.4

Problem 1.5

Problem 1.6

1.6.2 Solutions: Indeterminacy and Degrees-of-freedom

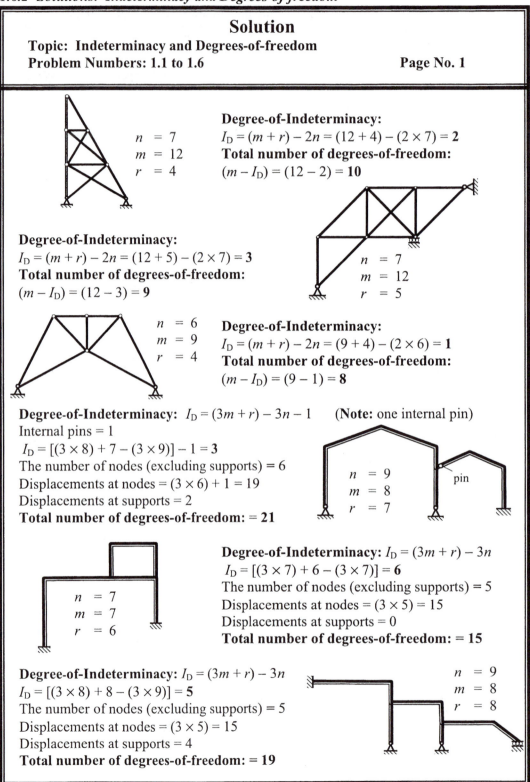

Solution

Topic: Indeterminacy and Degrees-of-freedom

Problem Numbers: 1.1 to 1.6   **Page No. 1**

$n = 7$
$m = 12$
$r = 4$

Degree-of-Indeterminacy:
$I_D = (m + r) - 2n = (12 + 4) - (2 \times 7) = \mathbf{2}$
Total number of degrees-of-freedom:
$(m - I_D) = (12 - 2) = \mathbf{10}$

Degree-of-Indeterminacy:
$I_D = (m + r) - 2n = (12 + 5) - (2 \times 7) = \mathbf{3}$
Total number of degrees-of-freedom:
$(m - I_D) = (12 - 3) = \mathbf{9}$

$n = 7$
$m = 12$
$r = 5$

$n = 6$
$m = 9$
$r = 4$

Degree-of-Indeterminacy:
$I_D = (m + r) - 2n = (9 + 4) - (2 \times 6) = \mathbf{1}$
Total number of degrees-of-freedom:
$(m - I_D) = (9 - 1) = \mathbf{8}$

Degree-of-Indeterminacy: $I_D = (3m + r) - 3n - 1$ **(Note:** one internal pin)
Internal pins = 1
$I_D = [(3 \times 8) + 7 - (3 \times 9)] - 1 = \mathbf{3}$
The number of nodes (excluding supports) = 6
Displacements at nodes = $(3 \times 6) + 1 = 19$
Displacements at supports = 2
Total number of degrees-of-freedom: = 21

$n = 9$
$m = 8$
$r = 7$
pin

Degree-of-Indeterminacy: $I_D = (3m + r) - 3n$
$I_D = [(3 \times 7) + 6 - (3 \times 7)] = \mathbf{6}$
The number of nodes (excluding supports) = 5
Displacements at nodes = $(3 \times 5) = 15$
Displacements at supports = 0
Total number of degrees-of-freedom: = 15

$n = 7$
$m = 7$
$r = 6$

Degree-of-Indeterminacy: $I_D = (3m + r) - 3n$
$I_D = [(3 \times 8) + 8 - (3 \times 9)] = \mathbf{5}$
The number of nodes (excluding supports) = 5
Displacements at nodes = $(3 \times 5) = 15$
Displacements at supports = 4
Total number of degrees-of-freedom: = 19

$n = 9$
$m = 8$
$r = 8$

2. Material and Section Properties

2.1 Introduction

Structural behaviour is dependent upon material characteristics such as elastic constants which describe the stress/strain relationships and the geometry of the cross-section of individual members. This section describes the principal characteristics and properties which must be considered and evaluated to enable mathematical modelling to be undertaken.

2.1.1 Simple Stress and Strain

The application of loads to structural members induces deformations and internal resisting forces within the materials. The intensity of these forces is known as the stress in the material and is measured as the force per unit area of the cross-sections which is normally given the symbol σ when it acts perpendicular to the surface of a cross-section and τ when it acts parallel to the surface. Different types of force cause different types and distributions of stress for example: axial stress, bending stress, shear stress, torsional stress and combined stress.

Consider the case of simple stress due to an axial load P which is supported by a column of cross-sectional area A and original length L as shown in Figure 2.1. The applied force induces an internal stress σ such that:

$P = (\sigma \times A)$ and hence $\sigma = P/A$ (i.e. load/unit area)

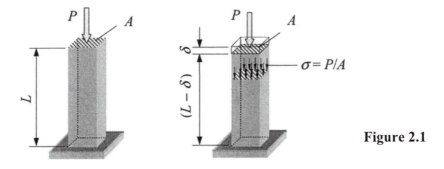

Figure 2.1

The deformation induced by the stress is quantified by relating the change in length to the original length and is known as the **strain** in the material normally given the symbol ε where:

$\varepsilon = $ (change in length/original length) $= (\delta/L)$

Note: the strain is dimensionless since the units of δ and L are the same.

The relationship between stress and strain was first established by Robert Hook in 1676 who determined that in an elastic material the strain is proportional to the stress. The general form of a stress/strain graph is shown in Figure 2.2.

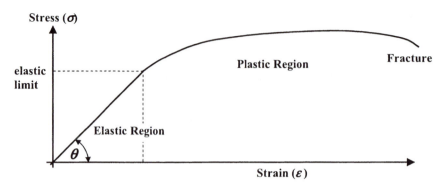

Figure 2.2

The point at which this graph ceases to obey Hook's Law and becomes non–linear is the '*elastic limit*' or '*proportional limit*'.

A typical stress-strain curve for concrete is shown in Figure 2.3(a). This is a non-linear curve in which the peak stress is developed at a compressive strain of approximately 0.002 (depending upon the strength of the concrete) with an ultimate strain of approximately 0.0035. There is no clearly defined elastic range over which the stress varies linearly with the strain. Such stress/strain curves are typical of brittle materials.

A typical stress-strain curve for hot-rolled mild steel is shown in Figure 2.3(b). When a test specimen of mild steel reinforcing bar is subjected to an axial tension in a testing machine, the stress/strain relationship is linearly elastic until the value of stress reaches a yield value, e.g. 250 N/mm^2 (MPa).

At this point an appreciable increase in the stretching of the sample occurs at constant load, this is known as **yielding**. During the process of yielding a molecular change takes place in the material which has the effect of hardening the steel. After approximately 5% strain has occurred sufficient *strain-hardening* will have developed to enable the steel to carry a further increase in load until a maximum load is reached.

The stress-strain curve falls after this point due to a local reduction in the diameter of the sample (known as *necking*) with a consequent smaller cross-sectional area and load carrying capacity. Eventually the sample fractures at approximately 35% strain.

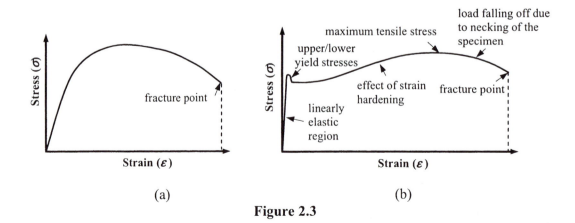

(a)   (b)

Figure 2.3

The characteristics of the stress/strain curves are fundamental to the development and use of structural analysis techniques. A number of frequently used material properties relating to these characteristics are defined in Sections 2.1.2 to 2.1.6.

2.1.2 Young's Modulus (Modulus of Elasticity) – E

From Hooke's Law (in the elastic region): stress ∝ strain ∴ stress = (constant × strain). The value of the constant is known as '*Young's Modulus*' and given the symbol '*E*'. Since strain is dimensionless the units of *E* are the same as those for stress. It represents a measure of material resistance to axial deformation. For some materials the value of Young's Modulus is different in tension than it is in compression. The numerical value of *E* is equal to the slope of the stress/strain curve in the elastic region, i.e. $\tan\theta$ in Figure 2.2.

2.1.3 Secant Modulus – E$_s$

The '*secant modulus*' is equal to the slope of a line drawn from the origin of the stress–strain graph to a point of interest beyond the elastic limit as shown in Figure 2.4.

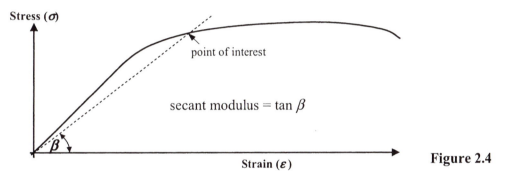

Figure 2.4

The secant modulus is used to describe the material resistance to deformation in the inelastic region of a stress/strain curve and is often expressed as a percentage of Young's Modulus, e.g. 75% – 0.75*E*.

2.1.4 Tangent Modulus – E$_t$

The '*tangent modulus*' is equal to the slope of a tangent line to the stress–strain graph at a point of interest beyond the elastic limit as shown in Figure 2.5.

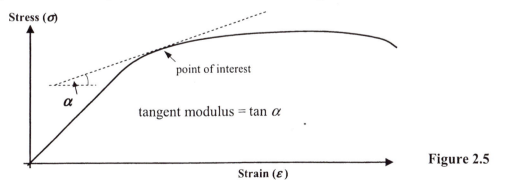

Figure 2.5

The tangent modulus can be used in inelastic buckling analysis of columns as shown in Section 6.3.6 of Chapter 6.

2.1.5 Shear Rigidity (Modulus of Rigidity) – G

The shear rigidity is used to describe the material resistance against shear deformation. similar to Young's Modulus for axial or normal stress/strain. The numerical value of G is equal to the slope of the shear stress/strain curve in the elastic region, where the shear strain is the change angle induced between two perpendicular surfaces subject to a shear stress.

2.1.6 Yield Strength

The yield strength corresponds with the point on the stress/strain graph where permanent deformation begins in the material. In some cases, e.g. in Figure 2.3(a) there is no distinct yield point whilst in others, such as in Figure 2.3(b) there is a well–defined yield region. In the former case a percentage offset is often used to obtain an approximate yield point, e.g. a 0.2% offset point can be determined by drawing a line parallel to the elastic linear line of the graph starting at a point 0.2% (0.002) along the strain axes as shown in Figure 2.6. The intersection of this line with the stress–strain curve defines the 0.2% yield point.

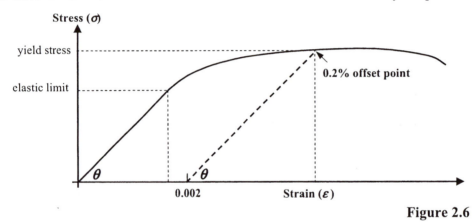

Figure 2.6

2.1.7 Ultimate Tensile Strength

The '*ultimate strength*' is the maximum stress which a material is capable of sustaining and corresponds to the highest point on the stress/strain curve; see Figure 2.3(b). In engineering terms this is normally the value adopted, however if a specimen undergoes considerable necking prior to fracture the true value will differ from this.

2.1.8 Modulus of Rupture in Bending

The '*modulus of rupture*' represents the ultimate strength in bending obtained during a bending test. It is determined by calculating the maximum bending stress in the extreme fibres in a member at failure.

2.1.9 Modulus of Rupture in Torsion

The '*modulus of rupture*' represents the ultimate strength in torsion obtained during torsion test. It is determined by calculating the maximum shear stress in the extreme fibres of a circular member at failure.

2.1.10   Poisson's Ratio – υ

The '*Poisson's Ratio*' for a material is a dimensionless constant representing the ratio of the lateral strain to the axial strain as shown in Figure 2.7.

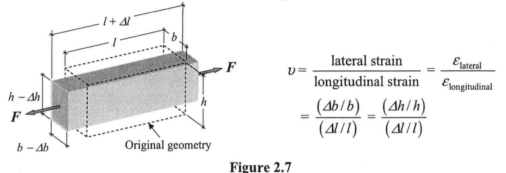

$$\upsilon = \frac{\text{lateral strain}}{\text{longitudinal strain}} = \frac{\varepsilon_{\text{lateral}}}{\varepsilon_{\text{longitudinal}}}$$

$$= \frac{(\Delta b / b)}{(\Delta l / l)} = \frac{(\Delta h / h)}{(\Delta l / l)}$$

Figure 2.7

2.1.11   Coefficient of Thermal Expansion – α

The linear coefficient of thermal expansion describes by how much a material will expand for each degree of temperature increase/decrease, e.g. the change in the length of a bar made from a particular material is given by:

$$\delta L = \alpha L \Delta_T$$

where
α is the coefficient of thermal expansion for the material,
L is the original length,
Δ_T is the change in temperature – a reduction being considered negative and an increase being positive.
The unit for coefficient of thermal expansion is typically °C^{-1}.

2.1.12   Elastic Assumptions

The laws of structural mechanics are well established in recognised *elastic theory* using the following assumptions:

- the material is *homogeneous* which implies its constituent parts have the same physical properties throughout its entire volume,
- the material is *isotropic* which implies that the elastic properties are the same in all directions,
- the material obeys *Hooke's Law*, i.e. when subjected to an external force system the deformations induced will be directly proportional to the magnitude of the applied force (i.e. $P \propto \delta$),
- the material is *elastic*, which implies that it will recover completely from any deformation after the removal of load,
- the *modulus of elasticity* is the same in tension and compression,
- *plane sections remain plane* during deformation. During bending this assumption is violated and is reflected in a non-linear bending stress diagram throughout cross-sections subject to a moment; in most cases this can be neglected.

2.2   *Elastic Cross-Section Properties*

An evaluation of the elastic section properties of a cross-section is fundamental to all structural analyses. These encompass a wide range of parameters such as the cross-sectional area, position of the centroid and the elastic neutral axes, the second moment of area about the centroidal axes and any parallel axes and the elastic section modulus (**Note:** not the Elastic Modulus of Elasticity which is discussed in Section 2.1). Each of these parameters is discussed separately in Sections 2.2.1 to 2.2.8.

Most structural elements have a cross-section for which standard properties are known, e.g. square, rectangle, triangle, trapezium, circle etc., or comprise a combination of one or more such shapes. If the properties of each shape which makes up a complete cross-section are known, this information can be used to determine the corresponding properties of the composite shape. A number of examples are given to illustrate this in the following sections.

In structural steelwork a variety of hot-rolled standard sections are available, the cross-sectional properties of which are given in published tables. A selection of the most commonly used ones are shown in Figure 2.8.

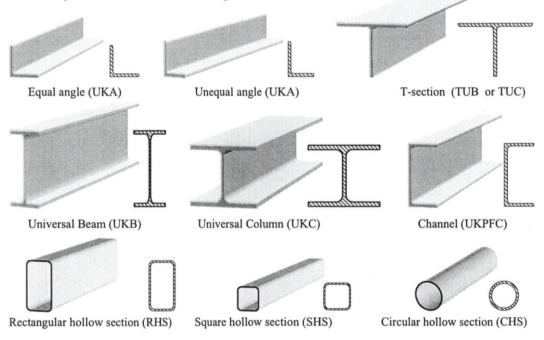

| Equal angle (UKA) | Unequal angle (UKA) | T-section (TUB or TUC) |

| Universal Beam (UKB) | Universal Column (UKC) | Channel (UKPFC) |

| Rectangular hollow section (RHS) | Square hollow section (SHS) | Circular hollow section (CHS) |

Figure 2.8

2.2.1 *Cross-sectional Area*

The cross-sectional area of a composite shape can be expressed as:

$$A_{total} = \sum_{i=1}^{number\ of\ parts} A_i$$

where:
A_{total} is the total area of the composite cross-section
A_i is the cross-sectional area of each component part

Consider the composite shapes indicated in (i) to (ix) and determine the value of A_{total}

(i)

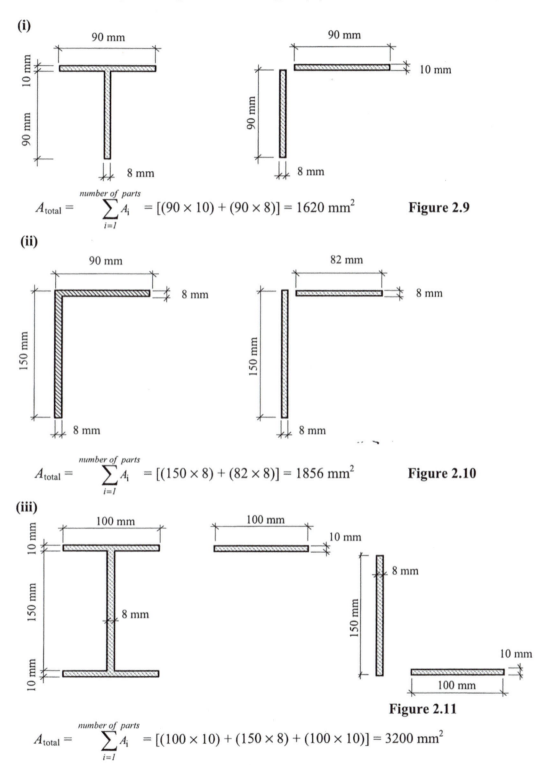

$$A_{total} = \sum_{i=1}^{number\ of\ parts} A_i = [(90 \times 10) + (90 \times 8)] = 1620\ mm^2$$ **Figure 2.9**

(ii)

$$A_{total} = \sum_{i=1}^{number\ of\ parts} A_i = [(150 \times 8) + (82 \times 8)] = 1856\ mm^2$$ **Figure 2.10**

(iii)

Figure 2.11

$$A_{total} = \sum_{i=1}^{number\ of\ parts} A_i = [(100 \times 10) + (150 \times 8) + (100 \times 10)] = 3200\ mm^2$$

(iv)

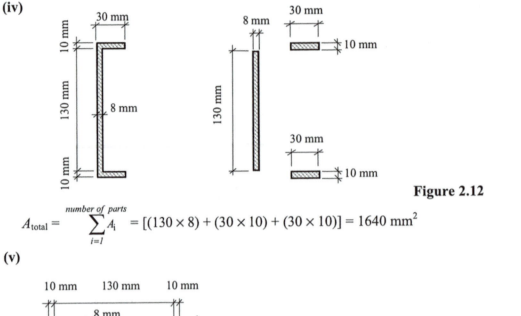

Figure 2.12

$$A_{\text{total}} = \sum_{i=1}^{number\ of\ parts} A_i = [(130 \times 8) + (30 \times 10) + (30 \times 10)] = 1640 \text{ mm}^2$$

(v)

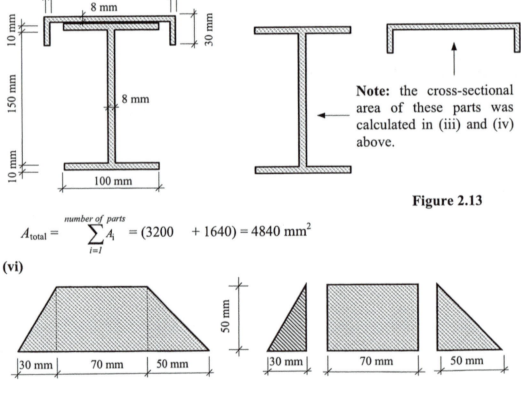

Note: the cross-sectional area of these parts was calculated in (iii) and (iv) above.

Figure 2.13

$$A_{\text{total}} = \sum_{i=1}^{number\ of\ parts} A_i = (3200 + 1640) = 4840 \text{ mm}^2$$

(vi)

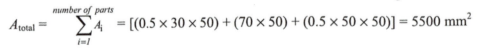

Figure 2.14

$$A_{\text{total}} = \sum_{i=1}^{number\ of\ parts} A_i = [(0.5 \times 30 \times 50) + (70 \times 50) + (0.5 \times 50 \times 50)] = 5500 \text{ mm}^2$$

Note: For a trapezium in general;

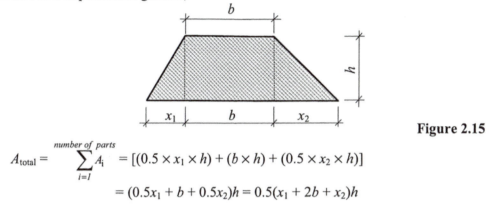

Figure 2.15

$$A_{\text{total}} = \sum_{i=1}^{\text{number of parts}} A_i = [(0.5 \times x_1 \times h) + (b \times h) + (0.5 \times x_2 \times h)]$$

$$= (0.5x_1 + b + 0.5x_2)h = 0.5(x_1 + 2b + x_2)h$$

$A_{\text{total}} = $ **[0.5 × (the sum of the lengths of the parallel sides) × (perpendicular height)]**

Check the area of the trapezium in **(vi)**: $A_{\text{total}} = [0.5 \times (70 + 150) \times 50] = 5500 \text{ mm}^2$
In a similar manner to adding the individual areas of component parts to obtain the total area, section properties can be evaluated by subtracting areas which do not exist, e.g. in hollow sections. Consider examples (vii) to (ix).

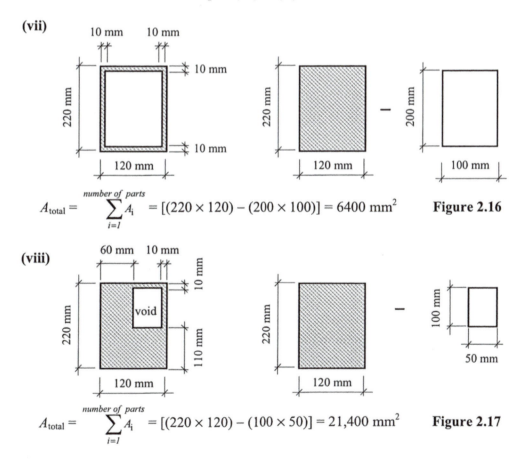

(vii)

$$A_{\text{total}} = \sum_{i=1}^{\text{number of parts}} A_i = [(220 \times 120) - (200 \times 100)] = 6400 \text{ mm}^2 \qquad \textbf{Figure 2.16}$$

(viii)

$$A_{\text{total}} = \sum_{i=1}^{\text{number of parts}} A_i = [(220 \times 120) - (100 \times 50)] = 21,400 \text{ mm}^2 \qquad \textbf{Figure 2.17}$$

(ix)

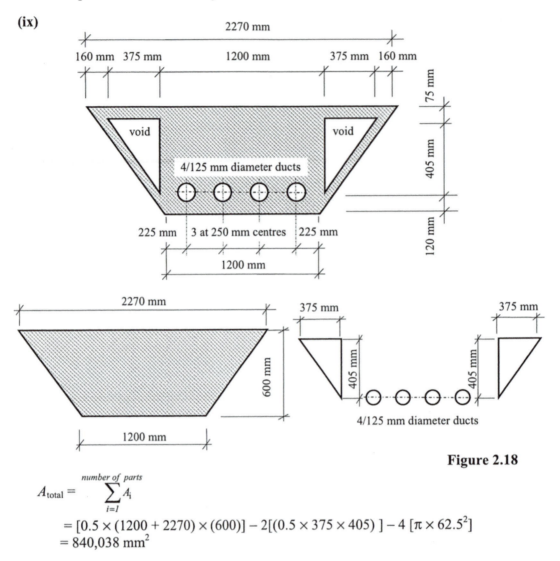

Figure 2.18

$$A_{\text{total}} = \sum_{i=1}^{\text{number of parts}} A_i$$

$$= [0.5 \times (1200 + 2270) \times (600)] - 2[(0.5 \times 375 \times 405)] - 4[\pi \times 62.5^2]$$
$$= 840,038 \text{ mm}^2$$

2.2.2 *Centre of Gravity and Centroid*

The centre of gravity of an object is the point through which the force due to gravity on the total mass of the object is considered to act. The corresponding position on a plane surface (i.e. relating to the cross-sectional area) is known as the **centroid**; both are indicated in Figure 2.19

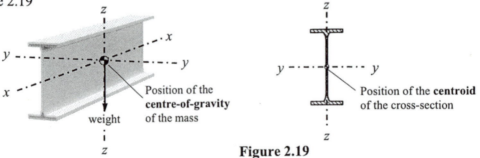

Figure 2.19

Consider the cross-section *A* shown in Figures 2.20(a) and (b) which can be considered to be an infinite number of elemental areas each equal to **δ𝐴**. The 1ˢᵗ moment of area (i.e. area × perpendicular lever arm) of the total area about any axis is equal to the sum of the 1ˢᵗ moments of area of each individual area about the same axis, i.e.

$$A \times \bar{y} = \sum(\delta A \times y) \qquad \therefore \ \bar{y} = \sum(\delta A \times y)/A$$
$$A \times \bar{z} = \sum(\delta A \times z) \qquad \therefore \ \bar{z} = \sum(\delta A \times z)/A$$

where:
A is the total area of the cross section
$\bar{y}$ is the distance in the y direction to the centroid for the total area
$\bar{z}$ is the distance in the z direction to the centroid for the total area
y is the distance in the y direction to the centroid of the elemental area
z is the distance in the z direction to the centroid of the elemental area

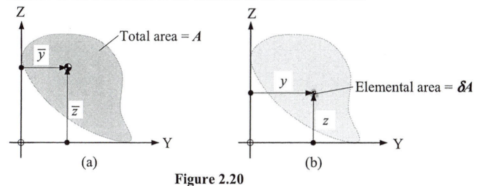

Figure 2.20

In precise terms, $\Sigma\delta Ay/A$ and $\Sigma\delta Az/A$ are the integrals for the shape being considered, however in most practical cases the cross-sectional area comprises a number of standard shapes (instead of the elemental area) i.e. rectangles, triangles, circles etc. in which the position of the centroid is known as shown in Figure 2.21

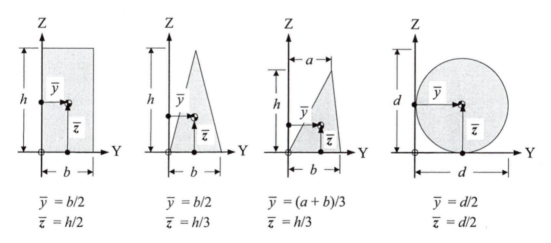

$$\bar{y} = b/2 \qquad\qquad \bar{y} = b/2 \qquad\qquad \bar{y} = (a+b)/3 \qquad\qquad \bar{y} = d/2$$
$$\bar{z} = h/2 \qquad\qquad \bar{z} = h/3 \qquad\qquad \bar{z} = h/3 \qquad\qquad\quad \bar{z} = d/2$$

Figure 2.21

Consider the composite shapes (i) to (ix) indicated previously to determine the co-ordinates of their centroids.

(i)

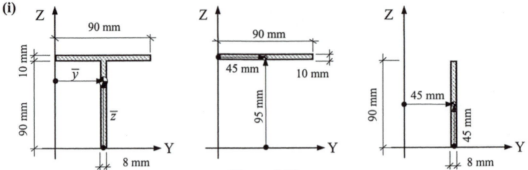

Figure 2.22

$\bar{y} = [(90 \times 10)(45) + (90 \times 8)(45)] / 1620 = 45$ mm   (Vertical axis of symmetry)

$\bar{z} = [(90 \times 10)(95) + (90 \times 8)(45)] / 1620 = 72.78$ mm

(ii)

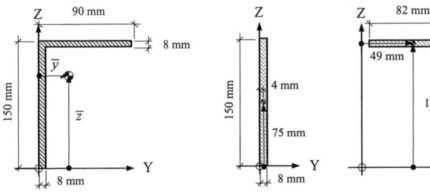

$\bar{y} = [(150 \times 8)(4) + (82 \times 8)(49)] / 1856 = 19.91$ mm

$\bar{z} = [(150 \times 8)(75) + (82 \times 8)(146)] / 1856 = 100.1$ mm

Figure 2.23

(iii)

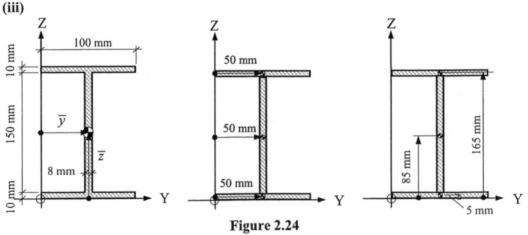

Figure 2.24

$\bar{y} = [(100 \times 10)(50) + (150 \times 8)(50) + (100 \times 10)(50)] / 3200 = 50.0$ mm

$\bar{z} = [(100 \times 10)(165) + (150 \times 8)(85) + (100 \times 10)(5)] / 3200 = 85.0$ mm

Note: If there are axes of symmetry then the centroid lies at the intersection point of the axes.

(iv)

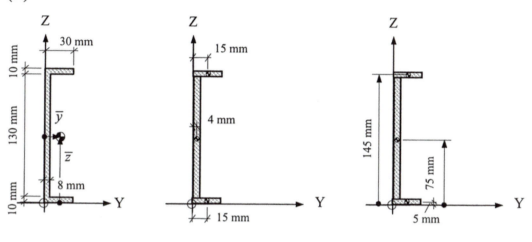

Figure 2.25

$\bar{y}$ = [(30 × 10)(15) + (130 × 8)(4) + (30 × 10)(15)] / 1640 = 8.02 mm
$\bar{z}$ = 75.0 mm   (Horizontal axis of symmetry)

(v)
The values of $\bar{y}$ and $\bar{z}$ for the sections in (iii) and (iv) are used in this calculation.

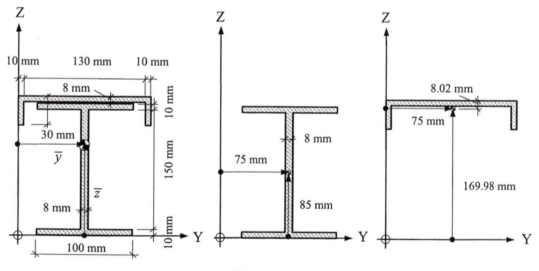

Figure 2.26

$\bar{y}$ = 75.0 mm   (Vertical axis of symmetry)
$\bar{z}$ = [(3200)(85) + (1640)(169.98)] / (3200 + 1640) = 113.80 mm

(vi)

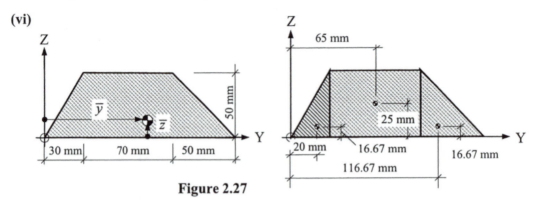

Figure 2.27

$\bar{y}$ = [(0.5 × 30 × 50)(20) + (70 × 50)(65) + (0.5 × 50 × 50)(116.67)] / 5500= 70.61 mm

$\bar{z}$ = [(0.5 × 30 × 50)(16.67) + (70 × 50)(25) + (0.5 × 50 × 50)(16.67)] / 5500 = 21.97 mm

(vii)

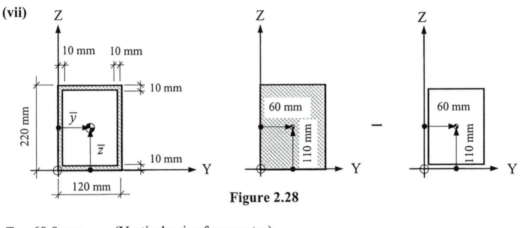

Figure 2.28

$\bar{y}$ = 60.0 mm   (Vertical axis of symmetry)

$\bar{z}$ = 110.0 mm   (Horizontal axis of symmetry)

(viii)

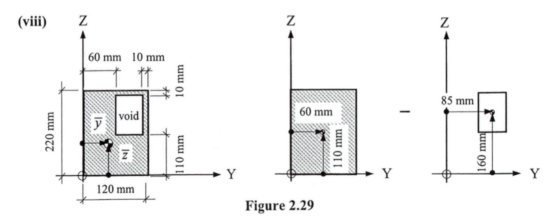

Figure 2.29

$\bar{y}$ = [(220 × 120)(60) − (100 × 50)(85)] / 21400 = 54.16 mm

$\bar{z}$ = [(220 × 120)(110) − (100 × 50)(160)] / 21400 = 98.32 mm

(ix)

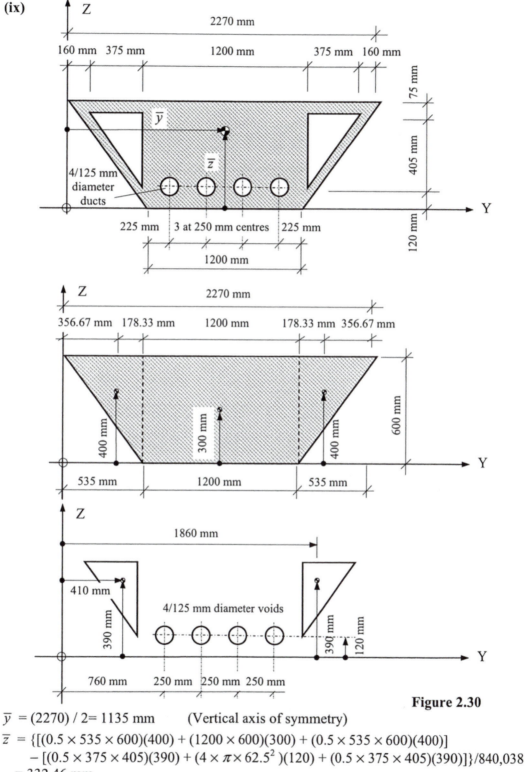

Figure 2.30

$\bar{y}$ = (2270) / 2= 1135 mm   (Vertical axis of symmetry)

$\bar{z}$ = {[(0.5 × 535 × 600)(400) + (1200 × 600)(300) + (0.5 × 535 × 600)(400)]
   − [(0.5 × 375 × 405)(390) + (4 × π × 62.5^2)(120) + (0.5 × 375 × 405)(390)]}/840,038
= 332.46 mm

2.2.3 Problems: Cross-sectional Area and Position of Centroid

Determine the cross-sectional area and the values of $\bar{y}$ and $\bar{z}$ to locate the position of the centroid for the sections shown in Problems 2.1 to 2.6. Assume the origin of the co-ordinate system to be at the bottom left-hand corner for each section.

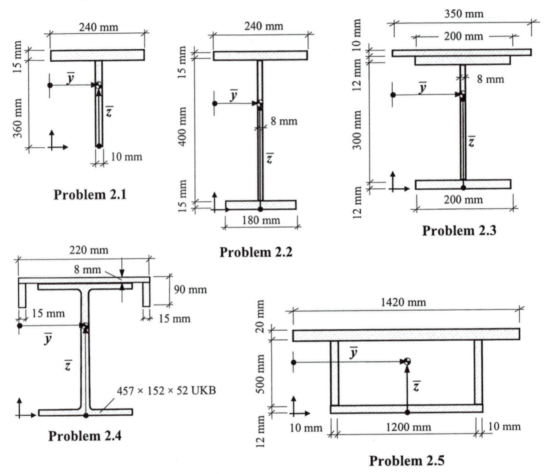

Problem 2.1

Problem 2.2

Problem 2.3

Problem 2.4

Problem 2.5

Section Properties for UKB sections:

457 × 152 × 52 UKB

Overall depth	$h = 449.8$ mm
Area	$A = 66.6$ cm^2
2nd Moment of area	$I_{yy} = 21400$ cm^4
2nd Moment of area	$I_{zz} = 645$ cm^4

533 × 210 × 82 UKB

Overall depth	$h = 528.3$ mm
Flange width	$b = 208.8$ mm
Area	$A = 105.0$ cm^2
Web thickness	$t_w = 9.6$ mm
2nd Moment of area	$I_{yy} = 47500$ cm^4
2nd Moment of area	$I_{zz} = 2010$ cm^4

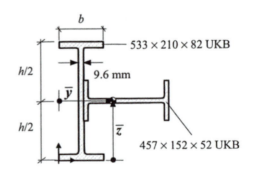

Problem 2.6

2.2.4 Solutions: Cross-sectional Area and Position of Centroid

---

Solution
Topic: Cross-sectional Area and Position of Centroid
Problem Numbers: 2.1 to 2.6   **Page No. 1**

Problem 2.1:
$A = [(240 \times 15) + (360 \times 10)] = 7200 \text{ mm}^2$
$\bar{y}$ lies on the vertical axis of symmetry
$\bar{y} = (240/2) = 120 \text{ mm}$
$\bar{z} = [(240 \times 15)(367.5) + (360 \times 10)(180)]/7200 = 273.75 \text{ mm}$

Problem 2.2:
$A = [(240 \times 15) + (400 \times 8) + (180 \times 15)] = 9500 \text{ mm}^2$
$\bar{y}$ lies on the vertical axis of symmetry
$\bar{y} = (240/2) = 120 \text{ mm}$
$\bar{z} = [(240 \times 15)(422.5) + (400 \times 8)(215) + (180 \times 15)(7.5)]/9500 = 234.66 \text{ mm}$

Problem 2.3:
$A = [(350 \times 10) + (200 \times 12) + (300 \times 8) + (200 \times 12)] = 10700 \text{ mm}^2$
$\bar{y}$ lies on the vertical axis of symmetry
$\bar{y} = (350/2) = 175 \text{ mm}$
$\bar{z} = [(350 \times 10)(329) + (200 \times 12)(318) + (300 \times 8)(162) + (200 \times 12)(6)]/10700$
$\quad = 216.63 \text{ mm}$

Problem 2.4:
$A = [(220 \times 8) + 2 (82 \times 15) + 6660] = 10880 \text{ mm}^2$
$\bar{y}$ lies on the vertical axis of symmetry
$\bar{y} = (220/2) = 110 \text{ mm}$
$\bar{z} = [(220 \times 8)(449.8 + 4) + 2(82 \times 15)(449.8 - 41) + (6660)(449.8/2)]/10880$
$\quad = 303.51 \text{ mm}$

Problem 2.5:
$A = [(1420 \times 20) + 2 (500 \times 10) + (1220 \times 12)] = 53040 \text{ mm}^2$
$\bar{y}$ lies on the vertical axis of symmetry
$\bar{y} = (1420/2) = 710 \text{ mm}$
$\bar{z} = [(1420 \times 20)(522) + 2(500 \times 10)(262) + (1220 \times 12)(6)]/53040 = 330.56 \text{ mm}$

Problem 2.6:
$A = [6660 + 10500] = 17160 \text{ mm}^2$
$\bar{y} = [(10500)(208.8/2) + (6660)(208.8/2 + 9.6/2 + 449.8/2)]/17160 = 193.55 \text{ mm}$
$\bar{z} = (528.3/2) = 264.15 \text{ mm}$
$\bar{z}$ lies on the horizontal axis of symmetry

2.2.5 Elastic Neutral Axes

Consider a beam of rectangular cross-section which is simply supported at the ends and carries a distributed load, as shown in Figure 2.31.

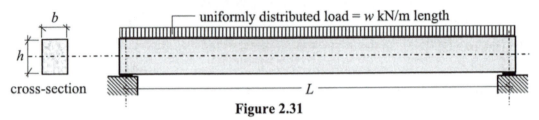

Figure 2.31

The beam will deflect due to the bending moments and shear forces induced by the applied loading, resulting in a curved shape as indicated in Figure 2.32.

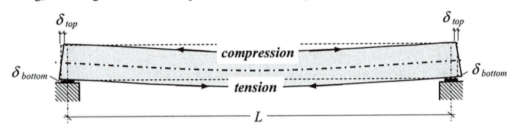

Original length of the beam before deformation = L
Final length of the top edge after deformation = $(L - 2\delta_{top})$ i.e. shortening
Final length of the bottom edge after deformation = $(L + 2\delta_{bottom})$ i.e. lengthening

Figure 2.32

Clearly if the ends of the beam are assumed to remain perpendicular to the longitudinal axis, then the material above this axis must be in compression, whilst that below it must be in tension. At a point between the top and the bottom of the beam a layer of fibres exist which remain at their original length and consequently do not have any bending stress in them. This layer of fibres forms the '*neutral surface*' and on a cross–section is indicated by the '*neutral axis*' as shown in Figure 2.33.

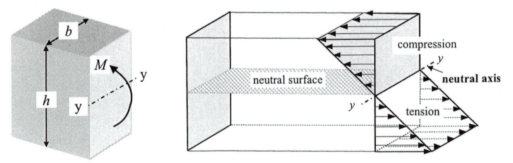

The neutral axis coincides with the centroidal axis discussed in Section 2.2.2

Figure 2.33

2.2.6 *Second Moment of Area – I and Radius of Gyration – i*

Two of the most important properties of a cross–section are the '*second moment of area*' and the '*radius of gyration*'. Consider the area shown in Figure 2.20(b). If the elemental area δA has its centroid at a perpendicular distance 'i' from a given axis, the second moment of area of the element about the given axis is the product of the area of the element and the square of the distance of the centroid from the axis, i.e.

Second moment of area $I = (\delta A \times i^2)$

The second moment of area of the total area A is equal to $\Sigma(\delta A i^2)$ over the whole area. It is convenient to consider two mutually perpendicular axes which intersect at the centroid of a cross–section and hence:

$$I_{yy} = Ai_{yy}^2 \qquad \text{and} \qquad I_{zz} = Ai_{zz}^2$$

Alternatively:

$$i_{yy} = \sqrt{\frac{I_{yy}}{A}} \qquad \text{and} \qquad i_{zz} = \sqrt{\frac{I_{zz}}{A}}$$

where i_{yy} and i_{zz} are known as the '*radii of gyration*' about the y–y and z–z axes respectively.
Consider the rectangular cross–section shown in Figure 2.34.

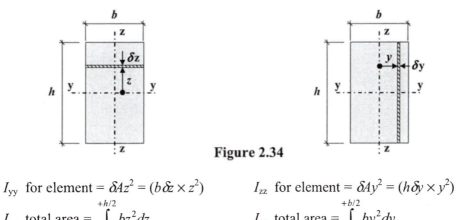

Figure 2.34

I_{yy} for element $= \delta A z^2 = (b \delta z \times z^2)$ I_{zz} for element $= \delta A y^2 = (h \delta y \times y^2)$

I_{yy} total area $= \displaystyle\int_{-h/2}^{+h/2} bz^2 dz$ I_{zz} total area $= \displaystyle\int_{-b/2}^{+b/2} by^2 dy$

$$I_{yy} = 2\left[\frac{bz^3}{3}\right]_0^{+h/2} = 2\left[\frac{b}{3}\times\left(\frac{h}{2}\right)^3\right] = \frac{bh^3}{12} \qquad I_{zz} = 2\left[\frac{hy^3}{3}\right]_0^{+b/2} = 2\left[\frac{h}{3}\times\left(\frac{b}{2}\right)^3\right] = \frac{hb^3}{12}$$

2.2.6.1 The Parallel Axis Theorem

It can also be shown that the second moment of area of a cross–sectional area A about an axis parallel to any other axis is equal to the second moment of area of A about that other axis plus the area multiplied by the square of the perpendicular distance between the axes. Consider the rectangular areas shown in Figure 2.35:

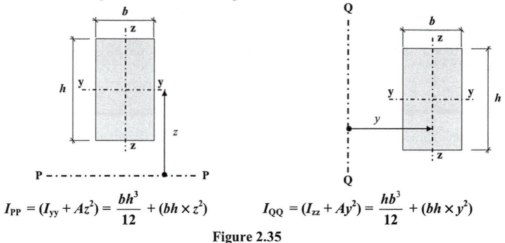

$$I_{PP} = (I_{yy} + Az^2) = \frac{bh^3}{12} + (bh \times z^2) \qquad I_{QQ} = (I_{zz} + Ay^2) = \frac{hb^3}{12} + (bh \times y^2)$$

Figure 2.35

These relationships are used extensively to determine the values of the second moment of area and radius of gyration of compound sections comprising defined areas such as rectangles, triangles circles etc.

Consider the cross–sectional area shown in Figures 2.24 and determine the values of the second moment of area and radius of gyration about the centroidal axes. Data from Figure 2.24 is indicated in Figure 2.36:

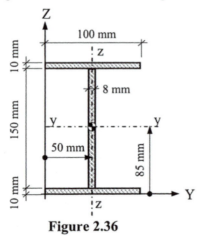

Figure 2.36

Area = 3200 mm² (see Figure 2.11)
$I_{PP} = (I_{yy} + Az^2)$ for each rectangle in which P–P is the y–y axis for the whole section.

$$I_{yy} = \sum \left(\frac{bh^3}{12} + bhz^2 \right) \text{ for each flange and the web}$$

$$= 2 \left(\frac{100 \times 10^3}{12} + 100 \times 10 \times 80^2 \right) + \frac{8 \times 150^3}{12}$$

(**Note:** the second term is zero for the web since the P–P axis coincides with its' centroidal axis.

$$I_{yy} = 15.07 \times 10^6 \text{ mm}^4$$

$I_{QQ} = (I_{zz} + Ay^2)$ for each rectangle in which Q–Q is the z–z axis for the whole section. In this case the second term for each rectangle is equal to zero since the Q–Q axis coincides with their centroidal axes.

$$I_{zz} = \frac{hb^3}{12} \text{ for each flange and the web} = 2 \left(\frac{10 \times 100^3}{12} \right) + \frac{150 \times 8^3}{12} = 1.67 \times 10^6 \text{ mm}^4$$

$$i_{yy} = \sqrt{\frac{I_{yy}}{A}} = \sqrt{\frac{15.07 \times 10^6}{3200}} = 68.63 \text{ mm}; \qquad i_{zz} = \sqrt{\frac{I_{zz}}{A}} = \sqrt{\frac{1.67 \times 10^6}{3200}} = 22.85 \text{ mm}$$

2.2.7 Elastic Section Modulus – W_{el}

The bending moments induced in a beam by an applied load system generate bending stresses in the material fibres which vary from a maximum in the extreme fibres to zero at the level of the neutral axis as shown in Figure 2.33 and Figure 2.37.

The magnitude of the bending stresses at any vertical cross-section can be determined using the simple theory of bending from which the following equation is derived:

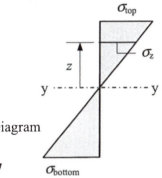

$$\frac{M}{I} = \frac{E}{R} = \frac{\sigma}{z} \quad \therefore \ \sigma = \frac{Mz}{I}$$ Bending Stress Diagram

Figure 2.37

where:
M is the applied bending moment at the section being considered,
E is the value of Young's modulus of elasticity,
R is the radius of curvature of the beam,
σ is the bending stress,
z is the distance measured from the elastic neutral axis to the level on the cross-section at which the stress is being evaluated,
I is the second moment of area of the full cross-section about the elastic neutral axis.

It is evident from the equation given above that for any specified cross-section in a beam subject to a known value of bending moment (i.e. M and I constant), the bending stress is directly proportional to the distance from the neutral axis; i.e.

$$\sigma = \text{constant} \times z \quad \therefore \ \sigma \ \alpha \ z$$

This is shown in Figure 2.37, in which the maximum bending stress occurs at the extreme fibres.

In design it is usually the extreme fibre stresses relating to the z_{maximum} values at the top and bottom which are critical. These can be determined using:

$$\sigma_{\text{top}} = \frac{M}{W_{el,y,top}} \quad \text{and} \quad \sigma_{\text{bottom}} = \frac{M}{W_{el,y,bottom}}$$

where σ and M are as before,

$W_{el,y,top}$ is the elastic section modulus relating to the top fibres and defined as $\dfrac{I_{yy}}{z_{top}}$

$W_{el,y,bottom}$ is the elastic section modulus relating to the bottom fibres and defined as $\dfrac{I_{yy}}{z_{bottom}}$

If a cross-section is symmetrical about the y–y axis then $W_{el,y,top} = W_{el,y,bottom}$. In asymmetric sections the maximum stress occurs in the fibres corresponding to the smallest W_{el} value. For a rectangular cross-section of breadth b and depth h subject to a bending moment M about the major y–y axis, the appropriate values of I, z and W_{el} are:

$$I_{yy} = \frac{bh^3}{12} \qquad z_{maximum} = \frac{h}{2} \qquad W_{el,y,minimum} = \frac{bh^2}{6}$$

In the case of bending about the minor z–z axis:

$$I_{zz} = \frac{hb^3}{12} \qquad y_{maximum} = \frac{b}{2} \qquad W_{el,z,minimum} = \frac{hb^2}{6}$$

Consider the cross–sectional area shown in Figures 2.29/2.38 and determine the values of the maximum and minimum elastic section modulii about the centroidal axes.

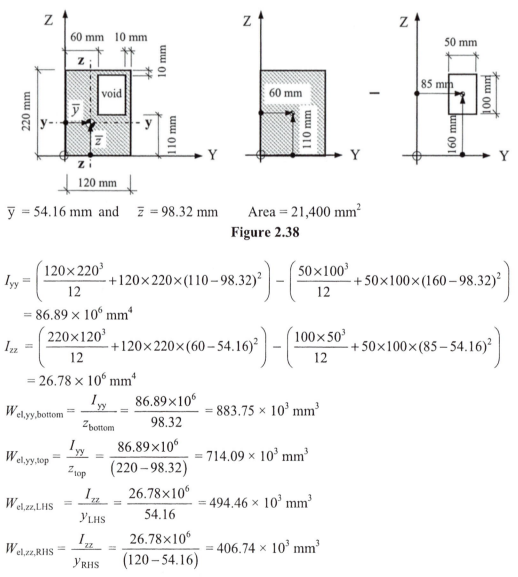

$\overline{y} = 54.16$ mm and $\overline{z} = 98.32$ mm   Area = 21,400 mm²

Figure 2.38

$$I_{yy} = \left(\frac{120 \times 220^3}{12} + 120 \times 220 \times (110 - 98.32)^2 \right) - \left(\frac{50 \times 100^3}{12} + 50 \times 100 \times (160 - 98.32)^2 \right)$$

$$= 86.89 \times 10^6 \text{ mm}^4$$

$$I_{zz} = \left(\frac{220 \times 120^3}{12} + 120 \times 220 \times (60 - 54.16)^2 \right) - \left(\frac{100 \times 50^3}{12} + 50 \times 100 \times (85 - 54.16)^2 \right)$$

$$= 26.78 \times 10^6 \text{ mm}^4$$

$$W_{el,yy,bottom} = \frac{I_{yy}}{z_{bottom}} = \frac{86.89 \times 10^6}{98.32} = 883.75 \times 10^3 \text{ mm}^3$$

$$W_{el,yy,top} = \frac{I_{yy}}{z_{top}} = \frac{86.89 \times 10^6}{(220 - 98.32)} = 714.09 \times 10^3 \text{ mm}^3$$

$$W_{el,zz,LHS} = \frac{I_{zz}}{y_{LHS}} = \frac{26.78 \times 10^6}{54.16} = 494.46 \times 10^3 \text{ mm}^3$$

$$W_{el,zz,RHS} = \frac{I_{zz}}{y_{RHS}} = \frac{26.78 \times 10^6}{(120 - 54.16)} = 406.74 \times 10^3 \text{ mm}^3$$

2.2.8 *Problems: Second Moments of Area and Elastic Section Modulii*

Determine the following values for the sections indicated in Problems 2.1 to 2.6.
(i)   the second moment of areas I_{yy} and I_{zz} and
(ii)   the elastic section modulii $W_{el,yy}$ and $W_{el,zz}$.

2.2.9 *Solutions: Second Moments of Area and Elastic Section Modulii*

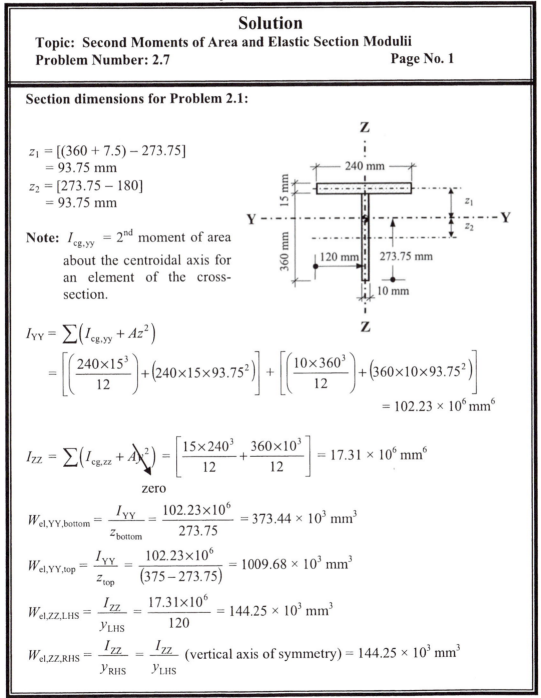

Solution

Topic: **Second Moments of Area and Elastic Section Modulii**

Problem Number: **2.7**   **Page No. 1**

Section dimensions for Problem 2.1:

$z_1 = [(360 + 7.5) - 273.75]$
    $= 93.75$ mm
$z_2 = [273.75 - 180]$
    $= 93.75$ mm

Note: $I_{cg,yy}$ = 2nd moment of area about the centroidal axis for an element of the cross-section.

$$I_{YY} = \sum\left(I_{cg,yy} + Az^2\right)$$

$$= \left[\left(\frac{240 \times 15^3}{12}\right) + \left(240 \times 15 \times 93.75^2\right)\right] + \left[\left(\frac{10 \times 360^3}{12}\right) + \left(360 \times 10 \times 93.75^2\right)\right]$$

$$= 102.23 \times 10^6 \, mm^6$$

$$I_{ZZ} = \sum\left(I_{cg,zz} + Ay^2\right) = \left[\frac{15 \times 240^3}{12} + \frac{360 \times 10^3}{12}\right] = 17.31 \times 10^6 \, mm^6$$

zero

$$W_{el,YY,bottom} = \frac{I_{YY}}{z_{bottom}} = \frac{102.23 \times 10^6}{273.75} = 373.44 \times 10^3 \, mm^3$$

$$W_{el,YY,top} = \frac{I_{YY}}{z_{top}} = \frac{102.23 \times 10^6}{(375 - 273.75)} = 1009.68 \times 10^3 \, mm^3$$

$$W_{el,ZZ,LHS} = \frac{I_{ZZ}}{y_{LHS}} = \frac{17.31 \times 10^6}{120} = 144.25 \times 10^3 \, mm^3$$

$$W_{el,ZZ,RHS} = \frac{I_{ZZ}}{y_{RHS}} = \frac{I_{ZZ}}{y_{LHS}} \text{ (vertical axis of symmetry)} = 144.25 \times 10^3 \, mm^3$$

Solution

Topic: Second Moments of Area and Elastic Section Modulii
Problem Number: 2.8   **Page No. 2**

Section dimensions for Problem 2.2:

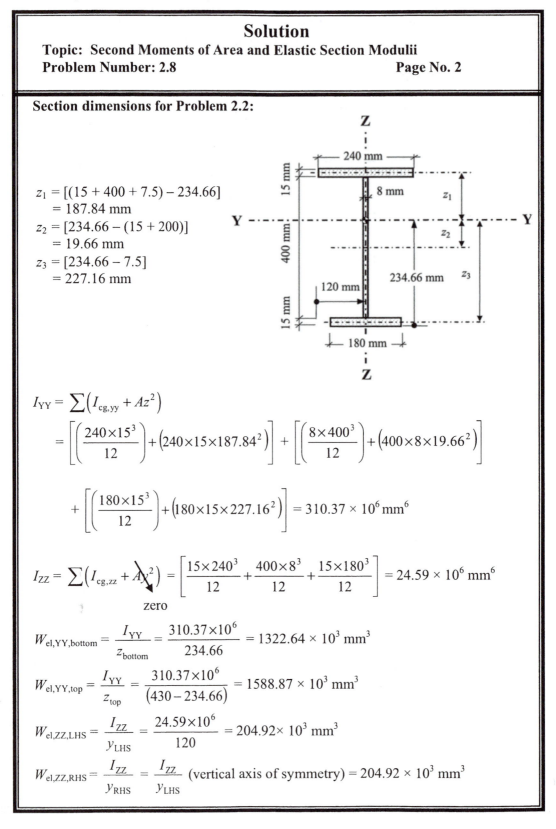

$z_1 = [(15 + 400 + 7.5) - 234.66]$
$\quad = 187.84 \text{ mm}$
$z_2 = [234.66 - (15 + 200)]$
$\quad = 19.66 \text{ mm}$
$z_3 = [234.66 - 7.5]$
$\quad = 227.16 \text{ mm}$

$$I_{YY} = \sum\left(I_{cg,yy} + Az^2\right)$$

$$= \left[\left(\frac{240 \times 15^3}{12}\right) + \left(240 \times 15 \times 187.84^2\right)\right] + \left[\left(\frac{8 \times 400^3}{12}\right) + \left(400 \times 8 \times 19.66^2\right)\right]$$

$$+ \left[\left(\frac{180 \times 15^3}{12}\right) + \left(180 \times 15 \times 227.16^2\right)\right] = 310.37 \times 10^6 \text{ mm}^6$$

$$I_{ZZ} = \sum\left(I_{cg,zz} + Ay^2\right) = \left[\frac{15 \times 240^3}{12} + \frac{400 \times 8^3}{12} + \frac{15 \times 180^3}{12}\right] = 24.59 \times 10^6 \text{ mm}^6$$

$$\text{zero}$$

$$W_{el,YY,bottom} = \frac{I_{YY}}{z_{bottom}} = \frac{310.37 \times 10^6}{234.66} = 1322.64 \times 10^3 \text{ mm}^3$$

$$W_{el,YY,top} = \frac{I_{YY}}{z_{top}} = \frac{310.37 \times 10^6}{(430 - 234.66)} = 1588.87 \times 10^3 \text{ mm}^3$$

$$W_{el,ZZ,LHS} = \frac{I_{ZZ}}{y_{LHS}} = \frac{24.59 \times 10^6}{120} = 204.92 \times 10^3 \text{ mm}^3$$

$$W_{el,ZZ,RHS} = \frac{I_{ZZ}}{y_{RHS}} = \frac{I_{ZZ}}{y_{LHS}} \text{ (vertical axis of symmetry)} = 204.92 \times 10^3 \text{ mm}^3$$

Solution
Topic: Second Moments of Area and Elastic Section Modulii
Problem Number: 2.9   **Page No. 3**

Section dimensions for Problem 2.3:

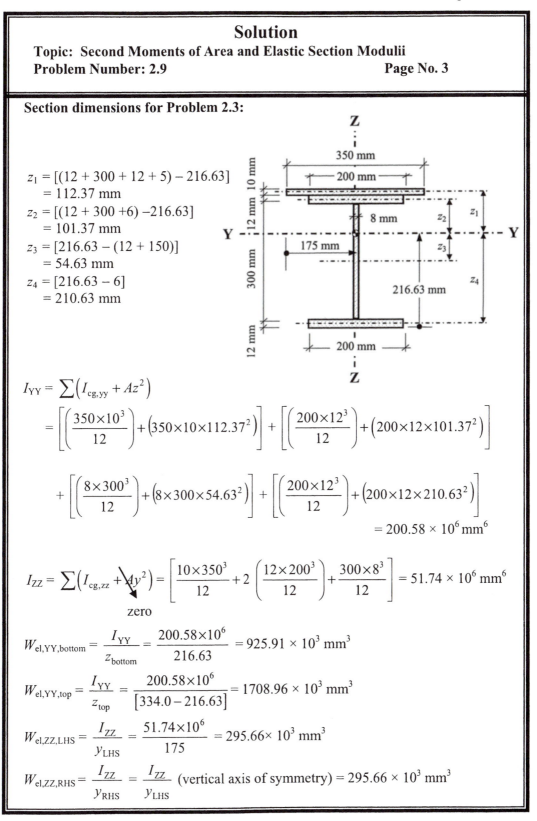

$z_1 = [(12 + 300 + 12 + 5) - 216.63]$
$\quad = 112.37 \text{ mm}$
$z_2 = [(12 + 300 + 6) - 216.63]$
$\quad = 101.37 \text{ mm}$
$z_3 = [216.63 - (12 + 150)]$
$\quad = 54.63 \text{ mm}$
$z_4 = [216.63 - 6]$
$\quad = 210.63 \text{ mm}$

$$I_{YY} = \sum\left(I_{cg,yy} + Az^2\right)$$

$$= \left[\left(\frac{350 \times 10^3}{12}\right) + \left(350 \times 10 \times 112.37^2\right)\right] + \left[\left(\frac{200 \times 12^3}{12}\right) + \left(200 \times 12 \times 101.37^2\right)\right]$$

$$+ \left[\left(\frac{8 \times 300^3}{12}\right) + \left(8 \times 300 \times 54.63^2\right)\right] + \left[\left(\frac{200 \times 12^3}{12}\right) + \left(200 \times 12 \times 210.63^2\right)\right]$$

$$= 200.58 \times 10^6 \text{ mm}^6$$

$$I_{ZZ} = \sum\left(I_{cg,zz} + \underset{\text{zero}}{Ay^2}\right) = \left[\frac{10 \times 350^3}{12} + 2\left(\frac{12 \times 200^3}{12}\right) + \frac{300 \times 8^3}{12}\right] = 51.74 \times 10^6 \text{ mm}^6$$

$$W_{el,YY,bottom} = \frac{I_{YY}}{z_{bottom}} = \frac{200.58 \times 10^6}{216.63} = 925.91 \times 10^3 \text{ mm}^3$$

$$W_{el,YY,top} = \frac{I_{YY}}{z_{top}} = \frac{200.58 \times 10^6}{[334.0 - 216.63]} = 1708.96 \times 10^3 \text{ mm}^3$$

$$W_{el,ZZ,LHS} = \frac{I_{ZZ}}{y_{LHS}} = \frac{51.74 \times 10^6}{175} = 295.66 \times 10^3 \text{ mm}^3$$

$$W_{el,ZZ,RHS} = \frac{I_{ZZ}}{y_{RHS}} = \frac{I_{ZZ}}{y_{LHS}} \text{ (vertical axis of symmetry)} = 295.66 \times 10^3 \text{ mm}^3$$

Solution

Topic: Second Moments of Area and Elastic Section Modulii
Problem Number: 2.10   **Page No. 4**

Section dimensions for Problem 2.4:
$y_1 = [110 - 7.5] = 102.5$ mm
$y_2 = [110 - 7.5] = 102.5$ mm

$z_1 = [(449.8 + 4) - 303.51]$
$\quad = 150.29$ mm
$z_2 = [(449.8 - 41) - 303.51]$
$\quad = 105.29$ mm
$z_3 = [303.51 - (449.8/2)]$
$\quad = 78.61$ mm

For 457 × 152 × 52 UKB:
$h = 449.8$ mm
$A = 66.6$ cm^2
$I_{yy} = 21400$ cm^4
$I_{zz} = 645$ cm^4

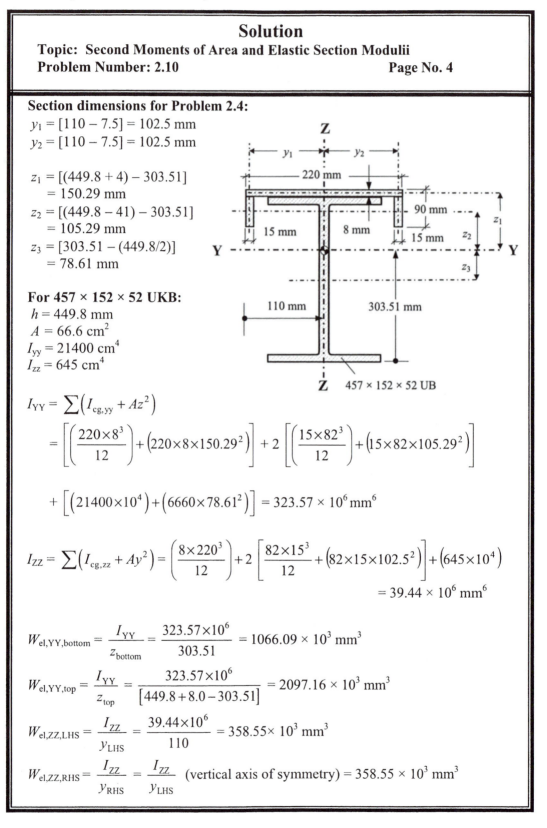

$I_{YY} = \sum \left(I_{cg,yy} + Az^2 \right)$

$= \left[\left(\frac{220 \times 8^3}{12} \right) + \left(220 \times 8 \times 150.29^2 \right) \right] + 2 \left[\left(\frac{15 \times 82^3}{12} \right) + \left(15 \times 82 \times 105.29^2 \right) \right]$

$+ \left[\left(21400 \times 10^4 \right) + \left(6660 \times 78.61^2 \right) \right] = 323.57 \times 10^6 \text{ mm}^6$

$I_{ZZ} = \sum \left(I_{cg,zz} + Ay^2 \right) = \left(\frac{8 \times 220^3}{12} \right) + 2 \left[\frac{82 \times 15^3}{12} + \left(82 \times 15 \times 102.5^2 \right) \right] + \left(645 \times 10^4 \right)$

$= 39.44 \times 10^6 \text{ mm}^6$

$W_{el,YY,bottom} = \dfrac{I_{YY}}{z_{bottom}} = \dfrac{323.57 \times 10^6}{303.51} = 1066.09 \times 10^3 \text{ mm}^3$

$W_{el,YY,top} = \dfrac{I_{YY}}{z_{top}} = \dfrac{323.57 \times 10^6}{[449.8 + 8.0 - 303.51]} = 2097.16 \times 10^3 \text{ mm}^3$

$W_{el,ZZ,LHS} = \dfrac{I_{ZZ}}{y_{LHS}} = \dfrac{39.44 \times 10^6}{110} = 358.55 \times 10^3 \text{ mm}^3$

$W_{el,ZZ,RHS} = \dfrac{I_{ZZ}}{y_{RHS}} = \dfrac{I_{ZZ}}{y_{LHS}}$ (vertical axis of symmetry) $= 358.55 \times 10^3 \text{ mm}^3$

Solution

Topic: Second Moments of Area and Elastic Section Modulii

Problem Number: 2.11   **Page No. 5**

Section dimensions for Problem 2.5:

$y_1 = [(1200/2) + 5] = 605$ mm
$y_2 = [(1200/2) + 5] = 605$ mm

$z_1 = [(12 + 500 + 10) - 330.56]$
    $= 191.44$ mm
$z_2 = [330.56 - (12 + 250)]$
    $= 68.56$ mm
$z_3 = [330.56 - 6]$
    $= 324.56$ mm

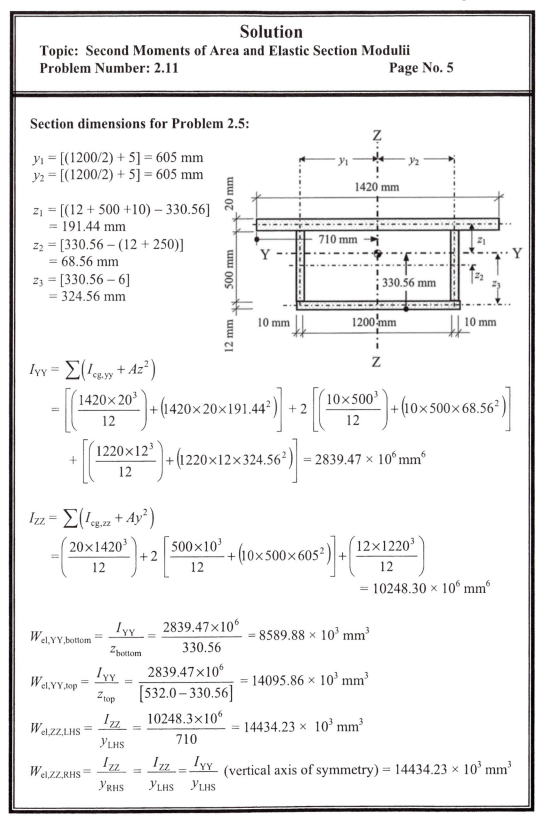

$$I_{YY} = \sum \left(I_{cg,yy} + Az^2 \right)$$

$$= \left[\left(\frac{1420 \times 20^3}{12} \right) + \left(1420 \times 20 \times 191.44^2 \right) \right] + 2 \left[\left(\frac{10 \times 500^3}{12} \right) + \left(10 \times 500 \times 68.56^2 \right) \right]$$

$$+ \left[\left(\frac{1220 \times 12^3}{12} \right) + \left(1220 \times 12 \times 324.56^2 \right) \right] = 2839.47 \times 10^6 \text{ mm}^6$$

$$I_{ZZ} = \sum \left(I_{cg,zz} + Ay^2 \right)$$

$$= \left(\frac{20 \times 1420^3}{12} \right) + 2 \left[\frac{500 \times 10^3}{12} + \left(10 \times 500 \times 605^2 \right) \right] + \left(\frac{12 \times 1220^3}{12} \right)$$

$$= 10248.30 \times 10^6 \text{ mm}^6$$

$$W_{el,YY,bottom} = \frac{I_{YY}}{z_{bottom}} = \frac{2839.47 \times 10^6}{330.56} = 8589.88 \times 10^3 \text{ mm}^3$$

$$W_{el,YY,top} = \frac{I_{YY}}{z_{top}} = \frac{2839.47 \times 10^6}{[532.0 - 330.56]} = 14095.86 \times 10^3 \text{ mm}^3$$

$$W_{el,ZZ,LHS} = \frac{I_{ZZ}}{y_{LHS}} = \frac{10248.3 \times 10^6}{710} = 14434.23 \times 10^3 \text{ mm}^3$$

$$W_{el,ZZ,RHS} = \frac{I_{ZZ}}{y_{RHS}} = \frac{I_{ZZ}}{y_{LHS}} = \frac{I_{YY}}{y_{LHS}} \text{ (vertical axis of symmetry)} = 14434.23 \times 10^3 \text{ mm}^3$$

Solution

Topic: Second Moments of Area and Elastic Section Modulii

Problem Number: 2.12   **Page No. 6**

Section dimensions for Problem 2.6:

$y_1 = [193.55 - (208.8/2)] = 89.15$ mm
$y_2 = [(208.8/2) + (9.6/2) + (449.8/2) - 193.55] = 140.55$ mm

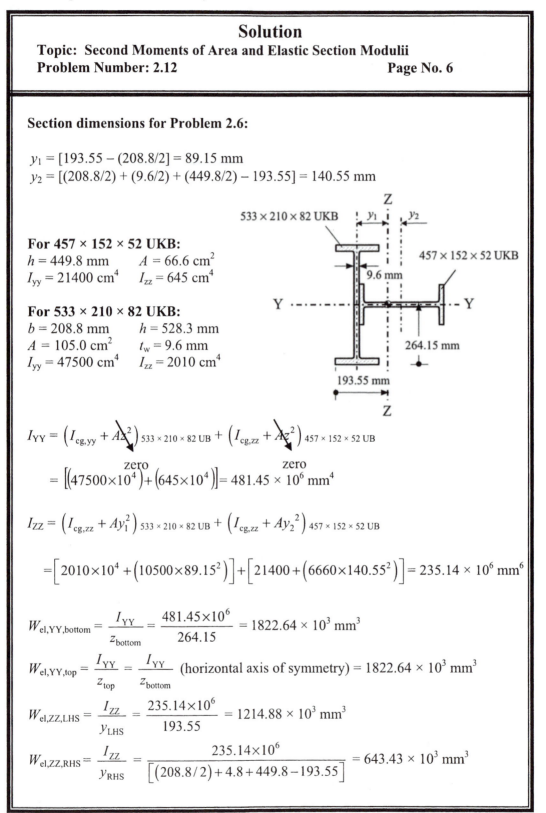

For 457 × 152 × 52 UKB:
$h = 449.8$ mm $A = 66.6$ cm^2
$I_{yy} = 21400$ cm^4 $I_{zz} = 645$ cm^4

For 533 × 210 × 82 UKB:
$b = 208.8$ mm $h = 528.3$ mm
$A = 105.0$ cm^2 $t_w = 9.6$ mm
$I_{yy} = 47500$ cm^4 $I_{zz} = 2010$ cm^4

$$I_{YY} = \left(I_{cg,yy} + A\!\!\diagdown\!\!z^2\right)_{533 \times 210 \times 82\,UB} + \left(I_{cg,zz} + A\!\!\diagdown\!\!z^2\right)_{457 \times 152 \times 52\,UB}$$

$$= \left[\left(47500 \times 10^4\right) + \left(645 \times 10^4\right)\right] = 481.45 \times 10^6 \text{ mm}^4$$

$$I_{ZZ} = \left(I_{cg,zz} + Ay_1^2\right)_{533 \times 210 \times 82\,UB} + \left(I_{cg,zz} + Ay_2^2\right)_{457 \times 152 \times 52\,UB}$$

$$= \left[2010 \times 10^4 + \left(10500 \times 89.15^2\right)\right] + \left[21400 + \left(6660 \times 140.55^2\right)\right] = 235.14 \times 10^6 \text{ mm}^6$$

$$W_{el,YY,bottom} = \frac{I_{YY}}{z_{bottom}} = \frac{481.45 \times 10^6}{264.15} = 1822.64 \times 10^3 \text{ mm}^3$$

$$W_{el,YY,top} = \frac{I_{YY}}{z_{top}} = \frac{I_{YY}}{z_{bottom}} \text{ (horizontal axis of symmetry)} = 1822.64 \times 10^3 \text{ mm}^3$$

$$W_{el,ZZ,LHS} = \frac{I_{ZZ}}{y_{LHS}} = \frac{235.14 \times 10^6}{193.55} = 1214.88 \times 10^3 \text{ mm}^3$$

$$W_{el,ZZ,RHS} = \frac{I_{ZZ}}{y_{RHS}} = \frac{235.14 \times 10^6}{\left[(208.8/2) + 4.8 + 449.8 - 193.55\right]} = 643.43 \times 10^3 \text{ mm}^3$$

2.3 *Plastic Cross-Section Properties*

When using elastic theory in design, the acceptance criterion can be based on "permissible" or "working" stresses. These are obtained by dividing the "yield stress" f_y of the material by a suitable factor of safety. The loads adopted to evaluate an actual working stress are "working loads".

In a structure fabricated from linearly elastic material, the factor of safety (F. of S.) can also be expressed in terms of the load required to produce yield stress and the working load. This is known as the Load Factor (λ).

$$\lambda = \frac{\text{Collapse load}}{\text{Working load}}$$

2.3.1 *Stress/Strain Relationship*

The plastic analysis and design of structures is based on collapse loads. A typical stress-strain curve for a ductile material having the characteristic of providing a large increase in strain beyond the yield point without any increase in stress, (e.g. steel) is given in Figure 2.39.

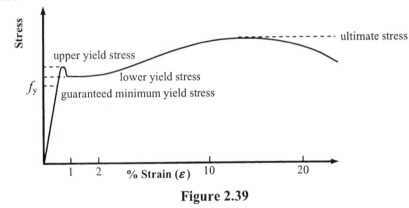

Figure 2.39

When adopting this curve for the theory of plasticity (see Chapter 8) it is idealised as indicated in Figure 2.40

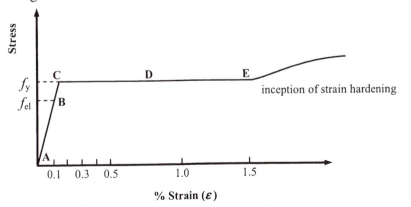

Figure 2.40

If a beam manufactured from material with a characteristic stress/strain curve as shown in Figure 2.39 has a rectangular cross section and is subjected to an increasing bending moment only, then the progression from elastic stress/strain distributions to plastic stress/strain distributions are as indicated in Figure 2.41.

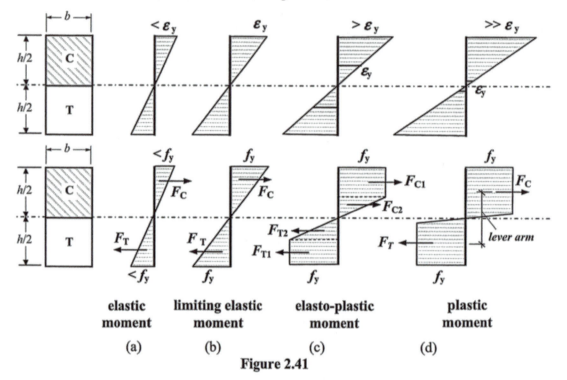

Figure 2.41

Initially at low values of applied moment (a) the maximum stress and strain values are less than the permissible working values as indicated in Figure 2.41 (i.e. between points A and B in Figure 2.40).

As the applied moment increases, then the stress and strain values increase until at stage (b), both attain the yield values ε_y and f_y. This corresponds to point C in Figure 2.40.

A further increase in the applied moment induces yield in some of the inner fibres of the material. Whilst the extreme fibre strains must now exceed ε_y, the stress must obviously remain at f_y. This corresponds to point D in Figure 2.40 and (c) in Figure 2.41.

As the applied moment increases still further, so the whole section eventually reaches the yield stress. (As indicated in (d) there is a very small region around the neutral axis which has not reached yield, but this can be ignored without any appreciable error). When the whole section has attained yield stress then the section cannot provide any further moment resistance and a **plastic hinge** is formed allowing the beam to rotate at the location of the beam. The value of the applied moment at which this occurs is known as the **Plastic Moment of Resistance** (M_{pl}).

2.3.2 Plastic Neutral Axis

At all stages of loading, the compression force (F_C) induced by the applied moment must equal the tension force (F_T). This being so, then at the formation of the **plastic hinge** where all the material is subjected to the same stress i.e. f_y, the **plastic neutral axis** must be that axis which equally divides the area into two separate parts, i.e.

The compression force $F_T = (A_C \times f_y)$ The tension force $F_T = (A_T \times f_y)$

where
A_C = Area in compression, A_T = Area in tension
f_y = yield stress

Force in compression = Force in tension
$F_C = F_T$
$(A_C \times f_y) = (A_T \times f_y)$
$\therefore A_C = A_T$ i.e.
The area of the cross-section in compression = The area of the cross-section in tension

In plastic analysis the neutral axis is the equal area axis.

2.3.3 Evaluation of Plastic Moment of Resistance and Plastic Section Modulus

In elastic analysis the limiting elastic moment can be expressed in terms of the yield stress and the *elastic* section modulus, at the limit of elasticity;

$M_{el} = (f_y \times W_{el})$ where W_{el} is the elastic section modulus

Similarly in plastic analysis, the plastic moment of resistance can be expressed in terms of the yield stress and the *plastic* section modulus.

$M_{pl} = (f_y \times W_{pl})$ where W_{pl} is the plastic section modulus

Consider the section shown in Figure 2.42.

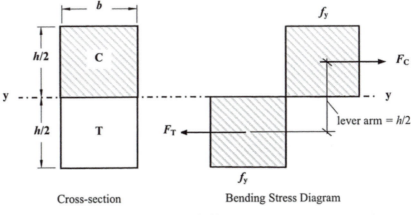

Cross-section   Bending Stress Diagram

Figure 2.42

If the rectangular section is subjected to a moment equal to the plastic moment of resistance M_{pl} of the section then we can determine a value for the plastic section modulus.

e.g. $M_{pl,yy} = f_y \times W_{pl,yy}$
$M_{pl,yy} = (F_C \times$ lever arm) or $[(F_T \times$ lever arm$)]$
$\therefore M_{pl,yy} =$ (stress $\times$ area $\times$ lever arm)
$= [f_y \times (b \times h/2) \times (h/2)] = f_y \, bh^2/4$

Hence for a rectangular section the Plastic Section Modulus $W_{pl,yy} = \dfrac{bh^2}{4}$

The Plastic Section Modulus $W_{pl} = 1^{st}$ moment of area about the equal area axis

2.3.4   *Shape Factor*

The ratio of the plastic modulus to the elastic modulus (or plastic moment to limiting elastic moment) is known as the **shape factor** given by the symbol ν.

For a rectangle $\nu = \dfrac{W_{pl}}{W_{el}} = \dfrac{bh^2/4}{bh^2/6} = 1.5;$ For most I sections $\nu \approx 1.15$

2.3.5 *Section Classification*

In design codes the compression elements of structural members are classified into four categories depending upon their resistance to local buckling effects which may influence their load carrying capacity. The compression may be due to direct axial forces, bending moments, or a combination of both. There are two distinct types of element in a cross-section identified in the code:

1. *Outstand elements*   – elements which are attached to an adjacent element at one edge only, the other edge being free, e.g. the flange of an **I**-section.

2. *Internal elements*   – elements which are attached to other elements on both longitudinal edges, including:
    – webs comprising the internal elements perpendicular to the axis of bending
    – flanges comprising the internal elements parallel to the axis of bending
    e.g. the webs and flanges of a rectangular hollow section.

The classifications specified in the Eurocode for structural steelwork (EN 1993-1-1) are:

- Class 1 $(M_{pl} = f_y \times W_{pl})$,
- Class 2 $(M_{pl} = f_y \times W_{pl})$,
- Class 3 $(M_{el} = f_y \times W_{el})$,
- Class 4 $(M_{eff} = f_y \times W_{effective})$.

Further explanation of local buckling and section classification is given in Chapter 6.

2.3.5.1   Aspect Ratio

The aspect ratio for various types of element can be determined using the variables indicated in the code for a wide range of cross-sections. A typical example is the hot–rolled **I**-section indicated in Figure 2.43.

Element	Aspect ratio
outstand of compression flange	c/t_f
web	c/t_w

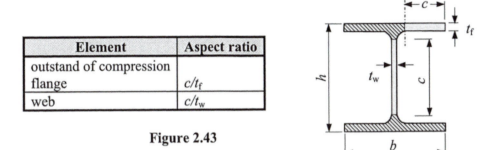

Figure 2.43

The limiting aspect ratios given must be modified to allow for the design strength f_y. This is done by multiplying each limiting ratio by ε which is defined as:

$$\varepsilon = \left(\frac{235}{f_y}\right)^2$$ in EN 1993-1-1. In the case of the web of a hybrid section ε should be based

on the design strength f_y of the flanges. In addition to ε, some limiting values also include parameters relating to stress ratios, these are not considered further here.

2.3.5.2   Type of Section

The type of section e.g. universal beam, universal column, circular hollow sections, welded tubes, hot finished rectangular hollow sections, cold formed rectangular hollow sections etc. also influences the classification.

The classifications given in codes indicate the moment/rotation characteristics of a section, as shown in Figure 2.44.

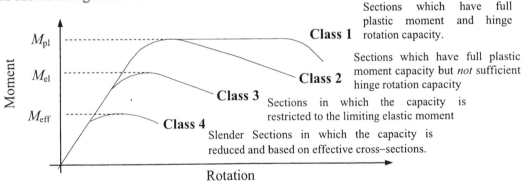

Class 1 Sections which have full plastic moment and hinge rotation capacity.

Class 2 Sections which have full plastic moment capacity but *not* sufficient hinge rotation capacity

Class 3 Sections in which the capacity is restricted to the limiting elastic moment

Class 4 Slender Sections in which the capacity is reduced and based on effective cross–sections.

Figure 2.44

where:
M_{pl} = the plastic moment of resistance,
M_{el} = the limiting elastic moment of resistance, (i.e the maximum stress = f_y).
M_{eff} = the elastic moment of resistance based on effective cross-section properties.

These characteristics determine whether or not a fully plastic moment can develop within a section and whether or not the section possesses sufficient rotational capacity to permit the section to be used in plastic design.

Consider a section subject to an increasing bending moment; the bending stress diagram changes from a linearly elastic condition with extreme fibre stresses less than the design strength (f_y), to one in which all of the fibres can be considered to have reached the design strength, as shown in Figure 2.45.

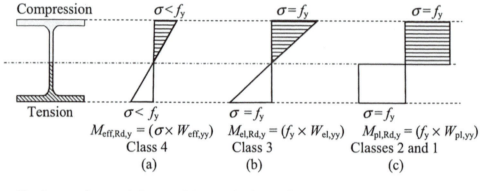

Compression $\sigma < f_y$	$\sigma = f_y$	$\sigma = f_y$
Tension　$\sigma < f_y$	$\sigma = f_y$	$\sigma = f_y$
$M_{eff,Rd,y} = (\sigma \times W_{eff,yy})$	$M_{el,Rd,y} = (f_y \times W_{el,yy})$	$M_{pl,Rd,y} = (f_y \times W_{pl,yy})$
Class 4	Class 3	Classes 2 and 1
(a)	(b)	(c)

where:
$W_{eff,yy}$ = effective section modulus;　$W_{el,yy}$ = elastic section modulus;
$W_{pl,yy}$ = plastic section modulus;　σ = bending stress;　f_y = design strength (yield stress)

Figure 2.45

2.4　*Example 2.1: Plastic Cross-section Properties – Section 1*

Determine the position of the plastic neutral axis $\overline{z}_{plastic}$, the plastic section modulus $W_{pl,yy}$ and the shape factor υ for the welded section indicated in Figure 2.46.

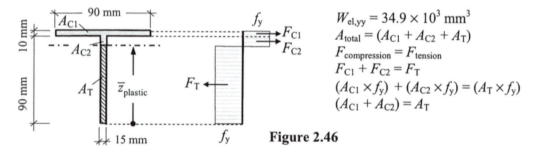

$W_{el,yy} = 34.9 \times 10^3$ mm³
$A_{total} = (A_{C1} + A_{C2} + A_T)$
$F_{compression} = F_{tension}$
$F_{C1} + F_{C2} = F_T$
$(A_{C1} \times f_y) + (A_{C2} \times f_y) = (A_T \times f_y)$
$(A_{C1} + A_{C2}) = A_T$

Figure 2.46

(i)　Position of plastic neutral axis $\left(\overline{z}_{plastic} \right)$
$A = [(90 \times 10) + (90 \times 15)] = 2250$ mm²　$A/2 = (2250/2) = 1125$ mm²

For equal area axis:
$\overline{z}_{plastic} = 1125/15 = 75$ mm

(ii)　Plastic section modulus: (1st moment of area about the plastic neutral axis)
$W_{pl,yy} = [(90 \times 10) \times 20] + [(15 \times 15) \times 7.5] + [(75 \times 15) \times 37.5] = 61.875 \times 10^3$ mm³

(iii)　Shape factor $\upsilon = \dfrac{W_{pl,yy}}{W_{el,yy}} = \left[\dfrac{61.875 \times 10^3}{34.9 \times 10^3} \right] = 1.77$

2.5   Problems: Plastic Cross-section Properties

Determine the following values for the welded sections indicated in Problems 2.13 to 2.16,

(i)   position of the plastic neutral axis $\overline{z}_{\text{plastic}}$,

(ii)   the plastic section modulus $W_{\text{pl,yy}}$ and

(iii)   the shape factor υ.

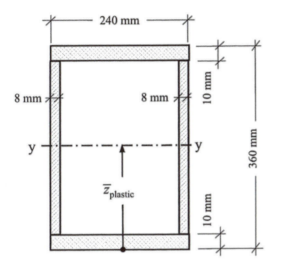

Problem 2.13

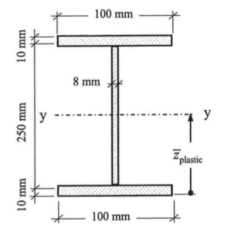

Problem 2.14

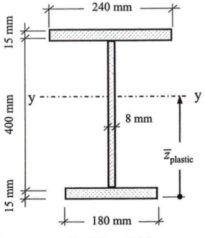

Problem 2.15

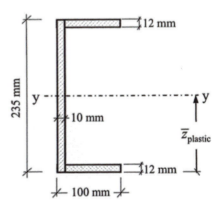

Problem 2.16

2.6   *Solutions: Plastic Cross-section Properties*

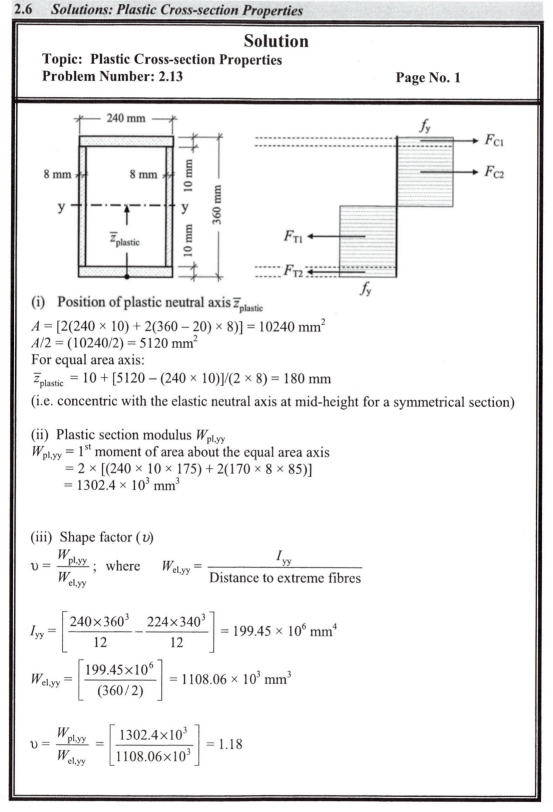

Solution
Topic: Plastic Cross-section Properties
Problem Number: 2.13   **Page No. 1**

(i)   Position of plastic neutral axis $\bar{z}_{plastic}$
$A = [2(240 \times 10) + 2(360 - 20) \times 8)] = 10240 \text{ mm}^2$
$A/2 = (10240/2) = 5120 \text{ mm}^2$
For equal area axis:
$\bar{z}_{plastic} = 10 + [5120 - (240 \times 10)]/(2 \times 8) = 180 \text{ mm}$

(i.e. concentric with the elastic neutral axis at mid-height for a symmetrical section)

(ii) Plastic section modulus $W_{pl,yy}$
$W_{pl,yy} = 1^{st}$ moment of area about the equal area axis
$= 2 \times [(240 \times 10 \times 175) + 2(170 \times 8 \times 85)]$
$= 1302.4 \times 10^3 \text{ mm}^3$

(iii) Shape factor (υ)
$\upsilon = \dfrac{W_{pl,yy}}{W_{el,yy}}$; where $W_{el,yy} = \dfrac{I_{yy}}{\text{Distance to extreme fibres}}$

$I_{yy} = \left[\dfrac{240 \times 360^3}{12} - \dfrac{224 \times 340^3}{12} \right] = 199.45 \times 10^6 \text{ mm}^4$

$W_{el,yy} = \left[\dfrac{199.45 \times 10^6}{(360/2)} \right] = 1108.06 \times 10^3 \text{ mm}^3$

$\upsilon = \dfrac{W_{pl,yy}}{W_{el,yy}} = \left[\dfrac{1302.4 \times 10^3}{1108.06 \times 10^3} \right] = 1.18$

Solution

Topic: Plastic Cross-section Properties
Problem Number: 2.14   **Page No. 1**

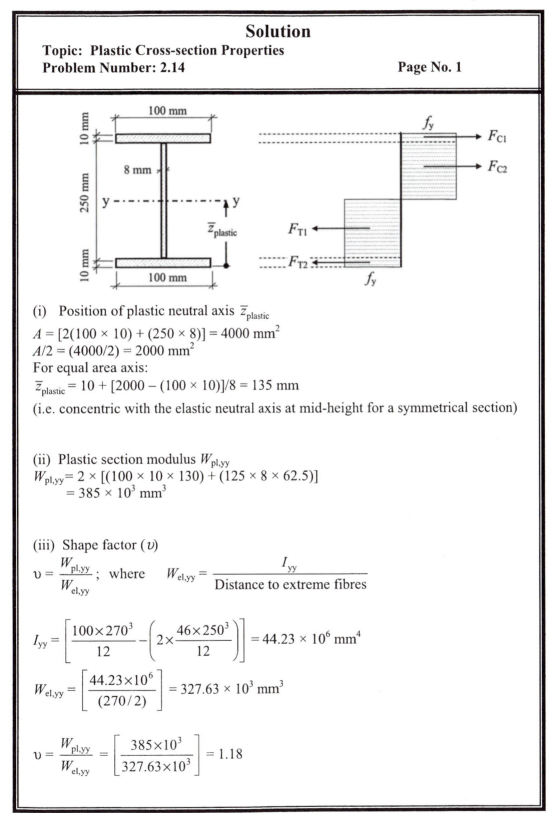

(i)   Position of plastic neutral axis $\overline{z}_{plastic}$

$A = [2(100 \times 10) + (250 \times 8)] = 4000$ mm^2
$A/2 = (4000/2) = 2000$ mm^2
For equal area axis:
$\overline{z}_{plastic} = 10 + [2000 - (100 \times 10)]/8 = 135$ mm

(i.e. concentric with the elastic neutral axis at mid-height for a symmetrical section)

(ii) Plastic section modulus $W_{pl,yy}$
$W_{pl,yy} = 2 \times [(100 \times 10 \times 130) + (125 \times 8 \times 62.5)]$
$\qquad = 385 \times 10^3$ mm^3

(iii) Shape factor (υ)

$\upsilon = \dfrac{W_{pl,yy}}{W_{el,yy}}$;   where $W_{el,yy} = \dfrac{I_{yy}}{\text{Distance to extreme fibres}}$

$I_{yy} = \left[\dfrac{100 \times 270^3}{12} - \left(2 \times \dfrac{46 \times 250^3}{12} \right) \right] = 44.23 \times 10^6$ mm^4

$W_{el,yy} = \left[\dfrac{44.23 \times 10^6}{(270/2)} \right] = 327.63 \times 10^3$ mm^3

$\upsilon = \dfrac{W_{pl,yy}}{W_{el,yy}} = \left[\dfrac{385 \times 10^3}{327.63 \times 10^3} \right] = 1.18$

Solution
Topic: Plastic Cross-section Properties
Problem Number: 2.15   **Page No. 1**

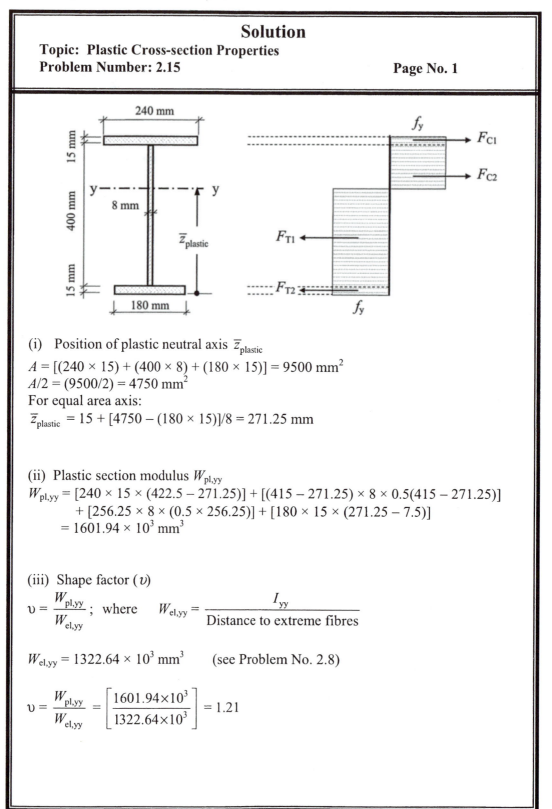

(i)   Position of plastic neutral axis $\bar{z}_{plastic}$

$A = [(240 \times 15) + (400 \times 8) + (180 \times 15)] = 9500 \text{ mm}^2$
$A/2 = (9500/2) = 4750 \text{ mm}^2$
For equal area axis:
$\bar{z}_{plastic} = 15 + [4750 - (180 \times 15)]/8 = 271.25 \text{ mm}$

(ii) Plastic section modulus $W_{pl,yy}$
$W_{pl,yy} = [240 \times 15 \times (422.5 - 271.25)] + [(415 - 271.25) \times 8 \times 0.5(415 - 271.25)]$
$\qquad + [256.25 \times 8 \times (0.5 \times 256.25)] + [180 \times 15 \times (271.25 - 7.5)]$
$\qquad = 1601.94 \times 10^3 \text{ mm}^3$

(iii) Shape factor (υ)

$\upsilon = \dfrac{W_{pl,yy}}{W_{el,yy}}$;   where $W_{el,yy} = \dfrac{I_{yy}}{\text{Distance to extreme fibres}}$

$W_{el,yy} = 1322.64 \times 10^3 \text{ mm}^3$ (see Problem No. 2.8)

$\upsilon = \dfrac{W_{pl,yy}}{W_{el,yy}} = \left[\dfrac{1601.94 \times 10^3}{1322.64 \times 10^3} \right] = 1.21$

Solution

Topic: Plastic Cross-section Properties
Problem Number: 2.16

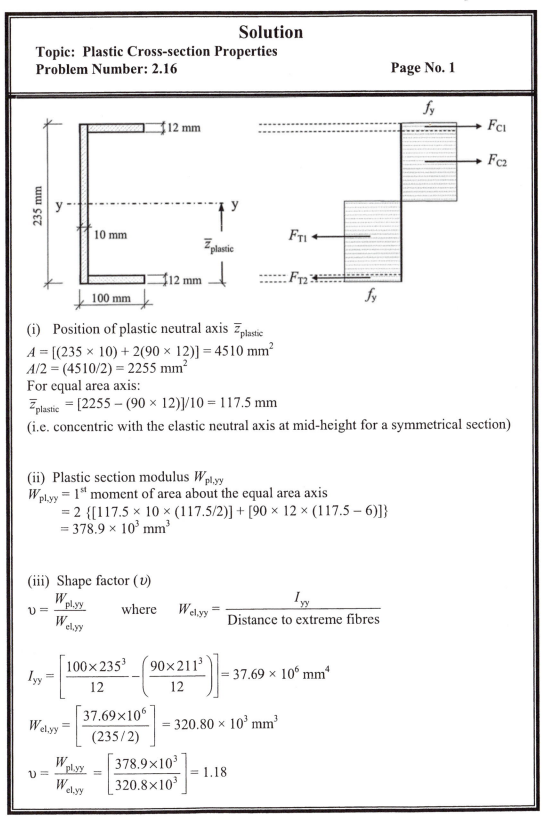

(i) Position of plastic neutral axis $\bar{z}_{\text{plastic}}$

$A = [(235 \times 10) + 2(90 \times 12)] = 4510 \text{ mm}^2$
$A/2 = (4510/2) = 2255 \text{ mm}^2$
For equal area axis:
$\bar{z}_{\text{plastic}} = [2255 - (90 \times 12)]/10 = 117.5 \text{ mm}$

(i.e. concentric with the elastic neutral axis at mid-height for a symmetrical section)

(ii) Plastic section modulus $W_{\text{pl,yy}}$

$W_{\text{pl,yy}}$ = 1st moment of area about the equal area axis
$\qquad = 2 \{[117.5 \times 10 \times (117.5/2)] + [90 \times 12 \times (117.5 - 6)]\}$
$\qquad = 378.9 \times 10^3 \text{ mm}^3$

(iii) Shape factor (υ)

$$\upsilon = \frac{W_{\text{pl,yy}}}{W_{\text{el,yy}}} \qquad \text{where} \qquad W_{\text{el,yy}} = \frac{I_{\text{yy}}}{\text{Distance to extreme fibres}}$$

$$I_{\text{yy}} = \left[\frac{100 \times 235^3}{12} - \left(\frac{90 \times 211^3}{12} \right) \right] = 37.69 \times 10^6 \text{ mm}^4$$

$$W_{\text{el,yy}} = \left[\frac{37.69 \times 10^6}{(235/2)} \right] = 320.80 \times 10^3 \text{ mm}^3$$

$$\upsilon = \frac{W_{\text{pl,yy}}}{W_{\text{el,yy}}} = \left[\frac{378.9 \times 10^3}{320.8 \times 10^3} \right] = 1.18$$

3. Pin-Jointed Frames

3.1 *Introduction*

The use of beams/plate-girders does not always provide the most economic or suitable structural solution when spanning large openings. In buildings which have lightly loaded, long span roofs, when large voids are required within the depth of roof structures for services, when plated structures are impractical, or for aesthetic/architectural reasons, the use of roof trusses, lattice girders or space-frames may be more appropriate.

Such trusses/girders/frames, generally, transfer their loads by inducing axial tension or compressive forces in the individual members. The magnitude and sense of these forces can be determined using standard methods of analysis such as '*the method of sections*', '*the method* of *joint-resolution*', '*the method of tension coefficients*' or the use of '*computer software*'. The first three methods indicated are summarized and illustrated in this Chapter.

3.2 *Method of Sections*

The *method of sections* involves the application of the three equations of static equilibrium to two-dimensional plane frames. The sign convention adopted to indicate ties (i.e. tension members) and struts (i.e. compression members) in frames is as shown in Figure 3.1.

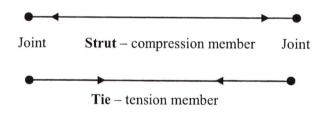

Figure 3.1

The method involves considering an imaginary section line which cuts the frame under consideration into two parts A and B as shown in Figure 3.4.

Since only three independent equations of equilibrium are available any section taken through a frame must not include more than three members for which the internal force is unknown.

Consideration of the equilibrium of the resulting force system enables the magnitude and sense (i.e. compression or tension) of the forces in the cut members to be determined.

3.2.1 *Example 3.1: Pin-Jointed Truss*

A pin-jointed truss supported by a pinned support at A and a roller support at G carries three loads at joints C, D and E as shown in Figure 3.2. Determine the magnitude and sense of the forces induced in members X, Y and Z as indicated.

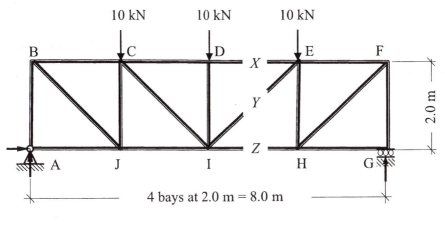

Figure 3.2

Step 1: Evaluate the support reactions. It is not necessary to know any information regarding the frame members at this stage other than dimensions as shown in Figure 3.3, since only **externally** applied loads and reactions are involved.

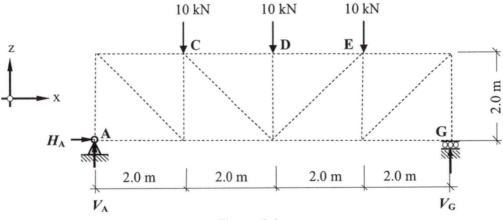

Figure 3.3

Apply the three equations of static equilibrium to the force system:

$+ve \uparrow \Sigma F_z = 0 \qquad V_A - (10 + 10 + 10) + V_G = 0 \qquad V_A + V_G = 30 \text{ kN}$

$+ve \longrightarrow \Sigma F_x = 0 \qquad\qquad\qquad\qquad\qquad\qquad\qquad\qquad \therefore H_A = 0$

$+ve \searrow \Sigma M_A = 0 \qquad (10 \times 2.0) + (10 \times 4.0) + (10 \times 6.0) - (V_G \times 8.0) = 0$

$$\therefore V_G = \mathbf{15 \text{ kN}} \uparrow$$

$$\text{Hence} \quad V_A = \mathbf{15 \text{ kN}} \uparrow$$

Step 2: Select a section through which the frame can be considered to be cut and using the same three equations of equilibrium determine the magnitude and sense of the unknown forces (i.e. the internal forces in the cut members).

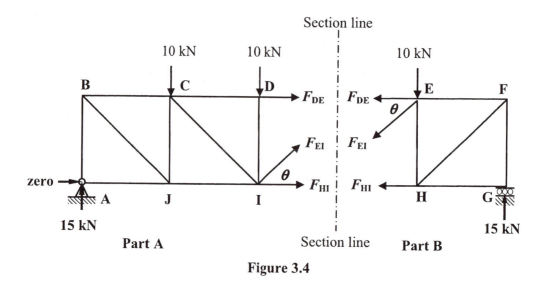

Figure 3.4

It is convenient to **assume** all unknown forces to be tensile and hence at the cut section their direction and lines of action are considered to be pointing away from the joints (refer to Figure 3.4). If the answer results in a negative force this means that the assumption of a tie was incorrect and the member is actually in compression, i.e. a strut.

The application of the equations of equilibrium to either part of the cut frame will enable the forces X (F_{DE}), Y (F_{EI}) and Z (F_{HI}) to be evaluated.

Note: the section considered must not cut through more than three members with unknown internal forces since only three equations of equilibrium are available.

Consider Part A:

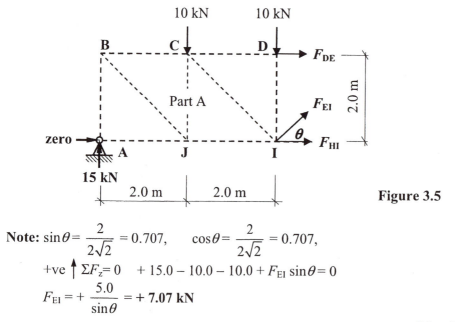

Figure 3.5

Note: $\sin\theta = \dfrac{2}{2\sqrt{2}} = 0.707,\qquad \cos\theta = \dfrac{2}{2\sqrt{2}} = 0.707,$

+ve $\uparrow$ $\Sigma F_z = 0$ $+15.0 - 10.0 - 10.0 + F_{EI}\sin\theta = 0$

$F_{EI} = +\dfrac{5.0}{\sin\theta} = +7.07$ **kN**

Member EI is a tie

$$+ve \longrightarrow \Sigma F_x = 0 \quad +F_{DE} + F_{HI} + F_{EI}\cos\theta = 0$$

$$+ve \; \rotatebox{-30}{$\curvearrowleft$} \; \Sigma M_I = 0 \quad + (15.0 \times 4.0) - (10.0 \times 2.0) + (F_{DE} \times 2.0) = 0$$

$$F_{DE} = -\textbf{20.0 kN}$$

Member DE is a strut

hence $F_{HI} = -F_{DE} - F_{EI}\cos\theta = -(-20.0) - (7.07 \times \cos\theta) = + \textbf{15.0 kN}$

Member HI is a tie

These answers can be confirmed by considering Part B of the structure and applying the equations as above.

3.3   *Method of Joint Resolution*

Considering the same frame using *joint resolution* highlights the advantage of the method of sections when only a few member forces are required.

In this technique (which can be considered as a special case of the method of sections), sections are taken which isolate each individual joint in turn in the frame, e.g.

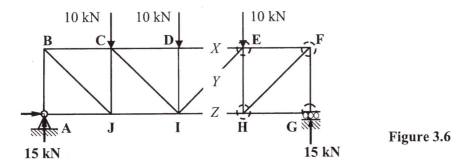

Figure 3.6

In Figure 3.6 four sections are shown, each of which isolates a joint in the structure as indicated in Figure 3.7.

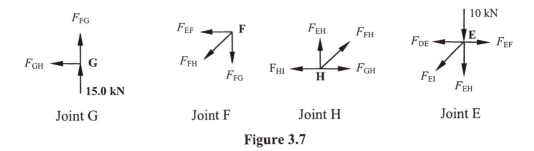

| Joint G | Joint F | Joint H | Joint E |

Figure 3.7

Since in each case the forces are coincident, the moment equation is of no value, hence only two independent equations are available. It is necessary when considering the equilibrium of each joint to do so in a sequence which ensures that there are no more than two unknown member forces in the joint under consideration. This can be carried out until all member forces in the structure have been determined.

Consider Joint G:

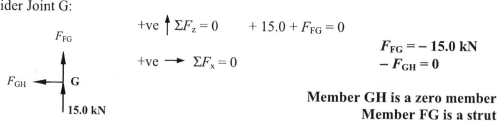

$$+ve \uparrow \Sigma F_z = 0 \qquad + 15.0 + F_{FG} = 0$$

$$F_{FG} = -15.0 \text{ kN}$$

$$+ve \longrightarrow \Sigma F_x = 0 \qquad\qquad - F_{GH} = 0$$

Member GH is a zero member
Member FG is a strut

Consider Joint F: substitute for calculated values, i.e. F_{FG} (direction of force is into the joint)

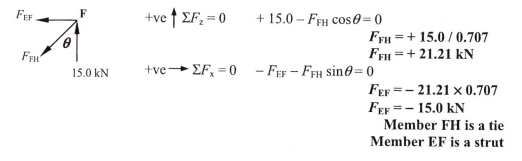

$$+ve \uparrow \Sigma F_z = 0 \qquad + 15.0 - F_{FH} \cos\theta = 0$$

$$F_{FH} = + 15.0 / 0.707$$
$$\boldsymbol{F_{FH} = + 21.21 \text{ kN}}$$

$$+ve \longrightarrow \Sigma F_x = 0 \quad - F_{EF} - F_{FH} \sin\theta = 0$$

$$F_{EF} = -21.21 \times 0.707$$
$$\boldsymbol{F_{EF} = -15.0 \text{ kN}}$$
Member FH is a tie
Member EF is a strut

Consider Joint H: substitute for calculated values, i.e. F_{GH} and F_{FH}

F_{EH}

21.21 kN

F_{HI} ← H → 0

$$+ve \uparrow \Sigma F_z = 0 \qquad + F_{EH} + 21.21 \sin\theta = 0$$
$$F_{EH} = -21.21 \times 0.707$$
$$\boldsymbol{F_{EH} = -15.0 \text{ kN}}$$

$$+ve \longrightarrow \Sigma F_x = 0 \quad - F_{HI} + 21.21 \cos\theta = 0$$
$$F_{HI} = + 21.21 \times 0.707$$
$$\boldsymbol{F_{HI} = + 15.0 \text{ kN}}$$
Member EH is a strut
Member HI is a tie

Consider Joint E: substitute for calculated values, i.e. F_{EF} and F_{EH}

10 kN

F_{DE} ← E ← 15 kN

θ

F_{EI}

15 kN

$$+ve \uparrow \Sigma F_z = 0 \qquad +15.0 - 10.0 - F_{EI} \cos\theta = 0$$
$$F_{EI} = + 5.0 / 0.707$$
$$\boldsymbol{F_{EI} = + 7.07 \text{ kN}}$$

$$+ve \longrightarrow \Sigma F_x = 0 \quad - F_{DE} - 15.0 - F_{EI} \sin\theta = 0$$
$$\boldsymbol{F_{DE} = -20.0 \text{ kN}}$$
Member EI is a tie
Member DE is a strut

3.3.1 Problems: Method of Sections and Joint Resolution

Determine the support reactions and the forces in the members of the pin-jointed frames indicated by the '*' in Problems 3.1 to 3.4 using the **method of sections**.

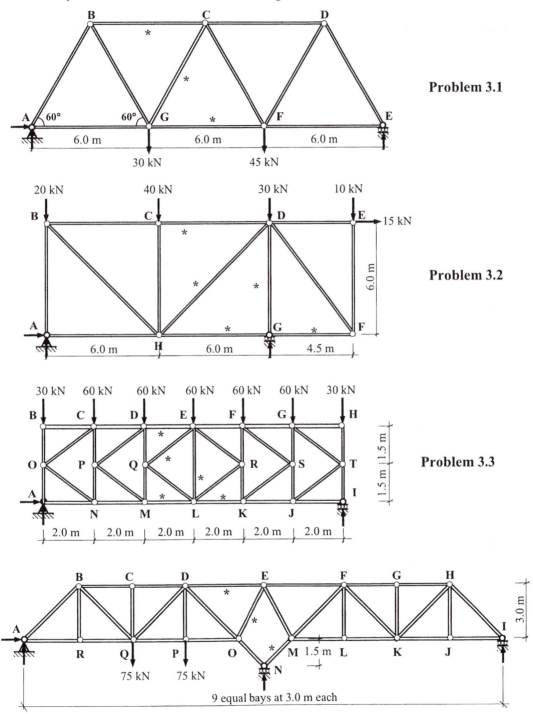

Problem 3.1

Problem 3.2

Problem 3.3

Problem 3.4

Determine the support reactions and the forces in the members of the pin-jointed frames indicated in Problems 3.5 to 3.10 using the **method of joint resolution**.

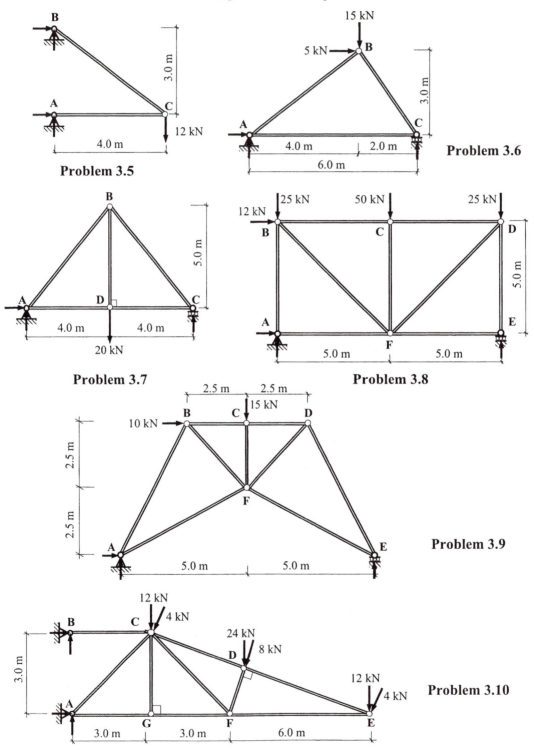

Problem 3.5

Problem 3.6

Problem 3.7

Problem 3.8

Problem 3.9

Problem 3.10

3.3.2 Solutions: Method of Sections and Joint Resolution

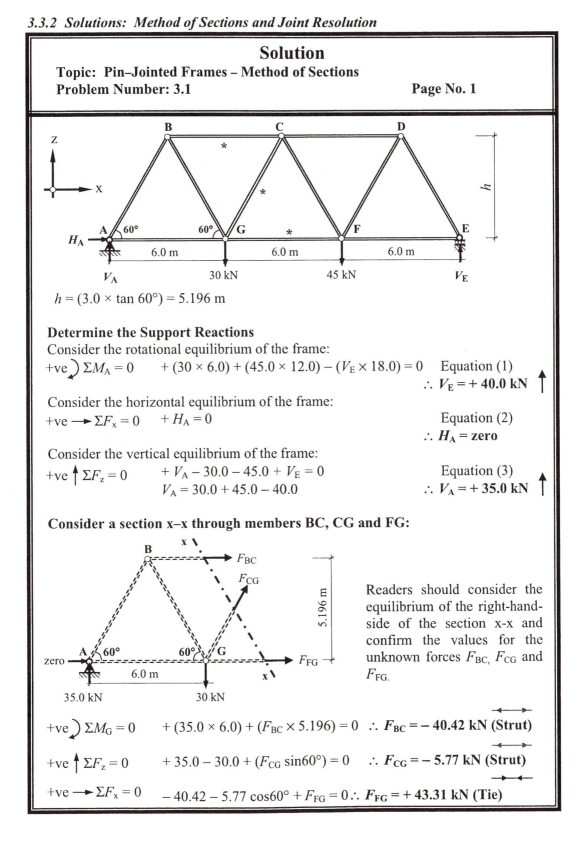

Solution

Topic: Pin–Jointed Frames – Method of Sections

Problem Number: 3.1   **Page No. 1**

$h = (3.0 \times \tan 60°) = 5.196$ m

Determine the Support Reactions

Consider the rotational equilibrium of the frame:

+ve $\curvearrowright$ $\Sigma M_A = 0$ $+ (30 \times 6.0) + (45.0 \times 12.0) - (V_E \times 18.0) = 0$ Equation (1)

$\therefore V_E = + \textbf{40.0 kN} \uparrow$

Consider the horizontal equilibrium of the frame:

+ve $\longrightarrow \Sigma F_x = 0$ $+ H_A = 0$ Equation (2)

$\therefore H_A = \textbf{zero}$

Consider the vertical equilibrium of the frame:

+ve $\uparrow \Sigma F_z = 0$ $+ V_A - 30.0 - 45.0 + V_E = 0$ Equation (3)

$V_A = 30.0 + 45.0 - 40.0$ $\therefore V_A = + \textbf{35.0 kN} \uparrow$

Consider a section x–x through members BC, CG and FG:

Readers should consider the equilibrium of the right-hand-side of the section x-x and confirm the values for the unknown forces F_{BC}, F_{CG} and F_{FG}.

+ve $\curvearrowright \Sigma M_G = 0$ $+ (35.0 \times 6.0) + (F_{BC} \times 5.196) = 0$ $\therefore F_{BC} = - \textbf{40.42 kN (Strut)}$

+ve $\uparrow \Sigma F_z = 0$ $+ 35.0 - 30.0 + (F_{CG} \sin 60°) = 0$ $\therefore F_{CG} = - \textbf{5.77 kN (Strut)}$

+ve $\longrightarrow \Sigma F_x = 0$ $- 40.42 - 5.77 \cos 60° + F_{FG} = 0 \therefore F_{FG} = + \textbf{43.31 kN (Tie)}$

Solution

Topic: Pin-Jointed Frames – Method of Sections

Problem Number: 3.2   **Page No. 1**

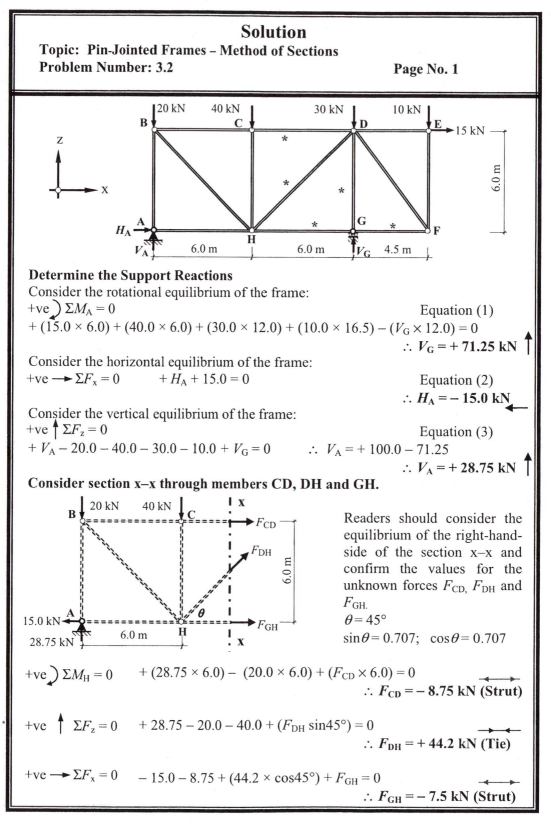

Determine the Support Reactions

Consider the rotational equilibrium of the frame:

$+ve \Big)\ \Sigma M_A = 0$ Equation (1)

$+ (15.0 \times 6.0) + (40.0 \times 6.0) + (30.0 \times 12.0) + (10.0 \times 16.5) - (V_G \times 12.0) = 0$

$$\therefore V_G = +\,\textbf{71.25 kN} \uparrow$$

Consider the horizontal equilibrium of the frame:

$+ve \longrightarrow \Sigma F_x = 0 \qquad + H_A + 15.0 = 0$ Equation (2)

$$\therefore H_A = -\,\textbf{15.0 kN} \longleftarrow$$

Consider the vertical equilibrium of the frame:

$+ve \uparrow \Sigma F_z = 0$ Equation (3)

$+ V_A - 20.0 - 40.0 - 30.0 - 10.0 + V_G = 0 \qquad \therefore V_A = +\,100.0 - 71.25$

$$\therefore V_A = +\,\textbf{28.75 kN} \uparrow$$

Consider section x–x through members CD, DH and GH.

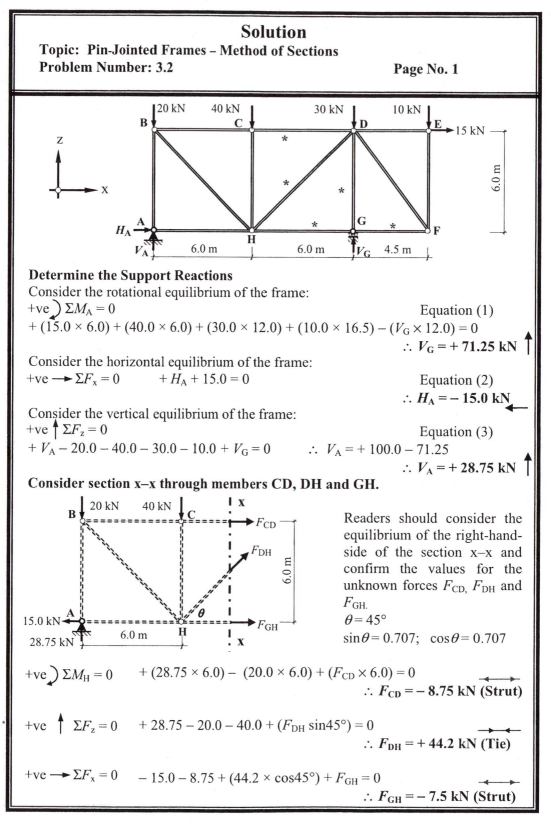

Readers should consider the equilibrium of the right-hand-side of the section x–x and confirm the values for the unknown forces F_{CD}, F_{DH} and F_{GH}.

$\theta = 45°$

$\sin\theta = 0.707; \quad \cos\theta = 0.707$

$+ve \Big)\ \Sigma M_H = 0 \qquad + (28.75 \times 6.0) - (20.0 \times 6.0) + (F_{CD} \times 6.0) = 0$

$$\therefore F_{CD} = -\,\textbf{8.75 kN (Strut)}$$

$+ve \uparrow \Sigma F_z = 0 \qquad + 28.75 - 20.0 - 40.0 + (F_{DH} \sin 45°) = 0$

$$\therefore F_{DH} = +\,\textbf{44.2 kN (Tie)}$$

$+ve \longrightarrow \Sigma F_x = 0 \qquad - 15.0 - 8.75 + (44.2 \times \cos 45°) + F_{GH} = 0$

$$\therefore F_{GH} = -\,\textbf{7.5 kN (Strut)}$$

Solution

Topic: Pin-Jointed Frames – Method of Sections

Problem Number: 3.2   **Page No. 2**

Consider section y–y through members CD, DH, DG and FG.

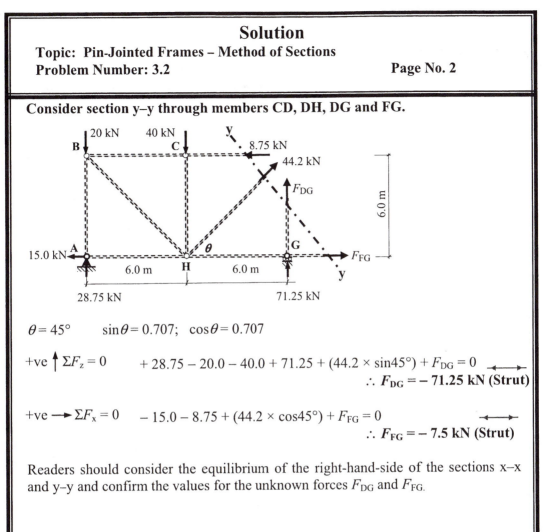

$\theta = 45°$ $\sin\theta = 0.707$; $\cos\theta = 0.707$

$+ve \uparrow \Sigma F_z = 0$ $+ 28.75 - 20.0 - 40.0 + 71.25 + (44.2 \times \sin45°) + F_{DG} = 0$

$$\therefore F_{DG} = -71.25 \text{ kN (Strut)}$$

$+ve \longrightarrow \Sigma F_x = 0$ $- 15.0 - 8.75 + (44.2 \times \cos45°) + F_{FG} = 0$

$$\therefore F_{FG} = -7.5 \text{ kN (Strut)}$$

Readers should consider the equilibrium of the right-hand-side of the sections x–x and y–y and confirm the values for the unknown forces F_{DG} and F_{FG}.

Solution
Topic: Pin-Jointed Frames – Method of Sections
Problem Number: 3.3   **Page No. 1**

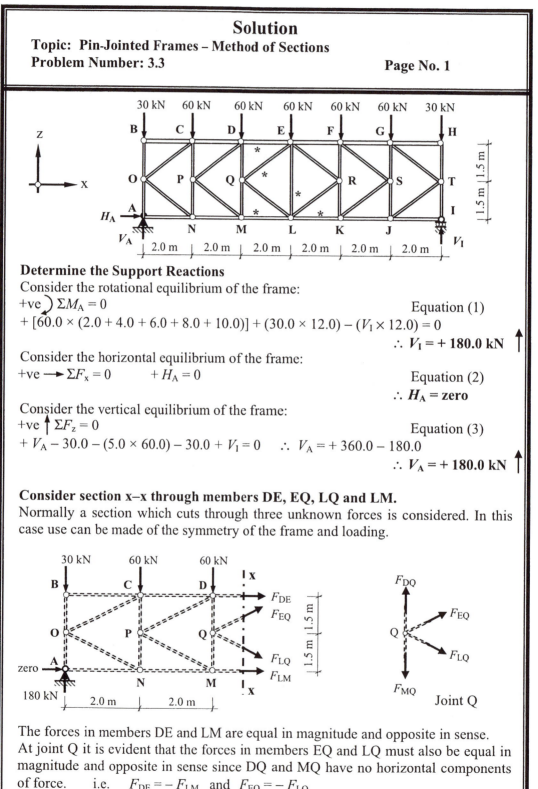

Determine the Support Reactions

Consider the rotational equilibrium of the frame:

+ve $\curvearrowright \Sigma M_A = 0$ Equation (1)

$+ [60.0 \times (2.0 + 4.0 + 6.0 + 8.0 + 10.0)] + (30.0 \times 12.0) - (V_I \times 12.0) = 0$

$\therefore V_I = + 180.0$ kN $\uparrow$

Consider the horizontal equilibrium of the frame:

+ve $\longrightarrow \Sigma F_x = 0$ $+ H_A = 0$ Equation (2)

$\therefore H_A =$ **zero**

Consider the vertical equilibrium of the frame:

+ve $\uparrow \Sigma F_z = 0$ Equation (3)

$+ V_A - 30.0 - (5.0 \times 60.0) - 30.0 + V_I = 0$ $\therefore V_A = + 360.0 - 180.0$

$\therefore V_A = + 180.0$ kN $\uparrow$

Consider section x–x through members DE, EQ, LQ and LM.

Normally a section which cuts through three unknown forces is considered. In this case use can be made of the symmetry of the frame and loading.

The forces in members DE and LM are equal in magnitude and opposite in sense. At joint Q it is evident that the forces in members EQ and LQ must also be equal in magnitude and opposite in sense since DQ and MQ have no horizontal components of force.   i.e. $F_{DE} = - F_{LM}$ and $F_{EQ} = - F_{LQ}$

Solution
Topic: Pin-Jointed Frames – Method of Sections
Problem Number: 3.3   **Page No. 2**

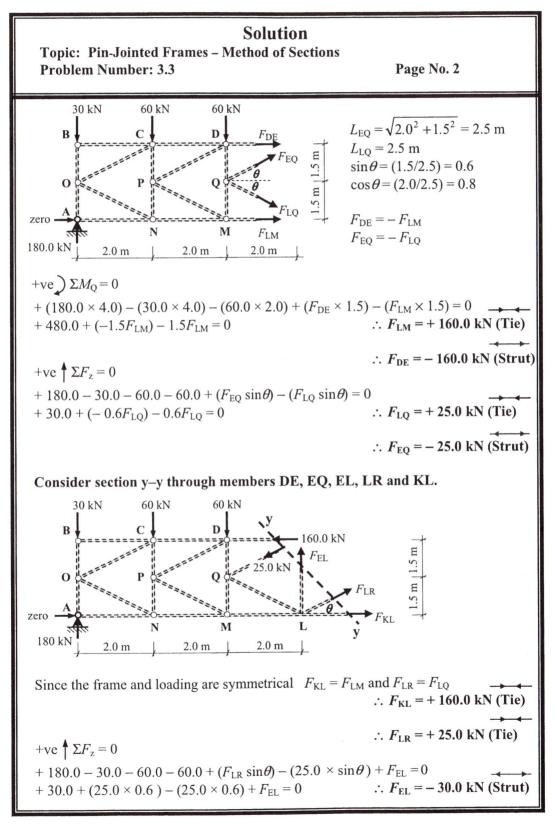

$L_{EQ} = \sqrt{2.0^2 + 1.5^2} = 2.5$ m

$L_{LQ} = 2.5$ m

$\sin\theta = (1.5/2.5) = 0.6$

$\cos\theta = (2.0/2.5) = 0.8$

$F_{DE} = -F_{LM}$

$F_{EQ} = -F_{LQ}$

+ve $\curvearrowright$ $\Sigma M_Q = 0$

$+ (180.0 \times 4.0) - (30.0 \times 4.0) - (60.0 \times 2.0) + (F_{DE} \times 1.5) - (F_{LM} \times 1.5) = 0$

$+ 480.0 + (-1.5 F_{LM}) - 1.5 F_{LM} = 0$ $\therefore F_{LM} = +160.0$ kN (Tie)

$\therefore F_{DE} = -160.0$ kN (Strut)

+ve $\uparrow$ $\Sigma F_z = 0$

$+ 180.0 - 30.0 - 60.0 - 60.0 + (F_{EQ} \sin\theta) - (F_{LQ} \sin\theta) = 0$

$+ 30.0 + (-0.6 F_{LQ}) - 0.6 F_{LQ} = 0$ $\therefore F_{LQ} = +25.0$ kN (Tie)

$\therefore F_{EQ} = -25.0$ kN (Strut)

Consider section y–y through members DE, EQ, EL, LR and KL.

Since the frame and loading are symmetrical $F_{KL} = F_{LM}$ and $F_{LR} = F_{LQ}$

$\therefore F_{KL} = +160.0$ kN (Tie)

$\therefore F_{LR} = +25.0$ kN (Tie)

+ve $\uparrow$ $\Sigma F_z = 0$

$+ 180.0 - 30.0 - 60.0 - 60.0 + (F_{LR} \sin\theta) - (25.0 \times \sin\theta) + F_{EL} = 0$

$+ 30.0 + (25.0 \times 0.6) - (25.0 \times 0.6) + F_{EL} = 0$ $\therefore F_{EL} = -30.0$ kN (Strut)

Solution

Topic: Pin-Jointed Frames – Method of Sections

Problem Number: 3.4

Page No. 1

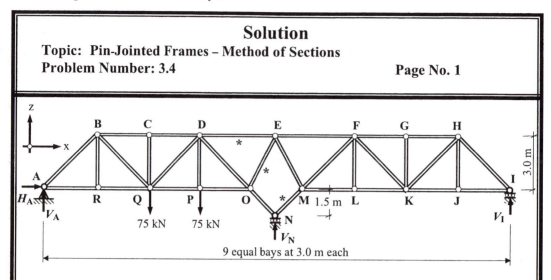

9 equal bays at 3.0 m each

This frame is similar to the frame given in Chapter 1: Figure 1.21 comprising two statically determinate frames.

There are four unknown reactions, however in addition to the three equations of static equilibrium, at support N the magnitude of the forces in members MN and NO are equal. (**Note:** the horizontal components must balance each other). This provides an additional equation which can be used to solve the problem.

Determine the Support Reactions

Consider the rotational equilibrium of the frame:

+ve $\curvearrowright \Sigma M_I = 0$

$+ (27.0 \times V_A) - (75.0 \times 21.0) - (75.0 \times 18.0) + (V_N \times 13.5) = 0$ Equation (1)

$+ 27.0V_A - 2925.0 + 13.5V_N$ $\therefore V_A = + 108.33 - 0.5V_N$

Consider the horizontal equilibrium of the frame:

+ve $\longrightarrow \Sigma F_x = 0$ $+ H_A = 0$ Equation (2)

$\therefore \boldsymbol{H_A = \text{zero}}$

Consider the vertical equilibrium of the frame:

+ve $\uparrow \Sigma F_z = 0$

$+ V_A - 75.0 - 75.0 + V_N + V_I = 0$ Equation (3)

$\therefore V_I = + 150.0 - V_A - V_N$

Consider section x–x at support N

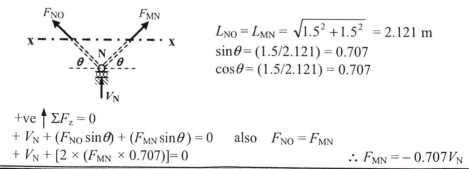

$L_{NO} = L_{MN} = \sqrt{1.5^2 + 1.5^2} = 2.121$ m

$\sin\theta = (1.5/2.121) = 0.707$

$\cos\theta = (1.5/2.121) = 0.707$

+ve $\uparrow \Sigma F_z = 0$

$+ V_N + (F_{NO} \sin\theta) + (F_{MN} \sin\theta) = 0$ also $F_{NO} = F_{MN}$

$+ V_N + [2 \times (F_{MN} \times 0.707)] = 0$ $\therefore F_{MN} = -0.707V_N$

Solution
Topic: **Pin-Jointed Frames – Method of Sections**
Problem Number: **3.4**   **Page No. 2**

Consider section y–y through members DE, EO and MN.

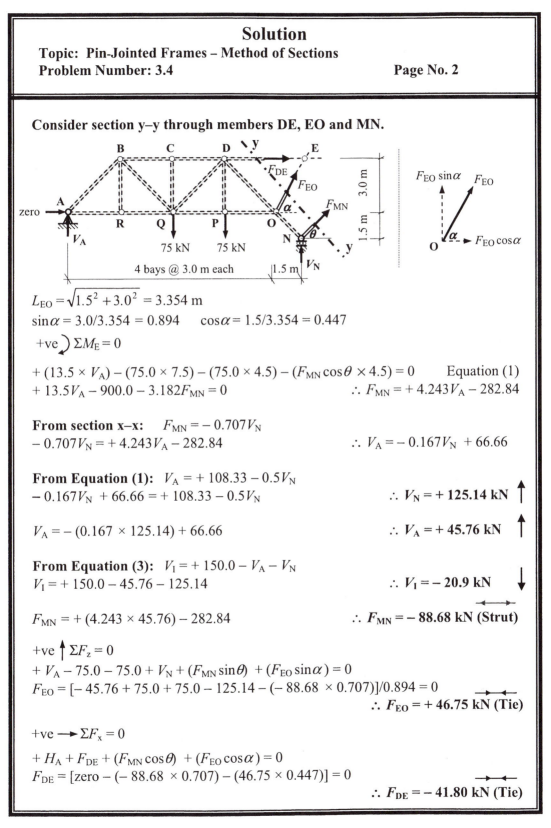

$L_{EO} = \sqrt{1.5^2 + 3.0^2} = 3.354 \text{ m}$

$\sin\alpha = 3.0/3.354 = 0.894 \qquad \cos\alpha = 1.5/3.354 = 0.447$

$+ve \,\circlearrowright\, \Sigma M_E = 0$

$+ (13.5 \times V_A) - (75.0 \times 7.5) - (75.0 \times 4.5) - (F_{MN}\cos\theta \times 4.5) = 0 \qquad$ Equation (1)

$+ 13.5V_A - 900.0 - 3.182F_{MN} = 0 \qquad\qquad \therefore F_{MN} = +4.243V_A - 282.84$

From section x–x: $\quad F_{MN} = -0.707V_N$

$-0.707V_N = +4.243V_A - 282.84 \qquad\qquad \therefore V_A = -0.167V_N + 66.66$

From Equation (1): $V_A = +108.33 - 0.5V_N$

$-0.167V_N + 66.66 = +108.33 - 0.5V_N \qquad\qquad \therefore V_N = +125.14 \text{ kN} \uparrow$

$V_A = -(0.167 \times 125.14) + 66.66 \qquad\qquad \therefore V_A = +45.76 \text{ kN} \uparrow$

From Equation (3): $V_I = +150.0 - V_A - V_N$

$V_I = +150.0 - 45.76 - 125.14 \qquad\qquad \therefore V_I = -20.9 \text{ kN} \downarrow$

$F_{MN} = +(4.243 \times 45.76) - 282.84 \qquad\qquad \therefore F_{MN} = -88.68 \text{ kN (Strut)}$

$+ve \,\uparrow\, \Sigma F_z = 0$

$+ V_A - 75.0 - 75.0 + V_N + (F_{MN}\sin\theta) + (F_{EO}\sin\alpha) = 0$

$F_{EO} = [-45.76 + 75.0 + 75.0 - 125.14 - (-88.68 \times 0.707)]/0.894 = 0$

$\therefore F_{EO} = +46.75 \text{ kN (Tie)}$

$+ve \longrightarrow \Sigma F_x = 0$

$+ H_A + F_{DE} + (F_{MN}\cos\theta) + (F_{EO}\cos\alpha) = 0$

$F_{DE} = [\text{zero} - (-88.68 \times 0.707) - (46.75 \times 0.447)] = 0$

$\therefore F_{DE} = -41.80 \text{ kN (Tie)}$

Solution
Topic: Pin-Jointed Frames – Joint Resolution
Problem Number: 3.5

Page No. 1

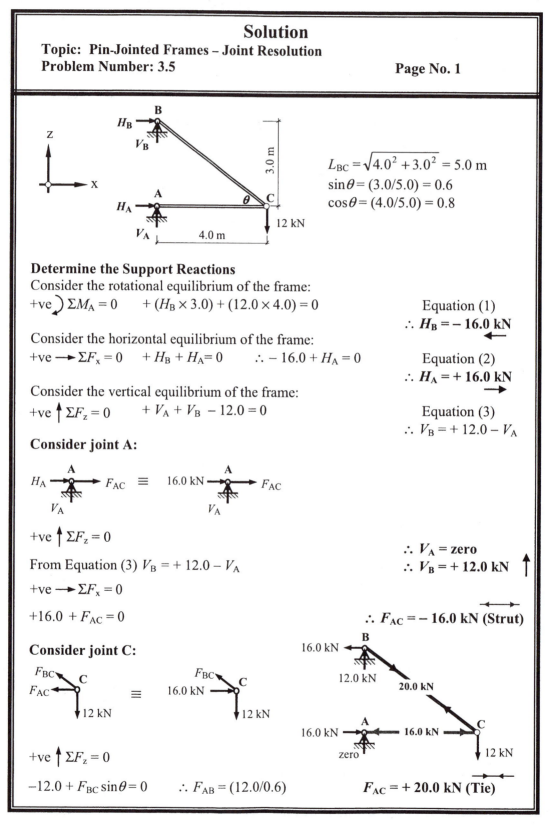

$L_{BC} = \sqrt{4.0^2 + 3.0^2} = 5.0$ m
$\sin\theta = (3.0/5.0) = 0.6$
$\cos\theta = (4.0/5.0) = 0.8$

Determine the Support Reactions
Consider the rotational equilibrium of the frame:
+ve $\curvearrowright$ $\Sigma M_A = 0$ $+ (H_B \times 3.0) + (12.0 \times 4.0) = 0$ Equation (1)
$\therefore H_B = -16.0$ kN

Consider the horizontal equilibrium of the frame:
+ve $\longrightarrow$ $\Sigma F_x = 0$ $+ H_B + H_A = 0$ $\therefore -16.0 + H_A = 0$ Equation (2)
$\therefore H_A = +16.0$ kN

Consider the vertical equilibrium of the frame:
+ve $\uparrow$ $\Sigma F_z = 0$ $+ V_A + V_B - 12.0 = 0$ Equation (3)
$\therefore V_B = +12.0 - V_A$

Consider joint A:

+ve $\uparrow$ $\Sigma F_z = 0$
$\therefore V_A$ = zero
From Equation (3) $V_B = +12.0 - V_A$
$\therefore V_B = +12.0$ kN $\uparrow$
+ve $\longrightarrow$ $\Sigma F_x = 0$
$+16.0 + F_{AC} = 0$
$\therefore F_{AC} = -16.0$ kN (Strut)

Consider joint C:

+ve $\uparrow$ $\Sigma F_z = 0$

$-12.0 + F_{BC} \sin\theta = 0$ $\therefore F_{AB} = (12.0/0.6)$
$F_{AC} = +20.0$ kN (Tie)

Solution

Topic: Pin–Jointed Frames – Joint Resolution
Problem Number: 3.6   **Page No. 1**

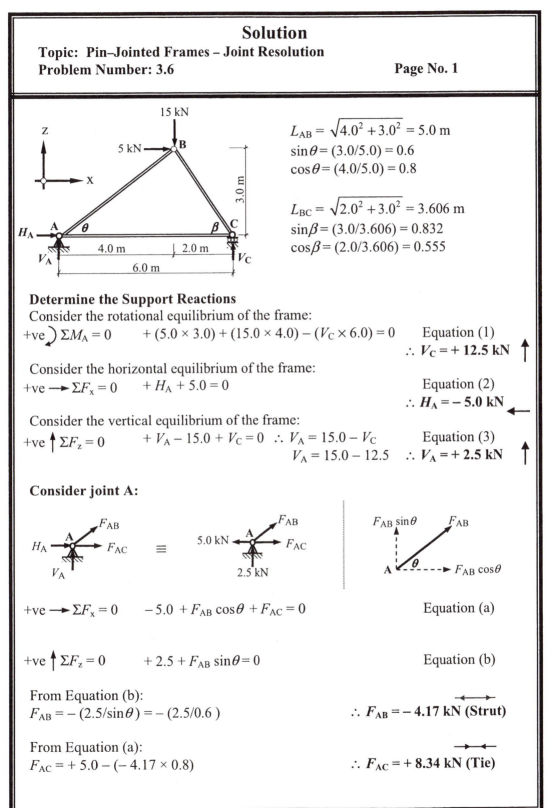

$$L_{AB} = \sqrt{4.0^2 + 3.0^2} = 5.0 \text{ m}$$
$$\sin\theta = (3.0/5.0) = 0.6$$
$$\cos\theta = (4.0/5.0) = 0.8$$

$$L_{BC} = \sqrt{2.0^2 + 3.0^2} = 3.606 \text{ m}$$
$$\sin\beta = (3.0/3.606) = 0.832$$
$$\cos\beta = (2.0/3.606) = 0.555$$

Determine the Support Reactions

Consider the rotational equilibrium of the frame:

$+ve \curvearrowright \Sigma M_A = 0 \qquad + (5.0 \times 3.0) + (15.0 \times 4.0) - (V_C \times 6.0) = 0$ Equation (1)
$$\therefore V_C = +\,12.5 \text{ kN} \uparrow$$

Consider the horizontal equilibrium of the frame:

$+ve \longrightarrow \Sigma F_x = 0 \qquad + H_A + 5.0 = 0$ Equation (2)
$$\therefore H_A = -\,5.0 \text{ kN} \longleftarrow$$

Consider the vertical equilibrium of the frame:

$+ve \uparrow \Sigma F_z = 0 \qquad + V_A - 15.0 + V_C = 0 \quad \therefore V_A = 15.0 - V_C$ Equation (3)
$$V_A = 15.0 - 12.5 \quad \therefore V_A = +\,2.5 \text{ kN} \uparrow$$

Consider joint A:

$+ve \longrightarrow \Sigma F_x = 0 \qquad -5.0 + F_{AB}\cos\theta + F_{AC} = 0$ Equation (a)

$+ve \uparrow \Sigma F_z = 0 \qquad + 2.5 + F_{AB}\sin\theta = 0$ Equation (b)

From Equation (b):
$F_{AB} = -(2.5/\sin\theta) = -(2.5/0.6)$ $\therefore F_{AB} = -\,4.17 \text{ kN (Strut)}$

From Equation (a):
$F_{AC} = +5.0 - (-4.17 \times 0.8)$ $\therefore F_{AC} = +\,8.34 \text{ kN (Tie)}$

Solution
Topic: Pin-Jointed Frames – Joint Resolution
Problem Number: 3.6   **Page No. 2**

Consider joint C:

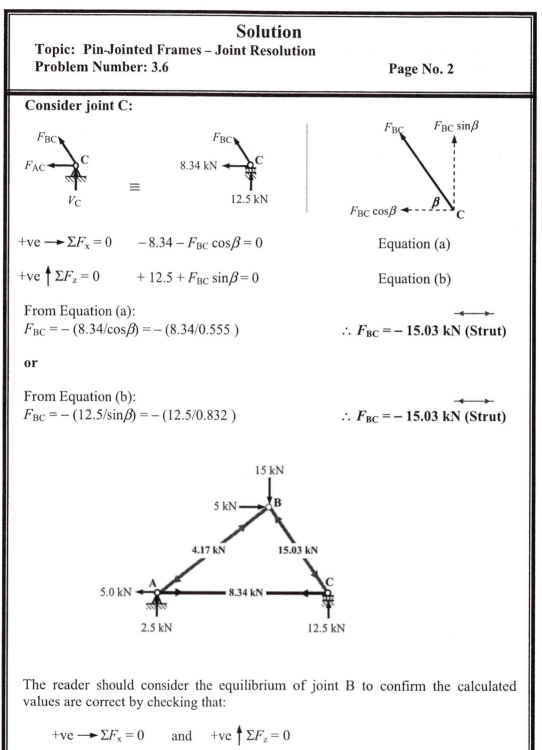

$+ve \longrightarrow \Sigma F_x = 0$ $-8.34 - F_{BC}\cos\beta = 0$ Equation (a)

$+ve \uparrow \Sigma F_z = 0$ $+12.5 + F_{BC}\sin\beta = 0$ Equation (b)

From Equation (a):
$F_{BC} = -(8.34/\cos\beta) = -(8.34/0.555)$ $\therefore F_{BC} = -15.03 \text{ kN (Strut)}$

or

From Equation (b):
$F_{BC} = -(12.5/\sin\beta) = -(12.5/0.832)$ $\therefore F_{BC} = -15.03 \text{ kN (Strut)}$

The reader should consider the equilibrium of joint B to confirm the calculated values are correct by checking that:

$+ve \longrightarrow \Sigma F_x = 0$ and $+ve \uparrow \Sigma F_z = 0$

Solution
Topic: Pin-Jointed Frames – Joint Resolution
Problem Number: 3.7   **Page No. 1**

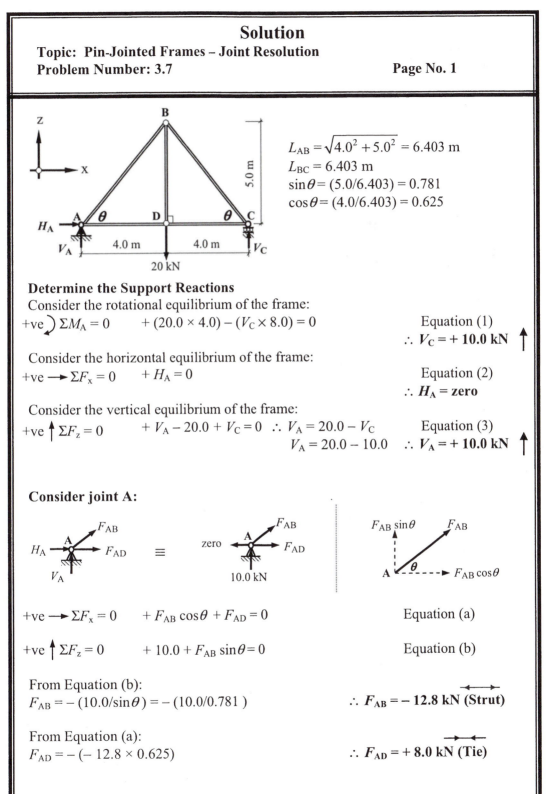

$$L_{AB} = \sqrt{4.0^2 + 5.0^2} = 6.403 \text{ m}$$
$$L_{BC} = 6.403 \text{ m}$$
$$\sin\theta = (5.0/6.403) = 0.781$$
$$\cos\theta = (4.0/6.403) = 0.625$$

Determine the Support Reactions

Consider the rotational equilibrium of the frame:

+ve $\curvearrowright$ $\Sigma M_A = 0$ $+ (20.0 \times 4.0) - (V_C \times 8.0) = 0$ Equation (1)
                                    $\therefore V_C = + 10.0 \text{ kN} \uparrow$

Consider the horizontal equilibrium of the frame:

+ve $\rightarrow \Sigma F_x = 0$ $+ H_A = 0$ Equation (2)
                              $\therefore H_A = \text{zero}$

Consider the vertical equilibrium of the frame:

+ve $\uparrow \Sigma F_z = 0$ $+ V_A - 20.0 + V_C = 0$ $\therefore V_A = 20.0 - V_C$ Equation (3)
                      $V_A = 20.0 - 10.0$ $\therefore V_A = + 10.0 \text{ kN} \uparrow$

Consider joint A:

$$+\text{ve} \rightarrow \Sigma F_x = 0 \qquad + F_{AB} \cos\theta + F_{AD} = 0 \qquad \text{Equation (a)}$$

$$+\text{ve} \uparrow \Sigma F_z = 0 \qquad + 10.0 + F_{AB} \sin\theta = 0 \qquad \text{Equation (b)}$$

From Equation (b):
$F_{AB} = - (10.0/\sin\theta) = - (10.0/0.781)$ $\therefore F_{AB} = - 12.8 \text{ kN (Strut)}$

From Equation (a):
$F_{AD} = - (- 12.8 \times 0.625)$ $\therefore F_{AD} = + 8.0 \text{ kN (Tie)}$

Solution

Topic: Pin-Jointed Frames – Joint Resolution
Problem Number: 3.7   **Page No. 2**

Consider joint D:

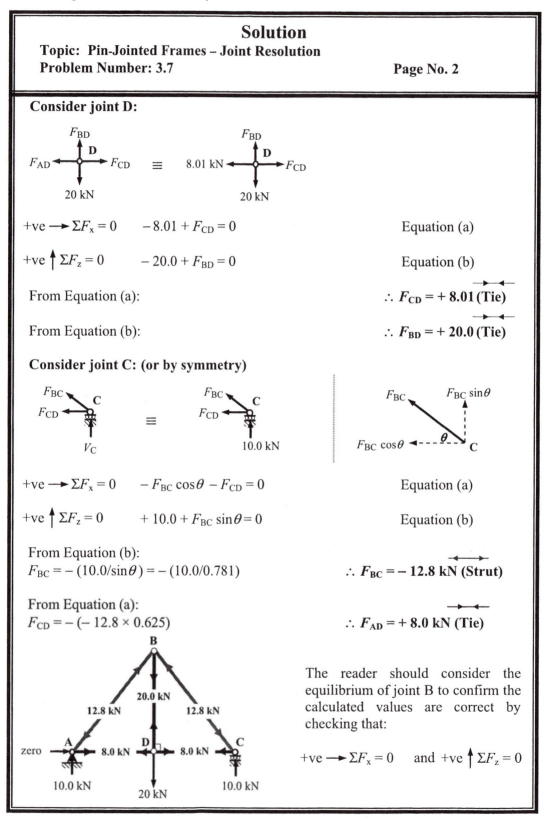

$+ve \longrightarrow \Sigma F_x = 0$ $-8.01 + F_{CD} = 0$ Equation (a)

$+ve \uparrow \Sigma F_z = 0$ $-20.0 + F_{BD} = 0$ Equation (b)

From Equation (a): $\therefore F_{CD} = +8.01 \text{ (Tie)}$

From Equation (b): $\therefore F_{BD} = +20.0 \text{ (Tie)}$

Consider joint C: (or by symmetry)

$+ve \longrightarrow \Sigma F_x = 0$ $-F_{BC}\cos\theta - F_{CD} = 0$ Equation (a)

$+ve \uparrow \Sigma F_z = 0$ $+10.0 + F_{BC}\sin\theta = 0$ Equation (b)

From Equation (b):
$F_{BC} = -(10.0/\sin\theta) = -(10.0/0.781)$ $\therefore F_{BC} = -12.8 \text{ kN (Strut)}$

From Equation (a):
$F_{CD} = -(-12.8 \times 0.625)$ $\therefore F_{AD} = +8.0 \text{ kN (Tie)}$

The reader should consider the equilibrium of joint B to confirm the calculated values are correct by checking that:

$+ve \longrightarrow \Sigma F_x = 0$ and $+ve \uparrow \Sigma F_z = 0$

Solution

Topic: Pin-Jointed Frames – Joint Resolution
Problem Number: 3.8   **Page No. 1**

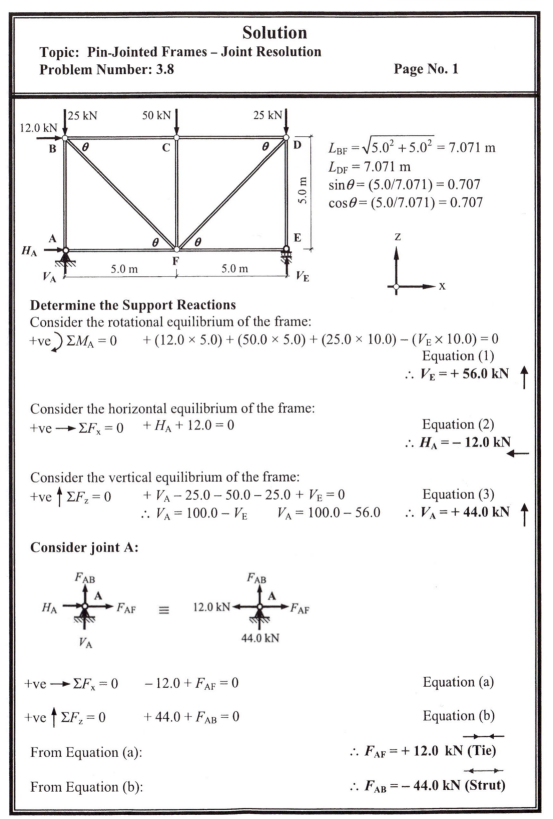

$L_{BF} = \sqrt{5.0^2 + 5.0^2} = 7.071 \text{ m}$

$L_{DF} = 7.071 \text{ m}$

$\sin\theta = (5.0/7.071) = 0.707$

$\cos\theta = (5.0/7.071) = 0.707$

Determine the Support Reactions

Consider the rotational equilibrium of the frame:

$+\text{ve} \; \curvearrowright \; \Sigma M_A = 0$ $+ (12.0 \times 5.0) + (50.0 \times 5.0) + (25.0 \times 10.0) - (V_E \times 10.0) = 0$
         Equation (1)
         $\therefore V_E = + 56.0 \text{ kN} \uparrow$

Consider the horizontal equilibrium of the frame:

$+\text{ve} \longrightarrow \Sigma F_x = 0$ $+ H_A + 12.0 = 0$ Equation (2)
         $\therefore H_A = - 12.0 \text{ kN} \leftarrow$

Consider the vertical equilibrium of the frame:

$+\text{ve} \uparrow \Sigma F_z = 0$ $+ V_A - 25.0 - 50.0 - 25.0 + V_E = 0$ Equation (3)
         $\therefore V_A = 100.0 - V_E$ $V_A = 100.0 - 56.0$ $\therefore V_A = + 44.0 \text{ kN} \uparrow$

Consider joint A:

$+\text{ve} \longrightarrow \Sigma F_x = 0$ $- 12.0 + F_{AF} = 0$ Equation (a)

$+\text{ve} \uparrow \Sigma F_z = 0$ $+ 44.0 + F_{AB} = 0$ Equation (b)

From Equation (a): $\therefore F_{AF} = + 12.0 \text{ kN (Tie)}$

From Equation (b): $\therefore F_{AB} = - 44.0 \text{ kN (Strut)}$

Solution
Topic: Pin-Jointed Frames – Joint Resolution
Problem Number: 3.8

Consider joint B:

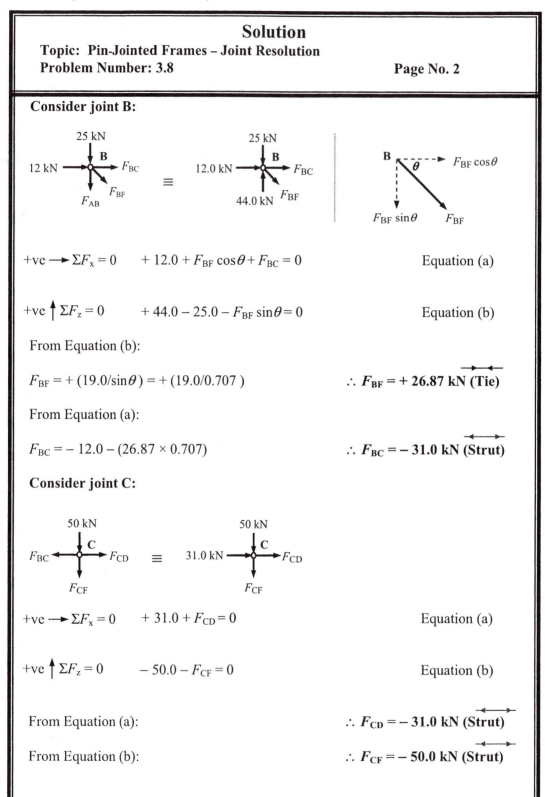

+ve $\longrightarrow \Sigma F_x = 0$ $+ 12.0 + F_{BF} \cos\theta + F_{BC} = 0$ Equation (a)

+ve $\uparrow \Sigma F_z = 0$ $+ 44.0 - 25.0 - F_{BF} \sin\theta = 0$ Equation (b)

From Equation (b):

$F_{BF} = + (19.0/\sin\theta) = + (19.0/0.707)$ $\therefore F_{BF} = + 26.87$ **kN (Tie)**

From Equation (a):

$F_{BC} = - 12.0 - (26.87 \times 0.707)$ $\therefore F_{BC} = - 31.0$ **kN (Strut)**

Consider joint C:

+ve $\longrightarrow \Sigma F_x = 0$ $+ 31.0 + F_{CD} = 0$ Equation (a)

+ve $\uparrow \Sigma F_z = 0$ $- 50.0 - F_{CF} = 0$ Equation (b)

From Equation (a): $\therefore F_{CD} = - 31.0$ **kN (Strut)**

From Equation (b): $\therefore F_{CF} = - 50.0$ **kN (Strut)**

Solution

Topic: Pin-Jointed Frames – Joint Resolution
Problem Number: 3.8

Consider joint D:

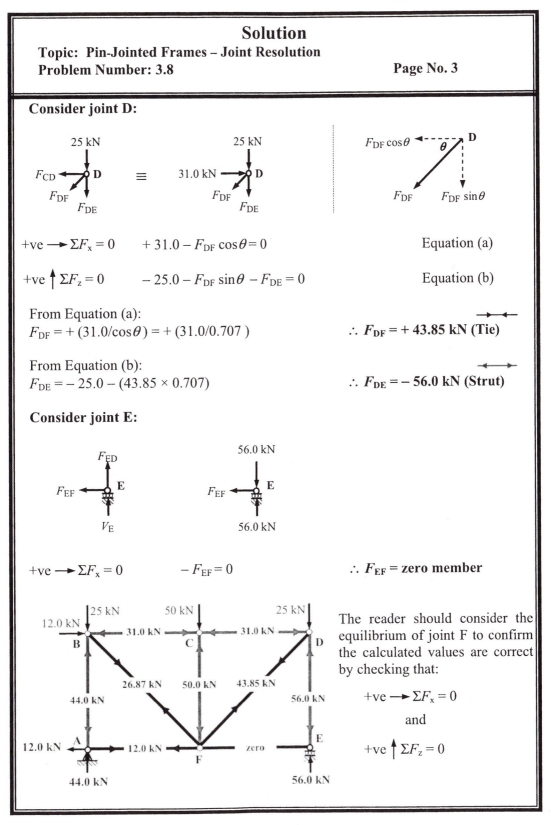

$+ve \longrightarrow \Sigma F_x = 0 \qquad + 31.0 - F_{DF} \cos\theta = 0$ 　　　　Equation (a)

$+ve \uparrow \Sigma F_z = 0 \qquad - 25.0 - F_{DF} \sin\theta - F_{DE} = 0$ 　　　　Equation (b)

From Equation (a):
$F_{DF} = + (31.0/\cos\theta) = + (31.0/0.707)$ 　　　$\therefore F_{DF} = + 43.85$ kN (Tie)

From Equation (b):
$F_{DE} = - 25.0 - (43.85 \times 0.707)$ 　　　$\therefore F_{DE} = - 56.0$ kN (Strut)

Consider joint E:

$+ve \longrightarrow \Sigma F_x = 0 \qquad - F_{EF} = 0$ 　　　$\therefore F_{EF} =$ **zero member**

The reader should consider the equilibrium of joint F to confirm the calculated values are correct by checking that:

$$+ve \longrightarrow \Sigma F_x = 0$$

and

$$+ve \uparrow \Sigma F_z = 0$$

Solution
Topic: Pin-Jointed Frames – Joint Resolution
Problem Number: 3.9

Page No. 1

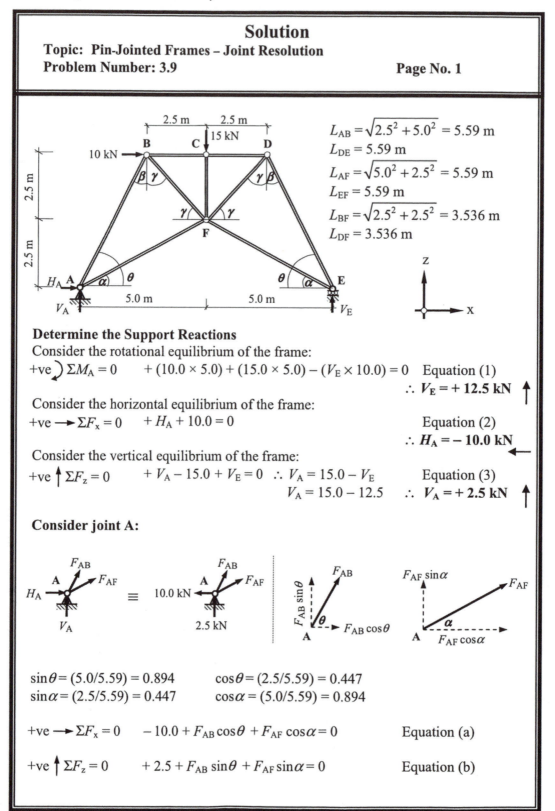

$$L_{AB} = \sqrt{2.5^2 + 5.0^2} = 5.59 \text{ m}$$
$$L_{DE} = 5.59 \text{ m}$$
$$L_{AF} = \sqrt{5.0^2 + 2.5^2} = 5.59 \text{ m}$$
$$L_{EF} = 5.59 \text{ m}$$
$$L_{BF} = \sqrt{2.5^2 + 2.5^2} = 3.536 \text{ m}$$
$$L_{DF} = 3.536 \text{ m}$$

Determine the Support Reactions
Consider the rotational equilibrium of the frame:

+ve $\curvearrowright$ $\Sigma M_A = 0$ $+ (10.0 \times 5.0) + (15.0 \times 5.0) - (V_E \times 10.0) = 0$ Equation (1)
$$\therefore V_E = +12.5 \text{ kN} \uparrow$$

Consider the horizontal equilibrium of the frame:

+ve $\longrightarrow \Sigma F_x = 0$ $+ H_A + 10.0 = 0$ Equation (2)
$$\therefore H_A = -10.0 \text{ kN} \longleftarrow$$

Consider the vertical equilibrium of the frame:

+ve $\uparrow \Sigma F_z = 0$ $+ V_A - 15.0 + V_E = 0$ $\therefore V_A = 15.0 - V_E$ Equation (3)
$$V_A = 15.0 - 12.5 \quad \therefore V_A = +2.5 \text{ kN} \uparrow$$

Consider joint A:

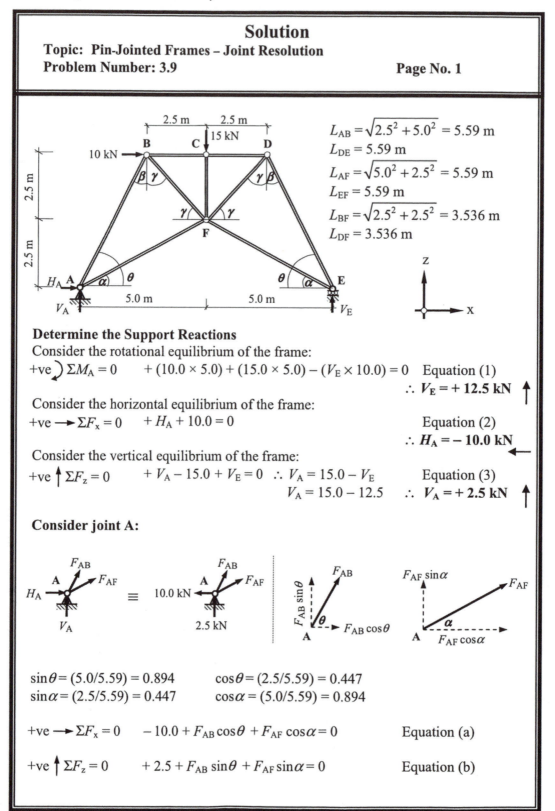

$\sin\theta = (5.0/5.59) = 0.894$ $\cos\theta = (2.5/5.59) = 0.447$
$\sin\alpha = (2.5/5.59) = 0.447$ $\cos\alpha = (5.0/5.59) = 0.894$

+ve $\longrightarrow \Sigma F_x = 0$ $- 10.0 + F_{AB}\cos\theta + F_{AF}\cos\alpha = 0$ Equation (a)

+ve $\uparrow \Sigma F_z = 0$ $+ 2.5 + F_{AB}\sin\theta + F_{AF}\sin\alpha = 0$ Equation (b)

Solution
Topic: Pin-Jointed Frames – Joint Resolution
Problem Number: 3.9   **Page No. 2**

From Equation (a):
$F_{AB} = [+10.0 - (F_{AF} \times 0.894)]/0.447$ $\therefore F_{AB} = +22.371 - 2.0F_{AF}$

Substitute for F_{AB} in Equation (b)
$+2.5 + (22.371 - 2.0F_{AF})\sin\theta + F_{AF}\sin\alpha = 0$
$+2.5 + [(22.371 \times 0.894) - (2.0F_{AF} \times 0.894) + (F_{AF} \times 0.447)] = 0$ →←
$+22.5 - 1.341F_{AF} = 0$ $\therefore F_{AF} = +16.78$ **kN**   **(Tie)**

$F_{AB} = +22.371 - (2.0 \times 16.78)$ $\therefore F_{AB} = -11.19$ **kN**   **(Strut)**

Consider joint B:

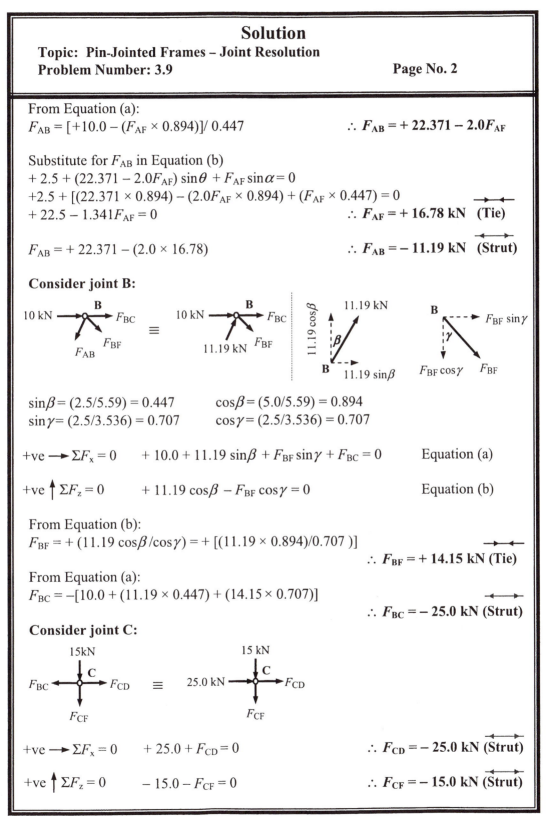

$\sin\beta = (2.5/5.59) = 0.447$ $\cos\beta = (5.0/5.59) = 0.894$
$\sin\gamma = (2.5/3.536) = 0.707$ $\cos\gamma = (2.5/3.536) = 0.707$

+ve → $\Sigma F_x = 0$ $+10.0 + 11.19\sin\beta + F_{BF}\sin\gamma + F_{BC} = 0$ Equation (a)

+ve ↑ $\Sigma F_z = 0$ $+11.19\cos\beta - F_{BF}\cos\gamma = 0$ Equation (b)

From Equation (b):
$F_{BF} = +(11.19\cos\beta/\cos\gamma) = +[(11.19 \times 0.894)/0.707)]$
                                 →← $\therefore F_{BF} = +14.15$ **kN (Tie)**

From Equation (a):
$F_{BC} = -[10.0 + (11.19 \times 0.447) + (14.15 \times 0.707)]$
                                 ←→ $\therefore F_{BC} = -25.0$ **kN (Strut)**

Consider joint C:

+ve → $\Sigma F_x = 0$ $+25.0 + F_{CD} = 0$ $\therefore F_{CD} = -25.0$ **kN (Strut)**

+ve ↑ $\Sigma F_z = 0$ $-15.0 - F_{CF} = 0$ $\therefore F_{CF} = -15.0$ **kN (Strut)**

Solution

Topic: Pin-Jointed Frames – Joint Resolution

Problem Number: 3.9

Consider joint D:

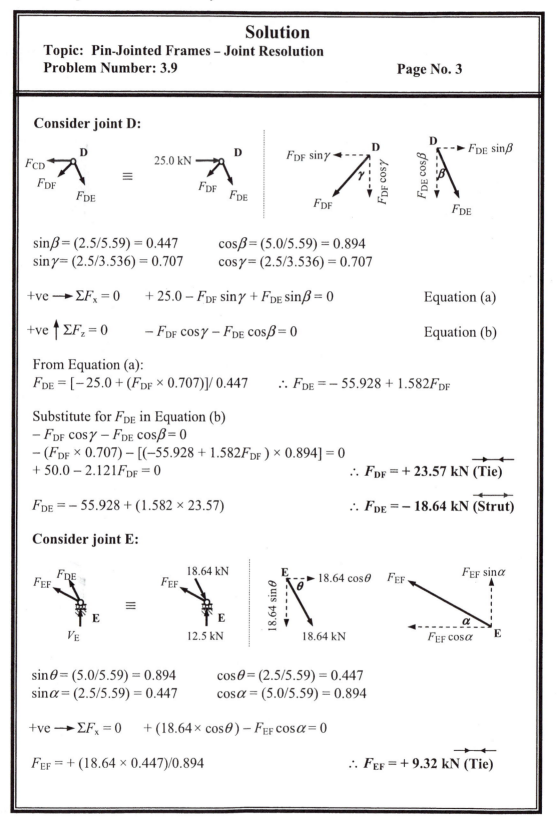

$\sin\beta = (2.5/5.59) = 0.447$ $\cos\beta = (5.0/5.59) = 0.894$

$\sin\gamma = (2.5/3.536) = 0.707$ $\cos\gamma = (2.5/3.536) = 0.707$

+ve $\longrightarrow \Sigma F_x = 0$ $+ 25.0 - F_{DF}\sin\gamma + F_{DE}\sin\beta = 0$ Equation (a)

+ve $\uparrow \Sigma F_z = 0$ $- F_{DF}\cos\gamma - F_{DE}\cos\beta = 0$ Equation (b)

From Equation (a):

$F_{DE} = [-25.0 + (F_{DF} \times 0.707)]/0.447$ $\therefore F_{DE} = -55.928 + 1.582F_{DF}$

Substitute for F_{DE} in Equation (b)

$- F_{DF}\cos\gamma - F_{DE}\cos\beta = 0$

$- (F_{DF} \times 0.707) - [(-55.928 + 1.582F_{DF}) \times 0.894] = 0$

$+ 50.0 - 2.121F_{DF} = 0$ $\therefore F_{DF} = + \textbf{23.57 kN (Tie)}$

$F_{DE} = -55.928 + (1.582 \times 23.57)$ $\therefore F_{DE} = - \textbf{18.64 kN (Strut)}$

Consider joint E:

$\sin\theta = (5.0/5.59) = 0.894$ $\cos\theta = (2.5/5.59) = 0.447$

$\sin\alpha = (2.5/5.59) = 0.447$ $\cos\alpha = (5.0/5.59) = 0.894$

+ve $\longrightarrow \Sigma F_x = 0$ $+ (18.64 \times \cos\theta) - F_{EF}\cos\alpha = 0$

$F_{EF} = + (18.64 \times 0.447)/0.894$ $\therefore F_{EF} = + \textbf{9.32 kN (Tie)}$

Solution
Topic: Pin-Jointed Frames – Joint Resolution
Problem Number: 3.9   **Page No. 4**

The values obtained above can be checked by confirming the horizontal and vertical equilibrium at joint F as follows:

Joint F:

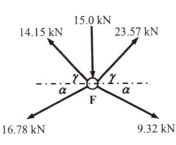

$$\sin\gamma = (2.5/3.536) = 0.707 \qquad \cos\gamma = (5.0/3.536) = 0.707$$
$$\sin\alpha = (2.5/5.59) = 0.447 \qquad \cos\alpha = (5.0/5.59) = 0.894$$

+ve $\longrightarrow \Sigma F_x$

$= -16.78 \cos\alpha - 14.15 \cos\gamma + 9.32 \cos\alpha + 23.57 \cos\gamma$
$= -(16.78 \times 0.894) - (14.15 \times 0.707) + (9.32 \times 0.894) + (23.57 \times 0.707)$
$= \text{zero}$

+ve $\uparrow \Sigma F_z = 0$

$= -16.78 \sin\alpha + 14.15 \sin\gamma - 9.32 \sin\alpha + 23.57 \sin\gamma - 15.0$
$= -(16.78 \times 0.447) + (14.15 \times 0.707) - (9.32 \times 0.447) + (23.57 \times 0.707) - 15.0$
$= \text{zero}$

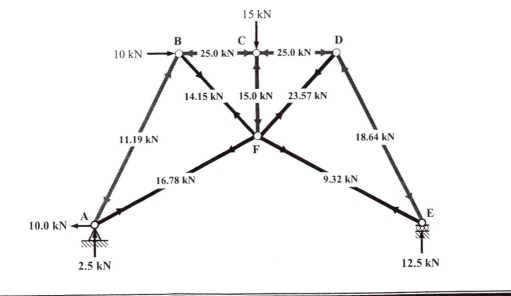

Solution

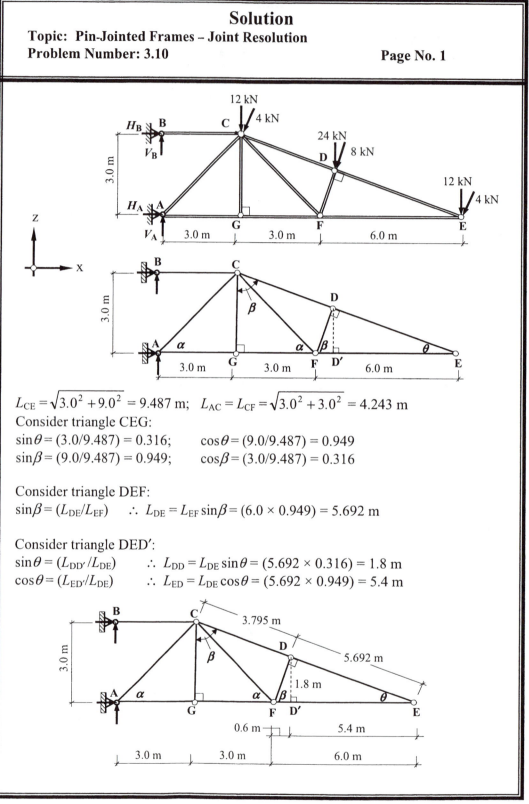

$L_{CE} = \sqrt{3.0^2 + 9.0^2} = 9.487$ m; $L_{AC} = L_{CF} = \sqrt{3.0^2 + 3.0^2} = 4.243$ m

Consider triangle CEG:

$\sin\theta = (3.0/9.487) = 0.316$; $\cos\theta = (9.0/9.487) = 0.949$
$\sin\beta = (9.0/9.487) = 0.949$; $\cos\beta = (3.0/9.487) = 0.316$

Consider triangle DEF:

$\sin\beta = (L_{DE}/L_{EF})$ $\therefore$ $L_{DE} = L_{EF}\sin\beta = (6.0 \times 0.949) = 5.692$ m

Consider triangle DED′:

$\sin\theta = (L_{DD'}/L_{DE})$ $\therefore$ $L_{DD} = L_{DE}\sin\theta = (5.692 \times 0.316) = 1.8$ m
$\cos\theta = (L_{ED'}/L_{DE})$ $\therefore$ $L_{ED} = L_{DE}\cos\theta = (5.692 \times 0.949) = 5.4$ m

Solution
Topic: **Pin-Jointed Frames – Joint Resolution**
Problem Number: **3.10**   Page No. 2

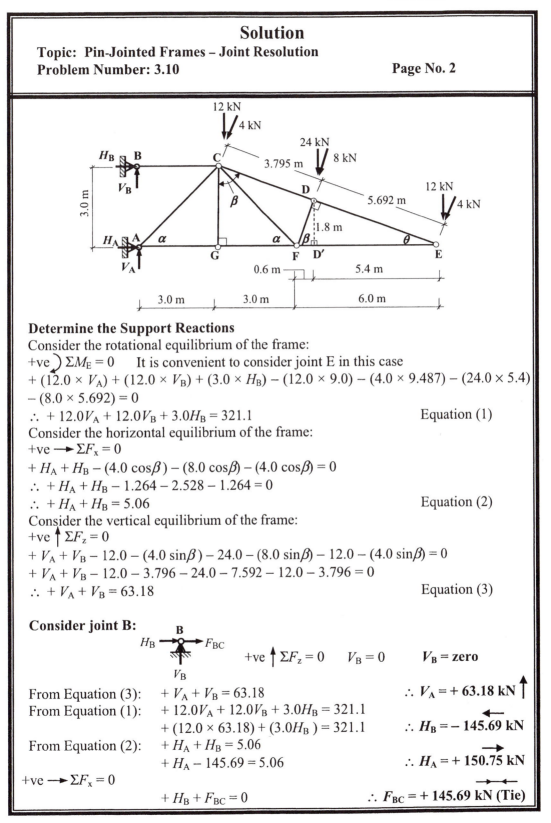

Determine the Support Reactions
Consider the rotational equilibrium of the frame:
+ve $\curvearrowright$ $\Sigma M_E = 0$ It is convenient to consider joint E in this case
+ $(12.0 \times V_A) + (12.0 \times V_B) + (3.0 \times H_B) - (12.0 \times 9.0) - (4.0 \times 9.487) - (24.0 \times 5.4)$
$- (8.0 \times 5.692) = 0$
$\therefore + 12.0 V_A + 12.0 V_B + 3.0 H_B = 321.1$ Equation (1)
Consider the horizontal equilibrium of the frame:
+ve $\longrightarrow \Sigma F_x = 0$
+ $H_A + H_B - (4.0 \cos\beta) - (8.0 \cos\beta) - (4.0 \cos\beta) = 0$
$\therefore + H_A + H_B - 1.264 - 2.528 - 1.264 = 0$
$\therefore + H_A + H_B = 5.06$ Equation (2)
Consider the vertical equilibrium of the frame:
+ve $\uparrow \Sigma F_z = 0$
+ $V_A + V_B - 12.0 - (4.0 \sin\beta) - 24.0 - (8.0 \sin\beta) - 12.0 - (4.0 \sin\beta) = 0$
+ $V_A + V_B - 12.0 - 3.796 - 24.0 - 7.592 - 12.0 - 3.796 = 0$
$\therefore + V_A + V_B = 63.18$ Equation (3)

Consider joint B:

$H_B \longrightarrow$ B $\longrightarrow F_{BC}$

V_B

+ve $\uparrow \Sigma F_z = 0$ $V_B = 0$ **V_B = zero**

From Equation (3):   + $V_A + V_B = 63.18$ $\therefore V_A = + 63.18$ kN $\uparrow$
From Equation (1):   + $12.0 V_A + 12.0 V_B + 3.0 H_B = 321.1$
                      + $(12.0 \times 63.18) + (3.0 H_B) = 321.1$ $\therefore H_B = -145.69$ kN $\longleftarrow$
From Equation (2):   + $H_A + H_B = 5.06$
                      + $H_A - 145.69 = 5.06$ $\therefore H_A = + 150.75$ kN $\longrightarrow$
+ve $\longrightarrow \Sigma F_x = 0$
                      + $H_B + F_{BC} = 0$ $\therefore F_{BC} = + 145.69$ kN (Tie)

Solution
Topic: Pin-Jointed Frames – Joint Resolution
Problem Number: 3.10

Consider joint A:

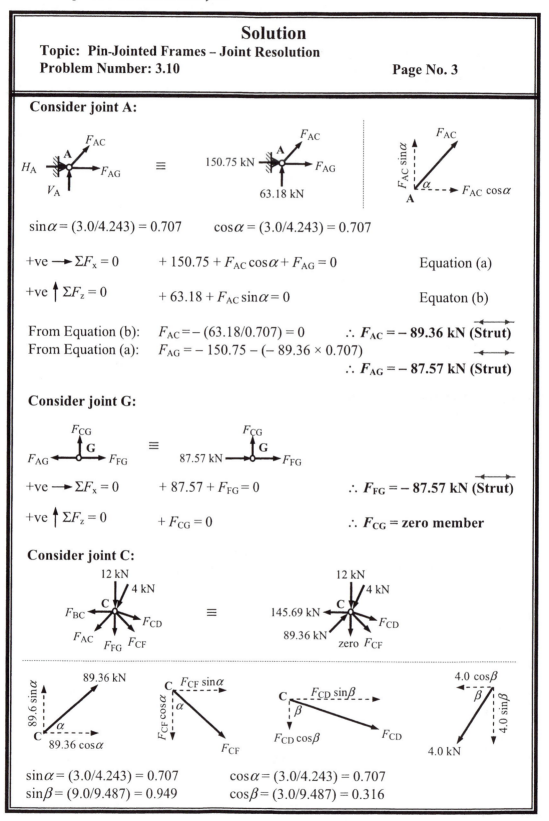

$\sin\alpha = (3.0/4.243) = 0.707$ $\cos\alpha = (3.0/4.243) = 0.707$

+ve $\longrightarrow \Sigma F_x = 0$ $+ 150.75 + F_{AC}\cos\alpha + F_{AG} = 0$ Equation (a)

+ve $\uparrow \Sigma F_z = 0$ $+ 63.18 + F_{AC}\sin\alpha = 0$ Equaton (b)

From Equation (b): $F_{AC} = - (63.18/0.707) = 0$ $\therefore \boldsymbol{F_{AC} = - 89.36 \text{ kN (Strut)}}$
From Equation (a): $F_{AG} = - 150.75 - (- 89.36 \times 0.707)$

$\therefore \boldsymbol{F_{AG} = - 87.57 \text{ kN (Strut)}}$

Consider joint G:

+ve $\longrightarrow \Sigma F_x = 0$ $+ 87.57 + F_{FG} = 0$ $\therefore \boldsymbol{F_{FG} = - 87.57 \text{ kN (Strut)}}$

+ve $\uparrow \Sigma F_z = 0$ $+ F_{CG} = 0$ $\therefore \boldsymbol{F_{CG} = \text{zero member}}$

Consider joint C:

$\sin\alpha = (3.0/4.243) = 0.707$ $\cos\alpha = (3.0/4.243) = 0.707$
$\sin\beta = (9.0/9.487) = 0.949$ $\cos\beta = (3.0/9.487) = 0.316$

Solution
Topic: Pin-Jointed Frames – Joint Resolution
Problem Number: 3.10   **Page No. 4**

+ve $\longrightarrow \Sigma F_x$

$-145.69 + 89.36\cos\alpha - 4.0\cos\beta + F_{CF}\sin\alpha + F_{CD}\sin\beta = 0$

$-145.69 + (89.36 \times 0.707) - (4.0 \times 0.316) + (F_{CF} \times 0.707) + (F_{CD} \times 0.949) = 0$

$-83.776 + 0.707F_{CF} + 0.949F_{CD} = 0$ Equation (a)

+ve $\uparrow \Sigma F_z = 0$

$-12.0 + 89.36\sin\alpha - 4.0\sin\beta - F_{CF}\cos\alpha - F_{CD}\cos\beta = 0$

$-12.0 + (89.36 \times 0.707) - (4.0 \times 0.949) - (F_{CF} \times 0.707) - (F_{CD} \times 0.316) = 0$

$+47.382 - 0.707F_{CF} - 0.316F_{CD} = 0$ Equation (b)

From Equation (a):
$F_{CF} = (+83.776 - 0.949F_{CD})/0.707$ $\therefore F_{CF} = +118.5 - 1.342F_{CD}$

Substitute for F_{CF} in Equation (b)
$+47.382 - 0.707F_{CF} - 0.316F_{CD} = 0$
$+47.382 - [0.707 \times (118.5 - 1.342F_{CD})] - 0.316F_{CD} = 0$
$+36.4 + 0.633F_{CD} = 0$ $\therefore \boldsymbol{F_{CD} = +57.50 \text{ kN (Tie)}}$

$\therefore F_{CF} = +118.5 - (1.342 \times 57.5)$ $\therefore \boldsymbol{F_{CF} = +41.34 \text{ kN (Tie)}}$

Consider joint F:

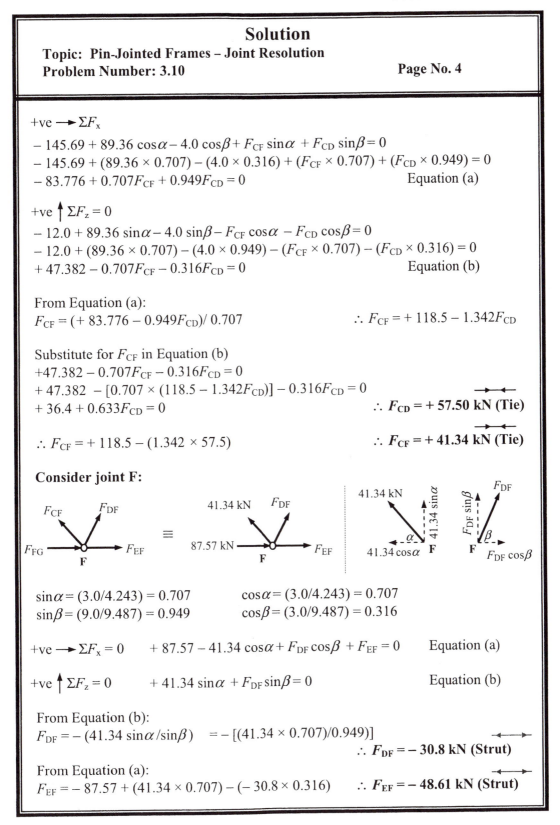

$\sin\alpha = (3.0/4.243) = 0.707$ $\cos\alpha = (3.0/4.243) = 0.707$
$\sin\beta = (9.0/9.487) = 0.949$ $\cos\beta = (3.0/9.487) = 0.316$

+ve $\longrightarrow \Sigma F_x = 0$ $+87.57 - 41.34\cos\alpha + F_{DF}\cos\beta + F_{EF} = 0$ Equation (a)

+ve $\uparrow \Sigma F_z = 0$ $+41.34\sin\alpha + F_{DF}\sin\beta = 0$ Equation (b)

From Equation (b):
$F_{DF} = -(41.34\sin\alpha/\sin\beta) = -[(41.34 \times 0.707)/0.949]$
$\therefore \boldsymbol{F_{DF} = -30.8 \text{ kN (Strut)}}$

From Equation (a):
$F_{EF} = -87.57 + (41.34 \times 0.707) - (-30.8 \times 0.316)$ $\therefore \boldsymbol{F_{EF} = -48.61 \text{ kN (Strut)}}$

Solution

Topic: Pin-Jointed Frames – Joint Resolution

Problem Number: 3.10 **Page No. 5**

Consider joint E:

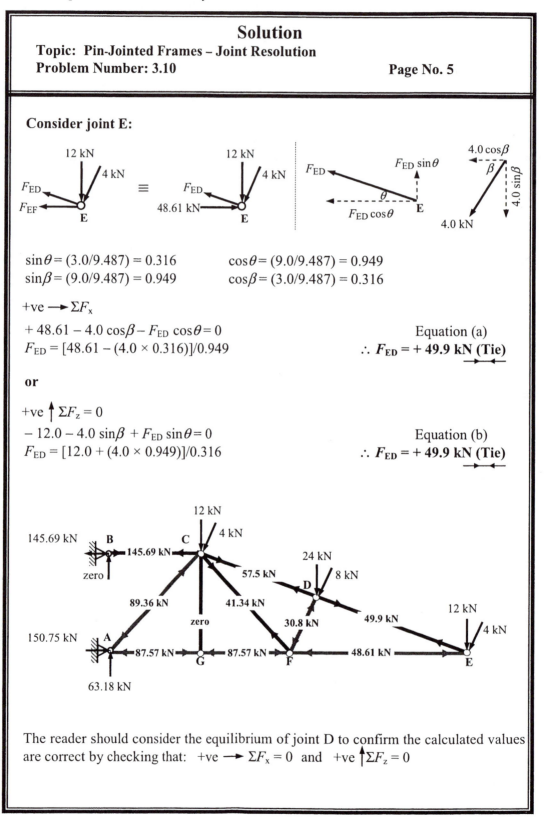

$\sin\theta = (3.0/9.487) = 0.316$ $\qquad$ $\cos\theta = (9.0/9.487) = 0.949$

$\sin\beta = (9.0/9.487) = 0.949$ $\qquad$ $\cos\beta = (3.0/9.487) = 0.316$

$+ve \longrightarrow \Sigma F_x$

$+ 48.61 - 4.0\cos\beta - F_{ED}\cos\theta = 0$ $\qquad\qquad$ Equation (a)

$F_{ED} = [48.61 - (4.0 \times 0.316)]/0.949$ $\qquad\qquad$ $\therefore F_{ED} = + \textbf{49.9 kN (Tie)}$

or

$+ve \uparrow \Sigma F_z = 0$

$- 12.0 - 4.0\sin\beta + F_{ED}\sin\theta = 0$ $\qquad\qquad$ Equation (b)

$F_{ED} = [12.0 + (4.0 \times 0.949)]/0.316$ $\qquad\qquad$ $\therefore F_{ED} = + \textbf{49.9 kN (Tie)}$

The reader should consider the equilibrium of joint D to confirm the calculated values are correct by checking that: $+ve \longrightarrow \Sigma F_x = 0$ and $+ve \uparrow\Sigma F_z = 0$

3.4   *Method of Tension Coefficients*

The method of tension coefficients is a tabular technique of carrying out joint resolution in either two or three dimensions. It is ideally suited to the analysis of pin-jointed space-frames.

Consider an individual member from a pin-jointed plane-frame, e.g. member AB shown in Figure 3.8 with reference to a particular X–Z co-ordinate system.

If AB is a member of length L_{AB} having a tensile force in it of T_{AB}, then the components of this force in the X and Z directions are $T_{AB} \cos\theta$ and $T_{AB} \sin\theta$ respectively.

If the co-ordinates of A and B are (X_A, Z_A) and (X_B, Z_B), then the component of T_{AB} in the x-direction is given by :

$$x\text{-component} = T_{AB} \frac{(X_B - X_A)}{L_{AB}} = t_{AB}(X_B - X_A)$$

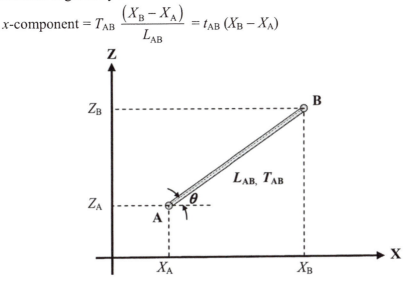

Figure 3.8

where $t_{AB} = \dfrac{T_{AB}}{L_{AB}}$ and is known as the **tension coefficient** of the bar. Similarly, the component of T_{AB} in the z-direction is given by:

$$z\text{-component} = T_{AB} = \frac{Z_B - Z_A}{L_{AB}} = t_{AB}(Z_B - Z_A)$$

If at joint A in the frame there are a number of bars, i.e. AB, AC ... AN, and external loads X_A and Z_A acting in the X and Z directions, then since the joint is in equilibrium the sum of the components of the external and internal forces must equal zero in each of those directions.

Expressing these conditions in terms of the components of each of the forces then gives:

$$t_{AB}(X_B - X_A) + t_{AC}(X_C - X_A) + \ldots \ldots \ldots \ldots t_{AN}(X_N - X_A) + X_A = 0 \qquad (1)$$

$$t_{AB}(Z_B - Z_A) + t_{AC}(Z_C - Z_A) + \ldots \ldots \ldots \ldots t_{AN}(Z_N - Z_A) + Z_A = 0 \qquad (2)$$

A similar pair of equations can be developed for each joint in the frame giving a total number of equation equal to $(2 \times$ number of joints$)$

In a statically determinate triangulated plane-frame the number of unknown member forces is equal to $[(2 \times$ number of joints$) - 3]$, hence there are three additional equations which can be used to determine the reactions or check the values of the tension coefficients.

Once a tension coefficient (e.g. t_{AB}) has been determined, the unknown member force is given by the product:

$$T_{AB} = t_{AB}L_{AB} \qquad (\textbf{Note: } T_{AB} \equiv T_{BA})$$

Note: A member which has a $-$ **ve** tension coefficient is in compression and is a strut.

3.4.1　Example 3.2:　Two-Dimensional Plane Truss

Consider the pin-jointed, plane-frame ABC loaded as shown in Figure 3.9.

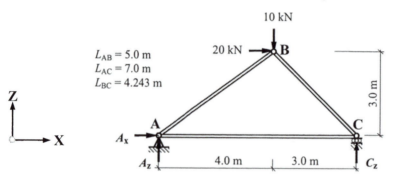

$L_{AB} = 5.0$ m
$L_{AC} = 7.0$ m
$L_{BC} = 4.243$ m

Figure 3.9

Construct a table in terms of tension coefficients and an X/Z co-ordinate system as shown in Table 3.1.

The equilibrium equations are solved in terms of the 't' values and hence the member forces and support reactions are evaluated and entered in the table as shown in Table 3.1.

Consider joint B:
There are only two unknowns and two equations, hence:
Adding both equations

$$-4t_{AB} + 3t_{BC} + 20 = 0$$
$$\underline{-3t_{AB} - 3t_{BC} - 10 = 0}$$
$$-7t_{AB} \qquad + 10 = 0 \qquad t_{AB} = +1.43$$

substitute for t_{AB} in the first equation $\qquad\qquad t_{BC} = -4.76$

Force in member　AB $= t_{AB} \times L_{AB} = +(1.43 \times 5.0) = +7.15$ kN　　**Tie**
Force in member　BC $= t_{BC} \times L_{BC} = -(4.76 \times 4.243) = -20.2$ kN　　**Strut**

Joints A and C can be considered in a similar manner until all unknown values, including reactions, have been determined.

The reader should complete this solution to obtain the following values: $F_{AC} = + 14.28$ kN
$A_x = + 20$ kN $A_z = - 4.29$ kN $C_z = + 14.28$ kN

Joint		Equilibrium Equations	Member	t	Length (m)	Force (kN)
A	X	$4t_{AB} + 7t_{AC} + \quad A_x = 0$	AB	+ 1.43	5.0	+ 7.15
	Z	$3t_{AB} \qquad\quad + \quad A_z = 0$	AC	?	7.0	?
			BC	– 4.76	4.243	– 20.20
B	X	$-4t_{AB} + 3t_{BC} + 20 \;= 0$	Support Reactions (kN)			
	Z	$-3t_{AB} - 3t_{BC} - 10 \;= 0$	Component		x	z
C	X	$-7t_{AC} - 3t_{BC} \qquad\quad = 0$	Support A			
	Z	$+3t_{BC} + C_z \qquad\qquad = 0$	Support C		zero	

Table 3.1

In the case of a space frame, each joint has three co-ordinates and the forces have components in the three orthogonal X, Z and Y directions. This leads to (3 × Number. of joints) equations which can be solved as above to determine the 't' values and subsequently the member forces and support reactions.

3.4.2 Example 3.3: Three-Dimensional Space Truss

The space frame shown in Figure 3.10 has three pinned supports at A, B and C, all of which lie on the same level as indicated. Member DE is horizontal and at a height of 10 m above the plane of the supports. The planar dimensions (z–x, x–y and z–y) of the frame are indicated in Figure 3.11.

Determine the forces in the members when the frame carries loads of 80 kN and 40 kN acting in a horizontal plane at joints E and D respectively as shown.

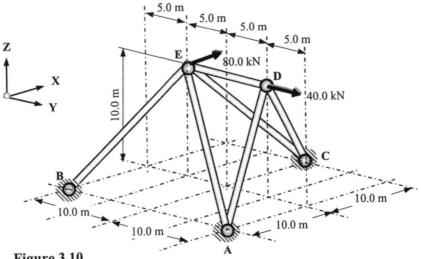

Figure 3.10

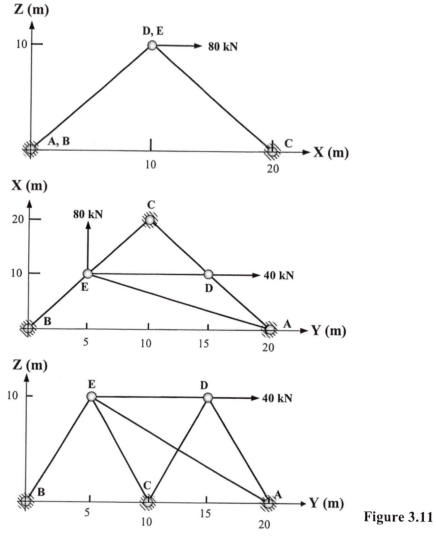

Figure 3.11

Solution:

Length of members: $L = \sqrt{\left(x^2 + y^2 + z^2\right)}$

$L_{DE} = 10.0$ m

$L_{AE} = \sqrt{\left(10.0^2 + 15.0^2 + 10.0^2\right)} = 20.62$ m

$L_{AD} = \sqrt{\left(10.0^2 + 5.0^2 + 10.0^2\right)} = 15.0$ m

$L_{BE} = \sqrt{\left(10.0^2 + 5.0^2 + 10.0^2\right)} = 15.0$ m

$L_{CD} = \sqrt{\left(10.0^2 + 5.0^2 + 10.0^2\right)} = 15.0$ m

$L_{CE} = \sqrt{\left(10.0^2 + 5.0^2 + 10.0^2\right)} = 15.0$ m

The equations from the Tension Coefficient Table are used to determine the 't' values. Since only three equations are available at any joint, only three unknowns can be determined at any one time, i.e. identify a joint with no more that three unknown member forces to begin the calculation; in this case the only suitable joint is D.

Solve the three simultaneous equations at joint D to determine the tension coefficients t_{AD}, t_{DE} and t_{CD}; i.e.

Consider Joint D: Equations (10), (11) and (12)

Equation (12)	$-10t_{AD} + 10t_{CD} = 0$	
Equation (11)	$+5t_{AD} - 5t_{CD} - 10t_{DE} + 40 = 0$	
Equation (10)	$-10t_{AD} - 10t_{CD} = 0$	

$$\left. \begin{array}{l} \\ \\ \\ \end{array} \right\} \Longrightarrow \begin{array}{l} t_{AD} = 0 \\ t_{DE} = +4.0 \\ t_{CD} = 0 \end{array}$$

Similarly for the next joint in which there are no more than three unknowns, i.e. Joint E

Consider Joint E: Equations (13), (14) and (15)

Equation (13)	$-10t_{AE} - 10t_{BE} + 10t_{CE} + 80 = 0$	
Equation (14)	$+15t_{AE} - 5t_{BE} + 5t_{CE} + 10t_{DE} = 0$	
Equation (15)	$-10t_{AE} - 10t_{BE} - 10t_{CE} = 0$	

$$\left. \begin{array}{l} \\ \\ \\ \end{array} \right\} \Longrightarrow \begin{array}{l} t_{AE} = 0 \\ t_{BE} = +4.0 \\ t_{CE} = -4.0 \end{array}$$

The support reactions can be determined after the tension coefficient values have been determined using Equations (1) to (9).

The sum of the reactions in the x, y and z directions should be checked by ensuring that they are equal and opposite to the applied load system.

Joint			Equilibrium Equations		Member	t	Length (m)	Force (kN)
1		X	$+10t_{AE} + 10t_{AD}$	$+ A_X = 0$	AD	0	15.0	0
2	A	Y	$-15t_{AE} - 5t_{AD}$	$+ A_Y = 0$	AE	0	20.62	0
3		Z	$+10t_{AE} + 10t_{AD}$	$+ A_Z = 0$	BE	$+4.0$	15.0	$+60.0$
4		X	$+ 10t_{BE}$	$+ B_X = 0$	CD	0	15.0	0
5	B	Y	$+ 5t_{BE}$	$+ B_Y = 0$	CE	-4.0	15.0	-60.0
6		Z	$+ 10t_{BE}$	$+ B_Z = 0$	DE	$+4.0$	10.0	$+40.0$
7		X	$- 10t_{CD} - 10t_{CE}$	$+ C_X = 0$	**Support Reactions (kN)**			
8	C	Y	$+ 5t_{CD} - 5t_{CE}$	$+ C_Y = 0$	**Component**	x	y	z
9		Z	$+10t_{CD} + 10t_{CE}$	$+ C_Z = 0$	**Support A**	zero	zero	zero
10		X	$-10t_{AD} + 10t_{CD}$	$= 0$	**Support B**	-40.0	-20.0	-40.0
11	D	Y	$+5t_{AD} - 5t_{CD} - 10t_{DE} + 40$	$= 0$	**Support C**	-40.0	-20.0	$+40.0$
12		Z	$-10t_{AD} - 10t_{CD}$	$= 0$	Σ Applied forces in X-direction $= +80$ kN			
13		X	$-10t_{AE} - 10t_{BE} + 10t_{CE} + 80 = 0$		Σ Applied forces in Y-direction $= +40$ kN			
14	E	Y	$+15t_{AE} - 5t_{BE} + 5t_{CE} + 10t_{DE} = 0$		Σ Applied forces in Z-direction $=$ zero			
15		Z	$-10t_{AE} - 10t_{BE} - 10t_{CE}$	$= 0$				

Table 3.2

3.4.3 Problems: Method of Tension Coefficients

The pin-jointed space-frames shown in Problems 3.11 to 3.16 have three pinned supports at A, B and C as indicated. In each case the supports A, B and C are in the same plane. Using the data given determine:

(i)   the member forces and
(ii)   the support reactions,

when the frames are subjected to the loading indicated.

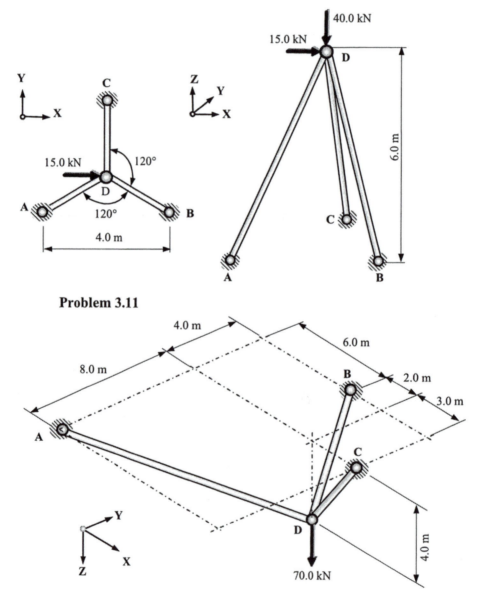

Problem 3.11

Problem 3.12

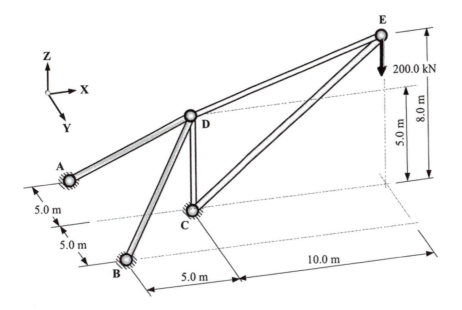

Problem 3.13

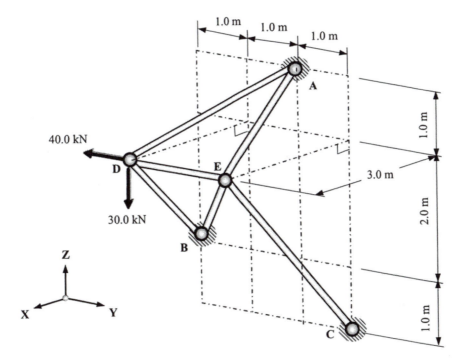

Problem 3.14

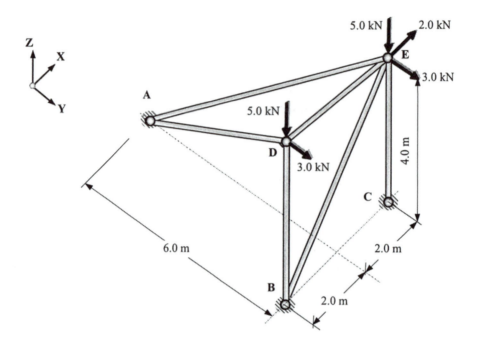

Problem 3.15

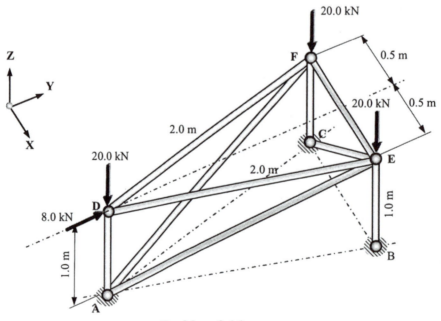

Problem 3.16

3.4.4 Solutions: Method of Tension Coefficients

Solution

Topic: Pin-Jointed Frames – Method of Tension Coefficients
Problem Number: 3.11   **Page No. 1**

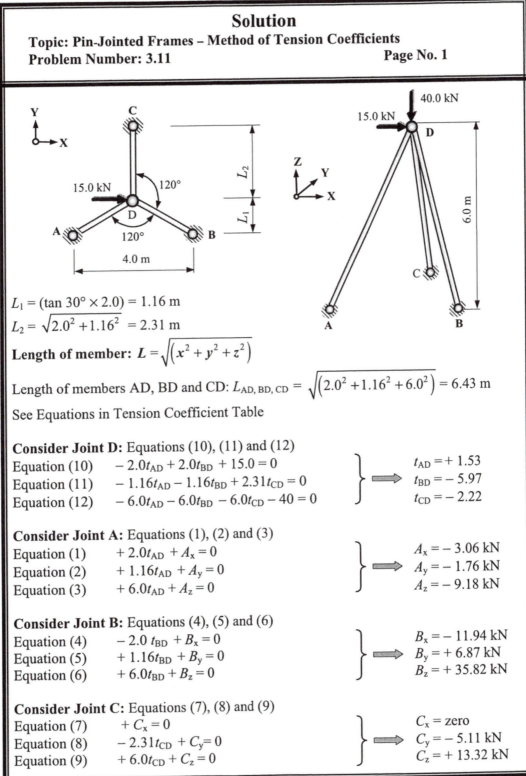

$L_1 = (\tan 30° \times 2.0) = 1.16$ m

$L_2 = \sqrt{2.0^2 + 1.16^2} = 2.31$ m

Length of member: $L = \sqrt{(x^2 + y^2 + z^2)}$

Length of members AD, BD and CD: $L_{AD,\ BD,\ CD} = \sqrt{(2.0^2 + 1.16^2 + 6.0^2)} = 6.43$ m

See Equations in Tension Coefficient Table

Consider Joint D: Equations (10), (11) and (12)

Equation (10) $-2.0t_{AD} + 2.0t_{BD} + 15.0 = 0$	$t_{AD} = +1.53$
Equation (11) $-1.16t_{AD} - 1.16t_{BD} + 2.31t_{CD} = 0$	$t_{BD} = -5.97$
Equation (12) $-6.0t_{AD} - 6.0t_{BD} - 6.0t_{CD} - 40 = 0$	$t_{CD} = -2.22$

Consider Joint A: Equations (1), (2) and (3)

Equation (1) $+2.0t_{AD} + A_x = 0$	$A_x = -3.06$ kN
Equation (2) $+1.16t_{AD} + A_y = 0$	$A_y = -1.76$ kN
Equation (3) $+6.0t_{AD} + A_z = 0$	$A_z = -9.18$ kN

Consider Joint B: Equations (4), (5) and (6)

Equation (4) $-2.0\ t_{BD} + B_x = 0$	$B_x = -11.94$ kN
Equation (5) $+1.16t_{BD} + B_y = 0$	$B_y = +6.87$ kN
Equation (6) $+6.0t_{BD} + B_z = 0$	$B_z = +35.82$ kN

Consider Joint C: Equations (7), (8) and (9)

Equation (7) $+C_x = 0$	$C_x = $ zero
Equation (8) $-2.31t_{CD} + C_y = 0$	$C_y = -5.11$ kN
Equation (9) $+6.0t_{CD} + C_z = 0$	$C_z = +13.32$ kN

Solution

Topic: Pin-Jointed Frames – Method of Tension Coefficients
Problem Number: 3.11   **Page No. 2**

Note: **+ve tension coefficient values indicate tension members**
−ve tension coefficient values indicate compression members

Joint			Equilibrium Equations		Member	t	Length (m)	Force (kN)
1		X	$+ 2.0\, t_{AD}$	$+ A_x = 0$	**AD**	$+ 1.53$	6.43	$+ 9.84$
2	A	Y	$+ 1.16\, t_{AD}$	$+ A_y = 0$	**BD**	$- 5.97$	6.43	$- 38.38$
3		Z	$+ 6.0\, t_{AD}$	$+ A_z = 0$	**CD**	$- 2.22$	6.43	$- 14.27$
4		X	$- 2.0\, t_{BD}$	$+ B_x = 0$				
5	B	Y	$+ 1.16\, t_{BD}$	$+ B_y = 0$				
6		Z	$+ 6.0\, t_{BD}$	$+ B_z = 0$				
7		X		$+ C_x = 0$				
8	C	Y	$- 2.31\, t_{CD}$	$+ C_y = 0$				
9		Z	$+ 6.0\, t_{CD}$	$+ C_z = 0$				
10		X	$- 2.0\, t_{AD} + 2.0\, t_{BD} + 15.0 \quad = 0$					
11	D	Y	$-1.16\, t_{AD} - 1.16\, t_{BD} + 2.31\, t_{CD} = 0$					
12		Z	$-6.0\, t_{AD} - 6.0\, t_{BD} - 6.0\, t_{CD} - 40 = 0$					

Support Reactions (kN)

Component	x	y	z
Support A	$- 3.06$	$- 1.76$	$- 9.18$
Support B	$- 11.94$	$+ 6.87$	$+ 35.82$
Support C	zero	$- 5.11$	$+ 13.32$

Σ **Applied forces in X-direction = + 15.0 kN**

Σ **Applied forces in Y-direction = zero**

Σ **Applied forces in Z-direction = − 40.0 kN**

Solution

Topic: Pin-Jointed Frames – Method of Tension Coefficients

Problem Number: 3.12   **Page No. 1**

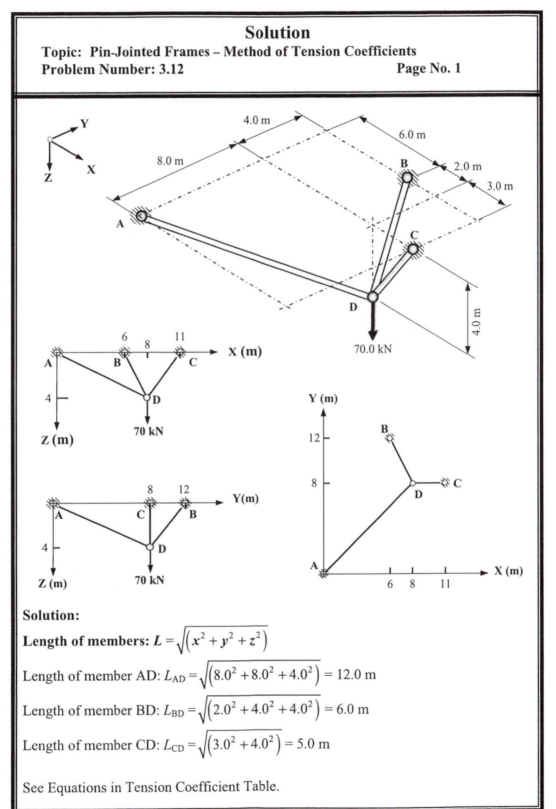

Solution:

Length of members: $L = \sqrt{\left(x^2 + y^2 + z^2\right)}$

Length of member AD: $L_{AD} = \sqrt{\left(8.0^2 + 8.0^2 + 4.0^2\right)} = 12.0$ m

Length of member BD: $L_{BD} = \sqrt{\left(2.0^2 + 4.0^2 + 4.0^2\right)} = 6.0$ m

Length of member CD: $L_{CD} = \sqrt{\left(3.0^2 + 4.0^2\right)} = 5.0$ m

See Equations in Tension Coefficient Table.

Solution
Topic: Pin-Jointed Frames – Method of Tension Coefficients
Problem Number: 3.12   **Page No. 2**

Consider Joint D: Equations (10), (11) and (12)

Equation (1) $-8.0t_{AD} - 2.0t_{BD} + 3t_{CD} = 0$

Equation (2) $-8.0t_{AD} + 4.0t_{BD} = 0$

Equation (3) $-4.0t_{AD} - 4.0t_{BD} - 4.0t_{CD} + 70.0 = 0$

$\Longrightarrow$ $t_{AD} = +2.5$ kN
$t_{BD} = +5.0$ kN
$t_{CD} = +10.0$ kN

Consider Joint A: Equations (1), (2) and (3)

Equation (1) $+8.0t_{AD} + A_x = 0$

Equation (2) $+8.0t_{AD} + A_y = 0$

Equation (3) $+4.0t_{AD} + A_z = 0$

$\Longrightarrow$ $A_x = -20.0$ kN
$A_y = -20.0$ kN
$A_z = -10.0$ kN

Consider Joint B: Equations (4), (5) and (6)

Equation (4) $+2.0t_{BD} + B_x = 0$

Equation (5) $-4.0t_{BD} + B_y = 0$

Equation (6) $+4.0t_{BD} + B_z = 0$

$\Longrightarrow$ $B_x = -10.0$ kN
$B_y = +20.0$ kN
$B_z = -20.0$ kN

Consider Joint C: Equations (7), (8) and (9)

Equation (7) $-3.0t_{CD} + C_x = 0$

Equation (8) $+C_y = 0$

Equation (9) $+4.0t_{CD} + C_z = 0$

$\Longrightarrow$ $C_x = +30.0$ kN
$C_y =$ zero
$C_z = -40.0$ kN

Note: +ve tension coefficient values indicate tension members
**   –ve tension coefficient values indicate compression members**

Joint			Equilibrium Equations		Member	t	Length (m)	Force (kN)
1		X	$+8.0t_{AD}$	$+A_x = 0$	**AD**	$+2.5$	12.0	$+30.0$
2	A	Y	$+8.0t_{AD}$	$+A_y = 0$	**BD**	$+5.0$	6.0	$+30.0$
3		Z	$+4.0t_{AD}$	$+A_z = 0$	**CD**	$+10.0$	5.0	$+50.0$
4		X	$+2.0t_{BD}$	$+B_x = 0$	**Support Reactions (kN)**			
5	B	Y	$-4.0t_{BD}$	$+B_y = 0$	**Component**	x	y	z
6		Z	$+4.0t_{BD}$	$+B_z = 0$	**Support A**	-20	-20	-10
7		X	$-3.0t_{CD}$	$+C_x = 0$	**Support B**	-10	$+20$	-20
8	C	Y		$+C_y = 0$	**Support C**	$+30$	zero	-40
9		Z	$+4.0t_{CD}$	$+C_z = 0$				
10		X	$-8.0t_{AD} - 2.0t_{BD} + 3t_{CD}$	$= 0$	Σ Applied forces in X-direction = zero			
11	D	Y	$-8.0t_{AD} + 4.0t_{BD}$	$= 0$	Σ Applied forces in Y-direction = zero			
12		Z	$-4.0t_{AD} - 4.0t_{BD} - 4.0t_{CD} + 70.0 = 0$		Σ Applied forces in Z-direction = $+70$ kN			

Solution

Topic: Pin-Jointed Frames – Method of Tension Coefficients

Problem Number: 3.13　　　　　　　　　　　　　　**Page No. 1**

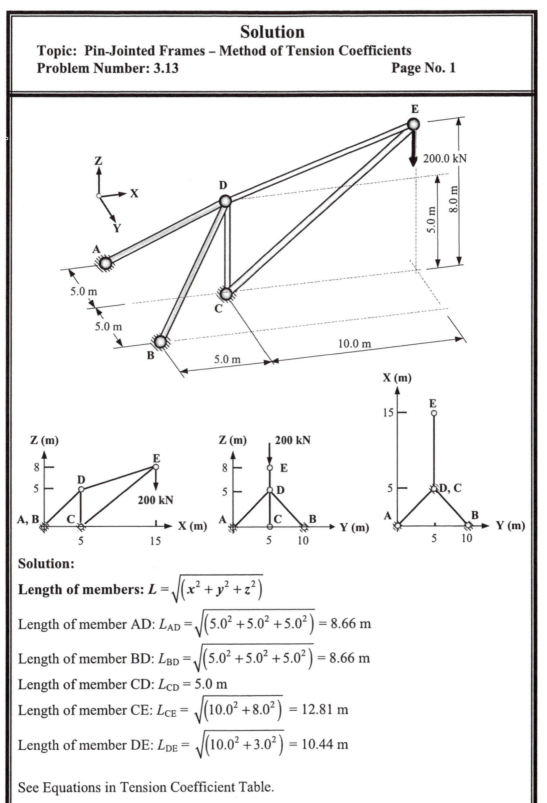

Solution:

Length of members: $L = \sqrt{(x^2 + y^2 + z^2)}$

Length of member AD: $L_{AD} = \sqrt{(5.0^2 + 5.0^2 + 5.0^2)} = 8.66$ m

Length of member BD: $L_{BD} = \sqrt{(5.0^2 + 5.0^2 + 5.0^2)} = 8.66$ m

Length of member CD: $L_{CD} = 5.0$ m

Length of member CE: $L_{CE} = \sqrt{(10.0^2 + 8.0^2)} = 12.81$ m

Length of member DE: $L_{DE} = \sqrt{(10.0^2 + 3.0^2)} = 10.44$ m

See Equations in Tension Coefficient Table.

Solution

Topic: Pin-Jointed Frames – Method of Tension Coefficients
Problem Number: 3.13   **Page No. 2**

Consider Joint E: Equations (13) and (15)
Equation (13) $-10.0t_{DE} - 10.0t_{CE} = 0$
Equation (15) $-3.0t_{DE} - 8.0t_{CE} - 200 = 0$

$\Rightarrow$ $t_{CE} = -40.0$
$t_{DE} = +40.0$

Consider Joint D: Equations (10), (11) and (12)
Equation (10) $-5.0t_{AD} - 5.0t_{BD} + 10.0t_{DE} = 0$
Equation (11) $-5.0t_{AD} + 5.0t_{BD} = 0$
Equation (12) $-5.0t_{AD} - 5.0t_{BD} + 3.0t_{DE} - 5.0t_{CD} = 0$

$\Rightarrow$ $t_{AD} = +40.0$
$t_{BD} = +40.0$
$t_{CD} = -56.0$

Similarly, the support reactions can be obtained by substituting the values of the tension coefficients in Equations (1) to (9).

Note: +ve tension coefficient values indicate tension members
            −ve tension coefficient values indicate compression members

Joint			Equilibrium Equations		Member	t	Length (m)	Force (kN)
1		X	$+5.0t_{AD}$	$+A_x = 0$	**AD**	+40.0	8.66	**+346.4**
2	A	Y	$+5.0t_{AD}$	$+A_y = 0$	**BD**	+40.0	8.66	**+346.6**
3		Z	$+5.0t_{AD}$	$+A_z = 0$	**CD**	−56.0	5.0	**−280.0**
4		X	$+5.0t_{BD}$	$+B_x = 0$	**CE**	−40.0	12.81	**−512.2**
5	B	Y	$-5.0t_{BD}$	$+B_y = 0$	**DE**	+40.0	10.44	**+417.6**
6		Z	$+5.0t_{BD}$	$+B_z = 0$	**Support Reactions (kN)**			
7		X	$+10.0t_{CE}$	$+C_x = 0$	**Component**	x	y	z
8	C	Y		$+C_y = 0$	**Support A**	−200	−200	−200
9		Z	$+5.0t_{CD} + 8.0t_{CE}$	$+C_z = 0$	**Support B**	−200	+200	−200
10		X	$-5.0t_{AD} - 5.0t_{BD} + 10.0t_{DE}$	$= 0$	**Support C**	+400	zero	+600
11	D	Y	$-5.0t_{AD} + 5.0t_{BD}$	$= 0$				
12		Z	$-5.0t_{AD} - 5.0t_{BD} + 3.0t_{DE} - 5.0t_{CD} = 0$		**Σ Applied forces in X-direction = zero**			
13		X	$-10.0t_{DE} - 10.0t_{CE}$	$= 0$	**Σ Applied forces in Y-direction = zero**			
14	E	Y			**Σ Applied forces in Z-direction = − 200 kN**			
15		Z	$-3.0t_{DE} - 8.0t_{CE} - 200$	$= 0$				

Solution

Topic: Pin-Jointed Frames – Method of Tension Coefficients
Problem Number: 3.14   **Page No. 1**

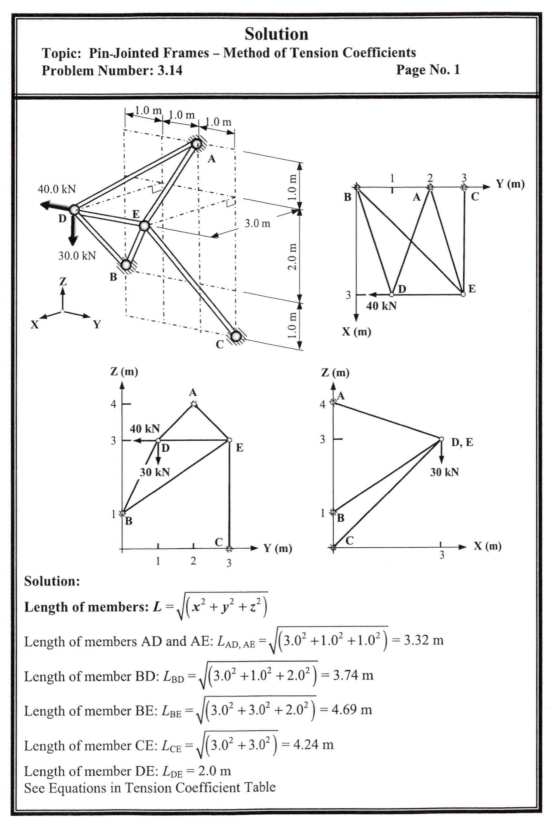

Solution:

Length of members: $L = \sqrt{(x^2 + y^2 + z^2)}$

Length of members AD and AE: $L_{AD, AE} = \sqrt{(3.0^2 + 1.0^2 + 1.0^2)} = 3.32$ m

Length of member BD: $L_{BD} = \sqrt{(3.0^2 + 1.0^2 + 2.0^2)} = 3.74$ m

Length of member BE: $L_{BE} = \sqrt{(3.0^2 + 3.0^2 + 2.0^2)} = 4.69$ m

Length of member CE: $L_{CE} = \sqrt{(3.0^2 + 3.0^2)} = 4.24$ m

Length of member DE: $L_{DE} = 2.0$ m
See Equations in Tension Coefficient Table

Solution
Topic: Pin-Jointed Frames – Method of Tension Coefficients
Problem Number: 3.14   **Page No. 2**

Consider Joint D: Equations (10), (11) and (12)

Equation (10) $-3.0t_{AD} - 3t_{BD} = 0$

Equation (11) $+1.0t_{AD} - 1.0t_{BD} + 2.0t_{DE} - 40.0 = 0$ $\Longrightarrow$ $t_{AD} = +10.0$
$t_{BD} = -10.0$
Equation (12) $+1.0t_{AD} - 2.0t_{BD} - 30.0 = 0$ $t_{DE} = +10.0$

Consider Joint E: Equations (13), (14) and (15)

Equation (13) $-3.0t_{AE} - 3.0t_{BE} - 3.0t_{CE} = 0$

Equation (14) $-1.0t_{AE} - 3.0t_{BE} - 2.0t_{DE} = 0$ $\Longrightarrow$ $t_{AE} = +1.82$
$t_{BE} = -7.28$
Equation (15) $+1.0t_{AE} - 2.0t_{BE} - 3.0t_{CE} = 0$ $t_{CE} = +5.46$

Similarly, the support reactions can be obtained by substituting the values of the tension coefficients in Equations (1) to (9).

Note: +ve tension coefficient values indicate tension members
−ve tension coefficient values indicate compression members

Joint			Equilibrium Equations			Member	t	Length (m)	Force (kN)
1		X	$+3.0t_{AD}$	$+3.0\,t_{AE}$	$+A_x = 0$	**AD**	+10.0	3.32	**+33.20**
2	A	Y	$-1.0t_{AD}$	$+1.0t_{AE}$	$+A_y = 0$	**AE**	+1.82	3.32	**+6.03**
3		Z	$-1.0t_{AD}$	$-1.0t_{AE}$	$+A_z = 0$	**BD**	−10.0	3.74	**−37.40**
4		X	$+3.0t_{BD}$	$+3.0t_{BE}$	$+B_x = 0$	**BE**	−7.28	4.69	**−34.14**
5	B	Y	$+1.0t_{BD}$	$+3.0t_{BE}$	$+B_y = 0$	**CE**	+5.46	4.24	**+23.17**
6		Z	$+2.0t_{BD}$	$+2.0t_{BE}$	$+B_z = 0$	**DE**	+10.0	2.0	**+20.0**
7		X	$+3.0t_{CE}$		$+C_x = 0$	**Support Reactions**			
8	C	Y			$+C_y = 0$	Component	x	y	z
9		Z	$+3.0t_{CE}$		$+C_z = 0$	**Support A**	− 35.5 kN	+ 8.2 kN	+ 11.8 kN
10		X	$-3.0t_{AD}$	$-3t_{BD}$	$= 0$	**Support B**	+ 51.8 kN	+ 31.8 kN	+ 34.6 kN
11	D	Y	$+1.0t_{AD}$	$-1.0t_{BD}$	$+2.0t_{DE} - 40.0 = 0$	**Support C**	− 16.4 kN	zero	− 16.4 kN
12		Z	$+1.0t_{AD}$	$-2.0t_{BD}$	$-30.0 = 0$	Σ Applied forces in X-direction = zero			
13		X	$-3.0t_{AE}$	$-3.0t_{BE}$	$-3.0t_{CE}$ $= 0$	Σ Applied forces in Y-direction = − 40 kN			
14	E	Y	$-1.0t_{AE}$	$-3.0t_{BE}$	$-2.0t_{DE} = 0$	Σ Applied forces in Z-direction = − 30 kN			
15		Z	$+1.0t_{AE}$	$-2.0t_{BE}$	$-3.0t_{CE}$ $= 0$				

Solution

Topic: Pin-Jointed Frames – Method of Tension Coefficients

Problem Number: 3.15   **Page No. 1**

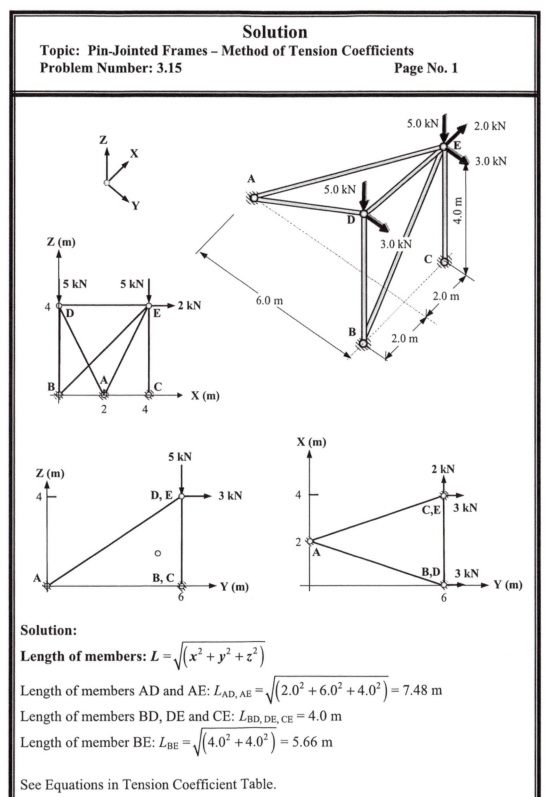

Solution:

Length of members: $L = \sqrt{\left(x^2 + y^2 + z^2\right)}$

Length of members AD and AE: $L_{AD,\,AE} = \sqrt{\left(2.0^2 + 6.0^2 + 4.0^2\right)} = 7.48$ m

Length of members BD, DE and CE: $L_{BD,\,DE,\,CE} = 4.0$ m

Length of member BE: $L_{BE} = \sqrt{\left(4.0^2 + 4.0^2\right)} = 5.66$ m

See Equations in Tension Coefficient Table.

Solution
Topic: Pin-Jointed Frames – Method of Tension Coefficients
Problem Number: 3.15   Page No. 2

Consider Joint D: Equations (10), (11) and (12)

Equation (10) $+ 2.0t_{AD} + 4.0t_{DE} = 0$

Equation (11) $- 6.0t_{AD} + 3.0 = 0$

Equation (12) $- 4.0t_{AD} - 4.0t_{BD} - 5.0 = 0$

$\left. \right\} \implies$
$t_{AD} = + 0.5$
$t_{BD} = - 1.75$
$t_{DE} = - 0.25$

Consider Joint E: Equations (13), (14) and (15)

Equation (13) $- 2.0t_{AE} - 4.0t_{DE} - 4.0t_{BE} + 2.0 = 0$

Equation (14) $- 6.0t_{AE} + 3.0 = 0$

Equation (15) $- 4.0t_{AE} - 4.0t_{BE} - 4.0t_{CE} - 5.0 = 0$

$\left. \right\} \implies$
$t_{AE} = + 0.5$
$t_{BE} = + 0.5$
$t_{CE} = - 2.25$

Similarly, the support reactions can be obtained by substituting the values of the tension coefficients in Equations (1) to (9).

Note: +ve tension coefficient values indicate tension members

          −ve tension coefficient values indicate compression members

Joint			Equilibrium Equations		Member	t	Length (m)	Force (kN)
1		X	$- 2.0t_{AD} + 2.0t_{AE}$	$+ A_x = 0$	AD	+ 0.5	7.48	+ 3.74
2	A	Y	$+ 6.0t_{AD} + 6.0t_{AE}$	$+ A_y = 0$	AE	+ 0.5	7.48	+ 3.74
3		Z	$+ 4.0t_{AD} + 4.0t_{AE}$	$+ A_z = 0$	BD	− 1.75	4.0	− 7.0
4		X	$+ 4.0t_{BE}$	$+ B_x = 0$	BE	+ 0.5	5.66	+ 2.83
5	B	Y		$+ B_y = 0$	CE	− 2.25	4.0	− 9.0
6		Z	$+ 4.0t_{BE} + 4.0t_{BD}$	$+ B_z = 0$	DE	− 0.25	4.0	− 1.0
7		X		$+ C_x = 0$	Support Reactions (kN)			
8	C	Y		$+ C_y = 0$	Component	x	y	z
9		Z	$+ 4.0t_{CE}$	$+ C_z = 0$	Support A	zero	− 6.0 kN	− 4.0 kN
10		X	$+ 2.0t_{AD} + 4.0t_{DE}$	$= 0$	Support B	− 2.0 kN	zero	+ 5.0 kN
11	D	Y	$- 6.0t_{AD}$	$+ 3.0 = 0$	Support C	zero	zero	+ 9.0 kN
12		Z	$- 4.0t_{AD} - 4.0t_{BD}$	$- 5.0 = 0$	Σ Applied forces in X-direction = + 2 kN			
13		X	$- 2.0t_{AE} - 4.0t_{DE} - 4.0t_{BE}$	$+ 2.0 = 0$	Σ Applied forces in Y-direction = + 6 kN			
14	E	Y	$- 6.0t_{AE}$	$+ 3.0 = 0$	Σ Applied forces in Z-direction = + 10 kN			
15		Z	$- 4.0t_{AE} - 4.0t_{BE} - 4.0t_{CE} - 5.0 = 0$					

Solution

Topic: Pin-Jointed Frames – Method of Tension Coefficients
Problem Number: 3.16   **Page No. 1**

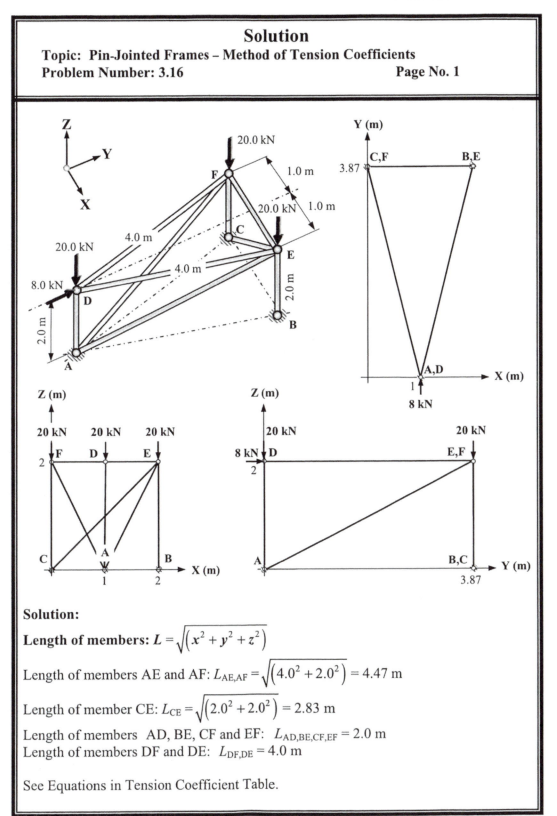

Solution:

Length of members: $L = \sqrt{\left(x^2 + y^2 + z^2\right)}$

Length of members AE and AF: $L_{AE,AF} = \sqrt{\left(4.0^2 + 2.0^2\right)} = 4.47$ m

Length of member CE: $L_{CE} = \sqrt{\left(2.0^2 + 2.0^2\right)} = 2.83$ m

Length of members AD, BE, CF and EF: $L_{AD,BE,CF,EF} = 2.0$ m
Length of members DF and DE: $L_{DF,DE} = 4.0$ m

See Equations in Tension Coefficient Table.

Solution

Topic: Pin-Jointed Frames – Method of Tension Coefficients
Problem Number: 3.16   **Page No. 2**

Consider Joint D: Equations (10), (11) and (12)

Equation (10) $+1.0t_{DE} - 1.0t_{DF} = 0$

Equation (11) $+3.87t_{DE} + 3.87t_{DF} + 8.0 = 0$

Equation (12) $-2.0t_{AD} - 20.0 = 0$

$\left.\begin{array}{l}\\\\\end{array}\right\} \Longrightarrow$ $t_{AD} = -10.0$, $t_{DE} = -1.03$, $t_{DF} = -1.03$

Consider Joint F: Equations (16), (17) and (18)

Equation (16) $+1.0t_{AF} + 1.0t_{DF} + 2.0t_{EF} = 0$

Equation (17) $-3.87t_{AF} - 3.87t_{DF} = 0$

Equation (18) $-2.0t_{CF} - 2.0t_{AF} - 20.0 = 0$

$\left.\begin{array}{l}\\\\\end{array}\right\} \Longrightarrow$ $t_{AF} = +1.03$, $t_{EF} = $ zero, $t_{CF} = -11.03$

Consider Joint E: Equations (13), (14) and (15)

Equation (13) $-1.0t_{AE} - 2.0t_{CE} - 1.0t_{DE} - 2.0t_{EF} = 0$

Equation (14) $-3.87t_{AE} - 3.87t_{DE} = 0$

Equation (15) $-2.0t_{AE} - 2.0t_{CE} - 2.0t_{BE} - 20.0 = 0$

$\left.\begin{array}{l}\\\\\end{array}\right\} \Longrightarrow$ $t_{AE} = +1.03$, $t_{BE} = -11.03$, $t_{CE} = $ zero

Similarly, the support reactions can be obtained by substituting the values of the tension coefficients in Equations (1) to (9).

Note: +ve tension coefficient values indicate tension members

−ve tension coefficient values indicate compression members

Joint			Equilibrium Equations		Member	t	Length (m)	Force (kN)
1		X	$+1.0t_{AE} - 1.0t_{AF}$	$+A_x = 0$	AD	−10.0	2.0	−20.0
2	A	Y	$+3.87t_{AE} + 3.87t_{AF}$	$+A_y = 0$	AE	+1.03	4.47	+4.61
3		Z	$+2.0t_{AE} + 2.0t_{AF} + 2.0t_{AD}$	$+A_z = 0$	AF	+1.03	4.47	+4.61
4		X		$+B_x = 0$	BE	−11.03	2.0	−22.06
5	B	Y		$+B_y = 0$	CE	zero	2.83	zero
6		Z	$+2.0t_{BE}$	$+B_z = 0$	CF	−11.03	2.0	−22.06
7		X	$+2.0t_{CE}$	$+C_x = 0$	DE	−1.03	4.0	−4.13
8	C	Y		$+C_y = 0$	DF	−1.03	4.0	−4.13
9		Z	$+2.0t_{CE} + 2.0t_{CF}$	$+C_z = 0$	EF	zero	2.0	zero

Joint			Equilibrium Equations		Support Reactions (kN)			
10		X	$+1.0t_{DE} - 1.0t_{DF}$	$= 0$	**Component**	**x**	**y**	**z**
11	D	Y	$+3.87t_{DE} + 3.87t_{DF} + 8.0$	$= 0$	Support A	zero	−8.0	+15.9
12		Z	$-2.0t_{AD}$ -20.0	$= 0$	Support B	zero	zero	+22.1
13		X	$-1.0t_{AE} - 2.0t_{CE} - 1.0t_{DE} - 2.0t_{EF} = 0$		Support C	zero	zero	+22.0
14	E	Y	$-3.87t_{AE}$ $-3.87t_{DE}$	$= 0$				
15		Z	$-2.0t_{AE} - 2.0t_{CE} - 2.0t_{BE} - 20.0 = 0$		Σ Applied forces in X-direction = zero			
16		X	$+1.0t_{AF} + 1.0t_{DF} + 2.0t_{EF}$	$= 0$	Σ Applied forces in Y-direction = +8 kN			
17	F	Y	$-3.87t_{AF} - 3.87t_{DF}$	$= 0$	Σ Applied forces in Z-direction = −60 kN			
18		Z	$-2.0t_{AF} - 2.0t_{CF}$ -20.0	$= 0$				

Pin-Jointed Frames 113

3.5   *Unit Load Method for Deflection*

The Unit Load Method of analysis is based on the principles of strain energy and Castigliano's 1st Theorem. When structures deflect under load the **work-done** by the displacement of the applied loads is stored in the members of the structure in the form of *strain energy*.

3.5.1 *Strain Energy (Axial Load Effects)*

Consider an axially loaded structural member of length '*L*', cross-sectional area '*A*', and of material with modulus of elasticity '*E*' as shown in Figure 3.12(a).

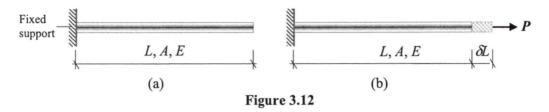

Fixed support

L, A, E   *L, A, E* δL

(a)   (b)

Figure 3.12

When an axial load '*P*' is applied as indicated, the member will increase in length by 'δL' as shown in Figure 3.12(b). Assuming linear elastic behaviour, $\delta L \propto P$, this relationship is represented graphically in Figure 3.13.

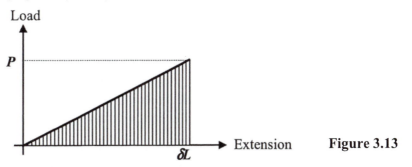

Load

P

δL Extension   **Figure 3.13**

The work-done by the externally applied load '*P*' is equal to:

(average value of the force × distance through which the force moves in its line of action)

i.e.   Work-done $= \left(\dfrac{P}{2} \times \delta L \right)$

For linearly elastic materials the relationship between the applied axial load and the change in length is:

$$\delta L = \frac{PL}{AE}$$

∴   Work-done $= \left(\dfrac{P}{2} \times \delta L \right) = \left(\dfrac{P}{2} \times \dfrac{PL}{AE} \right) = \dfrac{P^2 L}{2AE}$

This work-done by the externally applied load is equal to the 'energy' stored by the member when it changes length and is known as the strain energy, usually given the symbol 'U'. It is this energy which causes structural members to return to their original length when an applied load system is removed; (**Note:** it is assumed that the strains are within the elastic limits of the material.)

Strain energy = Work-done by the applied load system

$$U = \frac{P^2 L}{2AE}$$

(**Note:** the principles of strain energy also apply to members subject to shear, bending, torsion etc.)

3.5.2 Castigliano's 1st Theorem

Castigliano's 1st Theorem relating to strain energy and structural deformation can be expressed as follows:

'If the total strain energy in a structure is partially differentiated with respect to an applied load the result is equal to the displacement of that load in its line of action.'

In mathematical terms this is:

$$\Delta = \frac{\partial U}{\partial W}$$

where:
U is the total strain energy of the structure due to the applied load system,
W is the force acting at the point where the displacement is required,
Δ is the linear displacement in the direction of the line of action of W.

This form of the theorem is very useful in obtaining the deflection at joints in pin-jointed structures. Consider the pin-jointed frame shown in Figure 3.14 in which it is required to determine the vertical deflection of joint B.

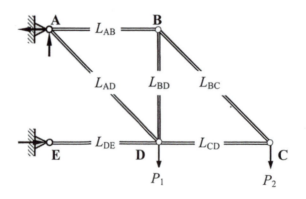

Figure 3.14

Step 1:
The member forces induced by the applied load system are calculated, in this case referred to as the '*P*' forces, as shown in Figure 3.15.

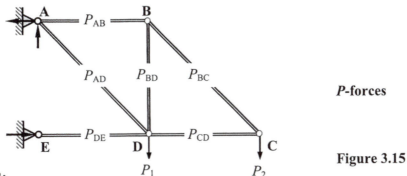

P-forces

Figure 3.15

Step 2:
The applied load system is removed from the structure and an imaginary Unit load is applied at the joint and in the direction of the required deflection, i.e. a vertical load equal to 1.0 at joint B. The resulting member forces due to the unit load are calculated and referred to as the '*u*' forces, as shown in Figure 3.16.

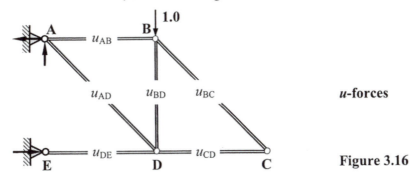

u-forces

Figure 3.16

If both the Step 1 and the Step 2 load systems are considered to act simultaneously, then by superposition the total force in each member is given by:

$$Q = (P + \beta u)$$

where:
P is the force due to the applied load system
u is the force due to the applied imaginary Unit load applied at B
β is a multiplying factor to reflect the value of the load applied at B (Since the unit load is an imaginary force the value of β = zero and is used here as a mathematical convenience.)

The total strain energy in the structure is equal to the sum of the energy stored in all the members:

$$U = \sum \frac{Q^2 L}{2AE}$$

Using Castigliano's 1st Theorem the deflection of joint B is given by:

$$\Delta = \frac{\partial U}{\partial W}$$

$$\therefore \; \Delta_\beta = \frac{\partial U}{\partial \beta} = \frac{\partial U}{\partial Q} \times \frac{\partial Q}{\partial \beta}$$

and

$$\frac{\partial U}{\partial Q} = \sum \frac{QL}{AE}; \quad \frac{\partial Q}{\partial \beta} = u$$

$$\therefore \; \Delta_\beta = \frac{\partial U}{\partial \beta} = \frac{\partial U}{\partial Q} \times \frac{\partial Q}{\partial \beta} = \sum \frac{QL}{AE} \times u = \sum \frac{(P + \beta u)L}{AE} \times u$$

Since $\beta =$ zero the vertical deflection at B (Δ_β) is given by:

$$\Delta_\beta = \sum \frac{PL}{AE} u$$

i.e. the deflection at any joint in a pin-jointed frame can be determined from:

$$\delta = \sum \frac{PL}{AE} u$$

where:
- δ is the displacement of the point of application of any load, along the line of action of that load,
- P is the force in a member due to the externally applied loading system,
- u is the force in a member due to a ***unit load*** acting at the position of, and in the direction of the desired displacement,
- L/A is the ratio of the length to the cross-sectional area of the members,
- E is the modulus of elasticity of the material for each member (i.e. Young's Modulus).

3.5.3 Example 3.4: Deflection of a Pin-Jointed Truss

A pin-jointed truss ABCD is shown in Figure 3.17 in which both a vertical and a horizontal load are applied at joint B as indicated. Determine the magnitude and direction of the resultant deflection at joint B and the vertical deflection at joint D.

Assume the cross-sectional area of all members is equal to A and all members are made from the same material, i.e. have the same modulus of elasticity E

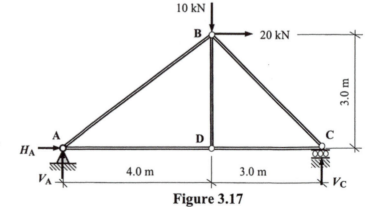

Figure 3.17

Step 1:   Evaluate the member forces. The reader should follow the procedure given in Example 3.1 to determine the following results:

Horizontal component of reaction at support A $H_A = -20.0$ kN
Vertical component of reaction at support A $V_A = -4.29$ kN
Vertical component of reaction at support C $V_C = +14.29$ kN

Use the *method of sections* or *joint resolution* as indicated in Section 3.2 and Section 3.3 respectively to determine the magnitude and sense of the unknown member forces (i.e. the **P** forces).
The reader should complete this calculation to determine the member forces as indicated in Figure 3.18.

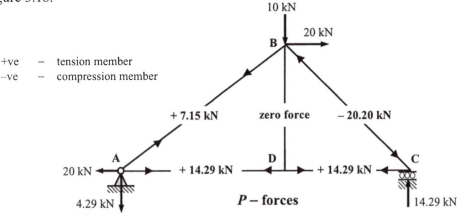

Figure 3.18

Step 2:   To determine the vertical deflection at joint B remove the externally applied load system and apply a **unit load only** in a vertical direction at joint B as shown in Figure 3.19. Use the *method of sections* or *joint resolution* as before to determine the magnitude and sense of the unknown member forces (i.e. the **u** forces).
The reader should complete this calculation to determine the member forces as indicated in Figure 3.19.

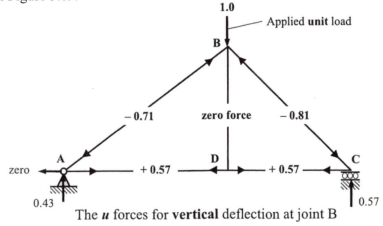

The **u** forces for **vertical** deflection at joint B

Figure 3.19

The vertical deflection $\delta_{V,B} = \sum \dfrac{PL}{AE} u$

This is better calculated in tabular form as shown in Table 3.3.

Member	Length (L)	Cross-section (A)	Modulus (E)	P forces (kN)	u forces	PL × u (kNm)
AB	5.0 m	A	E	+ 7.15	− 0.71	− 25.38
BC	4.24 m	A	E	− 20.20	− 0.81	+ 69.37
AD	4.0 m	A	E	+ 14.29	+ 0.57	+ 32.58
CD	3.0 m	A	E	+ 14.29	+ 0.57	+ 24.44
BD	3.0 m	A	E	0.0	0.0	0.0
					Σ	+ 101.01

Table 3.3

The +ve sign indicates that the deflection is in the same direction as the applied unit load.

Hence the vertical deflection $\delta_{V,B} = \sum \dfrac{PL}{AE} u = + (101.01/AE)$ ↓

Note: Where the members have different cross-sectional areas and/or moduli of elasticity each entry in the last column of the table should be based on $(PL \times u)/AE$ and not only $(PL \times u)$.

A similar calculation can be carried out to determine the horizontal deflection at joint B. The reader should complete this calculation to determine the member forces as indicated in Figure 3.20.

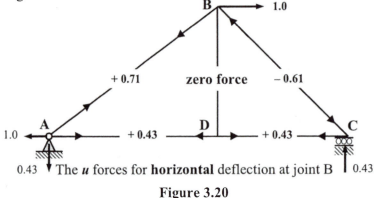

Figure 3.20

The horizontal deflection $\delta_{H,B} = \sum \dfrac{PL}{AE} u$

Member	Length (L)	Cross-section (A)	Modulus (E)	P forces (kN)	u forces	PL × u (kNm)
AB	5.0 m	A	E	+ 7.15	+ 0.71	+ 25.74
BC	4.24 m	A	E	− 20.20	− 0.61	+ 52.25
AD	4.0 m	A	E	+ 14.29	+ 0.43	+ 24.58
CD	3.0 m	A	E	+ 14.29	+ 0.43	+ 18.43
BD	3.0 m	A	E	0.0	0.0	0.0
					Σ	+ 121.00

Table 3.4

Hence the horizontal deflection $\delta_{\text{H,B}} = \sum \dfrac{PL}{AE}u = + (121.00/AE)$ →

The resultant deflection at joint B can be determined from the horizontal and vertical components evaluated above, i.e.

$R = \sqrt{\left(101.01^2 + 121.0^2\right)} / AE = 157.62/AE$

$\theta = \tan^{-1}(121.00/101.01) = 50.15°$

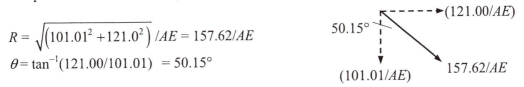

A similar calculation can be carried out to determine the vertical deflection at joint D. The reader should complete this calculation to determine the member forces as indicated in Figure 3.21.

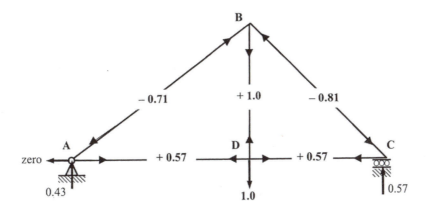

The member u forces for **vertical** deflection at joint D

Figure 3.21

The vertical deflection $\delta_{\text{V,D}} = \sum \dfrac{PL}{AE}u$

Member	Length (L)	Cross-section (A)	Modulus (E)	P forces (kN)	u forces	$PL \times u$ (kNm)
AB	5.0 m	A	E	+ 7.15	− 0.71	− 25.38
BC	4.24 m	A	E	− 20.20	− 0.81	+ 69.37
AD	4.0 m	A	E	+ 14.29	+ 0.57	+ 32.58
CD	3.0 m	A	E	+ 14.29	+ 0.57	+ 24.44
BD	3.0 m	A	E	0.0	+1.0	0.0
					Σ	+ 101.01

Table 3.5

Hence the vertical deflection $\delta_{\text{V,D}} = \sum \dfrac{PL}{AE}u = + (101.01/AE)$ ↓

3.5.3.1   Fabrication Errors – (Lack-of-fit)

During fabrication it is not unusual for a member length to be slightly too short or too long and assembly is achieved by forcing members in to place. The effect of this can be accommodated very easily in this method of analysis by adding additional terms relating to each member for which lack-of-fit applies. The δL term for the relevant members is equal to the magnitude of the error in length, i.e. Δ_L where negative values relate to members which are too short and positive values to members which are too long.

(**Note:** under normal applied loading the δL term $= \dfrac{PL}{AE}$).

3.5.3.2   Changes in Temperature

The effects of temperature change in members can also be accommodated in a similar manner; in this case the δL term is related to the coefficient of thermal expansion for the material, the change in temperature and the original length,

i.e. $\delta L = \alpha L \Delta_T$
where
α is the coefficient of thermal expansion,
L is the original length,
Δ_T is the change in temperature – a reduction being considered negative and an increase being positive.

Since this is an elastic analysis the principle of superposition can be used to obtain results when a combination of applied load, lack-of-fit and/or temperature difference occurs. This is illustrated in Example 3.5.

3.5.4 *Example 3.5: Lack-of-fit and Temperature Difference*

Consider the frame indicated in Example 3.4 and determine the vertical deflection at joint D assuming the existing loading and that member BC is too short by 2.0 mm, member CD is too long by 1.5 mm and that members AD and CD are both subject to an increase in temperature of 5°C. Assume $\alpha = 12.0 \times 10^{-6}/°C$ and $AE = 100 \times 10^3$ kN.

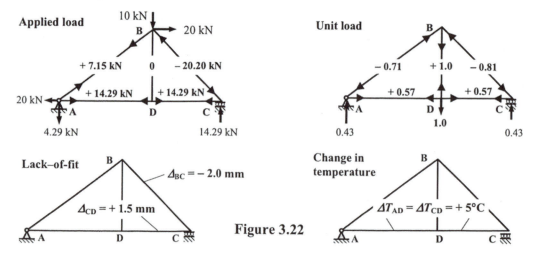

Figure 3.22

The δL value for member BC due to lack-of-fit $\Delta_L = -2.0$ mm
The δL value for member CD due to lack-of-fit $\Delta_L = +1.5$ mm

The δL value for member AD due to temperature change $= +\alpha L_{AD}\, \Delta_{T,AD}$
$$= +(12 \times 10^{-6} \times 4000 \times 5.0)$$
$$\Delta_T = +0.24 \text{ mm}$$

The δL value for member CD due to temperature change $= +\alpha L_{CD}\, \Delta_{T,CD}$
$$= +(12 \times 10^{-6} \times 3000 \times 5.0)$$
$$\Delta_T = +0.18 \text{ mm}$$

Member	Length (mm)	AE (kN)	P-force (kN)	PL/AE (mm)	Δ_L (mm)	Δ_T (mm)	u	$(PL/AE + \Delta_L + \Delta_T) \times u$ (mm)
AB	5000	100×10^3	+ 7.15	+ 0.36	0	0	− 0.71	− 0.26
BC	4243	100×10^3	− 20.20	− 0.86	− 2.0	0	− 0.81	+ 2.32
AD	4000	100×10^3	+ 14.29	+ 0.57	0	+ 0.24	+ 0.57	+ 0.46
CD	3000	100×10^3	+ 14.29	+ 0.43	+ 1.5	+ 0.18	+ 0.57	+ 1.20
BD	3000	100×10^3	0	0	0	0	1.0	0
								$\Sigma = + 3.72$

Table 3.6

The vertical deflection at joint D due to combined loading, lack-of-fit and temperature change is given by:

$$\delta_{V,D} = \sum \left(\frac{PL}{AE} + \Delta_l + \Delta_t \right) \times u = +3.72 \text{ mm} \downarrow$$

Note: Statically determinate, pin-jointed frames can accommodate small changes in geometry without any significant effect on the member forces induced by the applied load system, i.e. the member forces in Example 3.5 are the same as those in Example 3.4.

3.5.5 *Problems: Unit Load Method for Deflection of Pin-Jointed Frames*

A series of pin-jointed frames are shown in Problems 3.17 to 3.20. Using the applied load systems and data given in each case, determine the value of the deflections indicated. Assume $E = 205$ kN/mm^2 and $\alpha = 12 \times 10^{-6}$/°C where required.

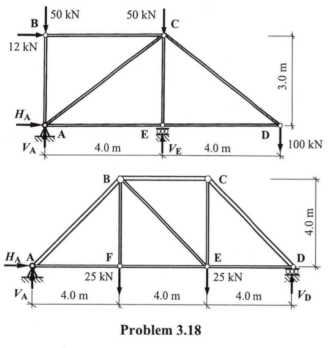

The cross-sectional area of all members is equal to 1500 mm^2.

Determine the value of the resultant deflection at joint D.

Problem 3.17

The   cross-sectional   area   of members AB, BC and CD is equal to 500 mm^2.
The cross-sectional area of all other members is equal to 250 mm^2.
Member BE is too short by 3.0 mm.

Determine the value of the vertical deflection at joint F and the horizontal deflection at joint B.

Problem 3.18

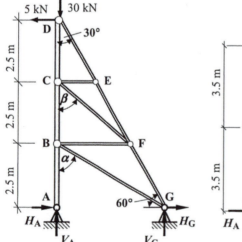

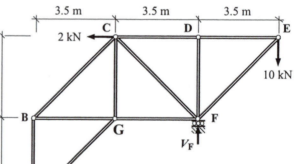

The cross-sectional area of all members is equal to 1200 mm^2.

Determine the value of the horizontal deflection at joint D.

Problem 3.19

The cross-sectional area of members AG, BG, CF CG, FG, and EF is equal to 400 mm^2.
 The cross-sectional area of all other members is equal to 100 mm^2.
All members are subjected to a decrease in temperature equal to 20°C.

Determine the horizontal deflection at joint F.

Problem 3.20

3.5.6 Solutions: Unit Load Method for Deflection of Pin-Jointed Frames

Solution
Topic: Unit Load Method for Deflection of Pin-Jointed Frames
Problem Number: 3.17   **Page No. 1**

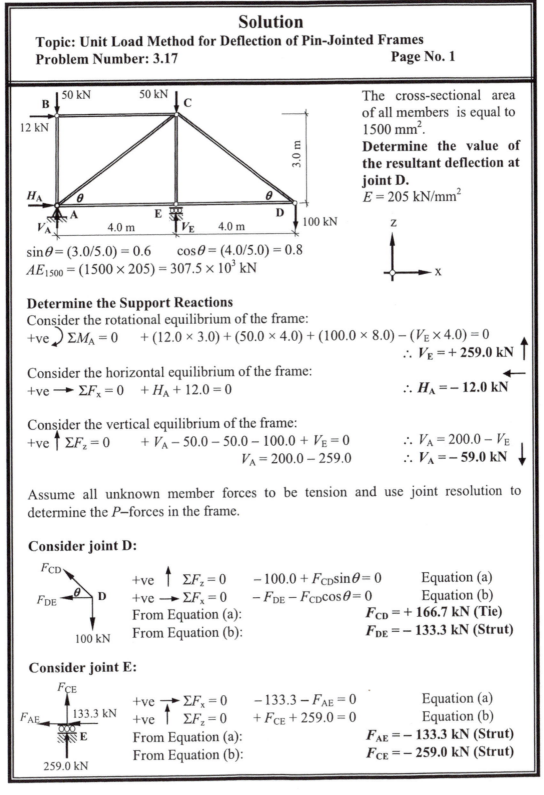

$\sin\theta = (3.0/5.0) = 0.6$ $\cos\theta = (4.0/5.0) = 0.8$
$AE_{1500} = (1500 \times 205) = 307.5 \times 10^3$ kN

The cross-sectional area of all members is equal to 1500 mm^2.
Determine the value of the resultant deflection at joint D.
$E = 205$ kN/mm^2

Determine the Support Reactions
Consider the rotational equilibrium of the frame:
+ve $\curvearrowright$ $\Sigma M_A = 0$ $+ (12.0 \times 3.0) + (50.0 \times 4.0) + (100.0 \times 8.0) - (V_E \times 4.0) = 0$
$\therefore V_E = + 259.0$ kN $\uparrow$

Consider the horizontal equilibrium of the frame:
+ve $\longrightarrow$ $\Sigma F_x = 0$ $+ H_A + 12.0 = 0$ $\therefore H_A = - 12.0$ kN

Consider the vertical equilibrium of the frame:
+ve $\uparrow$ $\Sigma F_z = 0$ $+ V_A - 50.0 - 50.0 - 100.0 + V_E = 0$ $\therefore V_A = 200.0 - V_E$
$V_A = 200.0 - 259.0$ $\therefore V_A = - 59.0$ kN $\downarrow$

Assume all unknown member forces to be tension and use joint resolution to determine the P–forces in the frame.

Consider joint D:

+ve $\uparrow$ $\Sigma F_z = 0$ $- 100.0 + F_{CD}\sin\theta = 0$ Equation (a)
+ve $\longrightarrow$ $\Sigma F_x = 0$ $- F_{DE} - F_{CD}\cos\theta = 0$ Equation (b)
From Equation (a): $F_{CD} = + 166.7$ kN (Tie)
From Equation (b): $F_{DE} = - 133.3$ kN (Strut)

Consider joint E:

+ve $\longrightarrow$ $\Sigma F_x = 0$ $- 133.3 - F_{AE} = 0$ Equation (a)
+ve $\uparrow$ $\Sigma F_z = 0$ $+ F_{CE} + 259.0 = 0$ Equation (b)
From Equation (a): $F_{AE} = - 133.3$ kN (Strut)
From Equation (b): $F_{CE} = - 259.0$ kN (Strut)

Solution

Topic: Unit Load Method for Deflection of Pin-Jointed Frames
Problem Number: 3.17   **Page No. 2**

Consider joint B:

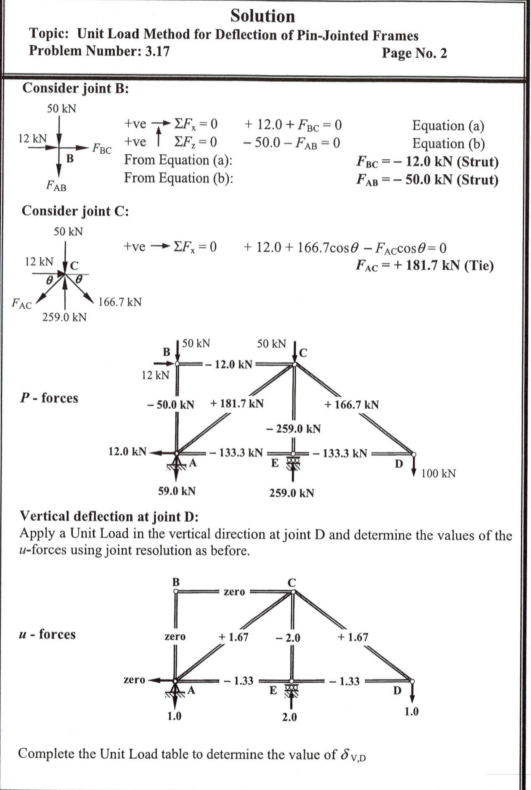

+ve $\longrightarrow \Sigma F_x = 0$ $+ 12.0 + F_{BC} = 0$ Equation (a)
+ve $\uparrow$ $\Sigma F_z = 0$ $- 50.0 - F_{AB} = 0$ Equation (b)
From Equation (a): $F_{BC} = - 12.0$ **kN (Strut)**
From Equation (b): $F_{AB} = - 50.0$ **kN (Strut)**

Consider joint C:

+ve $\longrightarrow \Sigma F_x = 0$ $+ 12.0 + 166.7\cos\theta - F_{AC}\cos\theta = 0$
$F_{AC} = + 181.7$ **kN (Tie)**

P - forces

Vertical deflection at joint D:
Apply a Unit Load in the vertical direction at joint D and determine the values of the
u-forces using joint resolution as before.

u - forces

Complete the Unit Load table to determine the value of $\delta_{V,D}$

Solution

Topic: Unit Load Method for Deflection of Pin-Jointed Frames
Problem Number: 3.17   **Page No. 3**

Member	Length (mm)	AE (kN)	P-force (kN)	PL/AE (mm)	u	$(PL/AE) \times u$
AB	3000	307.5×10^3	− 50.0	− 0.49	0	0
AC	5000	307.5×10^3	+ 181.7	+ 2.95	+ 1.67	+ 4.93
AE	4000	307.5×10^3	− 133.3	− 1.73	− 1.33	+ 2.31
BC	4000	307.5×10^3	− 12.0	− 0.16	0	0
CD	5000	307.5×10^3	+ 166.7	+ 2.71	+ 1.67	+ 4.53
CE	3000	307.5×10^3	− 259.0	− 2.53	− 2.0	+ 5.05
DE	4000	307.5×10^3	− 133.3	− 1.73	− 1.33	+ 2.31
						$\Sigma = + 19.13$

$$\delta_{V,D} = \sum \left(\frac{PL}{AE} \right) \times u = + 19.13 \text{ mm} \downarrow$$

Horizontal deflection at joint D:
Apply a Unit Load in the horizontal direction at joint D and determine the values of
the u-forces using joint resolution as before.

u - forces

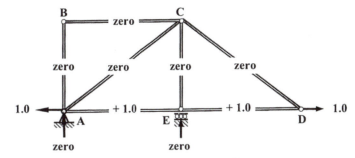

Complete the Unit Load table to determine the value of $\delta_{H,D}$

Member	Length (mm)	AE (kN)	P-force (kN)	PL/AE (mm)	u	$(PL/AE) \times u$
AB	3000	307.5×10^3	− 50.0	− 0.49	0	0
AC	5000	307.5×10^3	+ 181.7	+ 2.95	0	0
AE	4000	307.5×10^3	− 133.3	− 1.73	+ 1.0	− 1.73
BC	4000	307.5×10^3	− 12.0	− 0.16	0	0
CD	5000	307.5×10^3	+ 166.7	+ 2.71	0	0
CE	3000	307.5×10^3	− 259.0	− 2.53	0	0
DE	4000	307.5×10^3	− 133.3	− 1.73	+ 1.0	− 1.73
						$\Sigma = − 3.46$

$$\delta_{H,D} = \sum \left(\frac{PL}{AE} \right) \times u = − 3.46 \text{ mm} \leftarrow$$

Resultant deflection at joint D = $\delta_{R,D} = \sqrt{\left(19.13^2 + 3.46^2 \right)} = 19.44$ mm $\swarrow$ 10.3^0

Solution
Topic: Unit Load Method for Deflection of Pin-Jointed Frames
Problem Number: 3.18   **Page No. 1**

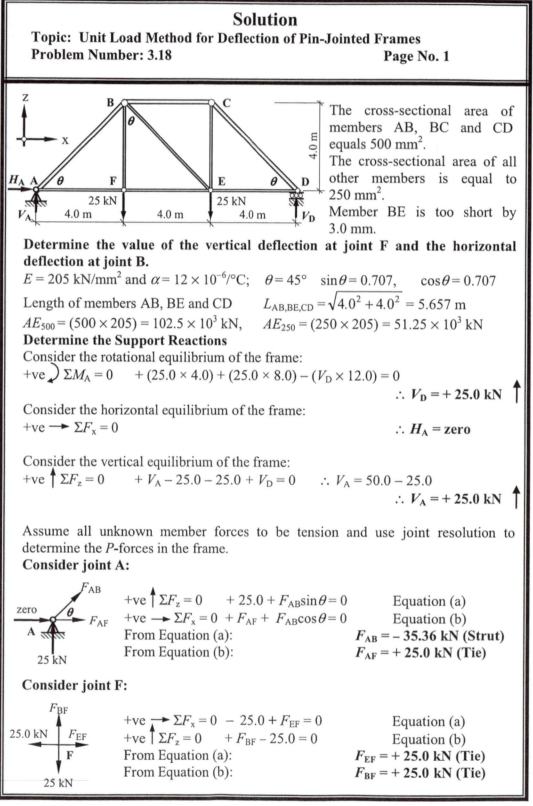

The cross-sectional area of members AB, BC and CD equals 500 mm^2.
The cross-sectional area of all other members is equal to 250 mm^2.
Member BE is too short by 3.0 mm.

Determine the value of the vertical deflection at joint F and the horizontal deflection at joint B.

$E = 205$ kN/mm^2 and $\alpha = 12 \times 10^{-6}$/°C; $\theta = 45°$ $\sin\theta = 0.707$, $\cos\theta = 0.707$

Length of members AB, BE and CD $L_{AB,BE,CD} = \sqrt{4.0^2 + 4.0^2} = 5.657$ m

$AE_{500} = (500 \times 205) = 102.5 \times 10^3$ kN, $AE_{250} = (250 \times 205) = 51.25 \times 10^3$ kN

Determine the Support Reactions

Consider the rotational equilibrium of the frame:

+ve $\curvearrowright$ $\Sigma M_A = 0$ $+ (25.0 \times 4.0) + (25.0 \times 8.0) - (V_D \times 12.0) = 0$

$$\therefore V_D = + 25.0 \text{ kN} \uparrow$$

Consider the horizontal equilibrium of the frame:

+ve $\longrightarrow$ $\Sigma F_x = 0$ $\therefore H_A = $ **zero**

Consider the vertical equilibrium of the frame:

+ve $\uparrow$ $\Sigma F_z = 0$ $+ V_A - 25.0 - 25.0 + V_D = 0$ $\therefore V_A = 50.0 - 25.0$

$$\therefore V_A = + 25.0 \text{ kN} \uparrow$$

Assume all unknown member forces to be tension and use joint resolution to determine the *P*-forces in the frame.

Consider joint A:

+ve $\uparrow$ $\Sigma F_z = 0$ $+ 25.0 + F_{AB}\sin\theta = 0$ Equation (a)
+ve $\longrightarrow$ $\Sigma F_x = 0$ $+ F_{AF} + F_{AB}\cos\theta = 0$ Equation (b)
From Equation (a): $F_{AB} = - 35.36$ kN (Strut)
From Equation (b): $F_{AF} = + 25.0$ kN (Tie)

Consider joint F:

+ve $\longrightarrow$ $\Sigma F_x = 0$ $- 25.0 + F_{EF} = 0$ Equation (a)
+ve $\uparrow$ $\Sigma F_z = 0$ $+ F_{BF} - 25.0 = 0$ Equation (b)
From Equation (a): $F_{EF} = + 25.0$ kN (Tie)
From Equation (b): $F_{BF} = + 25.0$ kN (Tie)

Solution

Topic: Unit Load Method for Deflection of Pin-Jointed Frames
Problem Number: 3.18   **Page No. 2**

Consider joint B:

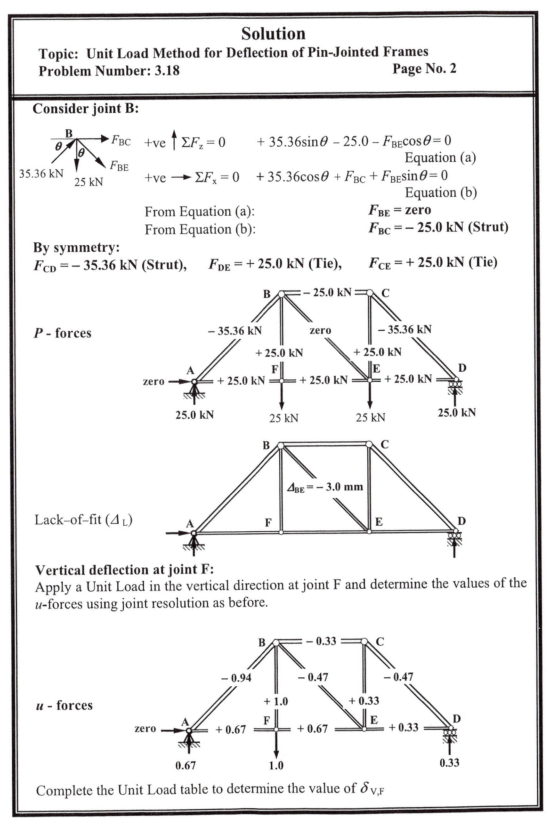

$+ve \uparrow \Sigma F_z = 0$ $+ 35.36\sin\theta - 25.0 - F_{BE}\cos\theta = 0$
                                                        Equation (a)

$+ve \rightarrow \Sigma F_x = 0$ $+ 35.36\cos\theta + F_{BC} + F_{BE}\sin\theta = 0$
                                                        Equation (b)

From Equation (a): F_{BE} = **zero**
From Equation (b): $F_{BC} = -25.0$ kN (Strut)

By symmetry:
$F_{CD} = -35.36$ kN (Strut), $F_{DE} = +25.0$ kN (Tie), $F_{CE} = +25.0$ kN (Tie)

Vertical deflection at joint F:
Apply a Unit Load in the vertical direction at joint F and determine the values of the *u*-forces using joint resolution as before.

Complete the Unit Load table to determine the value of $\delta_{V,F}$

Solution
Topic: Unit Load Method for Deflection of Pin-Jointed Frames
Problem Number: 3.18   **Page No. 3**

Member	Length (mm)	AE (kN)	P-force (kN)	PL/AE (mm)	Δ_L (mm)	u	$(PL/AE + \Delta_L) \times u$ (mm)
AB	5657	102.5×10^3	-35.36	-1.95	0	-0.94	$+1.83$
AF	4000	51.25×10^3	$+25.0$	$+1.95$	0	$+0.67$	$+1.31$
BC	4000	102.5×10^3	-25.0	-0.98	0	-0.33	$+0.32$
BE	5657	51.25×10^3	0	0	-3.0	-0.47	$+1.41$
BF	4000	51.25×10^3	$+25.0$	$+1.95$	0	$+1.0$	$+1.95$
CD	5657	102.5×10^3	-35.36	-1.95	0	-0.47	$+0.92$
CE	4000	51.25×10^3	$+25.0$	$+1.95$	0	$+0.33$	$+0.64$
DE	4000	51.25×10^3	$+25.0$	$+1.95$	0	$+0.33$	$+0.64$
EF	4000	51.25×10^3	$+25.0$	$+1.95$	0	$+0.67$	$+1.31$
							$\Sigma = +10.33$

$$\delta_{V,F} = \sum \left(\frac{PL}{AE} \right) \times u = +10.33 \text{ mm} \quad \downarrow$$

Horizontal deflection at joint B:
Apply a Unit Load in the horizontal direction at joint B and determine the values of the u-forces using joint resolution as before.

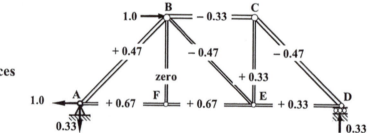

u - forces

Complete the Unit Load table to determine the value of $\delta_{H,B}$

Member	Length (mm)	AE (kN)	P-force (kN)	PL/AE (mm)	Δ_L (mm)	u	$(PL/AE + \Delta_L) \times u$ (mm)
AB	5657	102.5×10^3	-35.36	-1.95	0	$+0.47$	-0.92
AF	4000	51.25×10^3	$+25.0$	$+1.95$	0	$+0.67$	$+1.31$
BC	4000	102.5×10^3	-25.0	-0.98	0	-0.33	$+0.32$
BE	5657	51.25×10^3	0	0	-3.0	-0.47	$+1.41$
BF	4000	51.25×10^3	$+25.0$	$+1.95$	0	0	0
CD	5657	102.5×10^3	-35.36	-1.95	0	-0.47	$+0.92$
CE	4000	51.25×10^3	$+25.0$	$+1.95$	0	$+0.33$	$+0.64$
DE	4000	51.25×10^3	$+25.0$	$+1.95$	0	$+0.33$	$+0.64$
EF	4000	51.25×10^3	$+25.0$	$+1.95$	0	$+0.67$	$+1.31$
							$\Sigma = +5.63$

$$\delta_{H,B} = \sum \left(\frac{PL}{AE} \right) \times u = +5.63 \text{ mm} \quad \longrightarrow$$

Solution

Topic: Unit Load Method for Deflection of Pin-Jointed Frames
Problem Number: 3.19   **Page No. 1**

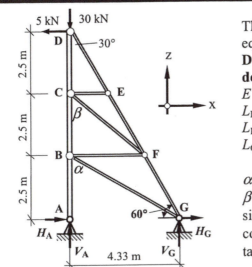

The cross-sectional area of all members is equal to 1200 mm^2.

Determine the value of the horizontal deflection at joint D.

$E = 205$ kN/mm^2

$L_{DE} = L_{EF} = L_{FG} = 2.887$ m

$L_{BF} = 2.887$ m $L_{CF} = 3.819$ m

$L_{CE} = 1.443$ m $L_{BG} = 5.0$ m

$\alpha = \tan^{-1}(4.33/2.5) = 60°$

$\beta = \tan^{-1}(2.887/2.5) = 49.11°$

$\sin\alpha = 0.866$ $\sin\beta = 0.756$

$\cos\alpha = 0.5$ $\cos\beta = 0.655$

$\tan\alpha = 1.732$ $\tan\beta = 1.155$

$AE_{1200} = (1200 \times 205) = 246.0 \times 10^3$ kN

Determine the Support Reactions

Consider the rotational equilibrium of the frame:

+ve $\curvearrowleft$ $\Sigma M_A = 0$ $-(5.0 \times 7.5) - (V_G \times 4.33) = 0$

$\therefore V_G = -8.66$ kN $\downarrow$

Consider the horizontal equilibrium of the frame:

+ve $\longrightarrow$ $\Sigma F_x = 0$ $+H_A + H_G - 5.0 = 0$ $\therefore H_G = 5.0 - H_A$

Consider the vertical equilibrium of the frame:

+ve $\uparrow$ $\Sigma F_z = 0$ $+V_A - 30.0 + V_G = 0$ $\therefore V_A = 30.0 + 8.66$

$\therefore V_A = +38.66$ kN $\uparrow$

Assume all unknown member forces to be tension and use joint resolution to determine the P-forces in the frame.

Consider joint A:

+ve $\uparrow$ $\Sigma F_z = 0$ $+38.66 + F_{AB} = 0$ Equation (a)

+ve $\longrightarrow$ $\Sigma F_x = 0$ $+H_A = 0$ Equation (b)

From Equation (a): $F_{AB} = -38.66$ kN (Strut)

From Equation (b): $H_A = $ zero

$\therefore$ $H_G = 5.0$ kN $\longrightarrow$

Solution
Topic: Unit Load Method for Deflection of Pin-Jointed Frames
Problem Number: 3.19   Page No. 2

Consider joint D:

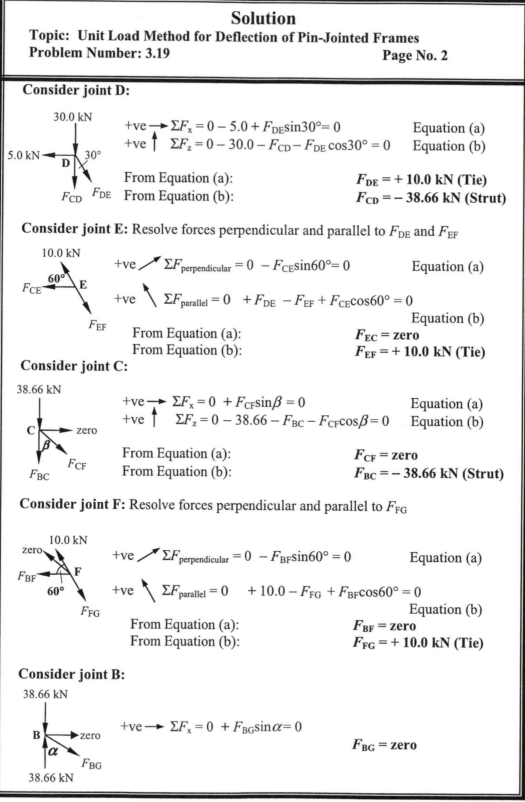

30.0 kN

5.0 kN $\longleftarrow$ D $\diagdown$ 30°

F_{CD} F_{DE}

+ve $\longrightarrow$ $\Sigma F_x = 0 - 5.0 + F_{DE}\sin30° = 0$ Equation (a)

+ve $\uparrow$ $\Sigma F_z = 0 - 30.0 - F_{CD} - F_{DE}\cos30° = 0$ Equation (b)

From Equation (a):   **$F_{DE} = + 10.0$ kN (Tie)**
From Equation (b):   **$F_{CD} = - 38.66$ kN (Strut)**

Consider joint E: Resolve forces perpendicular and parallel to F_{DE} and F_{EF}

10.0 kN

60°

$F_{CE} \longleftarrow$ E

F_{EF}

+ve $\nearrow$ $\Sigma F_{perpendicular} = 0 \ - F_{CE}\sin60° = 0$ Equation (a)

+ve $\nwarrow$ $\Sigma F_{parallel} = 0 \ + F_{DE} - F_{EF} + F_{CE}\cos60° = 0$

Equation (b)

From Equation (a):   **$F_{EC} = $ zero**
From Equation (b):   **$F_{EF} = + 10.0$ kN (Tie)**

Consider joint C:

38.66 kN

C $\longrightarrow$ zero

β

F_{CF}

F_{BC}

+ve $\longrightarrow$ $\Sigma F_x = 0 \ + F_{CF}\sin\beta = 0$ Equation (a)

+ve $\uparrow$ $\Sigma F_z = 0 - 38.66 - F_{BC} - F_{CF}\cos\beta = 0$ Equation (b)

From Equation (a):   **$F_{CF} = $ zero**
From Equation (b):   **$F_{BC} = - 38.66$ kN (Strut)**

Consider joint F: Resolve forces perpendicular and parallel to F_{FG}

zero

10.0 kN

$F_{BF} \longleftarrow$ F

60°

F_{FG}

+ve $\nearrow$ $\Sigma F_{perpendicular} = 0 \ - F_{BF}\sin60° = 0$ Equation (a)

+ve $\nwarrow$ $\Sigma F_{parallel} = 0 \ + 10.0 - F_{FG} + F_{BF}\cos60° = 0$

Equation (b)

From Equation (a):   **$F_{BF} = $ zero**
From Equation (b):   **$F_{FG} = + 10.0$ kN (Tie)**

Consider joint B:

38.66 kN

B $\longrightarrow$ zero

α

F_{BG}

38.66 kN

+ve $\longrightarrow$ $\Sigma F_x = 0 \ + F_{BG}\sin\alpha = 0$

$F_{BG} = $ zero

Solution
Topic: Unit Load Method for Deflection of Pin-Jointed Frames

Problem Number: 3.19   **Page No. 3**

Horizontal deflection at joint D:

Apply a Unit Load in the horizontal direction at joint D and determine the values of the *u*-forces using joint resolution as before.

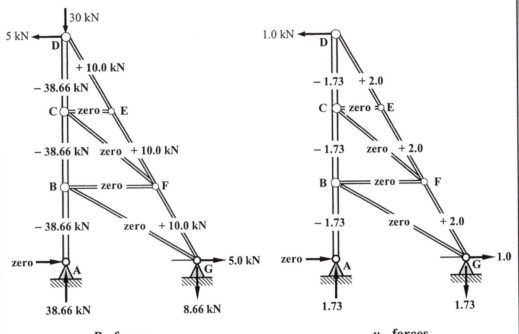

P - forces   *u* - forces

Complete the Unit Load table to determine the value of $\delta_{H,D}$

Member	Length (mm)	AE (kN)	P-force (kN)	PL/AE (mm)	u	(PL/AE) × u (mm)
AB	2500	246.0×10^3	− 38.66	− 0.39	− 1.73	+ 0.68
BC	2500	246.0×10^3	− 38.66	− 0.39	− 1.73	+ 0.68
BF	2887	246.0×10^3	0	0	0	0
BG	5000	246.0×10^3	0	0	0	0
CD	2500	246.0×10^3	− 38.66	− 0.39	− 1.73	+ 0.68
CE	1443	246.0×10^3	0	0	0	0
CF	3819	246.0×10^3	0	0	0	0
DE	2887	246.0×10^3	+ 10.0	+ 0.12	+ 2.0	+ 0.23
EF	2887	246.0×10^3	+ 10.0	+ 0.12	+ 2.0	+ 0.23
FG	2887	246.0×10^3	+ 10.0	+ 0.12	+ 2.0	+ 0.23
						$\Sigma = + 2.73$

$$\delta_{H,D} = \sum \left(\frac{PL}{AE} \right) \times u = + 2.73 \text{ mm} \quad \leftarrow$$

Solution

Topic: Unit Load Method for Deflection of Pin-Jointed Frames
Problem Number: 3.20   **Page No. 1**

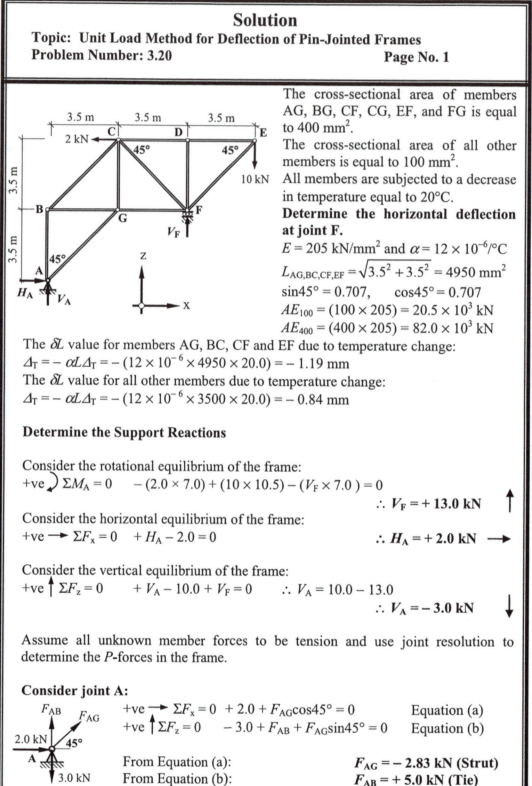

The cross-sectional area of members AG, BG, CF, CG, EF, and FG is equal to 400 mm^2.
The cross-sectional area of all other members is equal to 100 mm^2.
All members are subjected to a decrease in temperature equal to 20°C.
Determine the horizontal deflection at joint F.
$E = 205$ kN/mm^2 and $\alpha = 12 \times 10^{-6}/°$C

$L_{AG,BC,CF,EF} = \sqrt{3.5^2 + 3.5^2} = 4950$ mm^2
$\sin45° = 0.707, \quad \cos45° = 0.707$
$AE_{100} = (100 \times 205) = 20.5 \times 10^3$ kN
$AE_{400} = (400 \times 205) = 82.0 \times 10^3$ kN

The δL value for members AG, BC, CF and EF due to temperature change:
$\Delta_T = -\alpha L\Delta_T = -(12 \times 10^{-6} \times 4950 \times 20.0) = -1.19$ mm
The δL value for all other members due to temperature change:
$\Delta_T = -\alpha L\Delta_T = -(12 \times 10^{-6} \times 3500 \times 20.0) = -0.84$ mm

Determine the Support Reactions

Consider the rotational equilibrium of the frame:
+ve $\curvearrowright$ $\Sigma M_A = 0$ $-(2.0 \times 7.0) + (10 \times 10.5) - (V_F \times 7.0) = 0$
$$\therefore V_F = +13.0 \text{ kN} \uparrow$$

Consider the horizontal equilibrium of the frame:
+ve $\longrightarrow$ $\Sigma F_x = 0$ $+H_A - 2.0 = 0$
$$\therefore H_A = +2.0 \text{ kN} \longrightarrow$$

Consider the vertical equilibrium of the frame:
+ve $\uparrow$ $\Sigma F_z = 0$ $+V_A - 10.0 + V_F = 0$ $\therefore V_A = 10.0 - 13.0$
$$\therefore V_A = -3.0 \text{ kN} \downarrow$$

Assume all unknown member forces to be tension and use joint resolution to determine the *P*-forces in the frame.

Consider joint A:

F_{AB} F_{AG}
2.0 kN   45°
A
3.0 kN

+ve $\longrightarrow$ $\Sigma F_x = 0$ $+2.0 + F_{AG}\cos45° = 0$ Equation (a)
+ve $\uparrow$ $\Sigma F_z = 0$ $-3.0 + F_{AB} + F_{AG}\sin45° = 0$ Equation (b)

From Equation (a): $F_{AG} = -2.83$ kN (Strut)
From Equation (b): $F_{AB} = +5.0$ kN (Tie)

Solution

Topic: Unit Load Method for Deflection of Pin-Jointed Frames
Problem Number: 3.20   **Page No. 2**

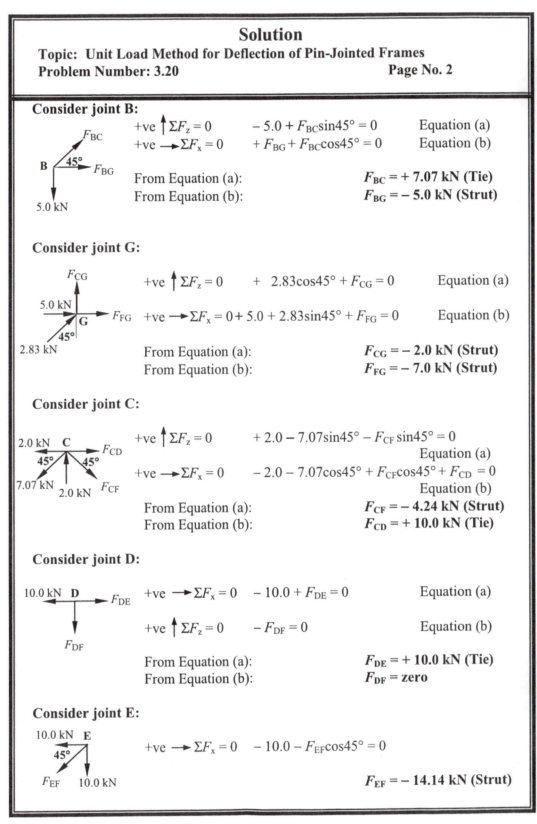

Consider joint B:

$+ve \uparrow \Sigma F_z = 0$ $- 5.0 + F_{BC}\sin45° = 0$ Equation (a)
$+ve \longrightarrow \Sigma F_x = 0$ $+ F_{BG} + F_{BC}\cos45° = 0$ Equation (b)

From Equation (a): $F_{BC} = + 7.07$ kN (Tie)
From Equation (b): $F_{BG} = - 5.0$ kN (Strut)

Consider joint G:

$+ve \uparrow \Sigma F_z = 0$ $+ 2.83\cos45° + F_{CG} = 0$ Equation (a)
$+ve \longrightarrow \Sigma F_x = 0 + 5.0 + 2.83\sin45° + F_{FG} = 0$ Equation (b)

From Equation (a): $F_{CG} = - 2.0$ kN (Strut)
From Equation (b): $F_{FG} = - 7.0$ kN (Strut)

Consider joint C:

$+ve \uparrow \Sigma F_z = 0$ $+ 2.0 - 7.07\sin45° - F_{CF}\sin45° = 0$
                                        Equation (a)
$+ve \longrightarrow \Sigma F_x = 0$ $- 2.0 - 7.07\cos45° + F_{CF}\cos45° + F_{CD} = 0$
                                        Equation (b)
From Equation (a): $F_{CF} = - 4.24$ kN (Strut)
From Equation (b): $F_{CD} = + 10.0$ kN (Tie)

Consider joint D:

$+ve \longrightarrow \Sigma F_x = 0$ $- 10.0 + F_{DE} = 0$ Equation (a)

$+ve \uparrow \Sigma F_z = 0$ $- F_{DF} = 0$ Equation (b)

From Equation (a): $F_{DE} = + 10.0$ kN (Tie)
From Equation (b): $F_{DF} = $ zero

Consider joint E:

$+ve \longrightarrow \Sigma F_x = 0$ $- 10.0 - F_{EF}\cos45° = 0$

$F_{EF} = - 14.14$ kN (Strut)

Solution

Topic: **Unit Load Method for Deflection of Pin-Jointed Frames**
Problem Number: 3.20　　　　　　　　　　　　　　　Page No. 3

Horizontal deflection at joint F:
Apply a Unit Load in the horizontal direction at joint F and determine the values of
the *u*-forces using joint resolution as before.

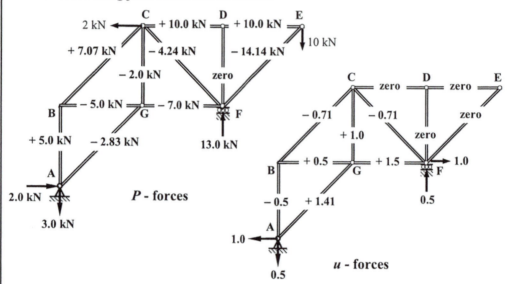

The δL value for members (AG, BC, CF and EF) due to temperature change:
$$\varDelta_T = -1.19 \text{ mm}$$

The δL value for all other members due to temperature change: $\quad \varDelta_T = -0.84$ mm
Complete the Unit Load table to determine the value of $\delta_{H,F}$

Member	Length (mm)	AE (kN)	P-force (kN)	PL/AE (mm)	$\varDelta_T$ (mm)	u	$(PL/AE + \varDelta_T) \times u$ (mm)
AB	3500	20.5×10^3	+ 5.0	+ 0.85	− 0.84	− 0.50	− 0.01
AG	4950	82.0×10^3	− 2.83	− 0.17	− 1.19	+ 1.41	− 1.92
BC	4950	20.5×10^3	+ 7.07	+ 1.71	− 1.19	− 0.71	− 0.37
BG	3500	82.0×10^3	− 5.0	− 0.21	− 0.84	+ 0.50	− 0.53
CD	3500	20.5×10^3	+ 10.0	+ 1.71	− 0.84	0	0
CF	4950	82.0×10^3	− 4.24	− 0.26	− 1.19	− 0.71	+ 1.02
CG	3500	82.0×10^3	− 2.0	− 0.09	− 0.84	+ 1.00	− 0.93
DE	3500	20.5×10^3	+ 10.0	+ 1.71	− 0.84	0	0
DF	3500	20.5×10^3	0	0	− 0.84	0	0
EF	4950	82.0×10^3	− 14.14	− 0.85	− 1.19	0	0
FG	3500	82.0×10^3	− 7.0	− 0.30	− 0.84	+ 1.50	− 1.71
							$\Sigma = -4.45$

$$\delta_{H,F} = \sum \left(\frac{PL}{AE} \right) \times u \ = -4.45 \text{ mm} \leftarrow$$

3.6 *Unit Load Method for Singly-Redundant Pin-Jointed Frames*

The method of analysis illustrated in Section 3.5 can also be adopted to determine the member forces in singly–redundant frames. Consider the frame shown in Example 3.6.

3.6.1 Example 3.6: Singly-Redundant Pin-Jointed Frame 1

Using the data given, determine the member forces and support reactions for the pin–jointed frame shown in Figure 3.23.

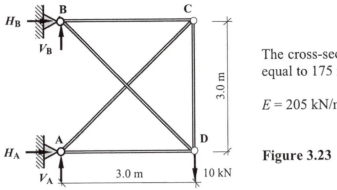

The cross-sectional area of all members is equal to 175 mm^2.

$E = 205$ kN/mm^2

Figure 3.23

The degree–of–indeterminacy $I_D = (m + r) - 2n = (5 + 4) - (2 \times 4) = 1$

Assume that member BD is a redundant member and consider the original frame to be the superposition of two structures as indicated in Figures 3.24(a) and (b). The frame in Figure 3.24(b) can be represented as shown in Figure 3.25.

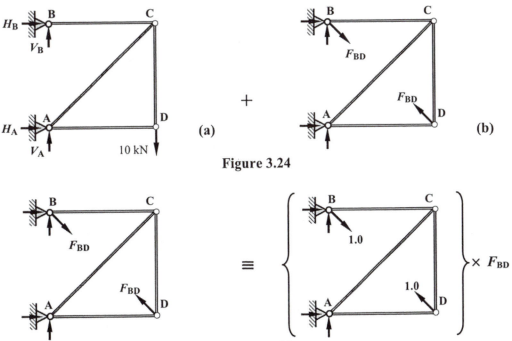

Figure 3.24

Figure 3.25

To maintain compatibility in the length of member BD in the original frame the change in length of the diagonal BD in Figure 3.24(a) must be equal and opposite to that in Figure 3.24(b) as shown in Figure 3.26.

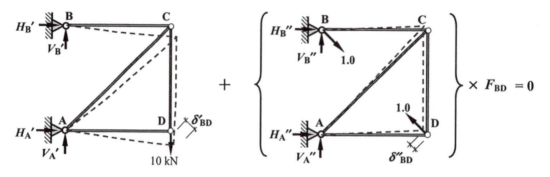

Figure 3.26

$(\delta'_{BD}$ due to P–forces$) + (\delta''_{BD}$ due to unit load forces$) \times F_{BD} = 0$

i.e. $\quad \sum \dfrac{PL}{AE} u + \left(\sum \dfrac{uL}{AE} u \right) \times F_{BD} = 0 \qquad \therefore \quad F_{BD} = -\sum \dfrac{PL}{AE} u \Big/ \sum \dfrac{uL}{AE} u$

Using joint resolution the P-forces and the u-forces can be determined as indicated in Figure 3.27.

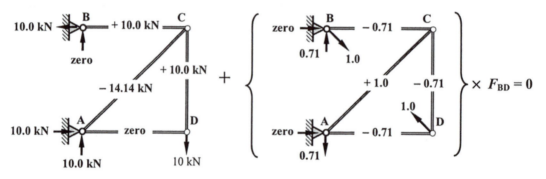

P - forces   **u - forces**

Figure 3.27

Member	Length (mm)	AE (kN)	P-force (kN)	PL/AE (mm)	u	(PL/AE) × u (mm)	(uL/AE) × u (mm)	Member forces
BC	3000	35.88 ×10³	+ 10.00	+ 0.84	− 0.71	− 0.59	0.04	+ 4.38
CD	3000	35.88 ×10³	+ 10.00	+ 0.84	− 0.71	− 0.59	0.04	+ 4.38
DA	3000	35.88 ×10³	0	0	− 0.71	0	0.04	− 5.62
AC	4243	35.88 ×10³	− 14.14	− 1.67	+ 1.00	− 1.67	0.12	− 6.23
BD	4243	35.88 ×10³	0	0	+ 1.00	0	0.12	+ 7.91
						Σ = − 2.85	Σ = + 0.36	

$$F_{BD} = -\sum \dfrac{PL}{AE} u \Big/ \sum \dfrac{uL}{AE} u = + 2.85/0.36 = + 7.91 \text{ kN (Tie)}$$

The final member forces = [*P*-forces + (*u*-forces × 7.91)] and are given in the last column of the table

$$V_A = + 10.0 - (0.71 \times 7.91) = + 4.38 \text{ kN} \quad \uparrow$$
$$H_A = + 10.0 + \text{zero} = + 10.0 \text{ kN} \quad \rightarrow$$
$$V_B = \text{zero} + (0.71 \times 7.91) = + 5.62 \text{ kN} \quad \uparrow$$
$$H_B = - 10.0 + \text{zero} = - 10.0 \text{ kN} \quad \leftarrow$$

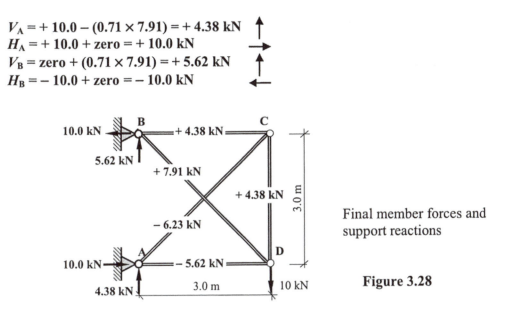

Final member forces and support reactions

Figure 3.28

3.6.2 Example 3.7: Singly-Redundant Pin-Jointed Frame 2

Using the data given, determine the member forces and support reactions for the pin-jointed frame shown in Figure 3.29.

The cross–sectional area of all members is equal to 140 mm². Assume $E = 205$ kN/mm²

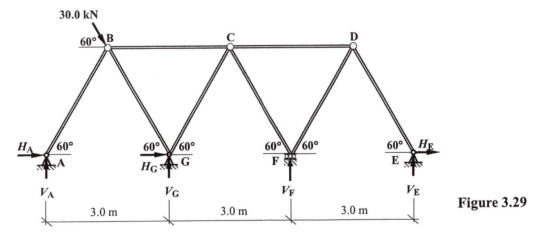

Figure 3.29

All member lengths $L = 3.0$ m
$AE = (140 \times 205) = 28.7 \times 10^3$ kN
$\sin 60° = 0.866 \qquad \cos 60° = 0.5$
Consider the applied load as two components $\quad 30.0\sin 60° = 25.98$ kN $\quad\downarrow$
$\qquad\qquad\qquad\qquad\qquad\qquad\qquad\qquad\qquad 30.0\cos 60° = 15.0$ kN $\quad\rightarrow$
The degree of indeterminacy $I_D = (m + r) - 2n = (8 + 7) - (2 \times 7) = 1$

Consider the vertical reaction at support F to be redundant. The equivalent system is the superposition of the statically determinate frame and the (unit load frame $\times V_F$) as shown in Figures 3.30 and 3.31.

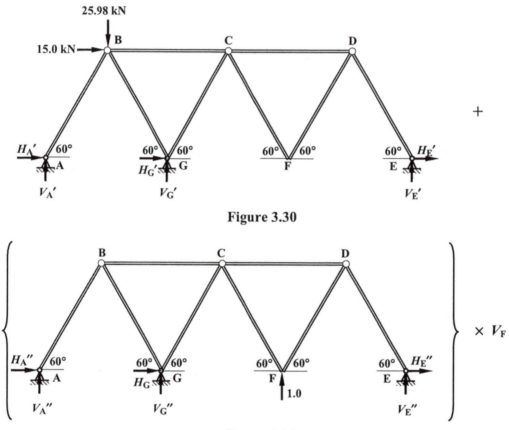

Figure 3.30

Figure 3.31

Using joint resolution the *P*-forces and the *u*-forces can be determined as indicated in Figures 3.32 and 3.33.

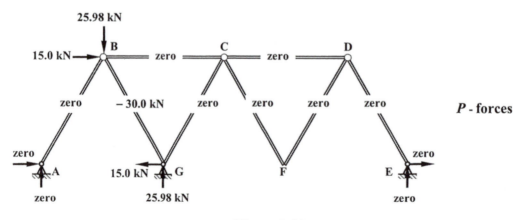

Figure 3.32

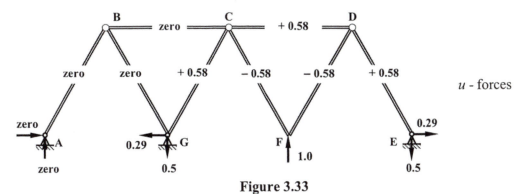

Figure 3.33

Member	Length (mm)	AE (kN)	P-force (kN)	PL/AE (mm)	u	(PL/AE) × u (mm)	(uL/AE) × u (mm)	Member forces
AB	3000	28.7×10^3	0	0	0	0	0	0
BC	3000	28.7×10^3	0	0	0	0	0	0
CD	3000	28.7×10^3	0	0	+ 0.58	0	0.035	0
DE	3000	28.7×10^3	0	0	+ 0.58	0	0.035	0
DF	3000	28.7×10^3	0	0	− 0.58	0	0.035	0
CF	3000	28.7×10^3	0	0	− 0.58	0	0.035	0
CG	3000	28.7×10^3	0	0	+ 0.58	0	0.035	0
BG	3000	28.7×10^3	− 30.00	− 3.14	0	0	0	− 30.00
						Σ = zero	Σ = + 0.18	

i.e. $\sum \dfrac{PL}{AE}u + \left(\sum \dfrac{uL}{AE}u\right) \times V_F = 0$

$V_F = -\sum \dfrac{PL}{AE}u \bigg/ \sum \dfrac{uL}{AE}u = 0/0.18 = $ zero

The final member forces = [P-forces + (u-forces × 0)] and are given in the last column of the table.

$V_G = + 25.98$ kN ↑
$H_G = - 15.0$ kN ←

All other reactions are equal to zero.

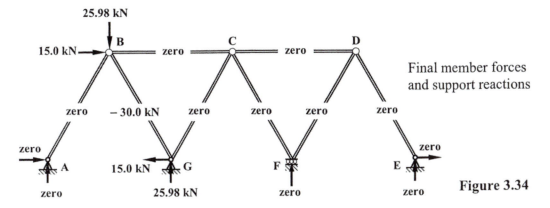

Final member forces and support reactions

Figure 3.34

3.6.3 Problems: Unit Load Method for Singly-Redundant Pin-Jointed Frames

Using the data given in the singly-redundant, pin-jointed frames shown in Problem 3.21 to Problem 3.24, determine the support reactions and the member forces due to the applied loads. Assume $E = 205$ kN/mm^2 and $\alpha = 12 \times 10^{-6}$/°C where required.

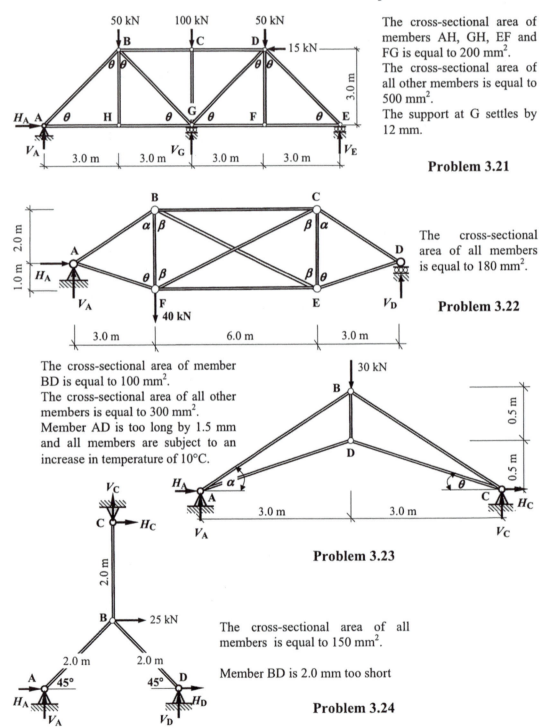

The cross-sectional area of members AH, GH, EF and FG is equal to 200 mm^2.
The cross-sectional area of all other members is equal to 500 mm^2.
The support at G settles by 12 mm.

Problem 3.21

The   cross-sectional area of all members is equal to 180 mm^2.

Problem 3.22

The cross-sectional area of member BD is equal to 100 mm^2.
The cross-sectional area of all other members is equal to 300 mm^2.
Member AD is too long by 1.5 mm and all members are subject to an increase in temperature of 10°C.

Problem 3.23

The   cross-sectional area of all members is equal to 150 mm^2.

Member BD is 2.0 mm too short

Problem 3.24

3.6.4 Solutions: Unit Load Method for Singly-Redundant Pin-Jointed Frames

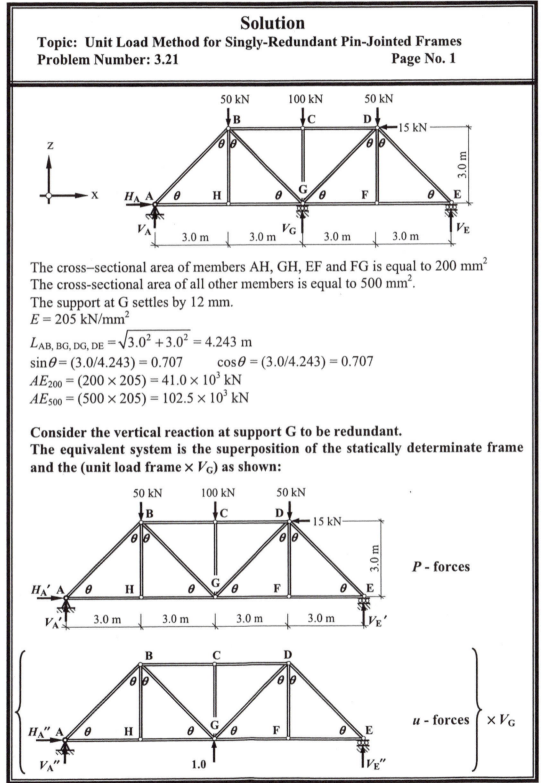

Solution

Topic: Unit Load Method for Singly-Redundant Pin-Jointed Frames
Problem Number: 3.21   **Page No. 1**

The cross–sectional area of members AH, GH, EF and FG is equal to 200 mm^2
The cross-sectional area of all other members is equal to 500 mm^2.
The support at G settles by 12 mm.
$E = 205$ kN/mm^2

$L_{AB, BG, DG, DE} = \sqrt{3.0^2 + 3.0^2} = 4.243$ m
$\sin\theta = (3.0/4.243) = 0.707$ $\cos\theta = (3.0/4.243) = 0.707$
$AE_{200} = (200 \times 205) = 41.0 \times 10^3$ kN
$AE_{500} = (500 \times 205) = 102.5 \times 10^3$ kN

Consider the vertical reaction at support G to be redundant.
The equivalent system is the superposition of the statically determinate frame
and the (unit load frame $\times V_G$) as shown:

Solution

Topic: Unit Load Method for Singly-Redundant Pin-Jointed Frames
Problem Number: 3.21   **Page No. 2**

Determine the Support Reactions for the statically determinate frame.

Consider the rotational equilibrium of the frame:
+ve $\curvearrowright$ $\Sigma M_A = 0$ $+ (50.0 \times 3.0) + (100 \times 6.0) + (50.0 \times 9.0) - (15.0 \times 3.0)$
$- (V_E' \times 12.0) = 0$ $\therefore V_E' = + 96.25$ kN $\uparrow$

Consider the horizontal equilibrium of the frame:
+ve $\longrightarrow$ $\Sigma F_x = 0$ $+ H_A' - 15.0 = 0$ $\therefore H_A' = + 15.0$ kN $\longrightarrow$

Consider the vertical equilibrium of the frame:
+ve $\uparrow$ $\Sigma F_z = 0$ $+ V_A' - 50.0 - 100.0 - 50.0 + V_E' = 0$ $\therefore V_A' = 200.0 - 96.25$
$\therefore V_A' = + 103.75$ kN $\uparrow$

Assume all unknown member forces to be tension and use joint resolution to determine the *P*-forces in the frame.

Consider joint A:

F_{AB}
15.0 kN θ
A $\quad F_{AH}$
103.75 kN

+ve $\uparrow$ $\Sigma F_z = 0$ $+ 103.75 + F_{AB}\sin\theta = 0$ Equation (a)
+ve $\longrightarrow$ $\Sigma F_x = 0$ $+ 15.0 + F_{AH} + F_{AB}\cos\theta = 0$ Equation (b)

From Equation (a): $F_{AB} = - 146.70$ kN (Strut)
From Equation (b): $F_{AH} = + 88.75$ kN (Tie)

Consider joint H:

F_{BH}
88.75 kN F_{GH}
H

+ve $\uparrow$ $\Sigma F_z = 0$ $+ F_{BH} = 0$ Equation (a)
+ve $\longrightarrow$ $\Sigma F_x = 0$ $- 88.75 + F_{GH} = 0$ Equation (b)

From Equation (a): $F_{BH} = $ **zero**
From Equation (b): $F_{GH} = + 88.75$ kN (Tie)

Consider joint B:

50 kN
B $\quad F_{BC}$
$\theta|\theta$
146.7 kN $\quad F_{BG}$
zero

+ve $\uparrow$ $\Sigma F_z = 0$ $- 50.0 + 146.7\cos\theta - F_{BG}\cos\theta = 0$
Equation (a)
+ve $\longrightarrow$ $\Sigma F_x = 0$ $+ 146.7\sin\theta + F_{BG}\sin\theta + F_{BC} = 0$
Equation (b)

From Equation (a): $F_{BG} = + 76.0$ kN (Tie)
From Equation (b): $F_{BC} = - 157.45$ kN (Strut)

Solution

Topic: Unit Load Method for Singly-Redundant Pin-Jointed Frames

Problem Number: 3.21   **Page No. 3**

Consider joint C:

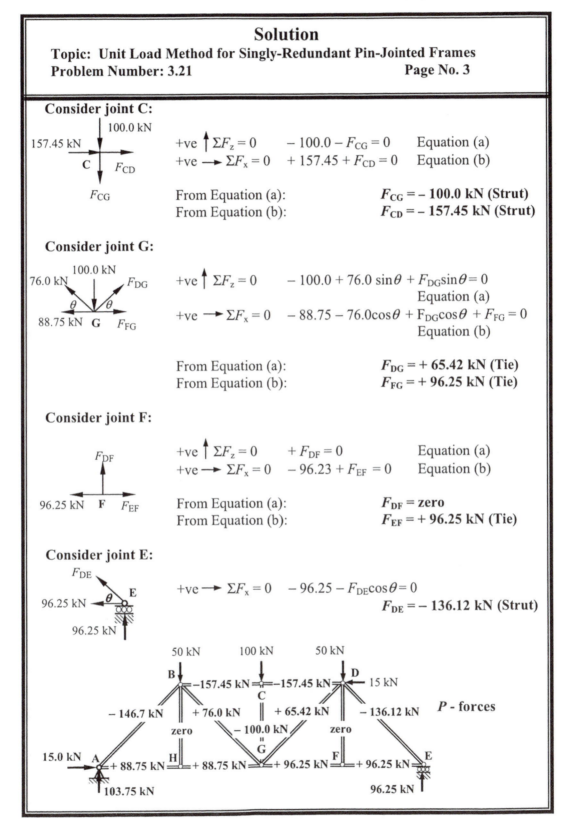

$$+ve \uparrow \Sigma F_z = 0 \qquad -100.0 - F_{CG} = 0 \qquad \text{Equation (a)}$$
$$+ve \longrightarrow \Sigma F_x = 0 \qquad +157.45 + F_{CD} = 0 \qquad \text{Equation (b)}$$

From Equation (a): $F_{CG} = -\textbf{100.0 kN (Strut)}$
From Equation (b): $F_{CD} = -\textbf{157.45 kN (Strut)}$

Consider joint G:

$$+ve \uparrow \Sigma F_z = 0 \qquad -100.0 + 76.0 \sin\theta + F_{DG}\sin\theta = 0$$
$$\text{Equation (a)}$$
$$+ve \longrightarrow \Sigma F_x = 0 \qquad -88.75 - 76.0\cos\theta + F_{DG}\cos\theta + F_{FG} = 0$$
$$\text{Equation (b)}$$

From Equation (a): $F_{DG} = +\textbf{65.42 kN (Tie)}$
From Equation (b): $F_{FG} = +\textbf{96.25 kN (Tie)}$

Consider joint F:

$$+ve \uparrow \Sigma F_z = 0 \qquad +F_{DF} = 0 \qquad \text{Equation (a)}$$
$$+ve \longrightarrow \Sigma F_x = 0 \qquad -96.23 + F_{EF} = 0 \qquad \text{Equation (b)}$$

From Equation (a): $F_{DF} = \textbf{zero}$
From Equation (b): $F_{EF} = +\textbf{96.25 kN (Tie)}$

Consider joint E:

$$+ve \longrightarrow \Sigma F_x = 0 \qquad -96.25 - F_{DE}\cos\theta = 0$$
$$F_{DE} = -\textbf{136.12 kN (Strut)}$$

P **- forces**

Solution
Topic: Unit Load Method for Singly-Redundant Pin-Jointed Frames
Problem Number: 3.21   **Page No. 4**

Apply a Unit Load in the vertical direction at support G and determine the values of the *u*-forces using joint resolution as before.

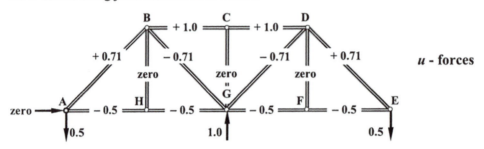

u - forces

$(\delta_{\text{VG}}$ due to *P*-forces$) + (\delta_{\text{VG}}$ due to unit forces$) \times V_{\text{G}} = - 12.0$ mm

i.e. $\sum \dfrac{PL}{AE} u + \left(\sum \dfrac{uL}{AE} u \right) \times V_{\text{G}} = - 12.0$

$\therefore V_{\text{G}} = \left(-12.0 - \sum \dfrac{PL}{AE} u \right) \Big/ \sum \dfrac{uL}{AE} u$

Complete the Unit Load table to determine the value of V_{G}

Member	Length (mm)	AE (kN)	P-force (kN)	PL/AE (mm)	u	(PL/AE) × u (mm)	(uL/AE) × u (mm)	Member forces
AB	4243	102.5 ×10³	−146.70	−6.07	+0.71	− 4.293	+ 0.021	− 70.40
AH	3000	41.0 ×10³	+88.75	+6.49	−0.50	− 3.247	+ 0.018	+ 34.79
BC	3000	102.5 ×10³	−157.45	−4.61	+1.00	− 4.608	+ 0.029	− 49.53
BG	4243	102.5 ×10³	+76.00	+3.15	−0.71	− 2.224	+ 0.021	− 0.30
BH	3000	102.5 ×10³	0	0	0	0	0	zero
CD	3000	102.5 ×10³	−157.45	−4.61	+1.00	− 4.608	+ 0.029	− 49.53
CG	3000	102.5 ×10³	−100.00	−2.93	0	0	0	− 100.00
DE	4243	102.5 ×10³	−136.12	−5.63	+0.71	− 3.984	+ 0.021	− 59.82
DG	4243	102.5 ×10³	+65.42	+ 2.71	−0.71	− 1.915	+ 0.021	− 10.88
DF	3000	102.5 ×10³	0	0	0	0	0	zero
EF	3000	41.0 ×10³	+96.25	+7.04	−0.50	− 3.521	+ 0.018	+ 42.29
FG	3000	41.0 ×10³	+96.25	+7.04	−0.50	− 3.521	+ 0.018	+ 42.29
GH	3000	41.0 ×10³	+88.75	+6.49	−0.50	− 3.247	+ 0.018	+ 34.79
						Σ = − 35.169	Σ = + 0.215	

$V_{\text{G}} = \left(-12.0 - \sum \dfrac{PL}{AE} u \right) \Big/ \sum \dfrac{uL}{AE} u = [- 12.0 - (- 35.169)]/0.215 = + \textbf{107.76 kN} \uparrow$

The final member forces = [*P*-forces + (*u*-forces × 107.76)] and are given in the last column of the table

$V_{\text{A}} = 103.75 - (0.5 \times 107.76) = + \textbf{49.87 kN} \uparrow \quad H_{\text{A}} = + \textbf{15.0 kN} \longrightarrow$
$V_{\text{E}} = 96.25 - (0.5 \times 107.76) = + \textbf{42.37 kN} \uparrow$

Solution

Topic: Unit Load Method for Singly-Redundant Pin-Jointed Frames
Problem Number: 3.22 Page No. 1

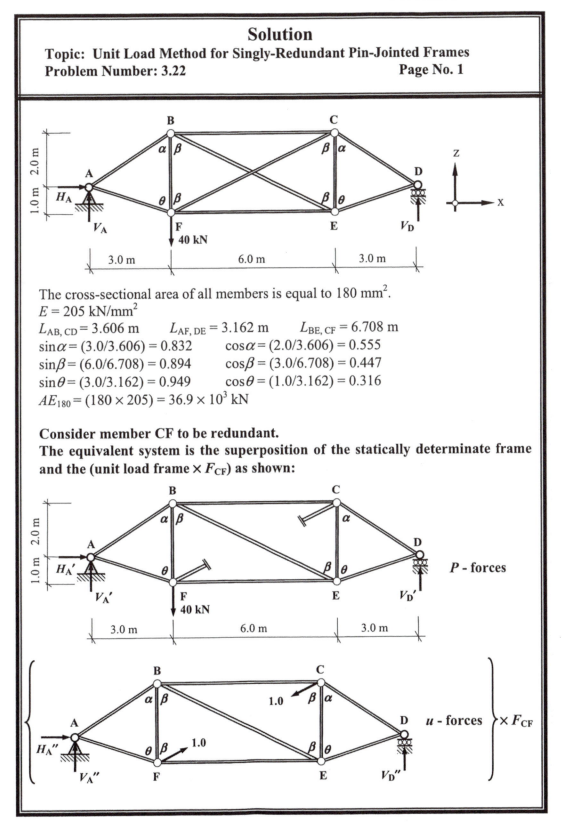

The cross-sectional area of all members is equal to 180 mm².
$E = 205 \text{ kN/mm}^2$
$L_{AB, CD} = 3.606 \text{ m}$ $L_{AF, DE} = 3.162 \text{ m}$ $L_{BE, CF} = 6.708 \text{ m}$
$\sin\alpha = (3.0/3.606) = 0.832$ $\cos\alpha = (2.0/3.606) = 0.555$
$\sin\beta = (6.0/6.708) = 0.894$ $\cos\beta = (3.0/6.708) = 0.447$
$\sin\theta = (3.0/3.162) = 0.949$ $\cos\theta = (1.0/3.162) = 0.316$
$AE_{180} = (180 \times 205) = 36.9 \times 10^3 \text{ kN}$

Consider member CF to be redundant.
The equivalent system is the superposition of the statically determinate frame and the (unit load frame × F_{CF}) as shown:

Solution
Topic: Unit Load Method for Singly-Redundant Pin-Jointed Frames
Problem Number: 3.22   **Page No. 2**

Determine the Support Reactions for the statically determinate frame.

Consider the rotational equilibrium of the frame:
+ve $\curvearrowright$ $\Sigma M_A = 0$ $+ (40.0 \times 3.0) - (V_D' \times 12.0) = 0$ $\therefore V_D' = +\textbf{10.0 kN}$ ↑

Consider the horizontal equilibrium of the frame:
+ve $\longrightarrow$ $\Sigma F_x = 0$ $+ H_A' = 0$ $\therefore H_A' = \textbf{zero}$

Consider the vertical equilibrium of the frame:
+ve ↑ $\Sigma F_z = 0$ $+ V_A' - 40.0 + V_D' = 0$ $\therefore V_A' = 40.0 - 10.0$
                                                          $\therefore V_A' = +\textbf{30.0 kN}$ ↑

Assume all unknown member forces to be tension and use joint resolution to determine the *P*-forces in the frame.

Consider joint A:

+ve ↑ $\Sigma F_z = 0$ $+ 30.0 + F_{AB} \cos\alpha - F_{AF}\cos\theta = 0$ Equation (a)
+ve $\longrightarrow$ $\Sigma F_x = 0$ $+ F_{AB}\sin\alpha + F_{AF} \sin\theta = 0$ Equation (b)

From Equation (a): $F_{AB} = -\textbf{36.06 kN (Strut)}$
From Equation (b): $F_{AF} = +\textbf{31.6 kN (Tie)}$

Consider joint F:

+ve ↑ $\Sigma F_z = 0$ $+ 31.6\cos\theta - 40.0 + F_{BF} = 0$ Equation (a)
+ve $\longrightarrow$ $\Sigma F_x = 0$ $- 31.6\sin\theta + F_{EF} = 0$ Equation (b)

From Equation (a): $F_{BF} = +\textbf{30.0 kN (Tie)}$
From Equation (b): $F_{EF} = +\textbf{30.0 kN (Tie)}$

Consider joint B:

+ve ↑ $\Sigma F_z = 0$ $- 30.0 + 36.06 \cos\alpha - F_{BE}\cos\beta = 0$
                                                                   Equation (a)
+ve $\longrightarrow$ $\Sigma F_x = 0$ $+ 36.06\sin\alpha + F_{BE} \sin\beta + F_{BC} = 0$
                                                                   Equation (b)

From Equation (a): $F_{BE} = -\textbf{22.34 kN (Strut)}$
From Equation (b): $F_{BC} = -\textbf{10.0 kN (Strut)}$

Solution
Topic: Unit Load Method for Singly-Redundant Pin-Jointed Frames
Problem Number: 3.22   **Page No. 3**

Consider joint E:

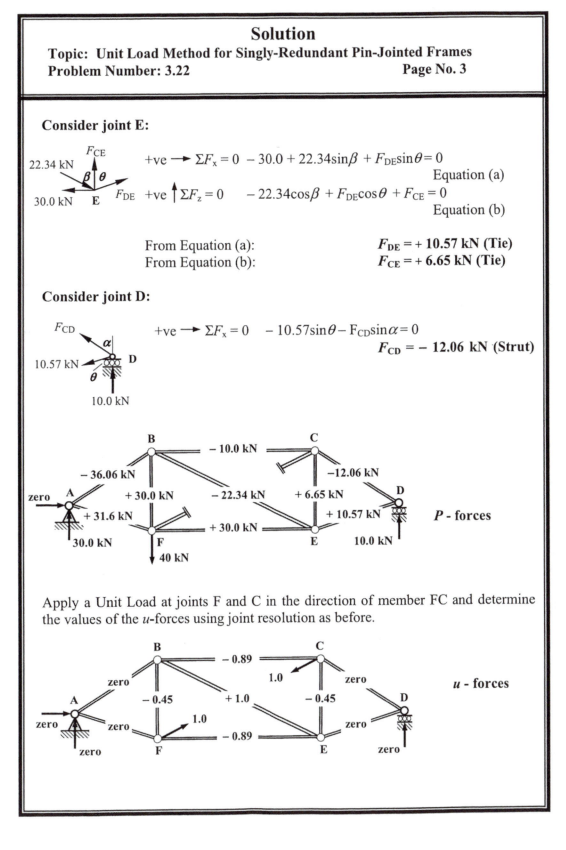

$+$ve $\longrightarrow \Sigma F_x = 0 \ -30.0 + 22.34\sin\beta + F_{DE}\sin\theta = 0$

Equation (a)

$+$ve $\uparrow \Sigma F_z = 0 \quad -22.34\cos\beta + F_{DE}\cos\theta + F_{CE} = 0$

Equation (b)

From Equation (a): $F_{DE} = +$ **10.57 kN (Tie)**
From Equation (b): $F_{CE} = +$ **6.65 kN (Tie)**

Consider joint D:

$+$ve $\longrightarrow \Sigma F_x = 0 \quad -10.57\sin\theta - F_{CD}\sin\alpha = 0$

$F_{CD} = -$ **12.06 kN (Strut)**

Apply a Unit Load at joints F and C in the direction of member FC and determine
the values of the *u*-forces using joint resolution as before.

Solution

Topic: Unit Load Method for Singly-Redundant Pin-Jointed Frames
Problem Number: 3.22   **Page No. 4**

$(\delta_{FC}$ due to P–forces$) + (\delta_{FC}$ due to unit forces$) \times F_{CF} = 0$

i.e. $\sum \dfrac{PL}{AE} u + \left(\sum \dfrac{uL}{AE} u \right) \times F_{CF} = 0$

$\therefore F_{CF} = -\sum \dfrac{PL}{AE} u \bigg/ \sum \dfrac{uL}{AE} u$

Complete the Unit Load table to determine the value of F_{CF}

Member	Length (mm)	AE (kN)	P-force (kN)	PL/AE (mm)	u	(PL/AE) × u (mm)	(uL/AE) × u (mm)	Member forces
AB	3606	36.9 ×10³	−36.06	− 3.52	0	0	0	− 36.06
AF	3162	36.9 ×10³	+ 31.60	+ 2.71	0	0	0	+ 31.60
BC	6000	36.9 ×10³	−10.00	− 1.63	− 0.89	+ 1.454	+ 0.130	− 21.31
BE	6708	36.9 ×10³	−22.34	− 4.06	1.00	− 4.061	+ 0.182	− 9.69
BF	3000	36.9 ×10³	+ 30.00	+ 2.44	− 0.45	− 1.090	+ 0.016	+ 24.35
CD	3606	36.9 ×10³	−12.06	− 1.18	0	0	0	− 12.06
CE	3000	36.9 ×10³	+ 6.65	+ 0.54	− 0.45	− 0.242	+ 0.016	+ 1.00
CF	6708	36.9 ×10³	0	0	1.00	0	+ 0.182	+ 12.65
DE	3162	36.9 ×10³	+ 10.57	+ 0.91	0	0	0	+ 10.57
EF	6000	36.9 ×10³	+ 30.00	+ 4.88	− 0.89	− 4.361	+ 0.130	+ 18.69
						Σ = − 8.30	Σ = + 0.656	

$F_{CF} = -\sum \dfrac{PL}{AE} u \bigg/ \sum \dfrac{uL}{AE} u = -(- 8.30)/0.656 = + 12.65$ kN **(Tie)**

The final member forces = [P-forces + (u-forces × 12.65)] and are given in the last column of the table

$V_A = + 30.0$ kN ↑ $H_A =$ zero $V_D = + 10.0$ kN ↑

Solution
Topic: Unit Load Method for Singly-Redundant Pin-Jointed Frames
Problem Number: 3.23   **Page No. 1**

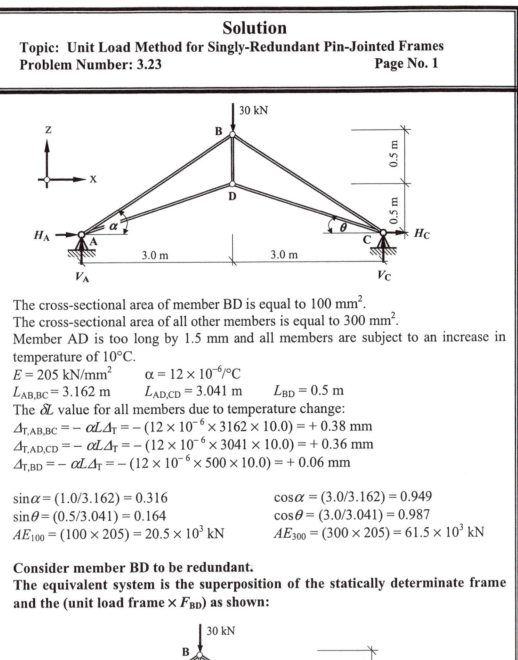

The cross-sectional area of member BD is equal to 100 mm².
The cross-sectional area of all other members is equal to 300 mm².
Member AD is too long by 1.5 mm and all members are subject to an increase in temperature of 10°C.

$E = 205$ kN/mm² $\alpha = 12 \times 10^{-6}/°C$
$L_{AB,BC} = 3.162$ m $L_{AD,CD} = 3.041$ m $L_{BD} = 0.5$ m

The δL value for all members due to temperature change:
$$\Delta_{T,AB,BC} = -\alpha L \Delta_T = -(12 \times 10^{-6} \times 3162 \times 10.0) = +0.38 \text{ mm}$$
$$\Delta_{T,AD,CD} = -\alpha L \Delta_T = -(12 \times 10^{-6} \times 3041 \times 10.0) = +0.36 \text{ mm}$$
$$\Delta_{T,BD} = -\alpha L \Delta_T = -(12 \times 10^{-6} \times 500 \times 10.0) = +0.06 \text{ mm}$$

$\sin\alpha = (1.0/3.162) = 0.316$ $\cos\alpha = (3.0/3.162) = 0.949$
$\sin\theta = (0.5/3.041) = 0.164$ $\cos\theta = (3.0/3.041) = 0.987$
$AE_{100} = (100 \times 205) = 20.5 \times 10^3$ kN $AE_{300} = (300 \times 205) = 61.5 \times 10^3$ kN

Consider member BD to be redundant.
The equivalent system is the superposition of the statically determinate frame and the (unit load frame $\times$ F_{BD}) as shown:

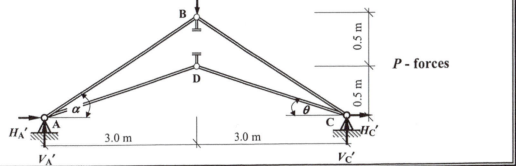

Solution
Topic: Unit Load Method for Singly-Redundant Pin-Jointed Frames
Problem Number: 3.23   **Page No. 2**

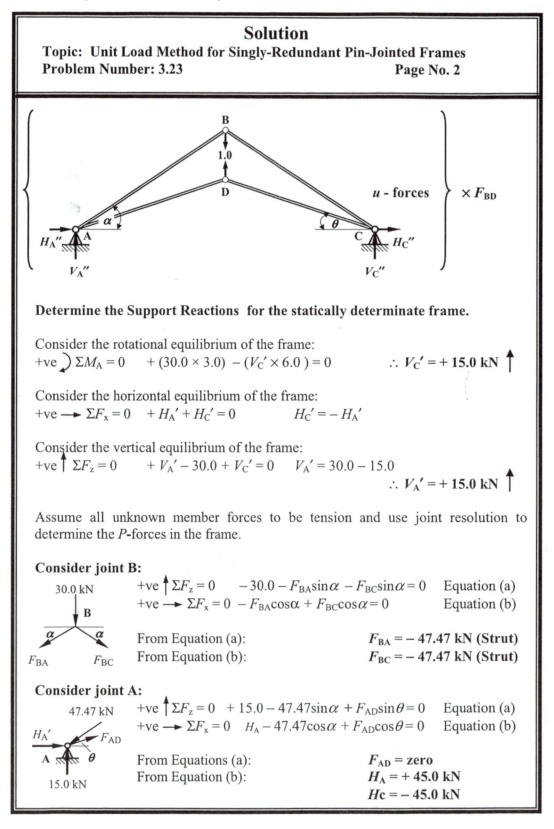

Determine the Support Reactions for the statically determinate frame.

Consider the rotational equilibrium of the frame:
+ve $\circlearrowright$ $\Sigma M_A = 0$ $+ (30.0 \times 3.0) - (V_C' \times 6.0) = 0$ $\therefore V_C' = + 15.0$ kN $\uparrow$

Consider the horizontal equilibrium of the frame:
+ve $\longrightarrow$ $\Sigma F_x = 0$ $+ H_A' + H_C' = 0$ $H_C' = - H_A'$

Consider the vertical equilibrium of the frame:
+ve $\uparrow$ $\Sigma F_z = 0$ $+ V_A' - 30.0 + V_C' = 0$ $V_A' = 30.0 - 15.0$
                                        $\therefore V_A' = + 15.0$ kN $\uparrow$

Assume all unknown member forces to be tension and use joint resolution to determine the *P*-forces in the frame.

Consider joint B:

30.0 kN

+ve $\uparrow$ $\Sigma F_z = 0$ $- 30.0 - F_{BA}\sin\alpha - F_{BC}\sin\alpha = 0$ Equation (a)
+ve $\longrightarrow$ $\Sigma F_x = 0$ $- F_{BA}\cos\alpha + F_{BC}\cos\alpha = 0$ Equation (b)

From Equation (a): $F_{BA} = - 47.47$ kN (Strut)
From Equation (b): $F_{BC} = - 47.47$ kN (Strut)

Consider joint A:

47.47 kN

+ve $\uparrow$ $\Sigma F_z = 0$ $+ 15.0 - 47.47\sin\alpha + F_{AD}\sin\theta = 0$ Equation (a)
+ve $\longrightarrow$ $\Sigma F_x = 0$ $H_A - 47.47\cos\alpha + F_{AD}\cos\theta = 0$ Equation (b)

From Equations (a): $F_{AD} = $ **zero**
From Equation (b): $H_A = + 45.0$ kN
                                            $H_C = - 45.0$ kN

Solution

Topic: Unit Load Method for Singly-Redundant Pin-Jointed Frames
Problem Number: 3.23   **Page No. 3**

Consider joint C:

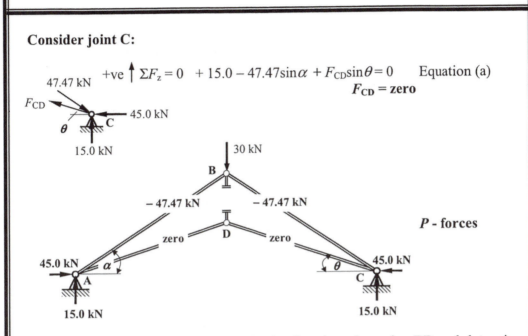

$$+\text{ve} \uparrow \Sigma F_z = 0 \quad +15.0 - 47.47\sin\alpha + F_{CD}\sin\theta = 0 \quad \text{Equation (a)}$$

$$F_{CD} = \text{zero}$$

P - **forces**

Apply a Unit Load at joints B and D in the direction of member BD and determine the values of the *u*-forces using joint resolution as before.

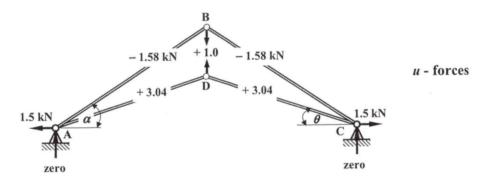

u - **forces**

$$(\delta_{BD} \text{ due to } P\text{-forces}) + (\delta_{BD} \text{ due to unit forces}) \times F_{BD} = 0$$

i.e. $\displaystyle\sum\left(\frac{PL}{AE} + \Delta_L + \Delta_T\right)u + \left(\sum\frac{uL}{AE}u\right) \times F_{BD} = 0$

$\therefore\ F_{BD} = -\displaystyle\sum\left(\frac{PL}{AE} + \Delta_L + \Delta_T\right)u \Big/ \sum\frac{uL}{AE}u$

Solution
Topic: Unit Load Method for Singly-Redundant Pin-Jointed Frames
Problem Number: 3.23   **Page No. 4**

The term $\sum\left(\dfrac{PL}{AE}+\varDelta_{L}+\varDelta_{T}\right)$ is evaluated separately here for convenience, normally this would be incorporated in one table.

Member	Length (mm)	AE (kN)	P-force (kN)	(PL/AE) (mm)	$\varDelta_L$	Temp. change	$\varDelta_T$	(PL/AE + $\varDelta_L$ + $\varDelta_T$) (mm)
AB	3162	61.5×10^3	-47.47	-2.44	0	$+10$	$+0.38$	-2.06
BC	3162	61.5×10^3	-47.47	-2.44	0	$+10$	$+0.38$	-2.06
BD	500	20.5×10^3	0	0	0	$+10$	$+0.06$	$+0.06$
CD	3041	61.5×10^3	0	0	0	$+10$	$+0.36$	$+0.36$
DA	3041	61.5×10^3	0	0	$+1.5$	$+10$	$+0.36$	$+1.86$

Complete the Unit Load table to determine the value of F_{BD}

Member	Length (mm)	AE (kN)	(PL/AE + $\varDelta_L$ + $\varDelta_T$) (mm)	u	(PL/AE + $\varDelta_L$ + $\varDelta_T$) $\times u$ (mm)	(uL/AE) $\times u$ (mm)	Member forces
AB	3162	61.5×10^3	-2.06	-1.58	$+3.261$	$+0.129$	-29.80
BC	3162	61.5×10^3	-2.06	-1.58	$+3.261$	$+0.129$	-29.80
BD	500	20.5×10^3	$+0.06$	$+1.00$	$+0.060$	$+0.024$	-11.17
CD	3041	61.5×10^3	$+0.36$	$+3.04$	$+1.110$	$+0.457$	-33.97
DA	3041	61.5×10^3	$+1.86$	$+3.04$	$+5.671$	$+0.457$	-33.97
					$\Sigma = +13.363$	$\Sigma = +1.196$	

$$F_{BD}=-\sum\left(\frac{PL}{AE}+\varDelta_{L}+\varDelta_{T}\right)u\bigg/\sum\frac{uL}{AE}u=-13.363/1.196=-11.17 \text{ kN (Strut)}$$

The final member forces = [P-forces + (u-forces × (−)11.17)] and are given in the last column of the table

$V_A = +15.0 + \text{zero} = +15.0$ kN   ↑

$H_A = +45.0 - (1.5 \times (-)11.17) = +61.76$ kN   →

$V_C = +15.0 + \text{zero} = +15.0$ kN   ↑

$H_C = -45.0 + (1.5 \times (-)11.17) = -61.76$ kN   ←

Solution

Topic: Unit Load Method for Singly-Redundant Pin-Jointed Frames
Problem Number: 3.24   **Page No. 1**

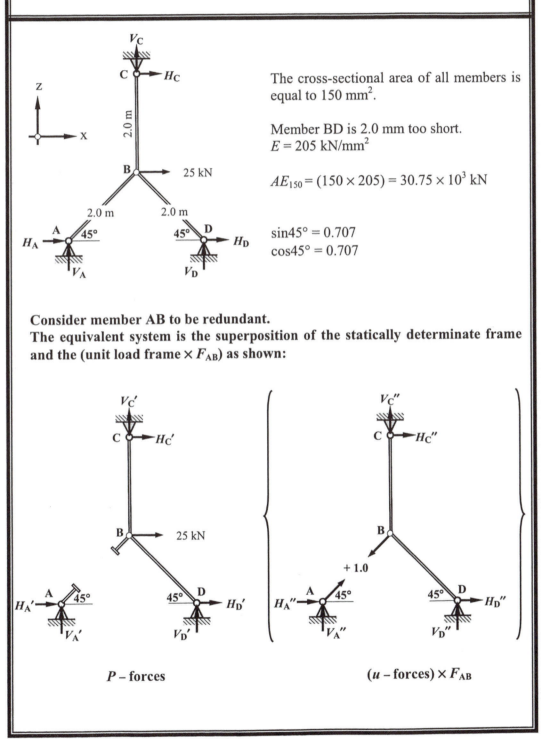

The cross-sectional area of all members is equal to 150 mm².

Member BD is 2.0 mm too short.
$E = 205$ kN/mm²

$AE_{150} = (150 \times 205) = 30.75 \times 10^3$ kN

$\sin 45° = 0.707$
$\cos 45° = 0.707$

Consider member AB to be redundant.
The equivalent system is the superposition of the statically determinate frame and the (unit load frame × F_{AB}) as shown:

P – forces   *(u – forces) × F_{AB}*

Solution

Topic: Unit Load Method for Singly-Redundant Pin-Jointed Frames
Problem Number: 3.24   **Page No. 2**

Assume all unknown member forces to be tension and use joint resolution to determine the *P*-forces in the frame.

Consider joint B:

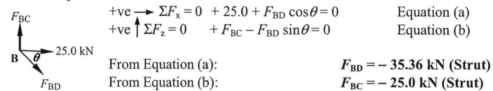

$+ve \longrightarrow \Sigma F_x = 0$ $+ 25.0 + F_{BD} \cos\theta = 0$ Equation (a)
$+ve \uparrow \Sigma F_z = 0$ $+ F_{BC} - F_{BD} \sin\theta = 0$ Equation (b)

From Equation (a): $F_{BD} = -\,35.36$ kN (Strut)
From Equation (b): $F_{BC} = -\,25.0$ kN (Strut)

Consider joint C:

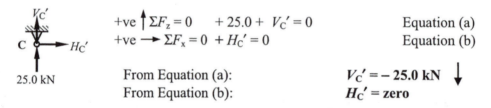

$+ve \uparrow \Sigma F_z = 0$ $+ 25.0 + V_C' = 0$ Equation (a)
$+ve \longrightarrow \Sigma F_x = 0$ $+ H_C' = 0$ Equation (b)

From Equation (a): $V_C' = -\,25.0$ kN $\downarrow$
From Equation (b): $H_C' = $ **zero**

Consider joint D:

$+ve \uparrow \Sigma F_z = 0$ $- 35.36\sin\theta + V_D' = 0$ Equation (a)
$+ve \longrightarrow \Sigma F_x = 0$ $+ 35.36\cos\theta + H_D' = 0$ Equation (b)

From Equation (a): $V_D' = +\,25.0$ kN $\uparrow$
From Equation (b): $H_D' = -\,25.0$ kN $\longleftarrow$

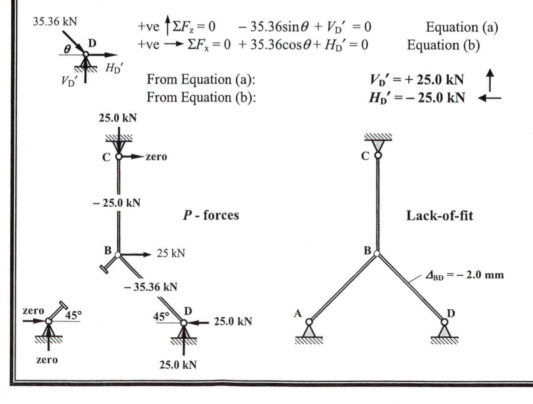

P - forces   Lack-of-fit

$\Delta_{BD} = -\,2.0$ mm

Solution

Topic: Unit Load Method for Singly-Redundant Pin-Jointed Frames

Problem Number: 3.24   **Page No. 3**

Apply a Unit Load at joints A and B in the direction of member AB and determine the values of the *u*-forces using joint resolution as before.

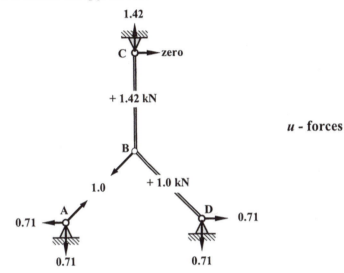

$(\delta_{AB}$ due to P-forces$) + (\delta_{AB}$ due to unit forces$) \times F_{AB} = 0$

i.e. $\sum \left(\dfrac{PL}{AE} + \varDelta_L \right) u + \left(\sum \dfrac{uL}{AE} u \right) \times F_{AB} = 0$

$\therefore F_{AB} = - \sum \left(\dfrac{PL}{AE} + \varDelta_L \right) u \Big/ \sum \dfrac{uL}{AE} u$

The term $\sum \left(\dfrac{PL}{AE} + \varDelta_L \right)$ is evaluated separately here for convenience, normally this would be incorporated in one table.

Member	Length (mm)	AE (kN)	P-force (kN)	(PL/AE) (mm)	$\varDelta_L$	(PL/AE +$\varDelta_L$) (mm)
AB	2000	30.75×10^3	0	0	0	0
BC	2000	30.75×10^3	− 25.00	− 1.63	0	− 1.63
BD	2000	30.75×10^3	− 35.36	− 2.30	− 2.0	− 4.30

Solution

Topic: Unit Load Method for Singly-Redundant Pin-Jointed Frames
Problem Number: 3.24   **Page No. 4**

Complete the Unit Load table to determine the value of F_{AB}

Member	Length (mm)	AE (kN)	$(PL/AE+\Delta_L)$ (mm)	u	$(PL/AE+\Delta_L) \times u$ (mm)	$(uL/AE) \times u$ (mm)	Member forces
AB	2000	30.75×10^3	0	+ 1.00	0	+ 0.065	+ 25.38
BC	2000	30.75×10^3	− 1.63	+ 1.42	− 2.315	+ 0.131	+ 10.87
BD	2000	30.75×10^3	− 4.30	+ 1.00	− 4.300	+ 0.065	− 9.99
					$\Sigma = - 6.615$	$\Sigma = + 0.261$	

$$F_{AB} = - \sum \left(\frac{PL}{AE} + \Delta_L \right) u \bigg/ \sum \frac{uL}{AE} u = +6.615/0.261 = 25.34 \text{ kN (Tie)}$$

The final member forces = [*P*-forces + (*u*-forces × 25.37)] and are given in the last column of the table

V_A = **zero** − (0.71 × 25.34) = − **17.99 kN** ↓
H_A = **zero** − (0.71 × 25.34) = − **17.99 kN** ←

V_C = − **25.0** + (1.42 × 25.34) = + **10.98 kN** ↑
H_C = **zero**

V_D = + **25.0** − (0.71 × 25.34) = + **7.01 kN** ↑
H_D = − **25.0** + (0.71 × 25.34) = − **7.01 kN** ←

4. Beams

4.1 Statically Determinate Beams

Two parameters which are fundamentally important to the design of beams are **shear force** and **bending moment**. These quantities are the result of internal forces acting on the material of a beam in response to an externally applied load system.

4.1.1 Example 4.1: Beam with Point Loads

Consider a simply-supported beam as shown in Figure 4.1 carrying a series of secondary beams each imposing a point load of 4 kN.

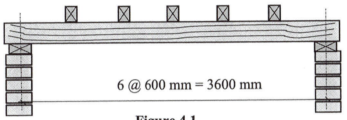

Figure 4.1

This structure can be represented as a line diagram as shown in Figure 4.2:

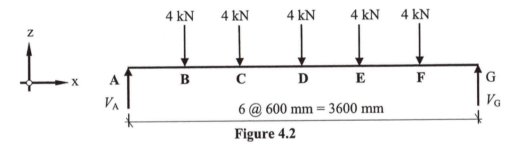

Figure 4.2

Since the externally applied force system is in equilibrium, the three equations of static equilibrium must be satisfied, i.e.

+ve $\uparrow \Sigma F_z = 0$ The sum of the vertical forces must equal zero.

+ve $\Sigma M = 0$ The sum of the moments of all forces about *any* point on the plane of the forces must equal zero.

+ve $\longrightarrow \Sigma F_x = 0$ The sum of the horizontal forces must equal zero.

The assumed positive directions are as indicated. In this particular problem there are no externally applied horizontal forces and consequently the third equation is not required.
(**Note:** It is still necessary to provide horizontal restraint to a structure since it can be subject to a variety of load cases, some of which may have a horizontal component.)

Consider the vertical equilibrium of the beam:

+ve $\uparrow \Sigma F_z = 0$

$+ V_A - (5 \times 4.0) + V_G = 0$ $\therefore V_A + V_G = 20$ kN   Equation (1)

Consider the rotational equilibrium of the beam:

+ve $\curvearrowleft \Sigma M_A = 0$

Note: The sum of the moments is taken about one end of the beam (end A) for convenience. Since one of the forces (V_A) passes through this point it does not produce a moment about A and hence does not appear in the equation. It should be recognised that the sum of the moments could have been considered about *any* known point in the same plane.

$+ (4.0 \times 0.6) + (4.0 \times 1.2) + (4.0 \times 1.8) + (4.0 \times 2.4) + (4.0 \times 3.0) - (V_G \times 3.6) = 0$

$\therefore V_G = 10$ kN   Equation (2)

Substituting into Equation (1) gives $V_A = 10$ kN

This calculation was carried out considering only the externally applied forces, i.e.

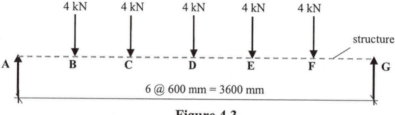

Figure 4.3

The structure itself was ignored, however the applied loads are transferred to the end supports through the material fibres of the beam. Consider the beam to be cut at section X–X producing two sections each of which is in equilibrium as shown in Figure 4.4.

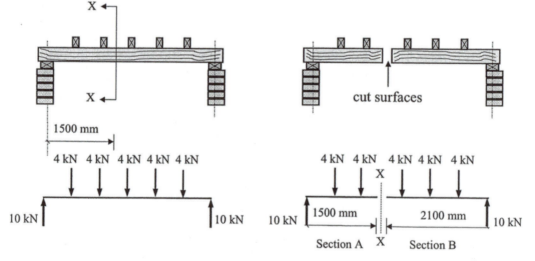

Figure 4.4

Clearly if the two sections are in equilibrium there must be internal forces acting on the cut surfaces to maintain this; these forces are known as the *shear force* and the *bending moment*, and are illustrated in Figure 4.5

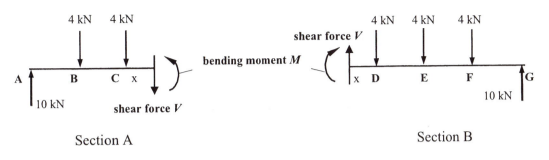

Section A   Section B

Figure 4.5

The force *V* and moment *M* are equal and opposite on each surface. The magnitude and direction of *V* and *M* can be determined by considering two equations of static equilibrium for either of the cut sections; both will give the same answer.

Consider the left-hand section with the 'assumed' directions of the internal forces *V* and *M* as shown in Figure 4.6.

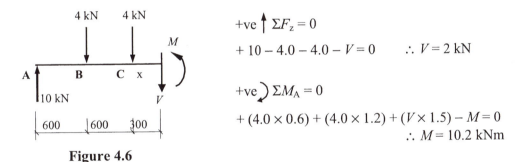

Figure 4.6

$$+ve \uparrow \Sigma F_z = 0$$
$$+ 10 - 4.0 - 4.0 - V = 0 \qquad \therefore V = 2 \text{ kN}$$

$$+ve \curvearrowright \Sigma M_A = 0$$
$$+ (4.0 \times 0.6) + (4.0 \times 1.2) + (V \times 1.5) - M = 0$$
$$\therefore M = 10.2 \text{ kNm}$$

4.1.2 Shear Force Diagrams

In a statically determinate beam, the numerical value of the shear force can be obtained by evaluating the algebraic sum of the vertical forces to one side of the section being considered. The convention adopted in this text to indicate positive and negative shear forces is shown in Figure 4.7.

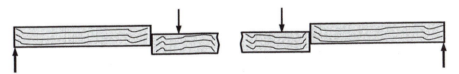

shear induced by a **+ve** shear force   shear induced by a **−ve** shear force

Figure 4.7

The calculation carried out to determine the shear force can be repeated at various locations along a beam and the values obtained plotted as a graph; this graph is known as the ***shear force diagram***. The shear force diagram indicates the variation of the shear force along a structural member.

Consider any section of the beam between A and B:

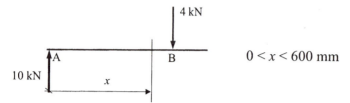

$0 < x < 600$ mm

Note: The value immediately under the point load at the cut section is not being considered.

The shear force at any position $x = \Sigma$ vertical forces to one side
$$= + 10.0 \text{ kN}$$

This value is a constant for all values of x between zero and 600 mm, the graph will therefore be a horizontal line equal to 10.0 kN. This force produces a +ve shear effect, i.e.

+ve shear effect

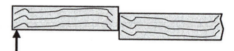

Consider any section of the beam between B and C:

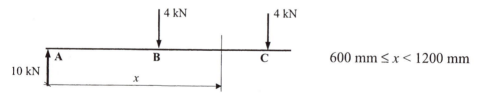

600 mm $\leq x < 1200$ mm

The shear force at any position $x = \Sigma$ vertical force to one side
$$= + 10.0 - 4.0 = 6.0 \text{ kN}$$

This value is a constant for all values of x between 600 mm and 1200 mm, the graph will therefore be a horizontal line equal to 6.0 kN. This force produces a +ve effect shear effect.

Similarly for any section between C and D:

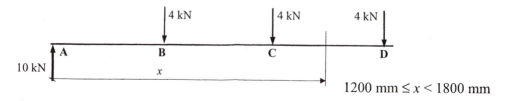

1200 mm $\leq x < 1800$ mm

The shear force at any position $x = \Sigma$ vertical forces to one side
$$= + 10.0 - 4.0 - 4.0 = 2.0 \text{ kN}$$

Consider any section of the beam between D and E:

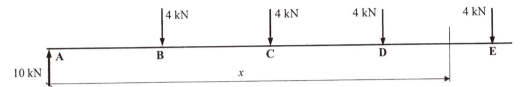

$$1800 \text{ mm} \leq x < 2400 \text{ mm}$$

The shear force at any position $x = \Sigma$ vertical forces to one side
$$= + 10.0 - 4.0 - 4.0 - 4.0 = - 2.0 \text{ kN}$$

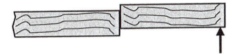

In this case the shear force is negative:

Similarly between E and F $2400 \text{ mm} < x < 3000 \text{ mm}$
    The shear force at any position $x = \Sigma$ vertical forces to one side
$$= + 10.0 - 4.0 - 4.0 - 4.0 - 4.0 = - 6.0 \text{ kN}$$

and

between F and G $3000 \text{ mm} < x < 3600 \text{ mm}$
    The shear force at any position $x = \Sigma$ vertical forces to one side
$$= + 10.0 - 4.0 - 4.0 - 4.0 - 4.0 - 4.0 = - 10.0 \text{ kN}$$

In each of the cases above the value has not been considered at the point of application of the load.
Consider the location of the applied load at B shown in Figure 4.8.

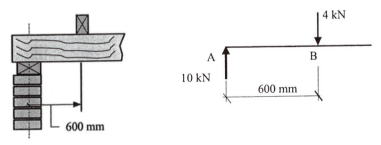

Figure 4.8

The 4.0 kN is not instantly transferred through the beam fibres at B but instead over the width of the actual secondary beam. The change in value of the shear force between $x < 600 \text{ mm}$ and $x > 600 \text{ mm}$ occurs over this width, as shown in Figure 4.9.

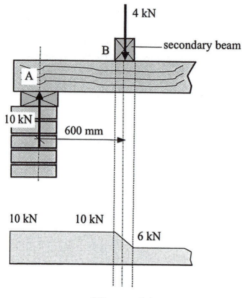

Figure 4.9

The width of the secondary beam is insignificant when compared with the overall span, and the shear force is assumed to change instantly at this point, producing a vertical line on the shear force diagram as shown in Figure 4.10.

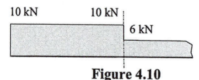

Figure 4.10

The full shear force diagram can therefore be drawn as shown in Figure 4.11.

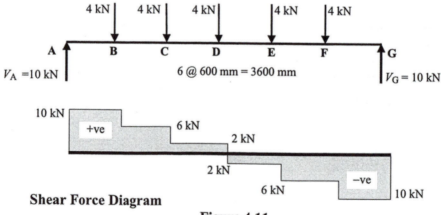

Shear Force Diagram

Figure 4.11

The same result can be obtained by considering sections from the right-hand side of the beam.

4.1.3 Bending Moment Diagrams

In a statically determinate beam the numerical value of the bending moment (i.e. moments caused by forces which tend to bend the beam) can be obtained by evaluating the algebraic sum of the moments of the forces to one side of a section. In the same manner as with shear forces either the left-hand or the right-hand side of the beam can be considered. The convention adopted in this text to indicate positive and negative bending moments is shown in Figures 4.12(a) and (b).

Bending inducing **tension on the underside** of a beam is considered **positive**.

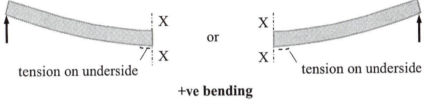

+ve **bending**

Figure 4.12 (a)

Bending inducing **tension on the top** of a beam is considered **negative**.

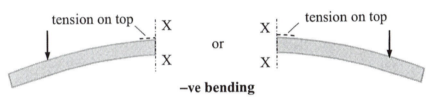

−ve **bending**

Figure 4.12 (b)

*Note: Clockwise/anti-clockwise moments do **not** define +ve or −ve **bending** moments. The sign of the bending moment is governed by the location of the tension surface at the point being considered.*

As with shear forces the calculation for bending moments can be carried out at various locations along a beam and the values plotted on a graph; this graph is known as the *'bending moment diagram'*. The bending moment diagram indicates the variation in the bending moment along a structural member.

Consider sections between A and B of the beam as before:

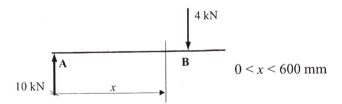

In this case when x = 600 mm the 4.0 kN load passes through the section being considered and does not produce a bending moment, and can therefore be ignored.

Bending moment = Σ algebraic sum of the moments of the forces to one side of a section.

$$= Σ \text{ (Force} × \text{lever arm)}$$
$$M_x = 10.0 × x = 10.0\,x \text{ kNm}$$

Unlike the shear force, this expression is not a constant and depends on the value of 'x' which varies between the limits given. This is a linear expression which should be reflected in the calculated values of the bending moment.

$x = 0$ $M_x = 10.0 × 0 = $ zero
$x = 200$ mm $M_x = 10.0 × 0.2 = 2.0$ kNm
$x = 400$ mm $M_x = 10.0 × 0.4 = 4.0$ kNm
$x = 600$ mm $M_x = 10.0 × 0.6 = 6.0$ kNm

Clearly the bending moment increases linearly from zero at the simply-supported end to a value of 6.0 kNm at point B.

Consider sections between B and C of the beam:

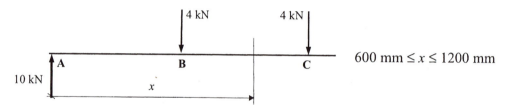

Bending moment = Σ algebraic sum of the moments of the forces to 'one' side of a section
$$M_x = + (10.0 × x) - (4.0 × [x - 0.6])$$

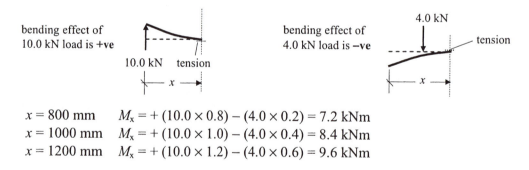

bending effect of
10.0 kN load is +ve

10.0 kN   tension

bending effect of
4.0 kN load is −ve

4.0 kN

tension

$x = 800$ mm $M_x = + (10.0 × 0.8) - (4.0 × 0.2) = 7.2$ kNm
$x = 1000$ mm $M_x = + (10.0 × 1.0) - (4.0 × 0.4) = 8.4$ kNm
$x = 1200$ mm $M_x = + (10.0 × 1.2) - (4.0 × 0.6) = 9.6$ kNm

As before the bending moment increases linearly, i.e. from 7.2 kNm at $x = 800$ mm to a value of 9.6 kNm at point C.

Since the variation is linear it is only necessary to evaluate the magnitude and sign of the bending moment at locations where the slope of the line changes, i.e. each of the point load locations.

Consider point D:

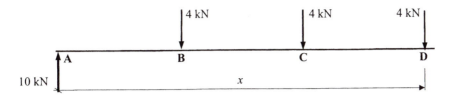

$x = 1800$ mm $M_x = (10.0 \times 1.8) - (4.0 \times 1.2) - (4.0 \times 0.6) = 10.8$ kNm

Consider point E:

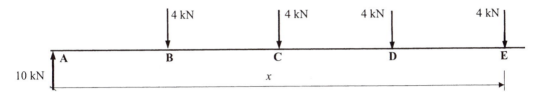

$x = 2400$ mm $M_x = (10.0 \times 2.4) - (4.0 \times 1.8) - (4.0 \times 1.2) - (4.0 \times 0.6) = 9.6$ kNm

Similarly at point F:

$x = 3000$ mm $M_x = (10.0 \times 3.0) - (4.0 \times 2.4) - (4.0 \times 1.8) - (4.0 \times 1.2) - (4.0 \times 0.6)$
           $= 6.0$ kNm

The full bending moment diagram can therefore be drawn as shown in Figure 4.13.

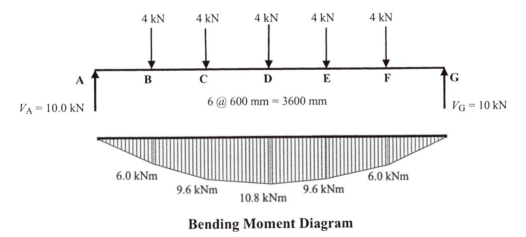

Bending Moment Diagram

Figure 4.13

The same result can be obtained by considering sections from the right-hand side of the beam. The value of the bending moment at any location can also be determined by evaluating the area under the shear force diagram.

Consider point B:

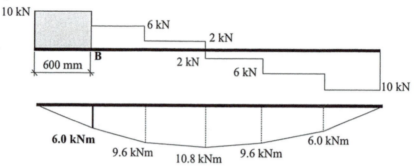

Bending moment at B = shaded area on the shear force diagram
$M_B = (10.0 \times 0.6) = 6.0$ kNm   as before

Consider a section at a distance of $x = 800$ mm along the beam between B and C:

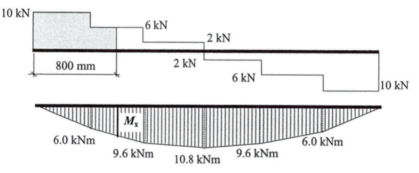

Bending moment at x = shaded area on the shear force diagram
$M_x = (10.0 \times 0.6) + (6.0 \times 0.2) = 7.2$ kNm as before

Consider a section at a distance of $x = 2100$ mm along the beam between D and E:

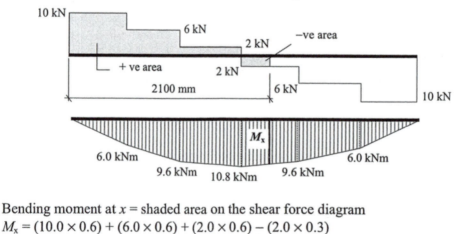

Bending moment at x = shaded area on the shear force diagram
$M_x = (10.0 \times 0.6) + (6.0 \times 0.6) + (2.0 \times 0.6) - (2.0 \times 0.3)$
   $= 10.2$ kNm

(**Note:** A maximum bending moment occurs at the same position as a zero shear force).

4.1.4 Example 4.2: Beam with a Uniformly Distributed Load (UDL)

Consider a simply-supported beam carrying a uniformly distributed load of 5 kN/m as shown in Figure 4.14.

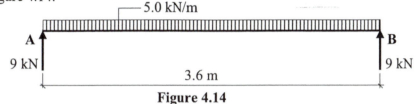

Figure 4.14

The shear force at any section a distance x from the support at A is given by:
V_x = algebraic sum of the vertical forces

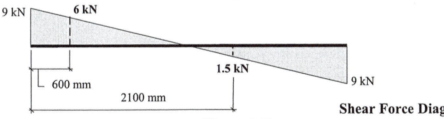

The force inducing +ve shear = 9.0 kN
The force inducing −ve shear = $(5.0 \times x) = 5.0x$ kN

$$V_x = +9.0 - 5.0x$$

This is a linear equation in which V_x decreases as x increases. The points of interest are at the supports where the maximum shear forces occur, and at the locations where the maximum bending moment occurs, i.e. the point of zero shear.

$V_x = 0$ when $+9.0 - 5.0x = 0$ $\therefore x = 1.8$ m

Any intermediate value can be found by substituting the appropriate value of 'x' in the equation for the shear force; e.g.

$x = 600$ mm $V_x = +9.0 - (5.0 \times 0.6) = +6.0$ kN
$x = 2100$ mm $V_x = +9.0 - (5.0 \times 2.1) = -1.5$ kN

The shear force can be drawn as shown in Figure 4.15.

Shear Force Diagram

Figure 4.15

The bending moment can be determined as before, **either** using an equation or evaluating the area under the shear force diagram.

Using an equation:

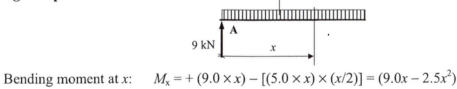

Bending moment at x: $M_x = +(9.0 \times x) - [(5.0 \times x) \times (x/2)] = (9.0x - 2.5x^2)$

In this case the equation is *not* linear, and the bending moment diagram will therefore be *curved*.

Consider several values:

$x = 0$ M_x = zero

$x = 600$ mm $M_x = + (9.0 \times 0.6) - (2.5 \times 0.6^2) = 4.5$ kNm

$x = 1800$ mm $M_x = + (9.0 \times 1.8) - (2.5 \times 1.8^2) = 8.1$ kNm

$x = 2100$ mm $M_x = + (9.0 \times 2.1) - (2.5 \times 2.1^2) = 7.88$ kNm

Using the shear force diagram:

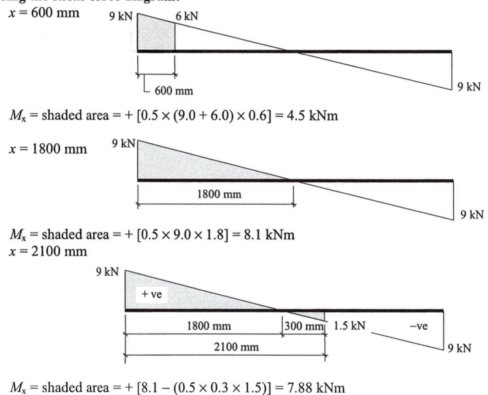

$x = 600$ mm

M_x = shaded area = $+ [0.5 \times (9.0 + 6.0) \times 0.6] = 4.5$ kNm

$x = 1800$ mm

M_x = shaded area = $+ [0.5 \times 9.0 \times 1.8] = 8.1$ kNm

$x = 2100$ mm

M_x = shaded area = $+ [8.1 - (0.5 \times 0.3 \times 1.5)] = 7.88$ kNm
The bending moment diagram is shown in Figure 4.16.

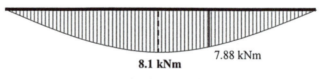

8.1 kNm   7.88 kNm

Bending Moment Diagram

Figure 4.16

The UDL loading is a 'standard' load case which occurs in numerous beam designs and can be expressed in general terms using L for the span and w for the applied load/metre or W_{total} $(= wL)$ for the total applied load, as shown in Figure 4.17.

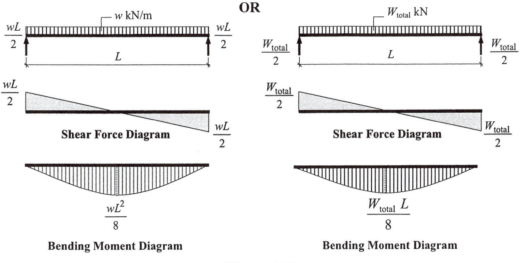

OR

Shear Force Diagram

$$\frac{wL^2}{8}$$

Bending Moment Diagram

Shear Force Diagram

$$\frac{W_{total}\ L}{8}$$

Bending Moment Diagram

Figure 4.17

Clearly both give the same magnitude of support reactions, shear forces and bending moments.

In cantilever beams, all support restraints are provided at one location, i.e. an 'encastré' or 'fixed' support as shown in Example 4.3.

4.1.5 Example 4.3: Cantilever Beam

Consider the cantilever beam shown in Figure 4.18 which is required to support a uniformly distributed load in addition to a mid-span point load as indicated.

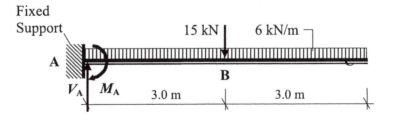

Fixed Support

15 kN   6 kN/m

A

B

V_A M_A 3.0 m   3.0 m

Figure 4.18

Support Reactions

Consider the rotational equilibrium of the beam: +ve $\circlearrowright$ $\Sigma M_A = 0$

$$M_A + (6.0 \times 6.0)(3.0) + (15.0 \times 3.0) = 0 \qquad \therefore M_A = -153.0 \text{ kNm} \circlearrowright$$

Consider the vertical equilibrium of the beam: +ve $\uparrow$ $\Sigma F_z = 0$

$$+ V_A - (6.0 \times 6.0) - 15.0 = 0 \qquad \therefore V_A = +51.0 \text{ kN} \uparrow$$

Shear force at B:

$V_B = [51.0 - (6.0 \times 3.0)] = \textbf{33.0 kN}$

and $= (33.0 - 15.0) = \textbf{18.0 kN}$

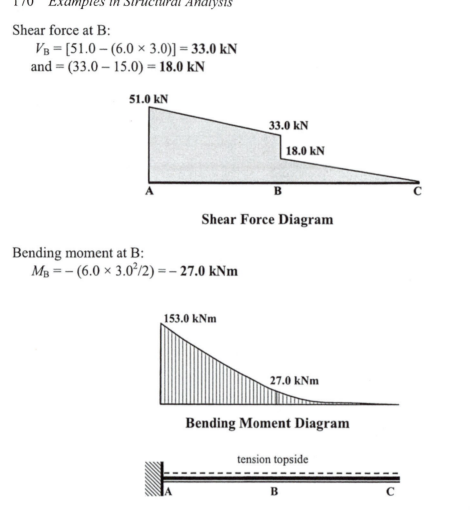

51.0 kN

33.0 kN

18.0 kN

A   B   C

Shear Force Diagram

Bending moment at B:

$M_B = - (6.0 \times 3.0^2/2) = - \textbf{27.0 kNm}$

153.0 kNm

27.0 kNm

Bending Moment Diagram

tension topside

A   B   C

4.1.6 Problems: Statically Determinate Beams – Shear Force and Bending Moment

A series of simply-supported beams are indicated in Problems 4.1 to 4.10. Using the applied loading given in each case:

i)   determine the support reactions,
ii) sketch the shear force diagram   and
iii) sketch the bending moment diagram indicating the maximum value(s).

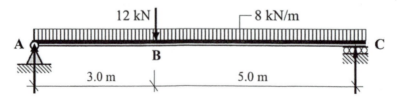

12 kN   ⌐ 8 kN/m

A   C
        B

3.0 m   5.0 m

Problem 4.1

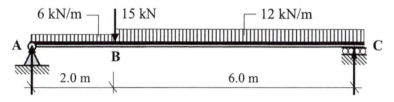

6 kN/m ⌐ ⏐ 15 kN ⌐ 12 kN/m

A   B   C

2.0 m   6.0 m

Problem 4.2

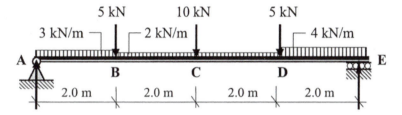

5 kN   10 kN   5 kN

3 kN/m ⌐ ⌐ 2 kN/m ⌐ 4 kN/m

A   B   C   D   E

2.0 m   2.0 m   2.0 m   2.0 m

Problem 4.3

8 kN   12 kN   6 kN/m

3 kN/m ⌐

A   B   C   D   E

2.0 m   1.0 m   3.0 m   3.0 m

Problem 4.4

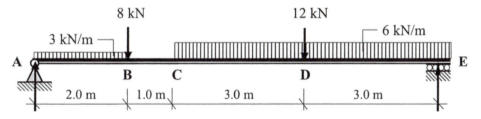

2 kN/m ⌐   6 kN/m ⌐   12 kN

A   B   C

4.0 m   2.0 m

Problem 4.5

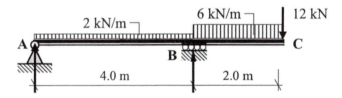

10 kN/m ⌐   8 kN ⏐ ⌐ 5 kN/m

A   B   C   D

6.0 m   1.0 m ⏐ 1.0 m

Problem 4.6

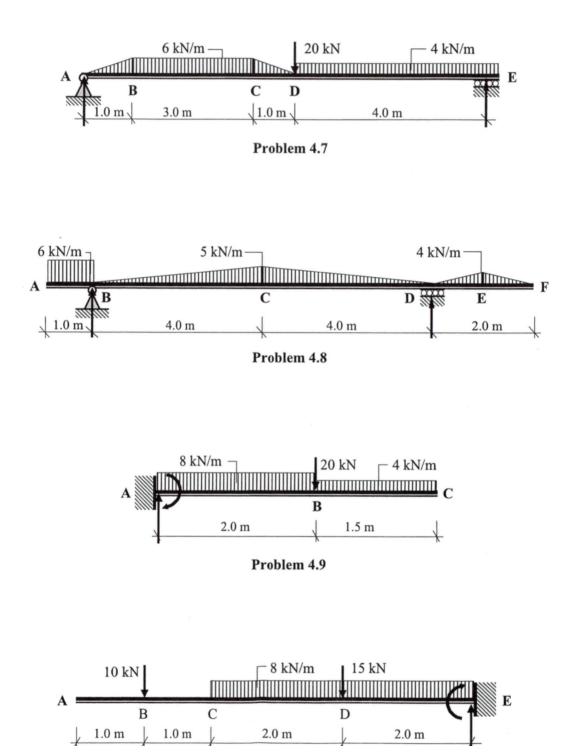

Problem 4.7

Problem 4.8

Problem 4.9

Problem 4.10

4.1.7 Solutions: Statically Determinate Beams – Shear Force and Bending Moment

---

Solution

Topic: Statically Determinate Beams – Shear Force and Bending Moment

Problem Number: 4.1   **Page No. 1**

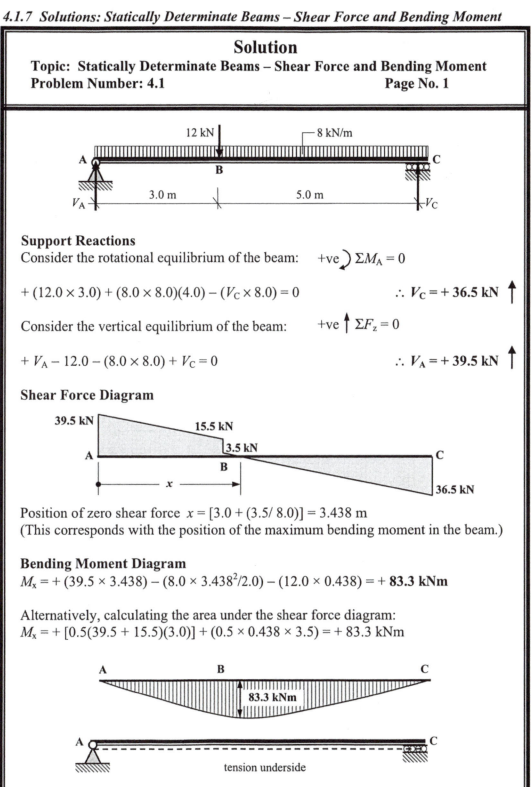

Support Reactions

Consider the rotational equilibrium of the beam: +ve $\circlearrowright$ $\Sigma M_A = 0$

$+ (12.0 \times 3.0) + (8.0 \times 8.0)(4.0) - (V_C \times 8.0) = 0$ $\therefore V_C = + 36.5$ kN $\uparrow$

Consider the vertical equilibrium of the beam: +ve $\uparrow$ $\Sigma F_z = 0$

$+ V_A - 12.0 - (8.0 \times 8.0) + V_C = 0$ $\therefore V_A = + 39.5$ kN $\uparrow$

Shear Force Diagram

Position of zero shear force $x = [3.0 + (3.5/8.0)] = 3.438$ m

(This corresponds with the position of the maximum bending moment in the beam.)

Bending Moment Diagram

$M_x = + (39.5 \times 3.438) - (8.0 \times 3.438^2/2.0) - (12.0 \times 0.438) = + \textbf{83.3 kNm}$

Alternatively, calculating the area under the shear force diagram:

$M_x = + [0.5(39.5 + 15.5)(3.0)] + (0.5 \times 0.438 \times 3.5) = + 83.3$ kNm

Solution
Topic: Statically Determinate Beams – Shear Force and Bending Moment
Problem Number: 4.2   **Page No. 1**

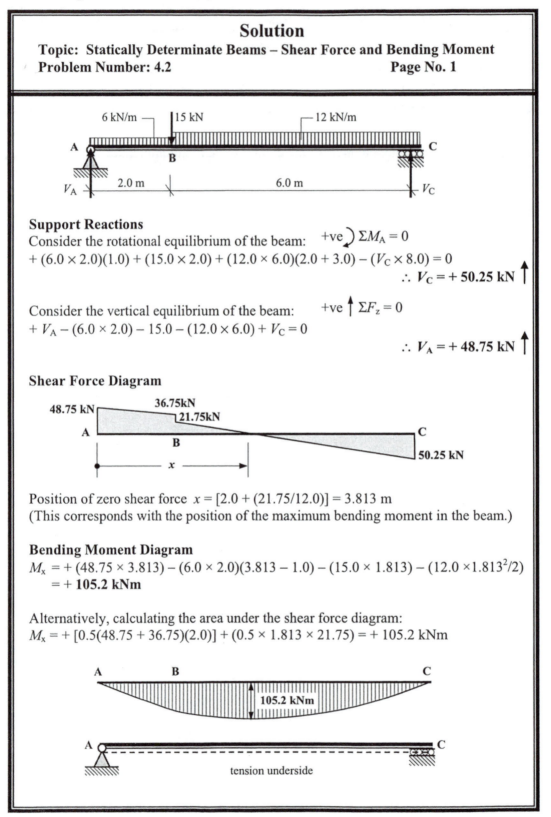

Support Reactions
Consider the rotational equilibrium of the beam: $+ve \curvearrowright \Sigma M_A = 0$
$+ (6.0 \times 2.0)(1.0) + (15.0 \times 2.0) + (12.0 \times 6.0)(2.0 + 3.0) - (V_C \times 8.0) = 0$
$$\therefore V_C = + 50.25 \text{ kN} \uparrow$$

Consider the vertical equilibrium of the beam: $+ve \uparrow \Sigma F_z = 0$
$+ V_A - (6.0 \times 2.0) - 15.0 - (12.0 \times 6.0) + V_C = 0$
$$\therefore V_A = + 48.75 \text{ kN} \uparrow$$

Shear Force Diagram

Position of zero shear force $x = [2.0 + (21.75/12.0)] = 3.813$ m
(This corresponds with the position of the maximum bending moment in the beam.)

Bending Moment Diagram
$M_x = + (48.75 \times 3.813) - (6.0 \times 2.0)(3.813 - 1.0) - (15.0 \times 1.813) - (12.0 \times 1.813^2/2)$
$\quad = + 105.2$ **kNm**

Alternatively, calculating the area under the shear force diagram:
$M_x = + [0.5(48.75 + 36.75)(2.0)] + (0.5 \times 1.813 \times 21.75) = + 105.2$ kNm

Solution
Topic: Statically Determinate Beams – Shear Force and Bending Moment
Problem Number: 4.3   **Page No. 1**

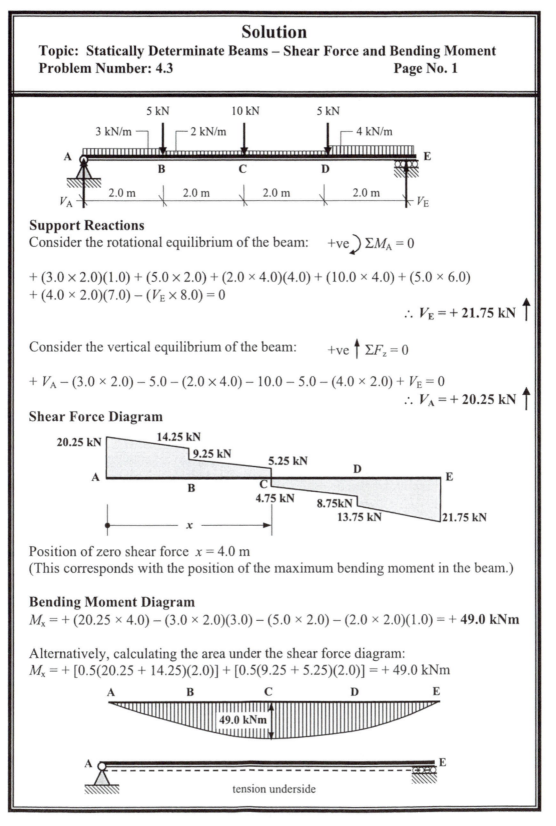

Support Reactions

Consider the rotational equilibrium of the beam: $+ve \circlearrowright \Sigma M_A = 0$

$+ (3.0 \times 2.0)(1.0) + (5.0 \times 2.0) + (2.0 \times 4.0)(4.0) + (10.0 \times 4.0) + (5.0 \times 6.0)$
$+ (4.0 \times 2.0)(7.0) - (V_E \times 8.0) = 0$

$$\therefore V_E = + 21.75 \text{ kN} \uparrow$$

Consider the vertical equilibrium of the beam: $+ve \uparrow \Sigma F_z = 0$

$+ V_A - (3.0 \times 2.0) - 5.0 - (2.0 \times 4.0) - 10.0 - 5.0 - (4.0 \times 2.0) + V_E = 0$
$$\therefore V_A = + 20.25 \text{ kN} \uparrow$$

Shear Force Diagram

Position of zero shear force $x = 4.0$ m
(This corresponds with the position of the maximum bending moment in the beam.)

Bending Moment Diagram

$M_x = + (20.25 \times 4.0) - (3.0 \times 2.0)(3.0) - (5.0 \times 2.0) - (2.0 \times 2.0)(1.0) = + 49.0 \text{ kNm}$

Alternatively, calculating the area under the shear force diagram:
$M_x = + [0.5(20.25 + 14.25)(2.0)] + [0.5(9.25 + 5.25)(2.0)] = + 49.0 \text{ kNm}$

Solution
Topic: Statically Determinate Beams – Shear Force and Bending Moment
Problem Number: 4.4   **Page No. 1**

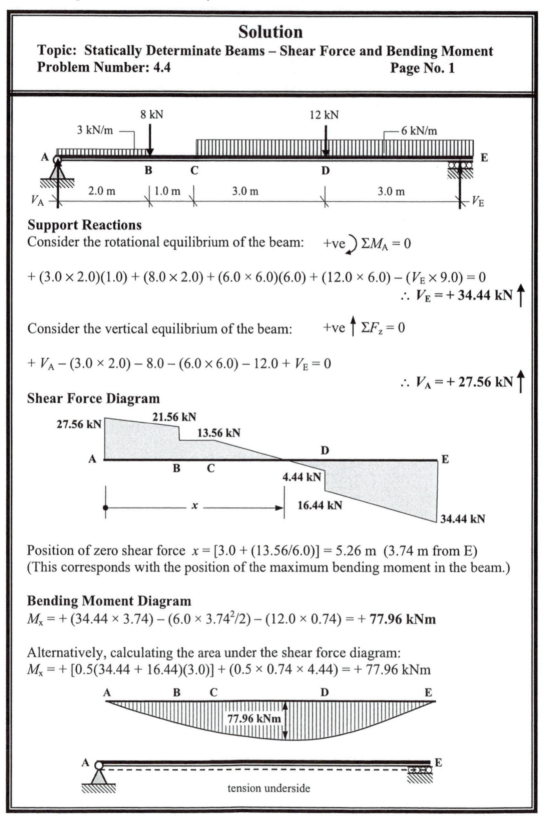

Support Reactions

Consider the rotational equilibrium of the beam: $+ve \curvearrowright \Sigma M_A = 0$

$$+ (3.0 \times 2.0)(1.0) + (8.0 \times 2.0) + (6.0 \times 6.0)(6.0) + (12.0 \times 6.0) - (V_E \times 9.0) = 0$$
$$\therefore V_E = + 34.44 \text{ kN} \uparrow$$

Consider the vertical equilibrium of the beam: $+ve \uparrow \Sigma F_z = 0$

$$+ V_A - (3.0 \times 2.0) - 8.0 - (6.0 \times 6.0) - 12.0 + V_E = 0$$
$$\therefore V_A = + 27.56 \text{ kN} \uparrow$$

Shear Force Diagram

Position of zero shear force $x = [3.0 + (13.56/6.0)] = 5.26$ m (3.74 m from E)
(This corresponds with the position of the maximum bending moment in the beam.)

Bending Moment Diagram
$$M_x = + (34.44 \times 3.74) - (6.0 \times 3.74^2/2) - (12.0 \times 0.74) = + \textbf{77.96 kNm}$$

Alternatively, calculating the area under the shear force diagram:
$$M_x = + [0.5(34.44 + 16.44)(3.0)] + (0.5 \times 0.74 \times 4.44) = + 77.96 \text{ kNm}$$

Solution

Topic: Statically Determinate Beams – Shear Force and Bending Moment

Problem Number: 4.5   Page No. 1

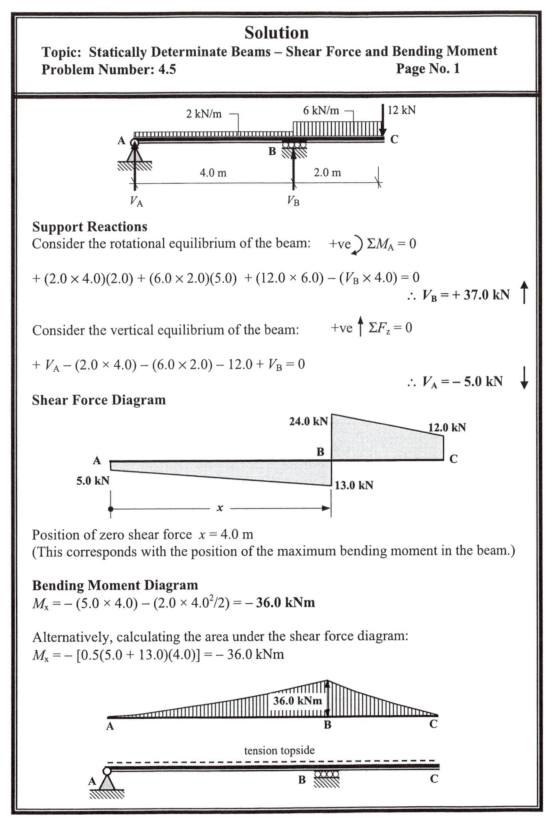

Support Reactions

Consider the rotational equilibrium of the beam: $+ve \curvearrowright \Sigma M_A = 0$

$$+ (2.0 \times 4.0)(2.0) + (6.0 \times 2.0)(5.0) + (12.0 \times 6.0) - (V_B \times 4.0) = 0$$

$$\therefore V_B = + 37.0 \text{ kN} \uparrow$$

Consider the vertical equilibrium of the beam: $+ve \uparrow \Sigma F_z = 0$

$$+ V_A - (2.0 \times 4.0) - (6.0 \times 2.0) - 12.0 + V_B = 0$$

$$\therefore V_A = - 5.0 \text{ kN} \downarrow$$

Shear Force Diagram

Position of zero shear force $x = 4.0$ m

(This corresponds with the position of the maximum bending moment in the beam.)

Bending Moment Diagram

$$M_x = - (5.0 \times 4.0) - (2.0 \times 4.0^2/2) = - 36.0 \text{ kNm}$$

Alternatively, calculating the area under the shear force diagram:

$$M_x = - [0.5(5.0 + 13.0)(4.0)] = - 36.0 \text{ kNm}$$

tension topside

Solution

Topic: Statically Determinate Beams – Shear Force and Bending Moment
Problem Number: 4.6   **Page No. 1**

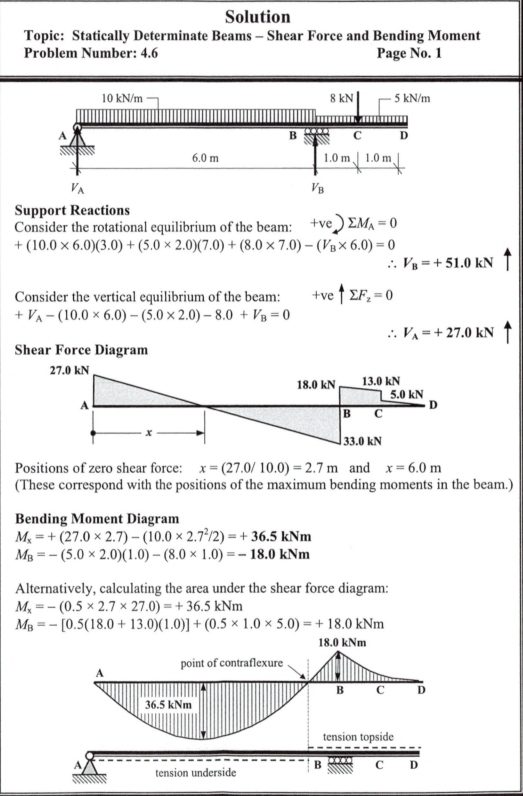

Support Reactions
Consider the rotational equilibrium of the beam: $+ve \;\; \sum M_A = 0$
$+ (10.0 \times 6.0)(3.0) + (5.0 \times 2.0)(7.0) + (8.0 \times 7.0) - (V_B \times 6.0) = 0$
$$\therefore V_B = + 51.0 \text{ kN} \uparrow$$

Consider the vertical equilibrium of the beam: $+ve \uparrow \sum F_z = 0$
$+ V_A - (10.0 \times 6.0) - (5.0 \times 2.0) - 8.0 + V_B = 0$
$$\therefore V_A = + 27.0 \text{ kN} \uparrow$$

Shear Force Diagram

Positions of zero shear force: $x = (27.0/\,10.0) = 2.7$ m   and $x = 6.0$ m
(These correspond with the positions of the maximum bending moments in the beam.)

Bending Moment Diagram
$M_x = + (27.0 \times 2.7) - (10.0 \times 2.7^2/2) = + \mathbf{36.5 \text{ kNm}}$
$M_B = - (5.0 \times 2.0)(1.0) - (8.0 \times 1.0) = - \mathbf{18.0 \text{ kNm}}$

Alternatively, calculating the area under the shear force diagram:
$M_x = - (0.5 \times 2.7 \times 27.0) = + 36.5 \text{ kNm}$
$M_B = - [0.5(18.0 + 13.0)(1.0)] + (0.5 \times 1.0 \times 5.0) = + 18.0 \text{ kNm}$

Solution
Topic: Statically Determinate Beams – Shear Force and Bending Moment
Problem Number: 4.7   **Page No. 1**

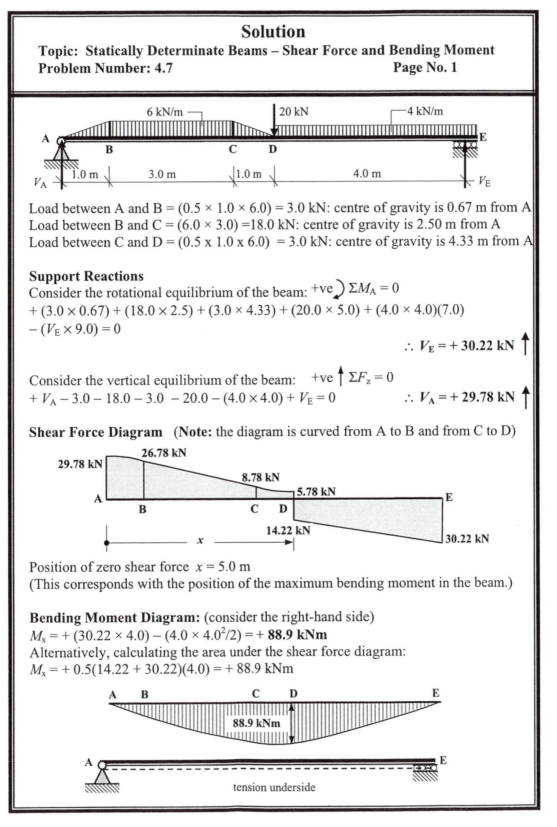

Load between A and B = $(0.5 \times 1.0 \times 6.0) = 3.0$ kN: centre of gravity is 0.67 m from A
Load between B and C = $(6.0 \times 3.0) = 18.0$ kN: centre of gravity is 2.50 m from A
Load between C and D = $(0.5 \times 1.0 \times 6.0) = 3.0$ kN: centre of gravity is 4.33 m from A

Support Reactions
Consider the rotational equilibrium of the beam: $+ve \; \curvearrowright \; \Sigma M_A = 0$
$+ (3.0 \times 0.67) + (18.0 \times 2.5) + (3.0 \times 4.33) + (20.0 \times 5.0) + (4.0 \times 4.0)(7.0)$
$- (V_E \times 9.0) = 0$

$$\therefore V_E = + \textbf{30.22 kN} \uparrow$$

Consider the vertical equilibrium of the beam: $+ve \uparrow \Sigma F_z = 0$
$+ V_A - 3.0 - 18.0 - 3.0 - 20.0 - (4.0 \times 4.0) + V_E = 0$ $\therefore V_A = + \textbf{29.78 kN} \uparrow$

Shear Force Diagram   (**Note:** the diagram is curved from A to B and from C to D)

Position of zero shear force $x = 5.0$ m
(This corresponds with the position of the maximum bending moment in the beam.)

Bending Moment Diagram: (consider the right-hand side)
$M_x = + (30.22 \times 4.0) - (4.0 \times 4.0^2/2) = + \textbf{88.9 kNm}$
Alternatively, calculating the area under the shear force diagram:
$M_x = + 0.5(14.22 + 30.22)(4.0) = + 88.9$ kNm

Solution
Topic: Statically Determinate Beams – Shear Force and Bending Moment
Problem Number: 4.8   **Page No. 1**

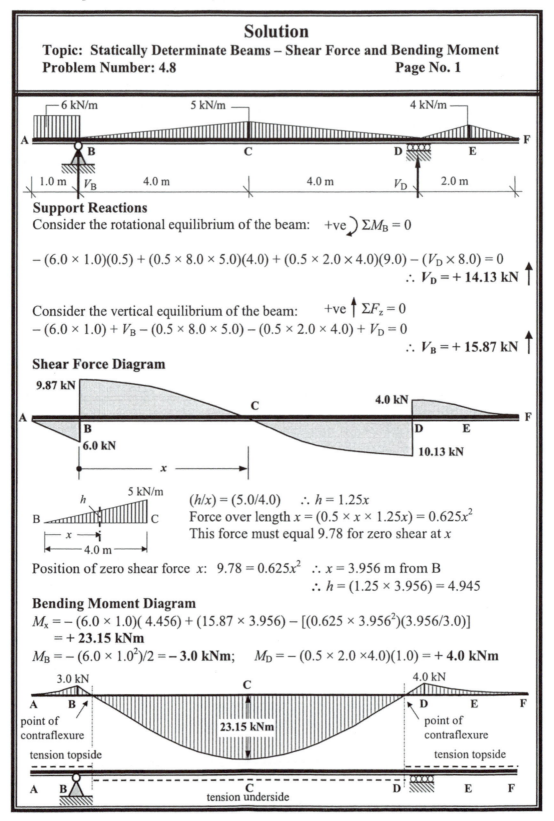

Support Reactions
Consider the rotational equilibrium of the beam: $+ve \curvearrowright \Sigma M_B = 0$

$-(6.0 \times 1.0)(0.5) + (0.5 \times 8.0 \times 5.0)(4.0) + (0.5 \times 2.0 \times 4.0)(9.0) - (V_D \times 8.0) = 0$
$$\therefore V_D = +\ 14.13 \text{ kN} \uparrow$$

Consider the vertical equilibrium of the beam: $+ve \uparrow \Sigma F_z = 0$
$-(6.0 \times 1.0) + V_B - (0.5 \times 8.0 \times 5.0) - (0.5 \times 2.0 \times 4.0) + V_D = 0$
$$\therefore V_B = +\ 15.87 \text{ kN} \uparrow$$

Shear Force Diagram

$(h/x) = (5.0/4.0) \quad \therefore h = 1.25x$
Force over length $x = (0.5 \times x \times 1.25x) = 0.625x^2$
This force must equal 9.78 for zero shear at x

Position of zero shear force x: $9.78 = 0.625x^2 \quad \therefore x = 3.956$ m from B
$$\therefore h = (1.25 \times 3.956) = 4.945$$

Bending Moment Diagram
$M_x = -(6.0 \times 1.0)(4.456) + (15.87 \times 3.956) - [(0.625 \times 3.956^2)(3.956/3.0)]$
$\quad = +\ \textbf{23.15 kNm}$
$M_B = -(6.0 \times 1.0^2)/2 = -\ \textbf{3.0 kNm}; \quad M_D = -(0.5 \times 2.0 \times 4.0)(1.0) = +\ \textbf{4.0 kNm}$

Solution

Topic: Statically Determinate Beams – Shear Force and Bending Moment

Problem Number: 4.9   **Page No. 1**

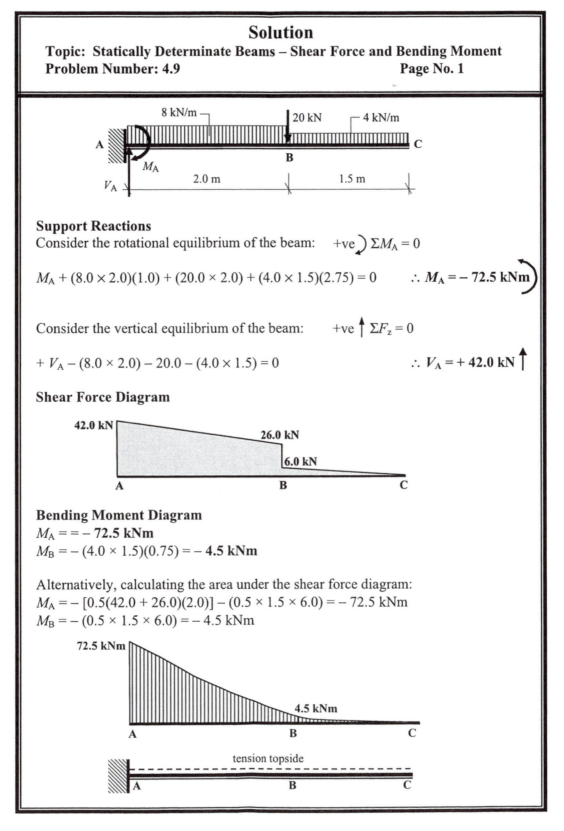

Support Reactions

Consider the rotational equilibrium of the beam: +ve $\curvearrowright$ $\Sigma M_A = 0$

$$M_A + (8.0 \times 2.0)(1.0) + (20.0 \times 2.0) + (4.0 \times 1.5)(2.75) = 0 \qquad \therefore M_A = -72.5 \text{ kNm} \curvearrowright$$

Consider the vertical equilibrium of the beam: +ve $\uparrow$ $\Sigma F_z = 0$

$$+ V_A - (8.0 \times 2.0) - 20.0 - (4.0 \times 1.5) = 0 \qquad \therefore V_A = +42.0 \text{ kN} \uparrow$$

Shear Force Diagram

Bending Moment Diagram

$M_A = = -72.5 \text{ kNm}$

$M_B = -(4.0 \times 1.5)(0.75) = -4.5 \text{ kNm}$

Alternatively, calculating the area under the shear force diagram:

$M_A = -[0.5(42.0 + 26.0)(2.0)] - (0.5 \times 1.5 \times 6.0) = -72.5 \text{ kNm}$

$M_B = -(0.5 \times 1.5 \times 6.0) = -4.5 \text{ kNm}$

Solution

Topic: Statically Determinate Beams – Shear Force and Bending Moment

Problem Number: 4.10   **Page No. 1**

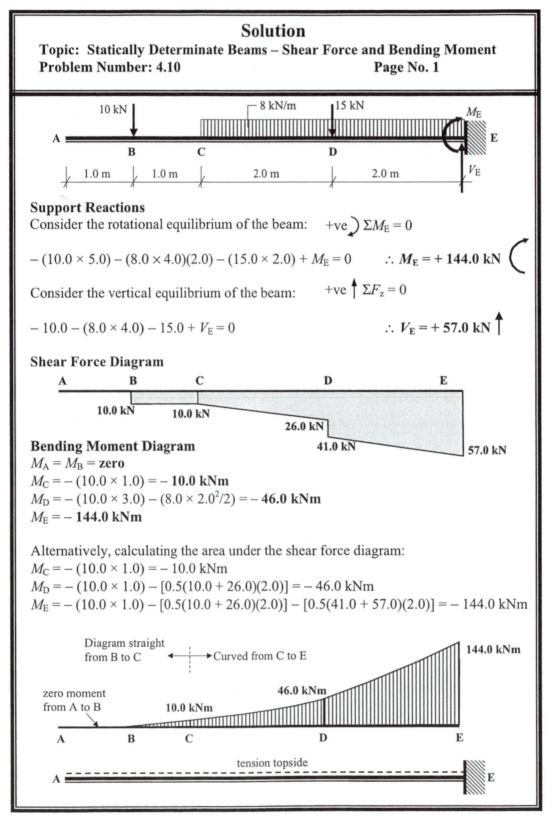

Support Reactions

Consider the rotational equilibrium of the beam: +ve $\curvearrowright \Sigma M_E = 0$

$- (10.0 \times 5.0) - (8.0 \times 4.0)(2.0) - (15.0 \times 2.0) + M_E = 0$ $\therefore M_E = + 144.0 \text{ kN}$

Consider the vertical equilibrium of the beam: +ve $\uparrow \Sigma F_z = 0$

$- 10.0 - (8.0 \times 4.0) - 15.0 + V_E = 0$ $\therefore V_E = + 57.0 \text{ kN} \uparrow$

Shear Force Diagram

Bending Moment Diagram

$M_A = M_B = \textbf{zero}$

$M_C = - (10.0 \times 1.0) = \textbf{- 10.0 kNm}$

$M_D = - (10.0 \times 3.0) - (8.0 \times 2.0^2/2) = \textbf{- 46.0 kNm}$

$M_E = \textbf{- 144.0 kNm}$

Alternatively, calculating the area under the shear force diagram:

$M_C = - (10.0 \times 1.0) = - 10.0 \text{ kNm}$

$M_D = - (10.0 \times 1.0) - [0.5(10.0 + 26.0)(2.0)] = - 46.0 \text{ kNm}$

$M_E = - (10.0 \times 1.0) - [0.5(10.0 + 26.0)(2.0)] - [0.5(41.0 + 57.0)(2.0)] = - 144.0 \text{ kNm}$

4.2 McCaulay's Method for the Deflection of Beams

In elastic analysis the deflected shape of a simply-supported beam is normally assumed to be a circular arc of radius R (R is known as the radius of curvature), as shown in Figure 4.19.

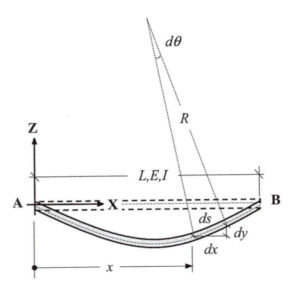

Consider the beam AB to be subject to a variable bending moment along its length. The beam is assumed to deflect as indicated.

R is the radius of curvature,
L is the span,
I is the second moment of area about the axis of bending,
E is the modulus of elasticity,
ds is an elemental length of beam measured a distance of x from the left-hand end
M is the value of the bending moment at position x.

Figure 4.19

The slope of the beam at position x is given by:

$$\text{slope} = \frac{dz}{dx} = \int \frac{M}{EI}dx$$

Differentiating the slope with respect to x gives:

$$\frac{d^2z}{dx^2} = \frac{M}{EI} \quad \text{and hence:}$$

$$EI\frac{d^2z}{dx^2} = M \qquad \text{Equation (1) – \textbf{bending moment} } (\boldsymbol{M_x})$$

Integrating Equation (1) with respect to x gives

$$EI\frac{dz}{dx} = \int Mdx \qquad \text{Equation (2) – } EI \times \textbf{slope } (\boldsymbol{EI\theta})$$

Integrating Equation (2) with respect to x gives

$$EIz = \iint (Mdx)dx \qquad \text{Equation (3) – } EI \times \textbf{deflection } (\boldsymbol{EI\delta})$$

Equations (1) and (2) result in two constants of integration A and B; these are determined by considering boundary conditions such as known values of slope and/or deflection at positions on the beam.

4.2.1 Example 4.4: Beam with Point Loads

Consider a beam supporting three point loads as shown in Figure 4.20.

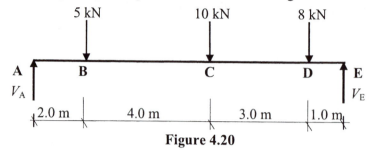

Figure 4.20

Step 1: Formulate an equation which represents the value of the bending moment at a position measured x from the left-hand end of the beam. This expression must include all of the loads and x should therefore be considered between points D and E.

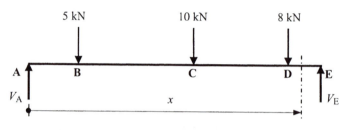

Figure 4.21

Consider the vertical equilibrium of the beam:

$+ve \uparrow \Sigma F_z = 0$

$V_A - 5.0 - 10.0 - 8.0 + V_E = 0$ $\therefore V_A + V_E = 23$ kN   (i)

Consider the rotational equilibrium of the beam:

$+ve \curvearrowright \Sigma M_A = 0$

$(5.0 \times 2.0) + (10.0 \times 6.0) + (8.0 \times 9.0) - (V_E \times 10.0) = 0$ (ii)

$\therefore V_E = 14.2$ kN

Substituting into equation (i) gives $V_A = 8.8$ kN

The equation for the **bending moment** at x:

$$EI\frac{d^2z}{dx^2} = M_x = +8.8x - 5.0[x-2] - 10.0[x-6] - 8.0[x-9]$$ Equation (1)

The equation for the **slope** (θ) at x:

$$EI\frac{dz}{dx} = \int Mdx = +\frac{8.8}{2}x^2 - \frac{5.0}{2}[x-2]^2 - \frac{10.0}{2}[x-6]^2 - \frac{8.0}{2}[x-9]^2 + A$$

Equation (2)

The equation for the **deflection** (δ) at x:

$$EIz = \iint (Mdx)\,dx = +\frac{8.8}{6}x^3 - \frac{5.0}{6}[x-2]^3 - \frac{10.0}{6}[x-6]^3 - \frac{8.0}{6}[x-9]^3 + Ax + B$$

Equation (3)

where A and B are constants of integration related to the boundary conditions.

Note: It is common practice to use square brackets, i.e. [ ], to enclose the lever arms for the forces as shown. These brackets are integrated as a unit and during the calculation for slope and deflection they are ignored if the contents are –ve, i.e. the position x being considered is to the left of the load associated with the bracket.

Boundary Conditions
The boundary conditions are known values associated with the slope and/or deflection. In this problem, assuming no settlement occurs at the supports then the **deflection** is equal to zero at these positions, i.e.

when $x = 0$, $z = 0$

$$+\frac{8.8}{6}x^3 - \frac{5.0}{6}[x-2]^3 - \frac{10.0}{6}[x-6]^3 - \frac{8.0}{6}[x-9]^3 + Ax + B = 0$$

               ignore   *ignore*   *ignore*

Substituting for x and z in equation (3) gives $B = 0$

when $x = 10.0$, $z = 0$

$$+\frac{8.8}{6}10^3 - \frac{5.0}{6}[10-2]^3 - \frac{10.0}{6}[10-6]^3 - \frac{8.0}{6}[10-9]^3 + (A \times 10) = 0$$

$$+ (1.466 \times 10^3) - (0.426 \times 10^3) - (0.106 \times 10^3) - 1.33 + 10A = 0$$

$$A = -93.265$$

The general equations for the slope and deflection at any point along the length of the beam are given by:
The equation for the **slope** at x:

$$EI\frac{dz}{dx} = EI\theta = +\frac{8.8}{2}x^2 - \frac{5.0}{2}[x-2]^2 - \frac{10.0}{2}[x-6]^2 - \frac{8.0}{2}[x-9]^2 - 93.265$$

Equation (4)

The equation for the **deflection** at x:

$$EIz = EI\delta = +\frac{8.8}{6}x^3 - \frac{5.0}{6}[x-2]^3 - \frac{10.0}{6}[x-6]^3 - \frac{8.0}{6}[x-9]^3 - 93.265x$$

Equation (5)

e.g. the deflection at the mid-span point can be determined from equation (5) by substituting the value of $x = 5.0$ and ignoring the [ ] when their contents are – ve, i.e.

$$EIz = +\frac{8.8}{6}5^3 - \frac{5.0}{6}[5-2]^3 - \frac{10.0}{6}[5-6]^3 - \frac{8.0}{6}[5-9]^3 - (93.265 \times 5)$$

                               ignore   *ignore*

$$EIz = + 183.33 - 22.5 - 466.325 \qquad \therefore z = -\frac{305.5}{EI} \text{ m} = -\left\{\frac{305.5 \times 10^3}{EI}\right\} \text{mm}$$

The maximum deflection can be determined by calculating the value of x when the slope, i.e. equation (4) is equal to zero and substituting the calculated value of x into equation (5) as above.

In most simply-supported spans the maximum deflection occurs near the mid-span point. This can be used to estimate the value of x in equation (4) and hence eliminate some of the [ ] brackets, e.g. if the maximum deflection is assumed to occur at a position less than 6.0 m from the left-hand end the last two terms in the [ ] brackets need not be used to determine the position of zero slope. This assumption can be checked and if incorrect a subsequent calculation carried out including an additional bracket until the correct answer is found.

Assume z_{maximum} occurs between 5.0 m and 6.0 m from the left-hand end of the beam, then:

The equation for the **slope** at x is:

$$EI\frac{dz}{dx} = +\frac{8.8}{2}x^2 - \frac{5.0}{2}[x-2]^2 - \underbrace{\frac{10.0}{2}[x-6]^2}_{ignore} - \underbrace{\frac{8.0}{2}[x-9]^2}_{ignore} - 93.265 = 0 \text{ for } z_{\text{maximum}}$$

This equation reduces to:

$$1.9x^2 + 10x - 103.265 = 0 \qquad \text{and hence} \qquad x = 5.2 \text{ m}$$

since x was assumed to lie between 5.0 m and 6.0 m ignoring the two [ ] terms was correct. The maximum deflection can be found by substituting the value of $x = 5.2$ m in equation (5) and ignoring the [ ] when their contents are −ve, i.e.

$$EIz_{\text{maximum}} = +\frac{8.8}{6}5.2^3 - \frac{5.0}{6}[5.2-2]^3 - \underbrace{\frac{10.0}{6}[5.2-6]^3}_{ignore} - \underbrace{\frac{8.0}{6}[5.2-9]^3}_{ignore} - (93.265 \times 5.2)$$

$$EIz_{\text{maximum}} = + 206.23 - 27.31 - 484.98 \qquad \therefore z_{\text{maximum}} = -\frac{306}{EI} \text{ m}$$

Note: There is no significant difference from the value calculated at mid-span.

4.2.2 *Example 4.5: Beam with Combined Point Loads and UDLs*

A simply-supported beam ABCD carries a uniformly distributed load of 3.0 kN/m between A and B, point loads of 4 kN and 6 kN at B and C respectively, and a uniformly distributed load of 5.0 kN/m between B and D as shown in Figure 4.22. Determine the position and magnitude of the maximum deflection.

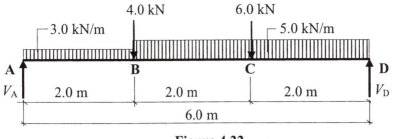

Figure 4.22

Consider the vertical equilibrium of the beam:

$+ve \uparrow \Sigma F_z = 0$

$V_A - (3.0 \times 2.0) - 4.0 - 6.0 - (5.0 \times 4.0) + V_D = 0$ $\qquad \therefore V_A + V_D = 36 \text{ kN}$ $\qquad$ (i)

Consider the rotational equilibrium of the beam:

$+ve \curvearrowright \Sigma M_A = 0$

$(3.0 \times 2.0 \times 1.0) + (4.0 \times 2.0) + (6.0 \times 4.0) + (5.0 \times 4.0 \times 4.0) - (V_D \times 6.0) = 0$ $\qquad$ (ii)
$$\therefore V_D = 19.67 \text{ kN}$$

Substituting into equation (i) gives $\qquad\qquad\qquad V_A = 16.33 \text{ kN}$

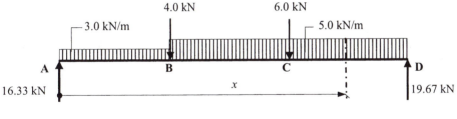

Figure 4.23

In the case of a **UDL** when a term is written in the moment equation in square brackets, **[ ]** this effectively applies the load for the full length of the beam. For example in Figure 4.23 the 3.0 kN/m load is assumed to apply from A to D and consequently only an additional 2.0 kN/m need be applied from position B onwards as shown in Figure 4.24.

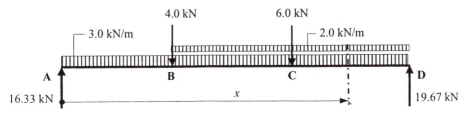

Figure 4.24

The equation for the **bending moment** at x is:

$$EI\frac{d^2z}{dx^2} = +16.33x - 3.0\frac{x^2}{2} - 4.0[x-2.0] - 2.0\frac{[x-2]^2}{2} - 6.0[x-4] \qquad \text{Equation (1)}$$

The equation for the **slope** at x is:

$$EI\frac{dz}{dx} = (\theta) = +16.33\frac{x^2}{2} - 3.0\frac{x^3}{6} - 4.0\frac{[x-2]^2}{2} - 2.0\frac{[x-2]^3}{6} - 6.0\frac{[x-4]^2}{2} + A$$

$$\text{Equation (2)}$$

The equation for the **deflection** at x is:

$$EIz = (\delta) = +16.33\frac{x^3}{6} - 3.0\frac{x^4}{24} - 4.0\frac{[x-2]^3}{6} - 2.0\frac{[x-2]^4}{24} - 6.0\frac{[x-4]^3}{6} + Ax + B$$

$$\text{Equation (3)}$$

where A and B are constants of integration related to the boundary conditions.

Boundary Conditions

In this problem, assuming no settlement occurs at the supports then the **deflection** is equal to zero at these positions, i.e.

when $x = 0$, $z = 0$

$$+16.33\frac{x^3}{6} - 3.0\frac{x^4}{24} - 4.0\cancel{\frac{[x-2]^3}{6}} - 2.0\cancel{\frac{[x-2]^4}{24}} - 6.0\cancel{\frac{[x-4]^3}{6}} + Ax + B$$

$$\qquad\qquad\qquad \textit{ignore} \qquad \textit{ignore} \qquad \textit{ignore}$$

Substituting for x and z in equation (3) $\therefore B = 0$

when $x = 6.0$, $z = 0$

$$+16.33\frac{x^3}{6} - 3.0\frac{x^4}{24} - 4.0\frac{[x-2]^3}{6} - 2.0\frac{[x-2]^4}{24} - 6.0\frac{[x-4]^3}{6} + Ax = 0$$

$$+16.33\frac{6.0^3}{6} - 3.0\frac{6.0^4}{24} - 4.0\frac{4.0^3}{6} - 2.0\frac{4.0^4}{24} - 6.0\frac{2.0^3}{6} + 6.0A = 0$$

$$\therefore A = -58.98$$

The general equations for the slope and bending moment at any point along the length of the beam are given by:

The equation for the **slope** at x:

$$EI\frac{dz}{dx} = +16.33\frac{x^2}{2} - 3.0\frac{x^3}{6} - 4.0\frac{[x-2]^2}{2} - 2.0\frac{[x-2]^3}{6} - 6.0\frac{[x-4]^2}{2} - 58.98$$

$$\text{Equation (4)}$$

The equation for the **deflection** at x:

$$EIz = +16.33\frac{x^3}{6} - 3.0\frac{x^4}{24} - 4.0\frac{[x-2]^3}{6} - 2.0\frac{[x-2]^4}{24} - 6.0\frac{[x-4]^3}{6} - 58.98x$$

$$\text{Equation (5)}$$

Assume $z_{maximum}$ occurs between 2.0 m and 4.0 m from the left-hand end of the beam, then:

The equation for the **slope** at 'x' is:

$$EI\frac{dz}{dx} = +16.33\frac{x^2}{2} - 3.0\frac{x^3}{6} - 4.0\frac{[x-2]^2}{2} - 2.0\frac{[x-2]^3}{6} - 6.0\frac{[x-4]^2}{2} - 58.98 = 0$$

$$\underset{ignore}{}$$

This cubic can be solved by iteration.

Guess a value for x, e.g. 3.1 m
$(16.33 \times 3.1^2)/2 - (3.0 \times 3.1^3)/6 - (4.0 \times 1.1^2)/2 - (2.0 \times 1.1^3)/6 - 58.98 = 1.73 > 0$

The assumed value of 3.1 is slightly high, try $x = 3.05$ m
$(16.33 \times 3.05^2)/2 - (3.0 \times 3.05^3)/6 - (4.0 \times 1.05^2)/2 - (2.0 \times 1.05^3)/6 - 58.98 = 0.20$

This value is close enough. $x = 3.05$ m and since x was assumed to lie between 2.0 m and 4.0 m, ignoring the $[x - 4]$ term was correct.

The maximum deflection can be found by substituting the value of $x = 3.05$ m in equation (5) and ignoring the [ ] when their contents are –ve, i.e.

$$EI\,z_{maximum} = +16.33\frac{x^3}{6} - 3.0\frac{x^4}{24} - 4.0\frac{[x-2]^3}{6} - 2.0\frac{[x-2]^4}{24} - 6.0\frac{[x-4]^3}{6} - 58.98\,x$$

$$\underset{ignore}{}$$

$$EI\,z_{maximum} = +77.22 - 10.82 - 0.77 - 0.1 - 179.89 \qquad \therefore z_{maximum} = -\frac{114.4}{EI}\ \text{m}$$

4.3 *Equivalent Uniformly Distributed Load Method for the Deflection of Beams*

In a simply-supported beam, the maximum deflection induced by the applied loading always approximates the mid-span value if it is not equal to it. A number of standard frequently used load cases for which the elastic deformation is required are given in Appendix 2 in this text.

In many cases beams support complex load arrangements which do not lend themselves either to an individual load case or to a combination of the load cases given in Appendix 2. Provided that deflection is not the governing design criterion, a calculation which gives an approximate answer is usually adequate. The equivalent UDL method is a useful tool for estimating the deflection in a simply-supported beam with a complex loading.

Consider a single-span, simply-supported beam carrying a *non-uniform* loading which induces a maximum bending moment of M as shown in Figure 4.25.

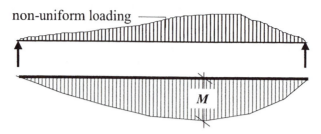

non-uniform loading

Bending Moment Diagram

Figure 4.25

The equivalent UDL (w_e) which would induce the same *magnitude* of maximum bending moment (but note that the position may be different) on a simply-supported span carrying a *uniform* loading can be determined from:

Maximum bending moment $M = \dfrac{w_e L^2}{8}$

$$\therefore\ w_e = \dfrac{8\,M}{L^2}$$

where w_e is the equivalent uniform distributed load.
The maximum deflection of the beam carrying the uniform loading will occur at the mid-span and will be equal to $\delta = \dfrac{5 w_e L^4}{384 EI}$ (see Appendix 2)

Using this expression, the maximum deflection of the beam carrying the non-uniform loading can be estimated by substituting for the w_e term, i.e.

$$\delta \approx \frac{5 w_e L^4}{384 EI}\ =\ \frac{5 \times \left(\dfrac{8M}{L^2}\right) L^4}{384\,EI}\ =\ \frac{0.104\,M\,L^2}{EI}$$

The maximum bending moments in Example 4.4 and Example 4.5 are 32.8 kNm and 30.67 kNm respectively (the reader should check these answers).
Using the equivalent UDL method to estimate the maximum deflection in each case gives:

Example 4.4 $\delta_{\text{maximum}} \approx \dfrac{0.104\,M\,L^2}{EI} = -\dfrac{341.1}{EI}\,\text{m}$ (actual value $= \dfrac{305.5}{EI}\,\text{m}$)

Example 4.5 $\delta_{\text{maximum}} \approx \dfrac{0.104\,M\,L^2}{EI} = -\dfrac{114.9}{EI}\,\text{m}$ (actual value $= \dfrac{114.4}{EI}\,\text{m}$)

Note: The estimated deflection is more accurate for beams which are predominantly loaded with distributed loads.

4.3.1 Problems: McCaulay's and Equivalent UDL Methods for Deflection of Beams

A series of simply-supported beams are indicated in Problems 4.11 to 4.15. Using the applied loading given in each case determine the maximum deflection. Assume all beams are uniform with Young's Modulus of Elasticity = E and Second Moment of Area = I

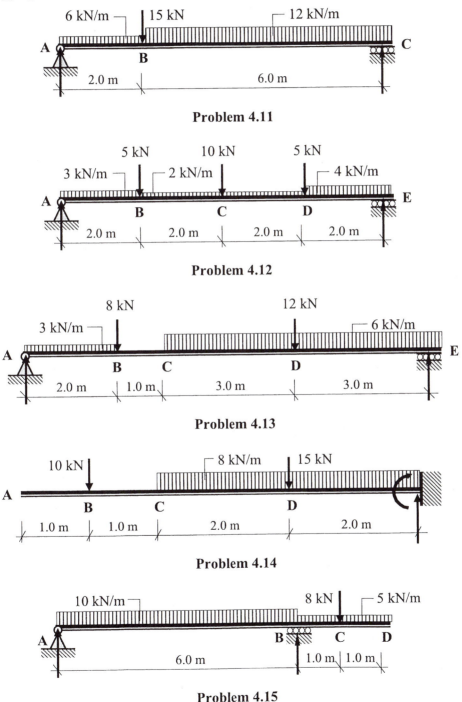

Problem 4.11

Problem 4.12

Problem 4.13

Problem 4.14

Problem 4.15

4.3.2 Solutions: McCaulay's and Equivalent UDL Methods for Deflection of Beams

---

Solution

Topic: Statically Determinate Beams – Deflection

Problem Number: 4.11   **Page No. 1**

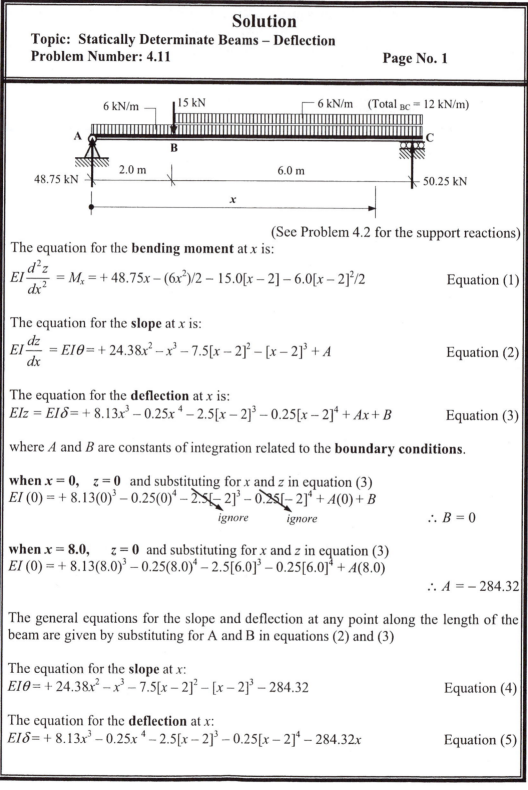

(See Problem 4.2 for the support reactions)

The equation for the **bending moment** at x is:

$$EI\frac{d^2z}{dx^2} = M_x = + 48.75x - (6x^2)/2 - 15.0[x-2] - 6.0[x-2]^2/2 \qquad \text{Equation (1)}$$

The equation for the **slope** at x is:

$$EI\frac{dz}{dx} = EI\theta = + 24.38x^2 - x^3 - 7.5[x-2]^2 - [x-2]^3 + A \qquad \text{Equation (2)}$$

The equation for the **deflection** at x is:

$$EIz = EI\delta = + 8.13x^3 - 0.25x^4 - 2.5[x-2]^3 - 0.25[x-2]^4 + Ax + B \qquad \text{Equation (3)}$$

where A and B are constants of integration related to the **boundary conditions**.

when $x = 0$, $z = 0$ and substituting for x and z in equation (3)

$$EI\,(0) = + 8.13(0)^3 - 0.25(0)^4 - 2.5[-2]^3 - 0.25[-2]^4 + A(0) + B$$

                                   ignore   *ignore* $\therefore B = 0$

when $x = 8.0$, $z = 0$ and substituting for x and z in equation (3)

$$EI\,(0) = + 8.13(8.0)^3 - 0.25(8.0)^4 - 2.5[6.0]^3 - 0.25[6.0]^4 + A(8.0)$$

$$\therefore A = -284.32$$

The general equations for the slope and deflection at any point along the length of the beam are given by substituting for A and B in equations (2) and (3)

The equation for the **slope** at x:

$$EI\theta = + 24.38x^2 - x^3 - 7.5[x-2]^2 - [x-2]^3 - 284.32 \qquad \text{Equation (4)}$$

The equation for the **deflection** at x:

$$EI\delta = + 8.13x^3 - 0.25x^4 - 2.5[x-2]^3 - 0.25[x-2]^4 - 284.32x \qquad \text{Equation (5)}$$

Solution

Topic: Statically Determinate Beams - Deflection
Problem Number: 4.11   **Page No. 2**

The position of the maximum deflection at the point of zero slope can be determined from equation (4) as follows:
Assume that zero slope occurs when $2.0 \leq x \leq 8.0$ and neglect [ ] when negative

$$EI\theta = 0 = + 24.38x^2 - x^3 - 7.5[x-2]^2 - [x-2]^3 - 284.32$$

Solve the resulting cubic equation by trial and error.
Guess $x = 3.9$ m   (i.e. slightly to the left of the mid-span)
$+ 24.38(3.9)^2 - 3.9^3 - 7.5(1.9)^2 - (1.9)^3 - 284.32 = -6.75$ Increase x
try $x = 3.95$
$+ 24.38(3.95)^2 - 3.95^3 - 7.5(1.95)^2 - (1.95)^3 - 284.32 = -1.49$ Increase x
try $x = 3.96$
$+ 24.38(3.96)^2 - 3.96^3 - 7.5(1.96)^2 - (1.96)^3 - 284.32 = -0.44$
Accept $x = 3.96$ m

The maximum deflection is given by:
$\delta_{max.} = \{+ 8.13(3.96)^3 - 0.25(3.96)^4 - 2.5(1.96)^3 - 0.25(1.96)^4 - 284.32(3.96)\}/EI$
$\boldsymbol{\delta_{max.} = -705.03/EI}$

Equivalent Uniformly Distributed Load Method:
$\delta_{max.} \approx - (0.104 M_{maximum}L^2)/EI$

The maximum bending moment = 105.2 kNm   (see Problem 4.2)

$\delta_{max.} \approx - (0.104 \times 105.2 \times 8.0^2)/EI = \boldsymbol{-700.2/EI}$

Solution

Topic: Statically Determinate Beams - Deflection
Problem Number: 4.12   **Page No. 1**

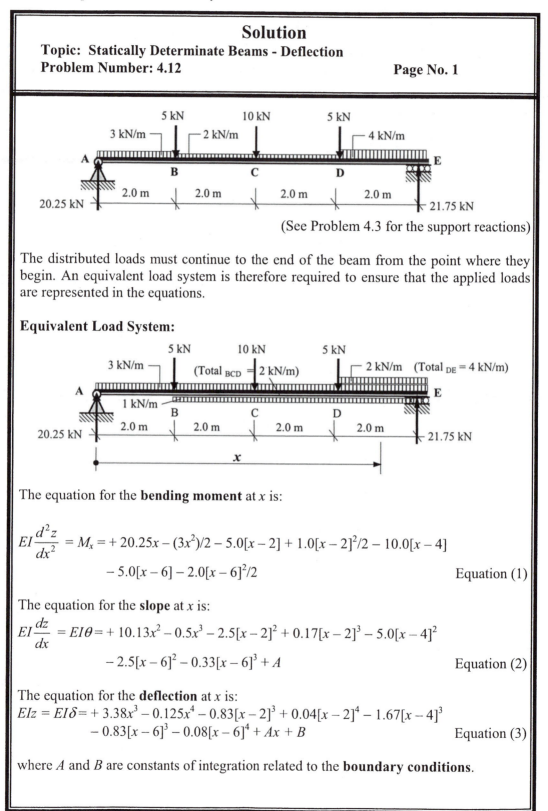

(See Problem 4.3 for the support reactions)

The distributed loads must continue to the end of the beam from the point where they begin. An equivalent load system is therefore required to ensure that the applied loads are represented in the equations.

Equivalent Load System:

The equation for the **bending moment** at x is:

$$EI\frac{d^2z}{dx^2} = M_x = +20.25x - (3x^2)/2 - 5.0[x-2] + 1.0[x-2]^2/2 - 10.0[x-4]$$
$$- 5.0[x-6] - 2.0[x-6]^2/2 \qquad \text{Equation (1)}$$

The equation for the **slope** at x is:

$$EI\frac{dz}{dx} = EI\theta = +10.13x^2 - 0.5x^3 - 2.5[x-2]^2 + 0.17[x-2]^3 - 5.0[x-4]^2$$
$$- 2.5[x-6]^2 - 0.33[x-6]^3 + A \qquad \text{Equation (2)}$$

The equation for the **deflection** at x is:
$$EIz = EI\delta = +3.38x^3 - 0.125x^4 - 0.83[x-2]^3 + 0.04[x-2]^4 - 1.67[x-4]^3$$
$$- 0.83[x-6]^3 - 0.08[x-6]^4 + Ax + B \qquad \text{Equation (3)}$$

where A and B are constants of integration related to the **boundary conditions**.

Solution

Topic: Statically Determinate Beams - Deflection

Problem Number: 4.12   **Page No. 2**

---

when $x = 0$, $z = 0$ and substituting for x and z in equation (3)

$EI(0) = + 3.38(0)^3 - 0.125(0)^4 - 0.83[-2.0]^3 + 0.04[-2.0]^4 - 1.67[-4.0]^3 - 0.83[-6.0]^3$
   ~~$- 0.08[-6.0]^4$~~ $+ A(0) + B$ *ignore*   *ignore*   *ignore*   *ignore*
      ignore $\therefore B = 0$

when $x = 8.0$, $z = 0$ and substituting for x and z in equation (3)

$EI(0) = + 3.38(8.0)^3 - 0.125(8.0)^4 - 0.83[6.0]^3 + 0.04[6.0]^4 - 1.67[4.0]^3$
   $- 0.83[2.0]^3 - 0.08[2.0]^4 + A(8.0)$ $\therefore A = -122.04$

The general equations for the slope and deflection at any point along the length of the beam are given by substituting for A and B in equations (2) and (3)

The equation for the **slope** at x:
$EI\theta = + 10.13x^2 - 0.5x^3 - 2.5[x-2]^2 + 0.17[x-2]^3 - 5.0[x-4]^2 - 2.5[x-6]^2$
   $- 0.33[x-6]^3 - 122.04$ Equation (4)

The equation for the **deflection** at x:
$EI\delta = + 3.38x^3 - 0.125x^4 - 0.83[x-2]^3 + 0.04[x-2]^4 - 1.67[x-4]^3 - 0.83[x-6]^3$
   $- 0.08[x-6]^4 - 122.04x$ Equation (5)

The position of the maximum deflection at the point of zero slope can be determined from equation (4) as follows:
Assume that zero slope occurs when $4.0 \leq x \leq 6.0$ and neglect [ ] when negative
$EI\theta = 0 = + 10.13x^2 - 0.5x^3 - 2.5[x-2]^2 + 0.17[x-2]^3 - 5.0[x-4]^2 - 2.5$~~$[x-6]^2$~~
   $- 0.33$~~$[x-6]^3$~~ $- 122.04$ *ignore*
      ignore

Solve the resulting cubic equation by trial and error.
Guess $x = 4.1$ m $EI\theta = + 4.33 > 0$ $\therefore$ reduce x
try $x = 4.05$ $EI\theta = + 1.86 > 0$ try $x = 4.02$ $EI\theta = + 0.38$
Accept $x = 4.02$ m

The maximum deflection is given by:
$\delta_{max.} = \{+ 3.38(4.02)^3 - 0.125(4.02)^4 - 0.83(2.02)^3 + 0.04(2.02)^4 - 1.67(0.02)^3$
   $- (122.04 \times 4.02)\}/EI$
$\delta_{max.} = -309.84/EI$

Equivalent Uniformly Distributed Load Method:
$\delta_{max.} \approx - (0.104 M_{maximum} L^2)/EI$
The maximum bending moment = 49.0 kNm   (see Problem 5.3)
$\delta_{max.} \approx - (0.104 \times 49.0 \times 8.0^2)/EI = -326.14/EI$

Solution
Topic: Statically Determinate Beams - Deflection
Problem Number: 4.13   **Page No. 1**

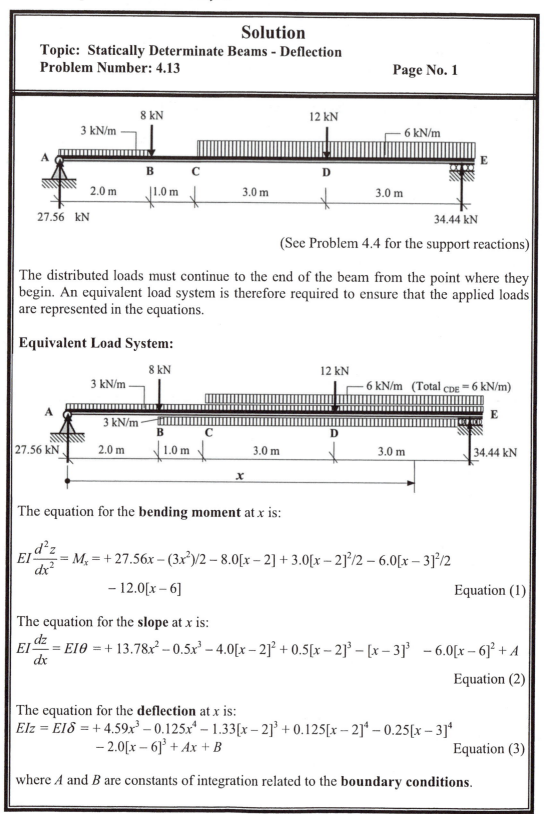

(See Problem 4.4 for the support reactions)

The distributed loads must continue to the end of the beam from the point where they begin. An equivalent load system is therefore required to ensure that the applied loads are represented in the equations.

Equivalent Load System:

The equation for the **bending moment** at x is:

$$EI\frac{d^2z}{dx^2} = M_x = +27.56x - (3x^2)/2 - 8.0[x-2] + 3.0[x-2]^2/2 - 6.0[x-3]^2/2$$
$$- 12.0[x-6] \qquad \text{Equation (1)}$$

The equation for the **slope** at x is:

$$EI\frac{dz}{dx} = EI\theta = +13.78x^2 - 0.5x^3 - 4.0[x-2]^2 + 0.5[x-2]^3 - [x-3]^3 - 6.0[x-6]^2 + A$$
$$\text{Equation (2)}$$

The equation for the **deflection** at x is:
$$EIz = EI\delta = +4.59x^3 - 0.125x^4 - 1.33[x-2]^3 + 0.125[x-2]^4 - 0.25[x-3]^4$$
$$- 2.0[x-6]^3 + Ax + B \qquad \text{Equation (3)}$$

where A and B are constants of integration related to the **boundary conditions**.

Solution

Topic: Statically Determinate Beams - Deflection

Problem Number: 4.13   **Page No. 2**

---

when $x = 0$, $z = 0$ and substituting for x and z in equation (3)

$EI(0) = +4.59(0)^3 - 0.125(0)^4 - 1.33[2.0]^3 + 0.125[2.0]^4 - 0.25[3.0]^4$
              $- 2.0[6.0]^3 - A(0) + B$ *ignore*   *ignore*   *ignore*

                          ignore $\therefore B = 0$

when $x = 9.0$, $z = 0$ and substituting for x and z in equation (3)

$EI(0) = +4.59(9.0)^3 - 0.125(9.0)^4 - 1.33[7.0]^3 + 0.125[7.0]^4 - 0.25[6.0]^4 - 2.0[3.0]^3$
              $- A(9.0)$ $\therefore A = -221.32$

The general equations for the slope and deflection at any point along the length of the beam are given by substituting for A and B in equations (2) and (3)

The equation for the **slope** at x:

$EI\theta = +13.78x^2 - 0.5x^3 - 4.0[x-2]^2 + 0.5[x-2]^3 - [x-3]^3 - 6.0[x-6]^2 - 221.32$

                                                                    Equation (4)

The equation for the **deflection** at x:

$EI\delta = +4.59x^3 - 0.125x^4 - 1.33[x-2]^3 + 0.125[x-2]^4 - 0.25[x-3]^4 - 2.0[x-6]^3$
              $- 221.32x$ Equation (5)

The position of the maximum deflection at the point of zero slope can be determined from equation (4) as follows:

Assume that zero slope occurs when $3.0 \le x \le 6.0$ and neglect [ ] when negative

$EI\theta = 0 = +13.78x^2 - 0.5x^3 - 4.0[x-2]^2 + 0.5[x-2]^3 - [x-3]^3 - 221.32$

Solve the resulting equation by trial and error.

Guess $x = 4.6$ m $EI\theta = -0.75 > 0$ $\therefore$ reduce x

try $x = 4.61$ m $EI\theta = +0.02 > 0$ Accept $x = 4.61$ m

The maximum deflection is given by:

$\delta_{max.} = \{+4.59(4.61)^3 - 0.125(4.61)^4 - 1.33(2.61)^3 + 0.125(2.61)^4 - 0.25(2.61)^4$
              $- (221.32 \times 4.61)\}/EI$

$\delta_{max.} = -656.5/EI$

Equivalent Uniformly Distributed Load Method:

$\delta_{max.} \approx -(0.104 M_{maximum} L^2)/EI$

The maximum bending moment = 78.0 kNm   (see Problem 5.4)

$\delta_{max.} \approx -(0.104 \times 78.0 \times 9.0^2)/EI = -657.4/EI$

Solution
Topic: Statically Determinate Beams - Deflection
Problem Number: 4.14   **Page No. 1**

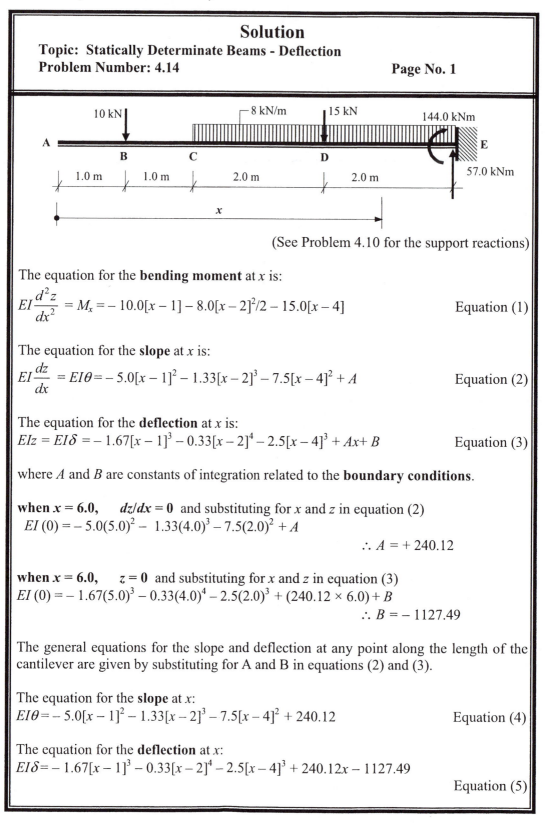

(See Problem 4.10 for the support reactions)

The equation for the **bending moment** at x is:

$$EI\frac{d^2z}{dx^2} = M_x = -10.0[x-1] - 8.0[x-2]^2/2 - 15.0[x-4] \qquad \text{Equation (1)}$$

The equation for the **slope** at x is:

$$EI\frac{dz}{dx} = EI\theta = -5.0[x-1]^2 - 1.33[x-2]^3 - 7.5[x-4]^2 + A \qquad \text{Equation (2)}$$

The equation for the **deflection** at x is:

$$EIz = EI\delta = -1.67[x-1]^3 - 0.33[x-2]^4 - 2.5[x-4]^3 + Ax + B \qquad \text{Equation (3)}$$

where A and B are constants of integration related to the **boundary conditions**.

when $x = 6.0$, $dz/dx = 0$ and substituting for x and z in equation (2)

$EI(0) = -5.0(5.0)^2 - 1.33(4.0)^3 - 7.5(2.0)^2 + A$

$$\therefore A = +240.12$$

when $x = 6.0$, $z = 0$ and substituting for x and z in equation (3)

$EI(0) = -1.67(5.0)^3 - 0.33(4.0)^4 - 2.5(2.0)^3 + (240.12 \times 6.0) + B$

$$\therefore B = -1127.49$$

The general equations for the slope and deflection at any point along the length of the cantilever are given by substituting for A and B in equations (2) and (3).

The equation for the **slope** at x:

$$EI\theta = -5.0[x-1]^2 - 1.33[x-2]^3 - 7.5[x-4]^2 + 240.12 \qquad \text{Equation (4)}$$

The equation for the **deflection** at x:

$$EI\delta = -1.67[x-1]^3 - 0.33[x-2]^4 - 2.5[x-4]^3 + 240.12x - 1127.49$$

$$\text{Equation (5)}$$

Solution
Topic: Statically Determinate Beams - Deflection
Problem Number: 4.14   **Page No. 2**

The maximum deflection occurs at the free end of the cantilever i.e. when $x = 0$ neglecting all [ ] which are negative.

$$\delta_{max.} = -1127.49 / EI$$

The deflection at any other location can be found by substituting the appropriate value of x, e.g.

At B: $x = 1.0$
$\delta_B = \{+ (240.12 \times 1.0) - 1127.49\}/EI$ $\delta_B = -887.4/EI$

At C: $x = 2.0$
$\delta_C = \{- 1.67(1)^3 + (240.12 \times 2.0) - 1127.49\}/EI$ $\delta_C = -648.9/EI$

At D: $x = 4.0$
$\delta_D = \{- 1.67(3.0)^3 - 0.33(2.0)^4 + (240.12 \times 4.0) - 1127.49\}/EI$ $\delta_D = -217.4/EI$

Note: The Equivalent Uniformly Distributed Load Method only applies to single-span beams.

Solution

Topic: Statically Determinate Beams - Deflection
Problem Number: 4.15

Page No. 1

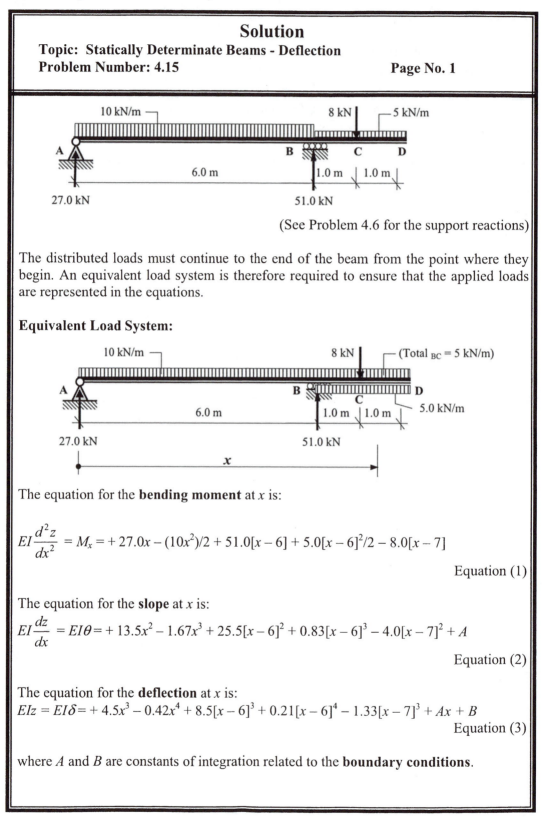

(See Problem 4.6 for the support reactions)

The distributed loads must continue to the end of the beam from the point where they begin. An equivalent load system is therefore required to ensure that the applied loads are represented in the equations.

Equivalent Load System:

The equation for the **bending moment** at x is:

$$EI\frac{d^2z}{dx^2} = M_x = + 27.0x - (10x^2)/2 + 51.0[x - 6] + 5.0[x - 6]^2/2 - 8.0[x - 7]$$

Equation (1)

The equation for the **slope** at x is:
$$EI\frac{dz}{dx} = EI\theta = + 13.5x^2 - 1.67x^3 + 25.5[x - 6]^2 + 0.83[x - 6]^3 - 4.0[x - 7]^2 + A$$

Equation (2)

The equation for the **deflection** at x is:
$$EIz = EI\delta = + 4.5x^3 - 0.42x^4 + 8.5[x - 6]^3 + 0.21[x - 6]^4 - 1.33[x - 7]^3 + Ax + B$$

Equation (3)

where A and B are constants of integration related to the **boundary conditions**.

Solution
Topic: Statically Determinate Beams - Deflection
Problem Number: 4.15

when $x = 0$, $z = 0$ and substituting for x and z in equation (3)

$EI(0) = + 4.5(0)^3 - 0.42(0)^4 + 8.5[-6.0]^3 + 0.21[-6.0]^4 - 1.33[-7.0]^3 + A(0) + B$

ignore   *ignore*   *ignore*

$\therefore B = 0$

when $x = 6.0$, $z = 0$ and substituting for x and z in equation (3)

$EI(0) = + 4.5(6.0)^3 - 0.42(6.0)^4 + A(6.0)$

$\therefore A = -71.28$

The general equations for the slope and deflection at any point along the length of the beam are given by substituting for A and B in equations (2) and (3)

The equation for the **slope** at x:

$EI\theta = + 13.5x^2 - 1.67x^3 + 25.5[x-6]^2 + 0.83[x-6]^3 - 4.0[x-7]^2 - 71.28$

Equation (4)

The equation for the **deflection** at x:

$EI\delta = + 4.5x^3 - 0.42x^4 + 8.5[x-6]^3 + 0.21[x-6]^4 - 1.33[x-7]^3 - 71.28x$

Equation (5)

The position of the maximum deflection between A and B at the point of zero slope can be determined from equation (4) as follows:

Assume that zero slope occurs when $3.0 \le x \le 6.0$ and neglect [ ] when negative

$EI\theta = 0 = + 13.5x^2 - 1.67x^3 - 71.28$

Solve the resulting equation by trial and error.

Guess $x = 2.9$ m $EI\theta = + 1.53 > 0$ $\therefore$ reduce x

try $x = 2.85$ m $EI\theta = -0.29 < 0$ Accept $x = 2.85$ m

The maximum deflection is given by:

$\delta_{AB\,max.} = \{+ 4.5(2.85)^3 - 0.42(2.85)^4 - (71.28 \times 2.85)\}/EI$ $\delta_{AB\,max.} = -\,\textbf{126.69}/\textbf{\textit{EI}}$

The maximum deflection of the cantilever occurs when $x = 8.0$ m

$\delta_D = \{+ 4.5(8.0)^3 - 0.42(8.0)^4 + 8.5(2.0)^3 + 0.21(2.0)^4 - 1.33(1.0)^3 - (71.28 \times 8.0)\}/EI$

$\delta_{D\,max.} = +\,\textbf{83.47}/\textbf{\textit{EI}}$

Equivalent Uniformly Distributed Load Method:

This can be used to give a conservative estimate of δ_{AB} assuming AB to be a simply supported 6.0 m span without the cantilever

$\delta_{max.} \approx -(0.104 M_{maximum} L^2)/EI$

The maximum bending moment in span AB = 36.5 kNm   (see Problem 5.6)

$\delta_{max.} \approx -(0.104 \times 36.5 \times 6.0^2)/EI = -\,\textbf{136.7}/\textbf{\textit{EI}}$

4.4   *The Principle of Superposition*

The Principle of Superposition can be stated as follows:

'*If the displacements at all points in a structure are proportional to the forces causing them, the effect produced on that structure by a number of forces applied simultaneously, is the same as the sum of the effects when each of the forces is applied individually.*'

This applies to any structure made from a material which has a linear load–displacement relationship. Consider the simply–supported beam ABCD shown in Figure 4.26 which carries two point loads at B and C as indicated.

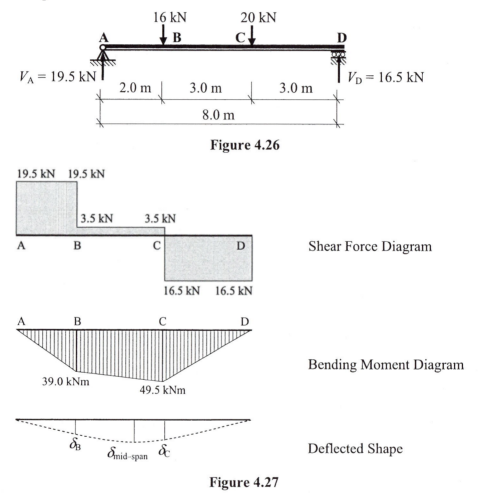

Figure 4.26

Figure 4.27

Note: the maximum deflection does not necessarily occur at the mid–span point.

When the loads are considered individually the corresponding functions are as indicated in Figure 4.28.

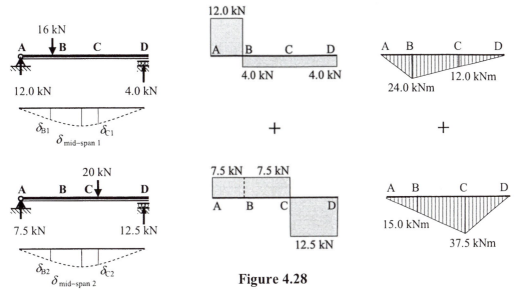

Figure 4.28

It is evident from Figure 4.28 that:

$V_A = (12.0 + 7.5) = 19.5$ kN;　　　　$V_D = (4.0 + 12.5) = 16.5$ kN

$\delta_B = (\delta_{B1} + \delta_{B2})$;　　　　$\delta_{mid-span} = (\delta_{mid-span1} + \delta_{mid-span2})$;　　　　$\delta_C = (\delta_{C1} + \delta_{C2})$

Shear Force at B $_{left-hand\ side} = (+12.0 + 7.5) = +19.5$ kN
Shear Force at B $_{right-hand\ side} = (-4.0 + 7.5) = +3.5$ kN
Shear Force at C $_{left-hand\ side} = (-4.0 + 7.5) = +3.5$ kN
Shear Force at C $_{right-hand\ side} = (-4.0 - 12.5) = -16.5$ kN
Bending Moment at B $= (+24.0 + 15.0) = +39.0$ kNm
Bending Moment at C $= (+12.0 + 37.5) = +49.5$ kNm

This Principle can be used very effectively when calculating the deflection of beams, (particularly non–uniform beams), as used in the Examples and Problems given in Section 4.5. Examples 5.6 to 5.10 illustrate the application of the Principle.

4.4.1 Example 4.6: Superposition–Beam 1

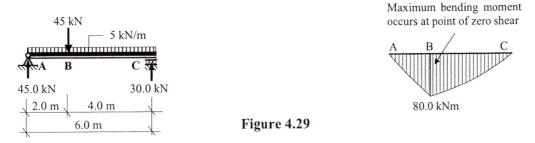

Figure 4.29

Using superposition this beam can be represented as the sum of the two load cases shown in Figure 4.30.

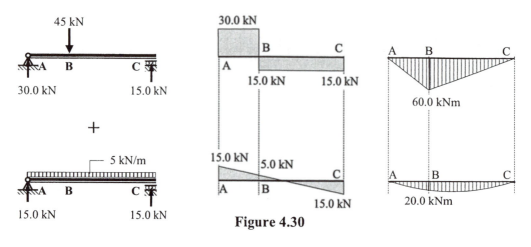

Figure 4.30

$V_A = (30.0 + 15.0) = 45.0$ kN; $V_C = (15.0 + 15.0) = 30.0$ kN

Shear Force at B $_{\text{left–hand side}} = (+ 30.0 + 5.0) = + 35.0$ kN

Shear Force at B $_{\text{right–hand side}} = (- 15.0 + 5.0) = + 10.0$ kN

Bending Moment at B $= (+ 60.0 + 20.0) = + 80.0$ kNm

4.4.2 Example 4.7: Superposition – Beam 2

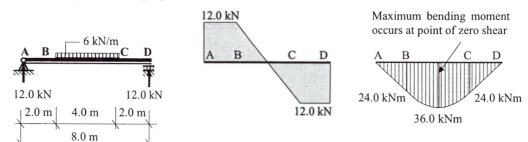

Figure 4.31

Using superposition this beam can be represented as the sum of:

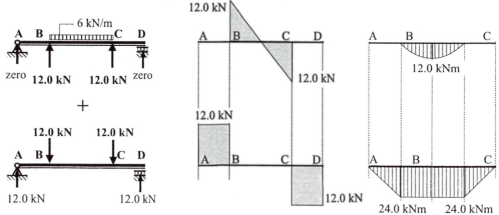

Figure 4.32

V_A = (zero + 12.0) = 12.0 kN; V_D = (zero + 12.0) = 12.0 kN;

Shear Force at B $_{left–hand\ side}$ = (zero + 12.0) = + 12.0 kN

Shear Force at B $_{right–hand\ side}$ = (+ 12.0 + zero) = + 12.0 kN

Shear Force at mid–span = zero

Shear Force at C $_{left–hand\ side}$ = (− 12.0 + zero) = − 12.0 kN

Shear Force at C $_{right–hand\ side}$ = (zero − 12.0) = − 12.0 kN

Bending Moment at B = (zero + 24.0) = + 24.0 kNm

Bending Moment at mid–span = (+ 12.0 + 24.0) = + 36.0 kNm

Bending Moment at C = (zero + 24.0) = + 24.0 kNm

4.4.3 Example 4.8: Superposition– Beam 3

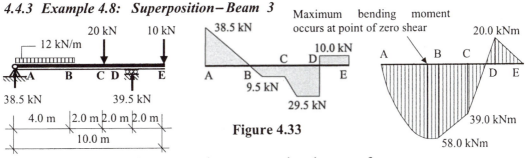

Figure 4.33

Using superposition this beam can be represented as the sum of:

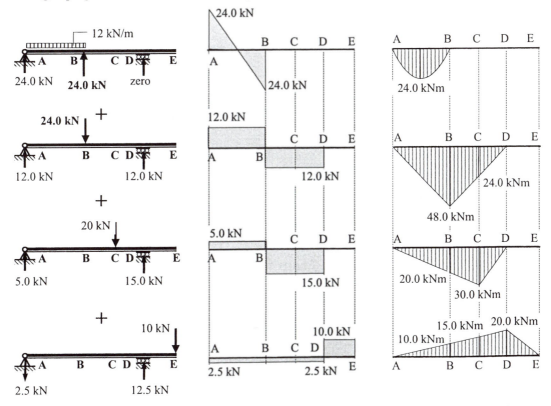

Figure 4.34

V_A = (24.0 + 12.0 + 5.0 − 2.5) = 38.5 kN;

V_D = (zero + 12.0 + 15.0 + 12.5) = 39.5 kN;
Shear Force at B $_{left-hand\ side}$ = (− 24.0 + 12.0 + 5.0 − 2.5) = − 9.5 kN
Shear Force at B $_{right-hand\ side}$ = (zero − 12.0 − 15.0 − 2.5) = − 29.5 kN
Shear Force at C $_{left-hand\ side}$ = (zero − 12.0 − 15.0 − 2.5) = − 29.5 kN
Shear Force at C $_{right-hand\ side}$ = (zero − 12.0 − 15.0 − 2.5) = − 29.5 kN
Shear Force at D $_{left-hand\ side}$ = (zero − 12.0 − 15.0 − 2.5) = − 29.5 kN
Shear Force at D $_{right-hand\ side}$ = + 10.0 kN
Shear Force at E = + 10.0 kN
Bending Moment at B = (zero + 48.0 + 20.0 − 10.0) = + 58.0 kNm
Bending Moment at C = (zero + 24.0 + 30.0 − 15.0) = + 39.0 kNm
Bending Moment at D = − 20.0 kNm

4.4.4 Example 4.9: Superposition− Beam 4

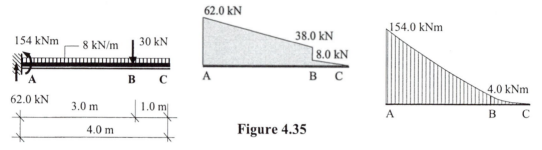

Figure 4.35

Using superposition this beam can be represented as the sum of:

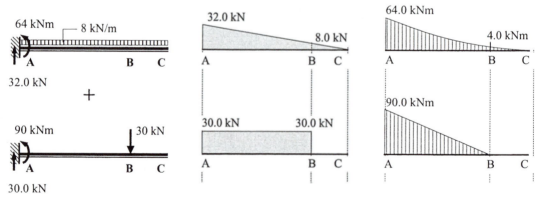

Figure 4.36

V_A = (32.0 + 30.0) = 62.0 kN
M_A = (− 64.0 − 90.0) = 154.0 kN
Shear Force at B $_{left-hand\ side}$ = (−8.0 − 30.0) = − 38.0 kN
Shear Force at B $_{right-hand\ side}$ = − 8.0 kN
Bending Moment at B = − 4.0 kNm

4.4.5 Example 4.10: Superposition – Beam 5

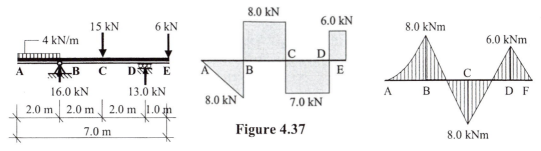

Figure 4.37

Using superposition this beam can be represented as the sum of:

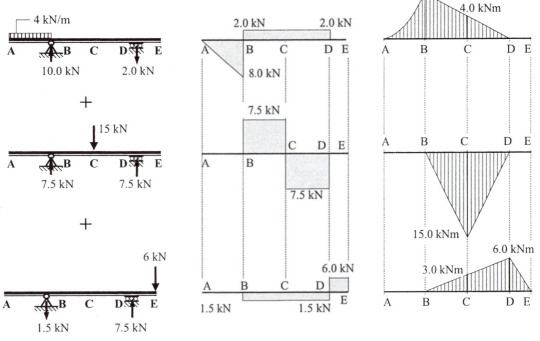

Figure 4.38

$V_B = (+\,10.0 + 7.5 - 1.5) = 16.0 \text{ kN};$

$V_D = (-\,2.0 + 7.5 + 7.5) = 13.0 \text{ kN};$

Shear Force at B $_{\text{left–hand side}} = -\,8.0 \text{ kN}$

Shear Force at B $_{\text{right–hand side}} = (+\,2.0 + 7.5 - 1.5) = +\,8.0 \text{ kN}$

Shear Force at C $_{\text{left–hand side}} = (+\,2.0 + 7.5 - 1.5) = +\,8.0 \text{ kN}$

Shear Force at C $_{\text{right–hand side}} = (+\,2.0 - 7.5 - 1.5) = -\,7.0 \text{ kN}$

Shear Force at D $_{\text{left–hand side}} = (+\,2.0 - 7.5 - 1.5) = -\,7.0 \text{ kN}$

Shear Force at D $_{\text{right–hand side}} = +\,6.0 \text{ kN}$

Shear Force at E $= +\,6.0 \text{ kN}$

Bending Moment at B $= -\,8.0 \text{ kNm}$

Bending Moment at C $= (-\,4.0 + 15.0 - 3.0) = +\,8.0 \text{ kNm}$

Bending Moment at D $= -\,6.0 \text{ kNm}$

4.5   *Unit Load Method for Deflection of Beams*

In Chapter 3, Section 3.5 the deflection of pin–jointed frames was calculated using the concept of strain energy and Castigliano's 1st. Theorem. This approach can also be applied to structures such as beams and rigid–jointed frames in which the members are primarily subject to bending effects.

In the case of pin–jointed frames the applied loads induce axial load effects and subsequent changes in the lengths of the members. In the case of beams and rigid–jointed frames, the corresponding applied loads induce bending moments and subsequent changes in the slope of the member.

Pin–jointed frames comprise discrete members with individual axial loads which are constant along the length of the member. In beams the bending moment generally varies along the length and consequently the summation of the bending effect for the entire beam is the integral of a function involving the bending moment.

4.5.1 *Strain Energy (Bending Load Effects)*

A simply–supported beam subjected to a single point load is shown in Figure 4.39. An incremental length of beam dx, over which the bending moment can be considered to be constant, is indicated a distance 'x' from the left–hand support.

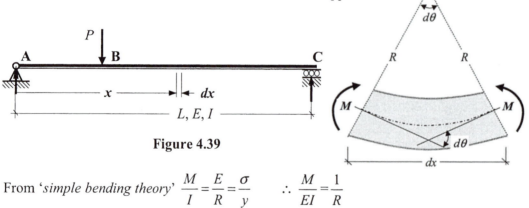

Figure 4.39

From '*simple bending theory*' $\dfrac{M}{I} = \dfrac{E}{R} = \dfrac{\sigma}{y}$ $\therefore \dfrac{M}{EI} = \dfrac{1}{R}$

where R is the radius of curvature and $1/R$ is the curvature of the beam, i.e. the rate of change of slope. $\therefore \dfrac{1}{R} = \dfrac{d\theta}{dx} = \dfrac{M}{EI}$

Assuming the moment is applied to the beam gradually, the relationship between the moment and the change in slope is as shown in Figure 4.40.

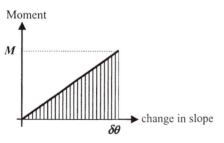

Figure 4.40

The external work–done on the member by the bending moment 'M' is equal to the strain energy stored and is given by the expression:

$$dU = \left(\frac{1}{2} M \times d\theta \right)$$

Differentiating the expression for strain energy with respect to x gives:

$$\frac{dU}{dx} = \left(\frac{1}{2}M \times \frac{d\theta}{dx}\right)$$

substituting for $\dfrac{d\theta}{dx}$ $\therefore \dfrac{dU}{dx} = \left(\dfrac{1}{2}M \times \dfrac{M}{EI}\right) = \dfrac{M^2}{2EI}$

Transposing dx in this equation $dU = \dfrac{M^2}{2EI}dx$

The total strain energy in the beam $U = \displaystyle\int_0^L \dfrac{M^2}{2EI}dx$

Using Castigliano's 1st Theorem relating to strain energy and structural deformation:

$$\varDelta = \frac{\partial U}{\partial W}$$

where:

U is the total strain energy of the structure due to the applied load system,

W is the force or moment acting at the point where the displacement or rotation is required,

$\varDelta$ is the linear displacement or rotation in the direction of the line of action of W.

Consider the simply–supported beam ABCD shown in Figure 4.41 in which it is required to determine the mid–span deflection at C due to an applied load P at position B.

Figure 4.41

Step 1:
The applied load bending moment diagram is determined as shown in Figure 4.42

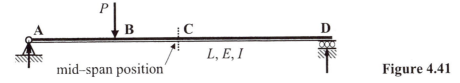

M applied load diagram   **Figure 4.42**

Step 2:
The applied load system is removed from the structure and an imaginary Unit load is applied at the position and in the direction of the required deflection, i.e. a vertical load equal to 1.0 at point C. The resulting bending moment diagram due to the unit load is indicated in Figure 4.43

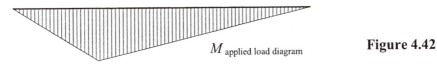

m unit load diagram   **Figure 4.43**

If both the Step 1 and the Step 2 load systems are considered to act simultaneously, then by superposition the bending moment in the beam is given by:

$$Q = (M + \beta m)$$

where:

M is the bending moment due to the applied load system

m is the bending moment due to the applied imaginary Unit load applied at C

β is a multiplying factor to reflect the value of the load applied at C, (since the unit load is an imaginary force the value of β = zero and is used here as a mathematical convenience.)

The strain energy in the structure is equal to the total energy stored along the full length of the beam:

$$U = \int_0^L \frac{Q^2}{2EI} dx$$

Using Castigliano's 1ˢᵗ Theorem the deflection of point C is given by:

$$\Delta = \frac{\partial U}{\partial W}$$

$$\therefore \Delta_\beta = \frac{\partial U}{\partial \beta} = \frac{\partial U}{\partial Q} \times \frac{\partial Q}{\partial \beta}$$

and $\dfrac{\partial U}{\partial Q} = \displaystyle\int_0^L \frac{Q}{EI} dx$; $\dfrac{\partial Q}{\partial \beta} = m$

$$\therefore \Delta_\beta = \frac{\partial U}{\partial \beta} = \frac{\partial U}{\partial Q} \times \frac{\partial Q}{\partial \beta} = \int_0^L \frac{Q}{EI} dx \times m = \int_0^L \frac{(M + \beta m)}{EI} dx \times m$$

Since β = zero the vertical deflection at C (Δ_β) is given by:

$$\Delta_\beta = \int_0^L \frac{Mm}{EI} dx$$

i.e. the deflection at any point in a beam can be determined from:

$$\delta = \int_0^L \frac{Mm}{EI} dx$$

where:

δ is the displacement of the point of application of any load, along the line of action of that load,

M is the bending in the member due to the externally applied load system,

m is the bending moment in member due to a **unit load** acting at the position of, and in the direction of the desired displacement,

I is the second–moment of area of the member,

E is the modulus of elasticity of the material for the member.

4.5.2 Example 4.11: Deflection and Slope of a Uniform Cantilever

A uniform cantilever beam is shown in Figure 4.44 in which a 20 kN is applied at B as indicated. Determine the magnitude and direction of the deflection and slope at B.

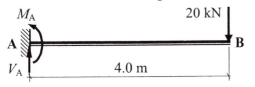

E and I are constant from A to B.

Figure 4.44

The bending moment diagrams for the applied load, a unit point load at B and a unit moment at B are shown in Figure 4.45.

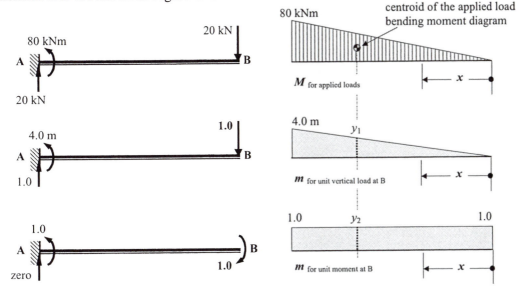

Figure 4.45

Solution:

$$\delta_B = \int_0^L \frac{Mm}{EI}\,dx$$

The bending moment at position 'x' due to the applied vertical load $M = -20.0x$

The bending moment at position 'x' due to the applied unit vertical load $m = -x$

$$Mm = +20x^2 \quad \therefore \delta_B = \int_{x=0}^{x=4} \frac{20x^2}{EI}\,dx = \left[\frac{20x^3}{3EI}\right]_0^4 = +\frac{426.67}{EI}\,\text{m} \downarrow$$

The bending moment at position 'x' due to the applied unit moment at B $m = -1.0$

$$Mm = +20x \quad \therefore \theta_B = \int_{x=0}^{x=4} \frac{20x}{EI}\,dx = \left[\frac{20x^2}{2EI}\right]_0^4 = +\frac{160}{EI}\,\text{rad.} \searrow$$

The product integral $\int_0^L Mm \, dx$ can be also be calculated as:

(***Area of the applied load bending moment diagram*** × ***the ordinate on the unit load bending moment diagram corresponding to the position of the centroid of the applied load bending moment diagram***), e.g. to determine the vertical deflection:

Area of the applied load bending moment diagram $\quad A = (0.5 \times 4.0 \times 80.0) = 160 \text{ kNm}^2$
Ordinate at the position of the centroid $\qquad\qquad y_1 = 2.67 \text{ m}$

$$\int_0^L Mm \, dx = (160 \times 2.67) = 426.67 \qquad \therefore \delta_B = \int_0^L \frac{Mm}{EI} dx = + \frac{426.67}{EI} \text{ m} \downarrow$$

To determine the slope:
Area of the applied load bending moment diagram $\quad A = (0.5 \times 4.0 \times 80.0) = 160 \text{ kNm}^2$
Ordinate at the position of the centroid $\qquad\qquad y_2 = 1.0$

$$\int_0^L Mm \, dx = (160 \times 1.0) = 160 \qquad \therefore \delta_B = \int_0^L \frac{Mm}{EI} dx = + \frac{160}{EI} \text{ rad.} \searrow$$

4.5.3 Example 4.12: Deflection and Slope of a Non–Uniform Cantilever

Consider the same problem as in Example 4.11 in which the cross–section of the cantilever has a variable *EI* value as indicated in Figure 4.46.

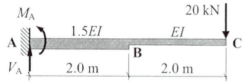

Figure 4.46

The bending moment diagrams for the applied load, a unit point load at C and a unit moment at C are shown in Figure 4.47.

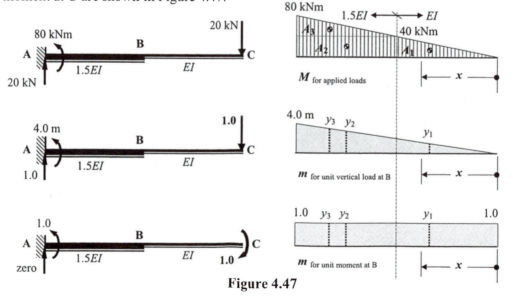

Figure 4.47

Solution:

$$\delta_C = \int_0^L \frac{Mm}{EI}\, dx$$

In this case since (Mm/EI) is not a continuous function the product integral must be evaluated between each of the discontinuities, i.e. C to B and B to A.

$$\delta_C = \int_0^L \frac{Mm}{EI}\, dx = \int_C^B \frac{Mm}{EI}\, dx + \int_B^A \frac{Mm}{1.5EI}\, dx$$

Consider the section from C to B: **$0 \le x \le 2.0$ m**

$$M = -20x \qquad m = -x \qquad \therefore Mm = +20x^2$$

$$\int_C^B \frac{Mm}{EI}\, dx = \int_0^2 \frac{20x^2}{EI}\, dx = \left[\frac{20x^3}{3EI}\right]_0^2 = +\frac{53.33}{EI}\ \text{m}$$

Consider the section from B to A: **$2.0 \le x \le 4.0$ m**

$$M = -20x \qquad m = -x \qquad \therefore Mm = +20x^2$$

$$\int_B^A \frac{Mm}{EI}\, dx = \int_2^4 \frac{20x^2}{1.5EI}\, dx = \left[\frac{20x^3}{4.5EI}\right]_2^4 = \left[\frac{20\times 4^3}{4.5EI} - \frac{20\times 2^3}{4.5EI}\right] = +\frac{248.89}{EI}\ \text{m}$$

$$\therefore \delta_C = +\frac{53.33}{EI} + \frac{248.89}{EI} = \frac{302.22}{EI}\ \text{m} \downarrow$$

Similarly to determine the slope:

$$\theta_C = \int_0^L \frac{Mm}{EI}\, dx = \int_C^B \frac{Mm}{EI}\, dx + \int_B^A \frac{Mm}{1.5EI}\, dx$$

Consider the section from C to B: **$0 \le x \le 2.0$ m**

$$M = -20x \qquad m = -1.0 \qquad \therefore Mm = 20x$$

$$\int_C^B \frac{Mm}{EI}\, dx = \int_0^2 \frac{20x}{EI}\, dx = \left[\frac{20x^2}{2EI}\right]_0^2 = +\frac{40.0}{EI}\ \text{rad.}$$

Consider the section from B to A: **$2.0 \le x \le 4.0$ m**

$$M = -20x \qquad m = -1.0 \qquad \therefore Mm = 20x$$

$$\int_B^A \frac{Mm}{EI}\, dx = \int_2^4 \frac{20x}{1.5EI}\, dx = \left[\frac{20x^2}{3.0EI}\right]_2^4 = \left[\frac{20\times 4^2}{3.0EI} - \frac{20\times 2^2}{3.0EI}\right] = +\frac{80.0}{EI}\ \text{rad.}$$

$$\therefore \theta_C = +\frac{40.0}{EI} + \frac{80.0}{EI} = +\frac{120.0}{EI}\ \text{rad.} \searrow$$

Alternatively, the applied bending moment diagram can be considered as a the sum of the areas created by the discontinuity. (In most cases this will result in a number of recognised shapes e.g. triangular, rectangular or parabolic, in which the areas and the position of the centroid can be easily calculated).

The deflection can then be determined by summing the products, i.e. (area × ordinate), for each of the shapes.

$A_1 = (0.5 \times 2.0 \times 40.0)$ kNm^2, $y_1 = 1.333$ m, $\therefore A_1 y_1 = 53.32$ kNm^3
$A_2 = (2.0 \times 40.0)$ kNm^2, $y_2 = 3.0$ m, $\therefore A_2 y_2 = 240.0$ kNm^3
$A_3 = (0.5 \times 2.0 \times 40.0)$ kNm^2, $y_3 = 3.333$ m, $\therefore A_3 y_3 = 133.32$ kNm^3

$$\delta_C = \int_0^L \frac{Mm}{EI} dx = (53.32/EI) + (240.0/1.5EI) + (133.32/1.5EI) = + (302.22/EI) \text{ m} \downarrow$$

The slope can then be determined by summing the products, i.e. (area × ordinate), for each of the shapes.

$A_1 = (0.5 \times 2.0 \times 40.0)$ kNm^2, $y_1 = 1.0$, $\therefore A_1 y_1 = 40.0$ kNm^3
$A_2 = (2.0 \times 40.0)$ kNm^2, $y_2 = 1.0$, $\therefore A_2 y_2 = 80.0$ kNm^3
$A_3 = (0.5 \times 2.0 \times 40.0)$ kNm^2, $y_3 = 1.0$, $\therefore A_3 y_3 = 40.0$ kNm^3

$$\delta_C = \int_0^L \frac{Mm}{EI} dx = (40.0/EI) + (80.0/1.5EI) + (40/1.5EI) = + (120.0/EI) \text{ rad.} \searrow$$

4.5.4 Example 4.13:   Deflection and Slope of a Linearly Varying Cantilever

Consider the same problem as in Example 4.11 in which the cross–section of the cantilever has an *I* which varies linearly from *I* at the free end to 2*I* at the fixed support at A as indicated in Figure 4.48. Determine the vertical displacement and the slope at point B for the loading indicated.

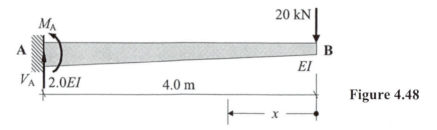

Figure 4.48

The value of *I* at position '*x*' along the beam is given by: $I + I(x/L) = I(L + x)/L$.

In this case since the *I* term is dependent on *x* it cannot be considered outside the integral as a constant. The displacement must be determined using integration and cannot be calculated using the sum of the (area × ordinate) as in Example 5.11 and Example 5.12.

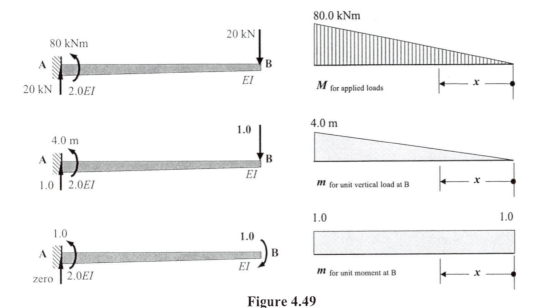

Figure 4.49

Solution:

The bending moment at position 'x' due to the applied vertical load $\qquad M = -20.0x$

The bending moment at position 'x' due to the applied unit vertical load $\quad m = -x$

$$Mm = +20x^2 \quad \therefore \quad \delta_B = \int_{x=0}^{x=4} \frac{20x^2 L}{EI(L+x)}\,dx = \frac{20L}{EI}\int_{x=0}^{x=4}\frac{x^2}{(L+x)}\,dx$$

Let $v = (L+x) \qquad \therefore x = (v-L) \quad dx = dv \qquad$ and $\qquad x^2 = (v-L)^2$

when $\quad x = 0 \qquad v = L = 4.0 \quad$ and $\quad$ when $\quad x = 4 \qquad v = (L+4.0) = 8.0$

$$\delta_B = \frac{20L}{EI}\int_{x=0}^{x=4}\frac{x^2}{(L+x)}\,dx = \frac{80.0}{EI}\int_{v=4}^{v=8}\frac{(v-4.0)^2}{v}\,dv = \frac{80.0}{EI}\int_{v=4}^{v=8}\frac{(v^2-8.0v+16.0)}{v}\,dv$$

$$= \frac{80.0}{EI}\int_{v=4}^{v=8}\left(v-8.0+\frac{16.0}{v}\right)dv = \frac{80.0}{EI}\left[\frac{v^2}{2}-8v+16.0\ln v\right]_{v=4.0}^{v=8.0}$$

$$= \frac{80.0}{EI}\left\{\left[\frac{8.0^2}{2}-(8\times8)+16.0\,ln\,8\right]-\left[\frac{4.0^2}{2}-(8\times4)+16.0\,ln\,4\right]\right\} = +\frac{247.20}{EI}\text{ m }\downarrow$$

The bending moment at position 'x' due to the applied unit moment at B $\quad m = -1.0$

$$Mm = +20x \quad \therefore \quad \theta_B = \int_{x=0}^{x=4}\frac{20xL}{EI(L+x)}\,dx = \frac{20L}{EI}\int_{x=0}^{x=4}\frac{x}{(L+x)}\,dx$$

$$\theta_B = \frac{20L}{EI}\int_{x=0}^{x=4}\frac{x}{(L+x)}\,dx = \frac{80.0}{EI}\int_{v=4}^{v=8}\frac{(v-4.0)}{v}\,dv = \frac{80.0}{EI}\int_{v=4}^{v=8}\left(1-\frac{4.0}{v}\right)dv$$

$$= \frac{80.0}{EI}\left[v-4.0\,lnv\right]_{v=4.0}^{v=8.0} = \frac{80.0}{EI}\left\{[8.0-4.0\,ln\,8]-[4.0-4.0\,ln\,4]\right\} = +\frac{98.19}{EI}\text{ rad. }\searrow$$

4.5.5 Example 4.14: Deflection of a Non-Uniform, Simply-Supported Beam

A non-uniform, single-span beam ABCD is simply-supported at A and D and carries loading as indicated in Figure 4.50. Determine the vertical displacement at point B.

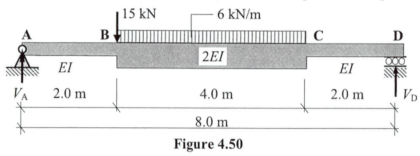

Figure 4.50

The bending moment diagrams for the applied load, a unit point load at B are shown in Figure 4.51.

The beam loading can be considered as the superposition of a number of load cases each of which produces a bending moment diagram with a standard shape. Since there are discontinuities in the bending moment diagrams the product integrals should be carried out for the three regions A to B, D to C and C to B.

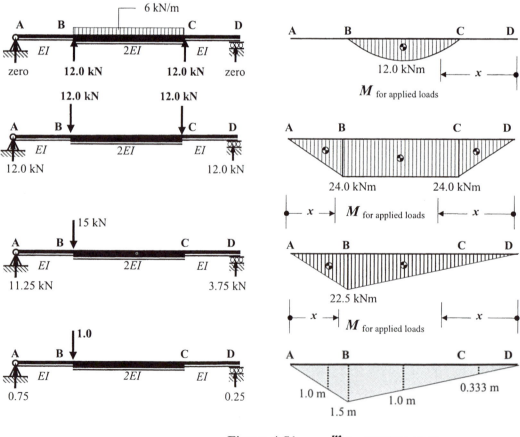

Figure 4.51 m for unit vertical load at B

Solution:

It is convenient in this problem to change the position of the origin from which '*x*' is measured for the different regions A–B, D–C and C–B as shown in Figure 4.51.

$$\delta_B = \int_0^L \frac{Mm}{EI}dx = \int_A^B \frac{Mm}{EI}dx + \int_D^C \frac{Mm}{EI}dx + \int_C^B \frac{Mm}{2EI}dx$$

Consider the section from A to B: **$0 \leq x \leq 2.0$ m**

$M = (12x + 11.25x) = 23.25x$ $\qquad$ $m = + 0.75\,x$ $\qquad$ $\therefore Mm = + 17.44\,x^2$

$$\int_A^B \frac{Mm}{EI}dx = \int_0^2 \frac{17.44x^2}{EI}dx = \left[\frac{17.44x^3}{3EI}\right]_0^2 = \left[\frac{17.44 \times 2^3}{3EI}\right] = + \frac{46.51}{EI}\text{m}$$

Consider the section from D to C: **$0 \leq x \leq 2.0$ m**

$M = (12x + 3.75x) = 15.75x$ $\qquad$ $m = + 0.25x$ $\qquad$ $\therefore Mm = + 3.94x^2$

$$\int_D^C \frac{Mm}{EI}dx = \int_0^2 \frac{3.94x^2}{EI}dx = \left[\frac{3.94x^3}{3EI}\right]_0^2 = \left[\frac{3.94 \times 2^3}{3EI}\right] = + \frac{10.51}{EI}\text{m}$$

Consider the section from C to B: **$2.0 \leq x \leq 6.0$ m**

$M = [12(x-2) - 6(x-2)^2/2] + [12x - 12(x-2)] + 3.75x = (27.75x - 3x^2 - 12)$

$m = + 0.25\,x$ $\qquad\qquad\qquad$ $\therefore Mm = (6.94x^2 - 0.75x^3 - 3x)$

$$\int_C^B \frac{Mm}{EI}dx = \int_2^6 \frac{(6.94x^2 - 0.75x^3 - 3x)}{2EI}dx = \left[\frac{6.94x^3}{6EI} - \frac{0.75x^4}{8EI} - \frac{3x^2}{4EI}\right]_2^6$$

$$= \left[\frac{6.94 \times 6^3}{6EI} - \frac{0.75 \times 6^4}{8EI} - \frac{3 \times 6^2}{4EI}\right] - \left[\frac{6.94 \times 2^3}{6EI} - \frac{0.75 \times 2^4}{8EI} - \frac{3 \times 2^2}{4EI}\right]$$

$$= + \frac{96.59}{EI}\text{m}$$

$$\therefore \delta_B = \left(\frac{46.51}{EI} + \frac{10.51}{EI} + \frac{96.59}{EI}\right) = \frac{153.61}{EI}\text{m} \downarrow$$

Alternatively: considering Σ (areas × ordinates)

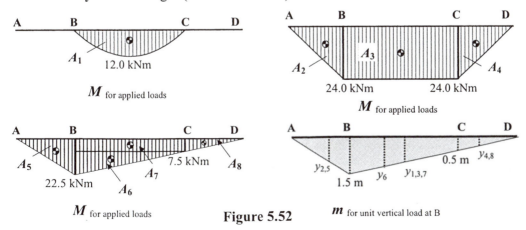

Figure 5.52

$A_1 = (0.667 \times 4.0 \times 12.0)$ kNm², $y_1 = 1.0$ m, $\therefore A_1 y_1 = 32.0$ kNm³
$A_2 = (0.5 \times 2.0 \times 24.0)$ kNm², $y_2 = 1.0$ m, $\therefore A_2 y_2 = 24.0$ kNm³
$A_3 = (4.0 \times 24.0)$ kNm², $y_3 = 1.0$ m, $\therefore A_3 y_3 = 96.0$ kNm³
$A_4 = (0.5 \times 2.0 \times 24.0)$ kNm², $y_4 = 0.333$ m, $\therefore A_4 y_4 = 8.0$ kNm³
$A_5 = (0.5 \times 2.0 \times 22.5)$ kNm², $y_5 = 1.0$ m, $\therefore A_5 y_5 = 22.5$ kNm³
$A_6 = (0.5 \times 4.0 \times 15.0)$ kNm², $y_6 = 1.167$ m, $\therefore A_6 y_6 = 35.0$ kNm³
$A_7 = (4.0 \times 7.5)$ kNm², $y_7 = 1.0$ m, $\therefore A_7 y_7 = 30.0$ kNm³
$A_8 = (0.5 \times 2.0 \times 7.5)$ kNm², $y_8 = 0.333$ m, $\therefore A_8 y_8 = 2.5$ kNm³

$$\delta_B = \int_0^L \frac{Mm}{EI} dx = (32.0/2EI) + (24.0/EI) + (96.0/2EI) + (8.0/EI) + (22.5/EI) + (35.0/2EI)$$

$$+ (30.0/2EI) + (2.5/EI) \qquad \therefore \delta_B = (153.5/EI) \text{ m} \downarrow$$

4.5.6 Example 4.15: Deflection of a Frame and Beam Structure

A uniform beam BCD is tied at B, supported on a roller at C and carries a vertical load at D as indicated in Figure 4.53. Using the data given determine the vertical displacement at point D.

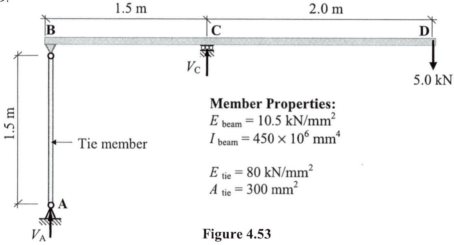

Member Properties:
$E_{beam} = 10.5$ kN/mm²
$I_{beam} = 450 \times 10^6$ mm⁴

$E_{tie} = 80$ kN/mm²
$A_{tie} = 300$ mm²

Figure 4.53

Solution:
Consider the rotational equilibrium of the beam:
$$+ve \, \curvearrowright \, \Sigma M_A = 0 \qquad -(V_C \times 1.5) + (5.0 \times 3.5) = 0 \qquad\qquad \therefore V_C = 11.67 \text{ kN} \uparrow$$

Consider the vertical equilibrium of the structure:
$$+ve \uparrow \Sigma F_z = 0 \qquad V_A + V_C - 5.0 = 0 \qquad\qquad \therefore V_A = -6.67 \text{ kN} \downarrow$$

Since the structure comprises both an axially loaded member and a flexural member the deflection at D is given by:

$$\delta_D = \left(\frac{PL}{AE} u \right)_{\text{Member AB}} + \left(\int_0^L \frac{Mm}{EI} dx \right)_{\text{Member BCD}}$$

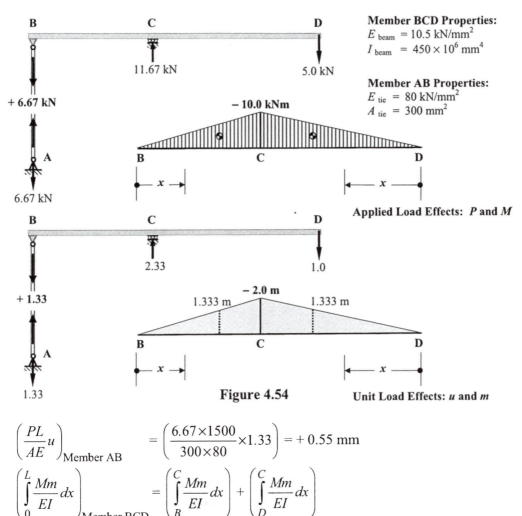

Figure 4.54

$$\left(\frac{PL}{AE}u\right)_{\text{Member AB}} = \left(\frac{6.67 \times 1500}{300 \times 80} \times 1.33\right) = +0.55 \text{ mm}$$

$$\left(\int_0^L \frac{Mm}{EI}dx\right)_{\text{Member BCD}} = \left(\int_B^C \frac{Mm}{EI}dx\right) + \left(\int_D^C \frac{Mm}{EI}dx\right)$$

Consider the section from B to C: **0 ≤ x ≤ 1.5 m**

$$M = -6.67x \qquad m = -1.33 x \qquad \therefore Mm = +8.87 x^2$$

$$\int_B^C \frac{Mm}{EI}dx = \int_0^{1.5} \frac{8.87x^2}{EI}dx = \left[\frac{8.87x^3}{3 \times EI}\right]_0^{1.5} = \left(\frac{29.94 \times 10^3}{3 \times 10.5 \times 450}\right) = +2.11 \text{ mm}$$

Consider the section from D to C: **0 ≤ x ≤ 2.0 m**

$$M = -5.0 x \qquad m = -1.0 x \qquad \therefore Mm = +5.0 x^2$$

$$\int_D^C \frac{Mm}{EI}dx = \int_0^2 \frac{5.0x^2}{EI}dx = \left[\frac{5.0x^3}{3EI}\right]_0^2 = \left(\frac{40.0 \times 10^3}{3 \times 10.5 \times 450}\right) = +2.82 \text{ mm}$$

$$\delta_D = \left(\frac{PL}{AE}u\right)_{\text{Member AB}} + \left(\int_0^L \frac{Mm}{EI}dx\right)_{\text{Member BCD}} = (0.55 + 2.11 + 2.82) = +\textbf{5.48 mm}\downarrow$$

In the previous examples the product integrals were also determined using:

(the area of the applied bending moment diagram × ordinate on the unit load bending moment diagram).

In Table 4.1 coefficients are given to enable the rapid evaluation of product integrals for standard cases along lengths of beam where the *EI* value is constant.

Product Integral $\int_0^L \dfrac{Mm}{EI}\,dx = [\text{Coefficient} \times a \times b \times L]/EI$

M \ m	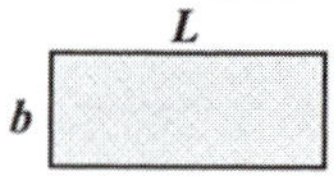 L (rectangle, b)	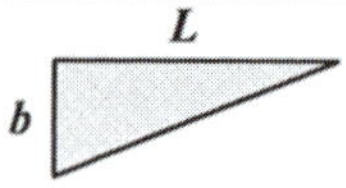 L (triangle, b)	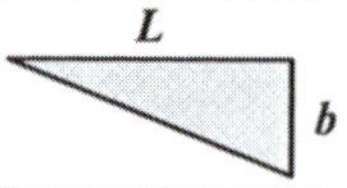 L (triangle, b)	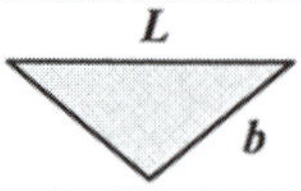 L (triangle, b)
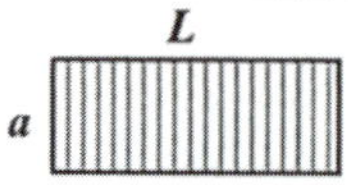 a (L, rectangle)	1.0	0.5	0.5	0.5
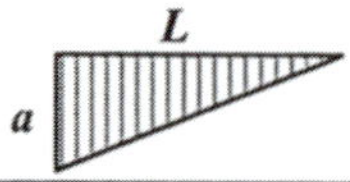 a (L, triangle)	0.5	**0.333**	0.167	0.25
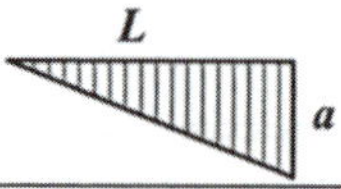 (L, triangle) a	0.5	0.167	**0.333**	0.25
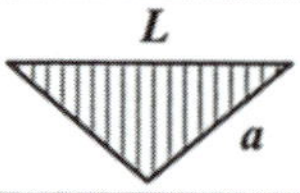 (L, triangle) a	0.5	0.25	0.25	0.333
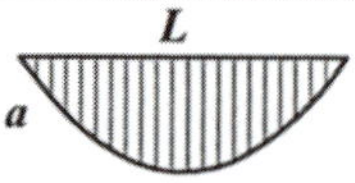 a (L, parabola)	0.667	0.333	0.333	0.417
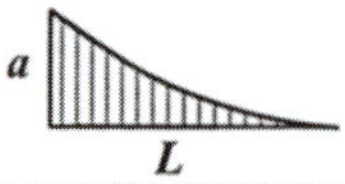 a (L, curve)	0.333	0.25	0.083	0.146
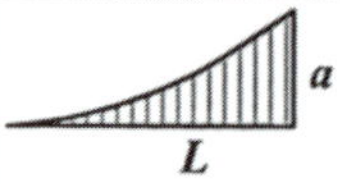 (L, curve) a	0.333	0.083	0.25	0.146

Table 4.1

Consider the contribution from the beam BCD to the vertical deflection at D in Example 4.15.

Product Integral $\int_0^L \dfrac{Mm}{EI}\,dx = \Sigma\,[\text{Coefficient} \times a \times b \times L]/EI$

From $(\text{B to C}) + (\text{D to C}) = [(0.333 \times 10.0 \times 2.0 \times 1.5) + (0.333 \times 10.0 \times 2.0 \times 2.0)]/EI$

$= +23.31/EI$ i.e. same as $[(2.11 + 2.82]$ calculated above.

4.5.7 *Example 4.16: Deflection of a Uniform Cantilever using Coefficients*

A uniform cantilever beam is shown in Figure 4.55 in which a uniformly distributed load and a vertical load is applied as indicated. Using the coefficients in Table 4.1 determine the magnitude and direction of the deflection at D.

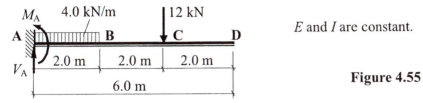

E and *I* are constant.

Figure 4.55

The bending moment diagrams for the applied loads and a unit point load at B are shown in Figure 4.56.

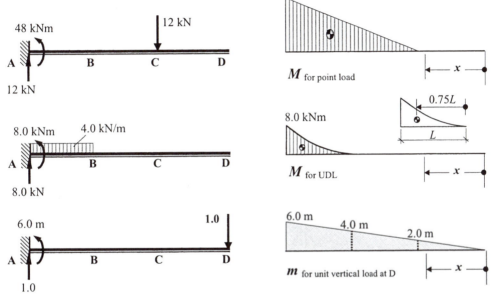

Figure 4.56

Solution:

Consider the unit load bending moment diagrams for both applied loads as the sum of rectangular and a triangular area as shown.

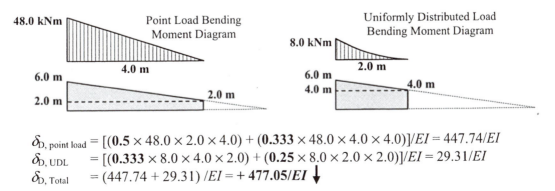

$$\delta_{D,\,point\,load} = [(\mathbf{0.5} \times 48.0 \times 2.0 \times 4.0) + (\mathbf{0.333} \times 48.0 \times 4.0 \times 4.0)]/EI = 447.74/EI$$
$$\delta_{D,\,UDL} = [(\mathbf{0.333} \times 8.0 \times 4.0 \times 2.0) + (\mathbf{0.25} \times 8.0 \times 2.0 \times 2.0)]/EI = 29.31/EI$$
$$\delta_{D,\,Total} = (447.74 + 29.31)\,/EI = +\,\mathbf{477.05/EI}\;\downarrow$$

4.5.8 Problems: Unit Load Method for Deflection of Beams / Frames

A series of statically–determinate beams/frames are indicated in Problems 4.16 to 4.23. Using the applied loading given in each case determine the deflections indicated. The relative values of Young's Modulus of Elasticity (E), Second Moment of Area (I) and Cross–sectional area (A) are given in each case.

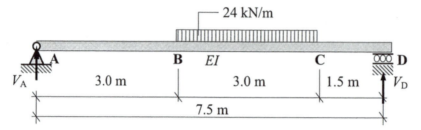

Determine the value of the vertical deflection at B given that $EI = 50.0 \times 10^3$ kNm^2

Problem 4.16

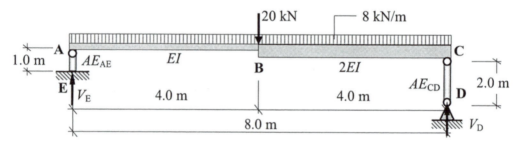

Determine the value of the vertical deflection at B given:
$E_{beam} = 9.0$ kN/mm^2 $I_{beam} = 14.6 \times 10^9$ mm^4
$E_{\text{AE and CD}} = 170$ kN/mm^2 $A_{AE} = 80$ mm^2 $A_{CD} = 120$ mm^2

Problem 4.17

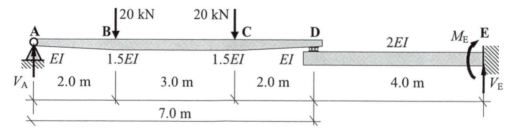

The EI value of the beam ABCD varies linearly from EI at the supports A and D to $1.5EI$ at B and C respectively and is constant between B and C.

Determine the value of the vertical deflection at B given that $EI = 15.0 \times 10^3$ kNm^2

Problem 4.18

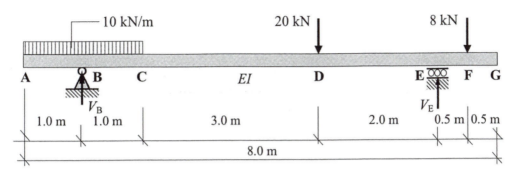

Determine the value of the vertical deflection at G given that $EI = 5.0 \times 10^3$ kNm^2

Problem 4.19

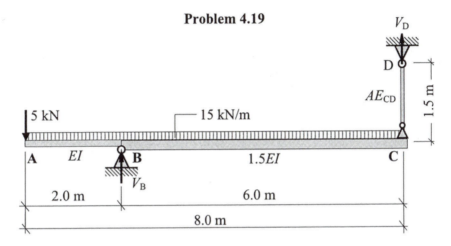

Determine the value of the vertical deflection at A given:

$E_{beam} = 205$ kN/mm^2 $I_{beam} = 60.0 \times 10^6$ mm^4
$E_{CD} = 205$ kN/mm^2 $A_{CD} = 50$ mm^2

Problem 4.20

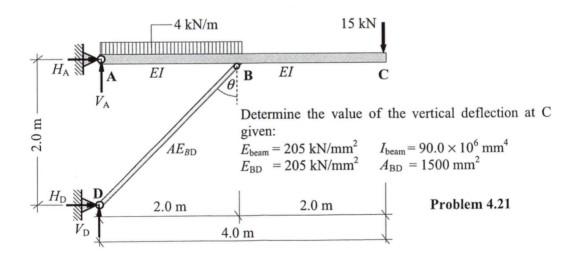

Determine the value of the vertical deflection at C given:

$E_{beam} = 205$ kN/mm^2 $I_{beam} = 90.0 \times 10^6$ mm^4
$E_{BD} = 205$ kN/mm^2 $A_{BD} = 1500$ mm^2

Problem 4.21

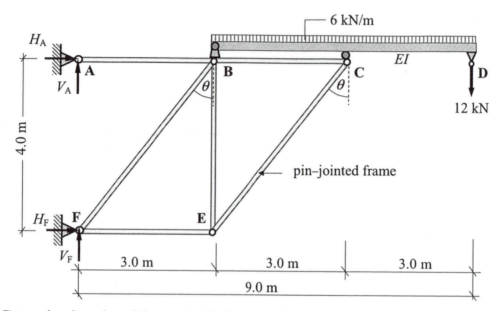

Determine the value of the vertical deflection at D given:

$E_{beam} = 205 \text{ kN/mm}^2$ $I_{beam} = 500.0 \times 10^6 \text{ mm}^4$

$E_{All \text{ frame members}} = 205 \text{ kN/mm}^2$ $A_{All \text{ frame members}} = 4000 \text{ mm}^2$

Problem 4.22

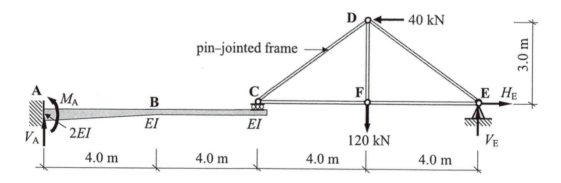

The *EI* value of the cantilever ABC varies linearly from 2*EI* at the fixed support to *EI* at B and is constant from B to C.

Determine the value of the vertical deflection at F and at C given:

$EI_{cantilever \text{ ABC}} = 1080 \times 10^3 \text{ kNm}^2$, $EA_{All \text{ frame members}} = 300 \times 10^3 \text{ kN}$

Problem 4.23

4.5.9 Solutions: Unit Load Method for Deflection of Beams / Frames

Solution

Topic: Statically Determinate Beams/Frames – Deflection Using Unit Load
Problem Number: 4.16   **Page No. 1**

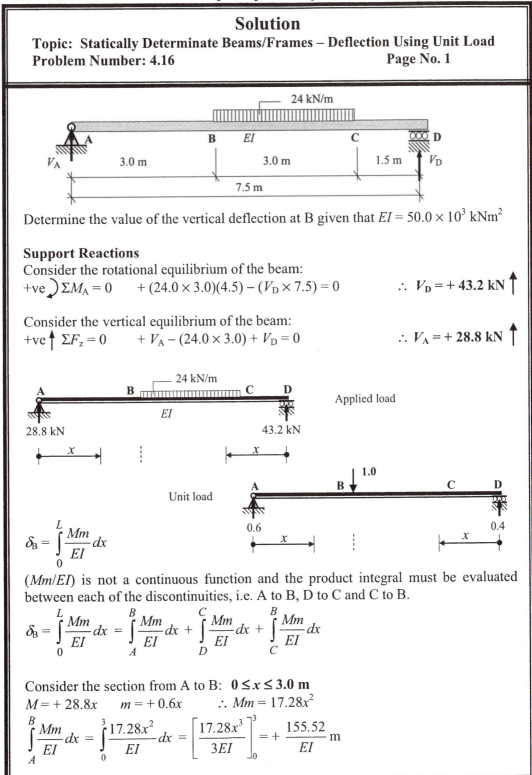

Determine the value of the vertical deflection at B given that $EI = 50.0 \times 10^3$ kNm^2

Support Reactions
Consider the rotational equilibrium of the beam:
+ve $\curvearrowright \Sigma M_A = 0$ $+ (24.0 \times 3.0)(4.5) - (V_D \times 7.5) = 0$ $\therefore V_D = + \textbf{43.2 kN} \uparrow$

Consider the vertical equilibrium of the beam:
+ve $\uparrow \Sigma F_z = 0$ $+ V_A - (24.0 \times 3.0) + V_D = 0$ $\therefore V_A = + \textbf{28.8 kN} \uparrow$

$$\delta_B = \int_0^L \frac{Mm}{EI} dx$$

(*Mm*/*EI*) is not a continuous function and the product integral must be evaluated between each of the discontinuities, i.e. A to B, D to C and C to B.

$$\delta_B = \int_0^L \frac{Mm}{EI} dx = \int_A^B \frac{Mm}{EI} dx + \int_D^C \frac{Mm}{EI} dx + \int_C^B \frac{Mm}{EI} dx$$

Consider the section from A to B: $\textbf{0} \leq x \leq \textbf{3.0 m}$
$M = + 28.8x$ $m = + 0.6x$ $\therefore Mm = 17.28x^2$

$$\int_A^B \frac{Mm}{EI} dx = \int_0^3 \frac{17.28x^2}{EI} dx = \left[\frac{17.28x^3}{3EI} \right]_0^3 = + \frac{155.52}{EI} \text{ m}$$

Solution
Topic: Determinate Beams/Frames – Deflection Using Unit Load
Problem Number: 4.16   **Page No. 2**

Consider the section from D to C: $\mathbf{0 \leq x \leq 1.5 \ m}$
$M = + 43.2x \qquad m = + 0.4x \qquad \therefore \ Mm = 17.28x^2$

$$\int_D^C \frac{Mm}{EI} dx = \int_0^{1.5} \frac{17.28x^2}{EI} dx = \left[\frac{17.28x^3}{3EI} \right]_0^{1.5} = + \frac{19.44}{EI} \ m$$

Consider the section from C to B: $\mathbf{1.5 \leq x \leq 4.5 \ m}$
$M = + 43.2x - 24(x - 1.5)^2/2 = 43.2x - 12(x^2 - 3x + 2.25)$
$\quad = - 12x^2 + 79.2x - 27.0$
$m = + 0.4x$
$Mm = - 4.8x^3 + 31.68x^2 - 10.8x$

$$\int_C^B \frac{Mm}{EI} dx = \int_{1.5}^{4.5} \frac{-4.8x^3 + 31.68x^2 - 10.8x}{EI} dx = \left[-\frac{4.8x^4}{4EI} + \frac{31.68x^3}{3EI} - \frac{10.8x^2}{2EI} \right]_{1.5}^{4.5}$$

$$= \left(+\frac{360.86}{EI} - \frac{17.42}{EI} \right) = +\frac{343.44}{EI}$$

$$\therefore \ \delta_B = \left(+\frac{155.52}{EI} + \frac{19.44}{EI} + \frac{343.44}{EI} \right) = \frac{518.4}{EI} = \frac{518.4}{50.0 \times 10^3} \ m = \mathbf{10.37 \ mm} \ \downarrow$$

Alternatively:

$\delta_B = \Sigma(\text{Area}_{\text{applied bending moment diagram}} \times \text{Ordinate}_{\text{unit load bending moment diagram}})$

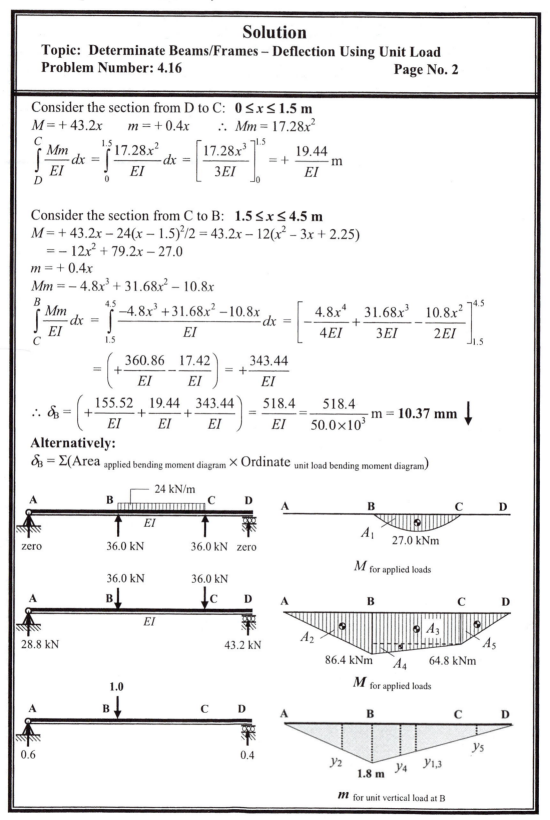

M for applied loads

M for applied loads

m for unit vertical load at B

Solution
Topic: Determinate Beams/Frames – Deflection Using Unit Load
Problem Number: 4.16   **Page No. 3**

$A_1 = (0.667 \times 3.0 \times 27.0)$ kNm^2, $y_1 = 1.2$ m, $\therefore A_1 y_1 = 64.83$ kNm^3

$A_2 = (0.5 \times 3.0 \times 86.4)$ kNm^2, $y_2 = 1.2$ m, $\therefore A_2 y_2 = 155.52$ kNm^3

$A_3 = (3.0 \times 64.8)$ kNm^2, $y_3 = 1.2$ m, $\therefore A_3 y_3 = 233.28$ kNm^3

$A_4 = (0.5 \times 3.0 \times 21.6)$ kNm^2, $y_4 = 1.4$ m, $\therefore A_4 y_4 = 45.36$ kNm^3

$A_5 = (0.5 \times 1.5 \times 64.8)$ kNm^2, $y_5 = 0.4$ m, $\therefore A_5 y_5 = 19.44$ kNm^3

$\delta_B = (64.83 + 155.52 + 233.28 + 45.36 + 19.44)/50.0 \times 10^3 = 0.0104$ m $=$ **10.37 mm** $\downarrow$

Using the coefficients given in Table 4.1:

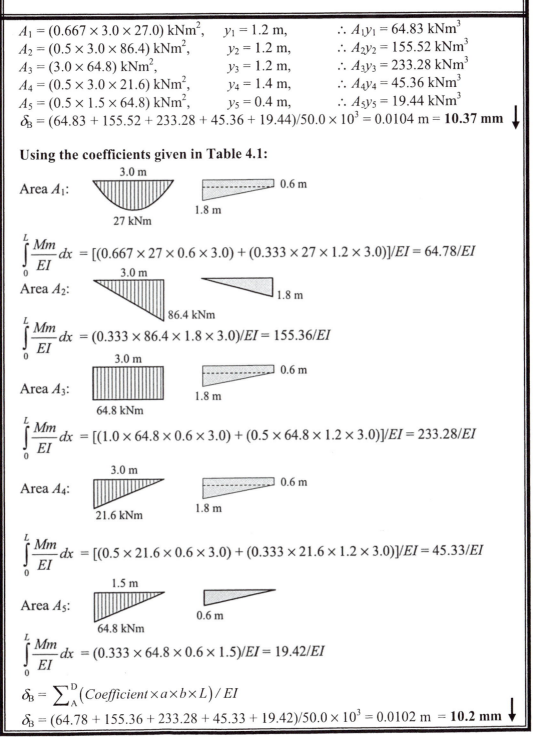

Area A_1:

3.0 m

27 kNm

0.6 m

1.8 m

$\int_0^L \dfrac{Mm}{EI} dx = [(0.667 \times 27 \times 0.6 \times 3.0) + (0.333 \times 27 \times 1.2 \times 3.0)]/EI = 64.78/EI$

Area A_2:

3.0 m

86.4 kNm

1.8 m

$\int_0^L \dfrac{Mm}{EI} dx = (0.333 \times 86.4 \times 1.8 \times 3.0)/EI = 155.36/EI$

Area A_3:

3.0 m

64.8 kNm

0.6 m

1.8 m

$\int_0^L \dfrac{Mm}{EI} dx = [(1.0 \times 64.8 \times 0.6 \times 3.0) + (0.5 \times 64.8 \times 1.2 \times 3.0)]/EI = 233.28/EI$

Area A_4:

3.0 m

21.6 kNm

0.6 m

1.8 m

$\int_0^L \dfrac{Mm}{EI} dx = [(0.5 \times 21.6 \times 0.6 \times 3.0) + (0.333 \times 21.6 \times 1.2 \times 3.0)]/EI = 45.33/EI$

Area A_5:

1.5 m

64.8 kNm

0.6 m

$\int_0^L \dfrac{Mm}{EI} dx = (0.333 \times 64.8 \times 0.6 \times 1.5)/EI = 19.42/EI$

$\delta_B = \sum_A^D \left(Coefficient \times a \times b \times L \right) / EI$

$\delta_B = (64.78 + 155.36 + 233.28 + 45.33 + 19.42)/50.0 \times 10^3 = 0.0102$ m $=$ **10.2 mm** $\downarrow$

Solution
Topic: Determinate Beams/Frames – Deflection Using Unit Load
Problem Number: 4.17   **Page No. 1**

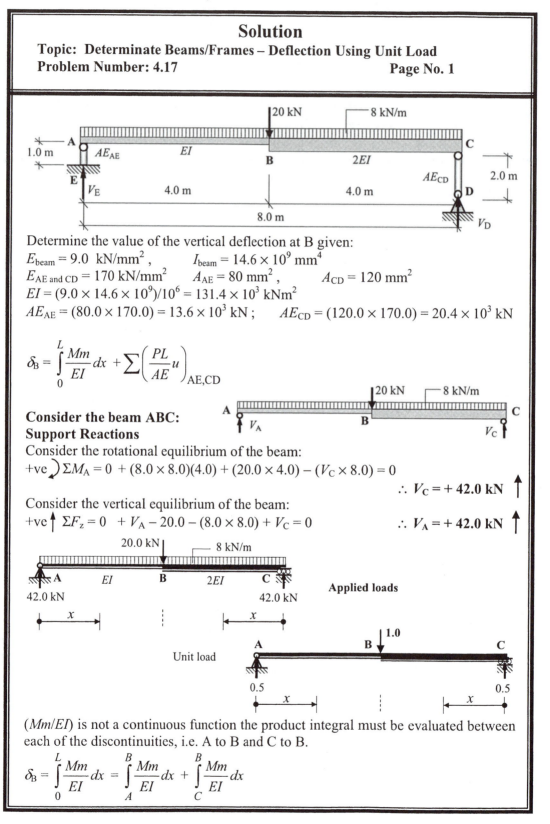

Determine the value of the vertical deflection at B given:

$E_{beam} = 9.0$ kN/mm^2 , $I_{beam} = 14.6 \times 10^9$ mm^4

$E_{AE \, and \, CD} = 170$ kN/mm^2 $A_{AE} = 80$ mm^2 , $A_{CD} = 120$ mm^2

$EI = (9.0 \times 14.6 \times 10^9)/10^6 = 131.4 \times 10^3$ kNm^2

$AE_{AE} = (80.0 \times 170.0) = 13.6 \times 10^3$ kN ; $AE_{CD} = (120.0 \times 170.0) = 20.4 \times 10^3$ kN

$$\delta_B = \int_0^L \frac{Mm}{EI} dx + \sum \left(\frac{PL}{AE} u \right)_{AE,CD}$$

Consider the beam ABC:
Support Reactions

Consider the rotational equilibrium of the beam:

+ve $\circlearrowright \Sigma M_A = 0$ + $(8.0 \times 8.0)(4.0) + (20.0 \times 4.0) - (V_C \times 8.0) = 0$

$\therefore V_C = + 42.0$ kN $\uparrow$

Consider the vertical equilibrium of the beam:

+ve $\uparrow \Sigma F_z = 0$ + $V_A - 20.0 - (8.0 \times 8.0) + V_C = 0$ $\therefore V_A = + 42.0$ kN $\uparrow$

(Mm/EI) is not a continuous function the product integral must be evaluated between each of the discontinuities, i.e. A to B and C to B.

$$\delta_B = \int_0^L \frac{Mm}{EI} dx = \int_A^B \frac{Mm}{EI} dx + \int_C^B \frac{Mm}{EI} dx$$

Solution

Topic: Determinate Beams/Frames – Deflection Using Unit Load
Problem Number: 4.17   **Page No. 2**

Consider the section from A to B: $0 \leq x \leq 4.0$ m
$M = +42.0x - 8.0x^2/2 = 42.0x - 4.0x^2$ $m = +0.5x$
$Mm = (42.0x - 4.0x^2)(0.5x) = 21.0x^2 - 2.0x^3$

$$\int_A^B \frac{Mm}{EI} dx = \int_0^4 \frac{21.0x^2 - 2.0x^3}{EI} dx = \left[\frac{21.0x^3}{3EI} - \frac{2.0x^4}{4EI}\right]_0^4 = +\frac{320.0}{EI} \text{ m}$$

Consider the section from C to B: $0 \leq x \leq 4.0$ m
$M = +42.0x - 8.0x^2/2 = 42.0x - 4.0x^2$ $m = +0.5x$
$Mm = (42.0x - 4.0x^2)(0.5x) = 21.0x^2 - 2.0x^3$

$$\int_C^B \frac{Mm}{EI} dx = \int_0^4 \frac{21.0x^2 - 2.0x^3}{2EI} dx = \left[\frac{21.0x^3}{6EI} - \frac{2.0x^4}{8EI}\right]_0^4 = +\frac{160.0}{EI} \text{ m}$$

$$\int_0^L \frac{Mm}{EI} dx = \left(+\frac{320.0}{EI} + \frac{160.0}{EI}\right) = \frac{480.0}{EI} = \frac{480.0}{131.4 \times 10^3} \text{ m} = 3.65 \text{ mm}$$

Consider the columns AE and CD:

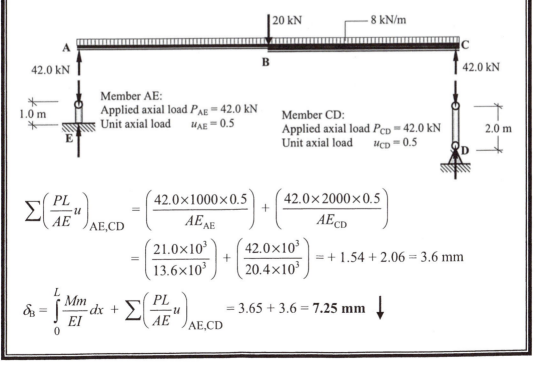

Member AE:
Applied axial load $P_{AE} = 42.0$ kN
Unit axial load $u_{AE} = 0.5$

Member CD:
Applied axial load $P_{CD} = 42.0$ kN
Unit axial load $u_{CD} = 0.5$

$$\sum\left(\frac{PL}{AE}u\right)_{AE,CD} = \left(\frac{42.0 \times 1000 \times 0.5}{AE_{AE}}\right) + \left(\frac{42.0 \times 2000 \times 0.5}{AE_{CD}}\right)$$

$$= \left(\frac{21.0 \times 10^3}{13.6 \times 10^3}\right) + \left(\frac{42.0 \times 10^3}{20.4 \times 10^3}\right) = +1.54 + 2.06 = 3.6 \text{ mm}$$

$$\delta_B = \int_0^L \frac{Mm}{EI} dx + \sum\left(\frac{PL}{AE}u\right)_{AE,CD} = 3.65 + 3.6 = \textbf{7.25 mm} \;\downarrow$$

Solution
Topic: Determinate Beams/Frames – Deflection Using Unit Load
Problem Number: 4.17   **Page No. 3**

Alternatively for the beam ABC:

$\delta_B = \Sigma(\text{Area }_{\text{applied bending moment diagram}} \times \text{Ordinate }_{\text{unit load bending moment diagram}})$

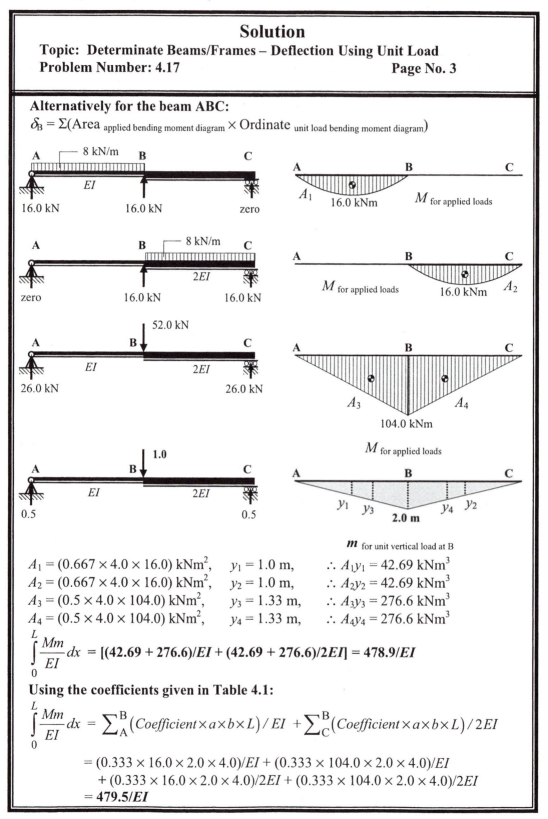

$A_1 = (0.667 \times 4.0 \times 16.0)\ \text{kNm}^2,\quad y_1 = 1.0\ \text{m},\quad \therefore A_1y_1 = 42.69\ \text{kNm}^3$

$A_2 = (0.667 \times 4.0 \times 16.0)\ \text{kNm}^2,\quad y_2 = 1.0\ \text{m},\quad \therefore A_2y_2 = 42.69\ \text{kNm}^3$

$A_3 = (0.5 \times 4.0 \times 104.0)\ \text{kNm}^2,\quad y_3 = 1.33\ \text{m},\quad \therefore A_3y_3 = 276.6\ \text{kNm}^3$

$A_4 = (0.5 \times 4.0 \times 104.0)\ \text{kNm}^2,\quad y_4 = 1.33\ \text{m},\quad \therefore A_4y_4 = 276.6\ \text{kNm}^3$

$$\int_0^L \frac{Mm}{EI}\,dx = [(42.69 + 276.6)/EI + (42.69 + 276.6)/2EI] = \mathbf{478.9}/\mathbf{EI}$$

Using the coefficients given in Table 4.1:

$$\int_0^L \frac{Mm}{EI}\,dx = \sum_A^B \left(Coefficient \times a \times b \times L\right)/EI + \sum_C^B \left(Coefficient \times a \times b \times L\right)/2EI$$

$$= (0.333 \times 16.0 \times 2.0 \times 4.0)/EI + (0.333 \times 104.0 \times 2.0 \times 4.0)/EI$$
$$+ (0.333 \times 16.0 \times 2.0 \times 4.0)/2EI + (0.333 \times 104.0 \times 2.0 \times 4.0)/2EI$$
$$= \mathbf{479.5}/\mathbf{EI}$$

Solution
Topic: Determinate Beams/Frames – Deflection Using Unit Load
Problem Number: 4.18   **Page No. 1**

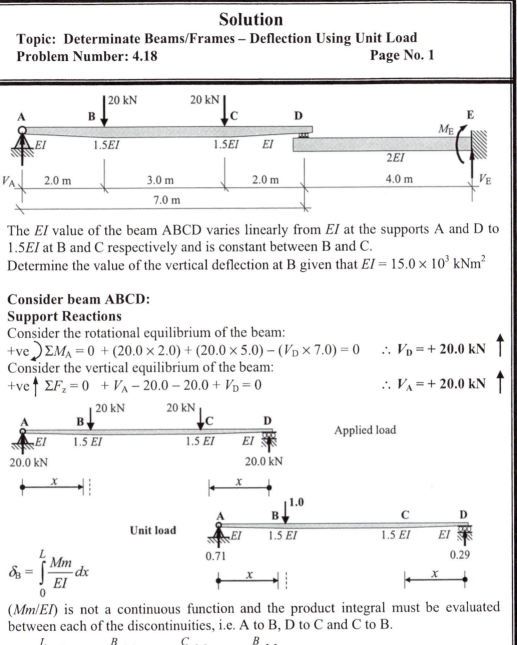

The *EI* value of the beam ABCD varies linearly from *EI* at the supports A and D to 1.5*EI* at B and C respectively and is constant between B and C.

Determine the value of the vertical deflection at B given that $EI = 15.0 \times 10^3 \text{ kNm}^2$

Consider beam ABCD:

Support Reactions

Consider the rotational equilibrium of the beam:

+ve $\curvearrowright \Sigma M_A = 0 \ + (20.0 \times 2.0) + (20.0 \times 5.0) - (V_D \times 7.0) = 0$ $\therefore V_D = + \textbf{20.0 kN} \uparrow$

Consider the vertical equilibrium of the beam:

+ve $\uparrow \Sigma F_z = 0 \ + V_A - 20.0 - 20.0 + V_D = 0$ $\therefore V_A = + \textbf{20.0 kN} \uparrow$

$$\delta_B = \int_0^L \frac{Mm}{EI} dx$$

(*Mm/EI*) is not a continuous function and the product integral must be evaluated between each of the discontinuities, i.e. A to B, D to C and C to B.

$$\delta_B = \int_0^L \frac{Mm}{EI} dx = \int_A^B \frac{Mm}{EI} dx + \int_D^C \frac{Mm}{EI} dx + \int_C^B \frac{Mm}{EI} dx$$

Consider the section from A to B: $\textbf{0} \le x \le \textbf{2.0 m}$

$M = + 20.0x$ $m = + 0.71x$ $\therefore Mm = 14.2x^2$

Also

The *EI* value varies linearly between A and B and at distance '*x*' from A is given by:

$EI (1 + 0.25x)$

Solution
Topic: Determinate Beams/Frames – Deflection Using Unit Load
Problem Number: 4.18   **Page No. 2**

$$\int_{A}^{B} \frac{Mm}{EI}dx = \int_{0}^{2} \frac{14.2x^2}{EI(1+0.25x)}dx$$

Let $v = (1 + 0.25x)$ $\therefore x = 4.0(v-1)$, $dx = 4.0dv$ and $x^2 = 16.0(v-1)^2$
when $x = 0$ $v = 1.0$ and when $x = 2$ $v = (1+0.5) = 1.5$

$Mm\ dx = 14.2x^2 = [14.2 \times 16.0(v-1)^2] \times 4.0dv = 908.8(v-1)^2\ dv$

$$= \int_{0}^{2} \frac{14.2x^2}{EI(1+0.25x)}dx = \frac{908.8}{EI}\int_{v=1.0}^{v=1.5} \frac{(v-1)^2}{v}dv = \frac{908.8}{EI}\int_{v=1.0}^{v=1.5} \frac{\left(v^2-2.0v+1.0\right)}{v}dv$$

$$= \frac{908.8}{EI}\int_{v=1.0}^{v=1.5}\left(v-2.0+\frac{1.0}{v}\right)dv = \frac{908.8}{EI}\left[\frac{v^2}{2}-2.0v+ln\,v\right]_{v=1.0}^{v=1.5}$$

$$= \frac{908.8}{EI}\left\{\left[\frac{1.5^2}{2}-(2.0\times1.5)+ln1.5\right]-\left[\frac{1.0^2}{2}-(2.0\times1.0)+ln1.0\right]\right\}$$

$$= +\frac{27.69}{EI}\,m$$

Consider the section from D to C: **$0 \le x \le 2.0$ m**
$M = + 20.0x$ $m = + 0.29x$ $\therefore Mm = 5.8x^2$
Also
The EI value varies linearly between D and C, and at distance x from A is given by:
$EI (1 + 0.25x)$

$$\int_{D}^{C} \frac{Mm}{EI}dx = \int_{0}^{2} \frac{5.8x^2}{EI(1+0.25x)}dx$$

Let $v = (1 + 0.25x)$ $\therefore x = 4.0(v-1)$, $dx = 4.0dv$ and $x^2 = 16.0(v-1)^2$
when $x = 0$ $v = 1.0$ and when $x = 2$ $v = (1+0.5) = 1.5$

$Mm\ dx = 5.8x^2 = [5.8 \times 16.0(v-1)^2] \times 4.0dv = 371.2(v-1)^2\ dv$

$$= \int_{0}^{2} \frac{14.2x^2}{EI(1+0.25x)}dx = \frac{371.2}{EI}\int_{v=1.0}^{v=1.5} \frac{(v-1)^2}{v}dv = \frac{371.2}{EI}\int_{v=1.0}^{v=1.5} \frac{\left(v^2-2.0v+1.0\right)}{v}dv$$

$$= \frac{371.2}{EI}\int_{v=1.0}^{v=1.5}\left(v-2.0+\frac{1.0}{v}\right)dv = \frac{371.2}{EI}\left[\frac{v^2}{2}-2.0v+ln\,v\right]_{v=1.0}^{v=1.5}$$

Solution
Topic: Determinate Beams/Frames – Deflection Using Unit Load
Problem Number: 4.18   **Page No. 3**

$$= \frac{371.2}{EI}\left\{\left[\frac{1.5^2}{2} - (2.0\times1.5) + ln1.5\right] - \left[\frac{1.0^2}{2} - (2.0\times1.0) + ln1.0\right]\right\}$$

$$= +\frac{11.31}{EI}\,\text{m}$$

Consider the section from C to B: **$2.0 \le x \le 5.0$ m**
$M = +20.0x - 20.0(x - 2.0) = 40.0 \qquad m = +0.29x \qquad \therefore Mm = 11.6x$

$$\int_{C}^{B}\frac{Mm}{1.5EI}\,dx \;=\; \int_{2}^{5}\frac{11.6x}{1.5EI}\,dx \;=\; \left[\frac{11.6x^2}{3.0EI}\right]_{2}^{5} \;=\; +\frac{81.2}{EI}\,\text{m}$$

Consider the cantilever beam DE:

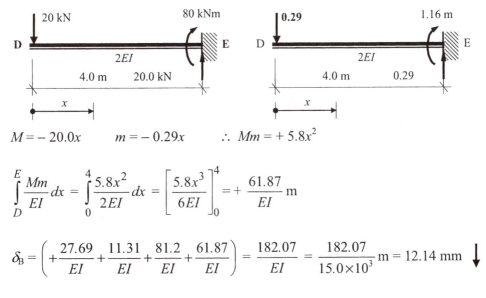

$M = -20.0x \qquad m = -0.29x \qquad \therefore Mm = +5.8x^2$

$$\int_{D}^{E}\frac{Mm}{EI}\,dx \;=\; \int_{0}^{4}\frac{5.8x^2}{2EI}\,dx \;=\; \left[\frac{5.8x^3}{6EI}\right]_{0}^{4} \;=\; +\frac{61.87}{EI}\,\text{m}$$

$$\delta_{B} = \left(+\frac{27.69}{EI} + \frac{11.31}{EI} + \frac{81.2}{EI} + \frac{61.87}{EI}\right) = \frac{182.07}{EI} = \frac{182.07}{15.0\times10^{3}}\,\text{m} = 12.14\,\text{mm}\;\downarrow$$

Alternatively:
Sections A to B and D to C must be carried out using the product integrals as above
The terms relating to the central section C to B and the cantilever beam D to E can
also be evaluated using the product (area × ordinate) or the Coefficients given in
Table 4.1 since the *EI* value is constant along these lengths.

The reader should carry out these calculations to confirm the results.

Solution
Topic: Determinate Beams/Frames – Deflection Using Unit Load
Problem Number: 4.19   **Page No. 1**

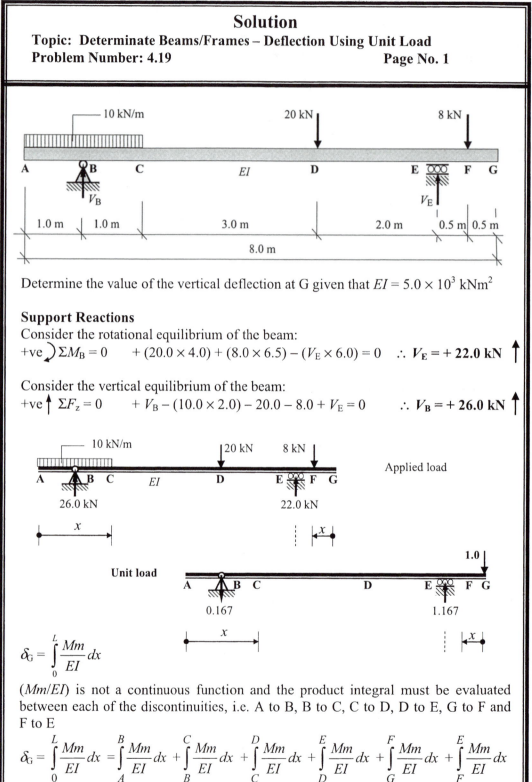

Determine the value of the vertical deflection at G given that $EI = 5.0 \times 10^3$ kNm^2

Support Reactions
Consider the rotational equilibrium of the beam:
$+ve \; \curvearrowright \Sigma M_B = 0 \qquad + (20.0 \times 4.0) + (8.0 \times 6.5) - (V_E \times 6.0) = 0 \quad \therefore \; V_E = + 22.0 \text{ kN} \uparrow$

Consider the vertical equilibrium of the beam:
$+ve \uparrow \; \Sigma F_z = 0 \qquad + V_B - (10.0 \times 2.0) - 20.0 - 8.0 + V_E = 0 \qquad \therefore \; V_B = + 26.0 \text{ kN} \uparrow$

$$\delta_G = \int_0^L \frac{Mm}{EI} dx$$

(*Mm*/*EI*) is not a continuous function and the product integral must be evaluated between each of the discontinuities, i.e. A to B, B to C, C to D, D to E, G to F and F to E

$$\delta_G = \int_0^L \frac{Mm}{EI} dx = \int_A^B \frac{Mm}{EI} dx + \int_B^C \frac{Mm}{EI} dx + \int_C^D \frac{Mm}{EI} dx + \int_D^E \frac{Mm}{EI} dx + \int_G^F \frac{Mm}{EI} dx + \int_F^E \frac{Mm}{EI} dx$$

Solution

Topic: Determinate Beams/Frames – Deflection Using Unit Load
Problem Number: 4.19   **Page No. 2**

Consider the section from A to B: **$0 \leq x \leq 1.0$ m**

$M = -10.0x^2/2 \qquad m = \text{zero} \qquad \therefore Mm = \text{zero}$

$$\int_A^B \frac{Mm}{EI}\,dx = \text{zero}$$

Consider the section from B to C: **$1.0 \leq x \leq 2.0$ m**

$M = -10.0x^2/2 + 26.0(x-1.0) = (-5.0x^2 + 26.0x - 26.0)$

$m = -0.167(x-1.0)$

$Mm = [(-5.0x^2 + 26.0x - 26.0)] \times [-0.167(x-1.0)]$

$\quad = (0.84x^3 - 5.18x^2 + 8.68x - 4.34)$

$$\int_B^C \frac{Mm}{EI}\,dx = \int_{1.0}^{2.0} \frac{0.84x^3 - 5.18x^2 + 8.68x - 4.34}{EI}\,dx$$

$$= \left[\frac{0.84x^4}{4EI} - \frac{5.18x^3}{3EI} + \frac{8.68x^2}{2EI} - \frac{4.34x}{EI} \right]_{1.0}^{2.0} = +\left[-\frac{1.77}{EI} - \left(-\frac{1.52}{EI} \right) \right] = -\frac{0.25}{EI}$$

Consider the section from C to D: **$2.0 \leq x \leq 5.0$ m**

$M = -(10.0 \times 2.0)(x - 1.0) + 26.0(x - 1.0) = +6.0(x-1.0) \qquad m = -0.167(x-1.0)$

$Mm = 6.0(x-1.0)(-0.167x + 0.167) = (-x^2 + 2.0x - 1.0)$

$$\int_C^D \frac{Mm}{EI}\,dx = \int_{2.0}^{5.0} \frac{-x^2 + 2.0x - 1.0}{EI}\,dx = \left[-\frac{x^3}{3EI} + \frac{2.0x^2}{2EI} - \frac{x}{EI} \right]_{2.0}^{5.0}$$

$$= \left(-\frac{21.67}{EI} - \frac{0.67}{EI} \right) = -\frac{22.34}{EI}$$

Consider the section from D to E: **$5.0 \leq x \leq 7.0$ m**

$M = -(10.0 \times 2.0)(x - 1.0) + 26.0(x - 1.0) - 20.0(x - 5.0) = (-14.0x + 94.0)$

$m = -0.167(x - 1.0)$

$Mm = (-14.0x + 94.0)(-0.167x + 0.167) = (2.34x^2 - 18.04x + 15.7)$

$$\int_D^E \frac{Mm}{EI}\,dx = \int_{5.0}^{7.0} \frac{2.34x^2 - 18.04x + 15.7}{EI}\,dx = \left[\frac{2.34x^3}{3EI} - \frac{18.04x^2}{2EI} + \frac{15.7x}{EI} \right]_{5.0}^{7.0}$$

$$= \left[-\frac{64.54}{EI} - \left(-\frac{49.5}{EI} \right) \right] = -\frac{15.04}{EI}$$

Solution
Topic: Determinate Beams/Frames – Deflection Using Unit Load
Problem Number: 4.19   Page No. 3

Consider the section from G to F: **$0 \le x \le 0.5$ m**

$M = $ zero $m = -x$ $\therefore Mm = $ zero

$\int_{G}^{F} \dfrac{Mm}{EI} dx = $ zero

Consider the section from F to E: **$0.5 \le x \le 1.0$ m**

$M = -8.0(x - 0.5)$ $m = -x$ $\therefore Mm = (8.0x^2 - 4.0x)$

$\int_{F}^{E} \dfrac{Mm}{EI} dx = \int_{0.5}^{1.0} \dfrac{8.0x^2 - 4.0x}{EI} dx = \left[\dfrac{8.0x^3}{3EI} - \dfrac{4.0x^2}{2EI} \right]_{0.5}^{1.0} = \left[+\dfrac{0.67}{EI} - \left(-\dfrac{0.17}{EI} \right) \right] = +\dfrac{0.84}{EI}$

$\delta_G = \left(-\dfrac{0.25}{EI} - \dfrac{22.34}{EI} - \dfrac{15.04}{EI} + \dfrac{0.84}{EI} \right) = -\dfrac{36.79}{EI} = -\dfrac{36.79}{5.0 \times 10^3}$ m $= -\,\textbf{7.36 mm}\uparrow$

Alternatively:

$\delta_G = \Sigma(\text{Area}_{\text{applied bending moment diagram}} \times \text{Ordinate}_{\text{unit load bending moment diagram}})$

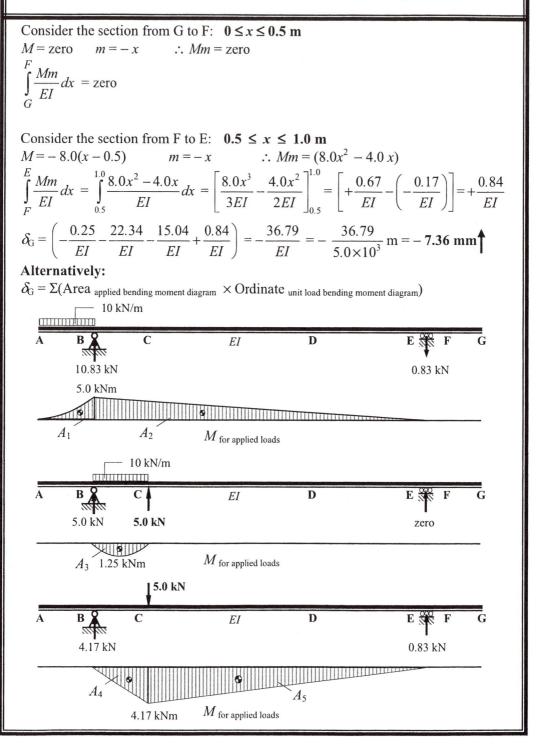

Solution

Topic: Determinate Beams/Frames – Deflection Using Unit Load

Problem Number: 4.19   **Page No. 4**

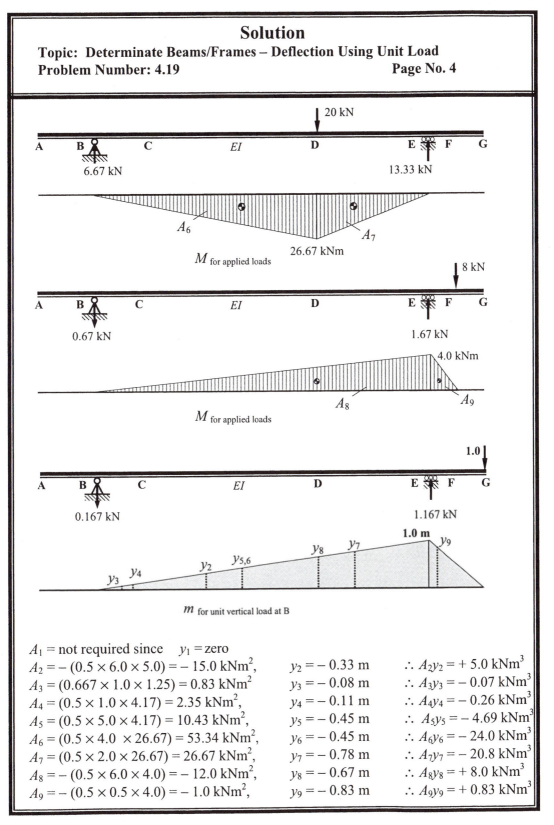

A_1 = not required since y_1 = zero

$A_2 = -(0.5 \times 6.0 \times 5.0) = -15.0$ kNm^2, $y_2 = -0.33$ m $\therefore A_2y_2 = +5.0$ kNm^3

$A_3 = (0.667 \times 1.0 \times 1.25) = 0.83$ kNm^2 $y_3 = -0.08$ m $\therefore A_3y_3 = -0.07$ kNm^3

$A_4 = (0.5 \times 1.0 \times 4.17) = 2.35$ kNm^2, $y_4 = -0.11$ m $\therefore A_4y_4 = -0.26$ kNm^3

$A_5 = (0.5 \times 5.0 \times 4.17) = 10.43$ kNm^2, $y_5 = -0.45$ m $\therefore A_5y_5 = -4.69$ kNm^3

$A_6 = (0.5 \times 4.0 \times 26.67) = 53.34$ kNm^2, $y_6 = -0.45$ m $\therefore A_6y_6 = -24.0$ kNm^3

$A_7 = (0.5 \times 2.0 \times 26.67) = 26.67$ kNm^2, $y_7 = -0.78$ m $\therefore A_7y_7 = -20.8$ kNm^3

$A_8 = -(0.5 \times 6.0 \times 4.0) = -12.0$ kNm^2, $y_8 = -0.67$ m $\therefore A_8y_8 = +8.0$ kNm^3

$A_9 = -(0.5 \times 0.5 \times 4.0) = -1.0$ kNm^2, $y_9 = -0.83$ m $\therefore A_9y_9 = +0.83$ kNm^3

Solution

Topic: Determinate Beams/Frames – Deflection Using Unit Load
Problem Number: 4.19   **Page No. 5**

$$\delta_G = \int_0^L \frac{Mm}{EI} dx = \Sigma(Ay)/EI$$

$$= (+\,5.0 - 0.07 - 0.26 - 4.69 - 24.0 - 20.8 + 8.0 + 0.83)/EI$$

$$\delta_G = -\,35.99/EI = -\,(35.99/5.0 \times 10^3)\text{ m} = -\,\textbf{7.20 mm} \uparrow$$

Using the coefficients given in Table 4.1: $\delta_G = \sum_0^L \left(Coefficient \times a \times b \times L\right)/EI$

Area A_2: $\int_0^L \dfrac{Mm}{EI} dx = +(0.167 \times 5.0 \times 1.0 \times 6.0)/EI = +5.0/EI$

Area A_3: $\int_0^L \dfrac{Mm}{EI} dx = -(0.333 \times 1.25 \times 0.167 \times 1.0)/EI = -0.07/EI$

Area A_4: $\int_0^L \dfrac{Mm}{EI} dx = -(0.333 \times 4.17 \times 0.167 \times 1.0)/EI = -0.23/EI$

Area A_5: $\int_0^L \dfrac{Mm}{EI} dx = -[(0.5 \times 4.17 \times 0.167 \times 5.0) + (0.167 \times 4.17 \times 0.83 \times 5.0)]/EI$

$$= -4.63/EI$$

Area A_6: $\int_0^L \dfrac{Mm}{EI} dx = -(0.333 \times 26.67 \times 0.67 \times 4.0)/EI = -23.80/EI$

Area A_7: $\int_0^L \dfrac{Mm}{EI} dx = -[(0.5 \times 26.67 \times 0.67 \times 2.0) + (0.167 \times 26.67 \times 0.33 \times 2.0)]/EI$

$$= -20.81/EI$$

Area A_8: $\int_0^L \dfrac{Mm}{EI} dx = +(0.333 \times 4.0 \times 1.0 \times 6.0)/EI = +8.0/EI$

Area A_9: $\int_0^L \dfrac{Mm}{EI} dx = +[(0.5 \times 4.0 \times 0.5 \times 0.5) + (0.333 \times 4.0 \times 0.5 \times 0.5)]/EI$

$$= +0.83/EI$$

$$\delta_G = (5.0 - 0.07 - 0.23 - 4.63 - 23.80 - 20.81 + 8.0 + 0.83)/EI = -35.71/EI$$
$$= -(35.71/5.0 \times 10^3)\text{ m} = -\,\textbf{7.14 mm}$$

Solution
Topic: Determinate Beams/Frames – Deflection Using Unit Load
Problem Number: 4.20   **Page No. 1**

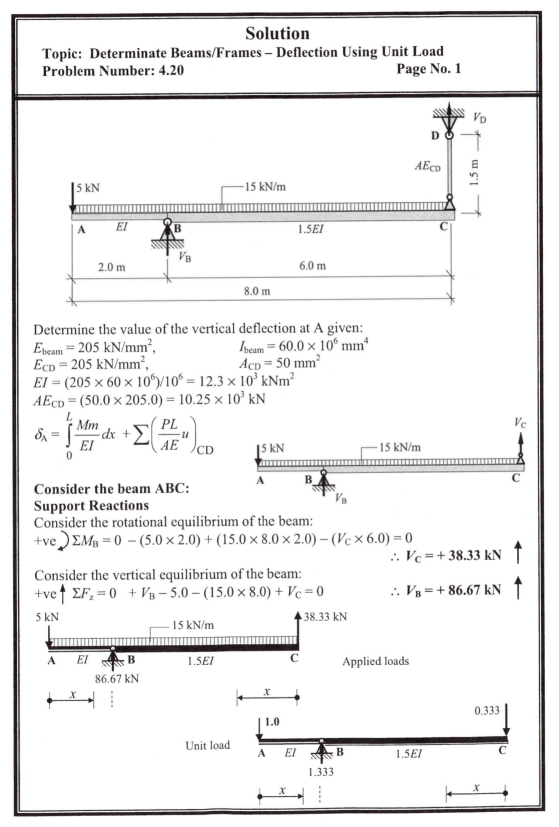

Determine the value of the vertical deflection at A given:

$E_{beam} = 205 \text{ kN/mm}^2$, $I_{beam} = 60.0 \times 10^6 \text{ mm}^4$

$E_{CD} = 205 \text{ kN/mm}^2$, $A_{CD} = 50 \text{ mm}^2$

$EI = (205 \times 60 \times 10^6)/10^6 = 12.3 \times 10^3 \text{ kNm}^2$

$AE_{CD} = (50.0 \times 205.0) = 10.25 \times 10^3 \text{ kN}$

$$\delta_A = \int_0^L \frac{Mm}{EI} dx + \sum \left(\frac{PL}{AE} u \right)_{CD}$$

Consider the beam ABC:
Support Reactions
Consider the rotational equilibrium of the beam:

+ve $\curvearrowright$ $\Sigma M_B = 0 \; -(5.0 \times 2.0) + (15.0 \times 8.0 \times 2.0) - (V_C \times 6.0) = 0$

$$\therefore V_C = +\textbf{38.33 kN} \uparrow$$

Consider the vertical equilibrium of the beam:

+ve $\uparrow$ $\Sigma F_z = 0 \; + V_B - 5.0 - (15.0 \times 8.0) + V_C = 0$ $\therefore V_B = +\textbf{86.67 kN} \uparrow$

Applied loads

Unit load

Solution
Topic: **Determinate Beams/Frames – Deflection Using Unit Load**
Problem Number: 4.20   Page No. 2

(Mm/EI) is not a continuous function and the product integral must be evaluated between each of the discontinuities, i.e. A to B and C to B.

$$\int_0^L \frac{Mm}{EI}\,dx = \int_A^B \frac{Mm}{EI}\,dx + \int_C^B \frac{Mm}{1.5EI}\,dx$$

Consider the section from A to B: **$0 \leq x \leq 2.0$ m**
$M = -5.0x - 15.0x^2/2 = -5.0x - 7.5x^2$ $\qquad$ $m = -x$
$Mm = (-5.0x - 7.5x^2)(x) = 5.0x^2 + 7.5x^3$

$$\int_A^B \frac{Mm}{EI}\,dx = \int_0^2 \frac{5.0x^2 + 7.5x^3}{EI}\,dx = \left[\frac{5.0x^3}{3EI} + \frac{7.5x^4}{4EI}\right]_0^2 = +\frac{43.33}{EI}\ \text{m}$$

Consider the section from C to B: **$0 \leq x \leq 6.0$ m**
$M = +38.33x - 15.0x^2/2 = +38.33x - 7.5x^2$ $\qquad$ $m = -0.333x$
$Mm = -(38.33x - 7.5x^2)(0.333x) = -12.77x^2 + 2.5x^3$

$$\int_C^B \frac{Mm}{EI}\,dx = \int_0^6 \frac{-12.77x^2 + 2.5x^3}{1.5EI}\,dx = \left[-\frac{12.77x^3}{4.5EI} + \frac{2.5x^4}{6EI}\right]_0^6 = -\frac{72.96}{EI}\ \text{m}$$

$$\int_0^L \frac{Mm}{EI}\,dx = \left(+\frac{43.33}{EI} - \frac{72.96}{EI}\right) = -\frac{29.63}{EI} = -\frac{29.63}{12.3\times10^3}\ \text{m} = -2.41\ \text{mm}$$

Consider member CD:
Applied axial load $\qquad P_{CD} = +58.33$ kN $\qquad$ (tension)

Unit axial load $\qquad\qquad u_{CD} = -0.333$ $\qquad\qquad$ (compression)

$$\sum\left(\frac{PL}{AE}u\right)_{CD} = -\left(\frac{38.33\times1500\times0.333}{AE_{CD}}\right) = -\left(\frac{19.146\times10^3}{10.25\times10^3}\right)\text{m} = -1.87\ \text{mm}$$

$$\delta_A = \int_0^L \frac{Mm}{EI}\,dx + \sum\left(\frac{PL}{AE}u\right)_{CD} = -2.41 - 1.87 = -\textbf{4.28 mm} \qquad \uparrow$$

Solution

Topic: Determinate Beams/Frames – Deflection Using Unit Load

Problem Number: 4.20   **Page No. 3**

Alternatively:

$$\delta_A = \Sigma(\text{Area}_{\text{applied bending moment diagram}} \times \text{Ordinate}_{\text{unit load bending moment diagram}})$$

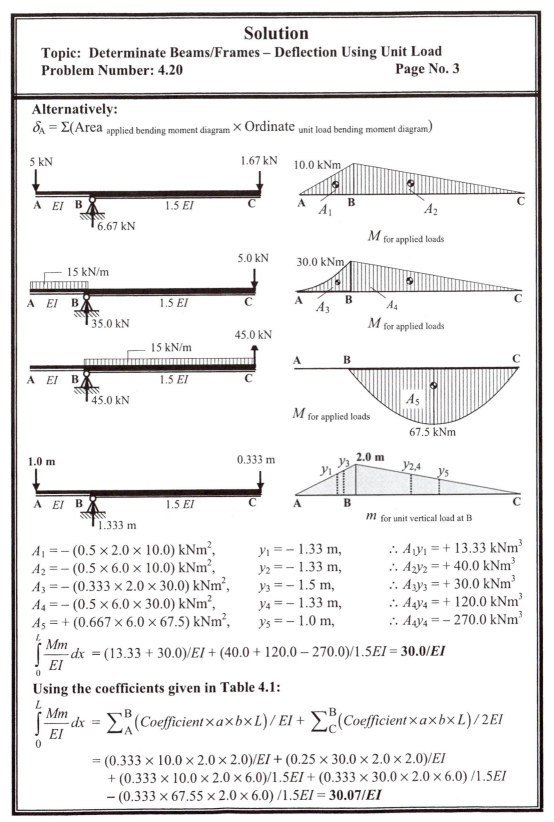

$A_1 = -(0.5 \times 2.0 \times 10.0)$ kNm^2, $y_1 = -1.33$ m, $\therefore A_1 y_1 = +13.33$ kNm^3

$A_2 = -(0.5 \times 6.0 \times 10.0)$ kNm^2, $y_2 = -1.33$ m, $\therefore A_2 y_2 = +40.0$ kNm^3

$A_3 = -(0.333 \times 2.0 \times 30.0)$ kNm^2, $y_3 = -1.5$ m, $\therefore A_3 y_3 = +30.0$ kNm^3

$A_4 = -(0.5 \times 6.0 \times 30.0)$ kNm^2, $y_4 = -1.33$ m, $\therefore A_4 y_4 = +120.0$ kNm^3

$A_5 = +(0.667 \times 6.0 \times 67.5)$ kNm^2, $y_5 = -1.0$ m, $\therefore A_4 y_4 = -270.0$ kNm^3

$$\int_0^L \frac{Mm}{EI}\,dx = (13.33 + 30.0)/EI + (40.0 + 120.0 - 270.0)/1.5EI = \mathbf{30.0}/\boldsymbol{EI}$$

Using the coefficients given in Table 4.1:

$$\int_0^L \frac{Mm}{EI}\,dx = \sum_A^B \left(Coefficient \times a \times b \times L\right)/EI + \sum_C^B \left(Coefficient \times a \times b \times L\right)/2EI$$

$$= (0.333 \times 10.0 \times 2.0 \times 2.0)/EI + (0.25 \times 30.0 \times 2.0 \times 2.0)/EI$$
$$+ (0.333 \times 10.0 \times 2.0 \times 6.0)/1.5EI + (0.333 \times 30.0 \times 2.0 \times 6.0)/1.5EI$$
$$- (0.333 \times 67.55 \times 2.0 \times 6.0)/1.5EI = \mathbf{30.07}/\boldsymbol{EI}$$

Solution
Topic: Determinate Beams/Frames – Deflection Using Unit Load
Problem Number: 4.21   **Page No. 1**

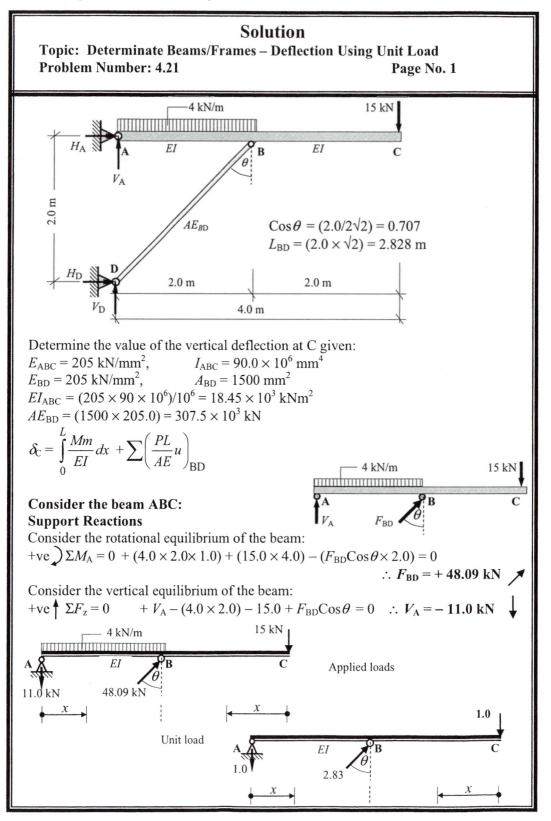

Determine the value of the vertical deflection at C given:
$E_{ABC} = 205$ kN/mm^2, $I_{ABC} = 90.0 \times 10^6$ mm^4
$E_{BD} = 205$ kN/mm^2, $A_{BD} = 1500$ mm^2
$EI_{ABC} = (205 \times 90 \times 10^6)/10^6 = 18.45 \times 10^3$ kNm^2
$AE_{BD} = (1500 \times 205.0) = 307.5 \times 10^3$ kN

$$\delta_C = \int_0^L \frac{Mm}{EI}\,dx + \sum\left(\frac{PL}{AE}u\right)_{BD}$$

Consider the beam ABC:
Support Reactions
Consider the rotational equilibrium of the beam:
$+ve \,\bigcirc\!\!\curvearrowright \Sigma M_A = 0 \ + (4.0 \times 2.0 \times 1.0) + (15.0 \times 4.0) - (F_{BD}\cos\theta \times 2.0) = 0$
$$\therefore F_{BD} = +\,48.09 \text{ kN} \ \nearrow$$

Consider the vertical equilibrium of the beam:
$+ve \uparrow \Sigma F_z = 0 \quad + V_A - (4.0 \times 2.0) - 15.0 + F_{BD}\cos\theta = 0 \quad \therefore V_A = -\,11.0 \text{ kN} \ \downarrow$

Solution

Topic: Determinate Beams/Frames – Deflection Using Unit Load
Problem Number: 4.21   **Page No. 2**

(Mm/EI) is not a continuous function and the product integral must be evaluated between each of the discontinuities, i.e. A to B and C to B.

$$\int_0^L \frac{Mm}{EI}\,dx = \int_A^B \frac{Mm}{EI}\,dx + \int_C^B \frac{Mm}{EI}\,dx$$

Consider the section from A to B: **$0 \le x \le 2.0$ m**

$M = -11.0x - 4.0x^2/2 = -11.0x - 2.0x^2$ $\qquad\qquad m = -x$

$Mm = -(-11.0x - 2.0x^2)(x) = 11.0x^2 + 2.0x^3$

$$\int_A^B \frac{Mm}{EI}\,dx = \int_0^2 \frac{11.0x^2 + 2.0x^3}{EI}\,dx = \left[\frac{11.0x^3}{3EI} + \frac{2.0x^4}{4EI}\right]_0^2 = +\frac{37.33}{EI}\,\text{m}$$

Consider the section from C to B: **$0 \le x \le 2.0$ m**

$M = -15.0x \qquad\qquad m = -x \qquad \therefore Mm = +15.0x^2$

$$\int_C^B \frac{Mm}{EI}\,dx = \int_0^2 \frac{15.0x^2}{EI}\,dx = \left[\frac{15.0x^3}{3.0EI}\right]_0^2 = +\frac{40.0}{EI}\,\text{m}$$

$$\int_0^L \frac{Mm}{EI}\,dx = \left(\frac{37.33}{EI} + \frac{40.0}{EI}\right) = +\frac{77.33}{EI} = +\frac{77.33}{18.45\times10^3}\,\text{m} = +4.19\,\text{mm}$$

Consider member BD:

Applied axial load $P_{BD} = -48.09$ kN   (compression)

Unit axial load $u_{BD} = -2.836$ (compression)

$$\sum\left(\frac{PL}{AE}u\right)_{BD} = +\left(\frac{48.09\times2828\times2.83}{AE_{BD}}\right) = +\left(\frac{384.88\times10^3}{307.5\times10^3}\right)\,\text{m} = +1.25\,\text{mm}$$

$$\delta_C = \int_0^L \frac{Mm}{EI}\,dx + \sum\left(\frac{PL}{AE}u\right)_{BD} = +4.19 + 1.25 = +5.44\,\text{mm}\;\downarrow$$

Solution
Topic: Determinate Beams/Frames – Deflection Using Unit Load
Problem Number: 4.21   **Page No. 3**

Alternatively:

$\delta_C = \Sigma(\text{Area}_{\text{applied bending moment diagram}} \times \text{Ordinate}_{\text{unit load bending moment diagram}})$

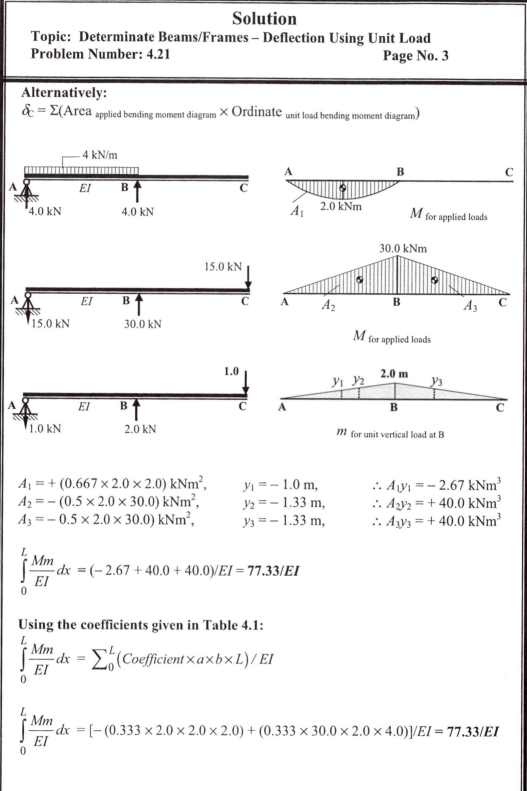

$A_1 = + (0.667 \times 2.0 \times 2.0) \text{ kNm}^2,$ $y_1 = -1.0 \text{ m},$ $\therefore A_1 y_1 = -2.67 \text{ kNm}^3$
$A_2 = -(0.5 \times 2.0 \times 30.0) \text{ kNm}^2,$ $y_2 = -1.33 \text{ m},$ $\therefore A_2 y_2 = +40.0 \text{ kNm}^3$
$A_3 = -0.5 \times 2.0 \times 30.0) \text{ kNm}^2,$ $y_3 = -1.33 \text{ m},$ $\therefore A_3 y_3 = +40.0 \text{ kNm}^3$

$$\int_0^L \frac{Mm}{EI}\,dx = (-2.67 + 40.0 + 40.0)/EI = \mathbf{77.33/\mathit{EI}}$$

Using the coefficients given in Table 4.1:

$$\int_0^L \frac{Mm}{EI}\,dx = \sum_0^L \left(\text{Coefficient} \times a \times b \times L\right)/EI$$

$$\int_0^L \frac{Mm}{EI}\,dx = [-(0.333 \times 2.0 \times 2.0 \times 2.0) + (0.333 \times 30.0 \times 2.0 \times 4.0)]/EI = \mathbf{77.33/\mathit{EI}}$$

Solution
Topic: Determinate Beams/Frames – Deflection Using Unit Load
Problem Number: 4.22   **Page No. 1**

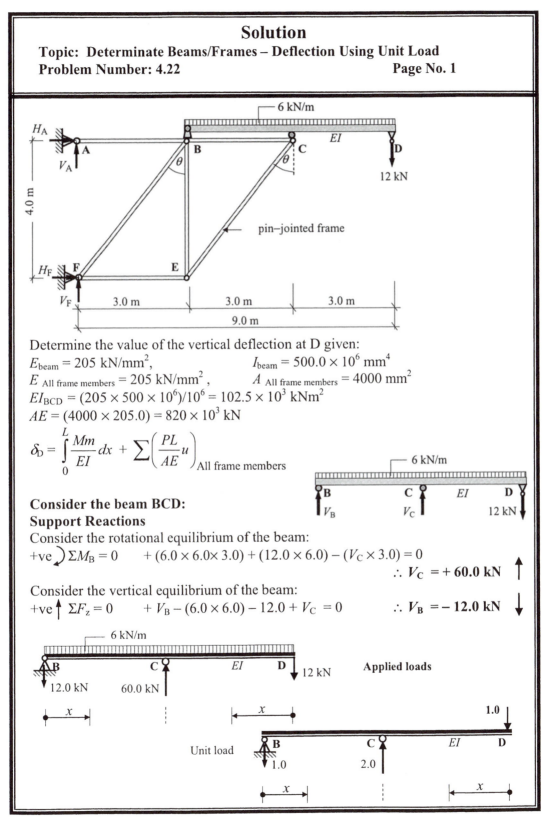

Determine the value of the vertical deflection at D given:

$E_{beam} = 205$ kN/mm^2, $I_{beam} = 500.0 \times 10^6$ mm^4

$E_{\text{All frame members}} = 205$ kN/mm^2 , $A_{\text{All frame members}} = 4000$ mm^2

$EI_{BCD} = (205 \times 500 \times 10^6)/10^6 = 102.5 \times 10^3$ kNm^2

$AE = (4000 \times 205.0) = 820 \times 10^3$ kN

$$\delta_D = \int_0^L \frac{Mm}{EI} dx + \sum \left(\frac{PL}{AE} u \right)_{\text{All frame members}}$$

Consider the beam BCD:
Support Reactions
Consider the rotational equilibrium of the beam:

+ve $\circlearrowright \Sigma M_B = 0$ $+ (6.0 \times 6.0 \times 3.0) + (12.0 \times 6.0) - (V_C \times 3.0) = 0$

$\therefore V_C = + 60.0$ kN $\uparrow$

Consider the vertical equilibrium of the beam:

+ve $\uparrow \Sigma F_z = 0$ $+ V_B - (6.0 \times 6.0) - 12.0 + V_C = 0$ $\therefore V_B = - 12.0$ kN $\downarrow$

Solution
Topic: Determinate Beams/Frames – Deflection Using Unit Load
Problem Number: 4.22   **Page No. 2**

(Mm/EI) is not a continuous function and the product integral must be evaluated between each of the discontinuities, i.e. B to C and D to C.

$$\int_0^L \frac{Mm}{EI}\,dx = \int_B^C \frac{Mm}{EI}\,dx + \int_D^C \frac{Mm}{EI}\,dx$$

Consider the section from B to C: **$0 \le x \le 3.0$ m**

$M = -12.0x - 6.0x^2/2 = -12.0x - 3.0x^2$ $m = -x$

$Mm = -(-12.0x - 3.0x^2)(x) = 12.0x^2 + 3.0x^3$

$$\int_B^C \frac{Mm}{EI}\,dx = \int_0^3 \frac{12.0x^2 + 3.0x^3}{EI}\,dx = \left[\frac{12.0x^3}{3EI} + \frac{3.0x^4}{4EI}\right]_0^3 = +\frac{168.75}{EI}\ \text{m}$$

Consider the section from D to C: **$0 \le x \le 3.0$ m**

$M = -12.0x - 6.0x^2/2 = -12.0x - 3.0x^2$ $m = -x$

$$\int_D^C \frac{Mm}{EI}\,dx = \int_0^3 \frac{12.0x^2 + 3.0x^3}{EI}\,dx = \left[\frac{12.0x^3}{3EI} + \frac{3.0x^4}{4EI}\right]_0^3 = +\frac{168.75}{EI}\ \text{m}$$

$$\int_0^L \frac{Mm}{EI}\,dx = \left(\frac{168.75}{EI} + \frac{168.75}{EI}\right) = +\frac{337.5}{EI} = +\frac{337.5}{102.5 \times 10^3}\ m = +3.29\ \text{mm}$$

Consider the pin–jointed frame:
The applied load axial effects (P–forces)and the unit load axial effects (u–forces) can be determined using joint resolution and/or the method of sections as indicated in Chapter 3.

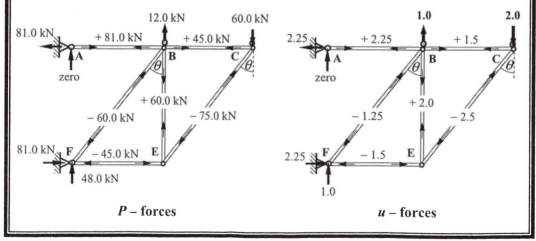

P – forces   **u – forces**

Solution

Topic: Determinate Beams/Frames – Deflection Using Unit Load
Problem Number: 4.22   **Page No. 3**

Member	Length (mm)	AE (kN)	P-force (kN)	PL/AE (mm)	u	$(PL/AE) \times u$
AB	3000	820.0×10^3	+ 81.0	+ 0.30	+ 2.25	+ 0.68
BC	3000	820.0×10^3	+ 45.0	+ 0.16	+ 1.50	+ 0.24
BF	5000	820.0×10^3	− 60.0	− 0.37	− 1.25	+ 0.46
BE	4000	820.0×10^3	+ 60.0	+ 0.29	+ 2.00	+ 0.58
CE	5000	820.0×10^3	− 75.0	− 0.46	− 2.50	+ 1.15
FE	3000	820.0×10^3	− 45.0	− 0.16	− 1.50	+ 0.24
						$\Sigma = + 3.35$

$$\sum \left(\frac{PL}{AE} u \right)_{\text{All frame members}} = 3.35 \text{ mm}$$

$$\delta_D = \int_0^L \frac{Mm}{EI} dx + \sum \left(\frac{PL}{AE} u \right)_{\text{All frame members}} = +3.29 + 3.35 = +6.64 \text{ mm} \quad \downarrow$$

Alternatively:

$$\delta_D = \Sigma (\text{Area}_{\text{applied bending moment diagram}} \times \text{Ordinate}_{\text{unit load bending moment diagram}})$$

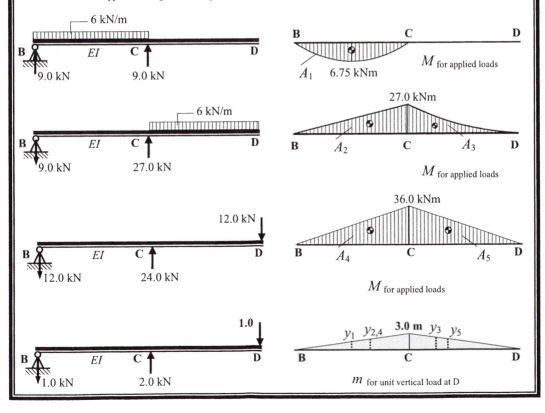

Solution

Topic: Determinate Beams/Frames – Deflection Using Unit Load

Problem Number: 4.22   **Page No. 4**

$A_1 = + (0.667 \times 3.0 \times 6.75) \text{ kNm}^2,$ $y_1 = -1.5 \text{ m},$ $\therefore A_1 y_1 = -20.26 \text{ kNm}^3$

$A_2 = - (0.5 \times 3.0 \times 27.0) \text{ kNm}^2,$ $y_2 = -2.0 \text{ m},$ $\therefore A_2 y_2 = +81.0 \text{ kNm}^3$

$A_3 = - (0.333 \times 3.0 \times 27.0) \text{ kNm}^2,$ $y_3 = -2.25 \text{ m},$ $\therefore A_3 y_3 = +60.69 \text{ kNm}^3$

$A_4 = - (0.5 \times 3.0 \times 36.0) \text{ kNm}^2,$ $y_4 = -2.0 \text{ m},$ $\therefore A_4 y_4 = +108.0 \text{ kNm}^3$

$A_5 = - (0.5 \times 3.0 \times 36.0) \text{ kNm}^2,$ $y_5 = -2.0 \text{ m},$ $\therefore A_5 y_5 = +108.0 \text{ kNm}^3$

$$\int_0^L \frac{Mm}{EI}\,dx = (-20.26 + 81.0 + 60.69 + 108.0 + 108.0)/EI = \mathbf{337.43}/\boldsymbol{EI}$$

Using the coefficients given in Table 4.1:

$$\int_0^L \frac{Mm}{EI}\,dx = \sum_0^L \left(Coefficient \times a \times b \times L \right)/EI$$

$$\int_0^L \frac{Mm}{EI}\,dx = [-(0.333 \times 6.75 \times 3.0 \times 3.0) + (0.333 \times 27.0 \times 3.0 \times 3.0)$$

$$+ (0.25 \times 27.0 \times 3.0 \times 3.0) + (0.333 \times 36.0 \times 3.0 \times 6.0)\,]/EI = \mathbf{337.22}/\boldsymbol{EI}$$

Solution

Topic: Determinate Beams/Frames – Deflection Using Unit Load
Problem Number: 4.23   **Page No. 1**

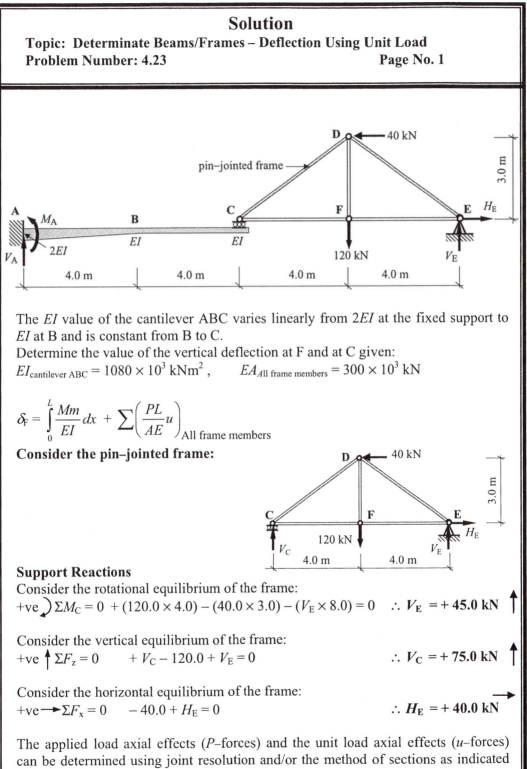

The *EI* value of the cantilever ABC varies linearly from 2*EI* at the fixed support to
EI at B and is constant from B to C.
Determine the value of the vertical deflection at F and at C given:

$$EI_{\text{cantilever ABC}} = 1080 \times 10^3 \text{ kNm}^2 , \qquad EA_{\text{All frame members}} = 300 \times 10^3 \text{ kN}$$

$$\delta_F = \int_0^L \frac{Mm}{EI} dx \; + \; \sum \left(\frac{PL}{AE} u \right)_{\text{All frame members}}$$

Consider the pin–jointed frame:

Support Reactions
Consider the rotational equilibrium of the frame:

$$+ve \; \curvearrowright \Sigma M_C = 0 \; + (120.0 \times 4.0) - (40.0 \times 3.0) - (V_E \times 8.0) = 0 \quad \therefore \; V_E = +45.0 \text{ kN} \uparrow$$

Consider the vertical equilibrium of the frame:

$$+ve \uparrow \Sigma F_z = 0 \qquad + V_C - 120.0 + V_E = 0 \qquad\qquad \therefore \; V_C = +75.0 \text{ kN} \uparrow$$

Consider the horizontal equilibrium of the frame:

$$+ve \longrightarrow \Sigma F_x = 0 \qquad -40.0 + H_E = 0 \qquad\qquad \therefore \; H_E = +40.0 \text{ kN} \longrightarrow$$

The applied load axial effects (*P*–forces) and the unit load axial effects (*u*–forces)
can be determined using joint resolution and/or the method of sections as indicated
in Chapter 3.

Solution

Topic: Determinate Beams/Frames – Deflection Using Unit Load

Problem Number: 4.23   **Page No. 2**

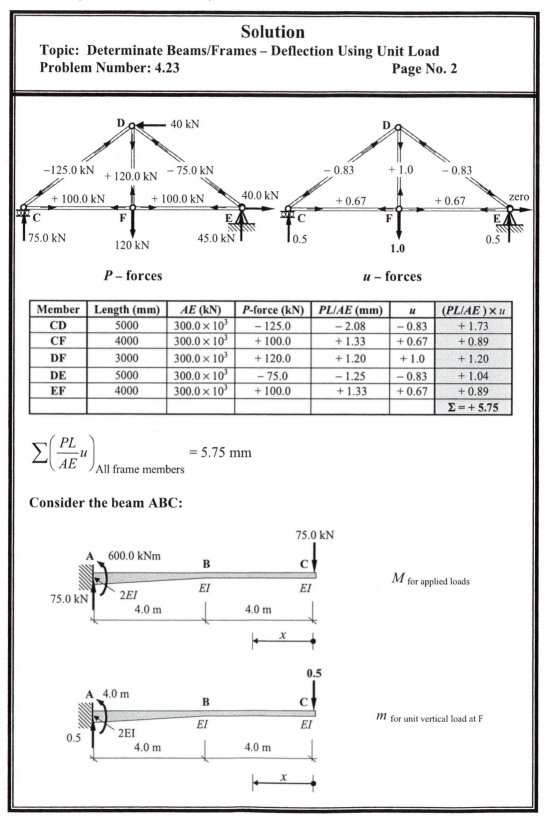

P – forces   ***u* – forces**

Member	Length (mm)	AE (kN)	P-force (kN)	PL/AE (mm)	u	$(PL/AE) \times u$
CD	5000	300.0×10^3	-125.0	-2.08	-0.83	$+1.73$
CF	4000	300.0×10^3	$+100.0$	$+1.33$	$+0.67$	$+0.89$
DF	3000	300.0×10^3	$+120.0$	$+1.20$	$+1.0$	$+1.20$
DE	5000	300.0×10^3	-75.0	-1.25	-0.83	$+1.04$
EF	4000	300.0×10^3	$+100.0$	$+1.33$	$+0.67$	$+0.89$
						$\Sigma = +5.75$

$$\sum \left(\frac{PL}{AE} u \right)_{\text{All frame members}} = 5.75 \text{ mm}$$

Consider the beam ABC:

M for applied loads

m for unit vertical load at F

Solution
Topic: **Determinate Beams/Frames – Deflection Using Unit Load**
Problem Number: 4.23   **Page No. 3**

(Mm/EI) is not a continuous function and the product integral must be evaluated between each of the discontinuities, i.e. C to B and B to A.

The value of EI at position 'x' along the beam between B and A is given by:
$EI + EI [(x - 4.0)/4] = 0.25EI\,x$

$$\int_0^L \frac{Mm}{EI}\,dx = \int_C^B \frac{Mm}{EI}\,dx + \int_B^A \frac{Mm}{0.25EIx}\,dx = \int_C^B \frac{Mm}{EI}\,dx + \frac{4.0}{EI}\int_B^A \frac{Mm}{x}\,dx$$

Consider the section from C to B: **$0 \leq x \leq 4.0$ m**
$$M = -75.0x \qquad\qquad m = -0.5x \qquad\qquad \therefore\; Mm = +37.5x^2$$

$$\int_C^B \frac{Mm}{EI}\,dx = \int_0^4 \frac{37.5x^2}{EI}\,dx = \left[\frac{37.5x^3}{3EI}\right]_0^4 = +\frac{800}{EI}\text{ m}$$

Consider the section from B to A: **$4.0 \leq x \leq 8.0$ m**
$$M = -75.0x \qquad\qquad m = -0.5x \qquad\qquad \therefore\; Mm = +37.5x^2$$

$$\frac{4.0}{EI}\int_B^A \frac{Mm}{x}\,dx = \frac{4.0}{EI}\int_{4.0}^{8.0} \frac{37.5x^2}{x}\,dx = \frac{150.0}{EI}\int_{4.0}^{8.0} x\,dx = \left[\frac{150.0x^2}{2EI}\right]_4^8 = +\frac{3600}{EI}\text{ m}$$

$$\int_0^L \frac{Mm}{EI}\,dx = \frac{800}{EI} + \frac{3600}{EI} = \frac{4400}{EI} = \frac{4400}{1080\times10^3}\text{ m} = +4.07\text{ mm}$$

$$\delta_F = \int_0^L \frac{Mm}{EI}\,dx + \sum\left(\frac{PL}{AE}u\right)_{\text{All frame members}} \qquad = +4.07 + 5.75 = +9.82\text{ mm} \quad \downarrow$$

Vertical deflection at C:
In this case when a unit load is applied at point C all of the u–forces for the pin–jointed frame are equal to zero.

$$\delta_C = \int_0^L \frac{Mm}{EI}\,dx + \sum\left(\frac{PL}{AE}u\right)_{\text{All frame members}}^{\text{zero}} \qquad \therefore\; \delta_C = \int_0^L \frac{Mm}{EI}\,dx$$

$$M = -75.0x \qquad\qquad m = -x \qquad\qquad \therefore\; Mm = +75.0x^2$$

$$\int_0^L \frac{Mm}{EI}\,dx = \int_0^4 \frac{75.0x^2}{EI}\,dx + \frac{4.0}{EI}\int_{4.0}^{8.0} \frac{75.0x^2}{x}\,dx = \frac{1600}{EI} + \frac{7200}{EI} = \frac{8800}{EI}$$

$$\delta_C = \frac{8800}{1080\times10^3}\text{ m} = +8.15\text{ mm} \quad \downarrow$$

4.6   *Statically Indeterminate Beams*

In many instances multi-span beams are used in design, and consequently it is necessary to consider the effects of the continuity on the support reactions and member forces. Such structures are ***indeterminate*** (see Chapter 1) and there are more unknown variables than can be solved using only the three equations of equilibrium. A few examples of such beams are shown in Figure 4.57 (a) to (d).

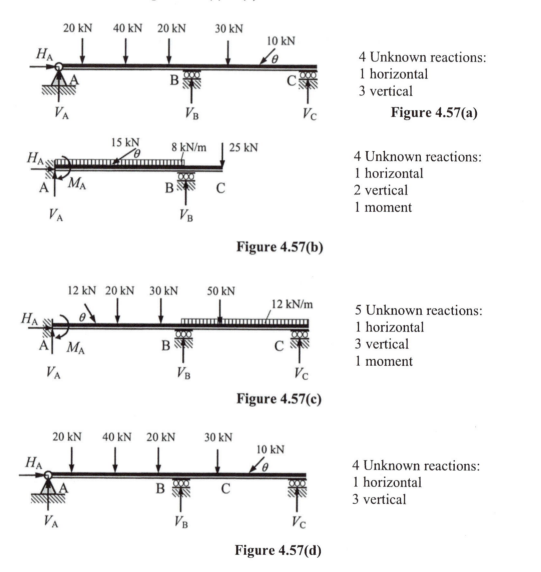

4 Unknown reactions:
1 horizontal
3 vertical

Figure 4.57(a)

4 Unknown reactions:
1 horizontal
2 vertical
1 moment

Figure 4.57(b)

5 Unknown reactions:
1 horizontal
3 vertical
1 moment

Figure 4.57(c)

4 Unknown reactions:
1 horizontal
3 vertical

Figure 4.57(d)

A number of analysis methods are available for determining the support reactions, and member forces in indeterminate beams. In the case of singly–redundant beams the '*unit–load method*' can be conveniently used to analyse the structure. In multi–redundant structures the method of '*moment distribution*' is a particularly useful hand–method of analysis. These methods are considered in Sections 4.6.1 and 4.6.2 respectively.

4.6.1 Unit Load Method for Singly–Redundant Beams

Using the method of analysis illustrated in Section 4.5 and considering the compatibility of displacements, member forces in singly–redundant beams can be determined as shown in Example 4.17 and Example 4.18 and in Problems 4.24 to 4.27.

4.6.2 Example 4.17: Singly–Redundant Beam 1

A propped cantilever ABC is fixed at A, supported on a roller at C and carries a mid–span point load of 15 kN as shown in Figure 4.58,

   (i)   determine the value of the support reactions and
  (ii)   sketch the shear force and bending moment diagram.

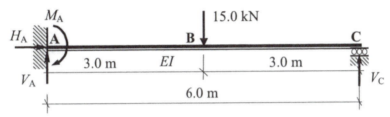

E and *I* are constant.

Figure 4.58

The degree–of–indeterminacy $I_D = [(3m + r)] - 3n = [(3 \times 1) + 4] - (3 \times 2) = 1$

Assume that the reaction at C is the redundant reaction and consider the original beam to be the superposition of two beams as indicated in Figures 4.59(a) and (b). The beam in Figure 4.59(b) can be represented as shown in Figure 4.60.   **(Note:** H_A = zero)

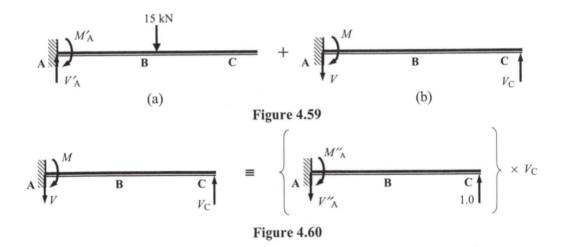

Figure 4.59

Figure 4.60

To maintain compatibility at the roller support, i.e. no resultant vertical displacement, the deformation of point C in Figure 4.59(a) must be equal and opposite to that in Figure 4.59(b) as shown in Figure 4.61.

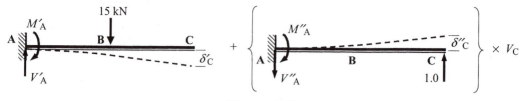

Figure 4.61

(δ' due to the applied load) + (δ'' due to the unit load) × V_C = 0

i.e. $\displaystyle\int_0^L \frac{Mm}{EI}dx + \left\{\int_0^L \frac{mm}{EI}dx\right\} \times V_C = 0$ ∴ $V_C = -\displaystyle\int_0^L \frac{Mm}{EI}dx \bigg/ \int_0^L \frac{m^2}{EI}dx$

The product integrals can be evaluated as before in Section 4.5, e.g. using the coefficients in Table 4.1.

Solution:
The bending moment diagrams for the applied loads and a unit point load at B are shown in Figure 4.62.

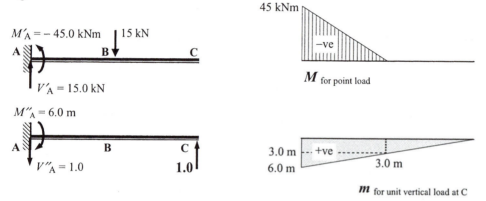

Figure 4.62

Using the coefficients given in Table 4.1:

$\displaystyle\delta'_{C,\ \text{point load}} = \int_0^L \frac{Mm}{EI}dx$

$\delta'_{C,\ \text{point load}} = [(0.5 \times 45.0 \times 3.0 \times 3.0) + (0.333 \times 45.0 \times 3.0 \times 3.0)]/EI = -337.5/EI$

$\displaystyle\delta''_{C,\ \text{unit load}} = \int_0^L \frac{m^2}{EI}dx$

$\delta''_{C,\ \text{unit load}} = (0.333 \times 6.0 \times 6.0 \times 6.0)/EI = +71.93/EI$

$V_C = -\displaystyle\int_0^L \frac{Mm}{EI}dx \bigg/ \int_0^L \frac{m^2}{EI}dx = -(-337.5/EI)/(71.93/EI) = 4.69 \text{ kN}$

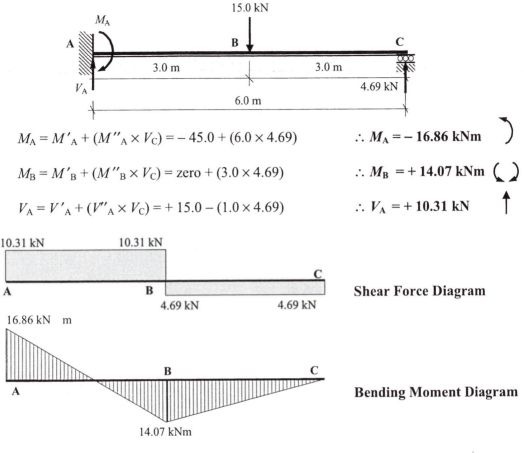

$$M_A = M'_A + (M''_A \times V_C) = -45.0 + (6.0 \times 4.69) \qquad \therefore M_A = -16.86 \text{ kNm}$$

$$M_B = M'_B + (M''_B \times V_C) = \text{zero} + (3.0 \times 4.69) \qquad \therefore M_B = +14.07 \text{ kNm}$$

$$V_A = V'_A + (V''_A \times V_C) = +15.0 - (1.0 \times 4.69) \qquad \therefore V_A = +10.31 \text{ kN}$$

Figure 4.63

4.6.3 Example 4.18: Singly–Redundant Beam 2

A non–uniform, two–span beam ABCD is simply-supported at A, B and D as shown in Figure 4.64. The beam carries a uniformly distributed load on span AB and a point at the mid–span point of BCD. Using the data given:

(i)   determine the value of the support reactions,
(ii)   sketch the shear force and bending moment diagrams.

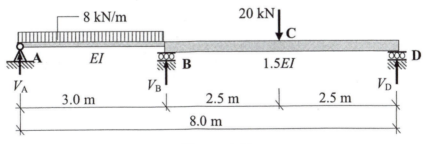

Figure 4.64

Solution:

Assume that the reaction at B is the redundant reaction. The bending moment diagrams for the applied loads and a unit point load at B are shown in Figure 4.65.

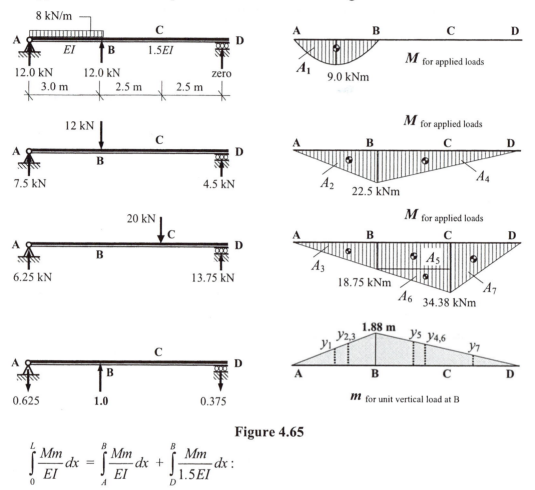

Figure 4.65

$$\int_0^L \frac{Mm}{EI}\,dx = \int_A^B \frac{Mm}{EI}\,dx + \int_D^B \frac{Mm}{1.5EI}\,dx :$$

$$A_1 = +\,(0.667 \times 3.0 \times 9.0) = +\,18.0 \text{ kNm}^2, \quad y_1 = -\,0.94 \text{ m}, \quad \therefore A_1 y_1 = -\,16.92 \text{ kNm}^3$$
$$A_2 = +\,(0.5 \times 3.0 \times 22.5) = +\,33.75 \text{ kNm}^2, \quad y_2 = -\,1.25 \text{ m}, \quad \therefore A_2 y_2 = -\,41.29 \text{ kNm}^3$$
$$A_3 = +\,(0.5 \times 3.0 \times 18.75) = +\,28.13 \text{ kNm}^2, \quad y_3 = -\,1.25 \text{ m}, \quad \therefore A_3 y_3 = -\,35.16 \text{ kNm}^3$$
$$A_4 = +\,(0.5 \times 5.0 \times 22.5) = +\,52.25 \text{ kNm}^2, \quad y_4 = -\,1.25 \text{ m}, \quad \therefore A_4 y_4 = -\,65.31 \text{ kNm}^3$$
$$A_5 = +\,(2.5 \times 18.75) = +\,46.88 \text{ kNm}^2, \qquad\qquad y_5 = -\,1.41 \text{ m}, \quad \therefore A_5 y_5 = -\,66.10 \text{ kNm}^3$$
$$A_6 = +\,(0.5 \times 2.5 \times 15.63) = +\,19.54 \text{ kNm}^2, \quad y_6 = -\,1.25 \text{ m}, \quad \therefore A_6 y_6 = -\,24.43 \text{ kNm}^3$$
$$A_7 = +\,(0.5 \times 2.5 \times 34.38) = +\,42.98 \text{ kNm}^2, \quad y_7 = -\,0.63 \text{ m}, \quad \therefore A_7 y_7 = -\,27.08 \text{ kNm}^3$$

$$\int_0^L \frac{Mm}{EI}\,dx = \sum_{n=1}^{n=7} \frac{\left(A_n y_n\right)}{EI}$$

$$= -\,[(16.92 + 41.29 + 35.16)/EI + (65.31 + 66.10 + 24.43 + 27.08)/1.5EI]$$
$$= -\,216.13/EI$$

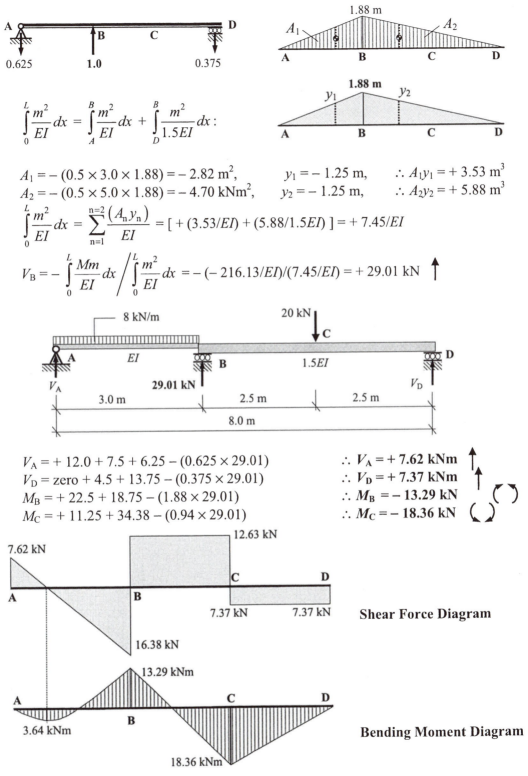

$$\int_0^L \frac{m^2}{EI}\,dx = \int_A^B \frac{m^2}{EI}\,dx + \int_D^B \frac{m^2}{1.5EI}\,dx :$$

$A_1 = -\,(0.5 \times 3.0 \times 1.88) = -\,2.82\ \text{m}^2,$ $y_1 = -\,1.25\ \text{m},$ $\therefore A_1 y_1 = +\,3.53\ \text{m}^3$

$A_2 = -\,(0.5 \times 5.0 \times 1.88) = -\,4.70\ \text{kNm}^2,$ $y_2 = -\,1.25\ \text{m},$ $\therefore A_2 y_2 = +\,5.88\ \text{m}^3$

$$\int_0^L \frac{m^2}{EI}\,dx = \sum_{n=1}^{n=2} \frac{(A_n y_n)}{EI} = [\,+(3.53/EI) + (5.88/1.5EI)\,] = +\,7.45/EI$$

$$V_B = -\int_0^L \frac{Mm}{EI}\,dx \Big/ \int_0^L \frac{m^2}{EI}\,dx = -(-216.13/EI)/(7.45/EI) = +\,29.01\ \text{kN} \uparrow$$

$V_A = +\,12.0 + 7.5 + 6.25 - (0.625 \times 29.01)$ $\therefore V_A = +\,7.62\ \text{kNm} \uparrow$

$V_D = \text{zero} + 4.5 + 13.75 - (0.375 \times 29.01)$ $\therefore V_D = +\,7.37\ \text{kNm} \uparrow$

$M_B = +\,22.5 + 18.75 - (1.88 \times 29.01)$ $\therefore M_B = -\,13.29\ \text{kN}$

$M_C = +\,11.25 + 34.38 - (0.94 \times 29.01)$ $\therefore M_C = -\,18.36\ \text{kN}$

Shear Force Diagram

Bending Moment Diagram

Figure 4.66

4.6.4 Problems: Unit Load Method for Singly-Redundant Beams

A series of singly-redundant beams are indicated in Problems 4.24 to 4.27. Using the applied loading given in each case:

    i)   determine the support reactions,
    ii) sketch the shear force diagram   and
    iii) sketch the bending moment diagram.

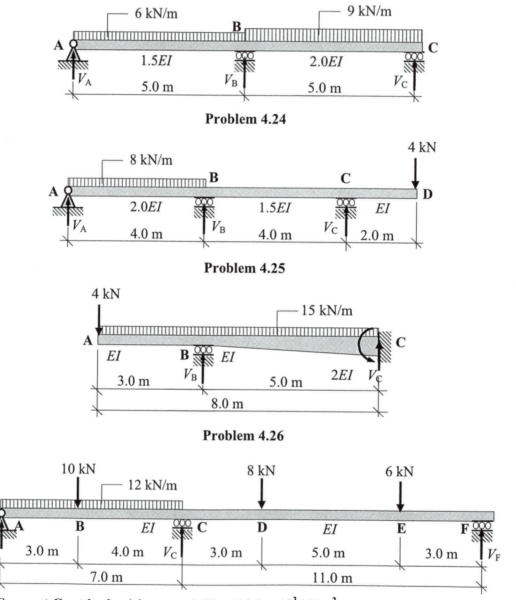

Problem 4.24

Problem 4.25

Problem 4.26

Support C settles by 4.0 mm and $EI = 100.0 \times 10^3$ kNm^2

Problem 4.27

4.6.5 Solutions: Unit Load Method for Singly-Redundant Beams

Solution

Topic: Unit Load – Singly-Redundant Beams
Problem Number: 4.24   **Page No. 1**

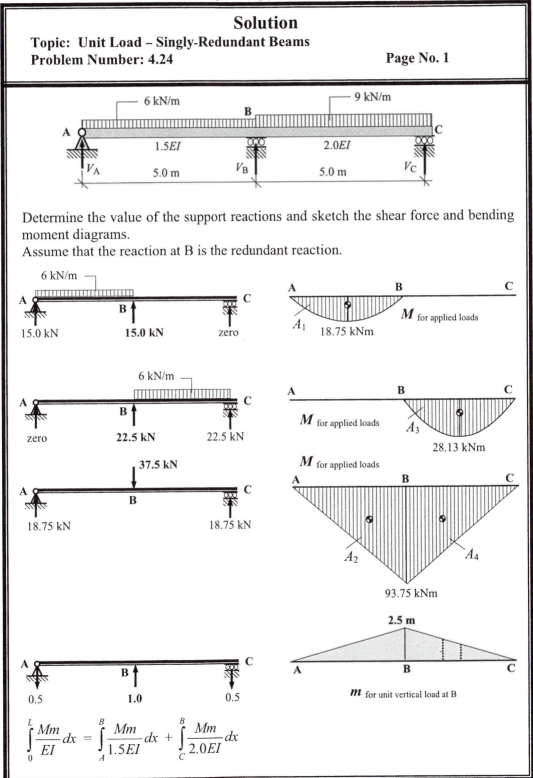

Determine the value of the support reactions and sketch the shear force and bending moment diagrams.
Assume that the reaction at B is the redundant reaction.

$$\int_0^L \frac{Mm}{EI}\,dx = \int_A^B \frac{Mm}{1.5EI}\,dx + \int_C^B \frac{Mm}{2.0EI}\,dx$$

Solution

Topic: Unit Load – Singly-Redundant Beams

Problem Number: 4.24   **Page No. 2**

Using the coefficients given in Table 4.1:

$$\int_0^L \frac{Mm}{EI}\,dx = \sum_0^L \left(Coefficient \times a \times b \times L\right)/EI$$

$$\int_0^L \frac{Mm}{EI}\,dx = [-(0.333 \times 18.75 \times 2.5 \times 5.0) - (0.333 \times 93.75 \times 2.5 \times 5.0)]/1.5EI$$

$$+ [-(0.333 \times 28.13 \times 2.5 \times 5.0) - (0.333 \times 93.75 \times 2.5 \times 5.0)\,]/2.0EI$$

$$= -565.85/EI$$

$$\int_0^L \frac{m^2}{EI}\,dx = +(0.333 \times 2.5 \times 2.5 \times 5.0)/1.5EI + (0.333 \times 2.5 \times 2.5 \times 5.0)/2.0EI$$

$$= +12.14/EI$$

$$V_B = -\int_0^L \frac{Mm}{EI}\,dx \bigg/ \int_0^L \frac{m^2}{EI}\,dx = -(-565.85/EI)/(12.14/EI) = +\mathbf{46.61}\ \mathbf{kN}\ \uparrow$$

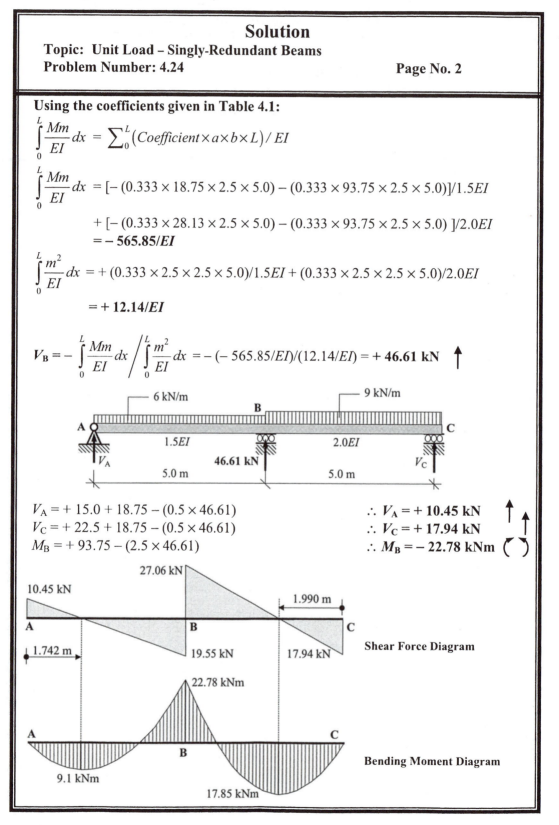

$V_A = +15.0 + 18.75 - (0.5 \times 46.61)$ $\therefore V_A = +\mathbf{10.45\ kN}\ \uparrow$

$V_C = +22.5 + 18.75 - (0.5 \times 46.61)$ $\therefore V_C = +\mathbf{17.94\ kN}$

$M_B = +93.75 - (2.5 \times 46.61)$ $\therefore M_B = -\mathbf{22.78\ kNm}$

Solution
Topic: Unit Load – Singly-Redundant Beams
Problem Number: 4.25   **Page No. 1**

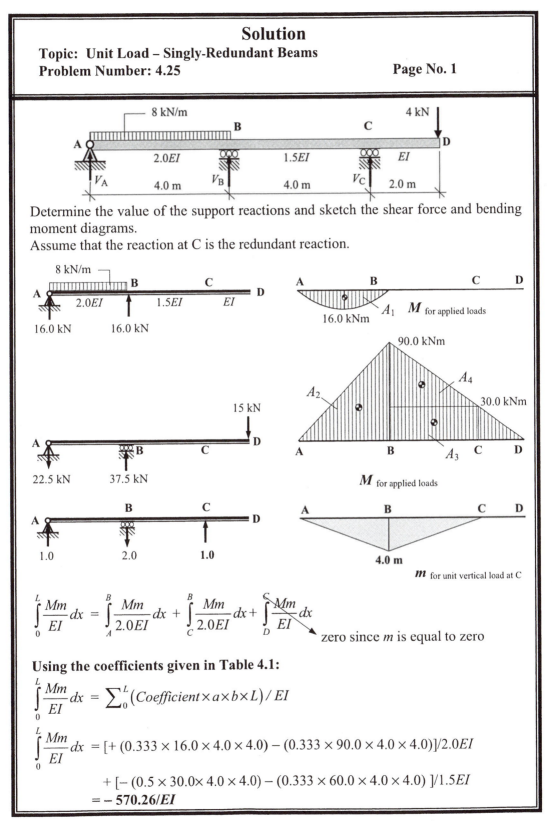

Determine the value of the support reactions and sketch the shear force and bending moment diagrams.

Assume that the reaction at C is the redundant reaction.

$$\int_0^L \frac{Mm}{EI}dx = \int_A^B \frac{Mm}{2.0EI}dx + \int_C^B \frac{Mm}{2.0EI}dx + \int_D^C \frac{Mm}{EI}dx$$

zero since m is equal to zero

Using the coefficients given in Table 4.1:

$$\int_0^L \frac{Mm}{EI}dx = \sum_0^L \left(Coefficient \times a \times b \times L\right)/EI$$

$$\int_0^L \frac{Mm}{EI}dx = [+(0.333 \times 16.0 \times 4.0 \times 4.0) - (0.333 \times 90.0 \times 4.0 \times 4.0)]/2.0EI$$

$$+ [-(0.5 \times 30.0 \times 4.0 \times 4.0) - (0.333 \times 60.0 \times 4.0 \times 4.0)]/1.5EI$$

$$= -570.26/EI$$

Solution

Topic: Unit Load – Singly-Redundant Beams
Problem Number: 4.25   **Page No. 2**

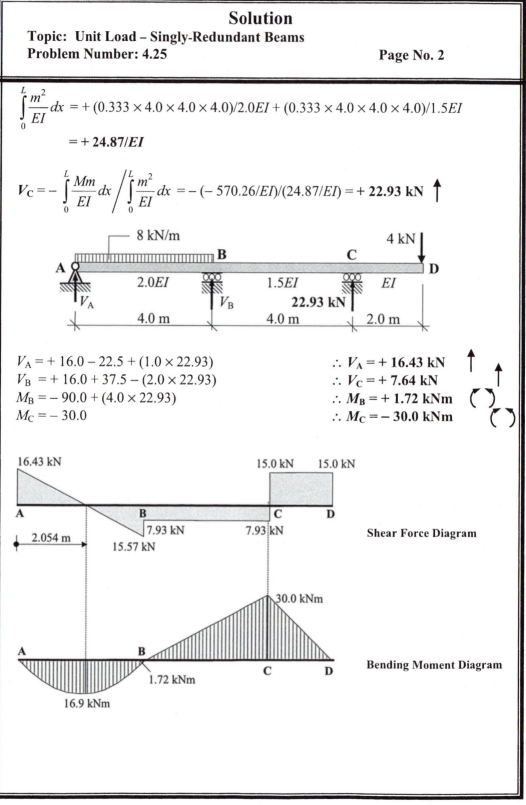

$$\int_0^L \frac{m^2}{EI}\,dx = +(0.333 \times 4.0 \times 4.0 \times 4.0)/2.0EI + (0.333 \times 4.0 \times 4.0 \times 4.0)/1.5EI$$

$$= +\mathbf{24.87}/\mathbf{EI}$$

$$V_C = -\int_0^L \frac{Mm}{EI}\,dx \Big/ \int_0^L \frac{m^2}{EI}\,dx = -(-570.26/EI)/(24.87/EI) = +\mathbf{22.93\ kN} \uparrow$$

$V_A = +16.0 - 22.5 + (1.0 \times 22.93)$ $\therefore V_A = +\mathbf{16.43\ kN}$
$V_B = +16.0 + 37.5 - (2.0 \times 22.93)$ $\therefore V_C = +\mathbf{7.64\ kN}$
$M_B = -90.0 + (4.0 \times 22.93)$ $\therefore M_B = +\mathbf{1.72\ kNm}$
$M_C = -30.0$ $\therefore M_C = -\mathbf{30.0\ kNm}$

Shear Force Diagram

Bending Moment Diagram

Solution

Topic: Unit Load – Singly-Redundant Beams

Problem Number: 4.26   **Page No. 1**

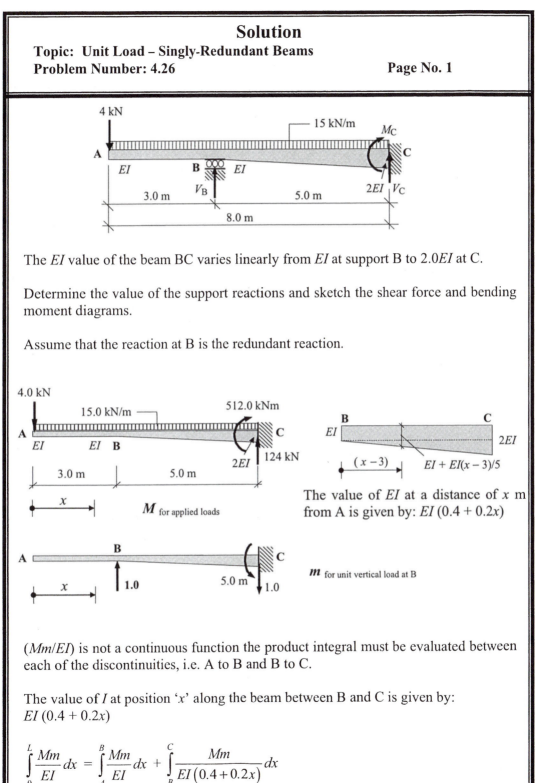

The *EI* value of the beam BC varies linearly from *EI* at support B to 2.0*EI* at C.

Determine the value of the support reactions and sketch the shear force and bending moment diagrams.

Assume that the reaction at B is the redundant reaction.

(*Mm/EI*) is not a continuous function the product integral must be evaluated between each of the discontinuities, i.e. A to B and B to C.

The value of *I* at position '*x*' along the beam between B and C is given by:
$EI (0.4 + 0.2x)$

$$\int_0^L \frac{Mm}{EI} dx = \int_A^B \frac{Mm}{EI} dx + \int_B^C \frac{Mm}{EI(0.4+0.2x)} dx$$

Solution

Topic: Unit Load – Singly-Redundant Beams

Problem Number: 4.26   **Page No. 2**

Consider the section from A to B: **$0 \leq x \leq 3.0$ m**

m = zero $\therefore Mm$ = zero

$$\int_A^B \frac{Mm}{EI} dx = \text{zero}$$

Consider the section from B to A: **$3.0 \leq x \leq 8.0$ m**

$M = -4.0x - 15.0x^2/2 = -4.0x - 7.5x^2$

$m = +1.0(x-3)$

$Mm = (x-3)(-4.0x - 7.5x^2) = 12.0x + 18.5x^2 - 7.5x^3$

$m^2 = +(x-3)^2$

$$\int_B^C \frac{Mm}{EI(0.4+0.2x)} dx = \int_{3.0}^{8.0} \frac{12.0x + 18.5x^2 - 7.5x^3}{EI(0.4+0.2x)} dx$$

Let $v = (0.4+0.2x)$ $\therefore x = (5v - 2)$ and $dx = 5dv$

$x^2 = (25v^2 - 20v + 4.0)$

$x^3 = (125v^3 - 150v^2 + 60v - 8.0)$

when $x = 3.0$ $v = 1.0$ and when $x = 8.0$ $v = 2.0$

$Mm = 12.0x + 18.5x^2 - 7.5x^3$

$\quad\quad = 12.0(5v - 2) + 18.5(25v^2 - 20v + 4.0) - 7.5(125v^3 - 150v^2 + 60v - 8.0)$

$\quad\quad = (-760v + 110 + 1587.5v^2 - 937.5v^3)$

$$\int_B^C \frac{Mm}{EI(0.4+0.2x)} dx = \int_{3.0}^{8.0} \frac{12.0x + 18.5x^2 - 7.5x^3}{EI(0.4+0.2x)} dx$$

$$= \int_{1.0}^{2.0} \frac{-760v + 110 + 1587.5v^2 - 937.5v^3}{EIv} 5.0dv$$

$$= \frac{5.0}{EI} \int_{1.0}^{2.0} \left(-760 + \frac{110}{v} + 1587.5v - 937.5v^2 \right) dv$$

$$= \frac{5.0}{EI} \left[-760v + 110 \ln v + \frac{1587.5v^2}{2.0} - \frac{937.5v^3}{3.0} \right]_{1.0}^{2.0}$$

$$= \frac{5.0}{EI} \left[(-768.8) - (-278.8) \right] = + \frac{2450.0}{EI} \text{ m}$$

Solution

Topic: Unit Load – Singly-Redundant Beams
Problem Number: 4.26

$$m^2 = + (x-3)^2 \;\; = x^2 - 6x + 9.0 \;\; = 25v^2 - 50v + 25.0$$

$$\int_B^C \frac{m^2}{EI(0.4+0.2x)}\,dx \; = \int_{3.0}^{8.0} \frac{x^2 - 6.0x + 9.0}{EI(0.4+0.2x)}\,dx \; = \int_{1.0}^{2.0} \frac{25.0v - 50.0v + 25.0}{EIv}\,5.0\,dv$$

$$= \frac{5.0}{EI} \int_{1.0}^{2.0} \left(25.0v - 50.0 + \frac{25.0}{v} \right)dv \; = \frac{5.0}{EI} \left[\frac{25.0v^2}{2.0} - 50.0v + 25.0\,lnv \right]_{1.0}^{2.0}$$

$$= \frac{5.0}{EI} \left[(-32.67) - (-37.5) \right] = + \frac{24.15}{EI}\,m$$

$$V_B = - \int_0^L \frac{Mm}{EI}\,dx \Big/ \int_0^L \frac{m^2}{EI}\,dx \; = -(-2450/EI)/(24.15/EI) = + \textbf{101.45 kN} \;\; \uparrow$$

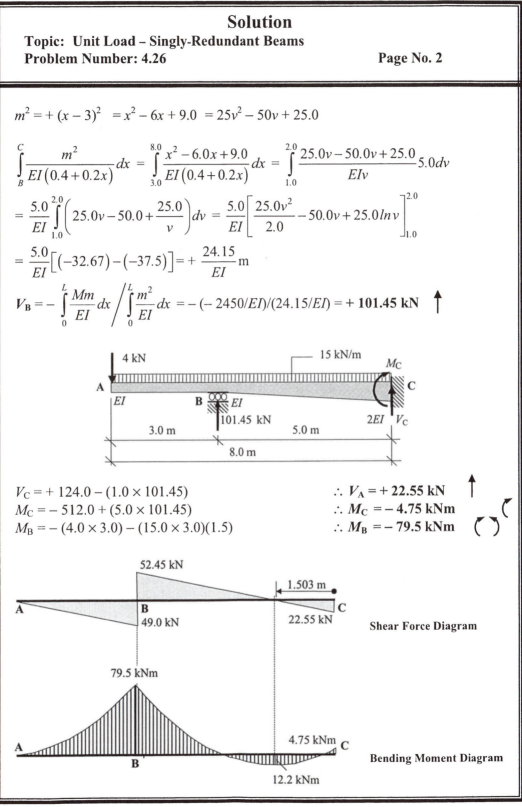

$$V_C = + 124.0 - (1.0 \times 101.45) \qquad\qquad \therefore V_A = + \textbf{22.55 kN} \;\; \uparrow$$
$$M_C = - 512.0 + (5.0 \times 101.45) \qquad\qquad \therefore M_C = - \textbf{4.75 kNm}$$
$$M_B = - (4.0 \times 3.0) - (15.0 \times 3.0)(1.5) \qquad \therefore M_B = - \textbf{79.5 kNm}$$

Shear Force Diagram

Bending Moment Diagram

Solution

Topic: Unit Load – Singly-Redundant Beams

Problem Number: 4.27

Page No. 1

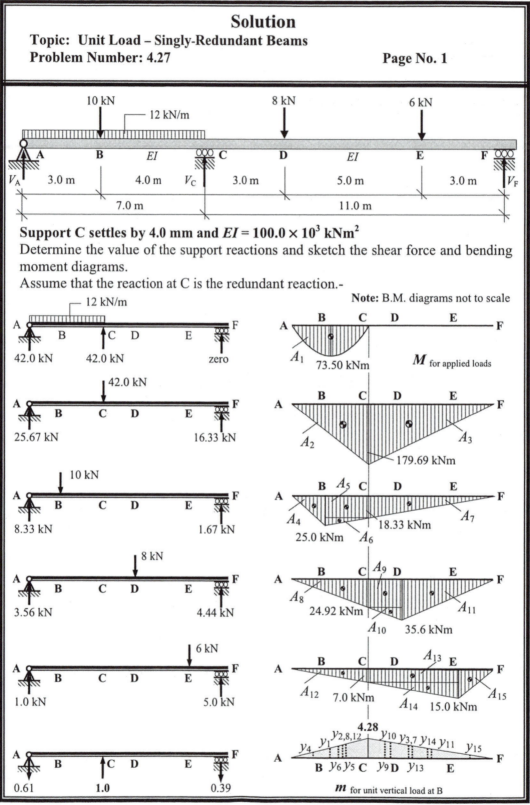

Support C settles by 4.0 mm and $EI = 100.0 \times 10^3$ kNm²

Determine the value of the support reactions and sketch the shear force and bending moment diagrams.

Assume that the reaction at C is the redundant reaction.-

Note: B.M. diagrams not to scale

Solution
Topic: Unit Load – Singly-Redundant Beams

Problem Number: 4.27   **Page No. 2**

$$\int_0^L \frac{Mm}{EI}dx + \left\{\int_0^L \frac{mm}{EI}dx\right\} \times V_C = -0.004 \qquad V_C = -\left(0.004 + \int_0^L \frac{Mm}{EI}dx\right)\bigg/\int_0^L \frac{m^2}{EI}dx$$

$$\int_0^L \frac{Mm}{EI}dx = \int_A^F \frac{Mm}{EI}dx$$

(**Note:** The reader should check this using the coefficients given in Table 4.1).

$A_1 = +(0.67 \times 7.0 \times 73.5) = +344.72 \text{ kNm}^2$

y_1 (3.5 m from A) $= -2.14$ m $\qquad\qquad \therefore A_1y_1 = -737.70 \text{ kNm}^3$

$A_2 = +(0.5 \times 7.0 \times 179.69) = +628.92 \text{ kNm}^2$

y_2 (4.67 m from A) $= -2.85$ m $\qquad\qquad \therefore A_2y_2 = -1792.42 \text{ kNm}^3$

$A_3 = +(0.5 \times 11.0 \times 179.69) = +988.30 \text{ kNm}^2$

y_3 (7.33 m from F) $= -2.85$ m $\qquad\qquad \therefore A_3y_3 = -2816.66 \text{ kNm}^3$

$A_4 = +(0.5 \times 3.0 \times 25.0) = +37.50 \text{ kNm}^2$

y_4 (2.0 m from A) $= -1.22$ m $\qquad\qquad \therefore A_4y_4 = -45.75 \text{ kNm}^3$

$A_5 = +(4.0 \times 18.33) = +73.32 \text{ kNm}^2$

y_5 (5.0 m from A) $= -3.05$ m $\qquad\qquad \therefore A_5y_5 = -223.63 \text{ kNm}^3$

$A_6 = +(0.5 \times 4.0 \times 6.67) = +13.34 \text{ kNm}^2$

y_6 (4.33 m from A) $= -2.64$ m $\qquad\qquad \therefore A_6y_6 = -35.22 \text{ kNm}^3$

$A_7 = +(0.5 \times 11.0 \times 18.33) = +100.82 \text{ kNm}^2$

y_7 (7.33 m from F) $= -2.85$ m $\qquad\qquad \therefore A_7y_7 = -287.34 \text{ kNm}^3$

$A_8 = +(0.5 \times 7.0 \times 24.92) = +87.22 \text{ kNm}^2$

y_8 (4.67 m from A) $= -2.85$ m $\qquad\qquad \therefore A_8y_8 = -248.58 \text{ kNm}^3$

$A_9 = +(3.0 \times 24.92) = +74.76 \text{ kNm}^2$

y_9 (9.50 m from F) $= -3.71$ m $\qquad\qquad \therefore A_9y_9 = -277.36 \text{ kNm}^3$

$A_{10} = +(0.5 \times 3.0 \times 10.68) = +16.02 \text{ kNm}^2$

y_{10} (9.0 m from F) $= -3.51$ m $\qquad\qquad \therefore A_{10}y_{10} = -56.23 \text{ kNm}^3$

$A_{11} = +(0.5 \times 8.0 \times 35.6) = +142.4 \text{ kNm}^2$

y_{11} (5.33 m from F) $= -2.08$ m $\qquad\qquad \therefore A_{11}y_{11} = -296.19 \text{ kNm}^3$

$A_{12} = +(0.5 \times 7.0 \times 7.0) = +24.50 \text{ kNm}^2$

y_{12} (4.67 m from A) $= -2.85$ m $\qquad\qquad \therefore A_{12}y_{12} = -69.83 \text{ kNm}^3$

$A_{13} = +(8.0 \times 7.0) = +56.0 \text{ kNm}^2$

y_{13} (7.0 m from F) $= -2.73$ m $\qquad\qquad \therefore A_{13}y_{13} = -152.88 \text{ kNm}^3$

$A_{14} = +(0.5 \times 8.0 \times 8.0) = +32.0 \text{ kNm}^2$

y_{14} (5.67 m from F) $= -2.21$ m $\qquad\qquad \therefore A_{14}y_{14} = -70.72 \text{ kNm}^3$

$A_{15} = +(0.5 \times 3.0 \times 15.0) = +22.5 \text{ kNm}^2$

y_{15} (2.0 m from F) $= -0.78$ m $\qquad\qquad \therefore A_{15}y_{15} = -17.55 \text{ kNm}^3$

$$\int_0^L \frac{Mm}{EI}dx = \sum_{n=1}^{n=15}\frac{(A_ny_n)}{EI} = -7128.06/EI \text{ m}$$

Solution

Topic: Unit Load – Singly-Redundant Beams
Problem Number: 4.27   **Page No. 3**

$$\int_0^L \frac{m^2}{EI}\,dx = \int_A^F \frac{m^2}{EI}\,dx$$

$A_1 = + (0.5 \times 7.0 \times 4.28) = - 14.98 \text{ kNm}^2$

$y_1 \ (4.67 \text{ m from A}) = - 2.85 \text{ m}$ $\therefore A_1 y_1 = + 42.69 \text{ kNm}^3$

$A_2 = + (0.5 \times 11.0 \times 4.28) = - 23.57 \text{ kNm}^2$

$y_2 \ (7.33 \text{ m from A}) = - 2.85 \text{ m}$ $\therefore A_2 y_2 = + 67.09 \text{ kNm}^3$

$$\int_0^L \frac{m^2}{EI}\,dx = \sum_{n=1}^{n=2} \frac{(A_n y_n)}{EI} = [+ (42.69/EI) + (67.09/EI)] = + 109.78/EI$$

$$V_C = - \left(0.004 + \int_0^L \frac{Mm}{EI}\,dx \right) \Bigg/ \int_0^L \frac{m^2}{EI}\,dx = - (0.004 - 7128.06/EI)/\,109.78/EI$$

$$= + 61.03 \text{ kN}$$

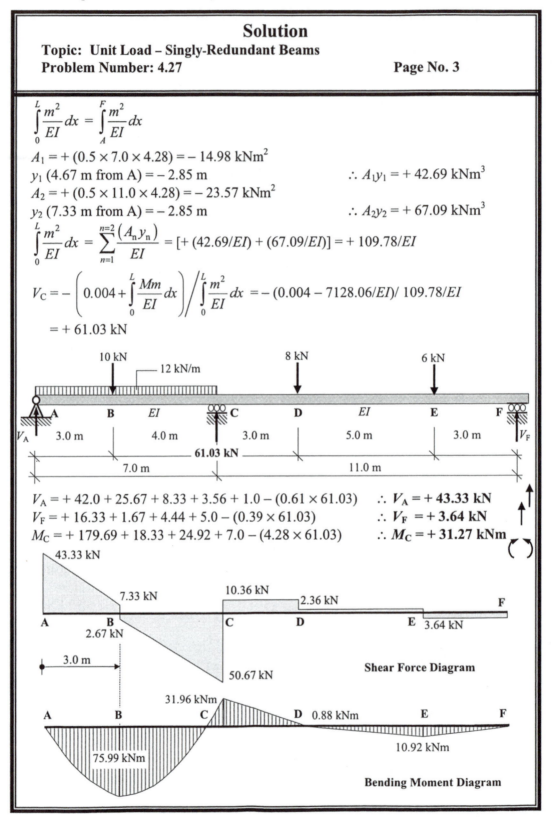

$V_A = + 42.0 + 25.67 + 8.33 + 3.56 + 1.0 - (0.61 \times 61.03)$ $\therefore V_A = + \textbf{43.33 kN}$

$V_F = + 16.33 + 1.67 + 4.44 + 5.0 - (0.39 \times 61.03)$ $\therefore V_F = + \textbf{3.64 kN}$

$M_C = + 179.69 + 18.33 + 24.92 + 7.0 - (4.28 \times 61.03)$ $\therefore M_C = + \textbf{31.27 kNm}$

Shear Force Diagram

Bending Moment Diagram

4.7   *Moment Distribution Method for Multi–Redundant Beams*

This section deals with continuous beams and propped cantilevers. An American engineer, Professor Hardy Cross, developed a very simple, elegant and practical method of analysis for such structures called *Moment Distribution*. This technique is one of developing successive approximations and is based on several basic concepts of structural behaviour which are illustrated in Sections 4.7.1 to 4.7.10.

4.7.1 *Bending (Rotational) Stiffness*

A fundamental relationship which exists in the elastic behaviour of structures and structural elements is that between an applied force system and the displacements which are induced by that system, i.e.

Force = Stiffness × Displacement   i.e. $P = k\delta$

where:
P is the applied force,
k is the stiffness,
δ is the displacement.

A definition of stiffness can be derived from this equation by rearranging it such that:

$k = P/\delta$

when $\delta = 1.0$ (i.e. unit displacement) the stiffness is: '*the force necessary to maintain a UNIT displacement, all other displacements being equal to zero.*'

The displacement can be a shear displacement, an axial displacement, a bending (rotational) displacement or a torsional displacement, each in turn producing the shear, axial, bending or torsional stiffness.
When considering beam elements in continuous structures using the moment distribution method of analysis, the bending stiffness is the principal characteristic which influences behaviour.
Consider the beam element AB shown in Figure 4.67 which is subject to a **UNIT** rotation at end A and is fixed at end B as indicated.

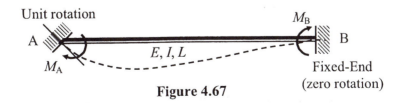

Figure 4.67

The force (M_A) necessary to maintain this displacement can be shown to be equal to ($4EI$)/L (see Chapter 7, Section 7.2.2). From the definition of stiffness given previously, the bending stiffness of the beam is equal to (Force/1.0), therefore $k = (4EI)/L$. This is known as the *absolute* bending stiffness of the element. Since most elements in continuous structures are made from the same material, the value of Young's Modulus (E) is constant throughout and $4E$ in the stiffness term is also a constant. This constant is normally ignored, to give $k = I/L$ which is known as the ***relative*** bending stiffness of the element. It is this value of stiffness which is normally used in the method of Moment Distribution.

It is evident from Figure 4.67 that when the beam element deforms due to the applied rotation at end A, an additional moment (M_B) is also transferred by the element to the remote end if it has zero slope (i.e. is fixed) The moment M_B is known as the ***carry-over*** moment.

4.7.2 Carry-Over Moment

Using the same analysis as that to determine M_A, it can be shown that $M_B = (2EI)/L$, i.e. ($\frac{1}{2} \times M_A$). It can therefore be stated that '***if a moment is applied to one end of a beam then a moment of the same sense and equal to half of its value will be transferred to the remote end provided that it is fixed.***'

If the remote end is '**pinned**', then the beam is less stiff, there is no carry-over moment and the value of M_A is smaller then when it is fixed as shown in Figure 4.69.

4.7.3 Pinned End

Consider the beam shown in Figure 4.68 in which a unit rotation is imposed at end A as before but the remote end B is pinned.

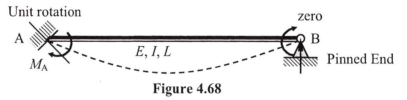

Figure 4.68

The force (M_A) necessary to maintain this displacement can be shown (e.g. using McCaulay's Method) to be equal to ($3EI)/L$, which represents the reduced absolute stiffness of a pin-ended beam. It can therefore be stated that '***the stiffness of a pin-ended beam is equal to ¾ × the stiffness of a fixed-end beam.***' In addition it can be shown that there is no carry-over moment to the remote end. These two cases are summarised in Figure 4.69.

Remote End Fixed:

$M_A = 4EI/L$

$k = (I/L)$

$M_B = 2EI/L$

Remote End Pinned:

$M_A = 3EI/L$

$k = ¾ (I/L)$

$M_B = $ zero

Figure 4.69

4.7.4 *Free and Fixed Bending Moments*

When a beam is free to rotate at both ends as shown in Figures 4.70(a) and (b) such that no bending moment can develop at the supports, then the bending moment diagram resulting from the applied loads on the beam is known as the ***Free Bending Moment Diagram***.

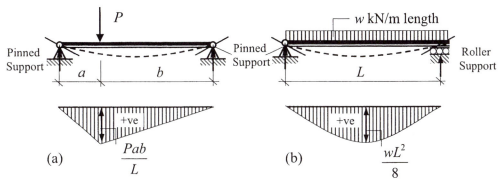

(a) $\dfrac{Pab}{L}$ (b) $\dfrac{wL^2}{8}$

Figure 4.70 – Free Bending Moment Diagrams

When a beam is fixed at the ends (encastré) such that it cannot rotate, i.e. zero slope at the supports, as shown in Figure 4.71, then bending moments are induced at the supports and are called *Fixed-End Moments*. The bending moment diagram associated **only** with the fixed-end moments is called the ***Fixed Bending Moment Diagram***.

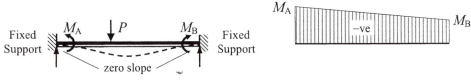

Figure 4.71 – Fixed Bending Moment Diagram

Using the principle of superposition, this beam can be considered in two parts in order to evaluate the support reactions and the **Final** bending moment diagram:

(i) *The fixed-reactions (moments and forces) at the supports*

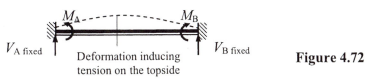

Figure 4.72

(ii) *The free reactions at the supports and the bending moments throughout the length due to the applied load, assuming the supports to be pinned*

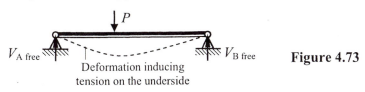

Figure 4.73

Combining (i) + (ii) gives the final bending moment diagram as shown in Figure 4.74:

$$V_A = (V_{A \text{ fixed}} + V_{A \text{ free}}); \qquad V_B = (V_{B \text{ fixed}} + V_{B \text{ free}})$$

$$M_A = (M_A + 0); \qquad\qquad M_B = (M_B + 0)$$

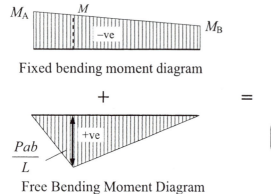

Fixed bending moment diagram

+

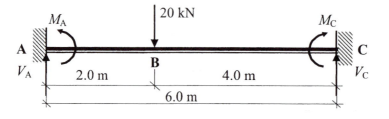

Free Bending Moment Diagram

Note: $M = -[M_B + (M_A - M_B)b/L]$

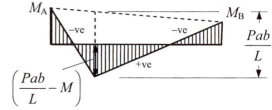

Final Bending Moment Diagram

Figure 4.74

The values of M_A and M_B for the most commonly applied load cases are given in Appendix 2. These are standard ***Fixed-End Moments*** relating to single-span encastré beams and are used extensively in structural analysis.

4.7.5 *Example 4.19: Single-span Encastré Beam*

Determine the support reactions and draw the bending moment diagram for the encastré beam loaded as shown in Figure 4.75.

Figure 4.75

Solution:
Consider the beam in two parts.

(i) *Fixed Support Reactions*
The values of the fixed-end moments are given in Appendix 2.

$$M_A = -\frac{Pab^2}{L^2} = -\frac{20 \times 2 \times 4^2}{6^2} = -17.78 \text{ kNm}$$

$$M_C = +\frac{Pa^2b}{L^2} = +\frac{20 \times 2^2 \times 4}{6^2} = +8.89 \text{ kNm}$$

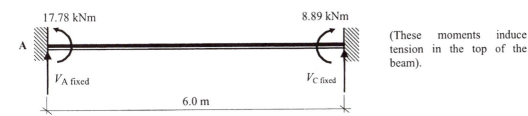

17.78 kNm   8.89 kNm

A

$V_{A \text{ fixed}}$ $V_{C \text{ fixed}}$

6.0 m

(These moments induce tension in the top of the beam).

Consider the rotational equilibrium of the beam:

+ve $\circlearrowright$ $\Sigma M_A = 0$

$- (17.78) + (8.89) - (6.0 \times V_{C \text{ fixed}}) = 0$ Equation (1)

$\therefore V_{C \text{ fixed}} = - 1.48 \text{ kN} \downarrow$

Consider the vertical equilibrium of the beam:

+ve $\uparrow$ $\Sigma F_z = 0$

$+ V_{A \text{ fixed}} + V_{C \text{ fixed}} = 0$ $\therefore V_{A \text{ fixed}} = - (- 1.48 \text{ kN}) = + 1.48 \text{ kN} \uparrow$ Equation (2)

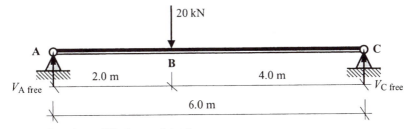

20 kN

A   C

                B

$V_{A \text{ free}}$ 2.0 m   4.0 m $V_{C \text{ free}}$

6.0 m

Consider the rotational equilibrium of the beam:

+ve $\circlearrowright$ $\Sigma M_A = 0$

$+ (20 \times 2.0) - (6.0 \times V_{C \text{ free}}) = 0$ $\therefore V_{C \text{ free}} = + 6.67 \text{ kN} \uparrow$ Equation (1)

Consider the vertical equilibrium of the beam:

+ve $\uparrow$ $\Sigma F_z = 0$

$+ V_{A \text{ free}} + V_{C \text{ free}} - 20 = 0$ $\therefore V_{A \text{ free}} = + 13.33 \text{ kN} \uparrow$ Equation (2)

Bending Moment under the point load $= (+ 13.33 \times 2.0) = + 26.67 \text{ kNm}$

(This induces tension in the bottom of the beam)

The final vertical support reactions are given by (i) + (ii):

$V_A = V_{A \text{ fixed}} + V_{A \text{ free}} = (+ 1.48 + 13.33) = + 14.81 \text{ kN} \uparrow$

$V_C = V_{C \text{ fixed}} + V_{C \text{ free}} = (- 1.48 + 6.67) = + 5.19 \text{ kN} \uparrow$

Check the vertical equilibrium: Total vertical force $= + 14.81 + 5.19 = + 20 \text{ kN} \uparrow$

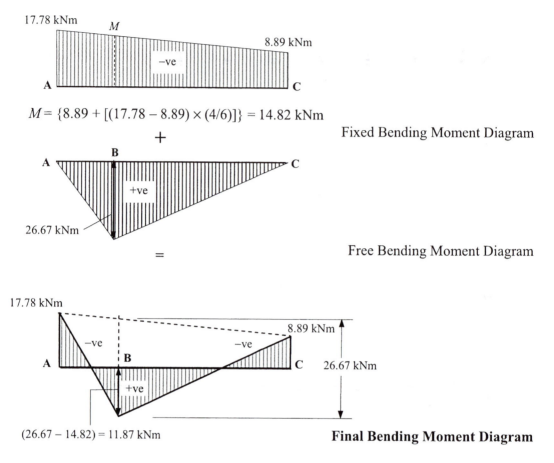

$M = \{8.89 + [(17.78 - 8.89) \times (4/6)]\} = 14.82$ kNm

Fixed Bending Moment Diagram

Free Bending Moment Diagram

Final Bending Moment Diagram

Figure 4.76

Note the similarity between the shape of the bending moment diagram and the final deflected shape as shown in Figure 4.77.

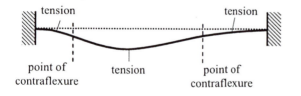

Deflected shape indicating tension zones and the similarity to the shape of the bending moment diagram

Figure 4.77

4.7.6 Propped Cantilevers

The fixed-end moment for propped cantilevers (i.e. one end fixed and the other end simply-supported) can be derived from the standard values given for encastré beams as follows. Consider the propped cantilever shown in Figure 4.78, which supports a uniformly distributed load as indicated.

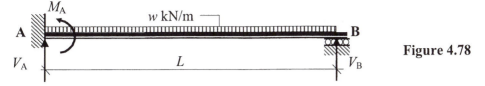

Figure 4.78

The structure can be considered to be the superposition of an encastré beam with the addition of an equal and opposite moment to M_B applied at B to ensure that the final moment at this support is equal to zero, as indicated in Figure 4.79.

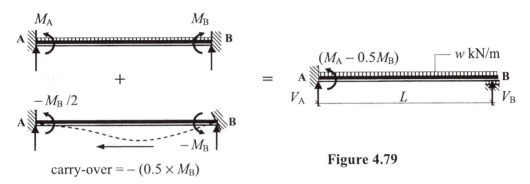

Figure 4.79

4.7.7 Example 4.20: Propped Cantilever

Determine the support reactions and draw the bending moment diagram for the propped cantilever shown in Figure 4.80.

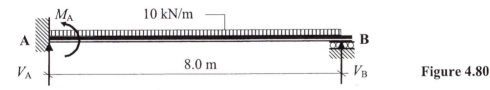

Figure 4.80

Solution
Fixed-End Moment for Propped Cantilever:
Consider the beam fixed at both supports.
The values of the fixed-end moments for encastre beams are given in Appendix 2.

$$M_A = - \frac{wL^2}{12} = - \frac{10 \times 8^2}{12} = - 53.33 \text{ kNm}$$

$$M_B = + \frac{wL^2}{12} = + \frac{10 \times 8^2}{12} = + 53.33 \text{ kNm}$$

The moment M_B must be cancelled out by applying an equal and opposite moment at B which in turn produces a carry-over moment equal to $- (0.5 \times M_B)$ at support A.

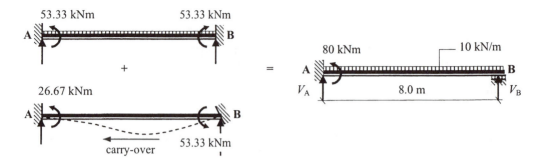

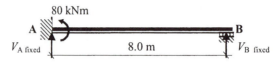

carry-over   53.33 kNm

(i) *Fixed Support Reactions*

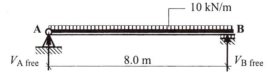

Consider the rotational equilibrium of the beam:

$$+ve \, \curvearrowright \Sigma M_A = 0$$

$$- (80) - (8.0 \times V_{B \text{ fixed}}) = 0 \qquad\qquad\qquad\qquad \text{Equation (1)}$$

$$\therefore V_{B \text{ fixed}} = - 10.0 \text{ kN} \downarrow$$

Consider the vertical equilibrium of the beam:

$$+ve \uparrow \Sigma F_z = 0$$

$$+ V_{A \text{ fixed}} + V_{B \text{ fixed}} = 0 \quad \therefore V_{A \text{ fixed}} = - (- 10.0 \text{ kN}) = + 10.0 \text{ kN} \uparrow \qquad \text{Equation (2)}$$

(ii) *Free Support Reactions*

Consider the rotational equilibrium of the beam:

$$+ve \, \curvearrowright \Sigma M_A = 0$$
$$+ (10 \times 8.0 \times 4.0) - (8.0 \times V_{B \text{ free}}) = 0 \quad \therefore V_{B \text{ free}} = + 40.0 \text{ kN} \uparrow \qquad \text{Equation (1)}$$

Consider the vertical equilibrium of the beam:

$$+ve \uparrow \Sigma F_z = 0$$
$$+ V_{A \text{ free}} + V_{B \text{ free}} - (10 \times 8.0) = 0 \qquad \therefore V_{A \text{ free}} = + 40.0 \text{ kN} \uparrow \qquad \text{Equation (2)}$$

The final vertical support reactions are given by (i) + (ii):

$$V_A = V_{A\ fixed} + V_{A\ free} = (+\ 10.0 + 40.0) = +\ 50.0\ \text{kN} \uparrow$$
$$V_B = V_{B\ fixed} + V_{B\ free} = (-\ 10.0 + 40.0) = +\ 30.0\ \text{kN} \uparrow$$

Check the vertical equilibrium: Total vertical force $= +\ 50.0 + 30.0 = +\ 80\ \text{kN} \uparrow$

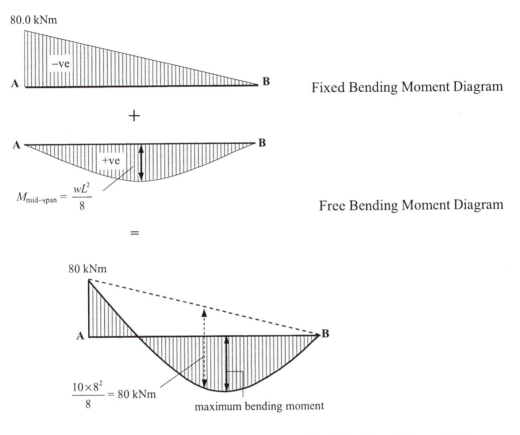

Fixed Bending Moment Diagram

Free Bending Moment Diagram

Final Bending Moment Diagram

Figure 4.81

Note the similarity between the shape of the bending moment diagram and the final deflected shape as shown in Figure 4.82.

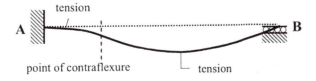

Deflected shape indicating tension zones and the similarity to the shape of the bending moment diagram

Figure 4.82

The position of the maximum bending moment can be determined by finding the point of zero shear force as shown in Figure 4.83.

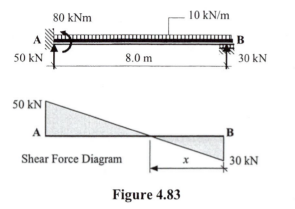

The position of zero shear:

$$x = \frac{30}{10} = 3.0 \text{ m}$$

Figure 4.83

Maximum bending moment:
$$M = [+ (30 \times 3.0) - (10 \times 3.0 \times 1.5)]$$
$$= + 45.0 \text{ kNm}$$
or
$$M = \text{shaded area over length '}x\text{'}$$
$$= (0.5 \times 30.0 \times 3.0) = 45.0 \text{ kNm}$$

4.7.8 Distribution Factors

Consider a uniform two-span continuous beam, as shown in Figure 4.84.

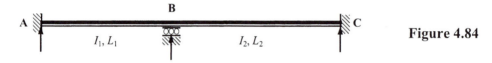

Figure 4.84

If an external moment M is applied to this structure at support B it will produce a rotation of the beam at the support; part of this moment is absorbed by each of the two spans BA and BC, as indicated in Figure 4.85.

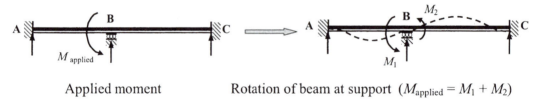

Applied moment   Rotation of beam at support $(M_{\text{applied}} = M_1 + M_2)$

Figure 4.85

The proportion of each moment induced in each span is directly proportional to the relative stiffnesses, e.g.

Stiffness of span BA = $k_{\text{BA}} = (I_1/L_1)$
Stiffness of span BC = $k_{\text{BC}} = (I_2/L_2)$

Total stiffness of the beam at the support = $k_{\text{total}} = (k_{\text{BA}} + k_{\text{BC}}) = [(I_1/L_1) + (I_2/L_2)]$

The moment absorbed by beam BA $M_1 = M_{\text{applied}} \times \left(\dfrac{k_{\text{BA}}}{k_{\text{total}}} \right)$

The moment absorbed by beam BC $M_2 = M_{\text{applied}} \times \left(\dfrac{k_{\text{BC}}}{k_{\text{total}}} \right)$

The ratio $\left(\dfrac{k}{k_{\text{total}}}\right)$ is known as the ***Distribution Factor*** for the member at the joint where the moment is applied.

As indicated in Section 4.7.2, when a moment (M) is applied to one end of a beam in which the other end is fixed, a carry-over moment equal to 50% of M is induced at the remote fixed-end and consequently moments equal to ½ M_1 and ½ M_2 will develop at supports A and C respectively, as shown in Figure 4.86.

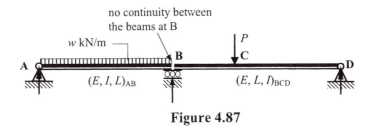

Figure 4.86

4.7.9 Application of the Method

All of the concepts outlined in Sections 4.7.1 to 4.7.8 are used when analysing indeterminate structures using the method of moment distribution. Consider the two *separate* beam spans indicated in Figure 4.87.

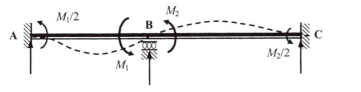

Figure 4.87

Since the beams are **not** connected at the support B they behave independently as simply-supported beams with separate reactions and bending moment diagrams, as shown in Figure 4.88.

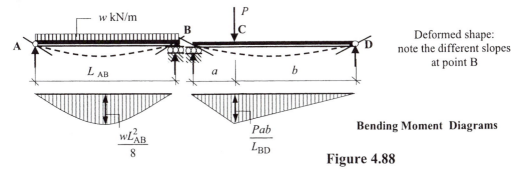

Figure 4.88

When the beams are continuous over support B as shown in Figure 4.89(a), a continuity moment develops for the continuous structure as shown in Figures 4.89(b) and (c). Note the similarity of the bending moment diagram for member AB to the propped cantilever in Figure 4.81. Both members AB and BD are similar to propped cantilevers in this structure.

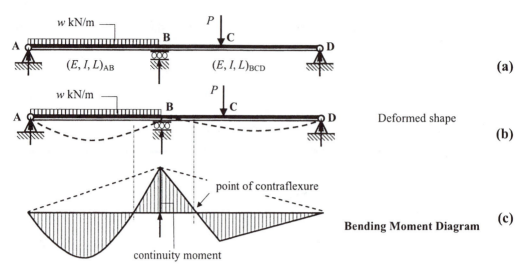

Figure 4.89

Moment distribution enables the evaluation of the continuity moments. The method is ideally suited to tabular representation and is illustrated in Example 4.21.

4.7.10   Example 4.21: Three-span Continuous Beam

A non-uniform, three span beam ABCDEF is fixed at support A and pinned at support F, as illustrated in Figure 4.90. Determine the support reactions and sketch the bending moment diagram for the applied loading indicated.

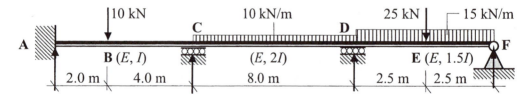

Figure 4.90

Solution:
Step 1
The first step is to assume that all supports are fixed against rotation and evaluate the 'fixed-end moments'.

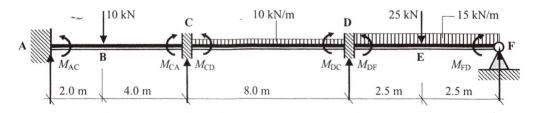

The values of the fixed-end moments for encastre beams are given in Appendix 2.

Span AC

$$M_{AC} = -\frac{Pab^2}{L^2} = -\frac{10 \times 2 \times 4^2}{6.0^2} = -8.89 \text{ kNm}$$

$$M_{CA} = +\frac{Pa^2b}{L^2} = +\frac{10 \times 2^2 \times 4}{6.0^2} = +4.44 \text{ kNm}$$

Span CD

$$M_{CD} = -\frac{wL^2}{12} = -\frac{10 \times 8^2}{12} = -53.33 \text{ kNm}$$

$$M_{DC} = +\frac{wL^2}{12} = +\frac{10 \times 8^2}{12} = +53.33 \text{ kNm}$$

Span DF*

$$M_{DF} = -\frac{wL^2}{12} - \frac{PL}{8} = -\frac{15 \times 5^2}{12} - \frac{25 \times 5}{8} = -46.89 \text{ kNm}$$

$$M_{FD} = +\frac{wL^2}{12} + \frac{PL}{8} = +\frac{15 \times 5^2}{12} + \frac{25 \times 5}{8} = +46.89 \text{ kNm}$$

* Since support F is pinned, the fixed-end moments are $(M_{DF} - 0.5M_{FD})$ at D and zero at F (see Figure 4.79): $(M_{DF} - 0.5M_{FD}) = [-46.89 - (0.5 \times 46.89)] = -70.34 \text{ kNm}.$

Step 2
The second step is to evaluate the member and total *stiffness* at each internal joint/support and determine the *distribution factors* at each support. Note that the applied force system is **not** required to do this.

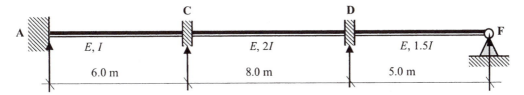

Support C
  Stiffness of CA = $k_{CA} = (I / 6.0) = 0.167I$
  Stiffness of CD = $k_{CD} = (2I / 8.0) = 0.25I$ } $k_{total} = (0.167 + 0.25)I = 0.417I$

  Distribution factor (DF) for CA = $\dfrac{k_{CA}}{k_{total}} = \dfrac{0.167I}{0.417I} = 0.4$

  Distribution factor (DF) for CD = $\dfrac{k_{CD}}{k_{total}} = \dfrac{0.25I}{0.417I} = 0.6$ } $\Sigma \text{ DF's} = 1.0$

Support D
  Stiffness of DC = $k_{DC} = k_{DC} = 0.25I$
  Stiffness of DF = $k_{DF} = \frac{3}{4} \times (1.5I / 5.0) = 0.225I$ } **Note:** the remote end F is pinned and $k = \frac{3}{4} (I/L)$

  $k_{total} = (0.25 + 0.225)I = 0.475I$

Distribution factor (DF) for DC $= \dfrac{k_{DC}}{k_{total}} = \dfrac{0.25I}{0.475I} = 0.53$

Distribution factor (DF) for DF $= \dfrac{k_{DF}}{k_{total}} = \dfrac{0.141I}{0.475I} = 0.47$

$\left.\vphantom{\dfrac{\dfrac{}{}}{\dfrac{}{}}}\right\} \Sigma$ DF's $= 1.0$

The structure and the distribution factors can be represented in tabular form, as shown in Figure 4.91.

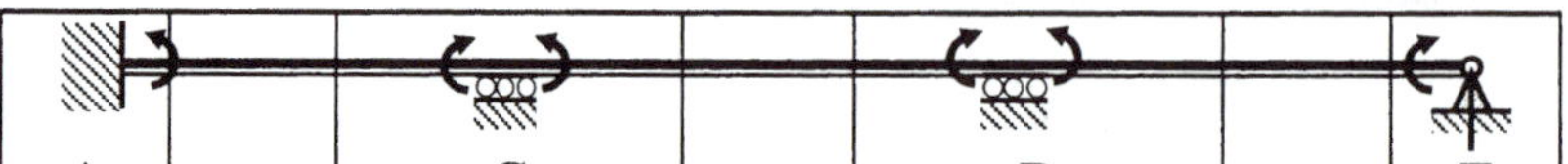

Joints/Support	A		C			D		F
Member	AC		CA	CD		DC	DF	FD
Distribution Factors	0		0.4	0.6		0.53	0.47	1.0

Figure 4.91

The distribution factor for fixed supports is equal to zero since any moment is resisted by an equal and opposite moment within the support and no balancing is required. In the case of pinned supports the distribution factor is equal to 1.0 since 100% of any applied moment, e.g. by a cantilever overhang, must be balanced and a carry-over of ½ × the balancing moment transferred to the remote end at the internal support.

Step 3
The fixed-end moments are now entered into the table at the appropriate locations, taking care to ensure that the signs are correct.

Joints/Support	A		C			D		F
Member	AC		CA	CD		DC	DF	FD
Distribution Factors	0		0.4	0.6		0.53	0.47	1.0
Fixed-End Moments	– 8.89		+ 4.44	– 53.33		+ 53.33	– 70.34	zero

Step 4
When the structure is restrained against rotation there is normally a resultant moment at a typical internal support. For example, consider the moments C:

$M_{CA} = + 4.44$ kNm $\circlearrowleft$ and $M_{CD} = - 53.33$ kNm $\circlearrowright$

The 'out-of-balance' moment is equal to the algebraic difference between the two:
The out-of-balance moment $= (+ 4.44 - 53.33) = - 48.89$ kNm $\circlearrowright$

If the imposed fixity at **one** support (all others remaining fixed), e.g. support C, is released, the beam will rotate sufficiently to induce a balancing moment such that equilibrium is achieved and the moments M_{CA} and M_{CD} are equal and opposite. The application of the balancing moment is distributed between CA and CD in proportion to the *distribution factors* calculated previously.

Moment applied to CA = + (48.89 × 0.4) = +19.56 kNm

Moment applied to CD = + (48.89 × 0.6) = + 29.33 kNm

Joints/Support	A		C			D			F
Member	AC		CA	CD		DC	DF		FD
Distribution Factors	0		0.4	0.6		0.53	0.47		1.0
Fixed-End Moments	− 8.89		+ 4.44	− 53.33		+ 53.33	− 70.34		zero
Balance Moment			+ 19.56	+ 29.33					

As indicated in Section 4.7.2, when a moment is applied to one end of a beam whilst the remote end is fixed, a carry-over moment equal to (½ × applied moment) and of the same sign is induced at the remote end. This is entered into the table as shown.

Joints/Support	A		C			D			F
Member	AC		CA	CD		DC	DF		FD
Distribution Factors	0		0.4	0.6		0.53	0.47		1.0
Fixed-End Moments	− 8.89		+ 4.44	− 53.33		+ 53.33	− 70.34		zero
Balance Moment			+ 19.56	+ 29.33					
Carry-over to Remote Ends	+ 9.78					+ 14.67			

Step 5

The procedure outline above is then carried out for each restrained support in turn. The reader should confirm the values given in the table for support D.

Joints/Support	A		C			D			F
Member	AC		CA	CD		DC	DF		FD
Distribution Factors	0		0.4	0.6		0.53	0.47		1.0
Fixed-End Moments	− 8.89		+ 4.44	− 53.33		+ 53.33	− 70.34		zero
Balance Moment			+ 19.56	+ 29.33					
Carry-over to Remote Ends	+ 9.78					+ 14.67			
Balance Moment						+ 1.27	+ 1.12		**Note:** No carry-over to the pinned end
Carry-over to Remote Ends				+ 0.64					

If the total moments at each internal support are now calculated they are:

$$M_{CA} = (+ 4.44 + 19.56) = + 24.0 \text{ kNm}$$
$$M_{CD} = (- 53.33 + 29.33 + 0.64) = - 23.36 \text{ kNm}$$

The difference = 0.64 kNm   i.e. the value of the carry-over moment

$$M_{DC} = (+ 53.33 + 14.67 + 1.27) = + 69.27 \text{ kNm}$$
$$M_{DF} = (- 70.34 + 1.12) = - 69.27 \text{ kNm}$$

The difference = 0

It is evident that after one iteration of each support moment the true values are nearer to 23.8 kNm and 69.0 kNm for C and D respectively. The existing out-of-balance moments which still exist, 0.64 kNm, can be distributed in the same manner as during the first iteration. This process is carried out until the desired level of accuracy has been achieved, normally after three or four iterations.

A slight modification to carrying out the distribution process which still results in the same answers is to carry out the balancing operation for **all** supports simultaneously and the carry-over operation likewise. This is quicker and requires less work. The reader should complete a further three/four iterations to the solution given above and compare the results with those shown in Figure 4.92.

Joints/Support	A		C			D			F
Member	AC		CA	CD		DC	DF		FD
Distribution Factors	0		0.4	0.6		0.53	0.47		1.0
Fixed-End Moments	− 8.89		+ 4.44	− 53.33		+ 53.33	− 70.34		zero
Balance Moment			+ 19.56	+ 29.33		+ 9.01	+ 7.99		
Carry-over to Remote Ends	+ 9.78			+ 4.50		+ 14.67			
Balance Moment			− 1.80	− 2.70		− 7.78	− 6.89		
Carry-over to Remote Ends	−0.91			− 3.89		− 1.35			
Balance Moment	+ 0.78	carry-over*	+ 1.56	+ 2.33		+ 0.72	+ 0.63		
Total	+ 0.76		+ 23.76	− 23.76		+ 68.60	− 68.61		zero

*The final carry-over, to the fixed support only, means that this value is one iteration more accurate than the internal joints.

Figure 4.92

The continuity moments are shown in Figure 4.93.

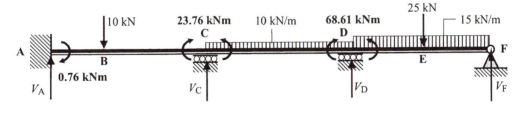

Figure 4.93

The support reactions and the bending moment diagrams for each span can be calculated using superposition as before by considering each span separately.

(i) *Fixed vertical reactions*

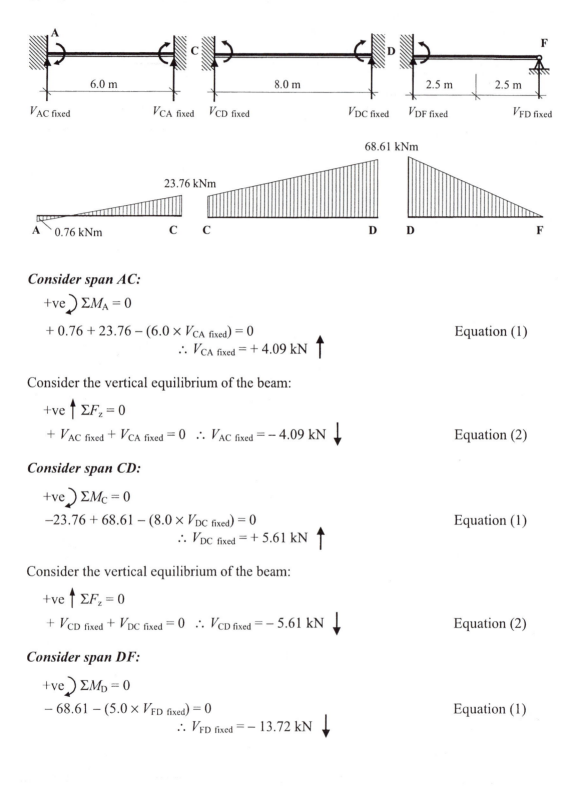

Consider span AC:

$$+ve \,\rotatebox{-20}{$\circlearrowright$}\, \Sigma M_A = 0$$

$$+ 0.76 + 23.76 - (6.0 \times V_{CA \text{ fixed}}) = 0 \qquad\qquad\qquad \text{Equation (1)}$$
$$\therefore V_{CA \text{ fixed}} = + 4.09 \text{ kN} \uparrow$$

Consider the vertical equilibrium of the beam:

$$+ve \uparrow \Sigma F_z = 0$$

$$+ V_{AC \text{ fixed}} + V_{CA \text{ fixed}} = 0 \;\; \therefore V_{AC \text{ fixed}} = - 4.09 \text{ kN} \downarrow \qquad\qquad \text{Equation (2)}$$

Consider span CD:

$$+ve \,\rotatebox{-20}{$\circlearrowright$}\, \Sigma M_C = 0$$

$$-23.76 + 68.61 - (8.0 \times V_{DC \text{ fixed}}) = 0 \qquad\qquad\qquad \text{Equation (1)}$$
$$\therefore V_{DC \text{ fixed}} = + 5.61 \text{ kN} \uparrow$$

Consider the vertical equilibrium of the beam:

$$+ve \uparrow \Sigma F_z = 0$$

$$+ V_{CD \text{ fixed}} + V_{DC \text{ fixed}} = 0 \;\; \therefore V_{CD \text{ fixed}} = - 5.61 \text{ kN} \downarrow \qquad\qquad \text{Equation (2)}$$

Consider span DF:

$$+ve \,\rotatebox{-20}{$\circlearrowright$}\, \Sigma M_D = 0$$

$$- 68.61 - (5.0 \times V_{FD \text{ fixed}}) = 0 \qquad\qquad\qquad \text{Equation (1)}$$
$$\therefore V_{FD \text{ fixed}} = - 13.72 \text{ kN} \downarrow$$

Consider the vertical equilibrium of the beam:

$$+ve \uparrow \Sigma F_z = 0$$

$$+ V_{\text{DF fixed}} + V_{\text{FD fixed}} = 0 \qquad \therefore V_{\text{DF fixed}} = + 13.72 \text{ kN} \uparrow \qquad \text{Equation (2)}$$

The total vertical reaction at each support due to the continuity moments is equal to the algebraic sum of the contributions from each beam at the support.

$$V_{\text{A fixed}} = V_{\text{AC fixed}} = - 4.09 \text{ kN}$$
$$V_{\text{C fixed}} = V_{\text{CA fixed}} + V_{\text{CD fixed}} = (+ 4.09 - 5.61) = - 1.52 \text{ kN}$$
$$V_{\text{D fixed}} = V_{\text{DC fixed}} + V_{\text{DF fixed}} = (+ 5.61 + 13.72) = + 19.33 \text{ kN}$$
$$V_{\text{F fixed}} = V_{\text{FD fixed}} = - 13.72 \text{ kN}$$

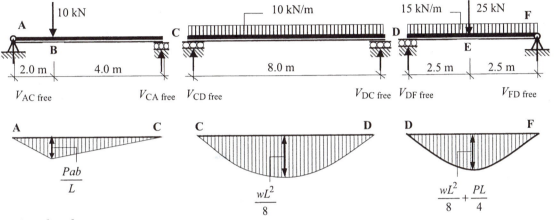

Free bending moments

Span AC $\dfrac{Pab}{L} = \dfrac{10 \times 2 \times 4}{6} = 13.3 \text{ kNm}$

Span CD $\dfrac{wL^2}{8} = \dfrac{10 \times 8^2}{8} = 80.0 \text{ kNm}$

Span DF $\left[\dfrac{wL^2}{8} + \dfrac{PL}{4} \right] = \left[\dfrac{15 \times 5^2}{8} + \dfrac{25 \times 5}{4} \right] = 78.13 \text{ kNm}$

(ii) *Free vertical reactions*

Consider span AC:

$$+ve \, \rotatebox{0}{$\curvearrowright$} \, \Sigma M_A = 0$$

$$+ (10 \times 2.0) - (6.0 \times V_{\text{CA free}}) = 0 \qquad \therefore V_{\text{CA free}} = + 3.33 \text{ kN} \uparrow \qquad \text{Equation (1)}$$

Consider the vertical equilibrium:

$$+ve \uparrow \Sigma F_z = 0$$

$$+ V_{\text{CA free}} + V_{\text{CA free}} - 10.0 = 0 \qquad \therefore V_{\text{AC free}} = + 6.67 \text{ kN} \uparrow \qquad \text{Equation (2)}$$

Consider span CD:

$$+ve \; \curvearrowright \; \Sigma M_C = 0$$

$+ (10 \times 8.0 \times 4.0) - (8.0 \times V_{DC \; free}) = 0$ $\quad \therefore V_{DC \; free} = + 40.0$ kN $\uparrow$ Equation (1)

Consider the vertical equilibrium:

$$+ve \; \uparrow \; \Sigma F_z = 0$$

$+ V_{CD \; free} + V_{DC \; free} - (10 \times 8.0) = 0$ $\quad \therefore V_{CD \; free} = + 40.0$ kN $\uparrow$ Equation (2)

Consider span DF:

$$+ve \; \curvearrowright \; \Sigma M_D = 0$$

$+ (25 \times 2.5) + (15 \times 5.0 \times 2.5) - (5.0 \times V_{FD \; free}) = 0$

$\therefore V_{FD \; free} = + 50.0$ kN $\uparrow$ Equation (1)

Consider the vertical equilibrium:

$$+ve \; \uparrow \; \Sigma F_z = 0$$

$+ V_{DF \; free} + V_{FD \; free} - 25.0 - (15 \times 5.0) = 0$ $\therefore V_{DF \; free} = + 50.0$ kN $\uparrow$ Equation (2)

$V_{A \; free} = V_{AC \; free} = + 6.67$ kN
$V_{C \; free} = V_{CA \; free} + V_{CD \; free} = (+ 3.33 + 40.0) = + 43.33$ kN
$V_{D \; free} = V_{DC \; free} + V_{DF \; free} = (+ 40.0 + 50.0) = + 90.0$ kN
$V_{F \; free} = V_{FD \; free} = + 50.0$ kN

The final vertical support reactions are given by (i) + (ii):
$V_A = V_{A \; fixed} + V_{A \; free} = (- 4.09 + 6.67) = + 2.58$ kN
$V_C = V_{C \; fixed} + V_{C \; free} = (- 1.58 + 43.33) = + 41.81$ kN
$V_D = V_{D \; fixed} + V_{D \; free} = (+ 19.33 + 90.0) = + 109.33$ kN
$V_F = V_{F \; fixed} + V_{F \; free} = (- 13.72 + 50.0) = + 36.28$ kN

Check the vertical equilibrium: Total vertical force $= + 2.58 + 41.81 + 109.33 + 36.28$
$= + 190$ kN (= total applied load)

The final bending moment diagram is shown in Figure 4.94.

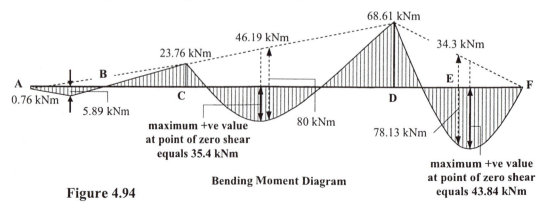

Bending Moment Diagram

Figure 4.94

4.7.11 Problems: Moment Distribution – Continuous Beams

A series of continuous beams are indicated in Problems **4.28** to **4.32** in which the relative *EI* values and the applied loading are given. In each case:

i)   determine the support reactions,
ii) sketch the shear force diagram   and
iii) sketch the bending moment diagram.

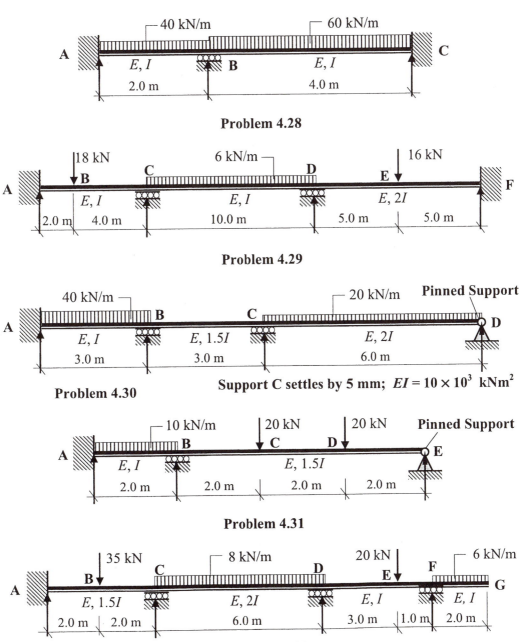

Problem 4.28

Problem 4.29

Problem 4.30   Support C settles by 5 mm; $EI = 10 \times 10^3$ kNm²

Problem 4.31

Problem 4.32

4.7.12 Solutions: Moment Distribution–Continuous Beams

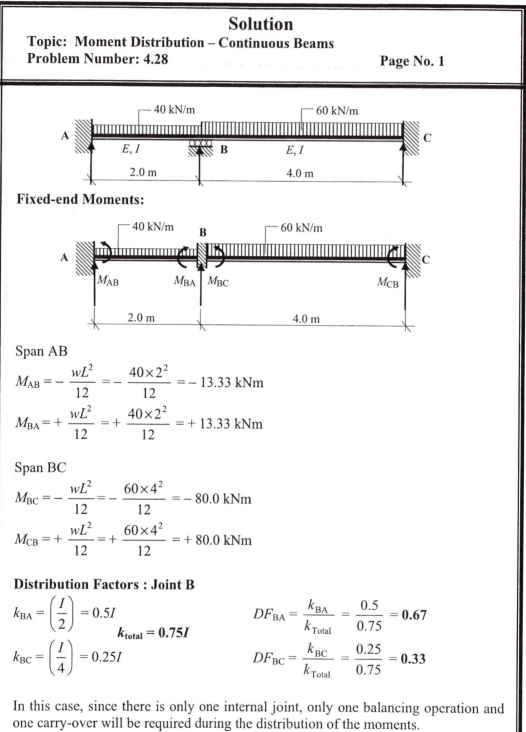

Solution

Topic: Moment Distribution – Continuous Beams

Problem Number: 4.28

Page No. 1

Fixed-end Moments:

Span AB

$$M_{AB} = -\frac{wL^2}{12} = -\frac{40 \times 2^2}{12} = -13.33 \text{ kNm}$$

$$M_{BA} = +\frac{wL^2}{12} = +\frac{40 \times 2^2}{12} = +13.33 \text{ kNm}$$

Span BC

$$M_{BC} = -\frac{wL^2}{12} = -\frac{60 \times 4^2}{12} = -80.0 \text{ kNm}$$

$$M_{CB} = +\frac{wL^2}{12} = +\frac{60 \times 4^2}{12} = +80.0 \text{ kNm}$$

Distribution Factors : Joint B

$$k_{BA} = \left(\frac{I}{2}\right) = 0.5I \qquad\qquad DF_{BA} = \frac{k_{BA}}{k_{Total}} = \frac{0.5}{0.75} = \mathbf{0.67}$$

$$k_{total} = \mathbf{0.75I}$$

$$k_{BC} = \left(\frac{I}{4}\right) = 0.25I \qquad\qquad DF_{BC} = \frac{k_{BC}}{k_{Total}} = \frac{0.25}{0.75} = \mathbf{0.33}$$

In this case, since there is only one internal joint, only one balancing operation and one carry-over will be required during the distribution of the moments.

Solution

Topic: Moment Distribution – Continuous Beams
Problem Number: 4.28   **Page No. 2**

Moment Distribution Table:

Joint	A		B			C
	AB		BA	BC		CB
Distribution Factors	0		0.67	0.33		0
Fixed-end Moments	− 13.33		+ 13.33	− 80.0		+ 80.0
Balance			+ 44.67	+ 22.0		
Carry-over	+ 22.33					+ 11.0
Total	+ 9.0		+ 58.0	− 58.0		+ 91.0

Continuity Moments:

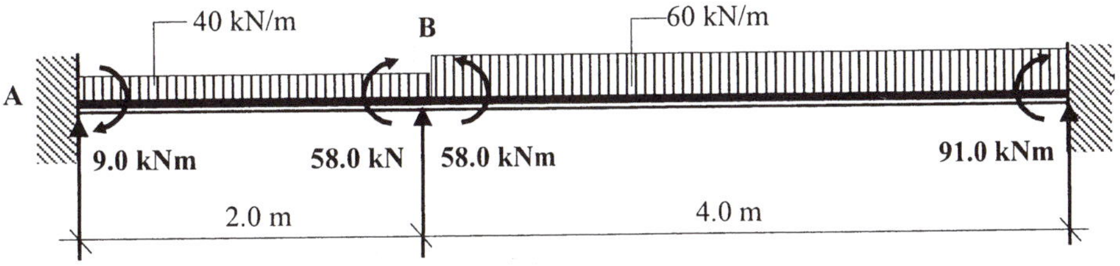

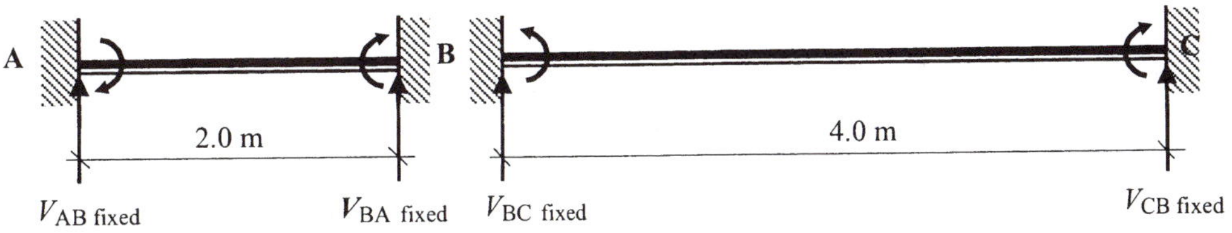

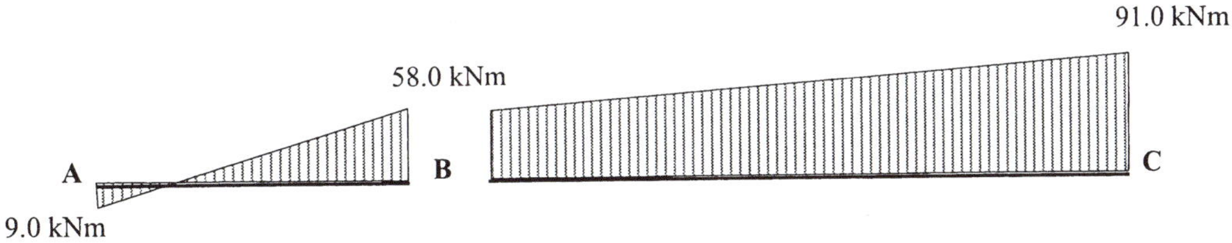

Fixed Bending Moment Diagrams

(i) Fixed vertical reactions:

Consider span AB: +ve $\circlearrowright \Sigma M_A = 0$

$+ 9.0 + 58.0 - (2.0 \times V_{\text{BA fixed}}) = 0$ $\therefore V_{\text{BA fixed}} = + 33.5$ kN ↑

Consider the vertical equilibrium of the beam: +ve ↑$\Sigma F_z = 0$

$+ V_{\text{AB fixed}} + V_{\text{BA fixed}} = 0$ $\therefore V_{\text{AB fixed}} = - 33.5$ kN ↓

Solution
Topic: Moment Distribution – Continuous Beams
Problem Number: 4.28

Page No. 3

Consider span BC: $+ve \downarrow \Sigma M_B = 0$
$-58.0 + 91.0 - (4.0 \times V_{CB\ fixed}) = 0$ $\qquad \therefore V_{CB\ fixed} = +8.25$ kN ↑

Consider the vertical equilibrium of the beam: $+ve \uparrow \Sigma F_z = 0$
$+ V_{BC\ fixed} + V_{CB\ fixed} = 0$ $\qquad \therefore V_{BC\ fixed} = -8.25$ kN ↓

The total vertical reaction at each support due to the continuity moments is equal to the algebraic sum of the contributions from each beam at the support.

$V_{A\ fixed} = V_{AB\ fixed} = -33.5$ kN
$V_{B\ fixed} = V_{BA\ fixed} + V_{BC\ fixed} = (+33.5 - 8.25) = +25.25$ kN
$V_{C\ fixed} = V_{CB\ fixed} = +8.25$ kN

Free bending moments:

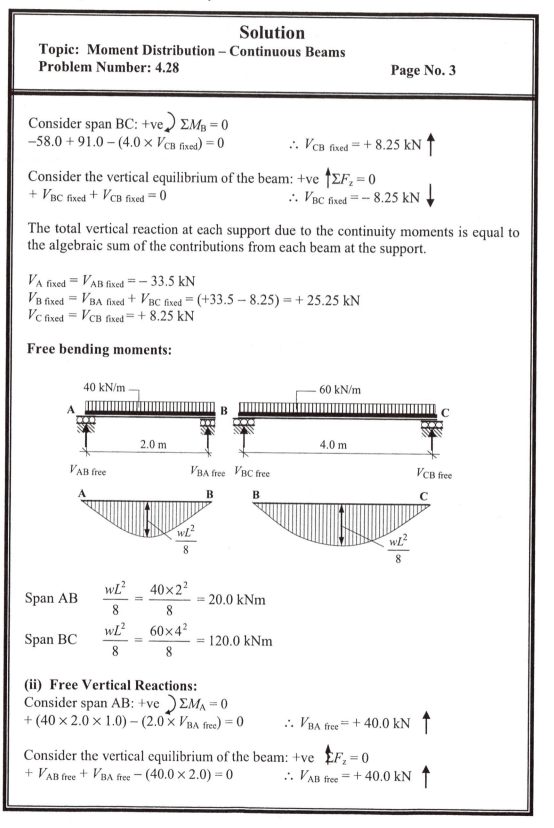

Span AB $\qquad \dfrac{wL^2}{8} = \dfrac{40 \times 2^2}{8} = 20.0$ kNm

Span BC $\qquad \dfrac{wL^2}{8} = \dfrac{60 \times 4^2}{8} = 120.0$ kNm

(ii) Free Vertical Reactions:
Consider span AB: $+ve \downarrow \Sigma M_A = 0$
$+ (40 \times 2.0 \times 1.0) - (2.0 \times V_{BA\ free}) = 0$ $\qquad \therefore V_{BA\ free} = +40.0$ kN ↑

Consider the vertical equilibrium of the beam: $+ve \uparrow \Sigma F_z = 0$
$+ V_{AB\ free} + V_{BA\ free} - (40.0 \times 2.0) = 0$ $\qquad \therefore V_{AB\ free} = +40.0$ kN ↑

Solution

Topic: Moment Distribution – Continuous Beams
Problem Number: 4.28   **Page No. 4**

Consider span BC: +ve $\circlearrowright$ $\Sigma M_B = 0$
$+ (60 \times 4.0 \times 2.0) - (4.0 \times V_{CB\ free}) = 0$ $\therefore V_{CB\ free} = + 120.0$ kN $\uparrow$

Consider the vertical equilibrium of the beam: +ve $\uparrow \Sigma F_z = 0$
$+ V_{BC\ free} + V_{CB\ free} - (60.0 \times 4.0) = 0$ $\therefore V_{BC\ free} = + 120.0$ kN $\uparrow$

$V_{A\ free} = V_{AB\ free} = + 40.0$ kN
$V_{B\ free} = V_{BA\ free} + V_{BC\ free} = (+ 40.0 + 120.0) = + 160.0$ kN
$V_{C\ free} = V_{CB\ free} = + 120.0$ kN

The final vertical support reactions are given by (i) + (ii):
$V_A = V_{A\ fixed} + V_{A\ free} = (- 33.5 + 40.0) = + 6.5$ kN $\uparrow$
$V_B = V_{B\ fixed} + V_{B\ free} = (+ 25.5 + 160.0) = + 185.25$ kN $\uparrow\uparrow$
$V_C = V_{C\ fixed} + V_{C\ free} = (+ 8.25 + 120.0) = + 128.25$ kN $\uparrow$

Check the vertical equilibrium:
Total vertical force $= + 6.5 + 185.25 + 128.25$
$\qquad\qquad\qquad = + 320.0$ kN (= total applied load)

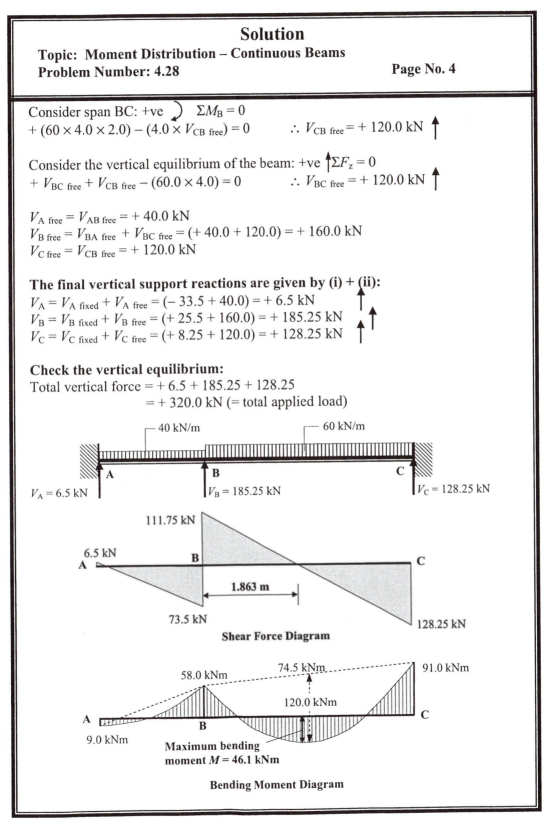

Shear Force Diagram

Bending Moment Diagram

Solution

Topic: Moment Distribution – Continuous Beams
Problem Number: 4.29

Page No. 1

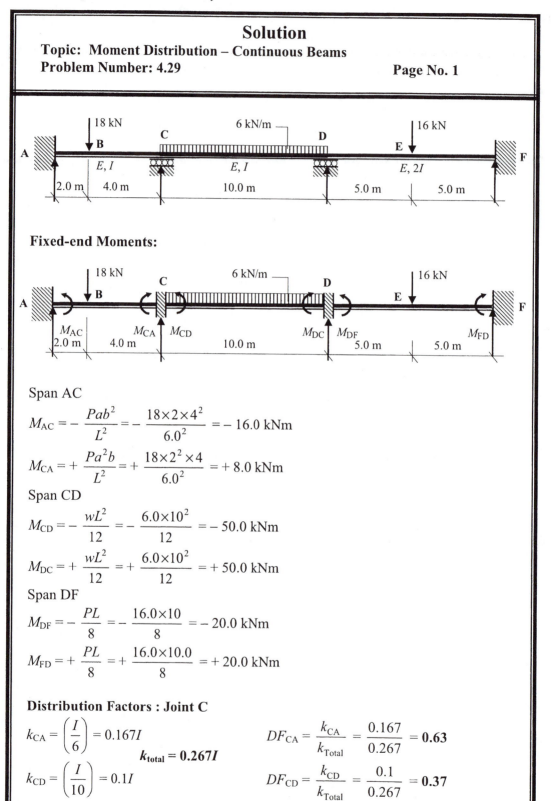

Fixed-end Moments:

Span AC

$$M_{AC} = -\frac{Pab^2}{L^2} = -\frac{18 \times 2 \times 4^2}{6.0^2} = -16.0 \text{ kNm}$$

$$M_{CA} = +\frac{Pa^2 b}{L^2} = +\frac{18 \times 2^2 \times 4}{6.0^2} = +8.0 \text{ kNm}$$

Span CD

$$M_{CD} = -\frac{wL^2}{12} = -\frac{6.0 \times 10^2}{12} = -50.0 \text{ kNm}$$

$$M_{DC} = +\frac{wL^2}{12} = +\frac{6.0 \times 10^2}{12} = +50.0 \text{ kNm}$$

Span DF

$$M_{DF} = -\frac{PL}{8} = -\frac{16.0 \times 10}{8} = -20.0 \text{ kNm}$$

$$M_{FD} = +\frac{PL}{8} = +\frac{16.0 \times 10.0}{8} = +20.0 \text{ kNm}$$

Distribution Factors : Joint C

$$k_{CA} = \left(\frac{I}{6}\right) = 0.167I \qquad\qquad DF_{CA} = \frac{k_{CA}}{k_{Total}} = \frac{0.167}{0.267} = \mathbf{0.63}$$

$$\mathbf{k_{total} = 0.267I}$$

$$k_{CD} = \left(\frac{I}{10}\right) = 0.1I \qquad\qquad DF_{CD} = \frac{k_{CD}}{k_{Total}} = \frac{0.1}{0.267} = \mathbf{0.37}$$

Solution

Topic: Moment Distribution – Continuous Beams

Problem Number: 4.29

Page No. 2

Distribution Factors : Joint D

$$k_{DC} = \left(\frac{I}{10}\right) = 0.1I \qquad\qquad DF_{DC} = \frac{k_{DC}}{k_{Total}} = \frac{0.1}{0.3} = \mathbf{0.33}$$

$$k_{total} = \mathbf{0.3I}$$

$$k_{DF} = \left(\frac{2I}{10}\right) = 0.2I \qquad\qquad DF_{DF} = \frac{k_{DF}}{k_{Total}} = \frac{0.2}{0.3} = \mathbf{0.67}$$

Moment Distribution Table:

Joint	A		C			D			F
	AC		CA	CD		DC	DF		FD
DF's	0		0.63	0.37		0.33	0.67		0
FEM's	− 16.0		+ 8.0	− 50.0		+ 50.0	− 20.0		+ 20.0
Balance			+ 26.5	+ 15.5		− 9.9	− 20.1		
Carry-over	+ 13.3			− 5.0		+ 7.8			− 10.1
Balance			+ 3.2	+ 1.8		− 2.6	− 5.2		
Carry-over	+ 1.6			− 1.3		+ 0.9			− 2.6
Balance			+ 0.8	+ 0.5		− 0.3	− 0.6		
Carry-over	+ 0.4								− 0.3
Total	**− 0.7**		**+ 38.5**	**− 38.5**		**+ 45.9**	**− 45.9**		**+ 7.0**

Continuity Moments:

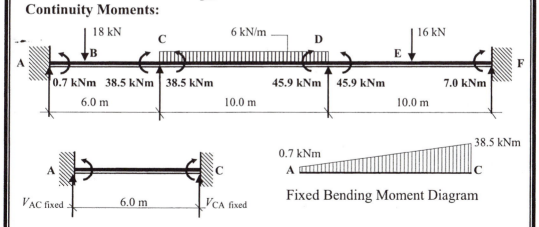

18 kN   6 kN/m   16 kN

A   B   C   D   E   F

0.7 kNm   38.5 kNm   38.5 kNm   45.9 kNm   45.9 kNm   7.0 kNm

6.0 m   10.0 m   10.0 m

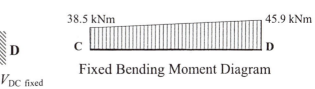

$V_{AC\ fixed}$ 6.0 m $V_{CA\ fixed}$

0.7 kNm   38.5 kNm

A   C

Fixed Bending Moment Diagram

$V_{CD\ fixed}$ 10.0 m $V_{DC\ fixed}$

38.5 kNm   45.9 kNm

C   D

Fixed Bending Moment Diagram

Solution
Topic: Moment Distribution – Continuous Beams
Problem Number: 4.29   **Page No. 3**

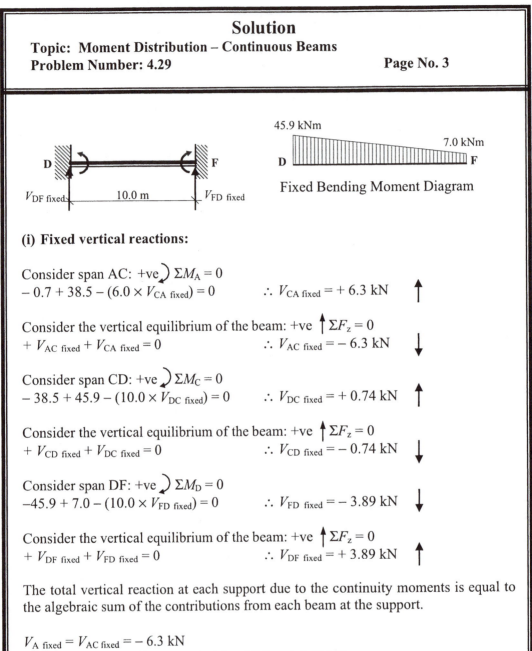

Fixed Bending Moment Diagram

(i) Fixed vertical reactions:

Consider span AC: $+ve \, \circlearrowright \, \Sigma M_A = 0$
$-0.7 + 38.5 - (6.0 \times V_{CA \text{ fixed}}) = 0$ $\therefore V_{CA \text{ fixed}} = +6.3 \text{ kN} \quad \uparrow$

Consider the vertical equilibrium of the beam: $+ve \, \uparrow \Sigma F_z = 0$
$+ V_{AC \text{ fixed}} + V_{CA \text{ fixed}} = 0$ $\therefore V_{AC \text{ fixed}} = -6.3 \text{ kN} \quad \downarrow$

Consider span CD: $+ve \, \circlearrowright \, \Sigma M_C = 0$
$-38.5 + 45.9 - (10.0 \times V_{DC \text{ fixed}}) = 0$ $\therefore V_{DC \text{ fixed}} = +0.74 \text{ kN} \quad \uparrow$

Consider the vertical equilibrium of the beam: $+ve \, \uparrow \Sigma F_z = 0$
$+ V_{CD \text{ fixed}} + V_{DC \text{ fixed}} = 0$ $\therefore V_{CD \text{ fixed}} = -0.74 \text{ kN} \quad \downarrow$

Consider span DF: $+ve \, \circlearrowright \, \Sigma M_D = 0$
$-45.9 + 7.0 - (10.0 \times V_{FD \text{ fixed}}) = 0$ $\therefore V_{FD \text{ fixed}} = -3.89 \text{ kN} \quad \downarrow$

Consider the vertical equilibrium of the beam: $+ve \, \uparrow \Sigma F_z = 0$
$+ V_{DF \text{ fixed}} + V_{FD \text{ fixed}} = 0$ $\therefore V_{DF \text{ fixed}} = +3.89 \text{ kN} \quad \uparrow$

The total vertical reaction at each support due to the continuity moments is equal to the algebraic sum of the contributions from each beam at the support.

$V_{A \text{ fixed}} = V_{AC \text{ fixed}} = -6.3 \text{ kN}$
$V_{C \text{ fixed}} = V_{CA \text{ fixed}} + V_{CD \text{ fixed}} = (+6.3 - 0.74) = +5.56 \text{ kN}$
$V_{D \text{ fixed}} = V_{DC \text{ fixed}} + V_{DF \text{ fixed}} = (+0.74 + 3.89) = +4.63 \text{ kN}$
$V_{F \text{ fixed}} = V_{FD \text{ fixed}} = -3.89 \text{ kN}$

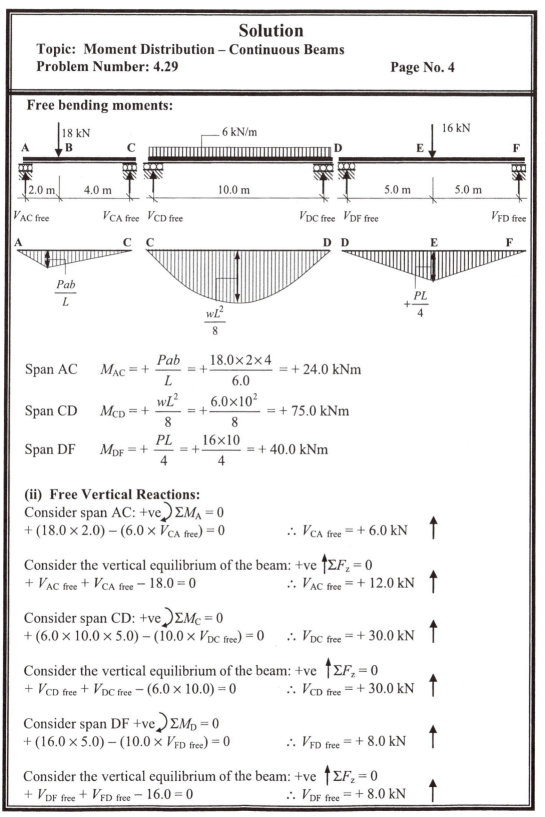

Solution

Topic: Moment Distribution – Continuous Beams
Problem Number: 4.29   **Page No. 4**

Free bending moments:

Span AC $M_{AC} = +\dfrac{Pab}{L} = +\dfrac{18.0 \times 2 \times 4}{6.0} = +24.0$ kNm

Span CD $M_{CD} = +\dfrac{wL^2}{8} = +\dfrac{6.0 \times 10^2}{8} = +75.0$ kNm

Span DF $M_{DF} = +\dfrac{PL}{4} = +\dfrac{16 \times 10}{4} = +40.0$ kNm

(ii) Free Vertical Reactions:
Consider span AC: $+ve \;\curvearrowright \Sigma M_A = 0$
$+ (18.0 \times 2.0) - (6.0 \times V_{CA\ free}) = 0$ $\therefore V_{CA\ free} = +6.0$ kN

Consider the vertical equilibrium of the beam: $+ve \;\uparrow \Sigma F_z = 0$
$+ V_{AC\ free} + V_{CA\ free} - 18.0 = 0$ $\therefore V_{AC\ free} = +12.0$ kN

Consider span CD: $+ve \;\curvearrowright \Sigma M_C = 0$
$+ (6.0 \times 10.0 \times 5.0) - (10.0 \times V_{DC\ free}) = 0$ $\therefore V_{DC\ free} = +30.0$ kN

Consider the vertical equilibrium of the beam: $+ve \;\uparrow \Sigma F_z = 0$
$+ V_{CD\ free} + V_{DC\ free} - (6.0 \times 10.0) = 0$ $\therefore V_{CD\ free} = +30.0$ kN

Consider span DF $+ve \;\curvearrowright \Sigma M_D = 0$
$+ (16.0 \times 5.0) - (10.0 \times V_{FD\ free}) = 0$ $\therefore V_{FD\ free} = +8.0$ kN

Consider the vertical equilibrium of the beam: $+ve \;\uparrow \Sigma F_z = 0$
$+ V_{DF\ free} + V_{FD\ free} - 16.0 = 0$ $\therefore V_{DF\ free} = +8.0$ kN

Solution

Topic: Moment Distribution – Continuous Beams
Problem Number: 4.29   **Page No. 5**

$V_{A\ free} = V_{AC\ free} = + 12.0\ kN$
$V_{C\ free} = V_{CA\ free} + V_{CD\ free} = (+ 6.0 + 30.0) = + 36.0\ kN$
$V_{D\ free} = V_{DC\ free} + V_{DF\ free} = (+ 30.0 + 8.0) = + 38.0\ kN$
$V_{F\ free} = V_{FD\ free} = + 8.0\ kN$

The final vertical support reactions are given by (i) + (ii):
$V_A = V_{A\ fixed} + V_{A\ free} = (- 6.3 + 12.0) = + 5.7\ kN$
$V_C = V_{C\ fixed} + V_{C\ free} = (+ 5.56 + 36.0) = + 41.56\ kN$
$V_D = V_{D\ fixed} + V_{D\ free} = (+ 4.63 + 38.0) = + 42.63\ kN$
$V_F = V_{F\ fixed} + V_{F\ free} = (- 3.89 + 8.0) = + 4.11\ kN$

Check the vertical equilibrium:
Total vertical force $= + 5.7 + 41.56 + 42.63 + 4.11$
$= + 94.0\ kN\ (= \text{total applied load})$

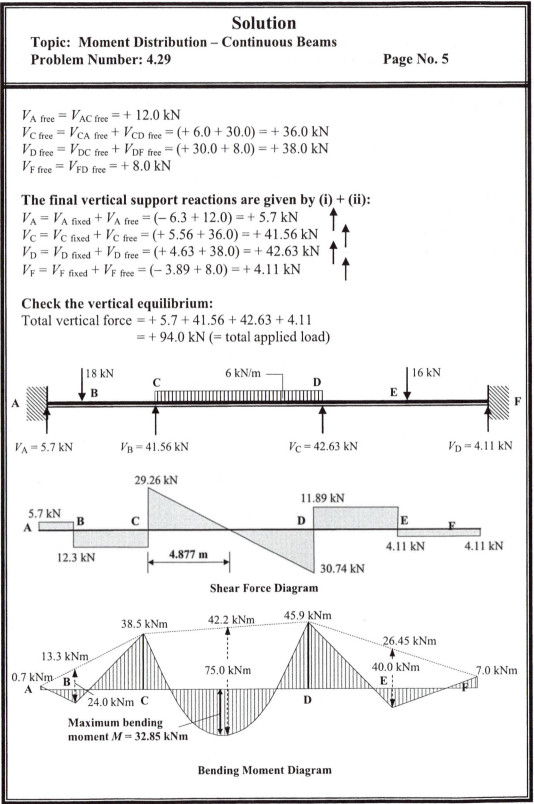

Shear Force Diagram

Bending Moment Diagram

Solution

Topic: Moment Distribution – Continuous Beams

Problem Number: 4.30   **Page No. 1**

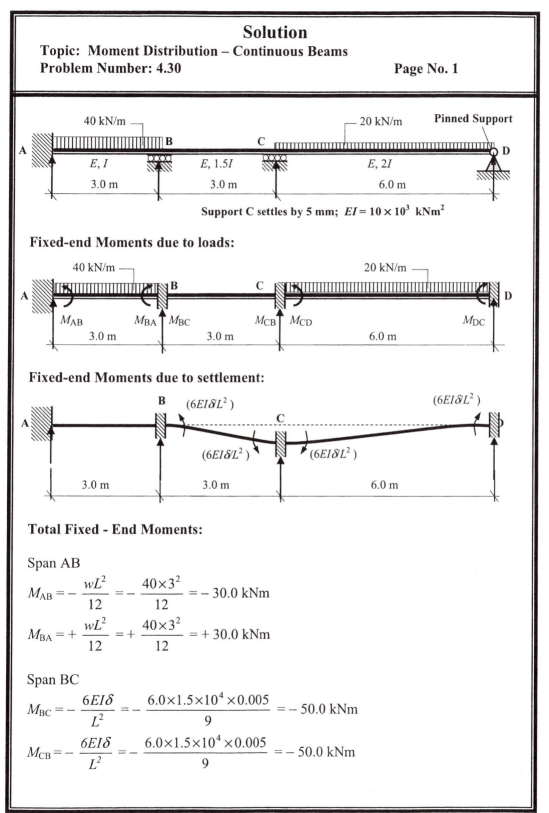

Support C settles by 5 mm; $EI = 10 \times 10^3$ kNm^2

Fixed-end Moments due to loads:

Fixed-end Moments due to settlement:

Total Fixed - End Moments:

Span AB

$$M_{AB} = -\frac{wL^2}{12} = -\frac{40 \times 3^2}{12} = -30.0 \text{ kNm}$$

$$M_{BA} = +\frac{wL^2}{12} = +\frac{40 \times 3^2}{12} = +30.0 \text{ kNm}$$

Span BC

$$M_{BC} = -\frac{6EI\delta}{L^2} = -\frac{6.0 \times 1.5 \times 10^4 \times 0.005}{9} = -50.0 \text{ kNm}$$

$$M_{CB} = -\frac{6EI\delta}{L^2} = -\frac{6.0 \times 1.5 \times 10^4 \times 0.005}{9} = -50.0 \text{ kNm}$$

Solution
Topic: Moment Distribution – Continuous Beams
Problem Number: 4.30   **Page No. 2**

Span CD *

$$M_{CD} = -\frac{wL^2}{12} + \frac{6EI\delta}{L^2} = -\frac{20.0 \times 6^2}{12} + \frac{6.0 \times 2.0 \times 10^4 \times 0.005}{36} = -43.33 \text{ kNm}$$

$$M_{DC} = +\frac{wL^2}{12} + \frac{6EI\delta}{L^2} = +\frac{20.0 \times 6^2}{12} + \frac{6.0 \times 2.0 \times 10^4 \times 0.005}{36} = +76.67 \text{ kNm}$$

* Since support D is pinned, the fixed-end moments are $(M_{CD} - 0.5M_{DC})$ at C and zero at D
$(M_{CD} - 0.5M_{DC}) = [-43.33 - (0.5 \times 76.67)] = -81.67$ kNm.

Distribution Factors : Joint B

$$k_{BA} = \left(\frac{I}{3}\right) = 0.333I$$

$$k_{BC} = \left(\frac{1.5I}{3}\right) = 0.5I$$

$$k_{total} = 0.833I$$

$$DF_{BA} = \frac{k_{BA}}{k_{Total}} = \frac{0.333}{0.833} = 0.4$$

$$DF_{BC} = \frac{k_{BC}}{k_{Total}} = \frac{0.5}{0.833} = 0.6$$

Distribution Factors : Joint C

$$k_{CB} = \left(\frac{1.5I}{3}\right) = 0.5I$$

$$k_{CD} = \left(\frac{3}{4} \times \frac{2I}{6}\right) = 0.25I$$

$$k_{total} = 0.75I$$

$$DF_{CB} = \frac{k_{CB}}{k_{Total}} = \frac{0.5}{0.75} = 0.67$$

$$DF_{CD} = \frac{k_{CD}}{k_{Total}} = \frac{0.25}{0.75} = 0.33$$

Moment Distribution Table:

Joint	A		B			C			D
	AB		BA	BC		CB	CD		DC
DF's	0		0.4	0.6		0.67	0.33		1.0
FEM's	− 30.0		+ 30.0	− 50.0		− 50.0	− 81.67		0
Balance			+ 8.0	+ 12.0		+ 88.22	+ 43.45		
Carry-over	+ 4.0			+ 44.1		+ 6.0			
Balance			− 17.6	− 26.5		− 4.0	− 2.0		
Carry-over	− 8.8			− 2.0		− 13.3			
Balance			+ 0.8	+ 1.2		+ 8.9	+ 4.4		
Carry-over	+ 0.4			+ 4.5		+ 0.6			
Balance			− 1.8	− 2.7		− 0.4	− 0.2		
Carry-over	− 0.9			− 0.2		− 1.2			
Balance			+ 0.1	+ 0.1		+ 0.8	+ 0.8		
Total	− 35.3		+ 19.5	− 19.5		+ 35.2	− 35.2		0

Solution

Topic: Moment Distribution – Continuous Beams

Problem Number: 4.30   **Page No. 3**

Continuity Moments:

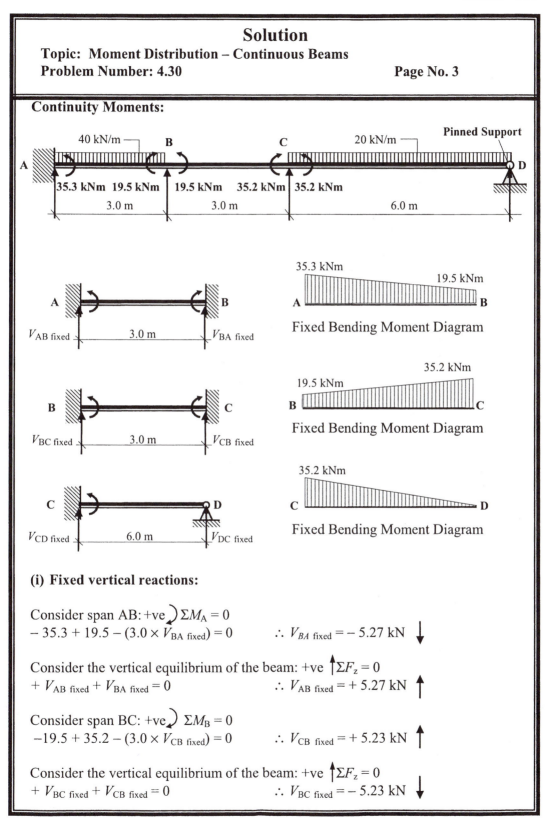

(i) Fixed vertical reactions:

Consider span AB: +ve $\curvearrowright \Sigma M_A = 0$

$-35.3 + 19.5 - (3.0 \times V_{BA \text{ fixed}}) = 0$ $\therefore V_{BA \text{ fixed}} = -5.27 \text{ kN} \downarrow$

Consider the vertical equilibrium of the beam: +ve $\uparrow \Sigma F_z = 0$

$+ V_{AB \text{ fixed}} + V_{BA \text{ fixed}} = 0$ $\therefore V_{AB \text{ fixed}} = +5.27 \text{ kN} \uparrow$

Consider span BC: +ve $\curvearrowright \Sigma M_B = 0$

$-19.5 + 35.2 - (3.0 \times V_{CB \text{ fixed}}) = 0$ $\therefore V_{CB \text{ fixed}} = +5.23 \text{ kN} \uparrow$

Consider the vertical equilibrium of the beam: +ve $\uparrow \Sigma F_z = 0$

$+ V_{BC \text{ fixed}} + V_{CB \text{ fixed}} = 0$ $\therefore V_{BC \text{ fixed}} = -5.23 \text{ kN} \downarrow$

Solution
Topic: **Moment Distribution – Continuous Beams**
Problem Number: **4.30**   Page No. 4

Consider span CD: +ve $\circlearrowright \Sigma M_C = 0$
$-35.2 - (6.0 \times V_{DC\ fixed}) = 0$ $\therefore V_{DC\ fixed} = -5.87$ kN $\downarrow$

Consider the vertical equilibrium of the beam: +ve $\uparrow \Sigma F_z = 0$
$+V_{CD\ fixed} + V_{DC\ fixed} = 0$ $\therefore V_{CD\ fixed} = +5.87$ kN $\uparrow$

The total vertical reaction at each support due to the continuity moments is equal to the algebraic sum of the contributions from each beam at the support.

$V_{A\ fixed} = V_{AB\ fixed} = +5.27$ kN
$V_{B\ fixed} = V_{BA\ fixed} + V_{BC\ fixed} = (-5.27 - 5.23) = -10.5$ kN
$V_{C\ fixed} = V_{CB\ fixed} + V_{CD\ fixed} = (+5.23 + 5.87) = +11.1$ kN
$V_{D\ fixed} = V_{DC\ fixed} = -5.87$

Free bending moments:

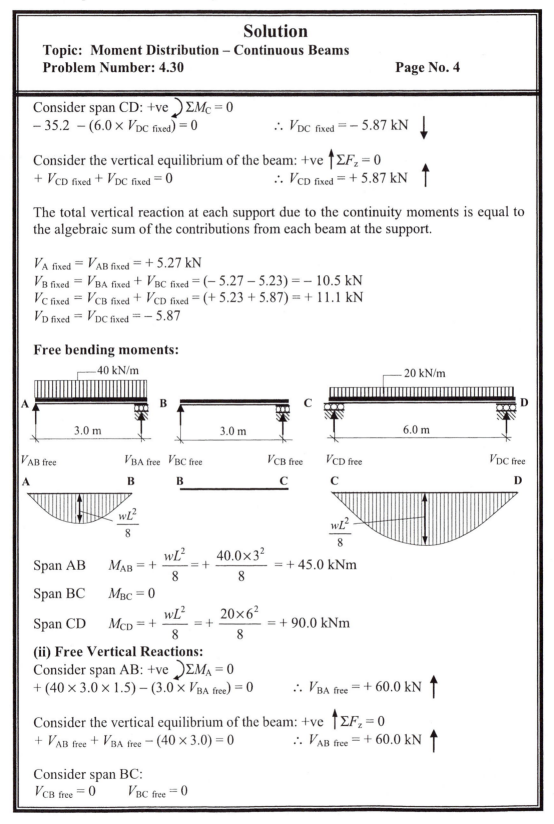

Span AB $M_{AB} = +\dfrac{wL^2}{8} = +\dfrac{40.0 \times 3^2}{8} = +45.0$ kNm

Span BC $M_{BC} = 0$

Span CD $M_{CD} = +\dfrac{wL^2}{8} = +\dfrac{20 \times 6^2}{8} = +90.0$ kNm

(ii) Free Vertical Reactions:
Consider span AB: +ve $\circlearrowright \Sigma M_A = 0$
$+(40 \times 3.0 \times 1.5) - (3.0 \times V_{BA\ free}) = 0$ $\therefore V_{BA\ free} = +60.0$ kN $\uparrow$

Consider the vertical equilibrium of the beam: +ve $\uparrow \Sigma F_z = 0$
$+V_{AB\ free} + V_{BA\ free} - (40 \times 3.0) = 0$ $\therefore V_{AB\ free} = +60.0$ kN $\uparrow$

Consider span BC:
$V_{CB\ free} = 0$ $V_{BC\ free} = 0$

Solution

Topic: Moment Distribution – Continuous Beams
Problem Number: 4.30

Consider span CD: $+ve \,\curvearrowright \Sigma M_C = 0$

$+ (20.0 \times 6.0 \times 3.0) - (6.0 \times V_{DC\ free}) = 0 \qquad \therefore V_{DC\ free} = +60.0 \text{ kN} \uparrow$

Consider the vertical equilibrium of the beam: $+ve \uparrow \Sigma F_z = 0$

$+ V_{CD\ free} + V_{DC\ free} - (60 \times 6.0) = 0 \qquad \therefore V_{CD\ free} = +60.0 \text{ kN} \uparrow$

$V_{A\ free} = V_{AB\ free} = +60.0 \text{ kN}$
$V_{B\ free} = V_{BA\ free} + V_{BC\ free} = (+60.0 + 0) = +60.0 \text{ kN}$
$V_{C\ free} = V_{CB\ free} + V_{CD\ free} = (0 + 60.0) = +60.0 \text{ kN}$
$V_{D\ free} = V_{DC\ free} = +60.0 \text{ kN}$

The final vertical support reactions are given by (i) + (ii):

$V_A = V_{A\ fixed} + V_{A\ free} = (+5.27 + 60.0) = +65.27 \text{ kN} \uparrow$
$V_B = V_{B\ fixed} + V_{B\ free} = (-10.5 + 60.0) = +49.5 \text{ kN}$
$V_C = V_{C\ fixed} + V_{C\ free} = (+11.1 + 60.0) = +71.1 \text{ kN}$
$V_D = V_{D\ fixed} + V_{D\ free} = (-5.87 + 60.0) = +54.13 \text{ kN}$

Check the vertical equilibrium:

Total vertical force $= +65.25 + 49.5 + 71.1 + 54.13$
$\qquad\qquad\qquad\quad = +240.0 \text{ kN}$ (= total applied load)

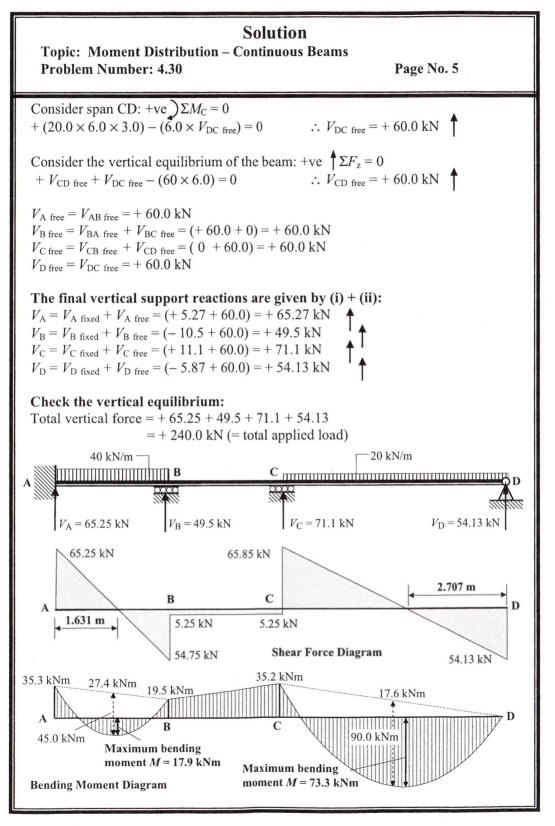

Shear Force Diagram

Maximum bending moment $M = 17.9$ kNm

Maximum bending moment $M = 73.3$ kNm

Bending Moment Diagram

Solution
Topic: Moment Distribution – Continuous Beams
Problem Number: 4.31   **Page No. 1**

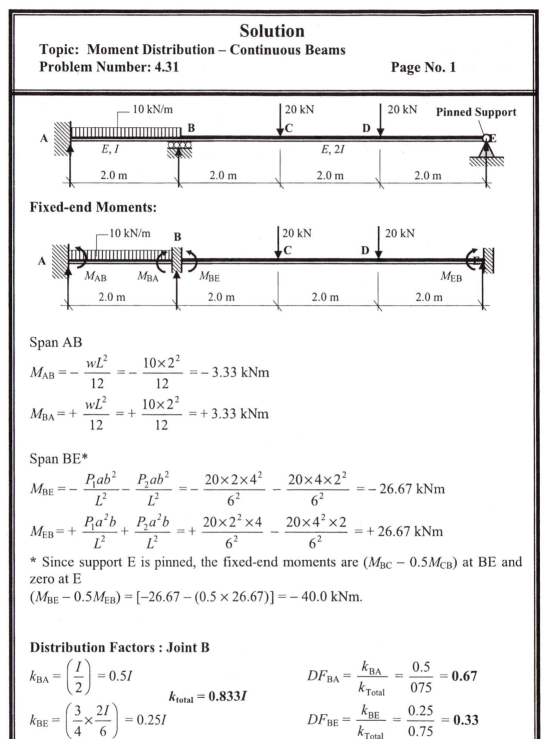

Fixed-end Moments:

Span AB

$$M_{AB} = -\frac{wL^2}{12} = -\frac{10 \times 2^2}{12} = -3.33 \text{ kNm}$$

$$M_{BA} = +\frac{wL^2}{12} = +\frac{10 \times 2^2}{12} = +3.33 \text{ kNm}$$

Span BE*

$$M_{BE} = -\frac{P_1ab^2}{L^2} - \frac{P_2ab^2}{L^2} = -\frac{20 \times 2 \times 4^2}{6^2} - \frac{20 \times 4 \times 2^2}{6^2} = -26.67 \text{ kNm}$$

$$M_{EB} = +\frac{P_1a^2b}{L^2} + \frac{P_2a^2b}{L^2} = +\frac{20 \times 2^2 \times 4}{6^2} - \frac{20 \times 4^2 \times 2}{6^2} = +26.67 \text{ kNm}$$

* Since support E is pinned, the fixed-end moments are ($M_{BC} - 0.5M_{CB}$) at BE and zero at E
($M_{BE} - 0.5M_{EB}$) = [−26.67 − (0.5 × 26.67)] = − 40.0 kNm.

Distribution Factors : Joint B

$$k_{BA} = \left(\frac{I}{2}\right) = 0.5I \qquad DF_{BA} = \frac{k_{BA}}{k_{Total}} = \frac{0.5}{075} = \mathbf{0.67}$$

$$k_{total} = \mathbf{0.833I}$$

$$k_{BE} = \left(\frac{3}{4} \times \frac{2I}{6}\right) = 0.25I \qquad DF_{BE} = \frac{k_{BE}}{k_{Total}} = \frac{0.25}{0.75} = \mathbf{0.33}$$

In this case, since there is only one internal joint, only one balancing operation and one carry-over will be required during the distribution of the moments.

Solution

Topic: Moment Distribution – Continuous Beams
Problem Number: 4.31   **Page No. 2**

Moment Distribution Table:

Joint	A		B			E
	AB		BA	BE		EB
Distribution Factors	0		0.67	0.33		1.0
Fixed-end Moments	− 3.33		+ 3.33	− 40.0		0
Balance			+ 24.57	+ 12.1		
Carry-over	+ 12.29					
Total	**+ 8.96**		**+ 27.9**	**− 27.9**		**0**

Continuity Moments:

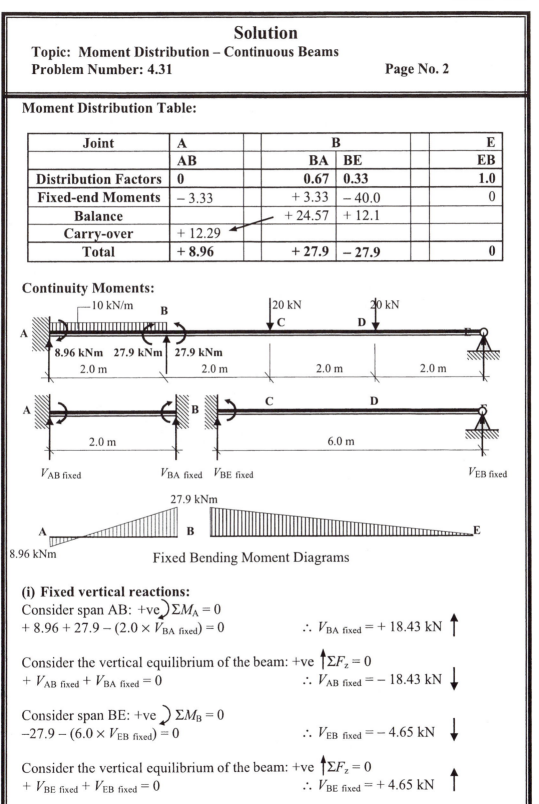

Fixed Bending Moment Diagrams

(i) Fixed vertical reactions:

Consider span AB: +ve $\circlearrowright \Sigma M_A = 0$

$+ 8.96 + 27.9 - (2.0 \times V_{BA \text{ fixed}}) = 0$ $\therefore V_{BA \text{ fixed}} = + 18.43$ kN ↑

Consider the vertical equilibrium of the beam: +ve ↑$\Sigma F_z = 0$

$+ V_{AB \text{ fixed}} + V_{BA \text{ fixed}} = 0$ $\therefore V_{AB \text{ fixed}} = − 18.43$ kN ↓

Consider span BE: +ve $\circlearrowright \Sigma M_B = 0$

$-27.9 - (6.0 \times V_{EB \text{ fixed}}) = 0$ $\therefore V_{EB \text{ fixed}} = − 4.65$ kN ↓

Consider the vertical equilibrium of the beam: +ve ↑$\Sigma F_z = 0$

$+ V_{BE \text{ fixed}} + V_{EB \text{ fixed}} = 0$ $\therefore V_{BE \text{ fixed}} = + 4.65$ kN ↑

Solution
Topic: Moment Distribution – Continuous Beams
Problem Number: 4.31   Page No. 3

The total vertical reaction at each support due to the continuity moments is equal to the algebraic sum of the contributions from each beam at the support.

$V_{\text{A fixed}} = V_{\text{AB fixed}} = -18.43$ kN
$V_{\text{B fixed}} = V_{\text{BA fixed}} + V_{\text{BE fixed}} = (+18.43 + 4.65) = +23.08$ kN
$V_{\text{E fixed}} = V_{\text{EB fixed}} = -4.65$ kN

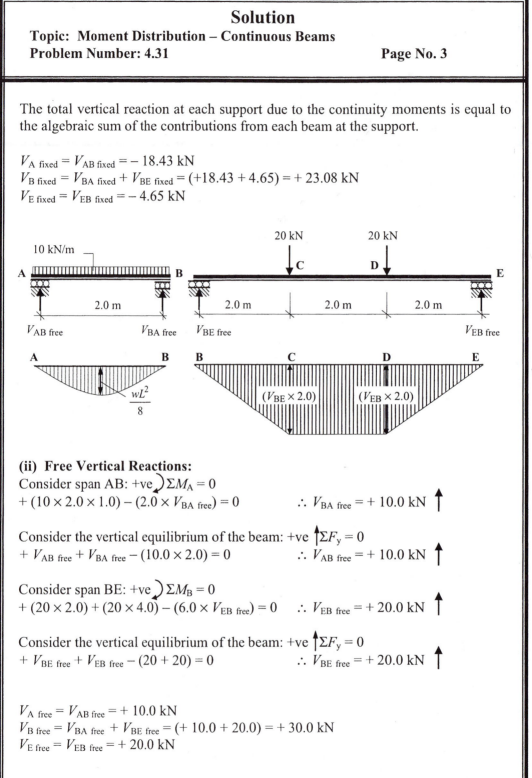

(ii) Free Vertical Reactions:
Consider span AB: $+\text{ve}\ \curvearrowright \Sigma M_{\text{A}} = 0$
$+ (10 \times 2.0 \times 1.0) - (2.0 \times V_{\text{BA free}}) = 0$ $\therefore V_{\text{BA free}} = +10.0$ kN $\uparrow$

Consider the vertical equilibrium of the beam: $+\text{ve} \uparrow \Sigma F_{\text{y}} = 0$
$+ V_{\text{AB free}} + V_{\text{BA free}} - (10.0 \times 2.0) = 0$ $\therefore V_{\text{AB free}} = +10.0$ kN $\uparrow$

Consider span BE: $+\text{ve}\ \curvearrowright \Sigma M_{\text{B}} = 0$
$+ (20 \times 2.0) + (20 \times 4.0) - (6.0 \times V_{\text{EB free}}) = 0$ $\therefore V_{\text{EB free}} = +20.0$ kN $\uparrow$

Consider the vertical equilibrium of the beam: $+\text{ve} \uparrow \Sigma F_{\text{y}} = 0$
$+ V_{\text{BE free}} + V_{\text{EB free}} - (20 + 20) = 0$ $\therefore V_{\text{BE free}} = +20.0$ kN $\uparrow$

$V_{\text{A free}} = V_{\text{AB free}} = +10.0$ kN
$V_{\text{B free}} = V_{\text{BA free}} + V_{\text{BE free}} = (+10.0 + 20.0) = +30.0$ kN
$V_{\text{E free}} = V_{\text{EB free}} = +20.0$ kN

Solution

Topic: Moment Distribution – Continuous Beams

Problem Number: 4.31   **Page No. 4**

Free bending moments:

Span AB $\dfrac{wL^2}{8} = \dfrac{10 \times 2^2}{8} = 5.0$ kNm

Span BE $(V_{\text{BE free}} \times 2.0) = (20 \times 2.0) = 40.0$ kNm

The final vertical support reactions are given by (i) + (ii):

$V_A = V_{A \text{ fixed}} + V_{A \text{ free}} = (-18.43 + 10.0) = -8.43$ kN

$V_B = V_{B \text{ fixed}} + V_{B \text{ free}} = (+23.08 + 30.0) = +53.08$ kN

$V_E = V_{E \text{ fixed}} + V_{E \text{ free}} = (-4.65 + 20.0) = +15.35$ kN

Check the vertical equilibrium:

Total vertical force $= -8.43 + 53.08 + 15.35$

$\qquad\qquad\qquad = +60.0$ kN (= total applied load)

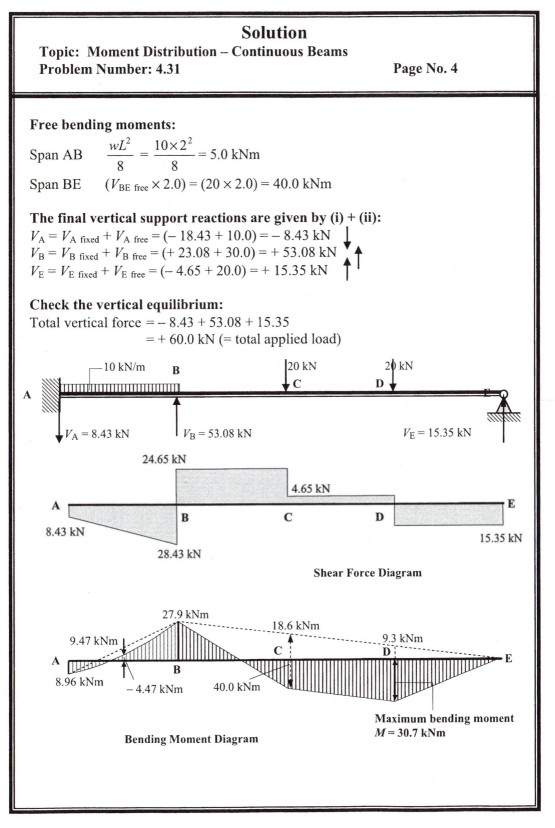

Shear Force Diagram

Bending Moment Diagram

Solution

Topic: Moment Distribution – Continuous Beams
Problem Number: 4.32   **Page No. 1**

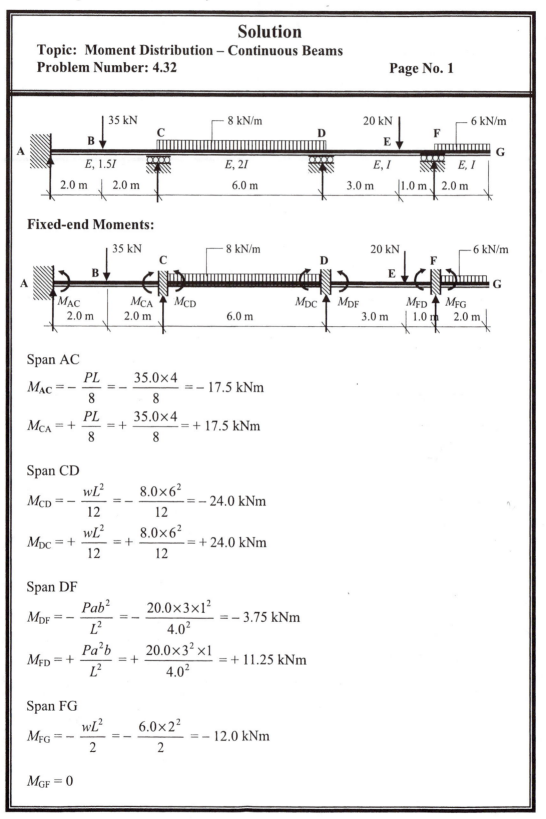

Fixed-end Moments:

Span AC

$$M_{AC} = -\frac{PL}{8} = -\frac{35.0 \times 4}{8} = -17.5 \text{ kNm}$$

$$M_{CA} = +\frac{PL}{8} = +\frac{35.0 \times 4}{8} = +17.5 \text{ kNm}$$

Span CD

$$M_{CD} = -\frac{wL^2}{12} = -\frac{8.0 \times 6^2}{12} = -24.0 \text{ kNm}$$

$$M_{DC} = +\frac{wL^2}{12} = +\frac{8.0 \times 6^2}{12} = +24.0 \text{ kNm}$$

Span DF

$$M_{DF} = -\frac{Pab^2}{L^2} = -\frac{20.0 \times 3 \times 1^2}{4.0^2} = -3.75 \text{ kNm}$$

$$M_{FD} = +\frac{Pa^2b}{L^2} = +\frac{20.0 \times 3^2 \times 1}{4.0^2} = +11.25 \text{ kNm}$$

Span FG

$$M_{FG} = -\frac{wL^2}{2} = -\frac{6.0 \times 2^2}{2} = -12.0 \text{ kNm}$$

$$M_{GF} = 0$$

Solution

Topic: Moment Distribution – Continuous Beams
Problem Number: 4.32   **Page No. 2**

Distribution Factors : Joint C

$$k_{CA} = \left(\frac{1.5I}{4}\right) = 0.375I \qquad\qquad DF_{CA} = \frac{k_{CA}}{k_{Total}} = \frac{0.375}{0.708} = \mathbf{0.53}$$

$$k_{total} = \mathbf{0.708}I$$

$$k_{CD} = \left(\frac{2I}{6}\right) = 0.333I \qquad\qquad DF_{CD} = \frac{k_{CD}}{k_{Total}} = \frac{0.333}{0.708} = \mathbf{0.47}$$

Distribution Factors : Joint D

Note: At joint D the stiffness of member DF is ($\frac{3}{4} \times I/L$) since support F is a simple support with a cantilever end, i.e. rotation can occur at this point.

$$k_{DC} = \left(\frac{2I}{6}\right) = 0.333I \qquad\qquad DF_{DC} = \frac{k_{DC}}{k_{Total}} = \frac{0.333}{0.521} = \mathbf{0.64}$$

$$k_{total} = \mathbf{0.521}I$$

$$k_{DF} = \left(\frac{3}{4} \times \frac{I}{4}\right) = 0.188I \qquad\qquad DF_{DF} = \frac{k_{DF}}{k_{Total}} = \frac{0.18}{0.521} = \mathbf{0.36}$$

Distribution Factors : Joint F

Note: At joint F the cantilever FG has zero stiffness.

$$k_{FD} = \left(\frac{I}{4}\right) = 0.25I \qquad\qquad DF_{FD} = \frac{k_{FD}}{k_{Total}} = \frac{0.25}{0.25} = \mathbf{1.0}$$

$$k_{total} = \mathbf{0.25}I$$

$$k_{FG} = 0 \qquad\qquad\qquad\qquad DF_{FG} = \mathbf{0}$$

Moment Distribution Table:

Joint	A		C			D			F		G
	AC		CA	CD		DC	DF		FD	FG	GF
DF's	0		0.53	0.47		0.64	0.36		1.0	0	0
FEM's	− 17.5		+ 17.5	− 24.0		+ 24.0	− 3.75		+ 11.25	− 12.0	0
Balance			+ 3.4	+ 3.1		− 13.0	− 7.25		+ 0.75		
Carry-over	+ 1.7			− 6.5		+ 1.6	+ 0.38				
Balance			+ 3.4	+ 3.1		− 1.27	− 0.71				
Carry-over	+ 1.7			− 0.63		+ 1.6					
Balance			+ 0.33	+ 0.30		− 1.0	− 0.6				
Carry-over	+ 0.2										
Total	**− 13.9**		**+ 24.6**	**− 24.6**		**+ 11.9**	**− 11.9**		**+ 12.0**	**− 12.0**	**0**

Note: The out-of-balance moment at joint F is balanced during the first balancing operation and ($\frac{1}{2} \times$ moment) carried-over to joint D. Since ($\frac{3}{4} \times$ stiffness) was used for k_{DF}, no carry-overs are made from D to F.

Solution

Topic: Moment Distribution – Continuous Beams
Problem Number: 4.32   **Page No. 3**

Continuity Moments:

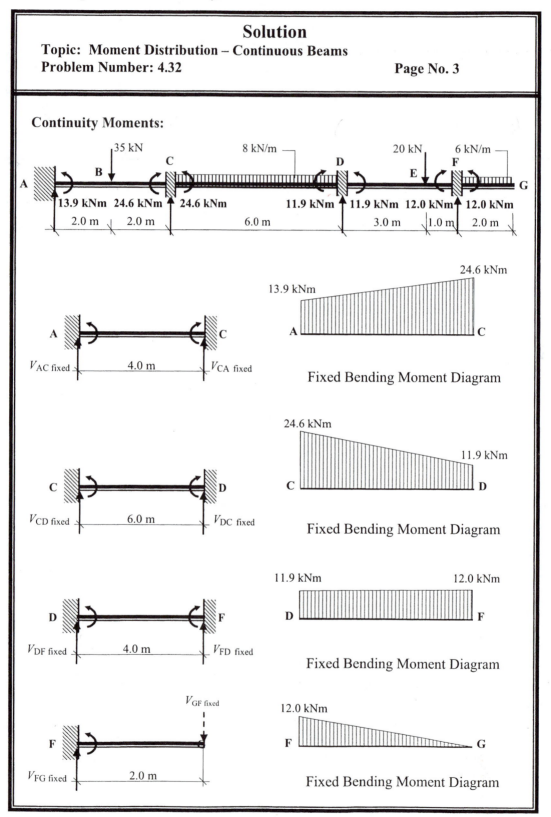

Fixed Bending Moment Diagram

Fixed Bending Moment Diagram

Fixed Bending Moment Diagram

Fixed Bending Moment Diagram

Solution

Topic: Moment Distribution – Continuous Beams
Problem Number: 4.32   **Page No. 4**

(i) Fixed vertical reactions:

Consider span AC: $+ve \curvearrowright \Sigma M_A = 0$
$-13.9 + 24.6 - (4.0 \times V_{CA \text{ fixed}}) = 0$ $\qquad$ $\therefore V_{CA \text{ fixed}} = +2.68 \text{ kN} \uparrow$

Consider the vertical equilibrium of the beam: $+ve \uparrow \Sigma F_z = 0$
$+V_{AC \text{ fixed}} + V_{CA \text{ fixed}} = 0$ $\qquad$ $\therefore V_{AC \text{ fixed}} = -2.68 \text{ kN} \downarrow$

Consider span CD: $+ve \curvearrowright \Sigma M_C = 0$
$-24.6 + 11.9 - (6.0 \times V_{DC \text{ fixed}}) = 0$ $\qquad$ $\therefore V_{DC \text{ fixed}} = -2.12 \text{ kN} \downarrow$

Consider the vertical equilibrium of the beam: $+ve \uparrow \Sigma F_z = 0$
$+V_{CD \text{ fixed}} + V_{DC \text{ fixed}} = 0$ $\qquad$ $\therefore V_{CD \text{ fixed}} = +2.12 \text{ kN} \uparrow$

Consider span DF: $+ve \curvearrowright \Sigma M_D = 0$
$-11.9 + 12.0 - (4.0 \times V_{FD \text{ fixed}}) = 0$ $\qquad$ $\therefore V_{FD \text{ fixed}} = +0.03 \text{ kN} \downarrow$

Consider the vertical equilibrium of the beam: $+ve \uparrow \Sigma F_z = 0$
$+V_{DF \text{ fixed}} + V_{FD \text{ fixed}} = 0$ $\qquad$ $\therefore V_{DF \text{ fixed}} = -0.03 \text{ kN} \uparrow$

Consider span FG: $+ve \curvearrowright \Sigma M_F = 0$
$-12.0 - (2.0 \times V_{GF \text{ fixed}}) = 0$ $\qquad$ $\therefore V_{GF \text{ fixed}} = -6.0 \text{ kN} \downarrow$

Consider the vertical equilibrium of the beam: $+ve \uparrow \Sigma F_z = 0$
$+V_{FG \text{ fixed}} + V_{GF \text{ fixed}} = 0$ $\qquad$ $\therefore V_{FG \text{ fixed}} = +6.0 \text{ kN} \uparrow$

The total vertical reaction at each support due to the continuity moments is equal to the algebraic sum of the contributions from each beam at the support.

$V_{A \text{ fixed}} = V_{AC \text{ fixed}} = -2.73 \text{ kN}$
$V_{C \text{ fixed}} = V_{CA \text{ fixed}} + V_{CD \text{ fixed}} = (+2.68 + 2.12) = +4.8 \text{ kN}$
$V_{D \text{ fixed}} = V_{DC \text{ fixed}} + V_{DF \text{ fixed}} = (-2.12 - 0.03) = -2.15 \text{ kN}$
$V_{F \text{ fixed}} = V_{FD \text{ fixed}} + V_{FG \text{ fixed}} = (+0.03 + 6.0) = +6.03 \text{ kN}$
$V_{G \text{ fixed}} = V_{GF \text{ fixed}} = -6.0 \text{ kN}$

Solution

Topic: Moment Distribution – Continuous Beams
Problem Number: 4.32   **Page No. 5**

Free bending moments:

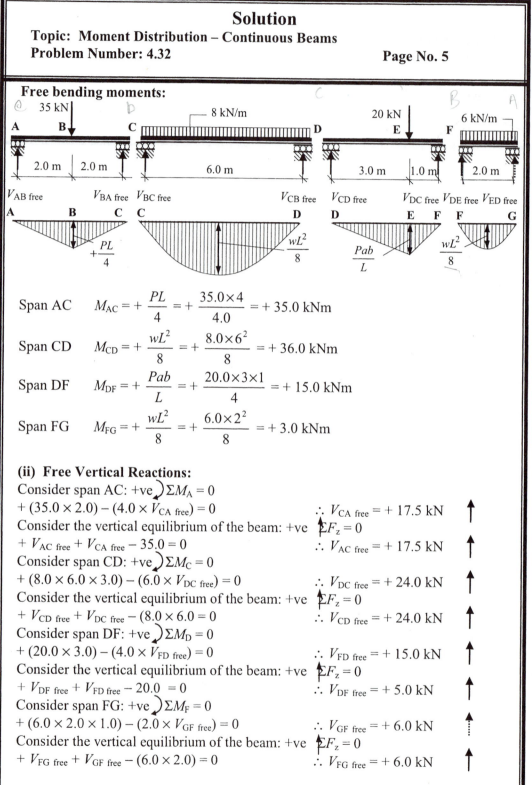

Span AC $M_{AC} = + \dfrac{PL}{4} = + \dfrac{35.0 \times 4}{4.0} = + 35.0 \text{ kNm}$

Span CD $M_{CD} = + \dfrac{wL^2}{8} = + \dfrac{8.0 \times 6^2}{8} = + 36.0 \text{ kNm}$

Span DF $M_{DF} = + \dfrac{Pab}{L} = + \dfrac{20.0 \times 3 \times 1}{4} = + 15.0 \text{ kNm}$

Span FG $M_{FG} = + \dfrac{wL^2}{8} = + \dfrac{6.0 \times 2^2}{8} = + 3.0 \text{ kNm}$

(ii) Free Vertical Reactions:
Consider span AC: $+ve \; \curvearrowright \Sigma M_A = 0$
$+ (35.0 \times 2.0) - (4.0 \times V_{CA \; free}) = 0$ $\therefore V_{CA \; free} = + 17.5 \text{ kN} \quad \uparrow$
Consider the vertical equilibrium of the beam: $+ve \; \uparrow \Sigma F_z = 0$
$+ V_{AC \; free} + V_{CA \; free} - 35.0 = 0$ $\therefore V_{AC \; free} = + 17.5 \text{ kN} \quad \uparrow$
Consider span CD: $+ve \; \curvearrowright \Sigma M_C = 0$
$+ (8.0 \times 6.0 \times 3.0) - (6.0 \times V_{DC \; free}) = 0$ $\therefore V_{DC \; free} = + 24.0 \text{ kN} \quad \uparrow$
Consider the vertical equilibrium of the beam: $+ve \; \uparrow \Sigma F_z = 0$
$+ V_{CD \; free} + V_{DC \; free} - (8.0 \times 6.0 = 0$ $\therefore V_{CD \; free} = + 24.0 \text{ kN} \quad \uparrow$
Consider span DF: $+ve \; \curvearrowright \Sigma M_D = 0$
$+ (20.0 \times 3.0) - (4.0 \times V_{FD \; free}) = 0$ $\therefore V_{FD \; free} = + 15.0 \text{ kN} \quad \uparrow$
Consider the vertical equilibrium of the beam: $+ve \; \uparrow \Sigma F_z = 0$
$+ V_{DF \; free} + V_{FD \; free} - 20.0 = 0$ $\therefore V_{DF \; free} = + 5.0 \text{ kN} \quad \uparrow$
Consider span FG: $+ve \; \curvearrowright \Sigma M_F = 0$
$+ (6.0 \times 2.0 \times 1.0) - (2.0 \times V_{GF \; free}) = 0$ $\therefore V_{GF \; free} = + 6.0 \text{ kN} \quad \uparrow$
Consider the vertical equilibrium of the beam: $+ve \; \uparrow \Sigma F_z = 0$
$+ V_{FG \; free} + V_{GF \; free} - (6.0 \times 2.0) = 0$ $\therefore V_{FG \; free} = + 6.0 \text{ kN} \quad \uparrow$

Solution

Topic: Moment Distribution – Continuous Beams

Problem Number: 4.32   **Page No. 6**

$V_{A \text{ free}} = V_{AC \text{ free}} = + 17.5 \text{ kN}$
$V_{C \text{ free}} = V_{CA \text{ free}} + V_{CD \text{ free}} = (+ 17.5 + 24.0) = + 41.5 \text{ kN}$
$V_{D \text{ free}} = V_{DC \text{ free}} + V_{DF \text{ free}} = (+ 24.0 + 5.0) = + 29.0 \text{ kN}$
$V_{F \text{ free}} = V_{FD \text{ free}} + V_{FG \text{ free}} = (+ 15.0 + 6.0) = + 21.0 \text{ kN}$
$V_{G \text{ free}} = V_{GF \text{ free}} = + 6.0 \text{ kN}$

The final vertical support reactions are given by (i) + (ii):
$V_A = V_{A \text{ fixed}} + V_{A \text{ free}} = (- 2.73 + 17.5) = + 14.77 \text{ kN}$ ↑
$V_C = V_{C \text{ fixed}} + V_{C \text{ free}} = (+ 4.8 + 41.5) = + 46.3 \text{ kN}$
$V_D = V_{D \text{ fixed}} + V_{D \text{ free}} = (- 2.15 + 29.0) = + 26.85 \text{ kN}$ ↑
$V_F = V_{F \text{ fixed}} + V_{F \text{ free}} = (+ 6.03 + 21.0) = + 27.03 \text{ kN}$ ↑
$V_G = V_{G \text{ fixed}} + V_{G \text{ free}} = (- 6.0 + 6.0) = + 0$

Check the vertical equilibrium:
Total vertical force = + 14.77 + 46.3 + 26.85 + 27.03
$= + 114.95 \text{ kN} (= \text{total applied load})$

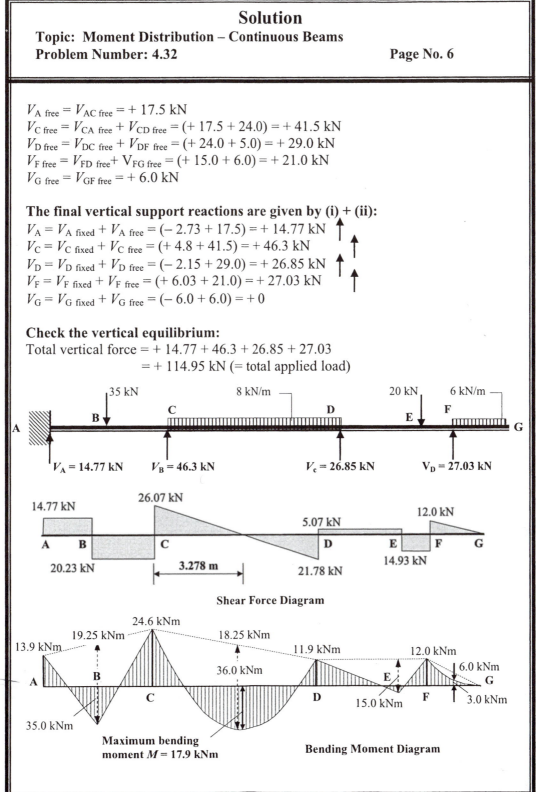

Shear Force Diagram

Bending Moment Diagram

Maximum bending moment M = 17.9 kNm

4.8   *Redistribution of Moments*

When continuous structures approach their failure load there is a redistribution of load as successive plastic hinges develop until failure occurs; this is dependent on the ductility of the material. Advantage can be taken of this behaviour to reduce the maximum moments whilst at the same time increasing others to maintain static equilibrium as shown in Example 4.22 below.

4.8.1   *Example 4.22: Redistribution of Moments in a Two-span Beam*

A two-span beam is required to support an ultimate design load of 150 kN/m as shown in Figure 4.95. Reduce the support moment by 20% and determine the redistributed bending moment diagram.

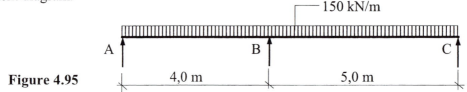

Figure 4.95

Use moment distribution to determine the moments over the supports and in the spans.

Fixed - end moments:

Span AB $M_{AB} = 0$ $M_{BA} = + \dfrac{wL^2}{8} = + \dfrac{150 \times 4,0^2}{8} = + 300,0$ kNm

Span BC $M_{BC} = - \dfrac{wL^2}{8} = - \dfrac{150 \times 5,0^2}{8} = - 469,0$ kNm $M_{CB} = 0$

Stiffnesses:

$K_{BA} = \dfrac{I}{4} = 0,25$ $DF_{BA} = \dfrac{0,25}{0,45} = 0,56$

$\qquad\qquad k_{total} = 0,45$

$K_{BC} = \dfrac{I}{5} = 0,2$ $DF_{Bc} = \dfrac{0,2}{0,45} = 0,44$

Moment distribution table:

Joint	A	B		C
	AB	BA	BC	CB
Distribution Factors	1,0	0,56	0,44	1,0
FEMs	0	+ 300,0	− 469,0	0
Balance		+ 94,6	+ 74,4	
Final Moments		+ 394,6	− 394,6	

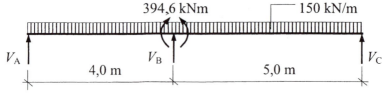

Figure 4.96

Consider span AB:

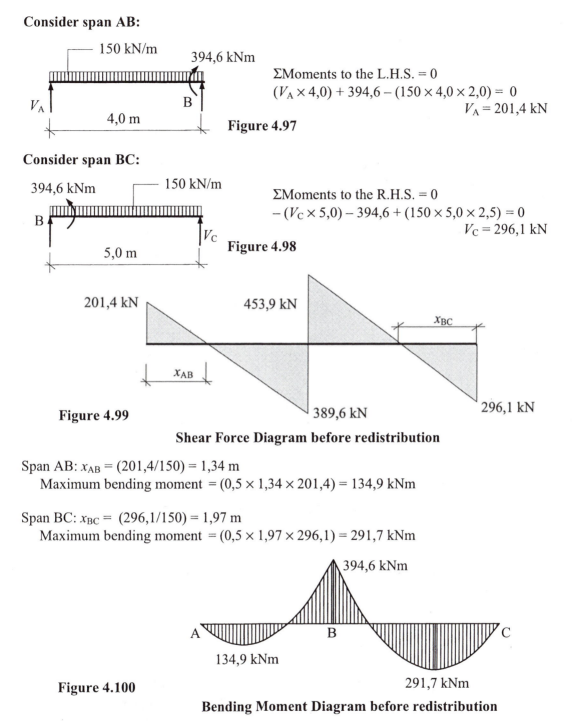

ΣMoments to the L.H.S. $= 0$

$(V_A \times 4,0) + 394,6 - (150 \times 4,0 \times 2,0) = 0$

$V_A = 201,4$ kN

Figure 4.97

Consider span BC:

ΣMoments to the R.H.S. $= 0$

$-(V_C \times 5,0) - 394,6 + (150 \times 5,0 \times 2,5) = 0$

$V_C = 296,1$ kN

Figure 4.98

Figure 4.99

Shear Force Diagram before redistribution

Span AB: $x_{AB} = (201,4/150) = 1,34$ m

Maximum bending moment $= (0,5 \times 1,34 \times 201,4) = 134,9$ kNm

Span BC: $x_{BC} = (296,1/150) = 1,97$ m

Maximum bending moment $= (0,5 \times 1,97 \times 296,1) = 291,7$ kNm

Figure 4.100

Bending Moment Diagram before redistribution

Allowing for 20% redistribution the reduced bending moment at support B is given by:

$M_{B,reduced} = 0,8 \times 394,6 = 315,7$ kNm.

The above calculation must be repeated with the reduced value of the support moment to determine the revised support reactions the redistributed maximum moments in the spans.

Consider span AB:

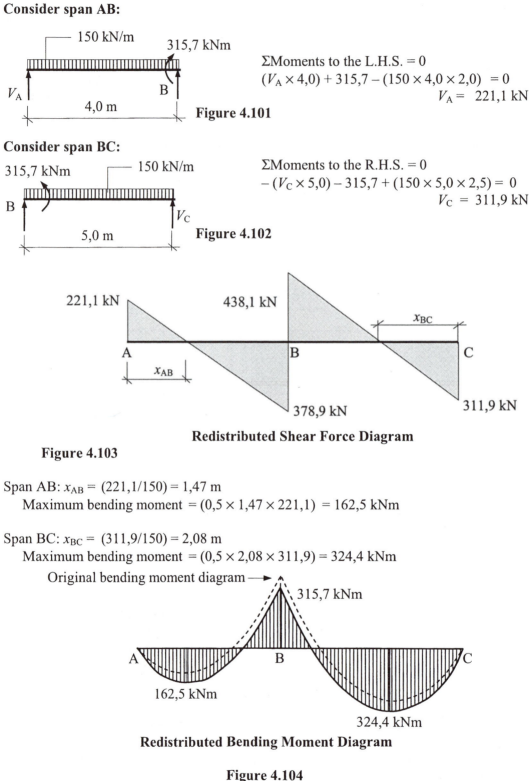

ΣMoments to the L.H.S. = 0
$(V_A \times 4,0) + 315,7 - (150 \times 4,0 \times 2,0) = 0$
$V_A = 221,1$ kN

Figure 4.101

Consider span BC:

ΣMoments to the R.H.S. = 0
$-(V_C \times 5,0) - 315,7 + (150 \times 5,0 \times 2,5) = 0$
$V_C = 311,9$ kN

Figure 4.102

Redistributed Shear Force Diagram
Figure 4.103

Span AB: $x_{AB} = (221,1/150) = 1,47$ m
   Maximum bending moment $= (0,5 \times 1,47 \times 221,1) = 162,5$ kNm

Span BC: $x_{BC} = (311,9/150) = 2,08$ m
   Maximum bending moment $= (0,5 \times 2,08 \times 311,9) = 324,4$ kNm

Original bending moment diagram ⟶

Redistributed Bending Moment Diagram

Figure 4.104

4.9   *Shear Force and Bending Moment Envelopes*

Shear force and bending moment envelopes are graphs which show the variation in the minimum and maximum values for the function along the structure due to the application of all the possible load cases or load combinations.

The diagrams are obtained by superimposing the separate diagrams for a function based on each load case or combination considered. The resulting diagram that shows the upper and lower bounds for the function along the structure due to the loading conditions is called the envelope.

A three-span beam with four separate load combinations and their associated bending moment diagrams, and the bending moment envelope encompassing all of the combinations considered indicating the positive and negative bending moments in each span, is shown in Figure 4.105. Note: the values of the bending moments are given for illustration only. A similar envelope can be drawn for the shear force diagrams.

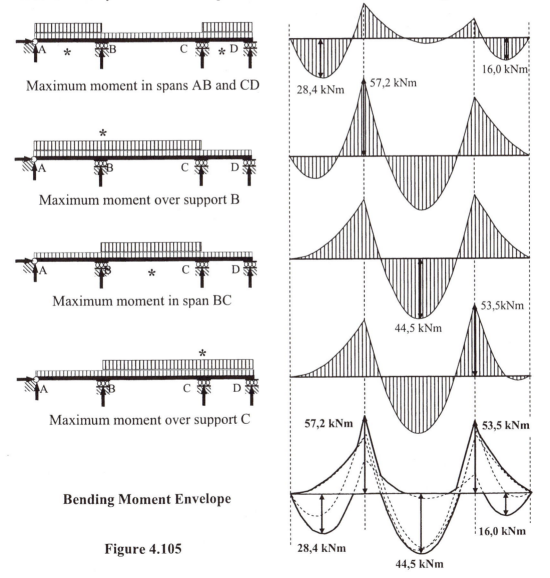

Maximum moment in spans AB and CD

Maximum moment over support B

Maximum moment in span BC

Maximum moment over support C

Bending Moment Envelope

Figure 4.105

5. Rigid-Jointed Frames

5.1 *Rigid–Jointed Frames*

Rigid–jointed frames are framed structures in which the members transmit applied loads by axial, shear and bending effects. There are basically two types of frame to consider;

(i)   statically determinate frames; see Figure 5.1(a) and

(ii)   statically indeterminate frames; see Figure 5.1(b).

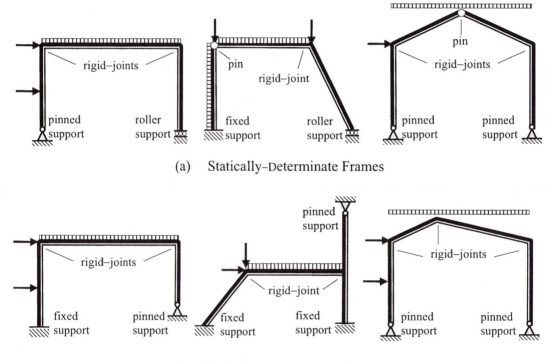

(a)   Statically-Determinate Frames

(b)   Statically-Indeterminate Frames

Figure 5.1

Rigid–joints (moment connections) are designed to transfer axial and shear forces in addition to bending moments between the connected members whilst pinned joints (simple connections) are designed to transfer axial and shear forces only. Typical moment and simple connections between steel members is illustrated in Figure 5.2.

In the case of statically determinate frames, only the equations of equilibrium are required to determine the member forces. They are often used where there is a possibility of support settlement since statically determinate frames can accommodate small changes of geometry without inducing significant secondary stresses. Analysis of such frames is illustrated in this Examples 5.1 and 5.2 and Problems 5.1 to 5.4.

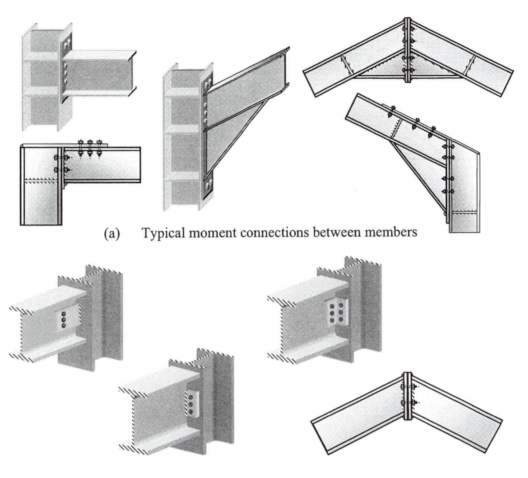

(a)　Typical moment connections between members

(b)　Typical simple connections between members

Figure 5.2

Statically indeterminate frames require consideration of compatibility when determining the member forces. The analysis of singly-redundant frames using the Unit Load method is illustrated in Example 5.3 and Problems 5.5 to 5.8 One of the most convenient and most versatile methods of analysis for such frames is moment distribution. When using this method there are two cases to consider; no–sway frames and sway frames. Analysis of the former is illustrated in Example 5.4 and Problems 5.9 to 5.16 and in the latter in Example 5.5 and Problems 5.17 to 5.22.

5.1.1 Example 5.1 Statically Determinate Rigid–Jointed Frame 1

An asymmetric portal frame is supported on a roller at A and pinned at support D as shown in Figure 5.3. For the loading indicated:

i) determine the support reactions and

ii) sketch the axial load, shear force and bending moment diagrams.

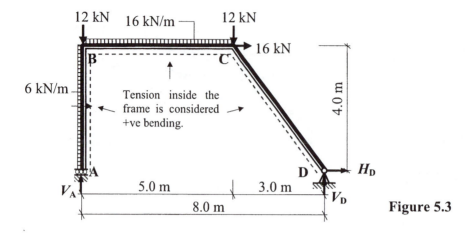

Figure 5.3

Solution:
Apply the three equations of static equilibrium to the force system

+ve ↑ $\Sigma F_z = 0$ $V_A - 12.0 - (16.0 \times 5.0) - 12.0 + V_D = 0$ Equation (1)

+ve → $\Sigma F_x = 0$ $(6.0 \times 4.0) + 16.0 + H_D = 0$ Equation (2)

+ve ⤸ $\Sigma M_A = 0$ $(6.0 \times 4.0)(2.0) + (16.0 \times 5.0)(2.5) + (12.0 \times 5.0) + (16.0 \times 4.0)$
$- (V_D \times 8.0) = 0$ Equation (3)

From equation (2): $40.0 + H_D = 0$ ∴ $H_D = -40.0$ **kN** ←

From equation (3): $372.0 - 8.0 V_D = 0$ ∴ $V_D = +46.5$ **kN** ↑

From equation (1): $V_A - 104.0 + 46.5 = 0$ ∴ $V_A = +57.5$ **kN** ↑

Assuming positive bending moments induce tension **inside** the frame:

$M_B = -(6.0 \times 4.0)(2.0) = -48.0$ kNm

$M_C = +(46.5 \times 3.0) - (40.0 \times 4.0) = -20.50$ kNm

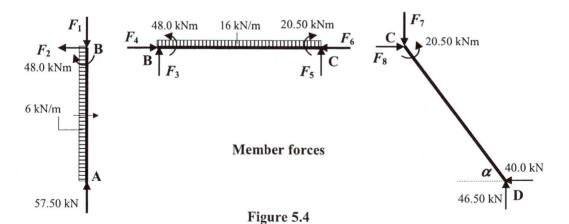

Member forces

Figure 5.4

The values of the end–forces F_1 to F_8 can be determined by considering the equilibrium of each member and joint in turn.

Consider member AB:

+ve $\uparrow \Sigma F_z = 0$ $+ 57.50 - F_1 = 0$ $\therefore F_1 = 57.50$ kN $\downarrow$

+ve $\longrightarrow \Sigma F_x = 0 + (6.0 \times 4.0) - F_2 = 0$ $\therefore F_2 = 24.0$ kN $\longleftarrow$

Consider joint B:

+ve $\uparrow \Sigma F_z = 0$ There is an applied vertical load at joint B = 12 kN $\downarrow$

$- F_1 + F_3 = - 12.0$ $\therefore F_3 = 45.50$ kN $\uparrow$

+ve $\longrightarrow \Sigma F_x = 0$

$- F_2 + F_4 = 0$ $\therefore F_4 = 24.0$ kN $\longrightarrow$

Consider member BC:

+ve $\uparrow \Sigma F_z = 0$ $+ 45.5 - (16.0 \times 5.0) + F_5 = 0$ $\therefore F_5 = 34.5$ kN $\uparrow$

+ve $\longrightarrow \Sigma F_x = 0$ $+ 24.0 - F_6 = 0$ $\therefore F_6 = 24.0$ kN $\longleftarrow$

Consider member CD:

+ve $\uparrow \Sigma F_z = 0$ $+ 46.5 - F_7 = 0$ $\therefore F_7 = 46.5$ kN $\downarrow$

+ve $\longrightarrow \Sigma F_x = 0$ $- 40.0 + F_8 = 0$ $\therefore F_8 = 40.0$ kN $\longrightarrow$

Check joint C:

+ve $\uparrow \Sigma F_z$ There is an applied vertical load at joint C = 12 kN $\downarrow$

$+ F_5 - F_7 = + 34.5 - 46.5 = - 12.0$

+ve $\longrightarrow \Sigma F_x$ There is an applied horizontal at joint C = 16 kN $\longrightarrow$

$- F_6 + F_8 = - 24.0 + 40.0 = + 16.0$

The axial force and shear force in member CD can be found from:

Axial load = +/− (horizontal force × cos α) +/− (vertical force × sin α)

Shear force = +/− (horizontal force × sin α) +/− (vertical force × cos α)

The signs are dependent on the directions of the respective forces.

Member CD:

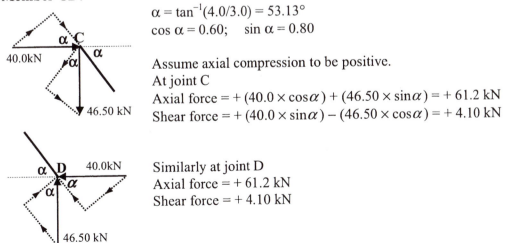

$\alpha = \tan^{-1}(4.0/3.0) = 53.13°$

$\cos \alpha = 0.60;$ $\sin \alpha = 0.80$

Assume axial compression to be positive.

At joint C

Axial force = $+ (40.0 \times \cos \alpha) + (46.50 \times \sin \alpha) = + 61.2$ kN

Shear force = $+ (40.0 \times \sin \alpha) - (46.50 \times \cos \alpha) = + 4.10$ kN

Similarly at joint D

Axial force = $+ 61.2$ kN

Shear force = $+ 4.10$ kN

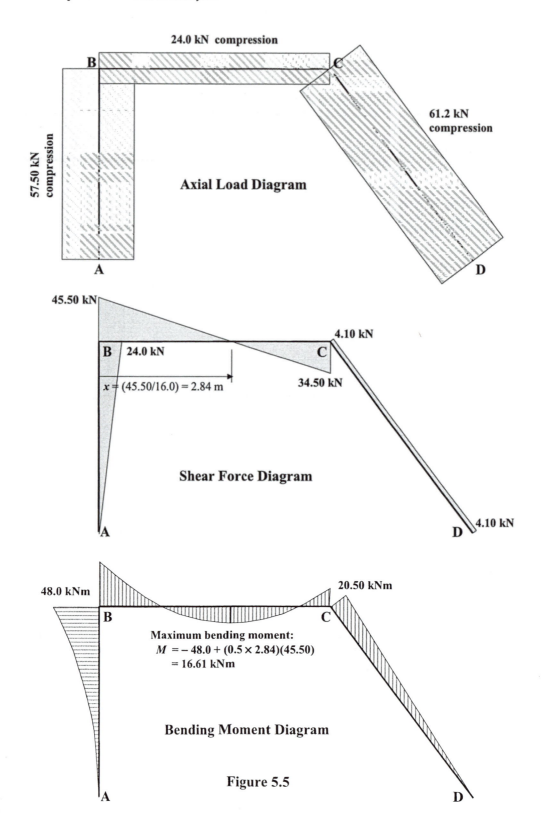

24.0 kN compression

B   **C**

61.2 kN
compression

57.50 kN
compression

Axial Load Diagram

A   **D**

45.50 kN

B   24.0 kN   4.10 kN

                                      C

$x = (45.50/16.0) = 2.84$ m   34.50 kN

Shear Force Diagram

A   **D** 4.10 kN

48.0 kNm   20.50 kNm

B   **C**

Maximum bending moment:
$M = -48.0 + (0.5 \times 2.84)(45.50)$
$= 16.61$ kNm

Bending Moment Diagram

Figure 5.5

A   **D**

5.1.2 Example 5.2 Statically Determinate Rigid-Jointed Frame 2

A pitched–roof portal frame is pinned at supports A and H and members CD and DEF are pinned at the ridge as shown in Figure 5.6. For the loading indicated:

i) determine the support reactions and

ii) sketch the axial load, shear force and bending moment diagrams.

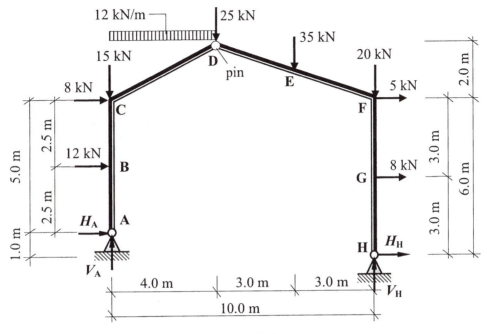

Figure 5.6

Apply the three equations of static equilibrium to the force system in addition to the Σ moments at the pin = 0:

$$+ve \uparrow \Sigma F_z = 0$$
$$V_A - 15.0 - (12.0 \times 4.0) - 25.0 - 35.0 - 20.0 + V_H = 0 \qquad \text{Equation (1)}$$
$$+ve \longrightarrow \Sigma F_x = 0$$
$$H_A + 12.0 + 8.0 + 5.0 + 8.0 + H_H = 0 \qquad \text{Equation (2)}$$
$$+ve \ \rotatebox{-20}{$\circlearrowright$} \ \Sigma M_A = 0$$
$$(12.0 \times 2.5) + (8.0 \times 5.0) + (12.0 \times 4.0)(2.0) + (25.0 \times 4.0) + (35.0 \times 7.0)$$
$$+ (20.0 \times 10.0) + (5.0 \times 5.0) + (8.0 \times 2.0) - (H_H \times 1.0) - (V_H \times 10.0) = 0$$
$$\text{Equation (3)}$$

$$+ve \ \rotatebox{-20}{$\circlearrowright$} \ \Sigma M_{pin} = 0 \quad \text{(right–hand side)}$$
$$+ (35.0 \times 3.0) + (20.0 \times 6.0) - (5.0 \times 2.0) - (8.0 \times 5.0) - (H_H \times 8.0) - (V_H \times 6.0) = 0$$
$$\text{Equation (4)}$$

From Equation (3): $+ 752.0 - H_H - 10.0V_H = 0$ Equation (3a)
From Equation (4): $+ 175.0 - 8.0H_H - 6.0V_H = 0$ Equation (3b)

Solve equations 3(a) and 3(b) simultaneously: $V_H = +\,78.93$ kN ↑ $H_H = -\,37.30$ kN ←
From Equation (2): $H_A + 33.0 + H_H = 0$ $H_A = +\,4.30$ kN →
From Equation (1): $V_A - 143.0 + V_H = 0$ $V_A = +\,64.07$ kN ↑

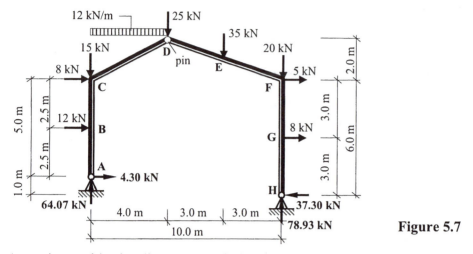

Figure 5.7

Assuming positive bending moments induce tension **inside** the frame:
$M_B = -\,(4.30 \times 2.5) = -\,10.75$ kNm
$M_C = -\,(4.30 \times 5.0) - (12.0 \times 2.5) = -\,51.50$ kNm
$M_D = $ zero (pin)
$M_E = -\,(20.0 \times 3.0) + (5.0 \times 1.0) + (8.0 \times 4.0) - (37.3 \times 7.0) + (78.93 \times 3.0)$
  $= -\,47.31$ kNm
$M_F = +\,(8.0 \times 3.0) - (37.30 \times 6.0) = -\,199.80$ kNm
$M_G = -\,(37.30 \times 3.0) = -\,111.90$ kNm

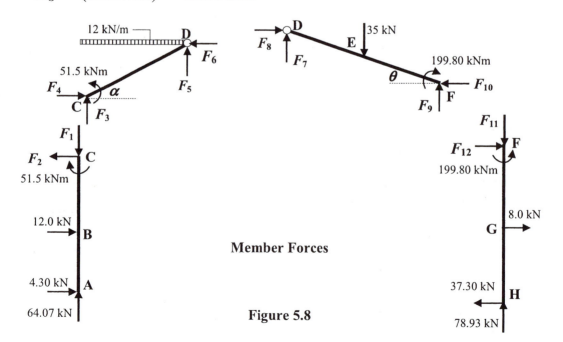

Member Forces

Figure 5.8

The values of the end–forces F_1 to F_{12} can be determined by considering the equilibrium of each member and joint in turn.

Consider member ABC:

$+ve \uparrow \Sigma F_z = 0 \qquad + 64.07 - F_1 = 0 \qquad\qquad\qquad \therefore F_1 = 64.07$ kN $\downarrow$

$+ve \longrightarrow \Sigma F_x = 0 \quad + 4.30 + 12.0 - F_2 = 0 \qquad\qquad \therefore F_2 = 16.30$ kN $\longleftarrow$

Consider joint C:

$+ve \uparrow \Sigma F_z = 0 \qquad$ There is an applied vertical load at joint C = 15 kN $\downarrow$

$- F_1 + F_3 = -15.0 \qquad\qquad\qquad\qquad\qquad\qquad \therefore F_3 = 49.07$ kN $\uparrow$

$+ve \longrightarrow \Sigma F_x = 0 \qquad$ There is an applied horizontal load at joint C = 8 kN $\longrightarrow$

$- F_2 + F_4 = + 8.0 \qquad\qquad\qquad\qquad\qquad\qquad \therefore F_4 = 24.30$ kN $\longrightarrow$

Consider member CD:

$+ve \uparrow \Sigma F_z = 0 \qquad + 49.07 - (12.0 \times 4.0) + F_5 = 0 \qquad \therefore F_5 = -1.07$ kN $\downarrow$

$+ve \longrightarrow \Sigma F_x = 0 \quad + 24.30 - F_6 = 0 \qquad\qquad\qquad \therefore F_6 = 24.30$ kN $\longleftarrow$

Consider member FGH:

$+ve \uparrow \Sigma F_z = 0 \qquad + 78.93 - F_{11} = 0 \qquad\qquad\qquad \therefore F_{11} = 78.93$ kN $\downarrow$

$+ve \longrightarrow \Sigma F_x = 0 \quad - 37.30 + 8.0 + F_{12} = 0 \qquad\qquad \therefore F_{12} = 29.30$ kN $\longrightarrow$

Consider joint F:

$+ve \uparrow \Sigma F_z = 0 \qquad$ There is an applied vertical load at joint F = 20 kN $\downarrow$

$F_{11} + F_9 = -20.0 \qquad\qquad\qquad\qquad\qquad\qquad \therefore F_9 = 58.93$ kN $\uparrow$

$+ve \longrightarrow \Sigma F_x = 0 \qquad$ There is an applied horizontal load at joint F = 5 kN $\longrightarrow$

$+ F_{12} - F_{10} = + 5.0 \qquad\qquad\qquad\qquad\qquad\qquad \therefore F_{10} = 24.30$ kN $\longleftarrow$

Consider member DF:

$+ve \uparrow \Sigma F_z = 0 \qquad + 58.93 - 35.0 + F_7 = 0 \qquad\qquad \therefore F_7 = 23.93$ kN $\downarrow$

$+ve \longrightarrow \Sigma F_x = 0 \quad - 24.30 + F_8 = 0 \qquad\qquad\qquad \therefore F_8 = 24.30$ kN $\longrightarrow$

The calculated values can be checked by considering the equilibrium at joint D.

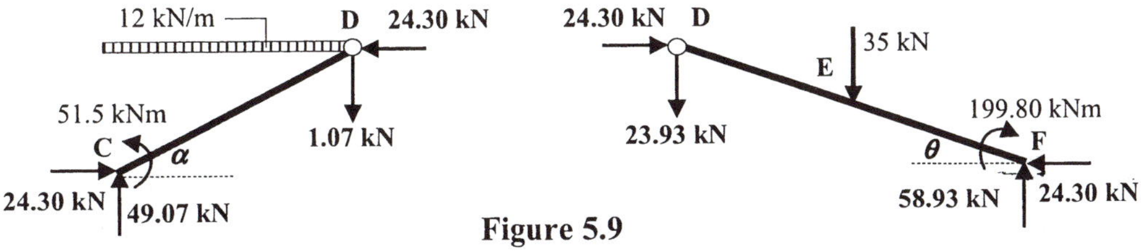

Figure 5.9

$+ve \longrightarrow \Sigma F_x \qquad - 24.30 + 24.30 = 0$

$+ve \uparrow \Sigma F_z \qquad - 1.07 - 23.93 = - 25.0$ kN (equal to the applied vertical load at D).

The axial force and shear force in member CD can be found from:

Axial load = +/− (horizontal force × cos α) +/− (vertical force × sin α)
Shear force = +/− (horizontal force × sin α) +/− (vertical force × cos α)
The signs are dependent on the directions of the respective forces.

Similarly with θ for member DEF.

Member CD:

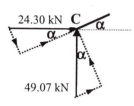

$\alpha = \tan^{-1}(2.0/4.0) = 26.565°$
$\cos \alpha = 0.894;$ $\sin \alpha = 0.447$

Assume axial compression to be positive.
At joint C
Axial force = + (24.30 × cos α) + (49.07× sin α) = + 43.66 kN
Shear force = − (24.30 × sin α) + (49.07× cos α) = + 33.01 kN

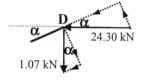

At joint D
Axial force = + (24.30 × cos α) + (1.07× sin α) = + 22.20 kN
Shear force = − (24.30 × sin α) + (49.07× cos α) = − 9.91 kN

Member DEF:

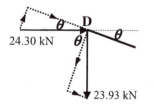

$\theta = \tan^{-1}(2.0/6.0) = 18.435°$
$\cos \theta = 0.947;$ $\sin \theta = 0.316$

Assume axial compression to be positive.
At joint D
Axial force = + (24.30 × cos θ) + (23.93× sin θ) = + 30.57 kN
Shear force = + (24.30 × sin θ) − (23.93 × cos θ) = + 14.98 kN

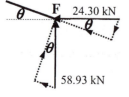

At joint F
Axial force = + (24.30 × cos θ) + (58.93× sin θ) = + 41.63 kN
Shear force = − (24.30 × sin θ) + (58.93× cos θ) = + 48.13 kN

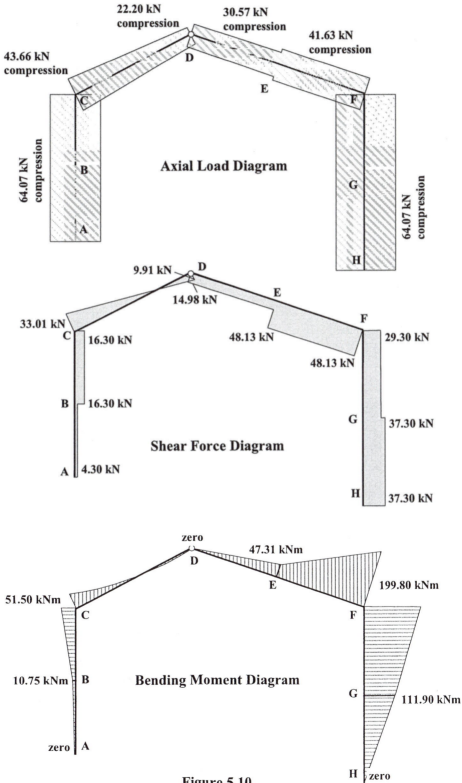

Figure 5.10

5.1.3 Problems: Statically Determinate Rigid-Jointed Frames

A series of statically determinate, rigid-jointed frames are indicated in Problems 5.1 to 5.4. In each case, for the loading given:

i) determine the support reactions and

ii) sketch the axial load, shear force and bending moment diagrams.

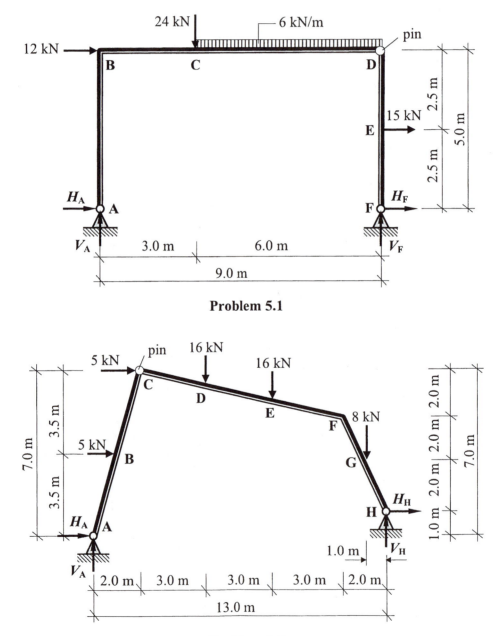

Problem 5.1

Problem 5.2

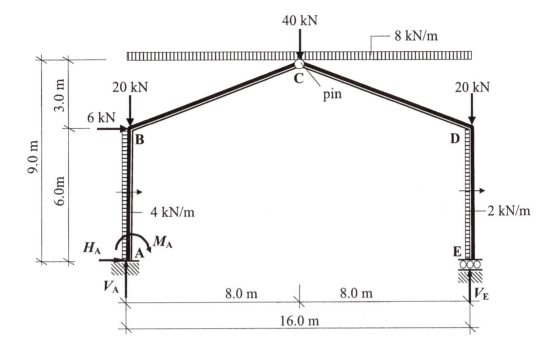

Problem 5.3

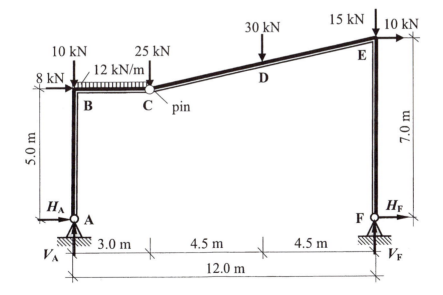

Problem 5.4

5.1.4 Solutions: Statically Determinate Rigid-Jointed Frames

Solution
Topic: Statically Determinate Rigid-Jointed Frames
Problem Number: 5.1   **Page No. 1**

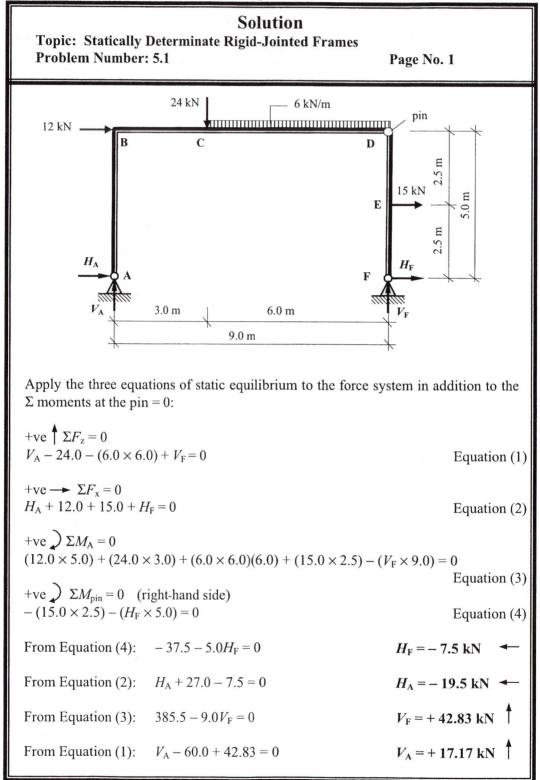

Apply the three equations of static equilibrium to the force system in addition to the Σ moments at the pin = 0:

$+ve \uparrow \Sigma F_z = 0$
$V_A - 24.0 - (6.0 \times 6.0) + V_F = 0$ Equation (1)

$+ve \longrightarrow \Sigma F_x = 0$
$H_A + 12.0 + 15.0 + H_F = 0$ Equation (2)

$+ve \,\rotatebox{0}{$\curvearrowright$}\, \Sigma M_A = 0$
$(12.0 \times 5.0) + (24.0 \times 3.0) + (6.0 \times 6.0)(6.0) + (15.0 \times 2.5) - (V_F \times 9.0) = 0$
                                                          Equation (3)

$+ve \,\rotatebox{0}{$\curvearrowright$}\, \Sigma M_{pin} = 0$ (right-hand side)
$- (15.0 \times 2.5) - (H_F \times 5.0) = 0$ Equation (4)

From Equation (4): $-37.5 - 5.0H_F = 0$ **$H_F = -7.5$ kN** $\longleftarrow$

From Equation (2): $H_A + 27.0 - 7.5 = 0$ **$H_A = -19.5$ kN** $\longleftarrow$

From Equation (3): $385.5 - 9.0V_F = 0$ **$V_F = +42.83$ kN** $\uparrow$

From Equation (1): $V_A - 60.0 + 42.83 = 0$ **$V_A = +17.17$ kN** $\uparrow$

Solution

Topic: Statically Determinate Rigid-Jointed Frames

Problem Number: 5.1

Page No. 2

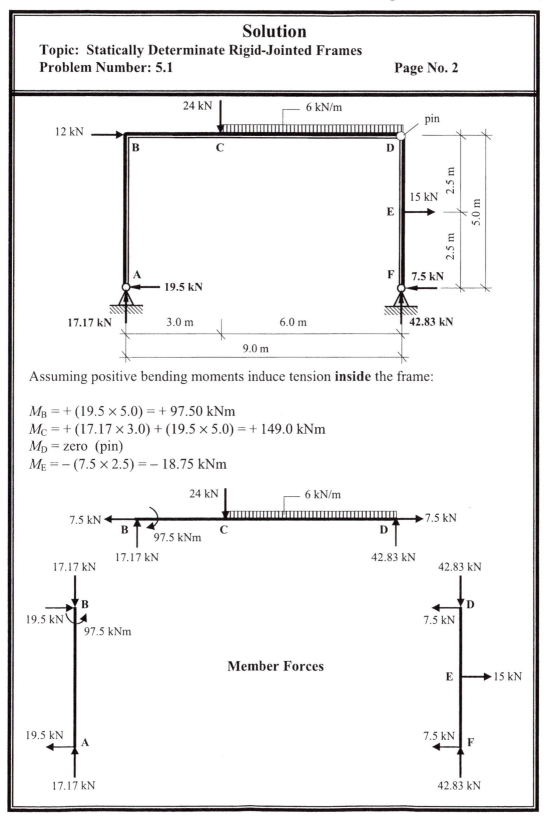

Assuming positive bending moments induce tension **inside** the frame:

$M_B = + (19.5 \times 5.0) = + 97.50$ kNm

$M_C = + (17.17 \times 3.0) + (19.5 \times 5.0) = + 149.0$ kNm

$M_D =$ zero (pin)

$M_E = - (7.5 \times 2.5) = - 18.75$ kNm

Member Forces

Solution

Topic: Statically Determinate Rigid-Jointed Frames
Problem Number: 5.1   **Page No. 3**

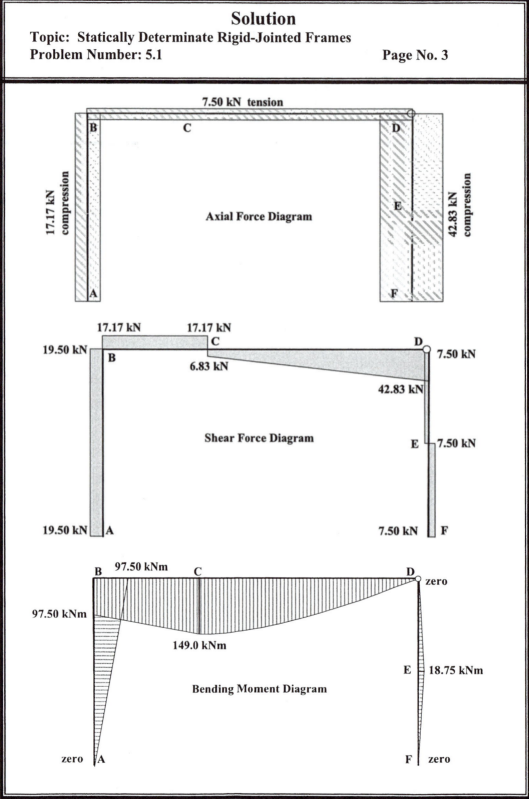

Solution

Topic: Statically Determinate Rigid-Jointed Frames
Problem Number: 5.2   **Page No. 1**

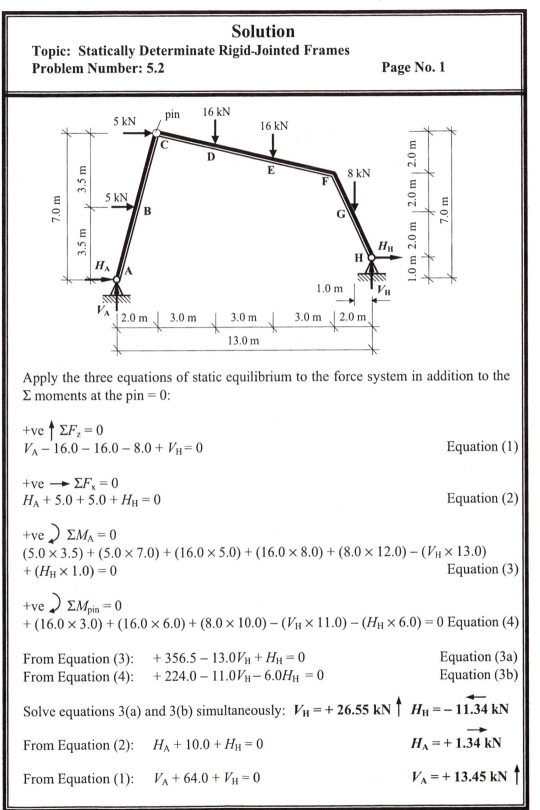

Apply the three equations of static equilibrium to the force system in addition to the Σ moments at the pin = 0:

+ve ↑ $\Sigma F_z = 0$
$V_A - 16.0 - 16.0 - 8.0 + V_H = 0$ Equation (1)

+ve → $\Sigma F_x = 0$
$H_A + 5.0 + 5.0 + H_H = 0$ Equation (2)

+ve ↻ $\Sigma M_A = 0$
$(5.0 \times 3.5) + (5.0 \times 7.0) + (16.0 \times 5.0) + (16.0 \times 8.0) + (8.0 \times 12.0) - (V_H \times 13.0)$
$+ (H_H \times 1.0) = 0$ Equation (3)

+ve ↻ $\Sigma M_{pin} = 0$
$+ (16.0 \times 3.0) + (16.0 \times 6.0) + (8.0 \times 10.0) - (V_H \times 11.0) - (H_H \times 6.0) = 0$ Equation (4)

From Equation (3): $+ 356.5 - 13.0V_H + H_H = 0$ Equation (3a)
From Equation (4): $+ 224.0 - 11.0V_H - 6.0H_H = 0$ Equation (3b)

Solve equations 3(a) and 3(b) simultaneously: $V_H = + 26.55$ kN ↑ $H_H = - 11.34$ kN ←

From Equation (2): $H_A + 10.0 + H_H = 0$ $H_A = + 1.34$ kN →

From Equation (1): $V_A + 64.0 + V_H = 0$ $V_A = + 13.45$ kN ↑

Solution

Topic: Statically Determinate Rigid-Jointed Frames
Problem Number: 5.2

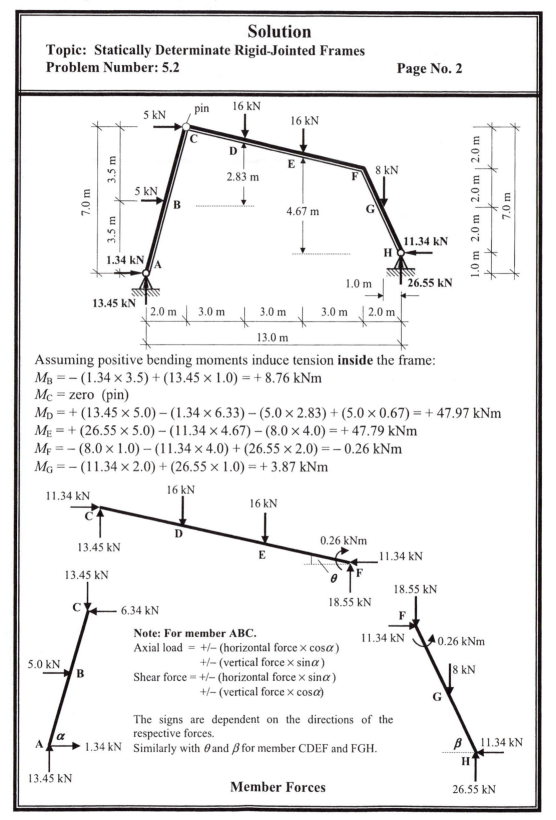

Assuming positive bending moments induce tension **inside** the frame:

$M_B = -(1.34 \times 3.5) + (13.45 \times 1.0) = +8.76$ kNm

$M_C =$ zero (pin)

$M_D = +(13.45 \times 5.0) - (1.34 \times 6.33) - (5.0 \times 2.83) + (5.0 \times 0.67) = +47.97$ kNm

$M_E = +(26.55 \times 5.0) - (11.34 \times 4.67) - (8.0 \times 4.0) = +47.79$ kNm

$M_F = -(8.0 \times 1.0) - (11.34 \times 4.0) + (26.55 \times 2.0) = -0.26$ kNm

$M_G = -(11.34 \times 2.0) + (26.55 \times 1.0) = +3.87$ kNm

Note: For member ABC.
Axial load $= +/-$ (horizontal force $\times \cos\alpha$)
　　　　　　$+/-$ (vertical force $\times \sin\alpha$)
Shear force $= +/-$ (horizontal force $\times \sin\alpha$)
　　　　　　$+/-$ (vertical force $\times \cos\alpha$)

The signs are dependent on the directions of the respective forces. Similarly with θ and β for member CDEF and FGH.

Member Forces

Solution

Topic: Statically Determinate Rigid-Jointed Frames

Problem Number: 5.2   Page No 3

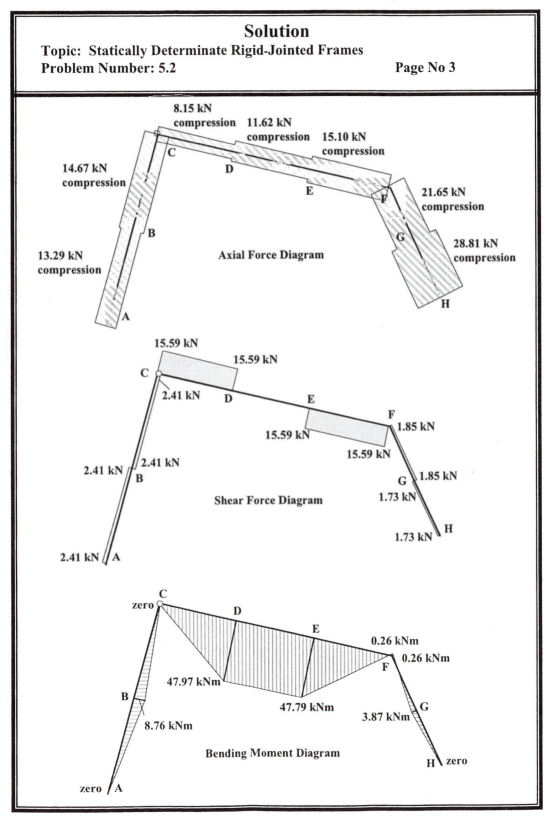

8.15 kN
compression   11.62 kN
compression   15.10 kN
compression

C

D

14.67 kN
compression   E

F   21.65 kN
compression

B

G   28.81 kN
compression

13.29 kN
compression   **Axial Force Diagram**

H

A

15.59 kN
15.59 kN

C

2.41 kN   D   E

F   1.85 kN

15.59 kN

15.59 kN

2.41 kN   2.41 kN
B   G   1.85 kN

Shear Force Diagram   1.73 kN

H

1.73 kN

2.41 kN A

C

zero   D

E

0.26 kNm

0.26 kNm

F

47.97 kNm

B

47.79 kNm

3.87 kNm   G

8.76 kNm

Bending Moment Diagram

H   zero

zero   A

Solution

Topic: Statically Determinate Rigid-Jointed Frames
Problem Number: 5.3   **Page No. 1**

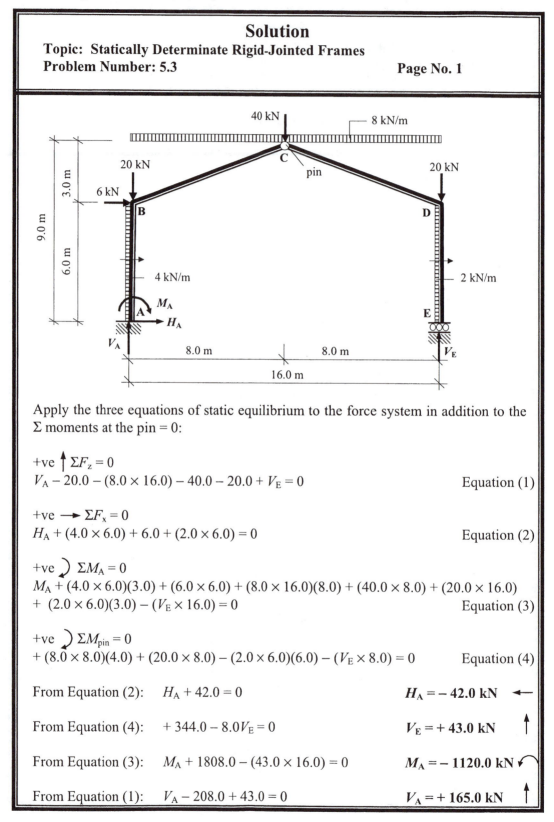

Apply the three equations of static equilibrium to the force system in addition to the Σ moments at the pin = 0:

+ve $\uparrow \Sigma F_z = 0$
$V_A - 20.0 - (8.0 \times 16.0) - 40.0 - 20.0 + V_E = 0$ Equation (1)

+ve $\longrightarrow \Sigma F_x = 0$
$H_A + (4.0 \times 6.0) + 6.0 + (2.0 \times 6.0) = 0$ Equation (2)

+ve $\circlearrowright \Sigma M_A = 0$
$M_A + (4.0 \times 6.0)(3.0) + (6.0 \times 6.0) + (8.0 \times 16.0)(8.0) + (40.0 \times 8.0) + (20.0 \times 16.0)$
$+ (2.0 \times 6.0)(3.0) - (V_E \times 16.0) = 0$ Equation (3)

+ve $\circlearrowright \Sigma M_{pin} = 0$
$+ (8.0 \times 8.0)(4.0) + (20.0 \times 8.0) - (2.0 \times 6.0)(6.0) - (V_E \times 8.0) = 0$ Equation (4)

From Equation (2): $H_A + 42.0 = 0$ $H_A = -42.0$ kN $\longleftarrow$

From Equation (4): $+ 344.0 - 8.0 V_E = 0$ $V_E = +43.0$ kN $\uparrow$

From Equation (3): $M_A + 1808.0 - (43.0 \times 16.0) = 0$ $M_A = -1120.0$ kN $\curvearrowleft$

From Equation (1): $V_A - 208.0 + 43.0 = 0$ $V_A = +165.0$ kN $\uparrow$

Solution

Topic: Statically Determinate Rigid-Jointed Frames
Problem Number: 5.3   Page No. 2

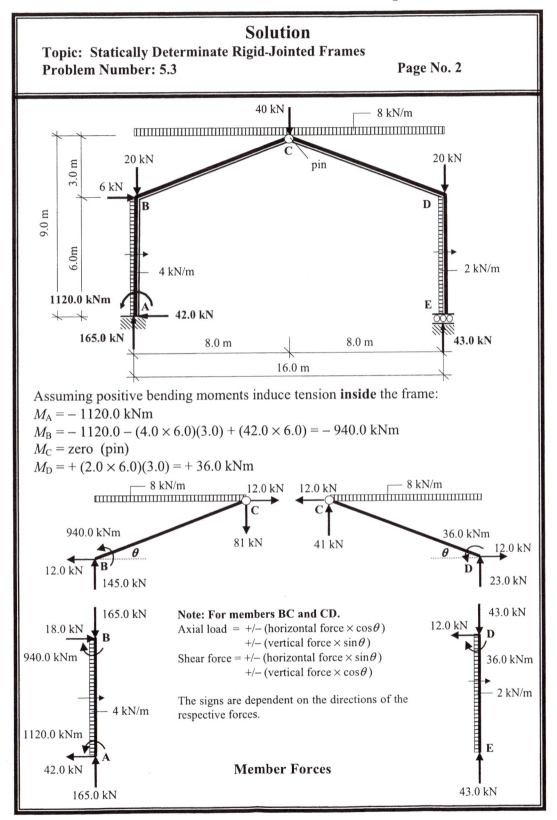

Assuming positive bending moments induce tension **inside** the frame:

$M_A = -1120.0$ kNm

$M_B = -1120.0 - (4.0 \times 6.0)(3.0) + (42.0 \times 6.0) = -940.0$ kNm

$M_C = $ zero (pin)

$M_D = + (2.0 \times 6.0)(3.0) = +36.0$ kNm

Note: For members BC and CD.
Axial load = +/– (horizontal force × cos θ)
                 +/– (vertical force × sin θ)
Shear force = +/– (horizontal force × sin θ)
                 +/– (vertical force × cos θ)

The signs are dependent on the directions of the
respective forces.

Member Forces

Solution

Topic: Statically Determinate Rigid-Jointed Frames

Problem Number: 5.3   Page No. 3

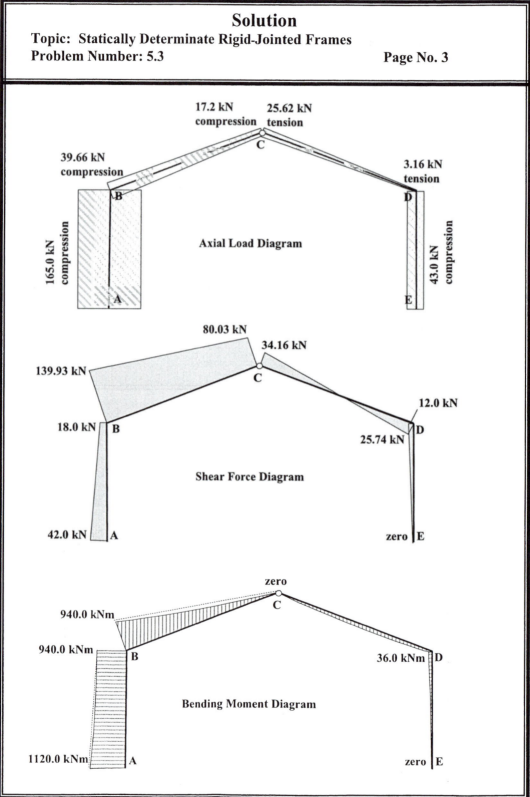

Axial Load Diagram

17.2 kN compression   25.62 kN tension

39.66 kN compression   3.16 kN tension

165.0 kN compression   43.0 kN compression

Shear Force Diagram

80.03 kN   34.16 kN

139.93 kN

18.0 kN   12.0 kN

25.74 kN

42.0 kN   zero

Bending Moment Diagram

zero

940.0 kNm

940.0 kNm   36.0 kNm

1120.0 kNm   zero

Solution

Topic: Statically Determinate Rigid-Jointed Frames
Problem Number: 5.4   Page No. 1

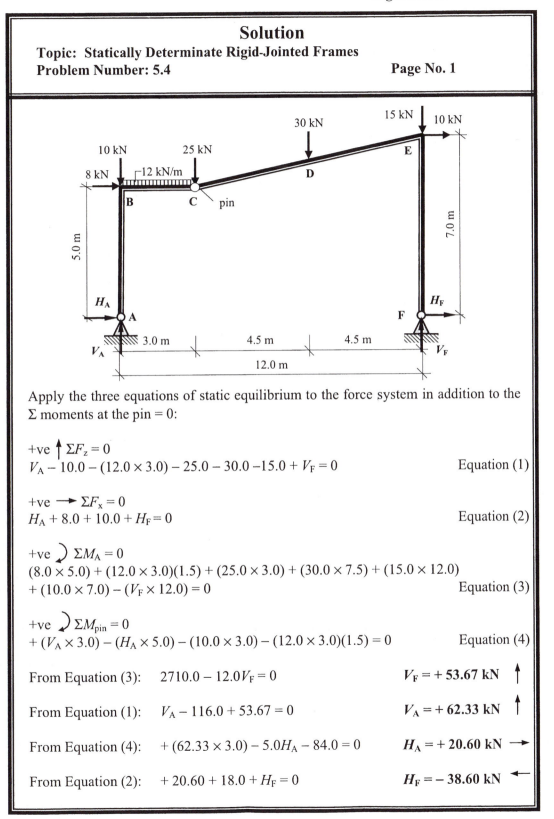

Apply the three equations of static equilibrium to the force system in addition to the Σ moments at the pin $= 0$:

+ve $\uparrow \Sigma F_z = 0$
$V_A - 10.0 - (12.0 \times 3.0) - 25.0 - 30.0 - 15.0 + V_F = 0$ Equation (1)

+ve $\longrightarrow \Sigma F_x = 0$
$H_A + 8.0 + 10.0 + H_F = 0$ Equation (2)

+ve $\circlearrowright \Sigma M_A = 0$
$(8.0 \times 5.0) + (12.0 \times 3.0)(1.5) + (25.0 \times 3.0) + (30.0 \times 7.5) + (15.0 \times 12.0)$
$+ (10.0 \times 7.0) - (V_F \times 12.0) = 0$ Equation (3)

+ve $\circlearrowright \Sigma M_{pin} = 0$
$+ (V_A \times 3.0) - (H_A \times 5.0) - (10.0 \times 3.0) - (12.0 \times 3.0)(1.5) = 0$ Equation (4)

From Equation (3): $2710.0 - 12.0 V_F = 0$ $V_F = + 53.67 \text{ kN} \uparrow$

From Equation (1): $V_A - 116.0 + 53.67 = 0$ $V_A = + 62.33 \text{ kN} \uparrow$

From Equation (4): $+ (62.33 \times 3.0) - 5.0 H_A - 84.0 = 0$ $H_A = + 20.60 \text{ kN} \longrightarrow$

From Equation (2): $+ 20.60 + 18.0 + H_F = 0$ $H_F = - 38.60 \text{ kN} \longleftarrow$

Solution
Topic: Statically Determinate Rigid-Jointed Frames
Problem Number: 5.4

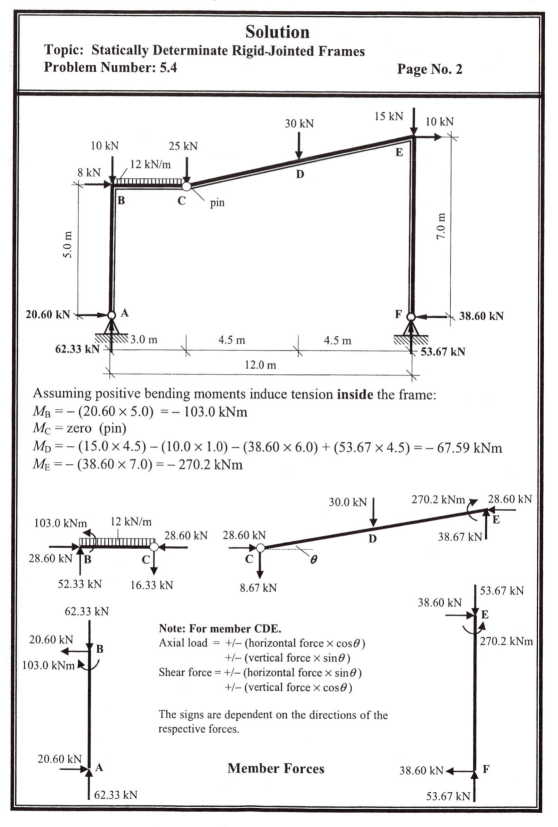

Assuming positive bending moments induce tension **inside** the frame:

$M_B = -(20.60 \times 5.0) = -103.0$ kNm

$M_C =$ zero (pin)

$M_D = -(15.0 \times 4.5) - (10.0 \times 1.0) - (38.60 \times 6.0) + (53.67 \times 4.5) = -67.59$ kNm

$M_E = -(38.60 \times 7.0) = -270.2$ kNm

Note: For member CDE.
Axial load = +/– (horizontal force × cosθ)
             +/– (vertical force × sinθ)
Shear force = +/– (horizontal force × sinθ)
             +/– (vertical force × cosθ)

The signs are dependent on the directions of the respective forces.

Member Forces

Solution

Topic: Statically Determinate Rigid-Jointed Frames
Problem Number: 5.4   **Page No. 3**

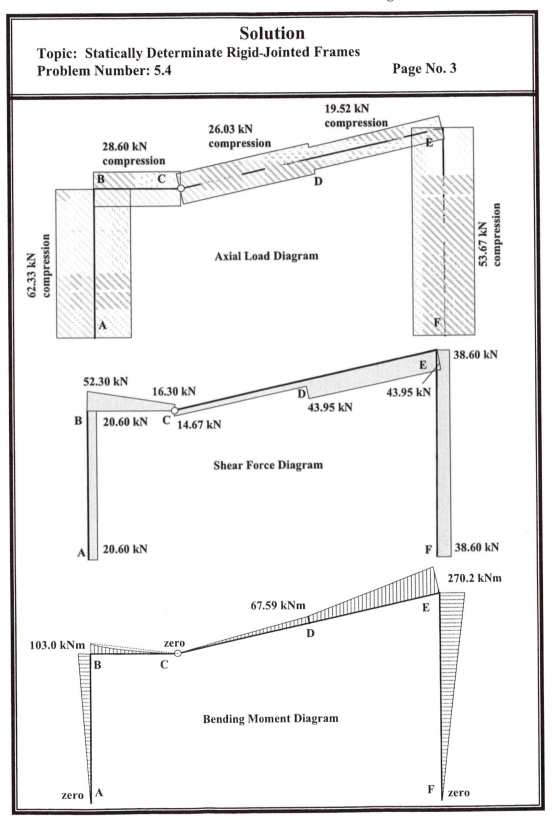

Axial Load Diagram

Shear Force Diagram

Bending Moment Diagram

5.2   *Unit Load Method for Singly-Redundant, Rigid-Jointed Frames*

The method of analysis illustrated in Chapter 4: Section 4.6.1 for singly redundant beams can be adopted for the analysis of singly-redundant, rigid-jointed frames. Consider the frame shown in Figure 5.11.

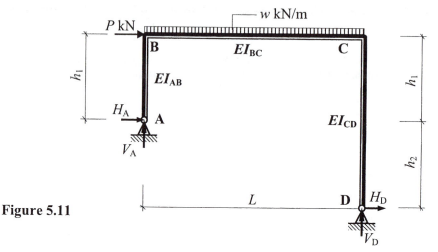

Figure 5.11

The frame is statically indeterminate where the number of degrees-of-indeterminacy is given by $I_D = [(3m + r) - 3n] = [(3 \times 3) + 4 - (3 \times 4)] = 1$

The frame shown in Figure 5.11 can be represented as the superposition of two separate structures, i.e.

Figure 5.12(a) in which a redundant reaction e.g. the horizontal component of reaction at support A is removed and all of the external loading is applied in addition to the components of reaction necessary to maintain equilibrium; they are V'_A, V'_D and H'_D and Figure 5.12(b) in which only the redundant reaction is applied in addition to the components of reaction necessary to maintain equilibrium as indicated in terms of a unit load, i.e V''_A, V''_D and H'_D.

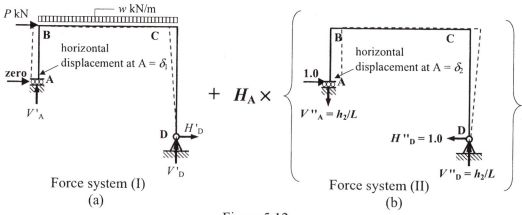

Figure 5.12

Assuming that there is no horizontal movement at support A, the following equation is valid: $\delta_A = (\delta_1 + H_A \delta_2) = 0$ where δ_1 and δ_2 are as indicated in Figures 12(a) and (b).

Considering force system (I) and using the Unit Load method considering bending effects only, the value of the displacement δ_1 can be determined as follows:

$$\delta_1 = \int_0^L \frac{Mm}{EI}dx = \int_A^B \frac{Mm}{EI}dx + \int_B^C \frac{Mm}{EI}dx + \int_C^D \frac{Mm}{1.5EI}dx$$

where:

M is the bending moment due to the applied load system,

m is the bending moment due to a horizontal unit load applied at support A as shown in Figure 5.12. (see Chapter 4: Section 4.5).

The product integral $\int_0^L \frac{Mm}{EI}dx$ can be be calculated as:

Σ (the **area of the applied load** bending moment diagram × the **ordinate on the unit load** bending moment diagram corresponding to the position of the centroid of the applied load bending moment diagram)/EI for each member.

Considering force system (II) and using the Unit Load method considering bending effects only, the value of the displacement δ_2 can be determined as follows:

$$\delta_2 = \int_0^L \frac{m^2}{EI}dx = \int_A^B \frac{m^2}{EI}dx + \int_B^C \frac{m^2}{EI}dx + \int_C^D \frac{m^2}{1.5EI}dx$$

The product integral $\int_0^L \frac{m^2}{EI}dx$ can be be calculated as:

Σ (the **area of the unit load** bending moment diagram × the **ordinate on the unit load** bending moment diagram corresponding to the position of the centroid of the applied load bending moment diagram)/EI for each member.

Considering the horizontal displacent at support A:

$$\delta_A = (\delta_1 + H_A\delta_2) = 0 \quad \text{gives} \quad \int_0^L \frac{Mm}{EI}dx + \left(H_A \times \int_0^L \frac{m^2}{EI}dx \right) = 0$$

and hence $H_A = -\dfrac{\displaystyle\int_0^L \frac{Mm}{EI}dx}{\displaystyle\int_0^L \frac{m^2}{EI}dx}$.

In the case of a horizontal movement δ_3 at support A then the above equation can be modified accordingly, i.e. $\delta_A = (\delta_1 + H_A\delta_2) = \delta_3$.

The magnitude and sense of the components of the support reactions and bending moments can be determined by superposition of the values determined from the two force systems as follows, (see Figure 5.12):

Final value = force system (I) value + H_A × force system (II) value   i.e.

Support reactions:
H_A = zero + H_A × (1.0)
$V_A = V'_A + H_A \times (V''_A)$
$H_D = H'_D + H_A \times (H''_D)$
$V_D = V'_D + H_A \times (V''_D)$

Bending moments at A, B, C and D:
M_A = zero
$M_B = M'_B + H_A \times (M''_B)$ i.e. = $M'_B + H_A \times (-1.0 \times h_1)$
$M_C = M'_C + H_A \times (M''_C)$ i.e. = $H'_D (h_1 + h_2) + H_A \times [-H''_D (h_1 + h_2)]$
V_D = zero

5.2.1 Example 5.3 Singly-Redundant, Rigid-Jointed Frame

A non-uniform, asymmetric pitched-roof portal frame ABCDE is pinned at supports A and E and subjected to the loading indicated in Figure 5.13.

i) determine the support reactions, and

ii) determine the change in the bending moment at B due to an outwards horizontal movement of support A equal to 15 mm ←

Note: $EI = 20.0 \times 10^3$ kNm²

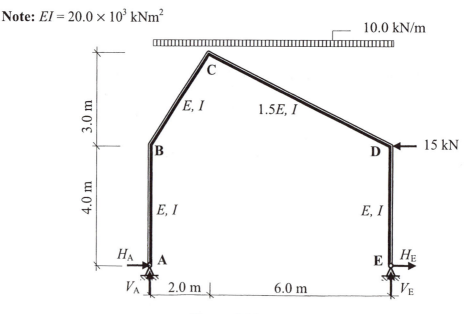

Figure 5.13

The reactions in force system (I) and force system (II) are as indicated in Figure 5.14.

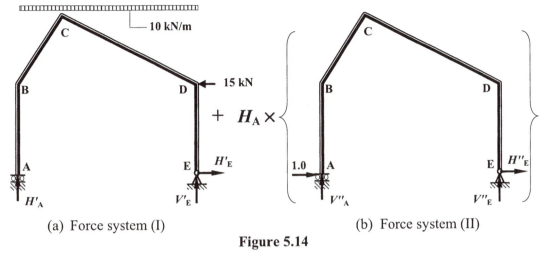

(a) Force system (I)   (b) Force system (II)

Figure 5.14

Apply the three equations of static equilibrium to force system (I) in Figure 5.14(a)

$+ve \uparrow \Sigma F_z = 0$ $V'_A - (10.0 \times 8.0) + V'_E = 0$ Equation (1)

$+ve \longrightarrow \Sigma F_x = 0$ $-15.0 + H'_E = 0$ Equation (2)

$+ve \,\rotatebox{0}{)}\, \Sigma M_A = 0$ $(10.0 \times 8.0)(4.0) - (15.0 \times 4.0) - (V'_E \times 8.0) = 0$ Equation (3)

From equation (2): $-15.0 + H'_E = 0$ $\therefore H'_E = + \mathbf{15.0 \ kN} \longrightarrow$

From equation (3): $260.0 - 8.0V'_E = 0$ $\therefore V'_E = + \mathbf{32.50 \ kN} \uparrow$

From equation (1): $V'_A - 80.0 + 32.5 = 0$ $\therefore V'_A = + \mathbf{47.50 \ kN} \uparrow$

Apply the three equations of static equilibrium to force system (II) in Figure 5.14(b)

$+ve \uparrow \Sigma F_z = 0$ $V''_A + V''_E = 0$ Equation (4)

$+ve \longrightarrow \Sigma F_x = 0$ $1.0 + H''_E = 0$ Equation (5)

$+ve \,\rotatebox{0}{)}\, \Sigma M_A = 0$ $-(V''_E \times 8.0) = 0$ Equation (6)

From equation (5): $1.0 + H''_E = 0$ $\therefore H''_E = - \mathbf{1.0 \ kN} \longleftarrow$

From equation (6): $-8.0V''_E = 0$ $\therefore V''_E = \mathbf{zero}$

From equation (4): $V''_A + V''_E = 0$ $\therefore V''_A = \mathbf{zero}$

Consider the frame shown in Figure 5.14.

Determine the values of M' and m at each of the node points for Figure 5.14(a) and Figure 5.14(b).

Node A: $M'_A = 0$ $m = 0$

Node B: $M'_B = 0$ $m = - (1.0 \times 4,0) = -4.0$

Node C: $M'_C = + (47.5 \times 2.0) - (10.0 \times 2.0 \times 1.0) = + 75.0 \ kNm$

                                                  $m = - (1.0 \times 7.0) = -7.0$

Node D: $M'_D = + (15.0 \times 4.0) = + 60.0 \ kNm$ $m = - (1.0 \times 4.0) = -4.0$

Node E: $M'_E = 0$
$m = 0$

The bending moment diagram for the frame in Figure 5.14(a) is given in Figure 5.15 and can be considered to be the sum of the fixed and free bending moment diagrams for each member as indicated in Figure 5.16.

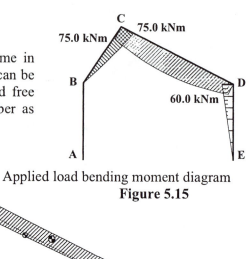

Applied load bending moment diagram
Figure 5.15

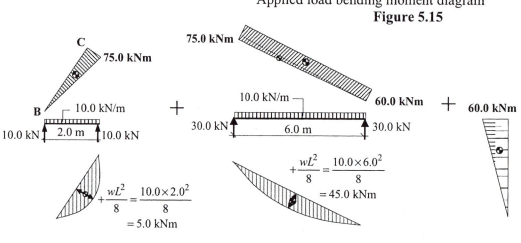

Fixed and free bending moment diagrams (*M*)
Figure 5.16

The unit load bending moment diagram for the frame shown in Figure 5.14(b) is indicated in Figure 5.17.

Length of BC = 3.61 m

Length of CD = 6.71 m

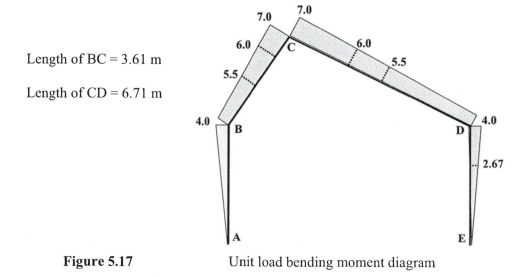

Figure 5.17   Unit load bending moment diagram

Using the unit load method $\delta_1 = \int_0^L \dfrac{Mm}{EI} dx = \int_A^B \dfrac{Mm}{EI} dx + \int_B^C \dfrac{Mm}{EI} dx + \int_C^D \dfrac{Mm}{1.5EI} dx + \int_D^E \dfrac{Mm}{EI} dx$

The product integral $\int_0^L Mm\ dx$ can be be calculated as:

Σ (Area of the applied load bending moment diagram $\times$ the ordinate on the unit load bending moment diagram corresponding to the position of the centroid of the applied load bending moment diagram) for each member.

$\int_A^B \dfrac{Mm}{EI} dx = 0$ since $M = 0$

$\int_B^C \dfrac{Mm}{EI} dx = -\dfrac{(0.5\times3.61\times75.0\times6.0)+(0.67\times3.61\times5.0\times5.5)}{EI} = -\dfrac{878.76}{EI}$

$\int_C^D \dfrac{Mm}{1.5EI} dx = -\dfrac{(0.5\times6.71\times15.0\times6.0)+(6.71\times60\times5.5)+(0.67\times6.71\times45.0\times5.5)}{1.5EI}$

$\qquad = -\dfrac{2419.29}{EI}$

$\int_D^E \dfrac{Mm}{EI} dx = -\dfrac{(0.5\times4.0\times60.0\times2.67)}{EI} = -\dfrac{320.40}{EI}$

$\therefore\ \delta_1 = \int_0^L \dfrac{Mm}{EI} dx = -(0 + 878.76 + 2419.29 + 320.40)/EI = -3618.45/EI$

Consider the frame shown in Figure 5.14(b) in which only the redundant reaction is applied in addition to the components of reaction necessary to maintain equilibrium.

$\delta_2 = \int_0^L \dfrac{m^2}{EI} dx = \int_A^B \dfrac{m^2}{EI} dx + \int_B^C \dfrac{m^2}{EI} dx + \int_C^D \dfrac{m^2}{1.5EI} dx + \int_D^E \dfrac{m^2}{EI} dx$

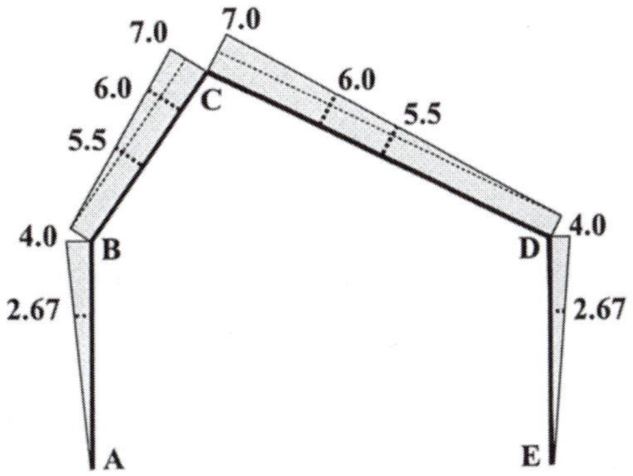

Figure 5.18   **Unit load bending moment diagrams (*m*)**

$$\int_{A}^{B} \frac{m^2}{EI} dx = \frac{(0.5 \times 4.0 \times 4.0 \times 2.67)}{EI} = \frac{21.36}{EI}$$

$$\int_{B}^{C} \frac{m^2}{EI} dx = \frac{(0.5 \times 3.61 \times 3.0 \times 6.0) + (3.61 \times 4.0 \times 5.5)}{EI} = \frac{111.91}{EI}$$

$$\int_{C}^{D} \frac{m^2}{1.5EI} dx = \frac{(0.5 \times 6.71 \times 3.0 \times 6.0) + (6.71 \times 4.0 \times 5.5)}{1.5EI} = \frac{138.67}{EI}$$

$$\int_{D}^{E} \frac{m^2}{EI} dx = \int_{A}^{B} \frac{m^2}{EI} dx = \frac{(0.5 \times 4.0 \times 4.0 \times 2.67)}{EI} = \frac{21.36}{EI}$$

$$\therefore \delta_2 = \int_{0}^{L} \frac{m^2}{EI} dx = (21.36 + 111.91 + 138.67 + 21.36)/EI = 293.54/EI$$

and

$$H_A \, \delta_2 = H_A \times \int_{0}^{L} \frac{m^2}{EI} dx = 293.54 H_A / EI$$

In case (i)

Considering the horizontal displacent at support A:

$$\delta_A = (\delta_1 + H_A \delta_2) = 0 \;\; \text{gives} \;\; \int_{0}^{L} \frac{Mm}{EI} dx + \left(H_A \times \int_{0}^{L} \frac{m^2}{EI} dx \right) = 0 \;\; \therefore \;\; H_A = - \frac{\displaystyle\int_{0}^{L} \frac{Mm}{EI} dx}{\displaystyle\int_{0}^{L} \frac{m^2}{EI} dx}$$

where:

$$\int_{0}^{L} \frac{Mm}{EI} dx = -3618.45/EI \quad \text{and} \quad \int_{0}^{L} \frac{m^2}{EI} dx = 293.54/EI$$

$$\therefore \; H_A = - \frac{\displaystyle\int_{0}^{L} \frac{Mm}{EI} dx}{\displaystyle\int_{0}^{L} \frac{m^2}{EI} dx} = -(-3618.45/293.54) = +12.33 \text{ kN} \quad \longrightarrow$$

i.e. in the same direction as the assumed unit load and, using superposition with force systems (I) and (II):

$$H_A = \text{zero} + H_A \times (1,0) = 0 + (12.33 \times 1.0) = +12.33 \text{ kN} \qquad \longrightarrow$$
$$V_A = V'_A + H_A \times (\text{zero}) = +47.5 + (12.33 \times 0) = +47.5 \text{ kN} \qquad \uparrow$$
$$M_A = \text{zero}$$

$H_E = H'_E + H_A \times (-1,0) = +15.0 + (12.33 \times -1.0) = +2.67 \text{ kN} \longrightarrow$
$V_E = V'_E + H_A \times (\text{zero}) = +32.5 + (12.33 \times 0) = +32.5 \text{ kN} \uparrow$
$M_E = \text{zero}$

Bending moments at B, C and D:
$M_B = M'_B + H_A \times (-4.0) = 0 - (12.33 \times 4.0) = -49.32 \text{ kNm}$
$M_C = M'_C + H_A \times (-7.0) = +75.0 - (12.33 \times 7.0) = -11.31 \text{ kNm}$
$M_D = M'_D + H_A \times (-4.0) = +60.0 - (12.33 \times 4.0) = +10.68 \text{ kNm}$

(Readers should complete the axial load and shear force diagrams indicating all the salient values.)

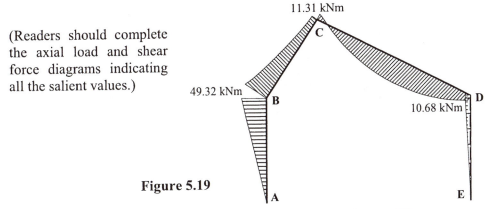

Figure 5.19

Bending Moment Diagram

In case (ii)
Considering the 15.0 mm horizontal displacement at support A:

$$\delta_A = (\delta_1 + H_A \delta_2) = -0.015 \text{ and hence } H_A = \dfrac{-0.015 - \displaystyle\int_0^L \dfrac{Mm}{EI}\,dx}{\displaystyle\int_0^L \dfrac{m^2}{EI}\,dx}$$

$$H_A = \dfrac{-0.015 + \dfrac{3618.45}{20.0 \times 10^3}}{\dfrac{293.54}{20.0 \times 10^3}} = +11.30 \text{ kN} \longrightarrow$$

i.e. in the same direction as the assumed unit load

The bending moment at B, $M_B = (H_A \times 4.0) = (11.30 \times 4.0) = 45.20 \text{ kNm}$

The change in the bending moment at B due to the horizontal displacement at A is given by $(49.32 - 45.20)$ i.e. a reduction of 4.12 kNm

5.2.2 Problems: Unit Load Method for Singly-Redundant, Rigid-Jointed Frames

A series of statically indeterminate, rigid-jointed frames are indicated in Problems 5.5 to 5.8. In each case, for the loading given:

i) **determine the support reactions and**

ii) **determine the magnitude and sense of the bending moments at the joints and under the point loads. (Assume tension inside the frame is positive bending).**

(Note: readers should complete the axial load, shear force and bending moment diagrams for each frame)

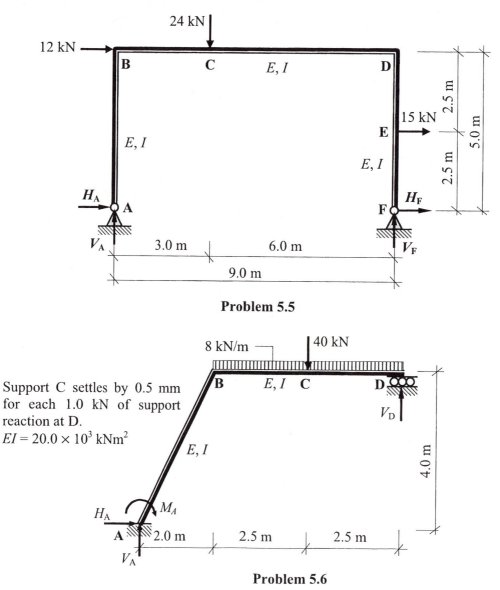

Problem 5.5

Problem 5.6

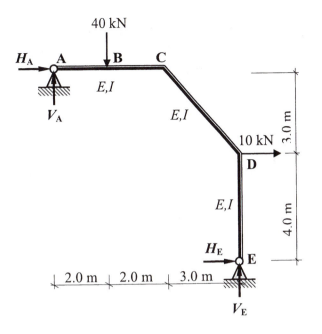

40 kN

H_A A B C

E,I

E,I

10 kN

D

3.0 m

E,I

4.0 m

H_E E

2.0 m | 2.0 m | 3.0 m

V_E

V_A

Problem 5.7

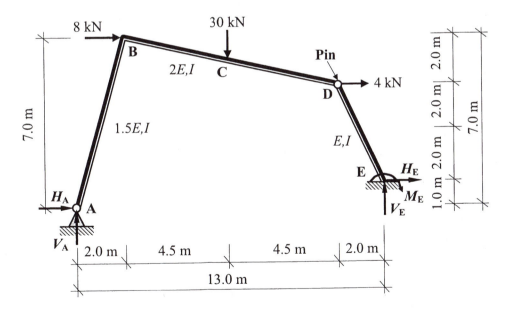

8 kN

30 kN

B

$2E,I$ C

Pin

D 4 kN

7.0 m

$1.5E,I$

E,I

E H_E

M_E

H_A A

V_E

V_A 2.0 m | 4.5 m | 4.5 m | 2.0 m

1.0 m 2.0 m | 2.0 m | 2.0 m

7.0 m

13.0 m

Problem 5.8

5.2.3 Solutions: Unit Load Method for Singly-Redundant, Rigid-Jointed Frames

Solution

Topic: Unit Load Method for Singly-Redundant, Rigid-Jointed Frames
Problem Number: 5.5   **Page No. 1**

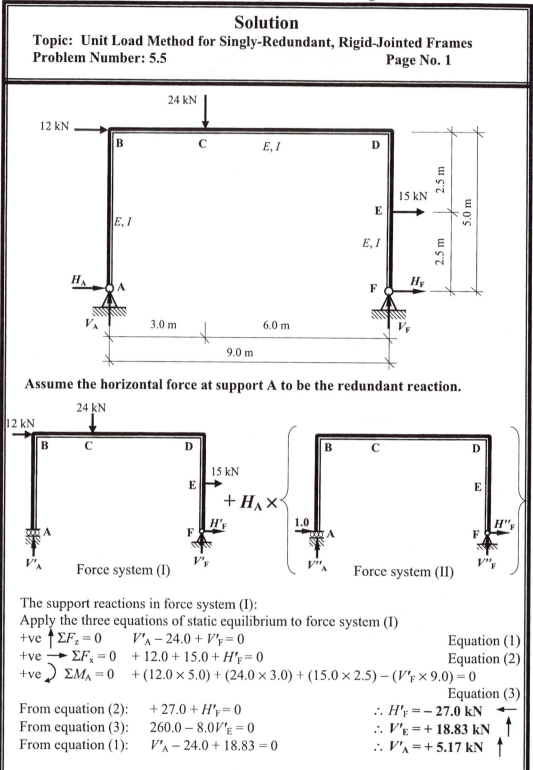

Assume the horizontal force at support A to be the redundant reaction.

Force system (I)

Force system (II)

The support reactions in force system (I):
Apply the three equations of static equilibrium to force system (I)

+ve $\uparrow \Sigma F_z = 0$ $V'_A - 24.0 + V'_F = 0$ Equation (1)

+ve $\longrightarrow \Sigma F_x = 0$ $+ 12.0 + 15.0 + H'_F = 0$ Equation (2)

+ve $\downarrow \Sigma M_A = 0$ $+ (12.0 \times 5.0) + (24.0 \times 3.0) + (15.0 \times 2.5) - (V'_F \times 9.0) = 0$

                                                                                    Equation (3)

From equation (2): $+ 27.0 + H'_F = 0$ $\therefore H'_F = -27.0$ kN $\longleftarrow$

From equation (3): $260.0 - 8.0 V'_E = 0$ $\therefore V'_E = +18.83$ kN $\uparrow$

From equation (1): $V'_A - 24.0 + 18.83 = 0$ $\therefore V'_A = +5.17$ kN $\uparrow$

Solution
Topic: Unit Load Method for Singly-Redundant, Rigid-Jointed Frames
Problem Number: 5.5   **Page No. 2**

The support reactions in force system (II):
Apply the three equations of static equilibrium to force system (II).

$+ve \uparrow \Sigma F_z = 0 \qquad V''_A + V''_F = 0$ Equation (4)

$+ve \longrightarrow \Sigma F_x = 0 \qquad 1.0 + H''_F = 0$ Equation (5)

$+ve \curvearrowright \Sigma M_A = 0 \qquad -(V''_F \times 9.0) = 0$ Equation (6)

From equation (5): $1.0 + H''_F = 0$ $\therefore H''_F = -1.0$ **kN** $\longleftarrow$

From equation (6): $-9.0V''_F = 0$ $\therefore V''_F = \textbf{zero}$

From equation (4): $V''_A + V''_F = 0$ $\therefore V''_A = \textbf{zero}$

Determine the values of M' and m at each of the node and load points:

Node A: $M' = 0$ $m = 0$

Node B: $M' = 0$ $m = -(1.0 \times 5,0) = -5.0$

Node D: $M' = +(15.0 \times 2.5) - (27.0 \times 5.0) = -97,50$ kNm

                                                   $m = -(1.0 \times 5.0) = -5.0$

Node F: $M' = 0$ $m = 0$

Point C: $M' = +(5.17 \times 3.0) = +15.51$ kNm $m = -(1.0 \times 5.0) = -5.0$

Point E: $M' = -(27.0 \times 2.5) = -67,50$ kNm $m = -(1.0 \times 2.5) = -2.5$

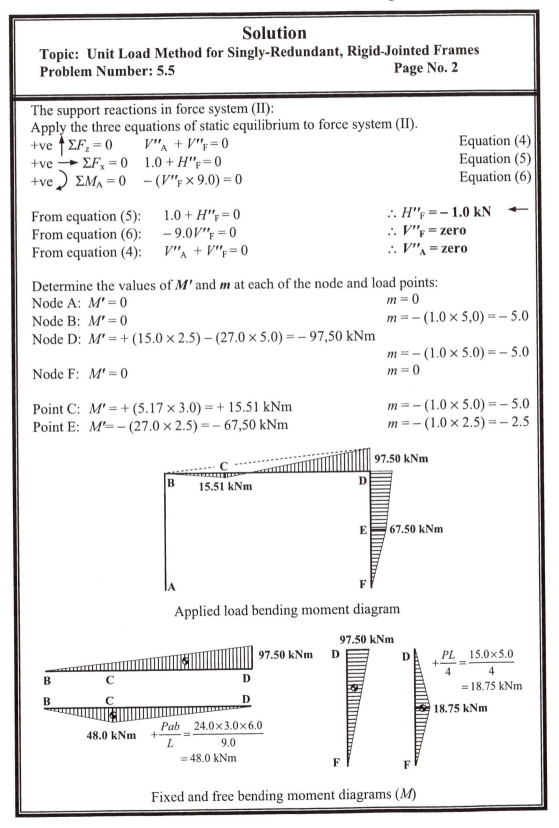

Applied load bending moment diagram

Fixed and free bending moment diagrams (M)

Solution

Topic: Unit Load Method for Singly-Redundant, Rigid-Jointed Frames
Problem Number: 5.5   **Page No. 3**

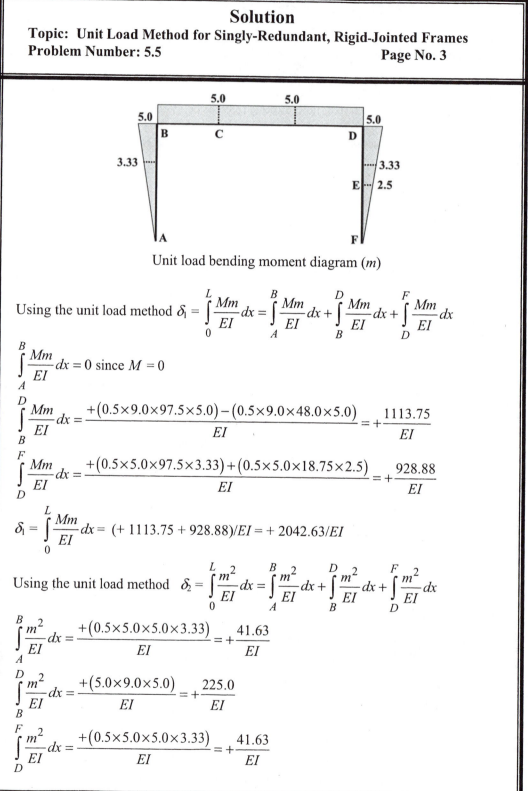

Unit load bending moment diagram (*m*)

Using the unit load method $\delta_1 = \int_0^L \dfrac{Mm}{EI}\,dx = \int_A^B \dfrac{Mm}{EI}\,dx + \int_B^D \dfrac{Mm}{EI}\,dx + \int_D^F \dfrac{Mm}{EI}\,dx$

$\int_A^B \dfrac{Mm}{EI}\,dx = 0$ since $M = 0$

$\int_B^D \dfrac{Mm}{EI}\,dx = \dfrac{+(0.5\times9.0\times97.5\times5.0)-(0.5\times9.0\times48.0\times5.0)}{EI} = +\dfrac{1113.75}{EI}$

$\int_D^F \dfrac{Mm}{EI}\,dx = \dfrac{+(0.5\times5.0\times97.5\times3.33)+(0.5\times5.0\times18.75\times2.5)}{EI} = +\dfrac{928.88}{EI}$

$\delta_1 = \int_0^L \dfrac{Mm}{EI}\,dx = (+1113.75 + 928.88)/EI = +2042.63/EI$

Using the unit load method $\delta_2 = \int_0^L \dfrac{m^2}{EI}\,dx = \int_A^B \dfrac{m^2}{EI}\,dx + \int_B^D \dfrac{m^2}{EI}\,dx + \int_D^F \dfrac{m^2}{EI}\,dx$

$\int_A^B \dfrac{m^2}{EI}\,dx = \dfrac{+(0.5\times5.0\times5.0\times3.33)}{EI} = +\dfrac{41.63}{EI}$

$\int_B^D \dfrac{m^2}{EI}\,dx = \dfrac{+(5.0\times9.0\times5.0)}{EI} = +\dfrac{225.0}{EI}$

$\int_D^F \dfrac{m^2}{EI}\,dx = \dfrac{+(0.5\times5.0\times5.0\times3.33)}{EI} = +\dfrac{41.63}{EI}$

Solution

Topic: **Unit Load Method for Singly-Redundant, Rigid-Jointed Frames**
Problem Number: **5.5**   Page No. **4**

$$\delta_2 = \int_0^L \frac{m^2}{EI} dx = (41.63 + 225.0 + 41.63)/EI = 308.26/EI$$

and

$$H_A \, \delta_2 = H_A \times \int_0^L \frac{m^2}{EI} dx = 308.26 H_A/EI$$

Considering the horizontal displacent at support A:

$$\delta_A = (\delta_1 + H_A\delta_2) = 0 \ \text{gives} \ \int_0^L \frac{Mm}{EI} dx + \left(H_A \times \int_0^L \frac{m^2}{EI} dx \right) = 0 \ \therefore \ H_A = -\frac{\displaystyle\int_0^L \frac{Mm}{EI} dx}{\displaystyle\int_0^L \frac{m^2}{EI} dx}$$

where:

$$\int_0^L \frac{Mm}{EI} dx = + 2042.63/EI \quad \text{and} \quad \int_0^L \frac{m^2}{EI} dx = + 308.26/EI$$

$$H_A = -\frac{\displaystyle\int_0^L \frac{Mm}{EI} dx}{\displaystyle\int_0^L \frac{m^2}{EI} dx} = - (2042.63/308.26) = - 6.63 \text{ kN} \quad \longleftarrow$$

i.e. in the opposite direction to the assumed unit load

and using superposition with force systems (I) and (II):

$H_A = \text{zero} + H_A \times (+ 1,0) = 0 + (6.63 \times 1.0) = - 6.63 \text{ kN} \qquad \longleftarrow$
$V_A = V'_A + H_A \times (\text{zero}) = + 5.17 + (6.63 \times 0) = + 5.17 \text{ kN} \qquad \uparrow$
$M_A = \text{zero}$
$H_F = H'_F + H_A \times (- 1,0) = - 27.0 + (6.63 \times - 1.0) = - 20.37 \text{ kN} \qquad \longleftarrow$
$V_E = V'_E + H_A \times (\text{zero}) = + 18.83 + (6.63 \times 0) = + 18.83 \text{ kN} \qquad \uparrow$
$M_F = \text{zero}$

Bending moments at B, C, D, and E :
$M_B = \text{zero} + H_A \times (- 5.0) = 0 + (6.63 \times 5.0) = + 33.15 \text{ kNm}$
$M_C = M'_C + H_A \times (- 5.0) = + 15.50 + (6.63 \times 5.0) = + 48.65 \text{ kNm}$
$M_D = M'_D + H_A \times (- 5.0) = - 97.50 + (6.63 \times 5.0) = - 64.35 \text{ kNm}$
$M_E = M'_E + H_A \times (- 2.5) = - 67.50 + (6.63 \times 2.5) = - 50.93 \text{ kNm}$

Solution
Topic: Unit Load Method for Singly-Redundant, Rigid-Jointed Frames
Problem Number: 5.6   **Page No. 1**

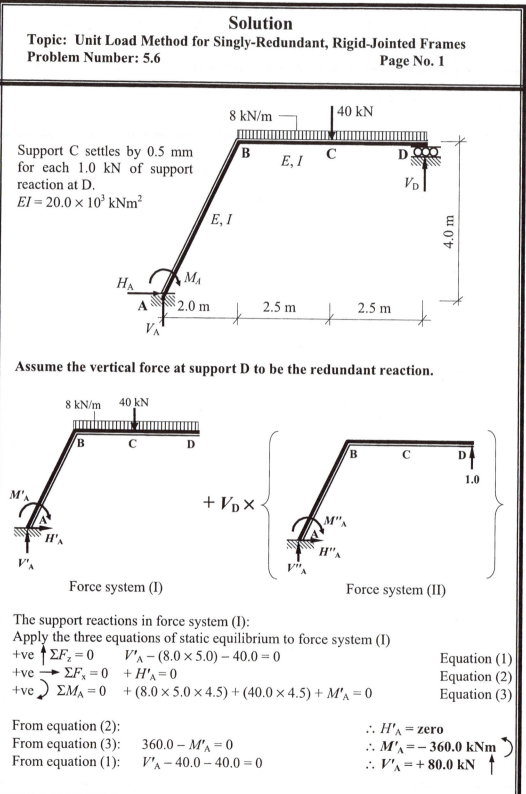

Assume the vertical force at support D to be the redundant reaction.

Force system (I)   Force system (II)

The support reactions in force system (I):
Apply the three equations of static equilibrium to force system (I)

+ve $\uparrow \Sigma F_z = 0$ $V'_A - (8.0 \times 5.0) - 40.0 = 0$ Equation (1)

+ve $\rightarrow \Sigma F_x = 0$ $+ H'_A = 0$ Equation (2)

+ve $\curvearrowright \Sigma M_A = 0$ $+ (8.0 \times 5.0 \times 4.5) + (40.0 \times 4.5) + M'_A = 0$ Equation (3)

From equation (2): $\therefore H'_A = \textbf{zero}$
From equation (3): $360.0 - M'_A = 0$ $\therefore M'_A = \textbf{– 360.0 kNm} \curvearrowright$
From equation (1): $V'_A - 40.0 - 40.0 = 0$ $\therefore V'_A = \textbf{+ 80.0 kN} \uparrow$

Solution

Topic: Unit Load Method for Singly-Redundant, Rigid-Jointed Frames
Problem Number: 5.6   **Page No. 2**

The support reactions in force system (II):
Apply the three equations of static equilibrium to force system (II).

+ve $\uparrow \Sigma F_z = 0$	$V''_A + 1.0 = 0$	Equation (4)
+ve $\longrightarrow \Sigma F_x = 0$	$+ H''_A = 0$	Equation (5)
+ve $\Sigma M_A = 0$	$M''_A - (1.0 \times 7.0) = 0$	Equation (6)

From equation (5): $H''_A = 0$ $\therefore H''_F = $ **zero**
From equation (6): $V''_A = -1.0$ $\therefore V''_A = -1.0 \downarrow$
From equation (4): $M''_A - 7.0 = 0$ $\therefore M''_A = +7.0$

Determine the values of M and m at each of the node and load points:

Node A: $M = -360.0$ kNm $m = +7.0$
Node B: $M = -360.0 + (80.0 \times 2.0) = -200.0$ kNm $m = -(1.0 \times 2,0) + 7.0 = +5.0$
Node D: $M = $ zero $m = $ zero

Point C: $M = -(8.0 \times 2.5 \times 1.25) = -25.0$ kNm $m = +(1.0 \times 2.5) = +2.5$

200.0 kNm   B   C   D   zero

$L_{AB} = (2.0^2 + 4.0^2)^{0.5} = 4.472$ m

360.0 kNm   A

Applied load bending moment diagram

200.0 kNm   B

200 kNm   B   C   D

B   C   D $+\dfrac{PL}{4} = \dfrac{40.0 \times 5.0}{4.0}$
50.0 kNm $= 50.0$ kNm

360.0 kNm   A

B   C   D $+\dfrac{wL^2}{8} = \dfrac{8.0 \times 5.0^2}{8}$
25.0 kNm $= 25.0$ kNm

Fixed and free bending moment diagrams (M)

Solution
Topic: Unit Load Method for Singly-Redundant, Rigid-Jointed Frames
Problem Number: 5.6   **Page No. 3**

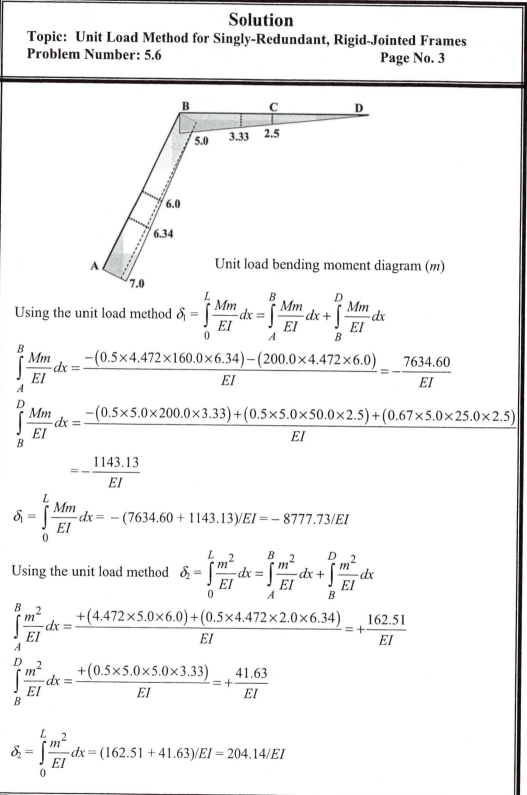

Unit load bending moment diagram (m)

Using the unit load method $\delta_1 = \int_0^L \dfrac{Mm}{EI}\,dx = \int_A^B \dfrac{Mm}{EI}\,dx + \int_B^D \dfrac{Mm}{EI}\,dx$

$$\int_A^B \frac{Mm}{EI}\,dx = \frac{-(0.5\times4.472\times160.0\times6.34)-(200.0\times4.472\times6.0)}{EI} = -\frac{7634.60}{EI}$$

$$\int_B^D \frac{Mm}{EI}\,dx = \frac{-(0.5\times5.0\times200.0\times3.33)+(0.5\times5.0\times50.0\times2.5)+(0.67\times5.0\times25.0\times2.5)}{EI}$$

$$= -\frac{1143.13}{EI}$$

$$\delta_1 = \int_0^L \frac{Mm}{EI}\,dx = -(7634.60 + 1143.13)/EI = -8777.73/EI$$

Using the unit load method $\delta_2 = \int_0^L \dfrac{m^2}{EI}\,dx = \int_A^B \dfrac{m^2}{EI}\,dx + \int_B^D \dfrac{m^2}{EI}\,dx$

$$\int_A^B \frac{m^2}{EI}\,dx = \frac{+(4.472\times5.0\times6.0)+(0.5\times4.472\times2.0\times6.34)}{EI} = +\frac{162.51}{EI}$$

$$\int_B^D \frac{m^2}{EI}\,dx = \frac{+(0.5\times5.0\times5.0\times3.33)}{EI} = +\frac{41.63}{EI}$$

$$\delta_2 = \int_0^L \frac{m^2}{EI}\,dx = (162.51 + 41.63)/EI = 204.14/EI$$

Solution
Topic: Unit Load Method for Singly-Redundant, Rigid-Jointed Frames
Problem Number: 5.6   **Page No. 4**

and

$$V_D \ \delta_2 = V_D \times \int_0^L \frac{m^2}{EI} dx \ = 204.14 V_D/EI$$

Considering the vertical displacent at support D:
$$\delta_D = (\delta_1 + V_D \delta_2) = - (V_D \times 0.0005)$$

$$\int_0^L \frac{Mm}{EI} \ dx + \left(V_D \times \int_0^L \frac{m^2}{EI} \ dx \right) = - 0.0005 V_D \ \therefore \ V_D = - \frac{\int_0^L \frac{Mm}{EI} dx}{\int_0^L \frac{m^2}{EI} dx + 0.0005}$$

where:
$$\int_0^L \frac{Mm}{EI} dx = - 8777.73/EI \quad \text{and} \quad \int_0^L \frac{m^2}{EI} dx = + 204.14/EI$$

$$V_D = - \frac{\int_0^L \frac{Mm}{EI} dx}{\int_0^L \frac{m^2}{EI} dx + 0.0005} = + 8777.73/[204.14 + (0.0005 \times 20 \times 10^3)] = + 41.0 \ \text{kN} \ \uparrow$$

    i.e. in the direction of the assumed unit load

and using superposition with force systems (I) and (II):

H_A = zero + $V_D \times$ (zero) = 0 + (41.0 × 0) = zero
$V_A = V'_A + V_D \times (- 1.0) = + 80.0 - (41.0 \times 1.0) = + 39.0 \ \text{kN}$ ↑
$M_A = M'_A + V_D \times (+ 7.0) = - 360.0 + (41.0 \times 7.0) = - 73.0 \ \text{kNm}$ ↶
$H_D = H'_D + V_D \times$ (zero) = 0 + (41.0 × 0) = zero
$V_D = V'_D + V_D \times (+ 1.0) = 0 + (41.0 \times 1.0) = + 41.0 \ \text{kN}$ ↑
M_D = zero

Bending moments at B and C:
$M_B = M'_B + V_D \times (+ 5.0) = - 200.0 + (41.0 \times 5.0) = + 5.0 \ \text{kNm}$
$M_C = M'_C + V_D \times (+ 2.5) = (- 100.0 + 50.0 + 25.0) + (41.0 \times 2.5) = + 77.5 \ \text{kNm}$

Solution

Topic: Unit Load Method for Singly-Redundant, Rigid-Jointed Frames

Problem Number: 5.7   **Page No. 1**

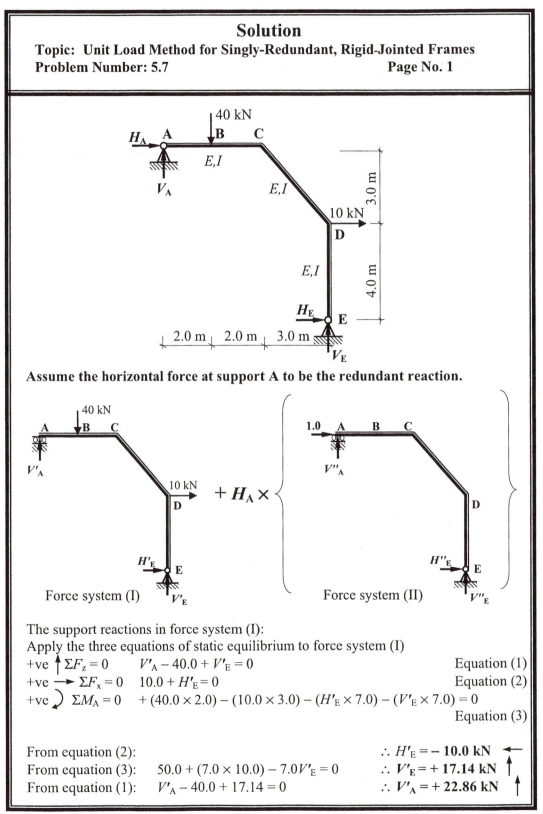

Assume the horizontal force at support A to be the redundant reaction.

Force system (I)   Force system (II)

The support reactions in force system (I):
Apply the three equations of static equilibrium to force system (I)

+ve $\uparrow \Sigma F_z = 0$ $V'_A - 40.0 + V'_E = 0$ Equation (1)

+ve $\longrightarrow \Sigma F_x = 0$ $10.0 + H'_E = 0$ Equation (2)

+ve $\Sigma M_A = 0$ $+ (40.0 \times 2.0) - (10.0 \times 3.0) - (H'_E \times 7.0) - (V'_E \times 7.0) = 0$

Equation (3)

From equation (2): $\therefore H'_E = -\textbf{10.0 kN}$ $\longleftarrow$

From equation (3): $50.0 + (7.0 \times 10.0) - 7.0V'_E = 0$ $\therefore V'_E = +\textbf{17.14 kN}$ $\uparrow$

From equation (1): $V'_A - 40.0 + 17.14 = 0$ $\therefore V'_A = +\textbf{22.86 kN}$ $\uparrow$

Solution
Topic: Unit Load Method for Singly-Redundant, Rigid-Jointed Frames
Problem Number: 5.7   **Page No. 2**

The support reactions in force system (II):
Apply the three equations of static equilibrium to force system (II).

$+ve \uparrow \Sigma F_z = 0 \qquad V''_A + V''_E = 0$ Equation (4)
$+ve \longrightarrow \Sigma F_x = 0 \quad 1.0 + H''_E = 0$ Equation (5)
$+ve \downarrow \Sigma M_A = 0 \quad -(H''_E \times 7.0) - (V''_E \times 7.0) = 0$ Equation (6)

From equation (5): $\quad H''_E = -1.0$ $\therefore H''_E = -1.0 \leftarrow$
From equation (6): $\quad +(1.0 \times 7.0) - (V''_E \times 7.0) = 0$ $\therefore V''_E = +1.0 \uparrow$
From equation (4): $\quad V''_A + 1.0 = 0$ $\therefore V''_A = -1.0 \downarrow$

Determine the values of **M** and **m** at each of the node and load points:
Node A: M = zero m = zero
Node C: $M = +(22.86 \times 4.0) - (40.0 \times 2.0) = +11.44$ kNm

$\qquad\qquad\qquad\qquad\qquad\qquad\qquad\qquad m = -(1.0 \times 4,0) = -4.0$
Node D: $M = -(10.0 \times 4.0) = -40.0$ kNm $m = -4.0$
Node E: M = zero m = zero

Point B: $M = +(22.86 \times 2.0) = +45.72$ kNm $m = -(1.0 \times 2,0) = -2.0$

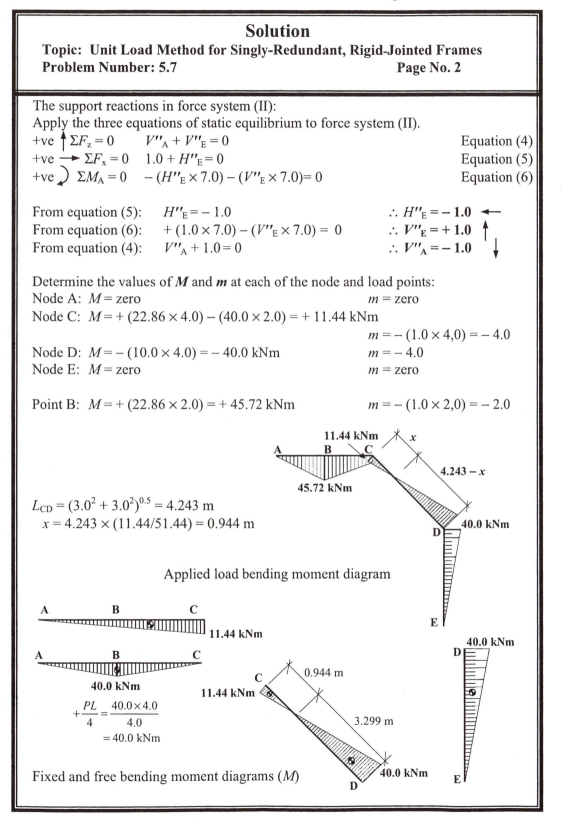

Applied load bending moment diagram

$L_{CD} = (3.0^2 + 3.0^2)^{0.5} = 4.243$ m
$\quad x = 4.243 \times (11.44/51.44) = 0.944$ m

$+\dfrac{PL}{4} = \dfrac{40.0 \times 4.0}{4.0}$
$\qquad = 40.0$ kNm

Fixed and free bending moment diagrams (M)

Solution

Topic: Unit Load Method for Singly-Redundant, Rigid-Jointed Frames

Problem Number: 5.7   **Page No. 3**

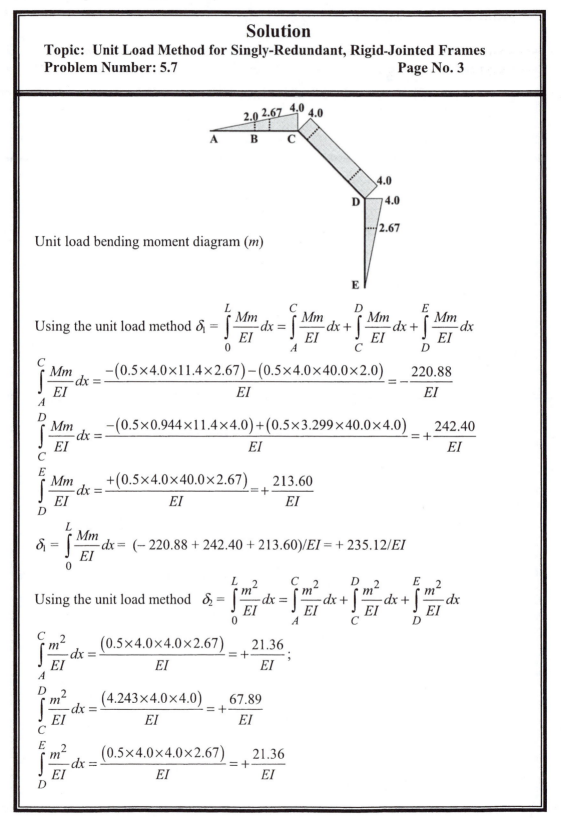

Unit load bending moment diagram (*m*)

Using the unit load method $\delta_1 = \int_0^L \dfrac{Mm}{EI}dx = \int_A^C \dfrac{Mm}{EI}dx + \int_C^D \dfrac{Mm}{EI}dx + \int_D^E \dfrac{Mm}{EI}dx$

$$\int_A^C \frac{Mm}{EI}dx = \frac{-(0.5\times4.0\times11.4\times2.67)-(0.5\times4.0\times40.0\times2.0)}{EI} = -\frac{220.88}{EI}$$

$$\int_C^D \frac{Mm}{EI}dx = \frac{-(0.5\times0.944\times11.4\times4.0)+(0.5\times3.299\times40.0\times4.0)}{EI} = +\frac{242.40}{EI}$$

$$\int_D^E \frac{Mm}{EI}dx = \frac{+(0.5\times4.0\times40.0\times2.67)}{EI} = +\frac{213.60}{EI}$$

$$\delta_1 = \int_0^L \frac{Mm}{EI}dx = (-220.88 + 242.40 + 213.60)/EI = +235.12/EI$$

Using the unit load method $\delta_2 = \int_0^L \dfrac{m^2}{EI}dx = \int_A^C \dfrac{m^2}{EI}dx + \int_C^D \dfrac{m^2}{EI}dx + \int_D^E \dfrac{m^2}{EI}dx$

$$\int_A^C \frac{m^2}{EI}dx = \frac{(0.5\times4.0\times4.0\times2.67)}{EI} = +\frac{21.36}{EI} \ ;$$

$$\int_C^D \frac{m^2}{EI}dx = \frac{(4.243\times4.0\times4.0)}{EI} = +\frac{67.89}{EI}$$

$$\int_D^E \frac{m^2}{EI}dx = \frac{(0.5\times4.0\times4.0\times2.67)}{EI} = +\frac{21.36}{EI}$$

Solution

Topic: Unit Load Method for Singly-Redundant, Rigid-Jointed Frames
Problem Number: 5.7   **Page No. 4**

$$\delta_2 = \int_0^L \frac{m^2}{EI} dx = (21.36 + 67.89 + 21.36)/EI = 110.61/EI$$

and

$$H_A \, \delta_2 = H_A \times \int_0^L \frac{m^2}{EI} dx = +110.61 H_A/EI$$

Considering the horizontal displacent at support A:
$$\delta_A = (\delta_1 + H_A \delta_2) = 0$$

$$\int_0^L \frac{Mm}{EI} dx + \left(H_D \times \int_0^L \frac{m^2}{EI} dx \right) = 0 \quad \therefore \quad H_A = -\frac{\int_0^L \dfrac{Mm}{EI} dx}{\int_0^L \dfrac{m^2}{EI} dx}$$

where:

$$\int_0^L \frac{Mm}{EI} dx = +235.12/EI \quad \text{and} \quad \int_0^L \frac{m^2}{EI} dx = +110.61/EI$$

$$H_A = -\frac{\int_0^L \dfrac{Mm}{EI} dx}{\int_0^L \dfrac{m^2}{EI} dx} = -235.12/110.61 = -2.13 \text{ kN} \quad \leftarrow$$

    i.e. opposite to the direction of the assumed unit load.

and using superposition with force systems (I) and (II):

$H_A = \text{zero} + H_A \times (+1,0) = 0 - (2.13 \times 1.0) = -2.13 \text{ kN} \quad \leftarrow$
$V_A = V'_A + H_A \times (-1.0) = +22.86 + (2.13 \times 1.0) = +24.99 \text{ kN} \quad \uparrow$
$M_A = \text{zero}$
$H_E = H'_E + H_A \times (-1.0) = -10.0 + (2.13 \times 0) = -7.87 \text{ kN} \quad \leftarrow$
$V_E = V'_E + H_A \times (+1.0) = +17.14 - (2.13 \times 1.0) = +15.01 \text{ kN} \quad \uparrow$
$M_E = \text{zero}$

Bending moments at B, C and D:
$M_B = M'_B + H_A \times (-2.0) = +45.72 + (2.13 \times 2.0) = +49.98 \text{ kNm}$
$M_C = M'_C + H_A \times (-4.0) = +11.44 + (2.13 \times 4.0) = +19.96 \text{ kNm}$
$M_D = M'_D + H_A \times (-4.0) = -40.0 + (2.13 + 4.0) = +31.48 \text{ kNm}$

Solution
Topic: Unit Load Method for Singly-Redundant, Rigid-Jointed Frames
Problem Number: 5.8   **Page No. 1**

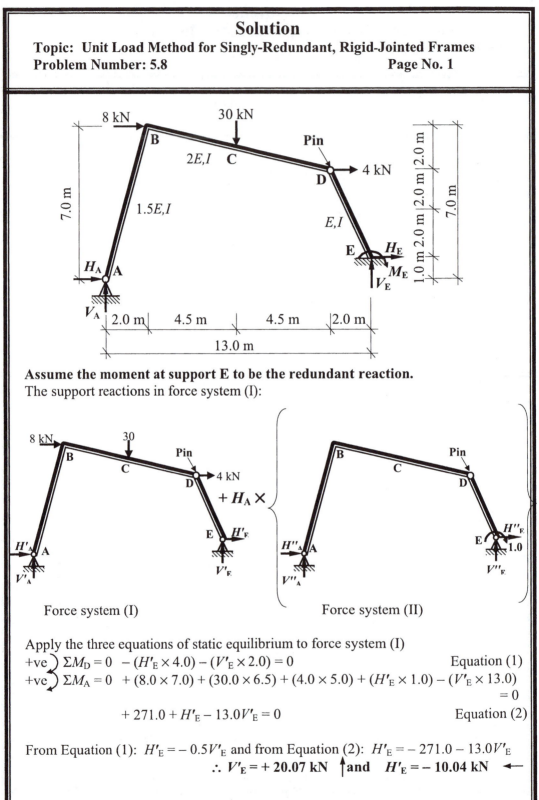

Assume the moment at support E to be the redundant reaction.
The support reactions in force system (I):

Force system (I)   Force system (II)

Apply the three equations of static equilibrium to force system (I)

+ve $\circlearrowright$ $\Sigma M_D = 0$ $-(H'_E \times 4.0) - (V'_E \times 2.0) = 0$ Equation (1)

+ve $\circlearrowleft$ $\Sigma M_A = 0$ $+(8.0 \times 7.0) + (30.0 \times 6.5) + (4.0 \times 5.0) + (H'_E \times 1.0) - (V'_E \times 13.0)$
$$= 0$$
$$+271.0 + H'_E - 13.0V'_E = 0 \qquad\qquad \text{Equation (2)}$$

From Equation (1): $H'_E = -0.5V'_E$ and from Equation (2): $H'_E = -271.0 - 13.0V'_E$
$$\therefore V'_E = +20.07 \text{ kN } \uparrow \text{and} \quad H'_E = -10.04 \text{ kN } \leftarrow$$

Solution

Topic: Unit Load Method for Singly-Redundant, Rigid-Jointed Frames

Problem Number: 5.8   **Page No. 2**

+ve ↑ $\Sigma F_z = 0$ $V'_A - 30.0 + 20.07 = 0$ $V'_A = +9.93$ kN ↑

+ve → $\Sigma F_x = 0$ $H'_A + 8.0 + 4.0 - 10.04 = 0$ $H'_A = -1.96$ kN ←

The support reactions in force system (II):

Apply the three equations of static equilibrium to force system (II).

+ve ↻ $\Sigma M_D = 0$ $+1.0 - (H''_E \times 4.0) - (V''_E \times 2.0) = 0$ Equation (3)

+ve ↻ $\Sigma M_A = 0$ $+1.0 + (H''_E \times 1.0) - (V''_E \times 13.0) = 0$ Equation (4)

From Equation (3): $H''_E = 0.25 - 0.5V''_E$

From Equation (4): $H''_E = -1.0 + 13.0V''_E$

$\therefore V''_E = +0.093$ kN →and $H''_E = +0.204$ kN ↑

+ve ↑ $\Sigma F_z = 0$ $V''_A - 0.093 = 0$ $V''_A = -0.093$ kN ↓

+ve → $\Sigma F_x = 0$ $H''_A + 0.204 = 0$ $H''_A = -0.204$ kN ←

Determine the values of **M** and **m** at each of the node and load points:

Node A: M = zero m = zero

Node B: $M = +(9.93 \times 2.0) + (1.96 \times 7.0) = +33.58$ kNm

$m = -(0.093 \times 2.0) + (0.204 \times 7.0) = +1.24$

Node C: $M = +(9.93 \times 6.5) + (1.96 \times 6.0) + (8.0 \times 1.0) = +84.31$ kNm

$m = -(0.093 \times 6.5) + (0.204 \times 6.0) = +0.62$

Node D: M = zero m = zero

Node E: M = zero $m = -1.0$

$L_{AB} = (7.0^2 + 2.0^2)^{0.5} = 7.280$ m

$L_{CD} = (9.0^2 + 2.0^2)^{0.5} = 9.220$ m

$L_{DE} = (4.0^2 + 2.0^2)^{0.5} = 4.472$ m

Applied load bending moment diagram

$+\dfrac{PL}{4} = \dfrac{30.0 \times 9.0}{4.0} = 67.5$ kNm

Fixed and free bending moment diagrams (*M*)

Solution

Topic: Unit Load Method for Singly-Redundant, Rigid-Jointed Frames
Problem Number: 5.8   **Page No. 3**

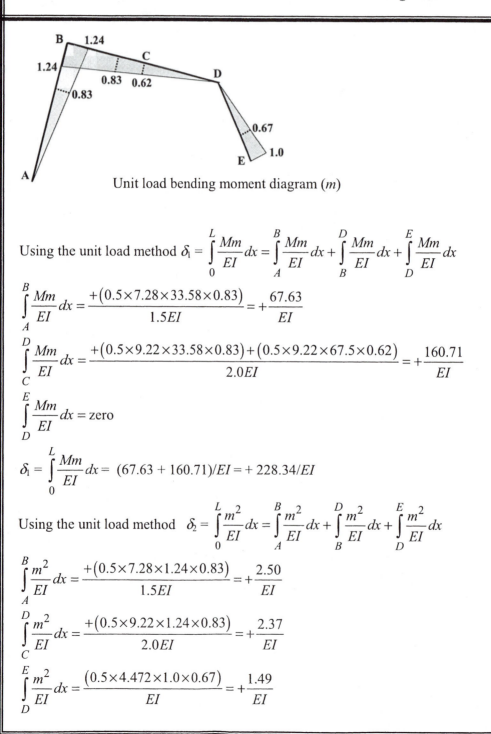

Unit load bending moment diagram (m)

Using the unit load method $\delta_1 = \int_0^L \dfrac{Mm}{EI}\,dx = \int_A^B \dfrac{Mm}{EI}\,dx + \int_B^D \dfrac{Mm}{EI}\,dx + \int_D^E \dfrac{Mm}{EI}\,dx$

$$\int_A^B \dfrac{Mm}{EI}\,dx = \dfrac{+(0.5\times7.28\times33.58\times0.83)}{1.5EI} = +\dfrac{67.63}{EI}$$

$$\int_C^D \dfrac{Mm}{EI}\,dx = \dfrac{+(0.5\times9.22\times33.58\times0.83)+(0.5\times9.22\times67.5\times0.62)}{2.0EI} = +\dfrac{160.71}{EI}$$

$$\int_D^E \dfrac{Mm}{EI}\,dx = \text{zero}$$

$$\delta_1 = \int_0^L \dfrac{Mm}{EI}\,dx = (67.63 + 160.71)/EI = +228.34/EI$$

Using the unit load method $\delta_2 = \int_0^L \dfrac{m^2}{EI}\,dx = \int_A^B \dfrac{m^2}{EI}\,dx + \int_B^D \dfrac{m^2}{EI}\,dx + \int_D^E \dfrac{m^2}{EI}\,dx$

$$\int_A^B \dfrac{m^2}{EI}\,dx = \dfrac{+(0.5\times7.28\times1.24\times0.83)}{1.5EI} = +\dfrac{2.50}{EI}$$

$$\int_C^D \dfrac{m^2}{EI}\,dx = \dfrac{+(0.5\times9.22\times1.24\times0.83)}{2.0EI} = +\dfrac{2.37}{EI}$$

$$\int_D^E \dfrac{m^2}{EI}\,dx = \dfrac{(0.5\times4.472\times1.0\times0.67)}{EI} = +\dfrac{1.49}{EI}$$

Solution
Topic: Unit Load Method for Singly-Redundant, Rigid-Jointed Frames
Problem Number: 5.8   **Page No. 4**

$$\delta_2 = \int_0^L \frac{m^2}{EI}\,dx = (2.50 + 2.37 + 1.49)/EI = 6.36/EI$$

and

$$M_E\,\delta_2 = M_E \times \int_0^L \frac{m^2}{EI}\,dx = +6.36 M_E/EI$$

Considering the rotational displacent at support E:
$$\delta_E = (\delta_1 + M_E\,\delta_2) = 0$$

$$\int_0^L \frac{Mm}{EI}\,dx + \left(M_E \times \int_0^L \frac{m^2}{EI}\,dx \right) = 0 \qquad \therefore \ M_E = - \frac{\displaystyle\int_0^L \frac{Mm}{EI}\,dx}{\displaystyle\int_0^L \frac{m^2}{EI}\,dx}$$

where:
$$\int_0^L \frac{Mm}{EI}\,dx = +228.34/EI \quad \text{and} \quad \int_0^L \frac{m^2}{EI}\,dx = +6.36/EI$$

$$M_E = - \frac{\displaystyle\int_0^L \frac{Mm}{EI}\,dx}{\displaystyle\int_0^L \frac{m^2}{EI}\,dx} = -228.34/6.36 = -35.90 \text{ kNm} \quad \curvearrowleft$$

i.e. opposite to the direction of the assumed unit moment.

and using superposition with force systems (I) and (II):

$$H_A = H'_A + M_E \times (-0.204) = -1.96 + (35.90 \times 0.204) = +5.36 \text{ kN} \qquad \rightarrow$$
$$V_A = V'_A + M_E \times (-0.093) = +9.93 + (35.90 \times 0.093) = +13.27 \text{ kN} \qquad \uparrow$$
$$M_A = \text{zero}$$
$$H_E = H'_E + M_E \times (+0.204) = -10.04 - (35.90 \times 0.204) = -17.32 \text{ kN} \qquad \leftarrow$$
$$V_E = V'_E + M_E \times (+0.093) = +20.07 - (35.90 \times 0.093) = +16.73 \text{ kN} \qquad \uparrow$$
$$M_E = -35.90 \text{ kNm} \qquad \curvearrowleft$$

Bending moments at B, C, D:
$$M_B = M'_B + M_E \times (+1.24) = +33.58 - (35.90 \times 1.24) = -10.94 \text{ kNm}$$
$$M_C = M'_C + M_E \times (+0.62) = +84.31 - (35.90 \times 0.62) = +62.05 \text{ kNm}$$
$$M_D = \text{zero}$$

5.3   *Moment Distribution for No-Sway Rigid-Jointed Frames*

The principles of moment distribution are explained in Chapter 4 in relation to the analysis of multi–span beams. In the case of rigid–jointed frames there are many instances where there more than two members meeting at a joint. This results in the out–of–balance moment induced by the fixed–end moments being distributed among several members. Consider the frame shown in Figure 5.20:

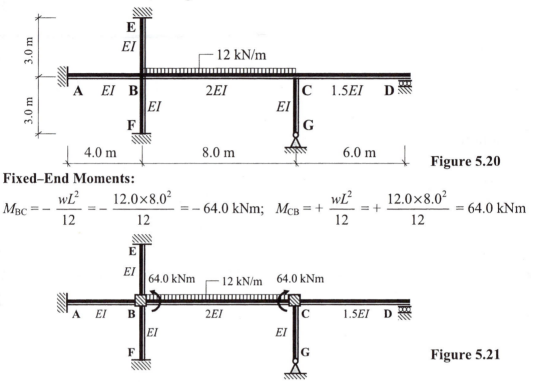

Figure 5.20

Fixed–End Moments:

$$M_{BC} = -\frac{wL^2}{12} = -\frac{12.0 \times 8.0^2}{12} = -64.0 \text{ kNm}; \quad M_{CB} = +\frac{wL^2}{12} = +\frac{12.0 \times 8.0^2}{12} = 64.0 \text{ kNm}$$

Figure 5.21

Distribution Factors:
At joint B there are four members contributing to the overall stiffness of the joint.

$$k_{BA} = \left(\frac{I}{L}\right) = \left(\frac{I}{4.0}\right) = 0.25I$$

$$k_{BC} = \left(\frac{I}{L}\right) = \left(\frac{2I}{8.0}\right) = 0.25I$$

$$k_{BE} = \left(\frac{I}{L}\right) = \left(\frac{I}{3.0}\right) = 0.33I$$

$$k_{BF} = \left(\frac{I}{L}\right) = \left(\frac{I}{3.0}\right) = 0.33I$$

$$k_{Total} = 1.16I$$

$$\text{D.F.}_{BA} = \frac{k_{BA}}{k_{Total}} = \left(\frac{0.25I}{1.16I}\right) = 0.22$$

$$\text{D.F.}_{BC} = \frac{k_{BC}}{k_{Total}} = \left(\frac{0.25I}{1.16I}\right) = 0.22$$

$$\text{D.F.}_{BE} = \frac{k_{BE}}{k_{Total}} = \left(\frac{0.33I}{1.16I}\right) = 0.28$$

$$\text{D.F.}_{BF} = \frac{k_{BF}}{k_{Total}} = \left(\frac{0.25I}{1.16I}\right) = 0.28$$

$$\sum \text{Distribution factors} = 1.0$$

The sum of the distribution factors is equal 1.0 since 100% of the out–of–balance moment must be distributed between the members.

At joint C there are three members contributing to the overall stiffness of the joint.

$$k_{CB} = \left(\frac{I}{L}\right) = \left(\frac{2I}{8.0}\right) = 0.25I \qquad\qquad \text{D.F.}_{CB} = \frac{k_{CB}}{k_{Total}} = \left(\frac{0.25I}{0.83I}\right) = 0.3$$

$$k_{CD} = \frac{3}{4} \times \left(\frac{I}{L}\right) = \left(\frac{1.5I}{6.0}\right) = 0.25I \quad\left.\begin{array}{l}\end{array}\right\} \quad k_{Total} = 0.83I \quad \text{D.F.}_{CD} = \frac{k_{CD}}{k_{Total}} = \left(\frac{0.25I}{0.83I}\right) = 0.3$$

$$k_{CG} = \frac{3}{4} \times \left(\frac{I}{L}\right) = \left(\frac{I}{3.0}\right) = 0.33I \qquad\qquad \text{D.F.}_{CG} = \frac{k_{CG}}{k_{Total}} = \left(\frac{0.33I}{0.83I}\right) = 0.4$$

The balancing moment at joint B = + 64.0 kNm
The balancing moment at joint C = − 64.0 kNm

At joint B:
Moment on BA = + (0.22 × 64.0) = + 14.08 kNm
Moment on BC = + (0.22 × 64.0) = + 14.08 kNm
Moment on BE = + (0.28 × 64.0) = + 17.92 kNm
Moment on BF = + (0.28 × 64.0) = + 17.92 kNm
At joint C:
Moment on CB = − (0.3 × 64.0) = − 19.20 kNm;
Moment on CD = − (0.3 × 64.0) = − 19.20 kNm
Moment on CG = − (0.4 × 64.0) = − 25.60 kNm

These balancing moments are indicated on the frame in Figure 5.22

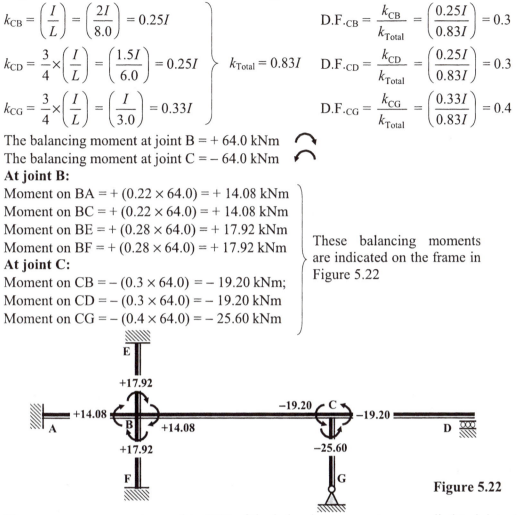

Figure 5.22

The carry–over moments equal to 50% of the balancing moments are applied to joints A, B, E, F and C.

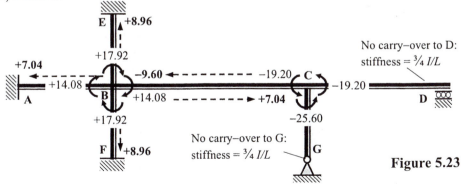

Figure 5.23

As before with beams, the above process is carried out until the required accuracy is obtained. This is illustrated in Example 5.4 and the solutions to Problems 5.9 to 5.16.

5.3.1 Example 5.4 No-Sway Rigid-Jointed Frame 1

A rigid-jointed, two-bay rectangular frame is pinned at supports A, D and E and carries loading as indicated in Figure 5.24. Given that supports D and E settle by 3 mm and 2 mm respectively and that $EI = 102.5 \times 10^3$ kNm^2;

    i) sketch the bending moment diagram and determine the support reactions,

    ii) sketch the deflected shape (assuming axially rigid members) and compare with the shape of the bending moment diagram, (the reader should check the answer using a computer analysis solution).

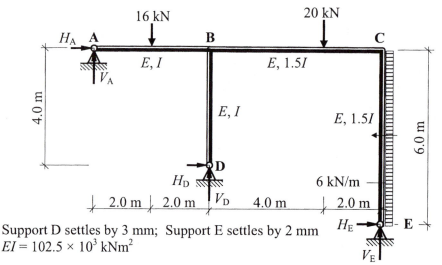

Figure 5.24

Fixed-end Moments:

The final fixed-end moments are due to the combined effects of the applied member loads and the settlement; consider the member loads,

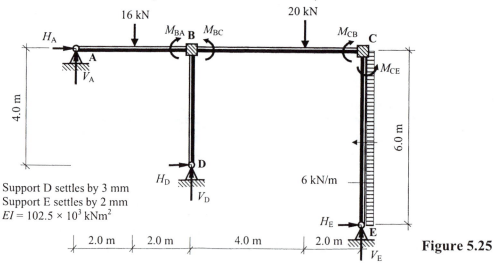

Figure 5.25

Member AB *

$$M_{AB} = -\frac{PL}{8} = -\frac{16.0 \times 4}{8} = -8.0 \text{ kNm}$$

$$M_{BA} = +\frac{PL}{8} = +\frac{16.0 \times 4}{8} = +8.0 \text{ kNm}$$

* Since support A is pinned, the fixed-end moments are $(M_{BA} - 0.5M_{AB})$ at B and zero at A.

$(M_{BA} - M_{AB}/2) = [+8.0 + (0.5 \times 8.0)] = +12.0$ kNm.

Member BC

$$M_{BC} = -\frac{Pab^2}{L^2} = -\left[-\left(\frac{20.0 \times 4.0 \times 2.0^2}{6^2}\right)\right] = -8.9 \text{ kNm}$$

$$M_{CB} = +\frac{Pa^2b}{L^2} = +\left[+\left(\frac{20.0 \times 4.0^2 \times 2.0}{6^2}\right)\right] = +17.8 \text{ kNm}$$

Member CE *

$$M_{CE} = -\frac{wL^2}{12} = -\frac{6.0 \times 6^2}{12} = -18.0 \text{ kNm};$$

$$M_{EC} = +\frac{wL^2}{12} = +\frac{6.0 \times 6^2}{12} = +18.0 \text{ kNm}$$

* Since support E is pinned, the fixed-end moments are $(M_{CE} - 0.5M_{EC})$ at C and zero at E.

$(M_{CE} - 0.5M_{EC}) = [-18.0 - (0.5 \times 18.0)] = -27.0$ kNm.

Consider the settlement of supports D and E: $\delta_{AB} = 3.0$ mm and $\delta_{BC} = 1.0$ mm

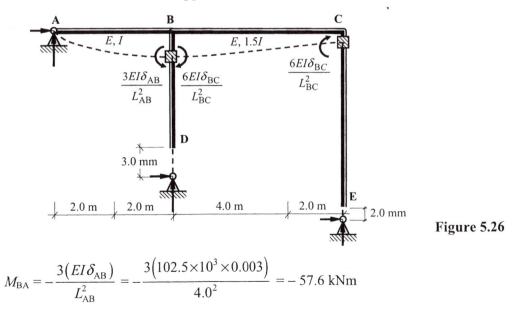

Figure 5.26

$$M_{BA} = -\frac{3(EI\delta_{AB})}{L^2_{AB}} = -\frac{3(102.5 \times 10^3 \times 0.003)}{4.0^2} = -57.6 \text{ kNm}$$

Note: the relative displacement between B and C i.e. $\delta_{BC} = (3.0 - 2.0) = 1.0$ mm

$$M_{BC} = + \frac{6\left(E1.5I\delta_{BC}\right)^*}{L_{BC}^2} = + \frac{6\left(1.5 \times 102.5 \times 10^3 \times 0.001\right)}{6.0^2} = + 25.6 \text{ kNm}$$

$M_{CB} = + 25.6$ kNm

Final Fixed-end Moments:

Member AB:	$M_{AB} = 0$	$M_{BA} = + 12.0 - 57.6 = - 45.6$ kNm
Member BC:	$M_{BC} = - 8.9 + 25.6 = + 16.7$ kNm	$M_{CB} = + 17.8 + 25.6 = + 43.4$ kNm
Member CE	$M_{CE} = - 27.0$ kNm	$M_{EC} = 0$

Distribution Factors : Joint B

$$k_{BA} = \left(\frac{3}{4} \times \frac{I}{4.0}\right) = 0.19I \qquad\qquad DF_{BA} = \frac{k_{BA}}{k_{Total}} = \frac{0.19}{0.63} = \mathbf{0.3}$$

$$k_{BC} = \left(\frac{1.5I}{6.0}\right) = 0.25I \qquad k_{total} = \mathbf{0.63I} \qquad DF_{BC} = \frac{k_{BC}}{k_{Total}} = \frac{0.25}{0.63} = \mathbf{0.4}$$

$$k_{BD} = \left(\frac{3}{4} \times \frac{I}{4.0}\right) = 0.19I \qquad\qquad DF_{BD} = \frac{k_{BD}}{k_{Total}} = \frac{0.19}{0.63} = \mathbf{0.3}$$

Distribution Factors : Joint C

$$k_{CB} = \left(\frac{1.5I}{6.0}\right) = 0.25I \qquad\qquad DF_{CB} = \frac{k_{CB}}{k_{Total}} = \frac{0.25}{0.44} = \mathbf{0.57}$$

$$k_{total} = \mathbf{0.44I}$$

$$k_{CE} = \left(\frac{3}{4} \times \frac{1.5I}{6.0}\right) = 0.19I \qquad\qquad DF_{CE} = \frac{k_{CE}}{k_{Total}} = \frac{0.19}{0.44} = \mathbf{0.43}$$

Moment Distribution Table:

Joint	A	D		B			C		E
	AB	DB		BA	BD	BC	CB	CE	EC
Distribution Factors	1.0	1.0		0.3	0.3	0.4	0.57	0.43	1.0
Fixed-end Moments				− 45.60		+ 16.7	+ 43.4	− 27.0	
Balance				+ 8.67	+ 8.67	+ 11.56	− 9.35	− 7.05	
Carry-over						− 4.67	+ 5.78		
Balance				+1.40	+ 1.40	+ 1.87	− 3.29	− 2.49	
Carry-over						− 1.65	+ 0.93		
Balance				+ 0.49	+ 0.49	+ 0.66	− 0.53	− 0.4	
Carry-over						− 0.27	+ 0.33		
Balance				0.08	+ 0.08	+ 0.11	− 0.19	− 0.14	
Total	0	0		−34.96	+ 10.65	+ 24.31	+ 37.08	− 37.08	0

Continuity Moments:

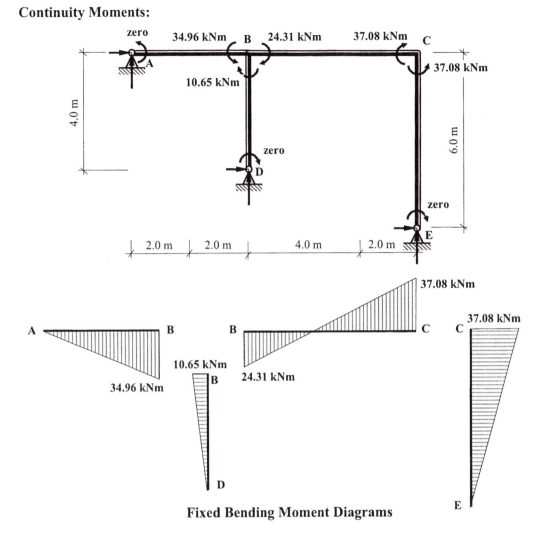

Fixed Bending Moment Diagrams

Free bending moments:

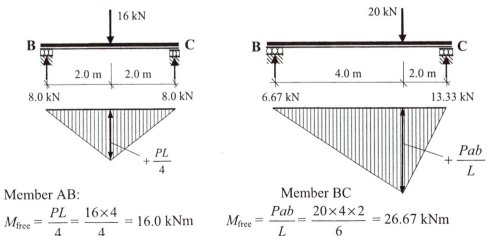

Member AB:

$$M_{\text{free}} = \frac{PL}{4} = \frac{16 \times 4}{4} = 16.0 \text{ kNm}$$

Member BC

$$M_{\text{free}} = \frac{Pab}{L} = \frac{20 \times 4 \times 2}{6} = 26.67 \text{ kNm}$$

Member CE:

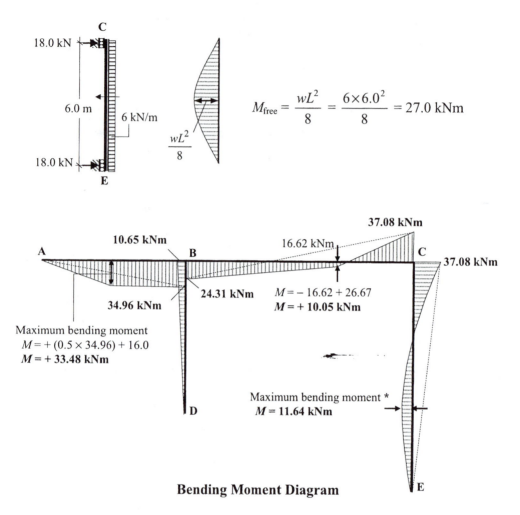

$$M_{free} = \frac{wL^2}{8} = \frac{6 \times 6.0^2}{8} = 27.0 \text{ kNm}$$

Bending Moment Diagram

The maximum value along the length of member CE can be found by identifying the point of zero shear as follows:

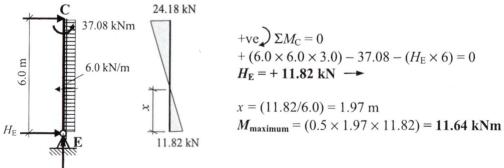

$+ve \curvearrowleft \Sigma M_C = 0$
$+ (6.0 \times 6.0 \times 3.0) - 37.08 - (H_E \times 6) = 0$
$H_E = + 11.82 \text{ kN} \longrightarrow$

$x = (11.82/6.0) = 1.97 \text{ m}$
$M_{maximum} = (0.5 \times 1.97 \times 11.82) = \textbf{11.64 kNm}$

Shear Force Diagram

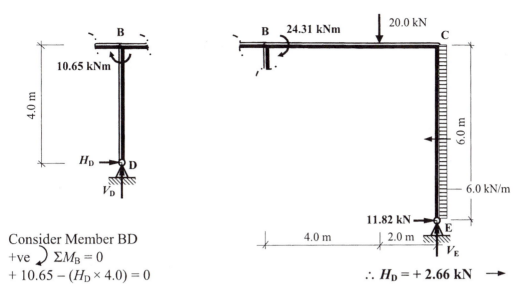

Consider Member BD

+ve $\curvearrowright$ $\Sigma M_B = 0$

$+ 10.65 - (H_D \times 4.0) = 0$

$\therefore H_D = + 2.66$ kN $\longrightarrow$

Consider a section at B

+ve $\curvearrowright$ $\Sigma M_B = 0$

$+ 24.31 + (20.0 \times 4.0) - (11.82 \times 6.0) + (6.0 \times 6.0 \times 3.0) - (V_E \times 6.0) = 0$

$\therefore V_E = + 23.57$ kN $\uparrow$

Consider Member AB:

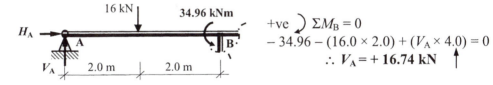

+ve $\curvearrowright$ $\Sigma M_B = 0$

$- 34.96 - (16.0 \times 2.0) + (V_A \times 4.0) = 0$

$\therefore V_A = + 16.74$ kN $\uparrow$

For the complete frame:

+ve $\uparrow$ $\Sigma F_z = 0$

$+ 16.74 - 16.0 - 20.0 + 23.57 + V_D = 0$

$\therefore V_D = - 4.31$ kN $\downarrow$

+ve $\longrightarrow$ $\Sigma F_x = 0$

$H_A + 11.82 + 2.66 - (6.0 \times 6.0) = 0$

$\therefore H_A = + 21.52$ kN $\longrightarrow$

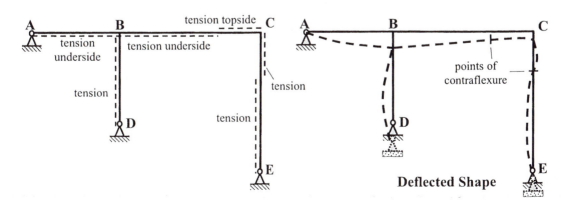

Deflected Shape

5.3.2 *Problems: Moment Distribution – No-Sway Rigid-Jointed Frames*

A series of rigid-jointed frames are indicated in Problems 5.9 to 5.16 in which the relative *EI* values and the applied loading are given. In each case:

i) sketch the bending moment diagram and determine the support reactions,

ii) sketch the deflected shape (assuming axially rigid members) and compare with the shape of the bending moment diagram, (check the answer using a computer analysis solution).

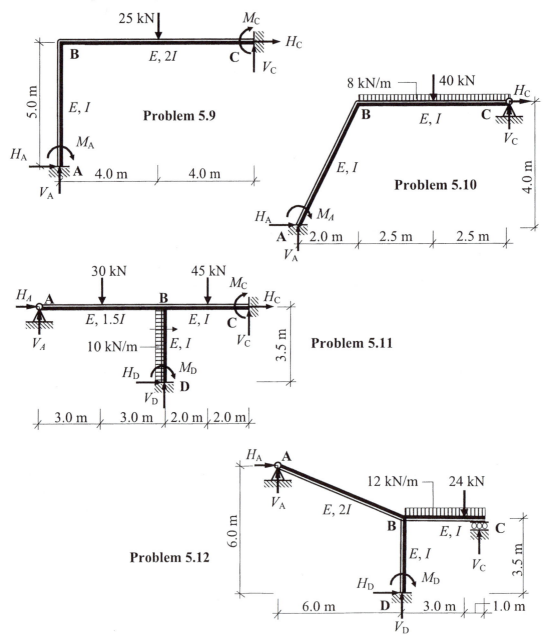

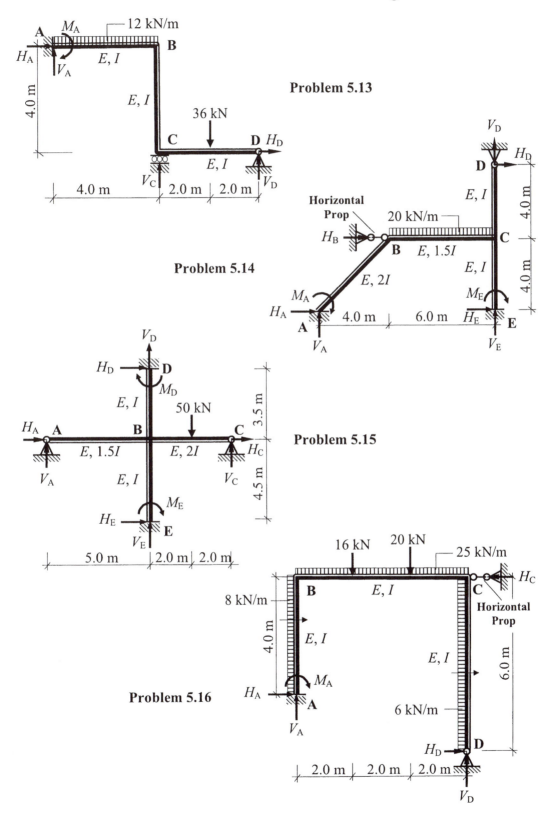

Problem 5.13

Problem 5.14

Problem 5.15

Problem 5.16

5.3.3 *Solutions: Moment Distribution – No-Sway Rigid-Jointed Frames*

Solution

Topic: Moment Distribution – No-Sway Rigid-Jointed Frames
Problem Number: 5.9   **Page No. 1**

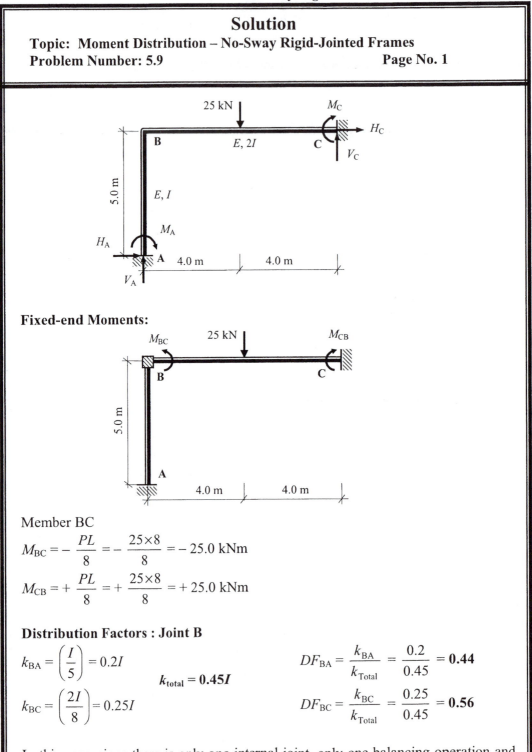

Fixed-end Moments:

Member BC

$$M_{BC} = - \frac{PL}{8} = - \frac{25 \times 8}{8} = -25.0 \text{ kNm}$$

$$M_{CB} = + \frac{PL}{8} = + \frac{25 \times 8}{8} = +25.0 \text{ kNm}$$

Distribution Factors : Joint B

$$k_{BA} = \left(\frac{I}{5} \right) = 0.2I \qquad\qquad DF_{BA} = \frac{k_{BA}}{k_{Total}} = \frac{0.2}{0.45} = \textbf{0.44}$$

$$k_{total} = \textbf{0.45}\boldsymbol{I}$$

$$k_{BC} = \left(\frac{2I}{8} \right) = 0.25I \qquad\qquad DF_{BC} = \frac{k_{BC}}{k_{Total}} = \frac{0.25}{0.45} = \textbf{0.56}$$

In this case, since there is only one internal joint, only one balancing operation and one carry-over will be required during the distribution of the moments.

Solution

Topic: Moment Distribution – No-Sway Rigid-Jointed Frames

Problem Number: 5.9   **Page No. 2**

Moment Distribution Table:

Joint	A		B		C
	AB		BA	BC	CB
Distribution Factors	0		0.44	0.56	0
Fixed-end Moments				− 25.0	+ 25.0
Balance			+ 11.0	+ 14.0	
Carry-over	+ 5.5				+ 7.0
Total	+ 5.5		+ 11.0	− 11.0	+ 32.0

Continuity Moments:

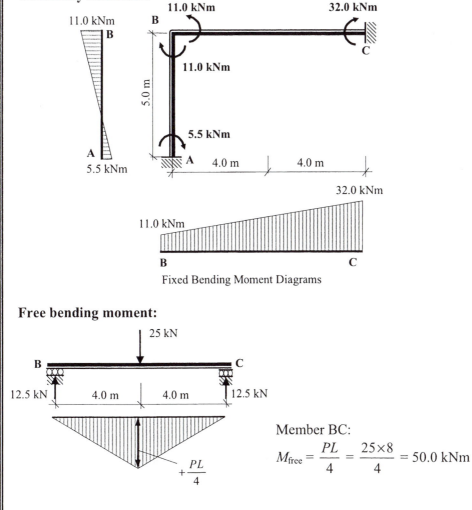

Fixed Bending Moment Diagrams

Free bending moment:

Member BC:

$$M_{free} = \frac{PL}{4} = \frac{25 \times 8}{4} = 50.0 \text{ kNm}$$

Solution
Topic: Moment Distribution – No-Sway Rigid-Jointed Frames
Problem Number: 5.9   **Page No. 3**

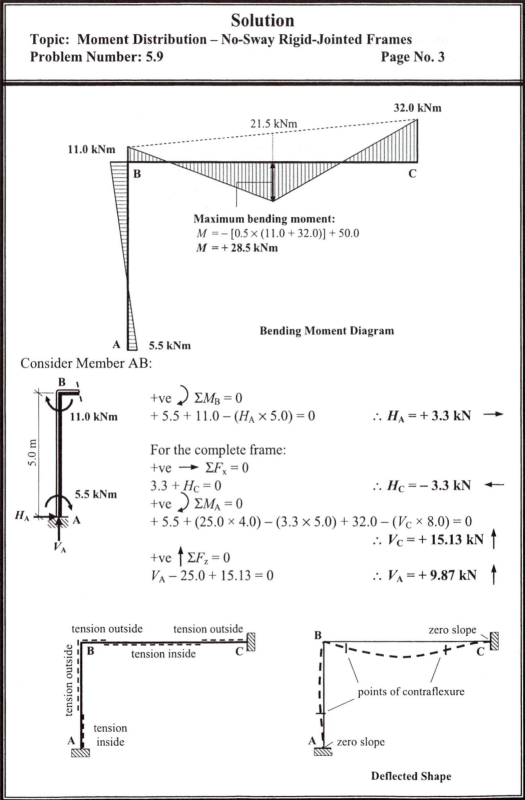

32.0 kNm

21.5 kNm

11.0 kNm

B

C

Maximum bending moment:
$M = -[0.5 \times (11.0 + 32.0)] + 50.0$
$M = +28.5 \text{ kNm}$

Bending Moment Diagram

A   5.5 kNm

Consider Member AB:

B

11.0 kNm

5.0 m

5.5 kNm

H_A A

V_A

$+ve \;\curvearrowright\; \Sigma M_B = 0$
$+ 5.5 + 11.0 - (H_A \times 5.0) = 0$ $\therefore H_A = +3.3 \text{ kN} \longrightarrow$

For the complete frame:
$+ve \;\longrightarrow\; \Sigma F_x = 0$
$3.3 + H_C = 0$ $\therefore H_C = -3.3 \text{ kN} \longleftarrow$
$+ve \;\curvearrowright\; \Sigma M_A = 0$
$+ 5.5 + (25.0 \times 4.0) - (3.3 \times 5.0) + 32.0 - (V_C \times 8.0) = 0$
        $\therefore V_C = +15.13 \text{ kN} \uparrow$
$+ve \;\uparrow\; \Sigma F_z = 0$
$V_A - 25.0 + 15.13 = 0$ $\therefore V_A = +9.87 \text{ kN} \uparrow$

tension outside   tension outside   zero slope

B   tension inside   C   B   C

tension outside

tension
inside

A

points of contraflexure

A   zero slope

Deflected Shape

Solution

Topic: Moment Distribution – No-Sway Rigid-Jointed Frames
Problem Number: 5.10   **Page No. 1**

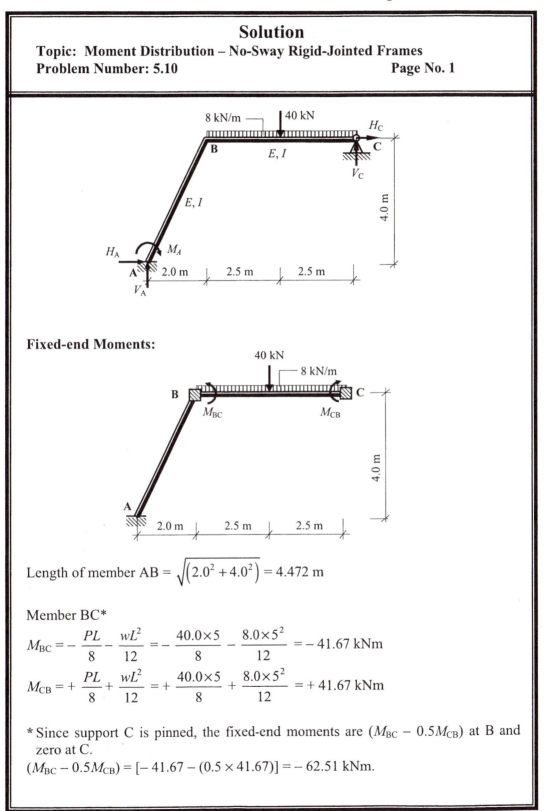

Length of member AB = $\sqrt{\left(2.0^2 + 4.0^2\right)}$ = 4.472 m

Member BC*

$$M_{BC} = -\frac{PL}{8} - \frac{wL^2}{12} = -\frac{40.0 \times 5}{8} - \frac{8.0 \times 5^2}{12} = -41.67 \text{ kNm}$$

$$M_{CB} = +\frac{PL}{8} + \frac{wL^2}{12} = +\frac{40.0 \times 5}{8} + \frac{8.0 \times 5^2}{12} = +41.67 \text{ kNm}$$

* Since support C is pinned, the fixed-end moments are $(M_{BC} - 0.5M_{CB})$ at B and zero at C.

$(M_{BC} - 0.5M_{CB}) = [-41.67 - (0.5 \times 41.67)] = -62.51$ kNm.

Solution

Topic: Moment Distribution – No-Sway Rigid-Jointed Frames
Problem Number: 5.10

Page No. 2

Distribution Factors : Joint B

$$k_{BA} = \left(\frac{I}{4.472} \right) = 0.22I$$

$$k_{total} = 0.37I$$

$$k_{BC} = \frac{3}{4} \times \left(\frac{I}{5} \right) = 0.15I$$

$$DF_{BA} = \frac{k_{BA}}{k_{Total}} = \frac{0.22}{0.37} = 0.59$$

$$DF_{BC} = \frac{k_{BC}}{k_{Total}} = \frac{0.15}{0.37} = 0.41$$

Moment Distribution Table:

Joint	A	B		C
	AB	BA	BC	CB
Distribution Factors	0	0.59	0.41	1.0
Fixed-end Moments			− 62.51	
Balance		+ 36.88	+ 25.63	
Carry-over	+ 18.44			
Total	+ 18.44	+ 36.88	− 36.88	0

Continuity Moments:

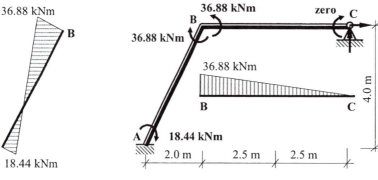

Fixed Bending Moment Diagrams

Free bending moment:

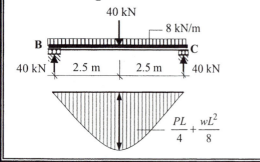

Member BC:

$$M_{free} = \frac{PL}{4} + \frac{wL^2}{8}$$

$$= \frac{40 \times 5}{4} + \frac{8.0 \times 5.0^2}{8}$$

$$= 75.0 \text{ kNm}$$

Solution

Topic: Moment Distribution – No-Sway Rigid-Jointed Frames
Problem Number: 5.10   **Page No. 3**

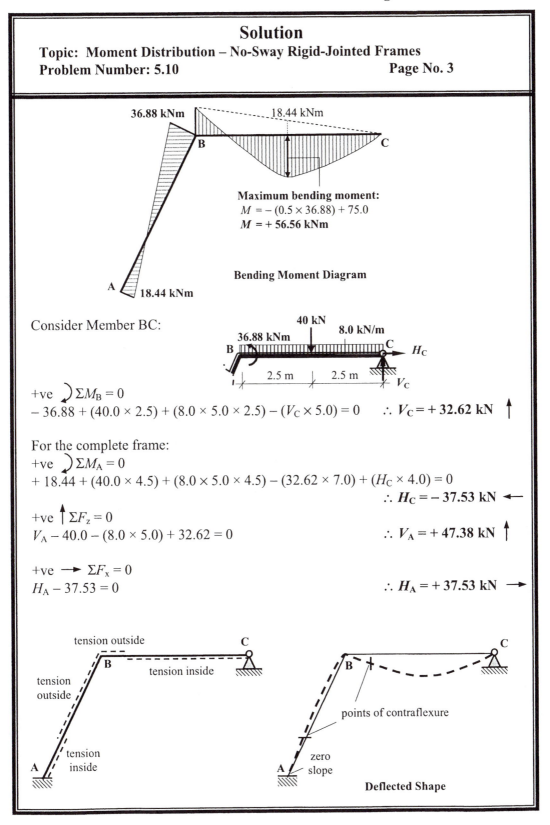

36.88 kNm   18.44 kNm

Maximum bending moment:
$M = -(0.5 \times 36.88) + 75.0$
$M = +56.56 \text{ kNm}$

Bending Moment Diagram

18.44 kNm

Consider Member BC:

36.88 kNm   40 kN   8.0 kN/m

2.5 m   2.5 m

$+ve \quad \circlearrowright \Sigma M_B = 0$
$-36.88 + (40.0 \times 2.5) + (8.0 \times 5.0 \times 2.5) - (V_C \times 5.0) = 0 \quad \therefore V_C = +32.62 \text{ kN} \uparrow$

For the complete frame:
$+ve \quad \circlearrowright \Sigma M_A = 0$
$+18.44 + (40.0 \times 4.5) + (8.0 \times 5.0 \times 4.5) - (32.62 \times 7.0) + (H_C \times 4.0) = 0$
$$\therefore H_C = -37.53 \text{ kN} \leftarrow$$

$+ve \uparrow \Sigma F_z = 0$
$V_A - 40.0 - (8.0 \times 5.0) + 32.62 = 0 \quad\quad\quad \therefore V_A = +47.38 \text{ kN} \uparrow$

$+ve \longrightarrow \Sigma F_x = 0$
$H_A - 37.53 = 0 \quad\quad\quad\quad\quad\quad\quad\quad\quad \therefore H_A = +37.53 \text{ kN} \longrightarrow$

tension outside

B   tension inside   C

tension
outside

tension
inside   A

points of contraflexure

zero
slope

Deflected Shape

Solution
Topic: Moment Distribution – No-Sway Rigid-Jointed Frames
Problem Number: 5.11   **Page No. 1**

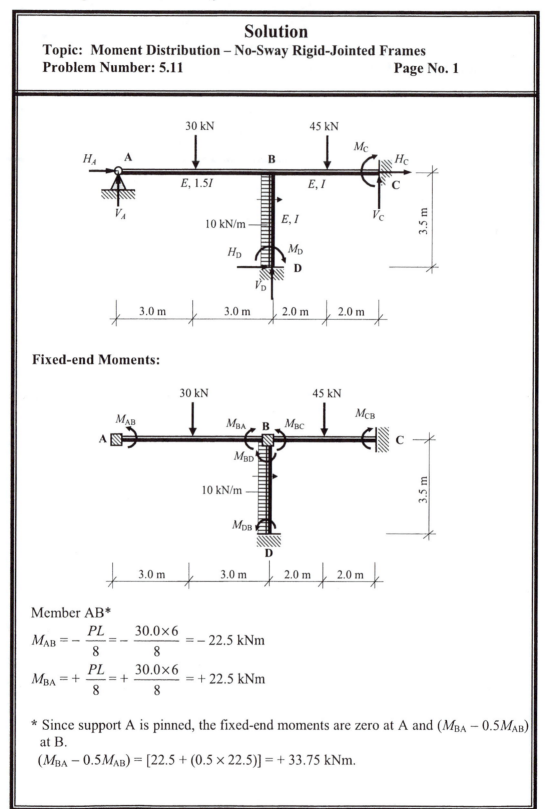

Fixed-end Moments:

Member AB*

$$M_{AB} = -\frac{PL}{8} = -\frac{30.0 \times 6}{8} = -22.5 \text{ kNm}$$

$$M_{BA} = +\frac{PL}{8} = +\frac{30.0 \times 6}{8} = +22.5 \text{ kNm}$$

* Since support A is pinned, the fixed-end moments are zero at A and $(M_{BA} - 0.5M_{AB})$ at B.

$(M_{BA} - 0.5M_{AB}) = [22.5 + (0.5 \times 22.5)] = +33.75 \text{ kNm}.$

Solution

Topic: Moment Distribution – No-Sway Rigid-Jointed Frames
Problem Number: 5.11   **Page No. 2**

Member BC

$$M_{BC} = -\frac{PL}{8} = -\frac{45.0 \times 4}{8} = -22.5 \text{ kNm}$$

$$M_{CB} = +\frac{PL}{8} = +\frac{45.0 \times 4}{8} = +22.5 \text{ kNm}$$

Member BD

$$M_{BD} = +\frac{wL^2}{12} = +\frac{10.0 \times 3.5^2}{12} = +10.21 \text{ kNm}$$

$$M_{DB} = -\frac{wL^2}{12} = -\frac{10.0 \times 3.5^2}{12} = -10.21 \text{ kNm}$$

Distribution Factors : Joint B

$$k_{BA} = \frac{3}{4} \times \left(\frac{1.5I}{6.0}\right) = 0.19I$$

$$k_{BC} = \left(\frac{I}{4}\right) = 0.25I$$

$$k_{BD} = \left(\frac{I}{3.5}\right) = 0.29I$$

$$k_{total} = 0.73I$$

$$DF_{BA} = \frac{k_{BA}}{k_{Total}} = \frac{0.19}{0.73} = 0.26$$

$$DF_{BC} = \frac{k_{BC}}{k_{Total}} = \frac{0.25}{0.73} = 0.34$$

$$DF_{BD} = \frac{k_{BD}}{k_{Total}} = \frac{0.29}{0.73} = 0.40$$

Moment Distribution Table:

Joint	A	B			C	D
	AB	BA	BD	BC	CB	DB
Distribution Factors	1.0	0.26	0.40	0.34	0	0
Fixed-end Moments		+ 33.75	+ 10.21	− 22.5	+ 22.5	− 10.21
Balance		− 5.58	− 8.58	− 7.3		
Carry-over					− 3.7	− 4.29
Total	0	+ 28.17	+ 1.63	− 29.8	+ 18.8	− 14.5

Note: the sum of the
moments at joint B = zero

Solution
Topic: Moment Distribution – No-Sway Rigid-Jointed Frames
Problem Number: 5.11   Page No. 3

Continuity Moments:

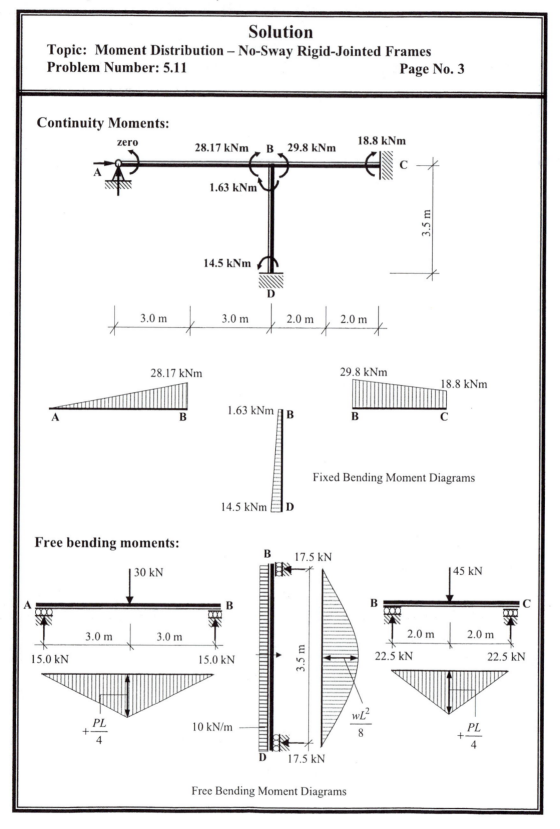

Fixed Bending Moment Diagrams

Free bending moments:

Free Bending Moment Diagrams

Solution

Topic: Moment Distribution – No-Sway Rigid-Jointed Frames
Problem Number: 5.11   **Page No. 4**

Member AB:

$$M_{free} = + \frac{PL}{4} = \frac{30.0 \times 6}{4} = 45.0 \text{ kNm}$$

Member BD:

$$M_{free} = + \frac{wL^2}{8} = \frac{10.0 \times 3.5^2}{8} = 15.31 \text{ kNm}$$

Member BC:

$$M_{free} = + \frac{PL}{4} = \frac{45.0 \times 4}{4} = 45.0 \text{ kNm}$$

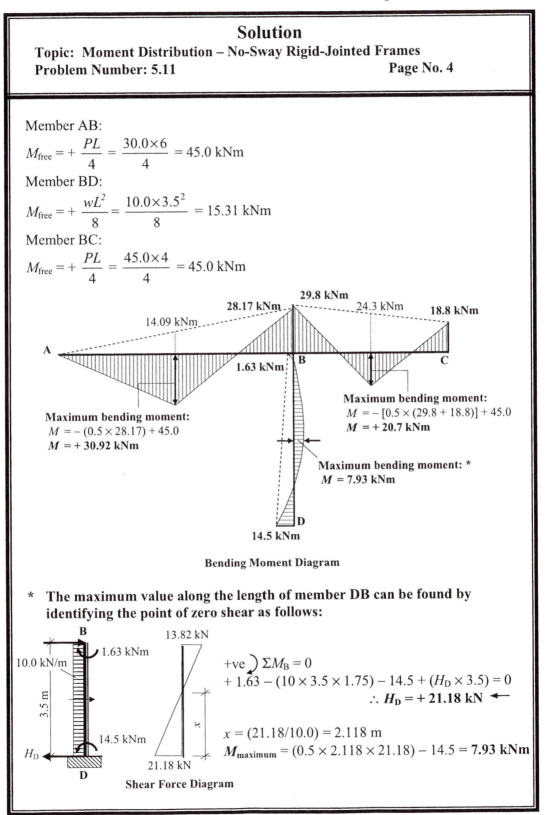

Bending Moment Diagram

* **The maximum value along the length of member DB can be found by identifying the point of zero shear as follows:**

Shear Force Diagram

$+ve \circlearrowright \Sigma M_B = 0$

$+ 1.63 - (10 \times 3.5 \times 1.75) - 14.5 + (H_D \times 3.5) = 0$

$\therefore H_D = + 21.18 \text{ kN} \leftarrow$

$x = (21.18/10.0) = 2.118 \text{ m}$

$M_{maximum} = (0.5 \times 2.118 \times 21.18) - 14.5 = 7.93 \text{ kNm}$

Solution

Topic: Moment Distribution – No-Sway Rigid-Jointed Frames
Problem Number: 5.11   **Page No. 4**

Consider Member AB:

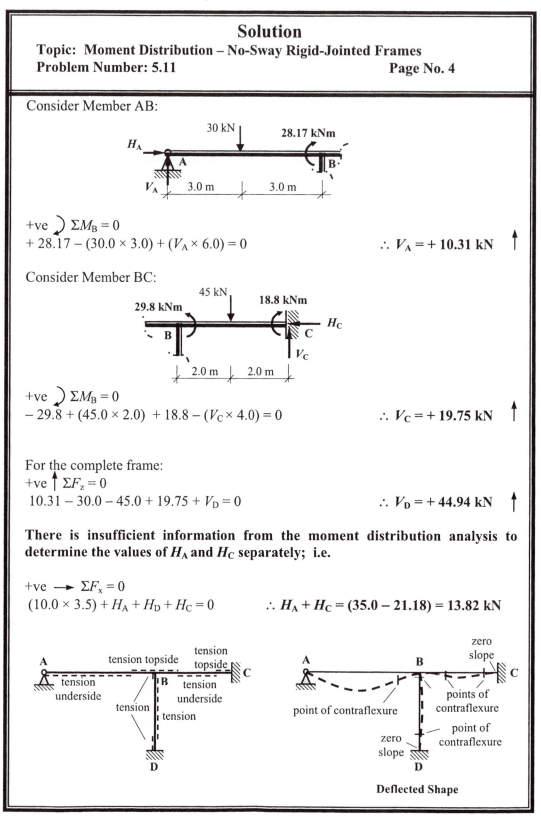

$+\text{ve}$ ⟳ $\Sigma M_B = 0$
$+ 28.17 - (30.0 \times 3.0) + (V_A \times 6.0) = 0$ $\therefore V_A = + 10.31$ kN ↑

Consider Member BC:

$+\text{ve}$ ⟳ $\Sigma M_B = 0$
$- 29.8 + (45.0 \times 2.0) + 18.8 - (V_C \times 4.0) = 0$ $\therefore V_C = + 19.75$ kN ↑

For the complete frame:
$+\text{ve}$ ↑ $\Sigma F_z = 0$
$10.31 - 30.0 - 45.0 + 19.75 + V_D = 0$ $\therefore V_D = + 44.94$ kN ↑

There is insufficient information from the moment distribution analysis to determine the values of H_A and H_C separately; i.e.

$+\text{ve}$ → $\Sigma F_x = 0$
$(10.0 \times 3.5) + H_A + H_D + H_C = 0$ $\therefore H_A + H_C = (35.0 - 21.18) = 13.82$ kN

Deflected Shape

Solution
Topic: Moment Distribution – No-Sway Rigid-Jointed Frames
Problem Number: 5.12   **Page No. 1**

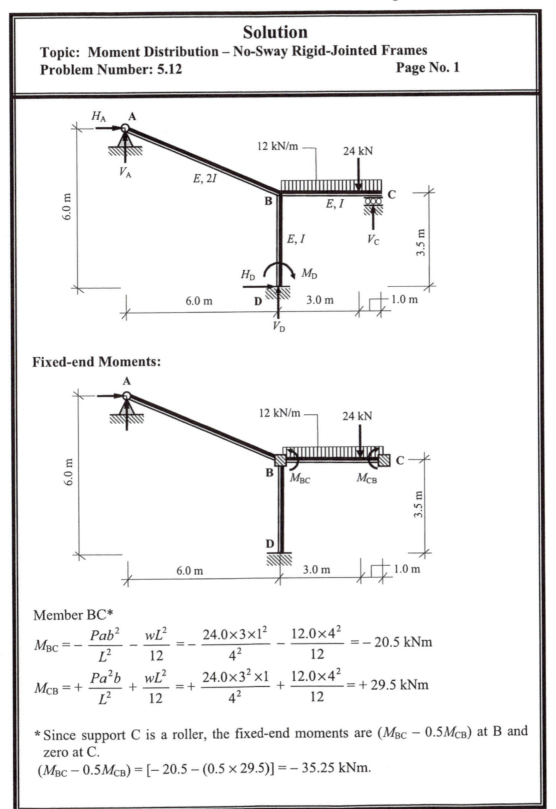

Fixed-end Moments:

Member BC*

$$M_{BC} = -\frac{Pab^2}{L^2} - \frac{wL^2}{12} = -\frac{24.0 \times 3 \times 1^2}{4^2} - \frac{12.0 \times 4^2}{12} = -20.5 \text{ kNm}$$

$$M_{CB} = +\frac{Pa^2b}{L^2} + \frac{wL^2}{12} = +\frac{24.0 \times 3^2 \times 1}{4^2} + \frac{12.0 \times 4^2}{12} = +29.5 \text{ kNm}$$

* Since support C is a roller, the fixed-end moments are $(M_{BC} - 0.5M_{CB})$ at B and zero at C.

$(M_{BC} - 0.5M_{CB}) = [-20.5 - (0.5 \times 29.5)] = -35.25 \text{ kNm}.$

Solution

Topic: Moment Distribution – No-Sway Rigid-Jointed Frames

Problem Number: 5.12   **Page No. 2**

Length of member AB = $\sqrt{\left(6.0^2 + 2.5^2\right)} = 6.5$ m

Distribution Factors : Joint B

$k_{BA} = \dfrac{3}{4} \times \left(\dfrac{2I}{6.5}\right) = 0.23I$

$k_{BC} = \dfrac{3}{4} \times \left(\dfrac{I}{4.0}\right) = 0.19I$ $\quad$ $k_{total} = \mathbf{0.71I}$

$k_{BD} = \left(\dfrac{I}{3.5}\right) = 0.29I$

$DF_{BA} = \dfrac{k_{BA}}{k_{Total}} = \dfrac{0.23}{0.71} = \mathbf{0.32}$

$DF_{BC} = \dfrac{k_{BC}}{k_{Total}} = \dfrac{0.19}{0.71} = \mathbf{0.27}$

$DF_{BD} = \dfrac{k_{BD}}{k_{Total}} = \dfrac{0.29}{0.71} = \mathbf{0.41}$

Moment Distribution Table:

Joint	A		B			C	D
	AB		**BA**	**BD**	**BC**	**CB**	**DB**
Distribution Factors	1.0		0.32	0.41	0.27	1.0	0
Fixed-end Moments					− 35.25		
Balance			+ 11.28	+ 14.45	+ 9.52		
Carry-over							+ 7.23
Total	0		+ 11.28	+ 14.45	− 25.73	0	+ 7.23

Note: the sum of the moments
at joint B = zero

Continuity Moments:

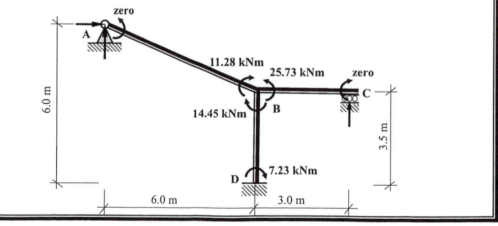

Solution

Topic: Moment Distribution – No-Sway Rigid-Jointed Frames
Problem Number: 5.12   **Page No. 3**

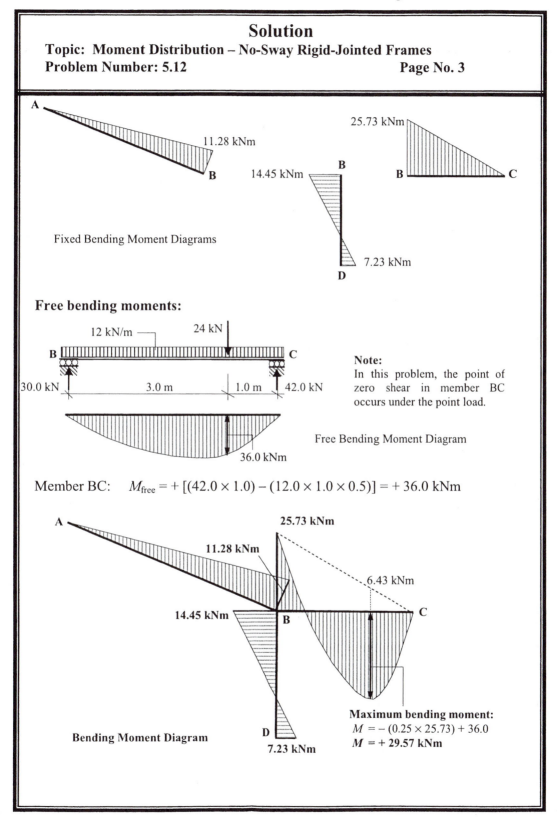

Fixed Bending Moment Diagrams

Free bending moments:

Note:
In this problem, the point of zero shear in member BC occurs under the point load.

Free Bending Moment Diagram

Member BC: $M_{\text{free}} = + [(42.0 \times 1.0) - (12.0 \times 1.0 \times 0.5)] = + 36.0 \text{ kNm}$

Maximum bending moment:
$M = -(0.25 \times 25.73) + 36.0$
$M = + 29.57 \text{ kNm}$

Bending Moment Diagram

Solution
Topic: Moment Distribution – No-Sway Rigid-Jointed Frames
Problem Number: 5.12   **Page No. 4**

Consider Member BC:

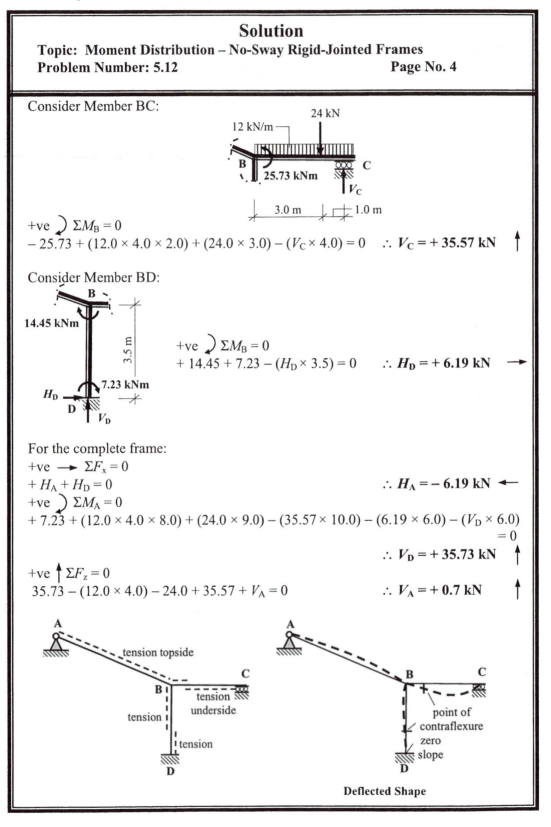

$+$ve $\circlearrowright$ $\Sigma M_B = 0$
$-25.73 + (12.0 \times 4.0 \times 2.0) + (24.0 \times 3.0) - (V_C \times 4.0) = 0$ $\therefore V_C = +\,35.57$ kN ↑

Consider Member BD:

$+$ve $\circlearrowright$ $\Sigma M_B = 0$
$+14.45 + 7.23 - (H_D \times 3.5) = 0$ $\therefore H_D = +\,6.19$ kN →

For the complete frame:
$+$ve → $\Sigma F_x = 0$
$+H_A + H_D = 0$ $\therefore H_A = -\,6.19$ kN ←
$+$ve $\circlearrowright$ $\Sigma M_A = 0$
$+7.23 + (12.0 \times 4.0 \times 8.0) + (24.0 \times 9.0) - (35.57 \times 10.0) - (6.19 \times 6.0) - (V_D \times 6.0)$
$= 0$
$\therefore V_D = +\,35.73$ kN ↑

$+$ve ↑ $\Sigma F_z = 0$
$35.73 - (12.0 \times 4.0) - 24.0 + 35.57 + V_A = 0$ $\therefore V_A = +\,0.7$ kN ↑

Deflected Shape

Solution

Topic: Moment Distribution – No-Sway Rigid-Jointed Frames
Problem Number: 5.13   **Page No. 1**

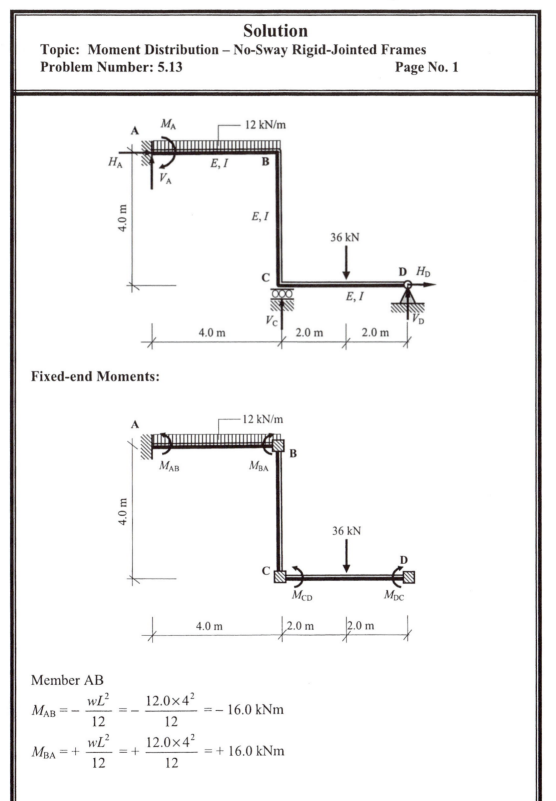

Fixed-end Moments:

Member AB

$$M_{AB} = -\frac{wL^2}{12} = -\frac{12.0 \times 4^2}{12} = -16.0 \text{ kNm}$$

$$M_{BA} = +\frac{wL^2}{12} = +\frac{12.0 \times 4^2}{12} = +16.0 \text{ kNm}$$

Solution

Topic: Moment Distribution – No-Sway Rigid-Jointed Frames
Problem Number: 5.13   **Page No. 2**

Member CD*

$$M_{CD} = -\frac{PL}{8} = -\frac{36.0 \times 4}{8} = -18.0 \text{ kNm}$$

$$M_{DC} = +\frac{PL}{8} = +\frac{36.0 \times 4}{8} = +18.0 \text{ kNm}$$

* Since support D is pinned, the fixed-end moments are $(M_{CD} - 0.5M_{DC})$ at C and zero at D.
$(M_{CD} - 0.5M_{DC}) = [-18.0 - (0.5 \times 18.0)] = -27.0 \text{ kNm}.$

Distribution Factors : Joint B

$$k_{BA} = \left(\frac{I}{4.0}\right) = 0.25I \qquad\qquad DF_{BA} = \frac{k_{BA}}{k_{Total}} = \frac{0.25}{0.5} = \mathbf{0.5}$$

$$k_{total} = \mathbf{0.51}I$$

$$k_{BC} = \left(\frac{I}{4.0}\right) = 0.25I \qquad\qquad DF_{BC} = \frac{k_{BC}}{k_{Total}} = \frac{0.25}{0.5} = \mathbf{0.5}$$

Distribution Factors : Joint C

$$k_{CB} = \left(\frac{I}{4.0}\right) = 0.25I \qquad\qquad DF_{CB} = \frac{k_{CB}}{k_{Total}} = \frac{0.25}{0.44} = \mathbf{0.57}$$

$$k_{total} = \mathbf{0.44}I$$

$$k_{CD} = \frac{3}{4} \times \left(\frac{I}{4.0}\right) = 0.19I \qquad\qquad DF_{CD} = \frac{k_{CD}}{k_{Total}} = \frac{0.19}{0.44} = \mathbf{0.43}$$

Moment Distribution Table:

Joint	A	B		C		D
	AB	BA	BC	CB	CD	DC
Distribution Factors	0	0.5	0.5	0.57	0.43	1.0
Fixed-end Moments	− 16.0	+ 16.0			− 27.0	
Balance		− 8.0	− 8.0	+ 15.39	+ 11.61	
Carry-over	− 4.0		+ 7.7	− 4.0		
Balance		− 3.85	− 3.85	+ 2.28	+ 1.72	
Carry-over	− 1.79		+ 1.14	− 1.93		
Balance		− 0.57	− 0.57	+ 1.1	+ 0.83	
Carry-over	− 0.29		+ 0.55	− 0.29		
Balance		− 0.27	− 0.27	+ 0.17	+ 0.12	
Carry-over	− 0.13					
Total	**− 22.35**	**+3.31**	**− 3.31**	**+ 12.72**	**− 12.72**	

Solution

Topic: Moment Distribution – No-Sway Rigid-Jointed Frames
Problem Number: 5.13   **Page No. 3**

Continuity Moments:

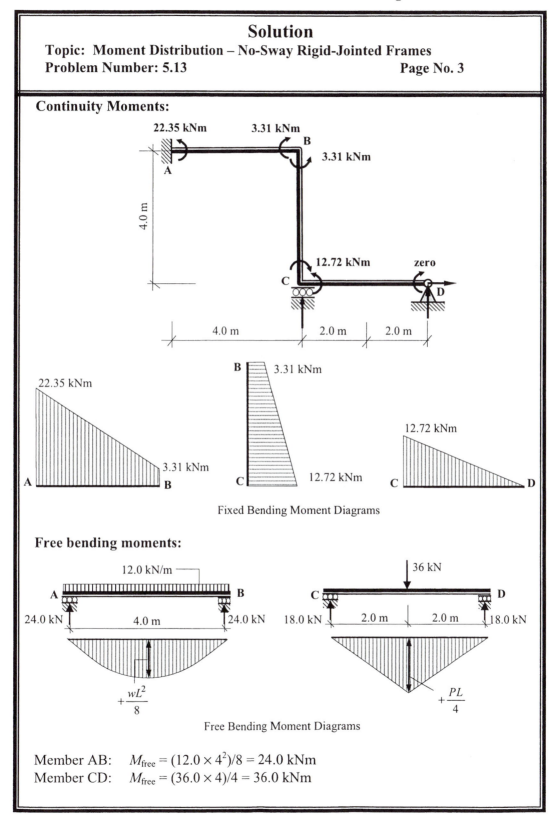

Fixed Bending Moment Diagrams

Free bending moments:

Free Bending Moment Diagrams

Member AB: $M_{\text{free}} = (12.0 \times 4^2)/8 = 24.0$ kNm
Member CD: $M_{\text{free}} = (36.0 \times 4)/4 = 36.0$ kNm

Solution
Topic: Moment Distribution – No-Sway Rigid-Jointed Frames
Problem Number: 5.13　　　　　　　　　　　　　**Page No. 4**

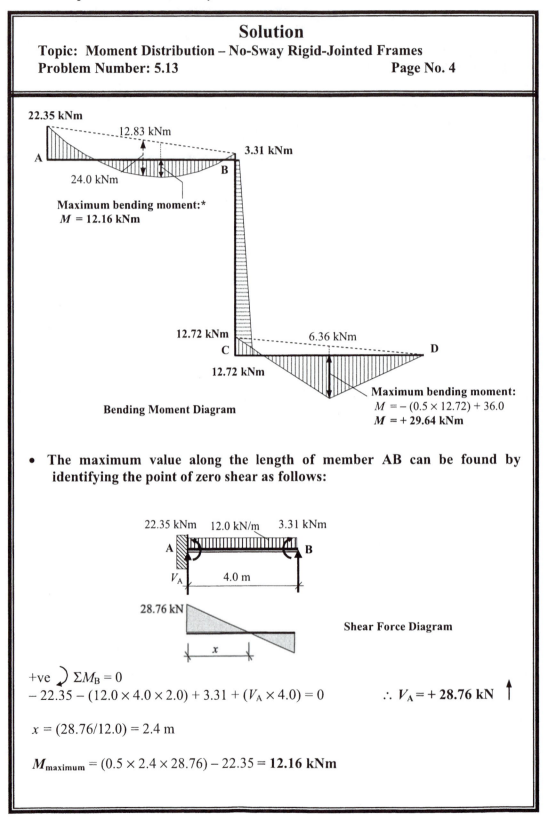

Bending Moment Diagram

22.35 kNm

12.83 kNm

A

3.31 kNm

B

24.0 kNm

Maximum bending moment:*
M = 12.16 kNm

12.72 kNm

6.36 kNm

C

D

12.72 kNm

Maximum bending moment:
$M = -(0.5 \times 12.72) + 36.0$
$M = +29.64$ kNm

- **The maximum value along the length of member AB can be found by identifying the point of zero shear as follows:**

22.35 kNm　　12.0 kN/m　　3.31 kNm

A

B

V_A　　　　4.0 m

28.76 kN

x

Shear Force Diagram

+ve $\sum M_B = 0$
$-22.35 - (12.0 \times 4.0 \times 2.0) + 3.31 + (V_A \times 4.0) = 0$　　　　$\therefore V_A = +\textbf{28.76 kN} \uparrow$

$x = (28.76/12.0) = 2.4$ m

$M_{\text{maximum}} = (0.5 \times 2.4 \times 28.76) - 22.35 = \textbf{12.16 kNm}$

Solution

Topic: Moment Distribution – No-Sway Rigid-Jointed Frames
Problem Number: 5.13   **Page No. 4**

Consider Member CD:

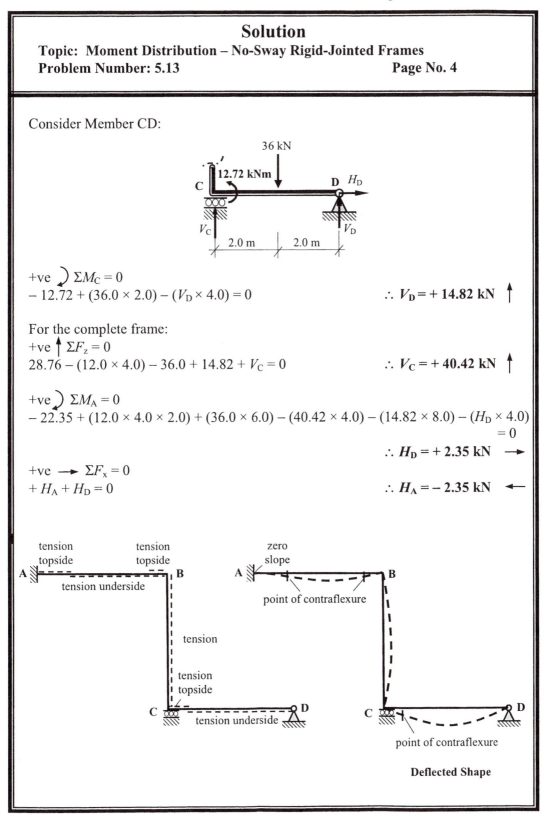

+ve $\circlearrowright$ $\Sigma M_C = 0$
$- 12.72 + (36.0 \times 2.0) - (V_D \times 4.0) = 0$ $\therefore V_D = + 14.82$ kN $\uparrow$

For the complete frame:
+ve $\uparrow$ $\Sigma F_z = 0$
$28.76 - (12.0 \times 4.0) - 36.0 + 14.82 + V_C = 0$ $\therefore V_C = + 40.42$ kN $\uparrow$

+ve $\circlearrowright$ $\Sigma M_A = 0$
$- 22.35 + (12.0 \times 4.0 \times 2.0) + (36.0 \times 6.0) - (40.42 \times 4.0) - (14.82 \times 8.0) - (H_D \times 4.0)$
$= 0$

$\therefore H_D = + 2.35$ kN $\longrightarrow$

+ve $\longrightarrow$ $\Sigma F_x = 0$
$+ H_A + H_D = 0$ $\therefore H_A = - 2.35$ kN $\longleftarrow$

tension topside   tension topside   zero slope

A ⊟- - - - - - - - - - B   A ⊟- - - - - - - - B
tension underside   point of contraflexure

tension

tension topside

C ⊟- - - - - - - - - - ○ D   C ⊟- - - - - - - - ○ D
tension underside   point of contraflexure

Deflected Shape

Solution
Topic: Moment Distribution – No-Sway Rigid-Jointed Frames
Problem Number: 5.14   **Page No. 1**

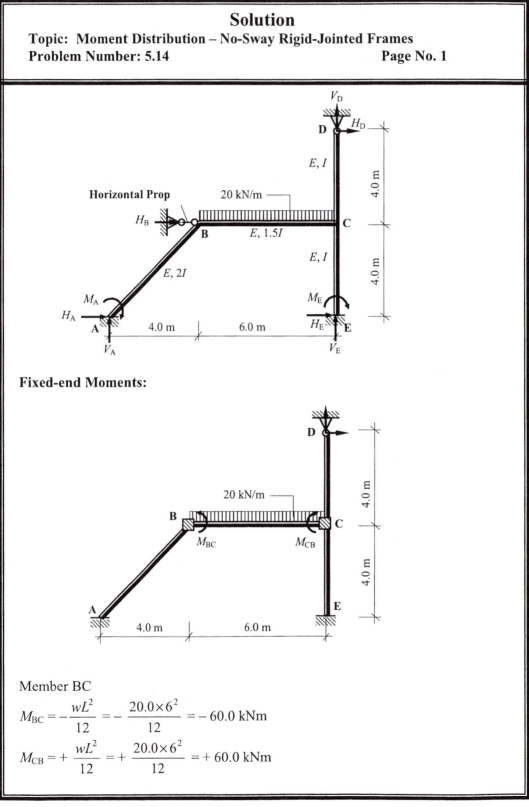

Fixed-end Moments:

Member BC

$$M_{BC} = -\frac{wL^2}{12} = -\frac{20.0 \times 6^2}{12} = -60.0 \text{ kNm}$$

$$M_{CB} = +\frac{wL^2}{12} = +\frac{20.0 \times 6^2}{12} = +60.0 \text{ kNm}$$

Solution
Topic: Moment Distribution – No-Sway Rigid-Jointed Frames
Problem Number: 5.14   **Page No. 2**

Length of member AB $= \sqrt{\left(4.0^2 + 4.0^2\right)} = 5.657$ m

Distribution Factors : Joint B

$k_{BA} = \left(\dfrac{2.0I}{5.657}\right) = 0.35I$ $\qquad DF_{BA} = \dfrac{k_{BA}}{k_{Total}} = \dfrac{0.35}{0.6} = \mathbf{0.58}$

$\qquad\qquad\qquad\qquad k_{total} = \mathbf{0.6I}$

$k_{BC} = \left(\dfrac{1.5I}{6.0}\right) = 0.25I$ $\qquad DF_{BC} = \dfrac{k_{BC}}{k_{Total}} = \dfrac{0.25}{0.6} = \mathbf{0.42}$

Distribution Factors : Joint C

$k_{CB} = \left(\dfrac{1.5I}{6.0}\right) = 0.25I$ $\qquad DF_{CB} = \dfrac{k_{CB}}{k_{Total}} = \dfrac{0.25}{0.69} = \mathbf{0.36}$

$k_{CD} = \dfrac{3}{4} \times \left(\dfrac{I}{4.0}\right) = 0.19I$ $\quad k_{total} = \mathbf{0.69I} \quad DF_{CD} = \dfrac{k_{CD}}{k_{Total}} = \dfrac{0.19}{0.69} = \mathbf{0.28}$

$k_{CE} = \left(\dfrac{I}{4.0}\right) = 0.25I$ $\qquad DF_{CE} = \dfrac{k_{CE}}{k_{Total}} = \dfrac{0.25}{0.69} = \mathbf{0.36}$

Moment Distribution Table:

Joint	A	B		C			E	D
	AB	BA	BC	CB	CD	CE	EC	DC
Distribution Factors	0	0.58	0.42	0.36	0.28	0.36	0	1.0
Fixed-end Moments			− 60.0	+ 60.0				
Balance		+ 34.8	+ 25.2	− 21.6	− 16.8	− 21.6		
Carry-over	+ 17.4		− 10.8	+ 12.6			− 10.8	
Balance		+ 6.26	+ 4.54	− 4.54	− 3.52	− 4.54		
Carry-over	+ 3.13		− 2.27	+ 2.27			− 2.27	
Balance		+ 1.32	+ 0.95	− 0.82	− 0.63	− 0.82		
Carry-over	+ 0.66		− 0.41	+ 0.48			0.41	
Balance		+ 0.24	+ 0.17	− 0.17	− 0.14	− 0.17		
Carry-over	+ 0.12						0.09	
Total	+ 21.3	+ 42.6	− 42.6	+ 48.2	− 21.1	− 27.1	− 13.6	0

Solution
Topic: Moment Distribution – No-Sway Rigid-Jointed Frames
Problem Number: 5.14   **Page No. 3**

Continuity Moments:

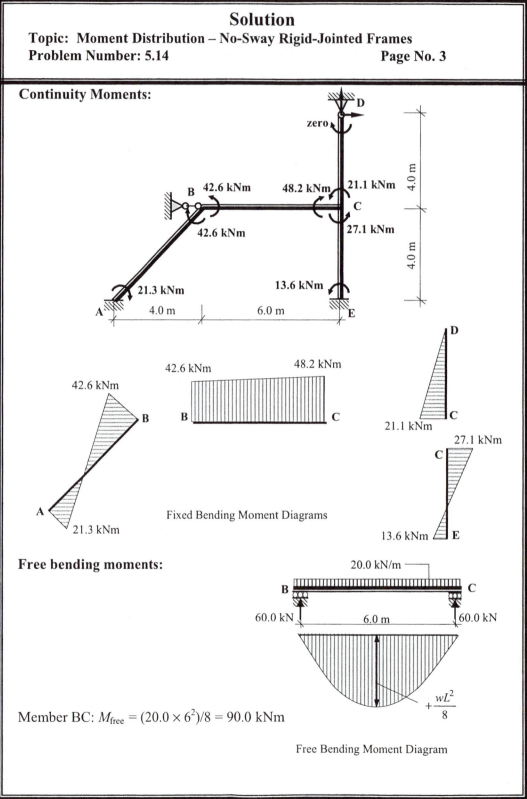

Fixed Bending Moment Diagrams

Free bending moments:

Member BC: $M_{free} = (20.0 \times 6^2)/8 = 90.0$ kNm

Free Bending Moment Diagram

Solution

Topic: Moment Distribution – No-Sway Rigid-Jointed Frames

Problem Number: 5.14   **Page No. 4**

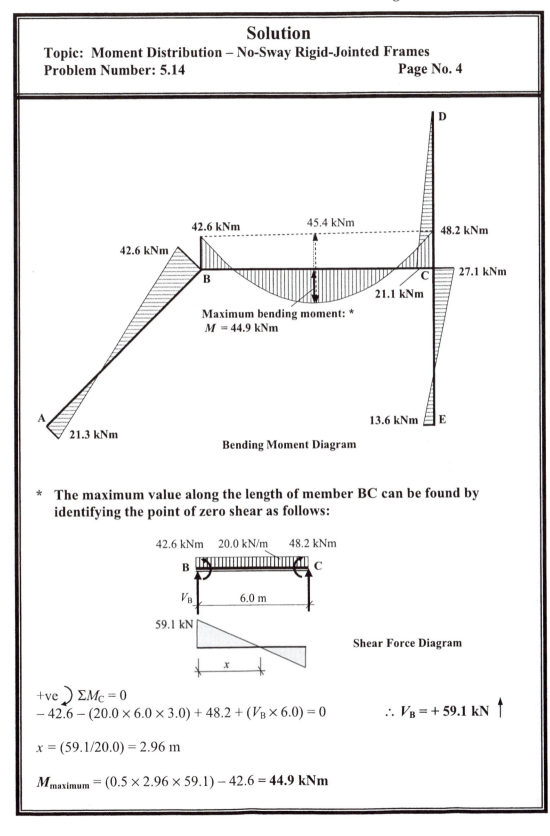

Bending Moment Diagram

*   **The maximum value along the length of member BC can be found by identifying the point of zero shear as follows:**

Shear Force Diagram

+ve $\curvearrowright$ $\Sigma M_C = 0$

$-42.6 - (20.0 \times 6.0 \times 3.0) + 48.2 + (V_B \times 6.0) = 0$ $\therefore V_B = +\,59.1 \text{ kN} \uparrow$

$x = (59.1/20.0) = 2.96 \text{ m}$

$M_{\text{maximum}} = (0.5 \times 2.96 \times 59.1) - 42.6 = \mathbf{44.9 \text{ kNm}}$

Solution

Topic: Moment Distribution – No-Sway Rigid-Jointed Frames
Problem Number: 5.14   **Page No. 4**

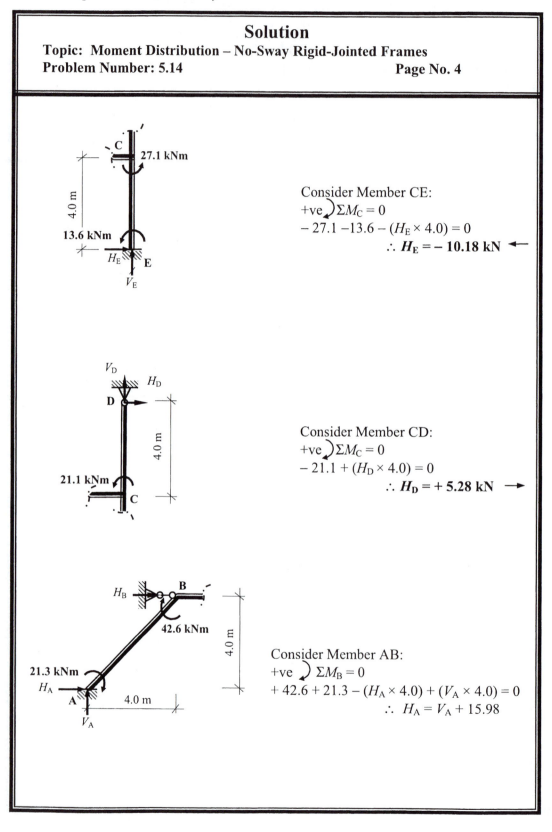

Consider Member CE:

$+ve\;\curvearrowright\Sigma M_{\mathrm{C}}=0$

$-27.1-13.6-(H_{\mathrm{E}}\times4.0)=0$

$\therefore\;\boldsymbol{H_{\mathrm{E}}=-10.18\;kN}\;\longleftarrow$

Consider Member CD:

$+ve\;\curvearrowright\Sigma M_{\mathrm{C}}=0$

$-21.1+(H_{\mathrm{D}}\times4.0)=0$

$\therefore\;\boldsymbol{H_{\mathrm{D}}=+5.28\;kN}\;\longrightarrow$

Consider Member AB:

$+ve\;\curvearrowright\Sigma M_{\mathrm{B}}=0$

$+42.6+21.3-(H_{\mathrm{A}}\times4.0)+(V_{\mathrm{A}}\times4.0)=0$

$\therefore\;\boldsymbol{H_{\mathrm{A}}=V_{\mathrm{A}}+15.98}$

Solution

Topic: Moment Distribution – No-Sway Rigid-Jointed Frames
Problem Number: 5.14 Page No. 5

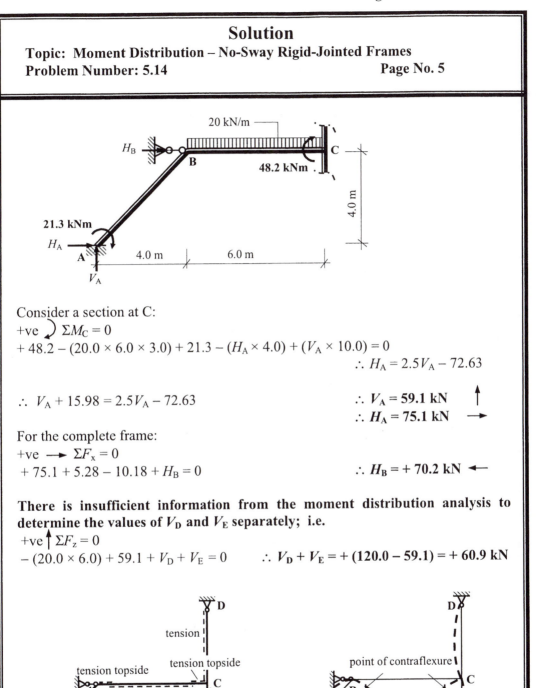

Consider a section at C:

+ve $\circlearrowright$ $\Sigma M_C = 0$

$+ 48.2 - (20.0 \times 6.0 \times 3.0) + 21.3 - (H_A \times 4.0) + (V_A \times 10.0) = 0$

$$\therefore H_A = 2.5 V_A - 72.63$$

$\therefore V_A + 15.98 = 2.5 V_A - 72.63$ $\qquad \therefore V_A = 59.1 \text{ kN}$ $\uparrow$

$\therefore H_A = 75.1 \text{ kN}$ $\rightarrow$

For the complete frame:

+ve $\rightarrow$ $\Sigma F_x = 0$

$+ 75.1 + 5.28 - 10.18 + H_B = 0$ $\qquad \therefore H_B = + 70.2 \text{ kN}$ $\leftarrow$

There is insufficient information from the moment distribution analysis to determine the values of V_D and V_E separately; i.e.

+ve $\uparrow$ $\Sigma F_z = 0$

$- (20.0 \times 6.0) + 59.1 + V_D + V_E = 0$ $\qquad \therefore V_D + V_E = + (120.0 - 59.1) = + 60.9 \text{ kN}$

Deflected Shape

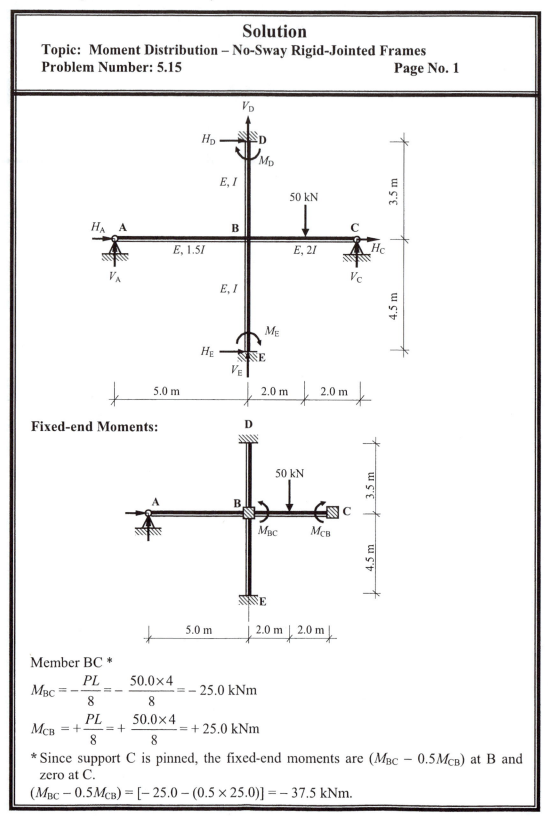

Solution

Topic:　Moment Distribution – No-Sway Rigid-Jointed Frames

Problem Number: 5.15　　　　　　　　　　　　　　　**Page No. 1**

Fixed-end Moments:

Member BC *

$$M_{BC} = -\frac{PL}{8} = -\frac{50.0 \times 4}{8} = -25.0 \text{ kNm}$$

$$M_{CB} = +\frac{PL}{8} = +\frac{50.0 \times 4}{8} = +25.0 \text{ kNm}$$

* Since support C is pinned, the fixed-end moments are $(M_{BC} - 0.5M_{CB})$ at B and zero at C.

$(M_{BC} - 0.5M_{CB}) = [-25.0 - (0.5 \times 25.0)] = -37.5 \text{ kNm}$.

Solution

Topic: Moment Distribution – No-Sway Rigid-Jointed Frames

Problem Number: 5.15   **Page No. 2**

Distribution Factors : Joint B

$$k_{BA} = \frac{3}{4} \times \left(\frac{1.5I}{5.0}\right) = 0.23I$$

$$k_{BC} = \frac{3}{4} \times \left(\frac{2I}{4.0}\right) = 0.38I$$

$$k_{BD} = \left(\frac{I}{3.5}\right) = 0.29I$$

$$k_{BE} = \left(\frac{I}{4.5}\right) = 0.22I$$

$$k_{total} = 1.12I$$

$$DF_{BA} = \frac{k_{BA}}{k_{Total}} = \frac{0.23}{1.12} = \mathbf{0.21}$$

$$DF_{BC} = \frac{k_{BC}}{k_{Total}} = \frac{0.38}{1.12} = \mathbf{0.33}$$

$$DF_{BD} = \frac{k_{BD}}{k_{Total}} = \frac{0.29}{1.12} = \mathbf{0.26}$$

$$DF_{BE} = \frac{k_{BE}}{k_{Total}} = \frac{0.22}{1.12} = \mathbf{0.20}$$

Moment Distribution Table:

Joint	A	D			B				E	C
	AB	DB		BD	BA	BC	BE		EB	CB
Distribution Factors	1.0	0		0.26	0.21	0.33	0.2		0	1.0
Fixed-end Moments						− 37.5				
Balance				+ 9.7	+ 7.9	+ 12.4	+ 7.5			
Carry-over		+ 4.9							+ 3.8	
Total	0	+ 4.9		+ 9.7	+ 7.9	− 25.1	+ 7.5		+ 3.8	0

Continuity Moments:

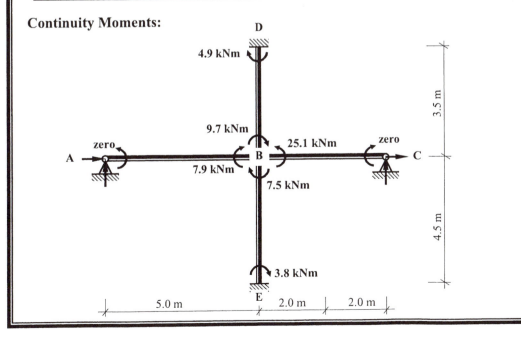

Solution

Topic: Moment Distribution – No-Sway Rigid-Jointed Frames
Problem Number: 5.15   **Page No. 3**

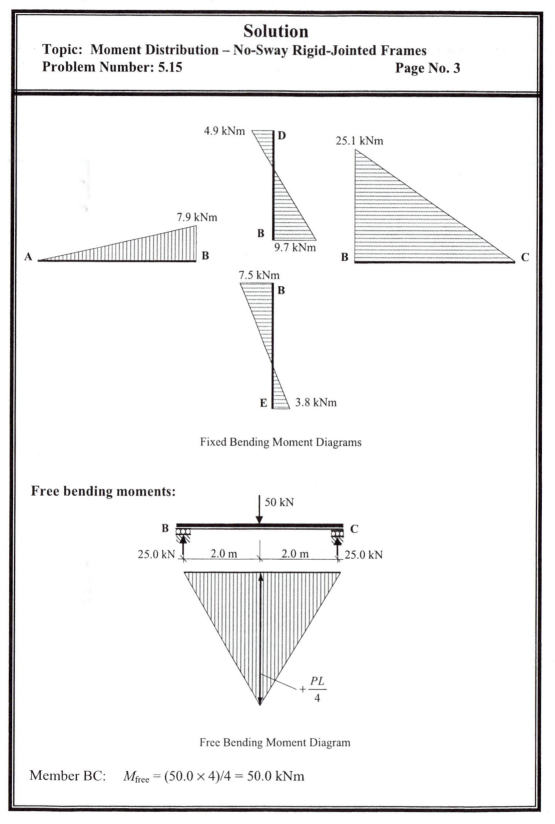

Fixed Bending Moment Diagrams

Free bending moments:

Free Bending Moment Diagram

Member BC: $M_{free} = (50.0 \times 4)/4 = 50.0$ kNm

Solution

Topic: Moment Distribution – No-Sway Rigid-Jointed Frames

Problem Number: 5.15   **Page No. 4**

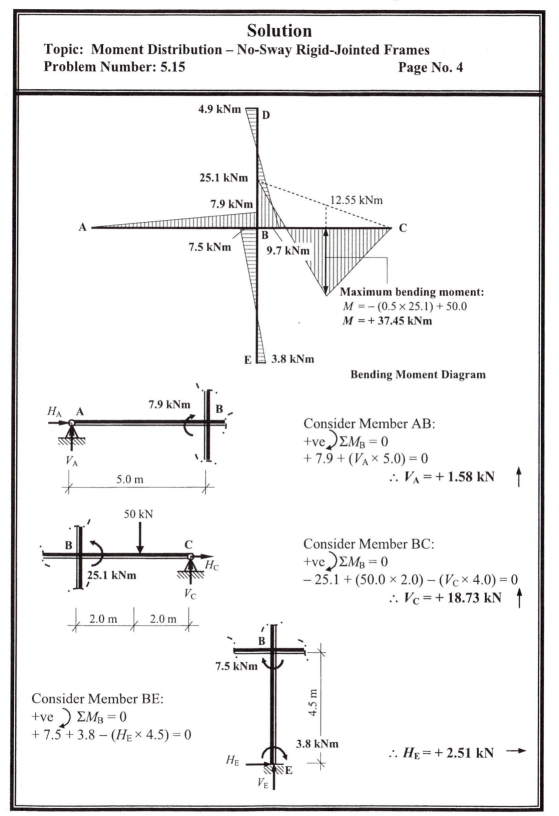

Bending Moment Diagram

Maximum bending moment:
$M = -(0.5 \times 25.1) + 50.0$
$M = +37.45$ kNm

Consider Member AB:
$+ve \; \curvearrowright \Sigma M_B = 0$
$+ 7.9 + (V_A \times 5.0) = 0$
$\therefore V_A = + 1.58$ kN $\uparrow$

Consider Member BC:
$+ve \; \curvearrowright \Sigma M_B = 0$
$- 25.1 + (50.0 \times 2.0) - (V_C \times 4.0) = 0$
$\therefore V_C = + 18.73$ kN $\uparrow$

Consider Member BE:
$+ve \; \curvearrowright \Sigma M_B = 0$
$+ 7.5 + 3.8 - (H_E \times 4.5) = 0$
$\therefore H_E = + 2.51$ kN $\rightarrow$

Solution
Topic: Moment Distribution – No-Sway Rigid-Jointed Frames
Problem Number: 5.15 Page No. 4

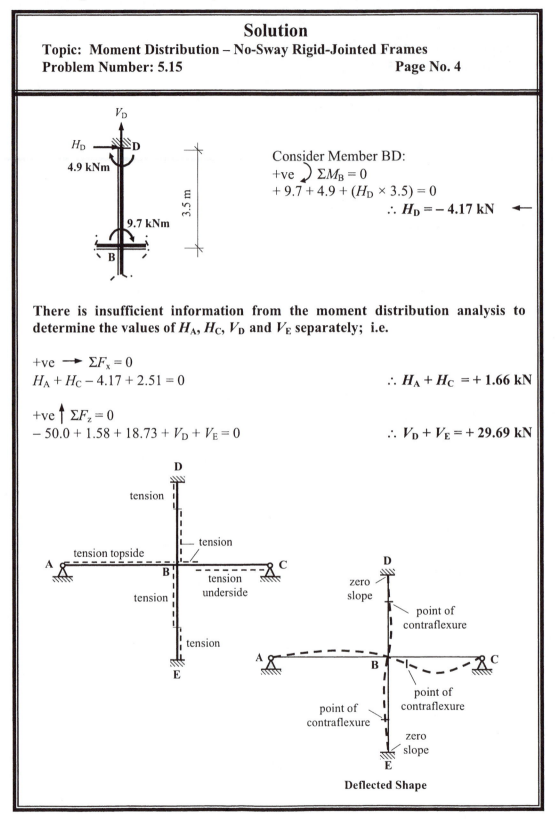

Consider Member BD:

+ve $\circlearrowleft$ $\Sigma M_B = 0$

$+ 9.7 + 4.9 + (H_D \times 3.5) = 0$

$\therefore H_D = - 4.17 \text{ kN}$ ←

There is insufficient information from the moment distribution analysis to determine the values of H_A, H_C, V_D and V_E separately; i.e.

+ve → $\Sigma F_x = 0$

$H_A + H_C - 4.17 + 2.51 = 0$ $\therefore H_A + H_C = + 1.66 \text{ kN}$

+ve ↑ $\Sigma F_z = 0$

$- 50.0 + 1.58 + 18.73 + V_D + V_E = 0$ $\therefore V_D + V_E = + 29.69 \text{ kN}$

Deflected Shape

Solution

Topic: Moment Distribution – No-Sway Rigid-Jointed Frames

Problem Number: 5.16   **Page No. 1**

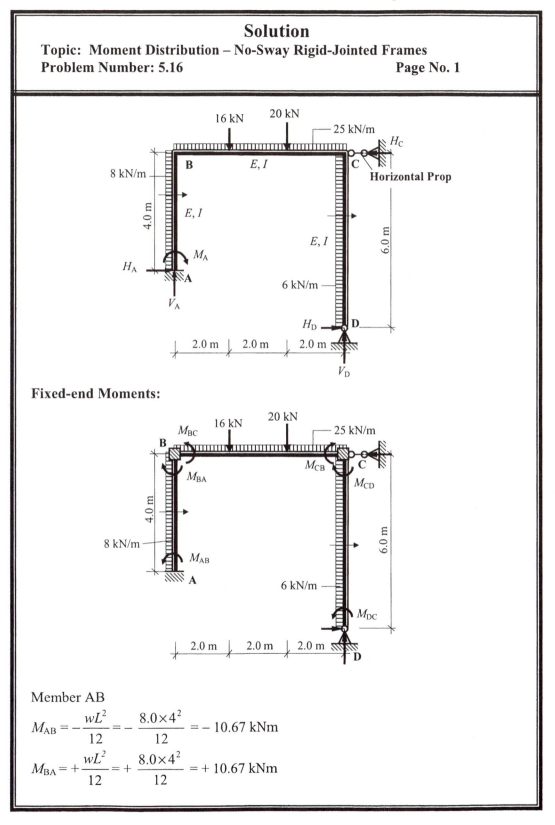

Fixed-end Moments:

Member AB

$$M_{AB} = -\frac{wL^2}{12} = -\frac{8.0 \times 4^2}{12} = -10.67 \text{ kNm}$$

$$M_{BA} = +\frac{wL^2}{12} = +\frac{8.0 \times 4^2}{12} = +10.67 \text{ kNm}$$

Solution

Topic: Moment Distribution – No-Sway Rigid-Jointed Frames
Problem Number: 5.16   **Page No. 2**

Member BC

$$M_{BC} = -\frac{wL^2}{12} - \frac{P_1ab^2}{L^2} - \frac{P_2ab^2}{L^2}$$

$$= -\left[\left(\frac{25.0 \times 6^2}{12}\right) + \left(\frac{16.0 \times 2.0 \times 4.0^2}{6^2}\right) + \left(\frac{20.0 \times 4.0 \times 2.0^2}{6^2}\right)\right] = -98.1 \text{ kNm}$$

$$M_{CB} = +\frac{wL^2}{12} + \frac{P_1a^2b}{L^2} + \frac{P_2a^2b}{L^2}$$

$$= +\left[\left(\frac{25.0 \times 6^2}{12}\right) + \left(\frac{16.0 \times 2.0^2 \times 4.0}{6^2}\right) + \left(\frac{20.0 \times 4.0^2 \times 2.0}{6^2}\right)\right] = +99.9 \text{ kNm}$$

Member CD *

$$M_{CD} = +\frac{wL^2}{12} = +\frac{6.0 \times 6^2}{12} = +18.0 \text{ kNm}$$

$$M_{DC} = -\frac{wL^2}{12} = \frac{6.0 \times 6^2}{12} = -18.0 \text{ kNm}$$

* Since support D is pinned, the fixed-end moments are ($M_{CD} - 0.5M_{DC}$) at C and zero at D.
($M_{CD} - 0.5M_{DC}$) = [+ 18.0 + (0.5 × 18.0)] = + 27.0 kNm.

Distribution Factors : Joint B

$$k_{BA} = \left(\frac{I}{4.0}\right) = 0.25I$$

$$DF_{BA} = \frac{k_{BA}}{k_{Total}} = \frac{0.25}{0.42} = \mathbf{0.6}$$

$$k_{total} = \mathbf{0.42I}$$

$$k_{BC} = \left(\frac{I}{6.0}\right) = 0.17I$$

$$DF_{BC} = \frac{k_{BC}}{k_{Total}} = \frac{0.17}{0.42} = \mathbf{0.4}$$

Distribution Factors : Joint C

$$k_{CB} = \left(\frac{I}{6.0}\right) = 0.17I$$

$$DF_{CB} = \frac{k_{CB}}{k_{Total}} = \frac{0.17}{0.3} = \mathbf{0.57}$$

$$k_{total} = \mathbf{0.3I}$$

$$k_{CD} = \frac{3}{4} \times \left(\frac{I}{6.0}\right) = 0.13I$$

$$DF_{CD} = \frac{k_{CD}}{k_{Total}} = \frac{0.13}{0.3} = \mathbf{0.43}$$

Solution

Topic: Moment Distribution – No-Sway Rigid-Jointed Frames

Problem Number: 5.16   **Page No. 3**

Moment Distribution Table:

Joint	A	B			C		D
	AB	BA	BC		CB	CD	DC
Distribution Factors	**0**	**0.6**	**0.4**		**0.57**	**0.43**	**1.0**
Fixed-end Moments	−10.67	+ 10.67	− 98.1		+ 99.9	+ 27.0	
Balance		+ 52.46	+ 34.97		− 72.3	− 54.6	
Carry-over	+ 26.23		− 36.2		+ 17.49		
Balance		+ 21.72	+ 14.48		− 9.97	− 7.52	
Carry-over	+ 10.86		− 4.99		+ 7.24		
Balance		+ 3.0	+ 1.99		− 4.13	− 3.11	
Carry-over	+ 1.5		− 2.07		+ 1.0		
Balance		+ 1.2	+ 0.87		− 0.57	− 0.43	
Carry-over	+ 0.6						
Total	**+ 28.52**	**+ 89.1**	**− 89.1**		**+ 38.66**	**− 38.66**	**0**

Continuity Moments:

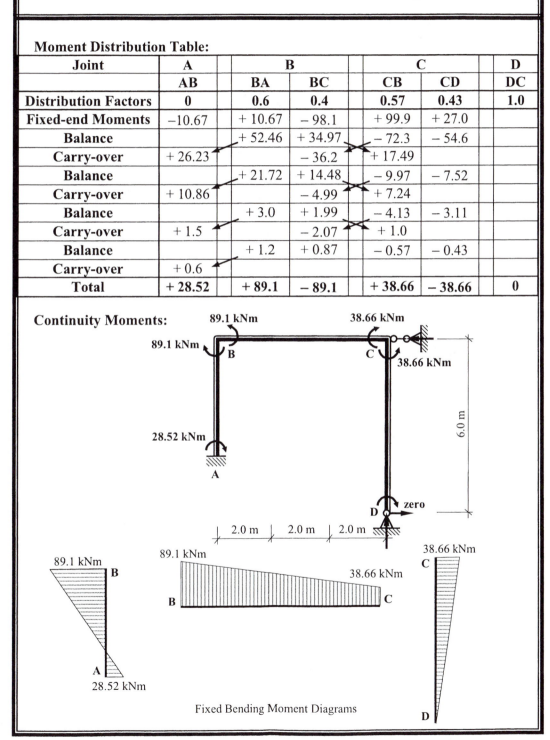

Fixed Bending Moment Diagrams

Solution

Topic: Moment Distribution – No-Sway Rigid-Jointed Frames
Problem Number: 5.16   **Page No. 4**

Free bending moments:

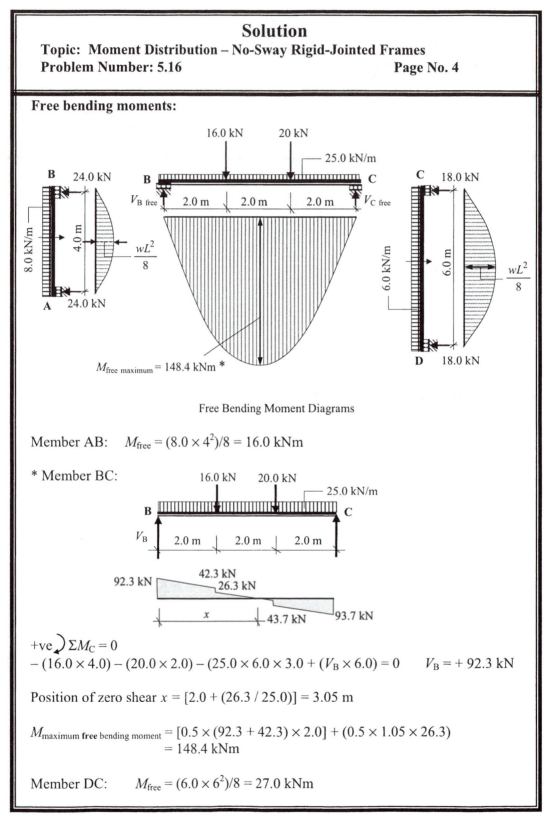

Free Bending Moment Diagrams

Member AB: $M_{free} = (8.0 \times 4^2)/8 = 16.0$ kNm

* Member BC:

$+ve \curvearrowright \Sigma M_C = 0$

$-(16.0 \times 4.0) - (20.0 \times 2.0) - (25.0 \times 6.0 \times 3.0 + (V_B \times 6.0) = 0$ $V_B = +92.3$ kN

Position of zero shear $x = [2.0 + (26.3 / 25.0)] = 3.05$ m

$M_{maximum\ free\ bending\ moment} = [0.5 \times (92.3 + 42.3) \times 2.0] + (0.5 \times 1.05 \times 26.3)$
$= 148.4$ kNm

Member DC: $M_{free} = (6.0 \times 6^2)/8 = 27.0$ kNm

Solution

Topic: **Moment Distribution – No-Sway Rigid-Jointed Frames**
Problem Number: **5.16**   **Page No. 5**

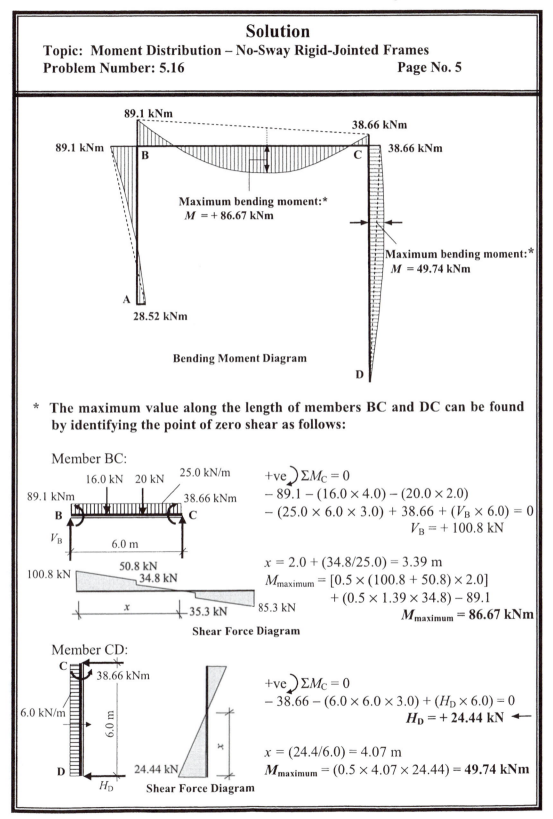

89.1 kNm

89.1 kNm

38.66 kNm

38.66 kNm

Maximum bending moment:*
$M = + 86.67$ kNm

Maximum bending moment:*
$M = 49.74$ kNm

28.52 kNm

Bending Moment Diagram

* The maximum value along the length of members BC and DC can be found by identifying the point of zero shear as follows:

Member BC:

16.0 kN   20 kN   25.0 kN/m

89.1 kNm

38.66 kNm

V_B

6.0 m

50.8 kN

34.8 kN

100.8 kN

x

35.3 kN

85.3 kN

Shear Force Diagram

$+ve \curvearrowright \Sigma M_C = 0$
$- 89.1 - (16.0 \times 4.0) - (20.0 \times 2.0)$
$- (25.0 \times 6.0 \times 3.0) + 38.66 + (V_B \times 6.0) = 0$
$V_B = + 100.8$ kN

$x = 2.0 + (34.8/25.0) = 3.39$ m
$M_{maximum} = [0.5 \times (100.8 + 50.8) \times 2.0]$
$+ (0.5 \times 1.39 \times 34.8) - 89.1$
$\boldsymbol{M_{maximum} = 86.67}$ **kNm**

Member CD:

C

38.66 kNm

6.0 kN/m

6.0 m

D

24.44 kN

x

H_D

Shear Force Diagram

$+ve \curvearrowright \Sigma M_C = 0$
$- 38.66 - (6.0 \times 6.0 \times 3.0) + (H_D \times 6.0) = 0$
$\boldsymbol{H_D = + 24.44}$ **kN** ←

$x = (24.4/6.0) = 4.07$ m
$M_{maximum} = (0.5 \times 4.07 \times 24.44) = \boldsymbol{49.74}$ **kNm**

Solution
Topic: Moment Distribution – No-Sway Rigid-Jointed Frames
Problem Number: 5.16   Page No. 6

Consider Member AB:

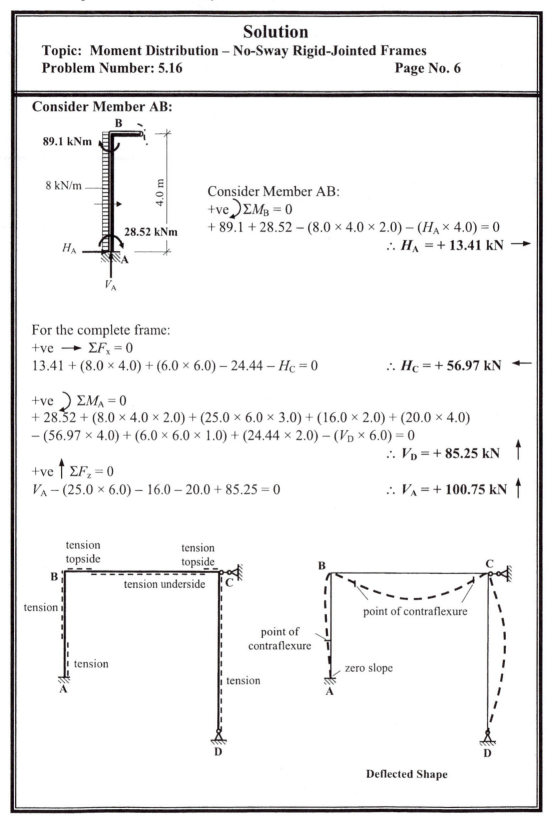

Consider Member AB:
$+ve \curvearrowright \Sigma M_B = 0$
$+ 89.1 + 28.52 - (8.0 \times 4.0 \times 2.0) - (H_A \times 4.0) = 0$
$\therefore H_A = + 13.41$ kN $\rightarrow$

For the complete frame:
$+ve \longrightarrow \Sigma F_x = 0$
$13.41 + (8.0 \times 4.0) + (6.0 \times 6.0) - 24.44 - H_C = 0$ $\qquad \therefore H_C = + 56.97$ kN $\leftarrow$

$+ve \curvearrowright \Sigma M_A = 0$
$+ 28.52 + (8.0 \times 4.0 \times 2.0) + (25.0 \times 6.0 \times 3.0) + (16.0 \times 2.0) + (20.0 \times 4.0)$
$- (56.97 \times 4.0) + (6.0 \times 6.0 \times 1.0) + (24.44 \times 2.0) - (V_D \times 6.0) = 0$
$\therefore V_D = + 85.25$ kN $\uparrow$

$+ve \uparrow \Sigma F_z = 0$
$V_A - (25.0 \times 6.0) - 16.0 - 20.0 + 85.25 = 0$ $\qquad \therefore V_A = + 100.75$ kN $\uparrow$

Deflected Shape

5.4 *Moment Distribution for Rigid-Jointed Frames with Sway*

The frames in Section 5.3 are prevented from any lateral movement by the support conditions. In frames where restraint against lateral movement is not provided at each level, unless the frame, the supports and the loading are symmetrical it will sway and consequently induce additional forces in the frame members.

Consider the frame indicated in Figure 5.27(a) in which the frame, supports and applied load are symmetrical.

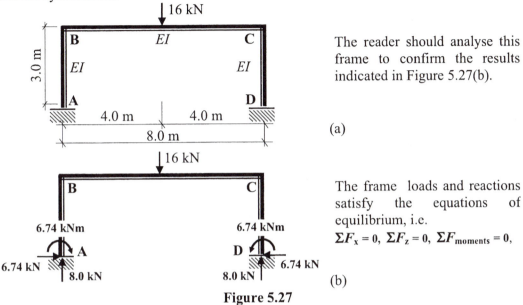

The reader should analyse this frame to confirm the results indicated in Figure 5.27(b).

(a)

The frame loads and reactions satisfy the equations of equilibrium, i.e.
$\Sigma F_x = 0$, $\Sigma F_z = 0$, $\Sigma F_{moments} = 0$,

(b)

Figure 5.27

Consider the same frame in which the load has been moved such that it is now asymmetric as indicated in Figure 5.28(a)

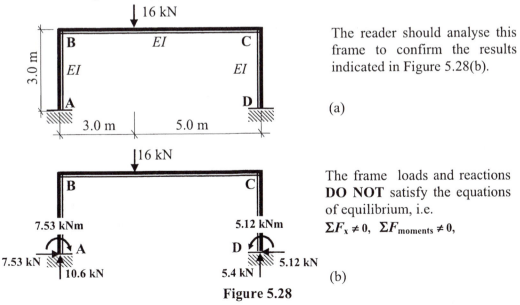

The reader should analyse this frame to confirm the results indicated in Figure 5.28(b).

(a)

The frame loads and reactions **DO NOT** satisfy the equations of equilibrium, i.e.
$\Sigma F_x \neq 0$, $\Sigma F_{moments} \neq 0$,

(b)

Figure 5.28

It is evident from Figure 5.28(b) that the solution to this problem is incomplete. Inspection of the deflected shapes of each of the frames in Figure 5.27(a) and 5.28(a) indicates the reason for the inconsistency in the asymmetric frame.

Consider the deflected shapes shown in Figures 5.29 (a) and (b):

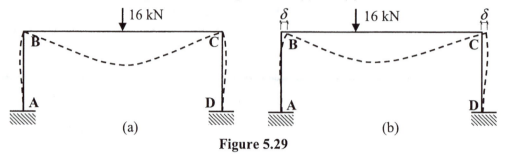

Figure 5.29

In case (a) the deflected shape indicates the equal rotations of the joints at B and C due to the balancing of the fixed–end moments induced by the load; note that there is no lateral movement at B and C.

In case (b) in addition to rotation due to the applied load there is also rotation of the joints due to the lateral movement 'δ' of B and C. The sway of the frame also induces forces in the members and this effect was **not** included in the results given in Figure 5.28(b). It is ignoring the '*sway*' of the frame which has resulted in the inconsistency. In effect, the frame which has been analysed is the one shown in Figure 5.30, i.e. including a prop force preventing sway. The value of the prop force 'P' is equal to the resultant horizontal force in Figure 5.28.

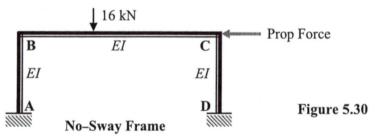

Figure 5.30

The complete analysis should include the effects of the sway and consequently an additional distribution must be carried out for sway–only and the effects added to the no–sway results, i.e. to cancel out the non–existent 'prop force' assumed in the no–sway frame.

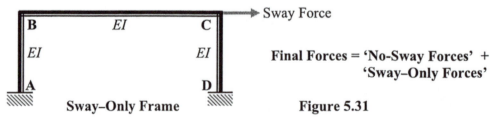

Final Forces = 'No–Sway Forces' + 'Sway–Only Forces'

Figure 5.31

The technique for completing this calculation including the sway effects is illustrated in Example 5.5 and the solutions to Problems 5.17 to 5.22.

5.4.1 *Example 5.5 Rigid–Jointed Frame with Sway– Frame 1*

A rigid–jointed frame is fixed at support A, pinned at support H and supported on a roller at F as shown in Figure 5.32. For the relative EI values and loading given:

i)   sketch the bending moment diagram,

ii) determine the support reactions and

iii) sketch the deflected shape (assuming axially rigid members) and compare with the shape of the bending moment diagram, (the reader should check the answer using a computer analysis solution). $EI = 10 \times 10^3$ kNm^2

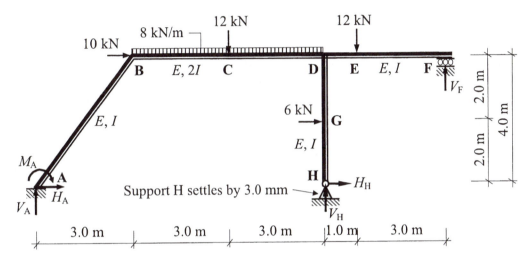

Figure 5.32

Consider the frame analysis as the superposition of two effects:

Final Forces = 'No-Sway Forces' + 'Sway Forces'

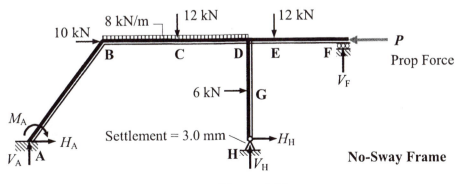

Figure 5.33

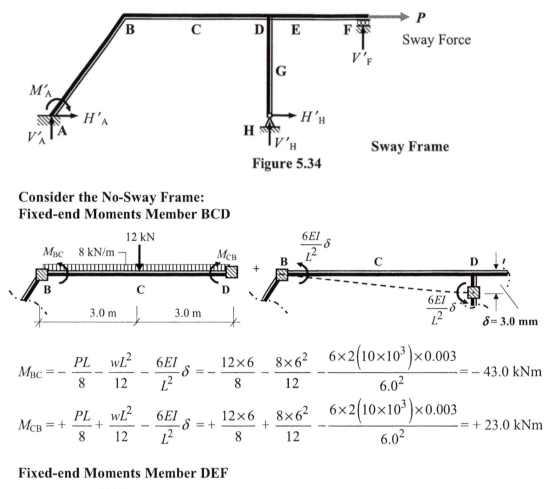

Figure 5.34

Consider the No-Sway Frame:
Fixed-end Moments Member BCD

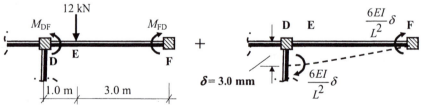

$$M_{BC} = -\frac{PL}{8} - \frac{wL^2}{12} - \frac{6EI}{L^2}\delta = -\frac{12\times6}{8} - \frac{8\times6^2}{12} - \frac{6\times2\left(10\times10^3\right)\times0.003}{6.0^2} = -43.0 \text{ kNm}$$

$$M_{CB} = +\frac{PL}{8} + \frac{wL^2}{12} - \frac{6EI}{L^2}\delta = +\frac{12\times6}{8} + \frac{8\times6^2}{12} - \frac{6\times2\left(10\times10^3\right)\times0.003}{6.0^2} = +23.0 \text{ kNm}$$

Fixed-end Moments Member DEF

Since F is a roller support, the fixed-end moments are $(M_{DF} - 0.5M_{FD})$ at D and zero at F.

$$M_{DF} = -\frac{Pab^2}{L^2} + \frac{6EI}{L^2}\delta = -\frac{12\times1.0\times3.0^2}{4.0^2} + \frac{6\times10\times10^3\times0.003}{4.0^2} = +4.5 \text{ kNm}$$

$$M_{FD} = +\frac{Pa^2b}{L^2} + \frac{6EI}{L^2}\delta = +\frac{12\times1.0^2\times3.0}{4.0^2} + \frac{6\times10\times10^3\times0.003}{4.0^2} = +13.5 \text{ kNm}$$

$(M_{DF} - 0.5M_{FD}) = [+4.5 - (0.5\times13.5)] = -2.25 \text{ kNm.}$

Fixed-end Moments Member DGH

Since support H pinned, the fixed-end moments are $(M_{DH} - 0.5M_{HD})$ at D and zero at H.

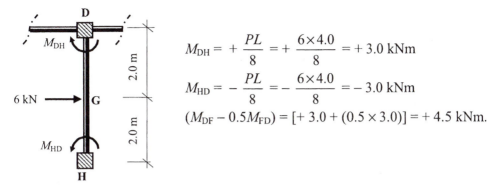

$$M_{DH} = +\frac{PL}{8} = +\frac{6 \times 4.0}{8} = +3.0 \text{ kNm}$$

$$M_{HD} = -\frac{PL}{8} = -\frac{6 \times 4.0}{8} = -3.0 \text{ kNm}$$

$$(M_{DF} - 0.5M_{FD}) = [+3.0 + (0.5 \times 3.0)] = +4.5 \text{ kNm}.$$

Distribution Factors : Joint B

$$k_{BA} = \left(\frac{I}{5}\right) = 0.2I \qquad\qquad DF_{BA} = \frac{k_{BA}}{k_{Total}} = \frac{0.2}{0.53} = \mathbf{0.38}$$

$$k_{total} = \mathbf{0.53}I$$

$$k_{BD} = \left(\frac{2I}{6}\right) = 0.33I \qquad\qquad DF_{BD} = \frac{k_{BD}}{k_{Total}} = \frac{0.33}{0.53} = \mathbf{0.62}$$

Distribution Factors : Joint D

$$k_{DB} = \left(\frac{2I}{6}\right) = 0.33I \qquad\qquad DF_{DB} = \frac{k_{DB}}{k_{Total}} = \frac{0.33}{0.71} = \mathbf{0.46}$$

$$k_{DH} = \frac{3}{4} \times \left(\frac{I}{4}\right) = 0.19I \qquad k_{total} = \mathbf{0.71}I \qquad DF_{DH} = \frac{k_{DH}}{k_{Total}} = \frac{0.19}{0.71} = \mathbf{0.27}$$

$$k_{DF} = \frac{3}{4} \times \left(\frac{I}{4}\right) = 0.19I \qquad\qquad DF_{DF} = \frac{k_{DF}}{k_{Total}} = \frac{0.19}{0.71} = \mathbf{0.27}$$

No-Sway Moment Distribution Table:

Joint	A		B			D			F	H
	AB		BA	BD		DB	DH	DF	FD	HD
Distribution Factors	0		0.38	0.62		0.46	0.27	0.27	1.0	1.0
Fixed-end Moments				− 43.0		+ 23.0	+ 4.5	− 2.25	0	0
Balance			+ 16.34	+ 26.66		− 11.62	− 6.82	− 6.82		
Carry-over	+ 8.17			− 5.81		+ 13.33				
Balance			+ 2.21	+ 3.60		− 6.13	− 3.60	− 3.60		
Carry-over	+ 1.10			− 3.07		+ 1.80				
Balance			+ 1.17	+ 1.90		− 0.83	− 0.49	− 0.48		
Carry-over	+ 0.58			− 0.41		+ 0.94				
Balance			+ 0.15	+ 0.26		− 0.44	− 0.25	− 0.25		
Carry-over	+ 0.08									
Total	+ 9.93		+ 19.87	− 19.87		+ 20.06	− 6.66	− 13.40	0	0

Determine the value of the reactions and prop force *P*:

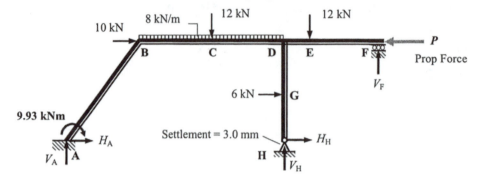

Consider member DEF:

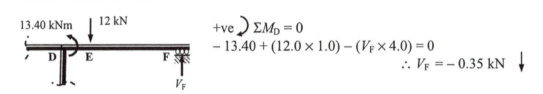

$+ve \curvearrowright \Sigma M_D = 0$

$-13.40 + (12.0 \times 1.0) - (V_F \times 4.0) = 0$

$\therefore V_F = -0.35 \text{ kN} \downarrow$

Consider member DGH:

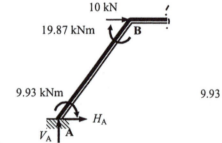

$+ve \curvearrowright \Sigma M_D = 0$

$-6.66 - (6.0 \times 2.0) - (H_H \times 4.0) = 0$

$\therefore H_H = -4.67 \text{ kN} \leftarrow$

Consider member BA and a section to the left of D:

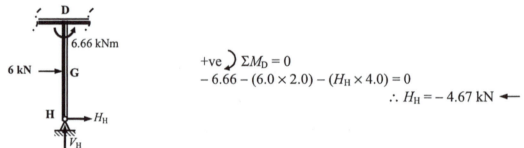

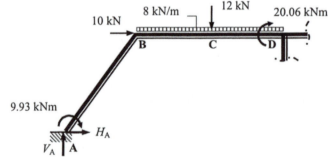

$+ve \curvearrowright \Sigma M_B = 0$

$+9.93 + (V_A \times 3.0) - (H_A \times 4.0) + 19.87 = 0 \quad \therefore V_A = -9.93 + 1.33 H_A \qquad \text{Equation (1)}$

$+ve \curvearrowright \Sigma M_D = 0$

$+9.93 + (V_A \times 9.0) - (H_A \times 4.0) - (8.0 \times 6.0)(3.0) - (12.0 \times 3.0) + 20.06 = 0$

$\therefore V_A = +16.67 + 0.44 H_A \qquad \text{Equation (2)}$

Solve equations (1) and (2) simultaneously:

$-9.93 + 1.33H_A = +16.67 + 0.44H_A$ $\therefore H_A = +29.89$ kN →

$V_A = +16.67 + (0.44 \times 29.89)$ $\therefore V_A = +29.82$ kN ↑

Consider the equilibrium of the complete frame:

+ve ↑ $\Sigma F_z = 0$

$V_A - (8.0 \times 6.0) - 12.0 + V_H - 12.0 + V_F = 0$

$+29.89 - 48.0 - 12.0 + V_H - 12.0 - 0.35 = 0$ $\therefore V_H = +42.46$ kN ↑

+ve → $\Sigma F_x = 0$

$H_A + 10.0 + 6.0 + H_H - P = 0$

$+29.89 + 16.0 - 4.67 - P = 0$ $\therefore P = +41.22$ kN ←

Since the direction of the prop force is right-to-left the sway of the frame is from left-to-right as shown.

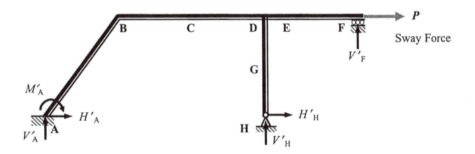

Apply an arbitrary sway force P' to determine the *ratios* of the fixed–end moments.

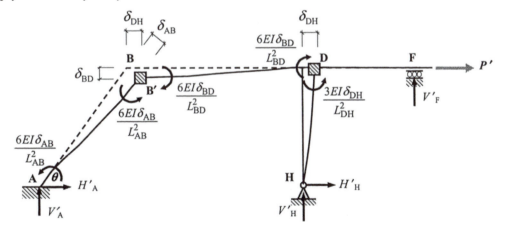

Fixed-end Moments due to Sway

The fixed–end moments in each member are related to the end–displacements (δ) in each case. The relationship between δ_{AB}, δ_{BD} and δ_{DH} can be determined by considering the displacement triangle at joint B and the geometry of the frame.

Displacement triangle:

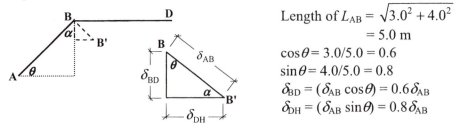

Length of $L_{AB} = \sqrt{3.0^2 + 4.0^2}$
$\qquad = 5.0$ m
$\cos\theta = 3.0/5.0 = 0.6$
$\sin\theta = 4.0/5.0 = 0.8$
$\delta_{BD} = (\delta_{AB}\cos\theta) = 0.6\delta_{AB}$
$\delta_{DH} = (\delta_{AB}\sin\theta) = 0.8\delta_{AB}$

Ratio of Fixed-end Moments: $\quad M_{AB} : M_{BA} : M_{BD} : M_{DB} : M_{DH}$

$$= -\frac{6(EI\delta_{AB})}{L_{AB}^2} : -\frac{6(EI\delta_{AB})}{L_{AB}^2} : +\frac{6(EI\delta_{BD})}{L_{BD}^2} : +\frac{6(EI\delta_{BD})}{L_{BD}^2} : -\frac{3(EI\delta_{DH})}{L_{DH}^2}$$

$$= -\frac{6(EI\delta_{AB})}{L_{AB}^2} : -\frac{6(EI\delta_{AB})}{L_{AB}^2} : +\frac{6(EI\times\delta_{AB}\cos\theta)}{L_{BD}^2} : +\frac{6(EI\times\delta_{AB}\cos\theta)}{L_{BD}^2} : -\frac{3(EI\times\delta_{AB}\sin\theta)}{L_{DH}^2}$$

$$= -\frac{6(EI\delta_{AB})}{5.0^2} : -\frac{6(EI\delta_{AB})}{5.0^2} : +\frac{6(2.0EI\delta_{AB}\times0.6)}{6.0^2} : +\frac{6(2.0EI\delta_{AB}\times0.6)}{6.0^2} : -\frac{3(EI\delta_{AB})\times0.8}{4.0^2}$$

$= \{- 0.24 : - 0.24 : + 0.20 : + 0.20 : - 0.15\} \times (EI\delta)_{AB}$

Assume arbitrary fixed-end moments equal to:
$\{- 24.0 : - 24.0 : + 20.0 : + 20.0 : - 15.0\} \times (EI\delta)_{AB}/100$

Sway-Only Moment Distribution Table:

Joint	A	B		D			F	H
	AB	BA	BD	DB	DH	DF	FD	HD
Distribution Factors	0	0.38	0.62	0.46	0.27	0.27	1.0	1.0
Fixed-end Moments	− 24.0	− 24.0	+ 20.0	+ 20.00	− 15.0		0	0
Balance		+ 1.52	+ 2.48	− 2.30	− 1.35	− 1.35		
Carry-over	+ 0.76		− 1.15	+ 1.24				
Balance		+ 0.44	+ 0.71	− 0.57	− 0.33	− 0.33		
Carry-over	+ 0.22		− 0.29	+ 0.36				
Balance		+ 0.11	+ 0.18	− 0.16	− 0.10	− 0.10		
Carry-over	+ 0.05		− 0.08	+ 0.09				
Balance		+ 0.03	+ 0.05	− 0.05	− 0.02	− 0.02		
Carry-over	+ 0.02							
Total	− 22.95	− 21.90	+ 21.90	+ 18.60	− 16.80	− 1.80	0	0

Determine the value of the arbitrary sway force P':

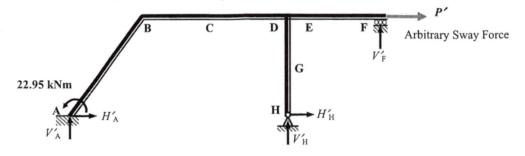

Consider member DEF:

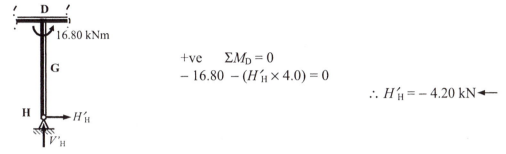

$+ve \;\curvearrowright\; \Sigma M_D = 0$

$-1.80 - (V'_F \times 4.0) = 0$

$\therefore V'_F = -0.45 \text{ kN} \;\downarrow$

Consider member DGH:

$+ve \quad \Sigma M_D = 0$

$-16.80 - (H'_H \times 4.0) = 0$

$\therefore H'_H = -4.20 \text{ kN} \;\leftarrow$

Consider member AB and a section to the left of D:

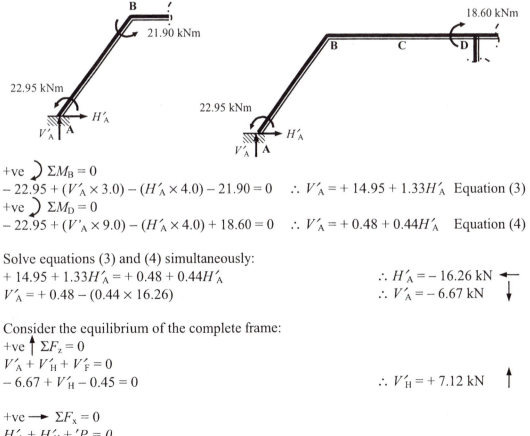

$+ve \;\curvearrowright\; \Sigma M_B = 0$

$-22.95 + (V'_A \times 3.0) - (H'_A \times 4.0) - 21.90 = 0 \quad \therefore V'_A = +14.95 + 1.33 H'_A \quad \text{Equation (3)}$

$+ve \;\curvearrowright\; \Sigma M_D = 0$

$-22.95 + (V'_A \times 9.0) - (H'_A \times 4.0) + 18.60 = 0 \quad \therefore V'_A = +0.48 + 0.44 H'_A \quad \text{Equation (4)}$

Solve equations (3) and (4) simultaneously:

$+14.95 + 1.33 H'_A = +0.48 + 0.44 H'_A$ $\therefore H'_A = -16.26 \text{ kN} \;\leftarrow$

$V'_A = +0.48 - (0.44 \times 16.26)$ $\therefore V'_A = -6.67 \text{ kN} \;\downarrow$

Consider the equilibrium of the complete frame:

$+ve \;\uparrow\; \Sigma F_z = 0$

$V'_A + V'_H + V'_F = 0$

$-6.67 + V'_H - 0.45 = 0$ $\therefore V'_H = +7.12 \text{ kN} \;\uparrow$

$+ve \;\rightarrow\; \Sigma F_x = 0$

$H'_A + H'_H + {}'P = 0$

$-16.26 - 4.20 + P' = 0$ $\therefore \boldsymbol{P' = +20.46 \text{ kN}} \;\dashrightarrow$

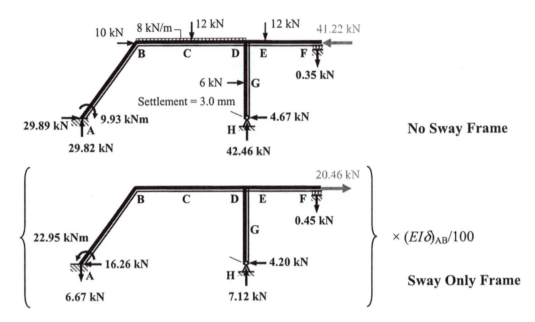

For the complete frame:

Final Forces = 'No-Sway Forces' + 'Sway Forces'

$P + P' = 0$
$-41.22 + [20.46 \times (EI\delta)_{AB}/100] = 0$ ∴ $(EI\delta)_{AB}/100 = +2.02$

The multiplying factor for the sway moments $= +2.02$

Final Moments Distribution Table:

Joint	A		B			D			F	H
	AB		BA	BD		DB	DH	DF	FD	HD
No-Sway Moments	+ 9.93		+ 19.87	− 19.87		+ 20.06	− 6.66	− 13.40	0	0
Sway Moments × 2.02	− 46.36		− 44.24	+ 44.24		+ 37.57	− 33.93	− 3.64	0	0
Final Moments (kNm)	− 36.43		− 24.37	+ 24.37		+ 57.63	− 40.59	− 17.04	0	0

$$\delta_{AB} = \left(\frac{2.02 \times 100}{EI}\right) = \left(\frac{2.02 \times 100}{10 \times 10^3}\right) = 0.02 \text{ m} = 20 \text{ mm}$$

The horizontal deflection at the rafter level $= \delta_{DH} = 0.8\delta_{AB} = (0.8 \times 20) = 16$ mm

Final values of support reactions:
$M_A = +9.93 - (22.95 \times 2.02) = -36.43$ kNm
$H_A = +29.89 - (16.26 \times 2.02) = -2.96$ kN
$V_A = +29.82 - (6.67 \times 2.02) = +16.35$ kN
$H_H = -4.67 - (4.20 \times 2.02) = -13.15$ kN
$V_H = +42.46 + (7.12 \times 2.02) = +56.84$ kN
$V_F = -0.35 - (0.45 \times 2.02) = -1.26$ kN

Continuity Moments:

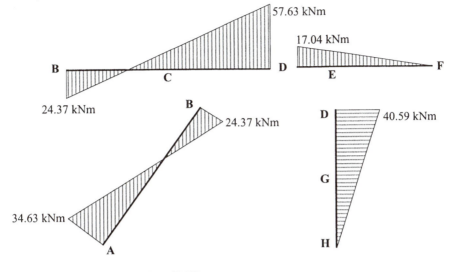

Free bending moment member BCD:

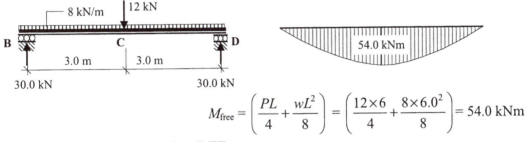

$$M_{\text{free}} = \left(\frac{PL}{4} + \frac{wL^2}{8} \right) = \left(\frac{12 \times 6}{4} + \frac{8 \times 6.0^2}{8} \right) = 54.0 \text{ kNm}$$

Free bending moment member DEF:

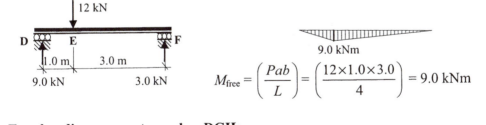

$$M_{\text{free}} = \left(\frac{Pab}{L} \right) = \left(\frac{12 \times 1.0 \times 3.0}{4} \right) = 9.0 \text{ kNm}$$

Free bending moment member DGH:

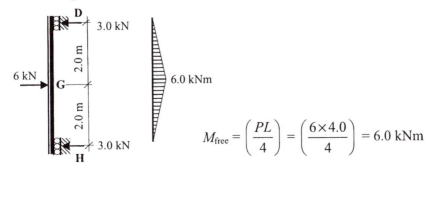

$$M_{\text{free}} = \left(\frac{PL}{4} \right) = \left(\frac{6 \times 4.0}{4} \right) = 6.0 \text{ kNm}$$

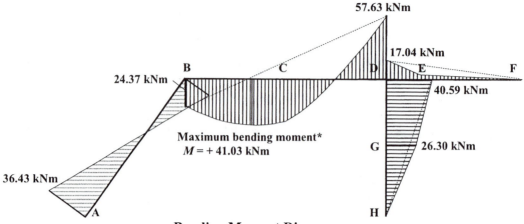

Bending Moment Diagram

* The maximum value along the length of members BCD can be found by identifying the point of zero shear as follows:

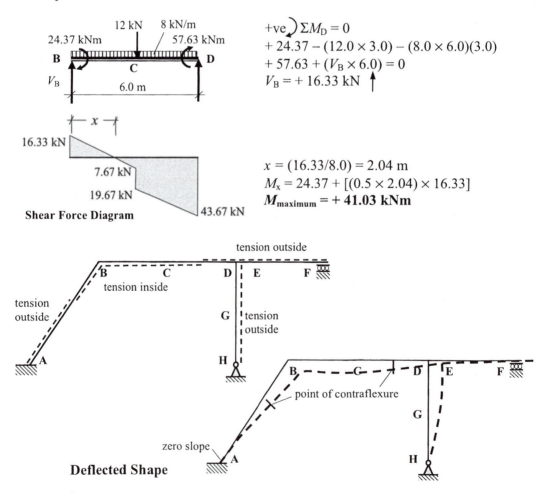

$+\text{ve} \circlearrowleft \Sigma M_D = 0$
$+ 24.37 - (12.0 \times 3.0) - (8.0 \times 6.0)(3.0)$
$+ 57.63 + (V_B \times 6.0) = 0$
$V_B = + 16.33 \text{ kN} \uparrow$

$x = (16.33/8.0) = 2.04 \text{ m}$
$M_x = 24.37 + [(0.5 \times 2.04) \times 16.33]$
$M_{\text{maximum}} = + \textbf{41.03 kNm}$

Shear Force Diagram

Deflected Shape

5.4.2 Problems: Moment Distribution – Rigid-Jointed Frames with Sway

A series of rigid-jointed frames are indicated in Problems 5.17 to 5.22 in which the relative *EI* values and the applied loading are given. In each case:

i) sketch the bending moment diagram   and

ii) sketch the deflected shape (assuming axially rigid members) and compare the shape of the bending moment diagram with a computer analysis solution of the deflected shape.

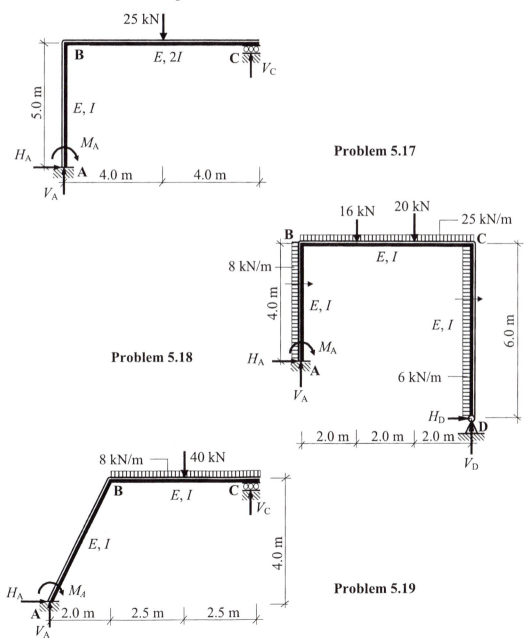

Problem 5.17

Problem 5.18

Problem 5.19

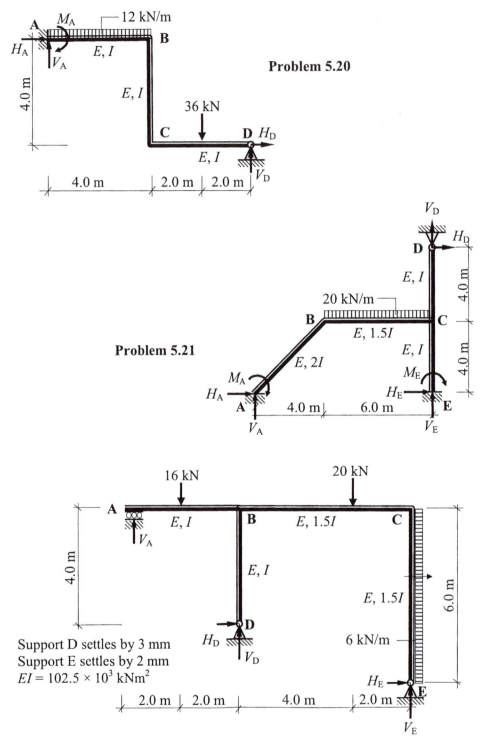

Problem 5.20

Problem 5.21

Support D settles by 3 mm
Support E settles by 2 mm
$EI = 102.5 \times 10^3$ kNm^2

Problem 5.22

5.4.3 Solutions: Moment Distribution – Rigid-Jointed Frames with Sway

Solution
Topic: Moment Distribution – Rigid-Jointed Frames with Sway

Problem Number: 5.17	Page No. 1

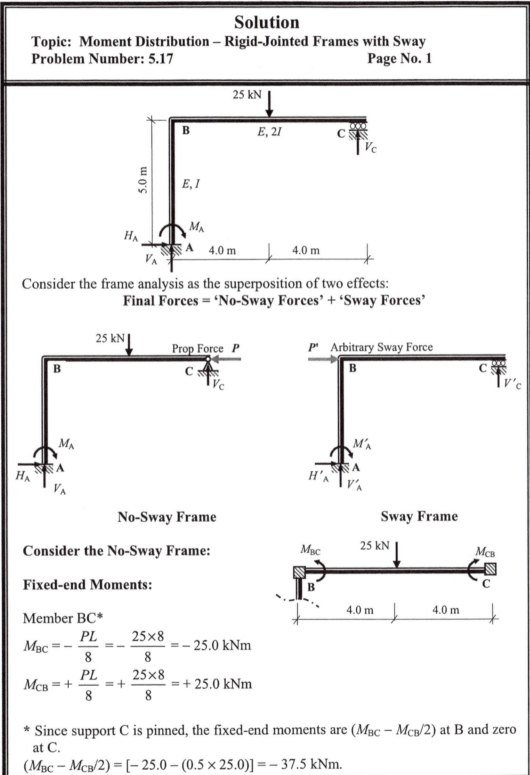

Consider the frame analysis as the superposition of two effects:

Final Forces = 'No-Sway Forces' + 'Sway Forces'

No-Sway Frame

Sway Frame

Consider the No-Sway Frame:

Fixed-end Moments:

Member BC*

$$M_{BC} = -\frac{PL}{8} = -\frac{25 \times 8}{8} = -25.0 \text{ kNm}$$

$$M_{CB} = +\frac{PL}{8} = +\frac{25 \times 8}{8} = +25.0 \text{ kNm}$$

* Since support C is pinned, the fixed-end moments are $(M_{BC} - M_{CB}/2)$ at B and zero at C.

$(M_{BC} - M_{CB}/2) = [-25.0 - (0.5 \times 25.0)] = -37.5 \text{ kNm}$.

Solution

Topic: Moment Distribution – Rigid-Jointed Frames with Sway
Problem Number: 5.17   **Page No. 2**

Distribution Factors : Joint B

$$k_{BA} = \left(\frac{I}{5}\right) = 0.2I$$

$$DF_{BA} = \frac{k_{BA}}{k_{Total}} = \frac{0.2}{0.39} = 0.51$$

$$k_{total} = 0.39I$$

$$k_{BC} = \frac{3}{4} \times \left(\frac{2I}{8}\right) = 0.19I$$

$$DF_{BC} = \frac{k_{BC}}{k_{Total}} = \frac{0.19}{0.39} = 0.49$$

In this case, since there is only one internal joint, only one balancing operation and one carry-over will be required during the distribution of the moments.

No-Sway Moment Distribution Table:

Joint	A		B		C
	AB	BA	BC		CB
Distribution Factors	0	0.51	0.49		0
Fixed-end Moments			− 37.5		0
Balance		+ 19.12	+ 18.38		
Carry-over	+ 9.56				
Total	+ 9.56	+ 19.12	− 19.12		0

Determine the value of the prop force *P*:

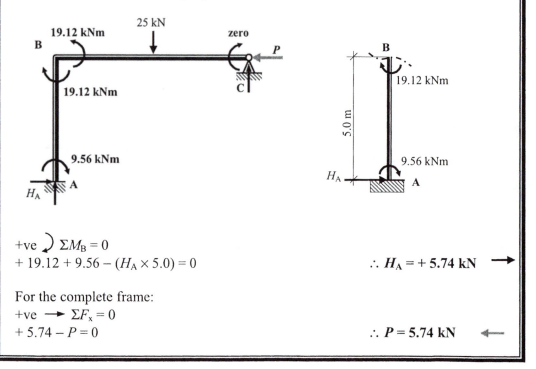

+ve $\circlearrowright$ $\Sigma M_B = 0$
+ 19.12 + 9.56 − ($H_A \times 5.0$) = 0 $\therefore H_A = + 5.74$ kN →

For the complete frame:
+ve → $\Sigma F_x = 0$
+ 5.74 − P = 0 $\therefore P = 5.74$ kN ←

Solution

Topic: Moment Distribution – Rigid-Jointed Frames with Sway
Problem Number: 5.17   **Page No. 3**

Since the direction of the prop force is right-to-left the sway of the frame is from left-to-right as shown.

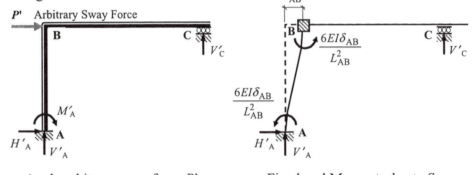

Apply arbitrary sway force P' Fixed-end Moments due to Sway

Ratio of Fixed-end Moments:

$$M_{AB} : M_{BA} = -\frac{6(EI\delta_{AB})}{L_{AB}^2} \; : \; -\frac{6(EI\delta_{AB})}{L_{AB}^2} \; = \; -\frac{6(EI\delta_{AB})}{25} \; : \; -\frac{6(EI\delta_{AB})}{25}$$

$$= \{-0.24 : -0.24\} \times (EI\delta)_{AB}$$

Assume arbitrary fixed-end moments equal to $\{-24.0 : -24.0\} \times (EI\delta)_{AB}/100$

Sway-Only Moment Distribution Table:

Joint	A		B			C
	AB		**BA**	**BC**		**CB**
Distribution Factors	0		0.51	0.49		0
Fixed-end Moments	− 24.0		− 24.0			0
Balance			+ 12.24	+ 11.76		
Carry-over	+ 6.12					
Total	− 17.88		− 11.76	+ 11.76		0

Determine the value of the arbitrary sway force P':

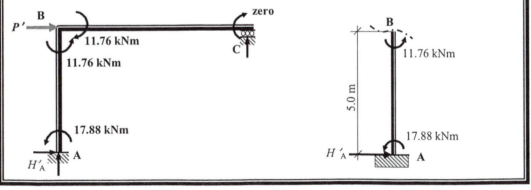

Solution

Topic: Moment Distribution – Rigid-Jointed Frames with Sway
Problem Number: 5.17   **Page No. 4**

+ve $\curvearrowright$ $\Sigma M_B = 0$
$- 11.76 - 17.88 - (H'_A \times 5.0) = 0$ $\therefore H'_A = -5.93 \text{ kN} \leftarrow$

For the complete frame:
+ve $\longrightarrow$ $\Sigma F_x = 0$
$- 5.93 + P' = 0$ $\therefore P' = +5.93 \text{ kN} \longrightarrow$

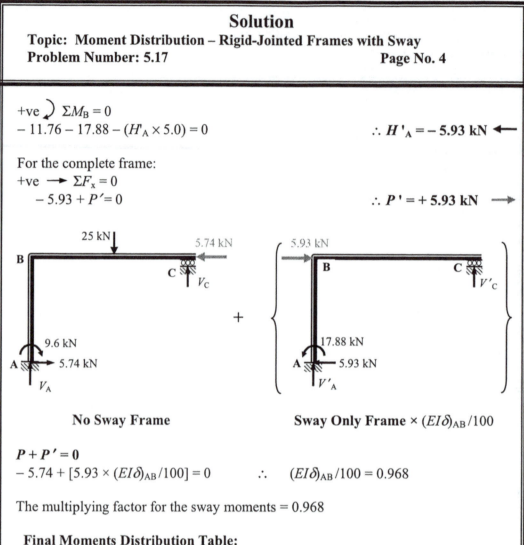

No Sway Frame   **Sway Only Frame** $\times (EI\delta)_{AB}/100$

$P + P' = 0$
$- 5.74 + [5.93 \times (EI\delta)_{AB}/100] = 0$ $\therefore$ $(EI\delta)_{AB}/100 = 0.968$

The multiplying factor for the sway moments = 0.968

Final Moments Distribution Table:

Joint	A		B		C
	AB		BA	BC		CB
No-Sway Moments	+ 9.56		+ 19.12	− 19.12		0
Sway Moments × 0.968	− 17.31		− 11.38	+ 11.38		0
Final Moments (kNm)	**− 7.75**		**+ 7.74***	**− 7.74**		**0**

The horizontal deflection at the rafter level = $\delta_{AB} = \left(\dfrac{0.968 \times 100}{EI} \right) = 96.8/EI$

For horizontal equilibrium at prop level: $- 5.74 + (5.93 \times 0.968) = 0$
Final value of $H_A = + 5.74 - (5.93 \times 0.968) = 0$

*   Since the horizontal reaction at A is equal to zero, the moment at the top of column AB is equal to M_A, i.e. approximately 7.75 kNm.

Solution
Topic: Moment Distribution – Rigid-Jointed Frames with Sway
Problem Number: 5.17   **Page No. 5**

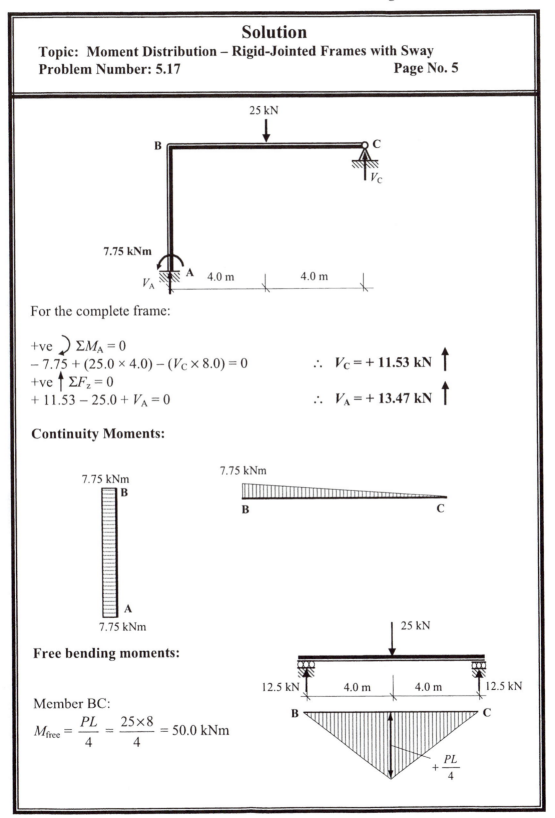

For the complete frame:

+ve $\circlearrowright$ $\Sigma M_A = 0$
$- 7.75 + (25.0 \times 4.0) - (V_C \times 8.0) = 0$ $\therefore$ $V_C = + 11.53$ kN $\uparrow$
+ve $\uparrow$ $\Sigma F_z = 0$
$+ 11.53 - 25.0 + V_A = 0$ $\therefore$ $V_A = + 13.47$ kN $\uparrow$

Continuity Moments:

7.75 kNm
B

A

7.75 kNm

7.75 kNm

B   C

Free bending moments:

Member BC:

$$M_{\text{free}} = \frac{PL}{4} = \frac{25 \times 8}{4} = 50.0 \text{ kNm}$$

25 kN

12.5 kN   4.0 m   4.0 m   12.5 kN

B   C

$+ \dfrac{PL}{4}$

Solution

Topic: Moment Distribution – Rigid-Jointed Frames with Sway

Problem Number: 5.17   **Page No. 6**

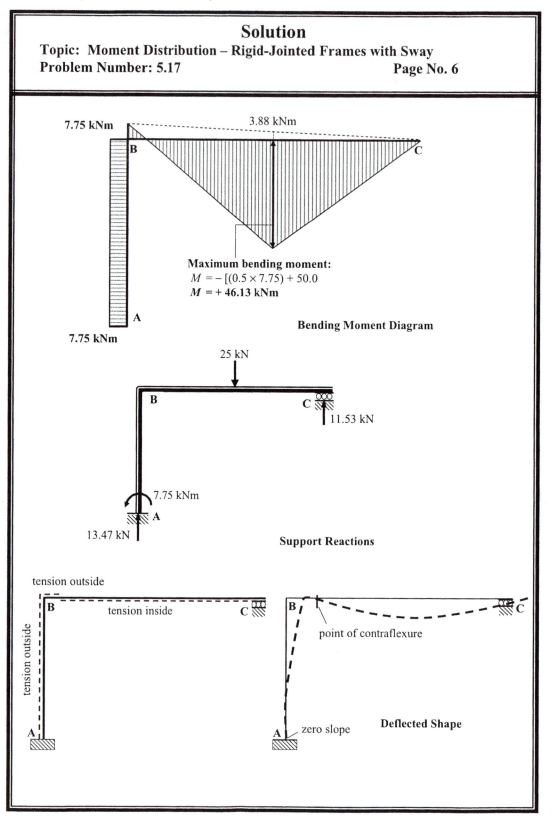

7.75 kNm

3.88 kNm

B   C

Maximum bending moment:
$M = -[(0.5 \times 7.75) + 50.0$
$M = +46.13$ kNm

A

7.75 kNm

Bending Moment Diagram

25 kN

B   C

11.53 kN

7.75 kNm

A

13.47 kN

Support Reactions

tension outside

B   tension inside   C   B   C

point of contraflexure

tension outside

A   A   zero slope

Deflected Shape

Solution

Topic: Moment Distribution – Rigid-Jointed Frames with Sway
Problem Number: 5.18   **Page No. 1**

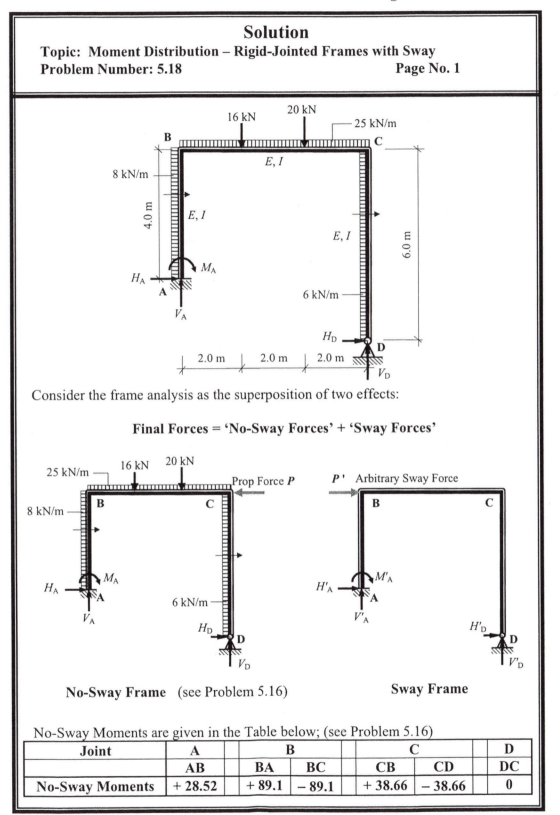

Consider the frame analysis as the superposition of two effects:

Final Forces = 'No-Sway Forces' + 'Sway Forces'

No-Sway Frame   (see Problem 5.16)   **Sway Frame**

No-Sway Moments are given in the Table below; (see Problem 5.16)

Joint	A	B		C		D
	AB	BA	BC	CB	CD	DC
No-Sway Moments	+ 28.52	+ 89.1	− 89.1	+ 38.66	− 38.66	0

Solution
Topic: Moment Distribution – Rigid-Jointed Frames with Sway
Problem Number: 5.18   **Page No. 2**

Determine the value of the prop force *P*:

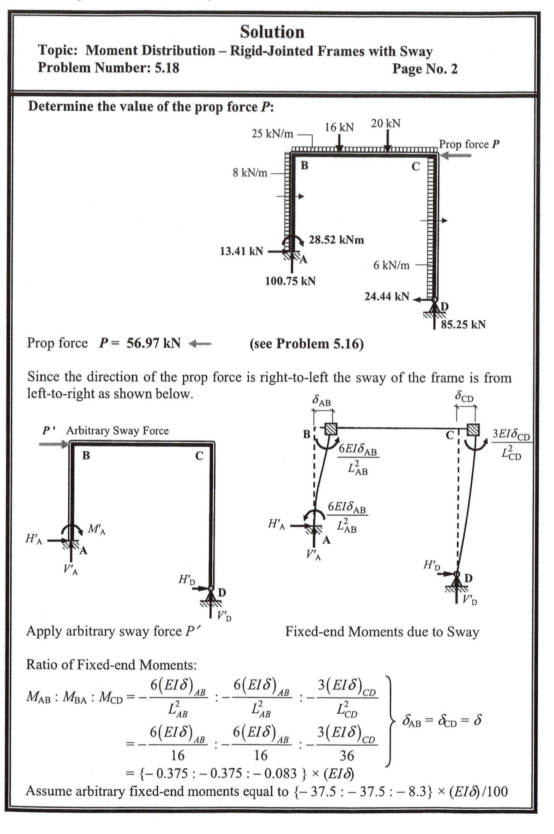

Prop force   *P* = **56.97 kN** ←   **(see Problem 5.16)**

Since the direction of the prop force is right-to-left the sway of the frame is from left-to-right as shown below.

Apply arbitrary sway force *P′*   Fixed-end Moments due to Sway

Ratio of Fixed-end Moments:

$$M_{AB} : M_{BA} : M_{CD} = -\frac{6(EI\delta)_{AB}}{L_{AB}^2} : -\frac{6(EI\delta)_{AB}}{L_{AB}^2} : -\frac{3(EI\delta)_{CD}}{L_{CD}^2} \Bigg\} \; \delta_{AB} = \delta_{CD} = \delta$$

$$= -\frac{6(EI\delta)_{AB}}{16} : -\frac{6(EI\delta)_{AB}}{16} : -\frac{3(EI\delta)_{CD}}{36}$$

$$= \{-0.375 : -0.375 : -0.083\} \times (EI\delta)$$

Assume arbitrary fixed-end moments equal to $\{-37.5 : -37.5 : -8.3\} \times (EI\delta)/100$

Solution

Topic: Moment Distribution – Rigid-Jointed Frames with Sway

Problem Number: 5.18 **Page No. 3**

Sway-Only Moment Distribution Table:

Joint	A		B		C		D
	AB		BA	BC	CB	CD	DC
Distribution Factors	0		0.6	0.4	0.57	0.43	1.0
Fixed-end Moments	−37.5		−37.5	0	0	−8.3	
Balance			+22.5	+15.0	+4.73	+3.57	
Carry-over	+11.25			+2.37	+7.5		
Balance			−1.42	−0.95	−4.28	−3.22	
Carry-over	−0.71			−2.14	−0.48		
Balance			+1.28	+0.86	+0.27	+0.21	
Carry-over	+0.64			+0.14	+0.43		
Balance			−0.08	−0.06	−0.25	−0.18	
Carry-over	−0.04						
Total	−26.36		−15.22	+15.22	+7.92	−7.92	0

Determine the value of the arbitrary sway force P':

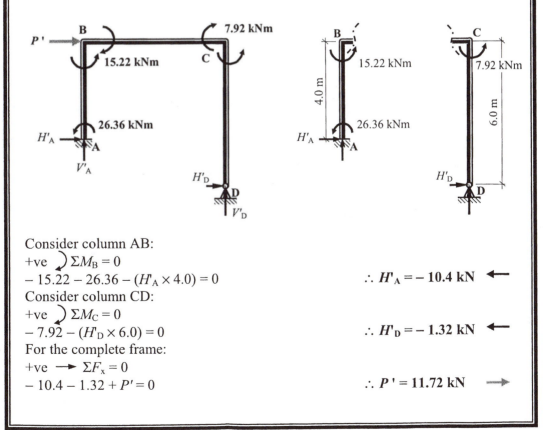

Consider column AB:

+ve $\circlearrowright \Sigma M_B = 0$

$-15.22 - 26.36 - (H'_A \times 4.0) = 0$ $\therefore H'_A = -10.4 \text{ kN} \longleftarrow$

Consider column CD:

+ve $\circlearrowright \Sigma M_C = 0$

$-7.92 - (H'_D \times 6.0) = 0$ $\therefore H'_D = -1.32 \text{ kN} \longleftarrow$

For the complete frame:

+ve $\longrightarrow \Sigma F_x = 0$

$-10.4 - 1.32 + P' = 0$ $\therefore P' = 11.72 \text{ kN} \longrightarrow$

Solution

Topic: Moment Distribution – Rigid-Jointed Frames with Sway

Problem Number: 5.18   **Page No. 4**

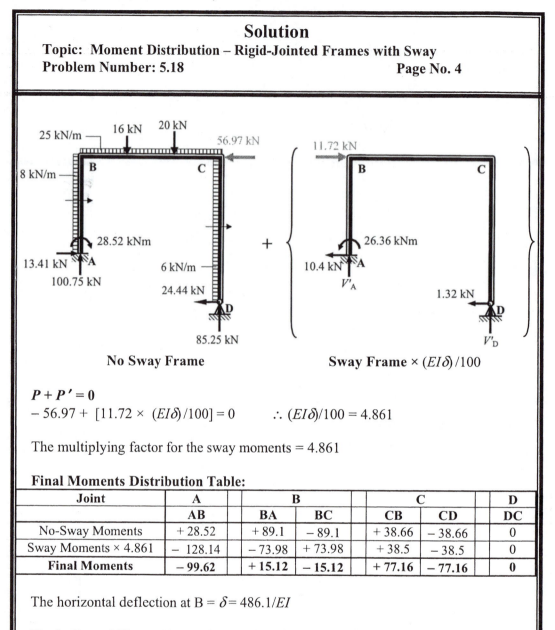

No Sway Frame   **Sway Frame × $(EI\delta)/100$**

$P + P' = 0$

$-56.97 + [11.72 \times (EI\delta)/100] = 0$ $\therefore (EI\delta)/100 = 4.861$

The multiplying factor for the sway moments = 4.861

Final Moments Distribution Table:

Joint	A	B		C		D
	AB	BA	BC	CB	CD	DC
No-Sway Moments	+ 28.52	+ 89.1	– 89.1	+ 38.66	– 38.66	0
Sway Moments × 4.861	– 128.14	– 73.98	+ 73.98	+ 38.5	– 38.5	0
Final Moments	**– 99.62**	**+ 15.12**	**– 15.12**	**+ 77.16**	**– 77.16**	**0**

The horizontal deflection at B = $\delta = 486.1/EI$

Final value of $H_A = + 13.41 - (10.4 \times 4.861) = -37.14$ kN ←

Final value of $H_D = -24.44 - (1.32 \times 4.861) = -30.86$ kN ←

Solution

Topic: Moment Distribution – Rigid-Jointed Frames with Sway
Problem Number: 5.18   **Page No. 5**

For addition of Continuity Moments and Free Bending Moments see Problem 5.16.

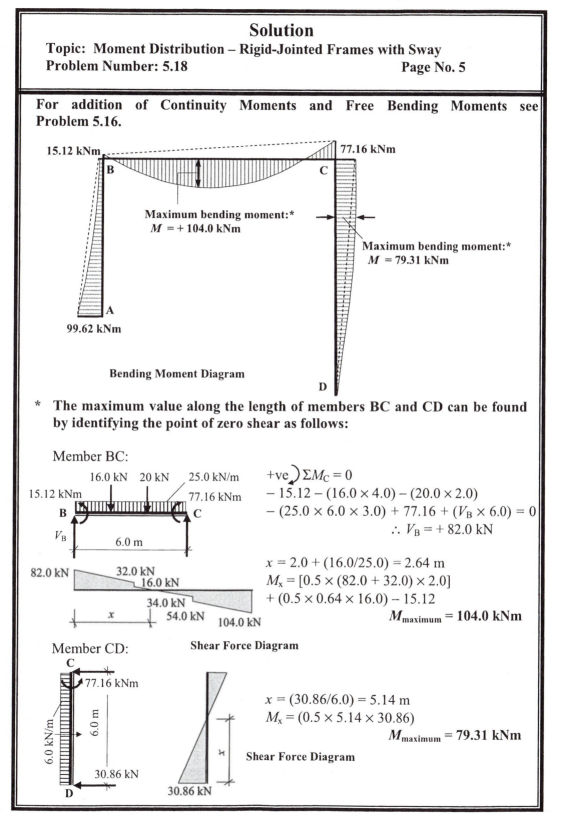

15.12 kNm

77.16 kNm

B

C

Maximum bending moment:*
M = + 104.0 kNm

Maximum bending moment:*
M = 79.31 kNm

A

99.62 kNm

Bending Moment Diagram

D

* **The maximum value along the length of members BC and CD can be found by identifying the point of zero shear as follows:**

Member BC:

16.0 kN   20 kN   25.0 kN/m

15.12 kNm   77.16 kNm

B   **C**

V_B

6.0 m

$+ve \curvearrowright \Sigma M_C = 0$

$-15.12 - (16.0 \times 4.0) - (20.0 \times 2.0)$
$- (25.0 \times 6.0 \times 3.0) + 77.16 + (V_B \times 6.0) = 0$
$\therefore V_B = + 82.0$ kN

82.0 kN   32.0 kN
                16.0 kN

                34.0 kN
            x 54.0 kN   104.0 kN

$x = 2.0 + (16.0/25.0) = 2.64$ m
$M_x = [0.5 \times (82.0 + 32.0) \times 2.0]$
$+ (0.5 \times 0.64 \times 16.0) - 15.12$
$M_{maximum} = 104.0$ kNm

Shear Force Diagram

Member CD:

C

77.16 kNm

6.0 kN/m

6.0 m

30.86 kN

D

x

30.86 kN

$x = (30.86/6.0) = 5.14$ m
$M_x = (0.5 \times 5.14 \times 30.86)$
$M_{maximum} = 79.31$ kNm

Shear Force Diagram

Solution
Topic: Moment Distribution – Rigid-Jointed Frames with Sway
Problem Number: 5.18   **Page No. 6**

Consider the complete frame:

+ve $\circlearrowright \Sigma M_A = 0$

$- 99.62 + (8.0 \times 4.0 \times 2.0) + (25.0 \times 6.0 \times 3.0) + (16.0 \times 2.0) + (20.0 \times 4.0)$
$+ (6.0 \times 6.0 \times 1.0) + (30.86 \times 1.0) - (V_D \times 6.0) = 0$

$\therefore V_D = + 104.0 \text{ kN} \uparrow$

+ve $\uparrow \Sigma F_z = 0$
$+ V_A - (25.0 \times 6.0) - 16.0 - 20.0 + 104.0 = 0$ $\therefore V_A = + 82.0 \text{ kN} \uparrow$

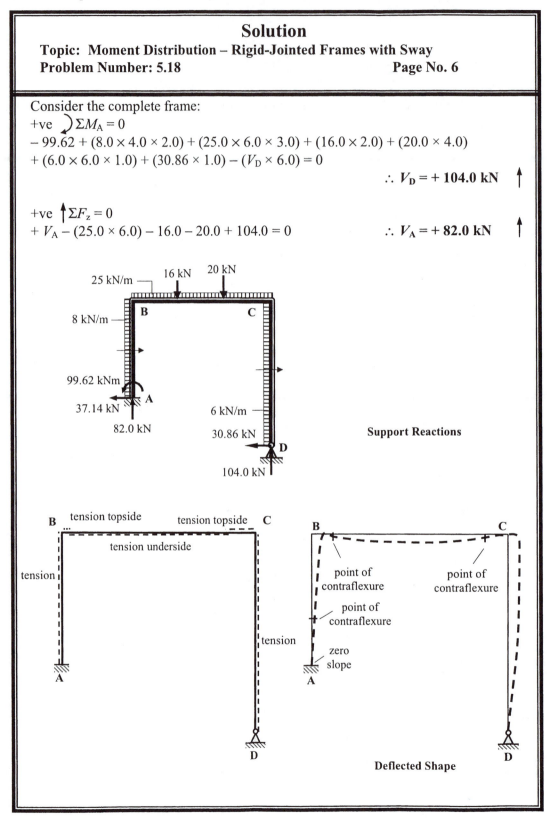

Solution

Topic: Moment Distribution – Rigid-Jointed Frames with Sway
Problem Number: 5.19   Page No. 1

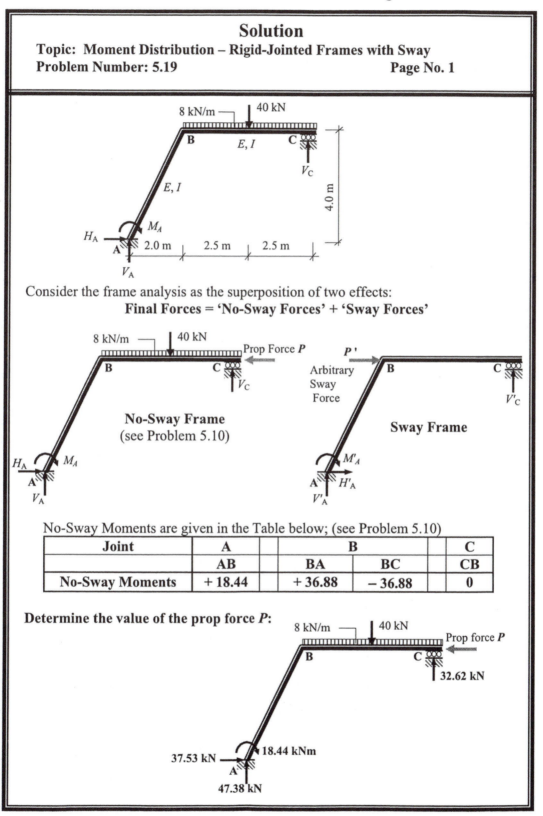

Consider the frame analysis as the superposition of two effects:

Final Forces = 'No-Sway Forces' + 'Sway Forces'

No-Sway Frame
(see Problem 5.10)

Sway Frame

No-Sway Moments are given in the Table below; (see Problem 5.10)

Joint	A		B		C
	AB		BA	BC	CB
No-Sway Moments	+ 18.44		+ 36.88	− 36.88	0

Determine the value of the prop force *P*:

Solution

Topic: Moment Distribution – Rigid-Jointed Frames with Sway

Problem Number: 5.19 **Page No. 2**

Prop force $P = 37.53$ kN ◄―― **(see Problem 5.10)**

Since the direction of the prop force is right-to-left the sway of the frame is from left-to-right as shown below.

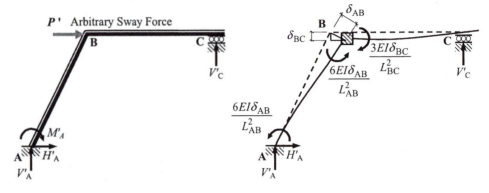

Apply arbitrary sway force P' Fixed-end Moments due to Sway

Displacement triangle:

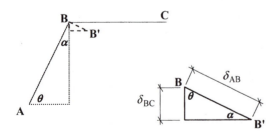

Length of $L_{AB} = \sqrt{2.0^2 + 4.0^2}$

$= 4.472$ m

$\cos\theta = 2.0/4.47 = 0.447$
$\sin\theta = 4.0/4.472 = 0.894$
$\delta_{BC} = (\delta_{AB}\cos\theta)$

Ratio of Fixed-end Moments:

$$M_{AB} : M_{BA} : M_{BC} = -\frac{6(EI\delta_{AB})}{L_{AB}^2} : -\frac{6(EI\delta_{AB})}{L_{AB}^2} : +\frac{3(EI\delta_{BC})}{L_{BC}^2}$$

$$= -\frac{6(EI\delta_{AB})}{4.472^2} : -\frac{6(EI\delta_{AB})}{4.472^2} : +\frac{3(EI\delta_{AB}\times0.447)}{5.0^2}$$

$$= \{-0.30 : -0.30 : +0.05\} \times (EI\delta_{AB})$$

Assume arbitrary fixed-end moments equal to $\{-30.0 : -30.0 : +5.0\} \times (EI\delta_{AB})/100$

Solution

Topic: Moment Distribution – Rigid-Jointed Frames with Sway
Problem Number: 5.19   **Page No. 3**

Sway- Only Moment Distribution Table:

Joint	A		B		C
	AB		BA	BC	CB
Distribution Factors	0		0.59	0.41	0
Fixed-end Moments	− 30.0		− 30.0	+ 5.0	0
Balance			+ 14.75	+ 10.25	
Carry-over	+ 7.38				
Total	− 22.62		− 15.25	+ 15.25	0

Determine the value of the arbitrary sway force *P'*:

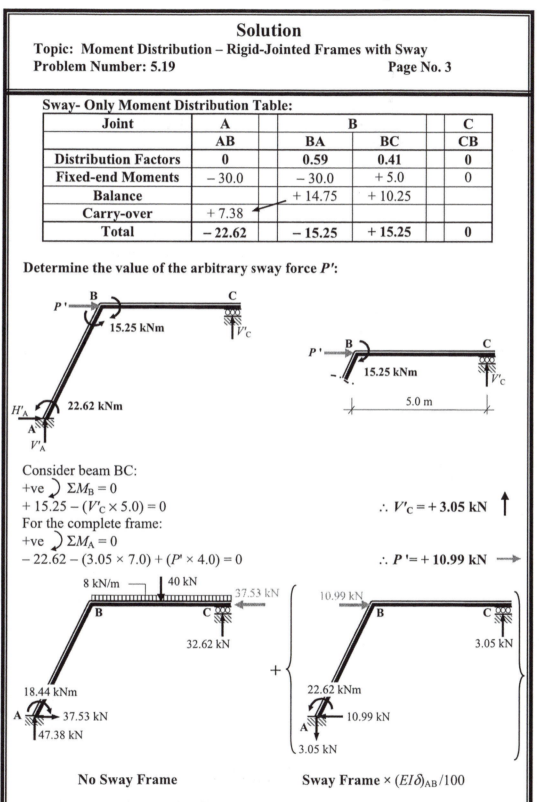

Consider beam BC:
+ve $\curvearrowright$ $\Sigma M_B = 0$
$+ 15.25 - (V'_C \times 5.0) = 0$ $\therefore V'_C = + 3.05$ kN ↑
For the complete frame:
+ve $\curvearrowright$ $\Sigma M_A = 0$
$- 22.62 - (3.05 \times 7.0) + (P' \times 4.0) = 0$ $\therefore P' = + 10.99$ kN →

No Sway Frame   **Sway Frame × $(EI\delta)_{AB}/100$**

Solution

Topic: Moment Distribution – Rigid-Jointed Frames with Sway
Problem Number: 5.19   **Page No. 4**

$P + P' = 0$
$-37.53 + [10.99 \times (EI\delta)_{AB}/100] = 0$ $\therefore (EI\delta)_{AB}/100 = 3.415$
The multiplying factor for the sway moments = 3.415

Final Moments Distribution Table:

Joint	A	B			C
	AB	**BA**	**BC**		**CB**
No-Sway Moments	+ 18.44	+ 36.88	− 36.88		0
Sway Moments × 3.415	− 77.25	− 52.08	+ 52.08		0
Final Moments	**− 58.81**	**− 15.20**	**+ 15.20**		**0**

The horizontal deflection at B = $(\delta_{AB} \sin\theta) = (341.5/EI) \times 0.894 = 305.3/EI$
The vertical deflection at B = $(\delta_{AB} \cos\theta) = (341.5/EI) \times 0.447 = 152.7/EI$

Final value of $H_A = + 37.53 - (10.99 \times 3.415) = \mathbf{0}$
Final value of $V_C = + 32.62 + (3.05 \times 3.415) = \mathbf{+ 43.0\ kN}$ ↑
Final value of $V_A = + 47.38 - (3.05 \times 3.415) = \mathbf{+ 37.0\ kN}$ ↑

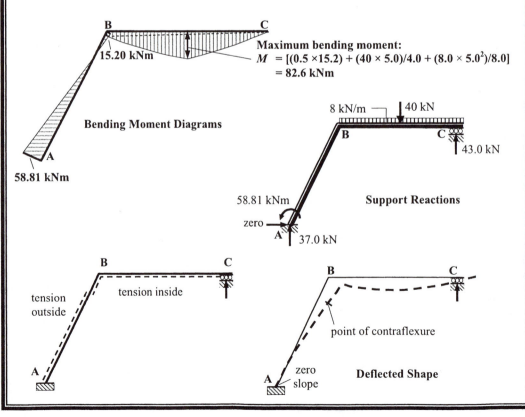

Maximum bending moment:
$M = [(0.5 \times 15.2) + (40 \times 5.0)/4.0 + (8.0 \times 5.0^2)/8.0]$
$= \mathbf{82.6\ kNm}$

Bending Moment Diagrams

15.20 kNm

58.81 kNm

8 kN/m   40 kN

43.0 kN

58.81 kNm

Support Reactions

zero

37.0 kN

tension
outside

tension inside

point of contraflexure

zero
slope

Deflected Shape

Solution

Topic: Moment Distribution – Rigid-Jointed Frames with Sway
Problem Number: 5.20 **Page No. 1**

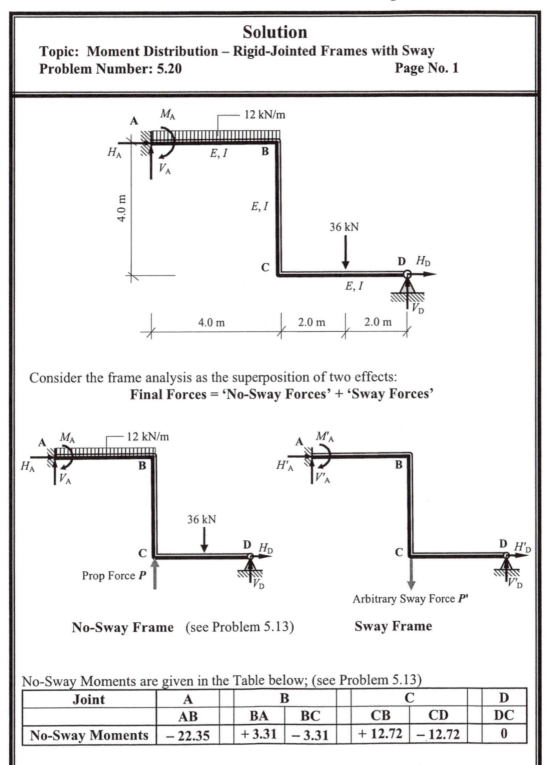

Consider the frame analysis as the superposition of two effects:

Final Forces = 'No-Sway Forces' + 'Sway Forces'

No-Sway Frame (see Problem 5.13) **Sway Frame**

No-Sway Moments are given in the Table below; (see Problem 5.13)

Joint	A	B		C		D
	AB	BA	BC	CB	CD	DC
No-Sway Moments	− 22.35	+ 3.31	− 3.31	+ 12.72	− 12.72	0

Solution

Topic: Moment Distribution – Rigid-Jointed Frames with Sway
Problem Number: 5.20　　　　　　　　　　　　　　　　**Page No. 2**

Determine the value of the prop force P:

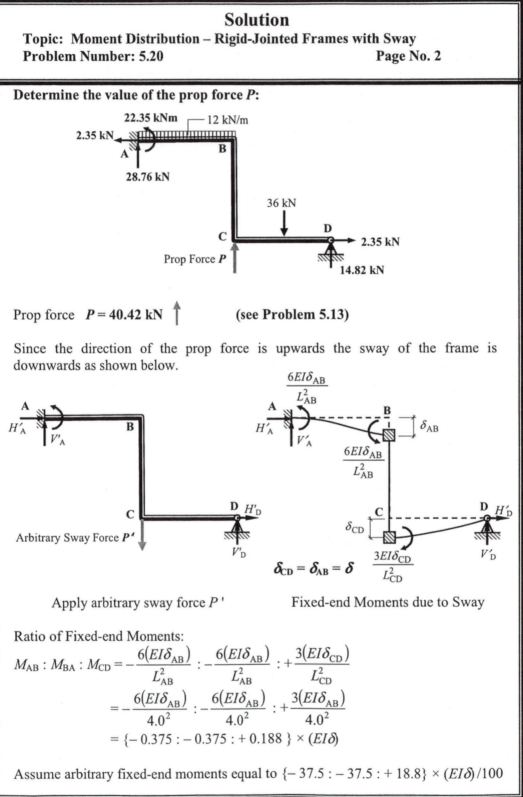

Prop force $P = 40.42$ kN $\uparrow$ 　　　　**(see Problem 5.13)**

Since the direction of the prop force is upwards the sway of the frame is downwards as shown below.

$$\delta_{CD} = \delta_{AB} = \delta$$

Apply arbitrary sway force P'　　　　　　Fixed-end Moments due to Sway

Ratio of Fixed-end Moments:

$$M_{AB} : M_{BA} : M_{CD} = -\frac{6(EI\delta_{AB})}{L_{AB}^2} : -\frac{6(EI\delta_{AB})}{L_{AB}^2} : +\frac{3(EI\delta_{CD})}{L_{CD}^2}$$

$$= -\frac{6(EI\delta_{AB})}{4.0^2} : -\frac{6(EI\delta_{AB})}{4.0^2} : +\frac{3(EI\delta_{AB})}{4.0^2}$$

$$= \{-0.375 : -0.375 : +0.188\} \times (EI\delta)$$

Assume arbitrary fixed-end moments equal to $\{-37.5 : -37.5 : +18.8\} \times (EI\delta)/100$

Solution
Topic: Moment Distribution – Rigid-Jointed Frames with Sway
Problem Number: 5.20   **Page No. 3**

Sway-Only Moment Distribution Table:

Joint	A	B			C		D
	AB	BA	BC		CB	CD	DC
Distribution Factors	1.0	0.5	0.5		0.57	0.43	1.0
Fixed-end Moments	− 37.5	− 37.5				+ 18.8	0
Balance		+ 18.75	+ 18.75		− 10.72	− 8.08	
Carry-over	+ 9.4		− 5.36		+ 9.38		
Balance		+ 2.68	+ 2.68		− 5.35	− 4.03	
Carry-over	+ 1.34		− 2.68		+ 1.34		
Balance		+ 1.34	+ 1.34		− 0.76	− 0.58	
Carry-over	+ 0.29		− 0.38		+ 0.67		
Balance		+ 0.19	+ 0.19		− 0.38	− 0.29	
Carry-over	+ 0.2						
Total	− 26.27	− 14.54	+ 14.54		− 5.82	+ 5.82	

Determine the value of the arbitrary sway force P'

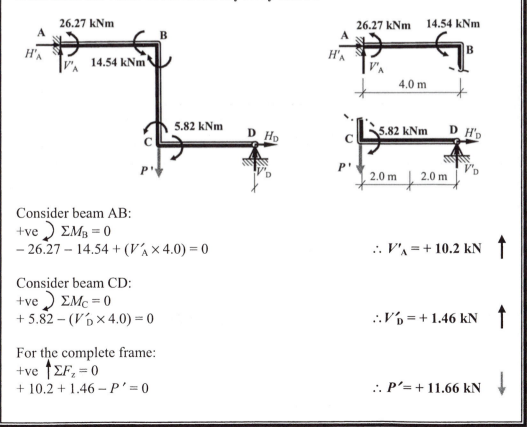

Consider beam AB:
+ve $\circlearrowright$ $\Sigma M_B = 0$
$- 26.27 - 14.54 + (V'_A \times 4.0) = 0$ $\therefore V'_A = + 10.2 \text{ kN}$ $\uparrow$

Consider beam CD:
+ve $\circlearrowright$ $\Sigma M_C = 0$
$+ 5.82 - (V'_D \times 4.0) = 0$ $\therefore V'_D = + 1.46 \text{ kN}$ $\uparrow$

For the complete frame:
+ve $\uparrow$ $\Sigma F_z = 0$
$+ 10.2 + 1.46 - P' = 0$ $\therefore P' = + 11.66 \text{ kN}$ $\downarrow$

Solution
Topic: Moment Distribution – Rigid-Jointed Frames with Sway
Problem Number: 5.20   **Page No. 4**

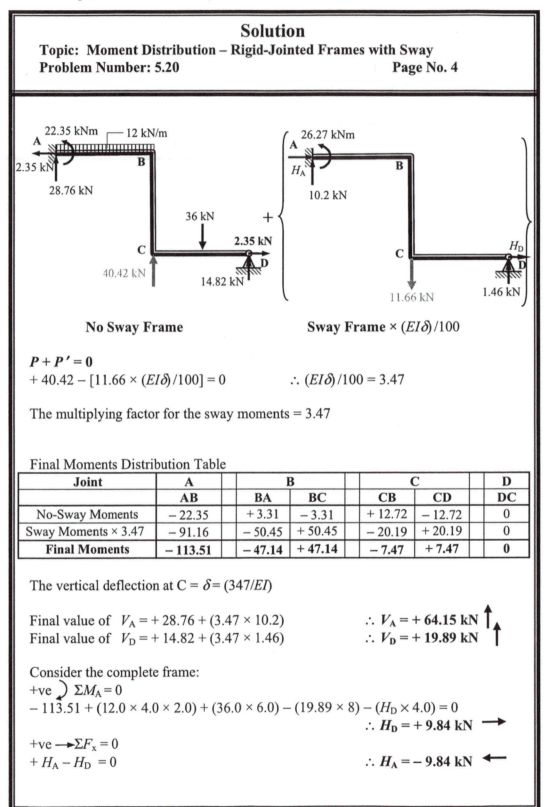

No Sway Frame   **Sway Frame × $(EI\delta)/100$**

$P + P' = 0$
$+ 40.42 - [11.66 \times (EI\delta)/100] = 0$ $\therefore (EI\delta)/100 = 3.47$

The multiplying factor for the sway moments = 3.47

Final Moments Distribution Table

Joint	A	B		C		D
	AB	BA	BC	CB	CD	DC
No-Sway Moments	− 22.35	+ 3.31	− 3.31	+ 12.72	− 12.72	0
Sway Moments × 3.47	− 91.16	− 50.45	+ 50.45	− 20.19	+ 20.19	0
Final Moments	**− 113.51**	**− 47.14**	**+ 47.14**	**− 7.47**	**+ 7.47**	**0**

The vertical deflection at C = $\delta = (347/EI)$

Final value of $V_A = + 28.76 + (3.47 \times 10.2)$ $\therefore V_A = + 64.15$ kN
Final value of $V_D = + 14.82 + (3.47 \times 1.46)$ $\therefore V_D = + 19.89$ kN

Consider the complete frame:
+ve $\circlearrowright$ $\Sigma M_A = 0$
$- 113.51 + (12.0 \times 4.0 \times 2.0) + (36.0 \times 6.0) - (19.89 \times 8) - (H_D \times 4.0) = 0$
$\therefore H_D = + 9.84$ kN $\longrightarrow$

+ve $\longrightarrow \Sigma F_x = 0$
$+ H_A - H_D = 0$ $\therefore H_A = - 9.84$ kN $\longleftarrow$

Solution

Topic: Moment Distribution – Rigid-Jointed Frames with Sway
Problem Number: 5.20   **Page No. 4**

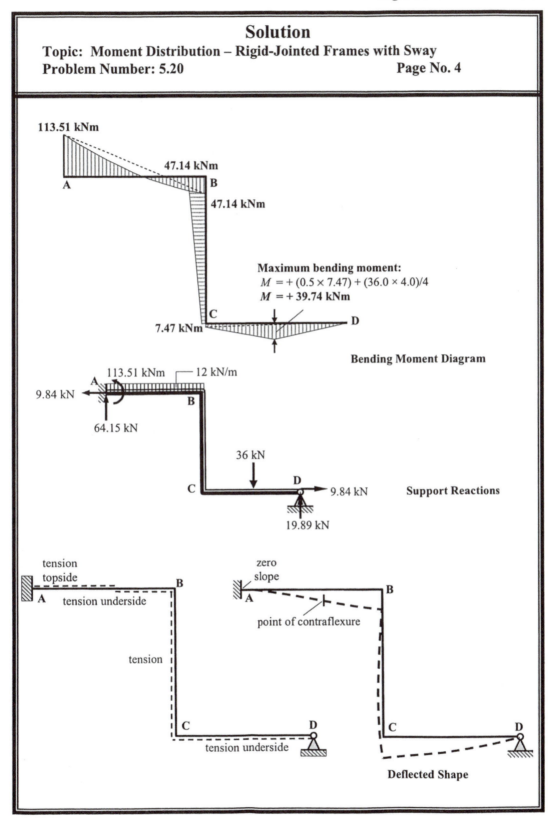

113.51 kNm

47.14 kNm

A   B

47.14 kNm

Maximum bending moment:
$M = +(0.5 \times 7.47) + (36.0 \times 4.0)/4$
$M = +39.74$ **kNm**

C   D

7.47 kNm

Bending Moment Diagram

113.51 kNm   12 kN/m

A

9.84 kN

B

64.15 kN

36 kN

C   D

9.84 kN   **Support Reactions**

19.89 kN

tension
topside

A   tension underside   B

zero
slope

A   B

point of contraflexure

tension

C   D

tension underside

C   D

Deflected Shape

Solution

Topic: Moment Distribution – Rigid-Jointed Frames with Sway
Problem Number: 5.21   **Page No. 1**

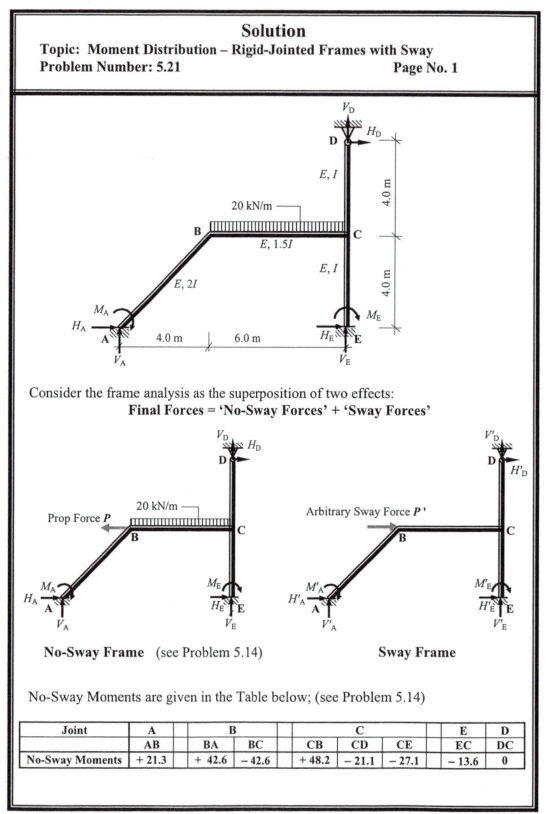

Consider the frame analysis as the superposition of two effects:
Final Forces = 'No-Sway Forces' + 'Sway Forces'

No-Sway Frame   (see Problem 5.14)   **Sway Frame**

No-Sway Moments are given in the Table below; (see Problem 5.14)

Joint	A	B		C			E	D
	AB	BA	BC	CB	CD	CE	EC	DC
No-Sway Moments	+ 21.3	+ 42.6	− 42.6	+ 48.2	− 21.1	− 27.1	− 13.6	0

Solution
Topic: Moment Distribution – Rigid-Jointed Frames with Sway
Problem Number: 5.21   **Page No. 2**

Determine the value of the prop force P:

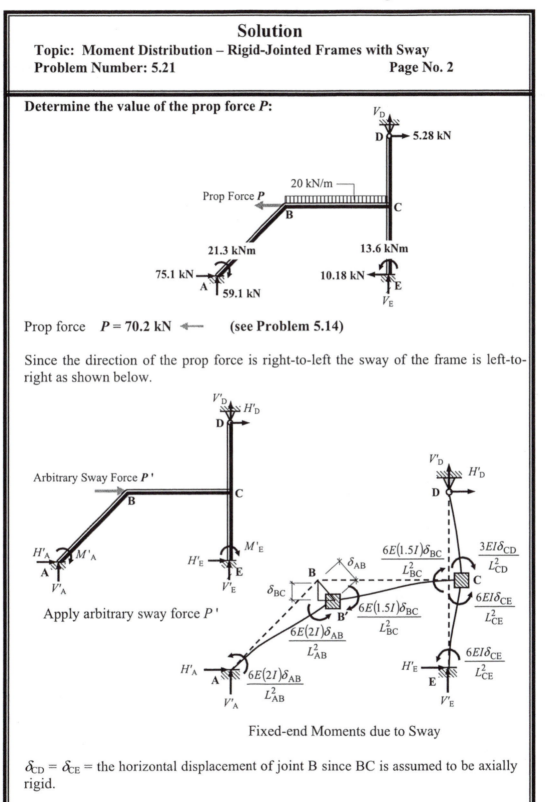

Prop force $P = 70.2$ kN $\longleftarrow$ **(see Problem 5.14)**

Since the direction of the prop force is right-to-left the sway of the frame is left-to-right as shown below.

Apply arbitrary sway force P'

Fixed-end Moments due to Sway

$\delta_{CD} = \delta_{CE}$ = the horizontal displacement of joint B since BC is assumed to be axially rigid.

Solution

Topic: Moment Distribution – Rigid-Jointed Frames with Sway
Problem Number: 5.21   **Page No. 3**

Displacement triangle:

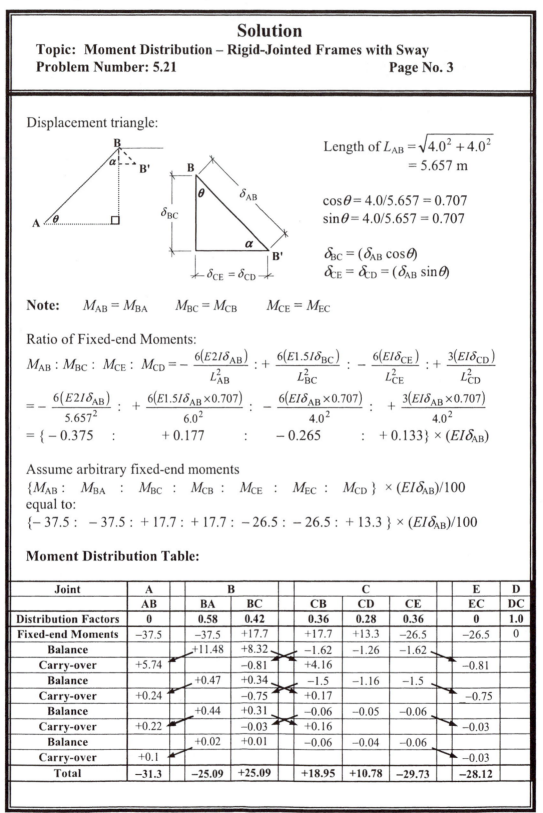

Length of $L_{AB} = \sqrt{4.0^2 + 4.0^2}$
                    $= 5.657$ m

$\cos\theta = 4.0/5.657 = 0.707$
$\sin\theta = 4.0/5.657 = 0.707$

$\delta_{BC} = (\delta_{AB}\cos\theta)$
$\delta_{CE} = \delta_{CD} = (\delta_{AB}\sin\theta)$

Note: $M_{AB} = M_{BA}$ $M_{BC} = M_{CB}$ $M_{CE} = M_{EC}$

Ratio of Fixed-end Moments:

$$M_{AB} : M_{BC} : M_{CE} : M_{CD} = -\frac{6(E2I\delta_{AB})}{L_{AB}^2} : +\frac{6(E1.5I\delta_{BC})}{L_{BC}^2} : -\frac{6(EI\delta_{CE})}{L_{CE}^2} : +\frac{3(EI\delta_{CD})}{L_{CD}^2}$$

$$= -\frac{6(E2I\delta_{AB})}{5.657^2} : +\frac{6(E1.5I\delta_{AB}\times0.707)}{6.0^2} : -\frac{6(EI\delta_{AB}\times0.707)}{4.0^2} : +\frac{3(EI\delta_{AB}\times0.707)}{4.0^2}$$

$$= \{-0.375 : +0.177 : -0.265 : +0.133\} \times (EI\delta_{AB})$$

Assume arbitrary fixed-end moments
$\{M_{AB} :   M_{BA} :   M_{BC} : M_{CB} :   M_{CE} :   M_{EC} :   M_{CD}\} \times (EI\delta_{AB})/100$
equal to:
$\{-37.5 : -37.5 : +17.7 : +17.7 : -26.5 : -26.5 : +13.3\} \times (EI\delta_{AB})/100$

Moment Distribution Table:

Joint	A	B		C			E	D
	AB	BA	BC	CB	CD	CE	EC	DC
Distribution Factors	0	0.58	0.42	0.36	0.28	0.36	0	1.0
Fixed-end Moments	−37.5	−37.5	+17.7	+17.7	+13.3	−26.5	−26.5	0
Balance		+11.48	+8.32	−1.62	−1.26	−1.62		
Carry-over	+5.74		−0.81	+4.16			−0.81	
Balance		+0.47	+0.34	−1.5	−1.16	−1.5		
Carry-over	+0.24		−0.75	+0.17			−0.75	
Balance		+0.44	+0.31	−0.06	−0.05	−0.06		
Carry-over	+0.22		−0.03	+0.16			−0.03	
Balance		+0.02	+0.01	−0.06	−0.04	−0.06		
Carry-over	+0.1						−0.03	
Total	−31.3	−25.09	+25.09	+18.95	+10.78	−29.73	−28.12	

Solution
Topic: Moment Distribution – Rigid-Jointed Frames with Sway
Problem Number: 5.21   **Page No. 4**

Determine the value of the arbitrary sway force P'

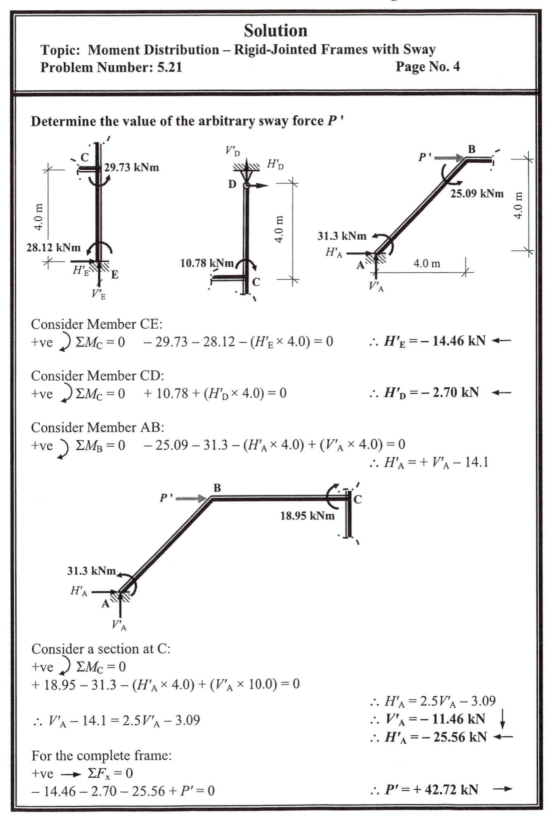

Consider Member CE:

+ve $\circlearrowright \Sigma M_C = 0$ $-29.73 - 28.12 - (H'_E \times 4.0) = 0$ $\therefore H'_E = -14.46 \text{ kN} \leftarrow$

Consider Member CD:

+ve $\circlearrowright \Sigma M_C = 0$ $+10.78 + (H'_D \times 4.0) = 0$ $\therefore H'_D = -2.70 \text{ kN} \leftarrow$

Consider Member AB:

+ve $\circlearrowright \Sigma M_B = 0$ $-25.09 - 31.3 - (H'_A \times 4.0) + (V'_A \times 4.0) = 0$

$$\therefore H'_A = +V'_A - 14.1$$

Consider a section at C:

+ve $\circlearrowright \Sigma M_C = 0$
$+18.95 - 31.3 - (H'_A \times 4.0) + (V'_A \times 10.0) = 0$

$\therefore H'_A = 2.5V'_A - 3.09$

$\therefore V'_A - 14.1 = 2.5V'_A - 3.09$

$\therefore V'_A = -11.46 \text{ kN} \downarrow$

$\therefore H'_A = -25.56 \text{ kN} \leftarrow$

For the complete frame:

+ve $\longrightarrow \Sigma F_x = 0$
$-14.46 - 2.70 - 25.56 + P' = 0$ $\therefore P' = +42.72 \text{ kN} \longrightarrow$

Solution

Topic: Moment Distribution – Rigid-Jointed Frames with Sway
Problem Number: 5.21   **Page No. 5**

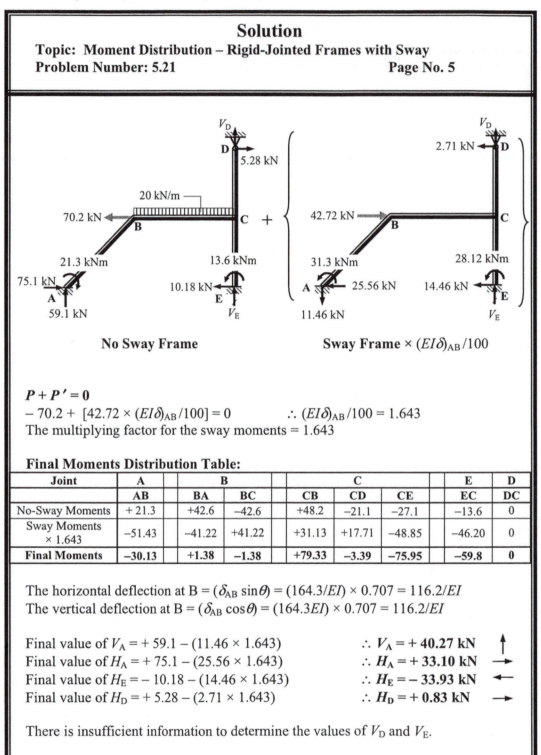

No Sway Frame   **Sway Frame** $\times (EI\delta)_{AB}/100$

$P + P' = 0$

$-70.2 + [42.72 \times (EI\delta)_{AB}/100] = 0$ $\therefore (EI\delta)_{AB}/100 = 1.643$

The multiplying factor for the sway moments = 1.643

Final Moments Distribution Table:

Joint	A	B		C			E	D
	AB	BA	BC	CB	CD	CE	EC	DC
No-Sway Moments	+ 21.3	+42.6	−42.6	+48.2	−21.1	−27.1	−13.6	0
Sway Moments × 1.643	−51.43	−41.22	+41.22	+31.13	+17.71	−48.85	−46.20	0
Final Moments	−30.13	+1.38	−1.38	+79.33	−3.39	−75.95	−59.8	0

The horizontal deflection at B = $(\delta_{AB} \sin\theta) = (164.3/EI) \times 0.707 = 116.2/EI$
The vertical deflection at B = $(\delta_{AB} \cos\theta) = (164.3EI) \times 0.707 = 116.2/EI$

Final value of $V_A = + 59.1 - (11.46 \times 1.643)$ $\therefore V_A = + 40.27$ kN ↑
Final value of $H_A = + 75.1 - (25.56 \times 1.643)$ $\therefore H_A = + 33.10$ kN →
Final value of $H_E = - 10.18 - (14.46 \times 1.643)$ $\therefore H_E = - 33.93$ kN ←
Final value of $H_D = + 5.28 - (2.71 \times 1.643)$ $\therefore H_D = + 0.83$ kN →

There is insufficient information to determine the values of V_D and V_E.

Solution

Topic: Moment Distribution – Rigid-Jointed Frames with Sway
Problem Number: 5.21   **Page No. 6**

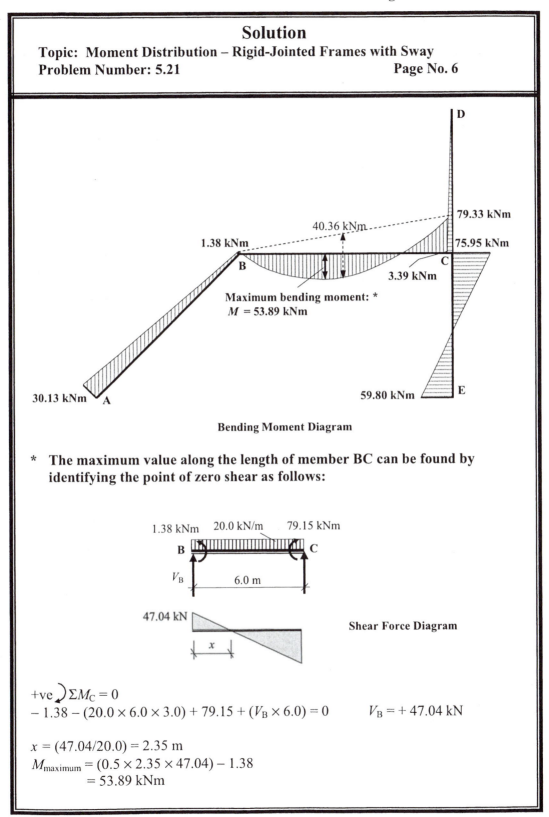

Bending Moment Diagram

* The maximum value along the length of member BC can be found by identifying the point of zero shear as follows:

Shear Force Diagram

$+ve \left.\right\rangle \Sigma M_C = 0$

$-1.38 - (20.0 \times 6.0 \times 3.0) + 79.15 + (V_B \times 6.0) = 0 \qquad V_B = +47.04 \text{ kN}$

$x = (47.04/20.0) = 2.35 \text{ m}$

$M_{maximum} = (0.5 \times 2.35 \times 47.04) - 1.38$

$\qquad\qquad = 53.89 \text{ kNm}$

Solution

Topic: Moment Distribution – Rigid-Jointed Frames with Sway
Problem Number: 5.21   **Page No. 7**

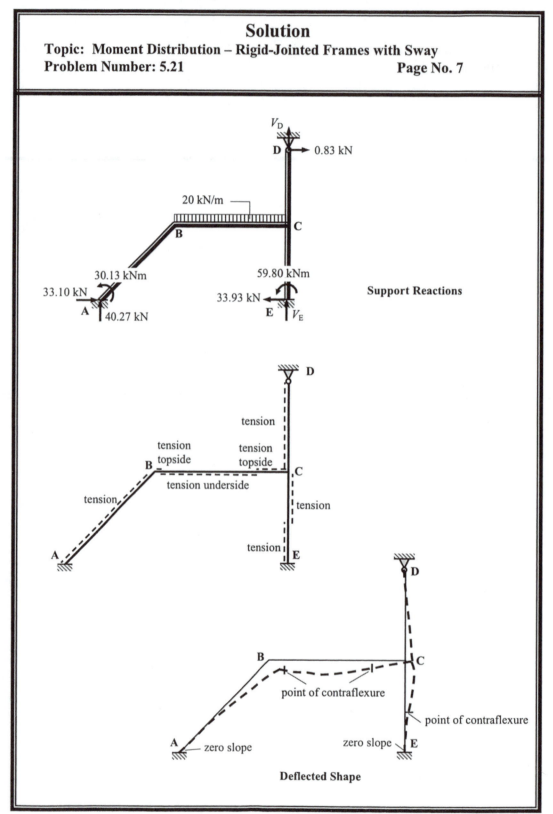

Support Reactions

Deflected Shape

Solution

Topic: Moment Distribution – Rigid-Jointed Frames with Sway
Problem Number: 5.22   **Page No. 1**

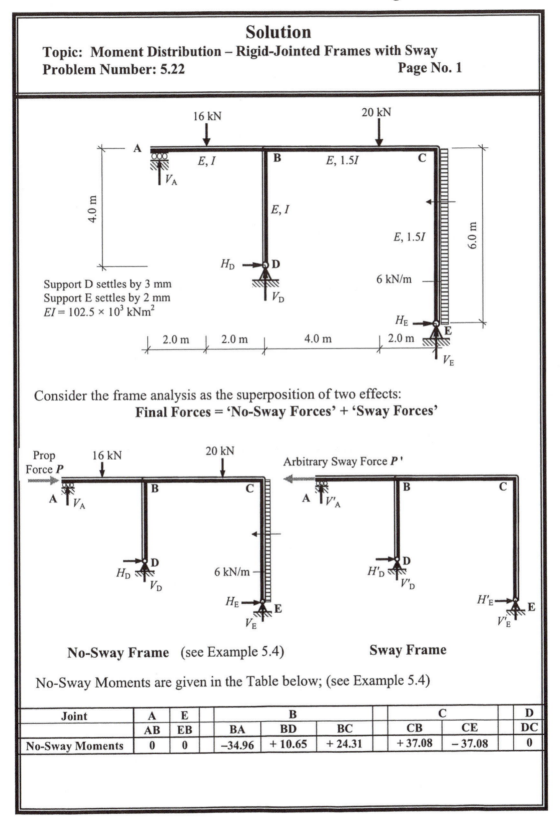

Consider the frame analysis as the superposition of two effects:
Final Forces = 'No-Sway Forces' + 'Sway Forces'

No-Sway Frame   (see Example 5.4)   **Sway Frame**

No-Sway Moments are given in the Table below; (see Example 5.4)

Joint	A	E		B				C		D
	AB	EB		BA	BD	BC		CB	CE	DC
No-Sway Moments	0	0		−34.96	+ 10.65	+ 24.31		+ 37.08	− 37.08	0

Solution

Topic: Moment Distribution – Rigid-Jointed Frames with Sway
Problem Number: 5.22   **Page No. 2**

Determine the value of the prop force P:

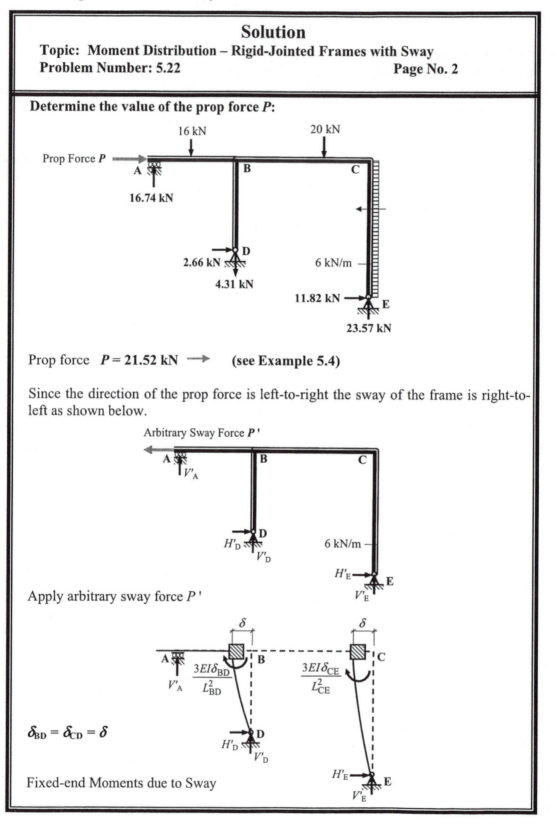

Prop force $P = 21.52$ kN $\longrightarrow$ **(see Example 5.4)**

Since the direction of the prop force is left-to-right the sway of the frame is right-to-left as shown below.

Apply arbitrary sway force P'

$\delta_{BD} = \delta_{CD} = \delta$

Fixed-end Moments due to Sway

Solution

Topic: Moment Distribution – Rigid-Jointed Frames with Sway

Problem Number: 5.22   **Page No. 3**

Ratio of Fixed-end Moments:

$$M_{BD} : M_{CE} = + \frac{3(EI\delta_{BD})}{L_{BD}^2} : + \frac{3(EI\delta_{CE})}{L_{CE}^2} = + \frac{3(EI\delta)}{4.0^2} : + \frac{3(E1.5I\delta)}{6.0^2}$$

$$= \{+ 0.188 : + 0.125\} \times (EI\delta)$$

Assume arbitrary fixed-end moments

$$\{M_{BD} : M_{CE}\} \times (EI\delta)/100 = \{+ 188 : + 125\} \times (EI\delta)/1000$$

Moment Distribution Table:

Joint	A	E		B				C			D
	AB	EB		BA	BD	BC		CB	CE		DC
Distribution Factors	1.0	1.0		0.3	0.3	0.4		0.57	0.43		1.0
Fixed-end Moments					+ 188.0				+125.0		
Balance				− 56.4	− 56.4	− 75.2		− 71.25	− 53.75		
Carry-over						−35.63		− 37.6			
Balance				+ 10.69	+ 10.69	+ 14.25		+ 21.43	+16.17		
Carry-over						+ 10.72		+ 7.13			
Balance				− 3.21	− 3.21	− 4.29		− 4.06	− 3.06		
Carry-over						− 2.03		− 2.14			
Balance				+ 0.61	+ 0.61	+ 0.81		+ 1.22	+ 0.92		
Total	0	0		− 48.32	+ 139.68	− 91.36		− 85.28	+ 85.28		0

Determine the value of the arbitrary sway force *P'*

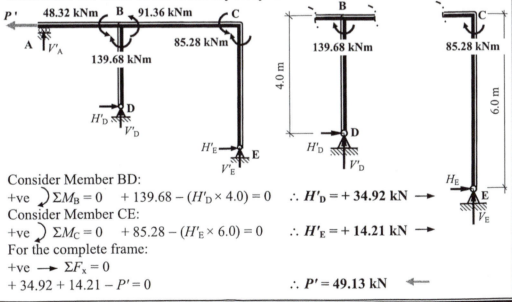

Consider Member BD:

+ve $\circlearrowright \Sigma M_B = 0$ $+ 139.68 − (H'_D \times 4.0) = 0$ $\therefore H'_D = + 34.92$ kN $\longrightarrow$

Consider Member CE:

+ve $\circlearrowright \Sigma M_C = 0$ $+ 85.28 − (H'_E \times 6.0) = 0$ $\therefore H'_E = + 14.21$ kN $\longrightarrow$

For the complete frame:

+ve $\longrightarrow \Sigma F_x = 0$

$+ 34.92 + 14.21 − P' = 0$ $\therefore P' = 49.13$ kN $\longleftarrow$

Solution
Topic: Moment Distribution – Rigid-Jointed Frames with Sway
Problem Number: 5.22　　　　　　　　　　　　　　　　　**Page No. 4**

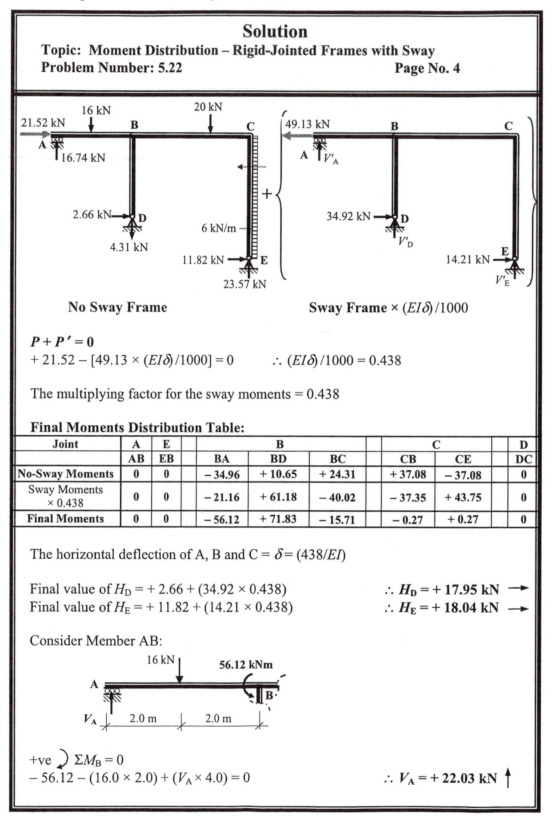

No Sway Frame　　　　　　　　　　　**Sway Frame × $(EI\delta)/1000$**

$P + P' = 0$
$+ 21.52 - [49.13 \times (EI\delta)/1000] = 0$ 　　　 $\therefore (EI\delta)/1000 = 0.438$

The multiplying factor for the sway moments = 0.438

Final Moments Distribution Table:

Joint	A	E		B			C		D
	AB	EB		BA	BD	BC	CB	CE	DC
No-Sway Moments	0	0		− 34.96	+ 10.65	+ 24.31	+ 37.08	− 37.08	0
Sway Moments × 0.438	0	0		− 21.16	+ 61.18	− 40.02	− 37.35	+ 43.75	0
Final Moments	0	0		− 56.12	+ 71.83	− 15.71	− 0.27	+ 0.27	0

The horizontal deflection of A, B and C = $\delta = (438/EI)$

Final value of $H_D = + 2.66 + (34.92 \times 0.438)$ 　　　　$\therefore H_D = + 17.95$ kN $\rightarrow$
Final value of $H_E = + 11.82 + (14.21 \times 0.438)$ 　　　　$\therefore H_E = + 18.04$ kN $\rightarrow$

Consider Member AB:

$+ve \;\curvearrowright \; \Sigma M_B = 0$
$- 56.12 - (16.0 \times 2.0) + (V_A \times 4.0) = 0$ 　　　　$\therefore V_A = + 22.03$ kN $\uparrow$

Solution
Topic: Moment Distribution – Rigid-Jointed Frames with Sway
Problem Number: 5.22   **Page No. 5**

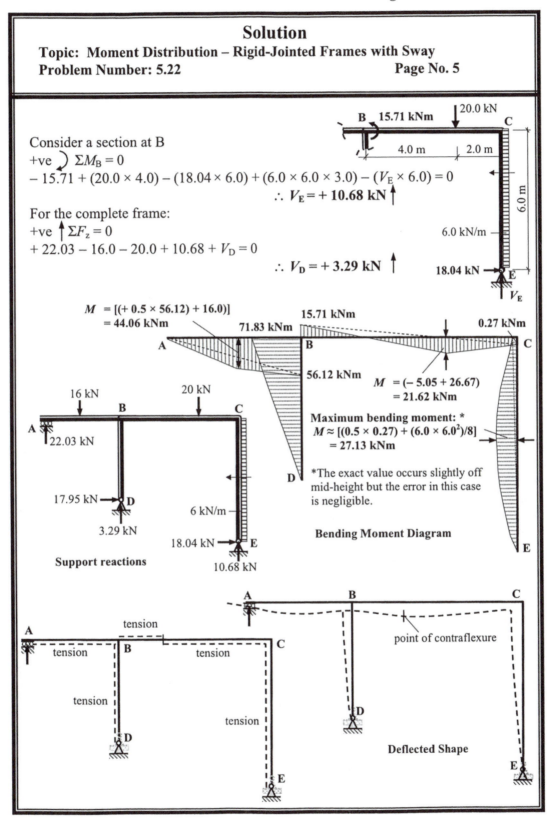

Consider a section at B

+ve $\curvearrowright$ $\Sigma M_B = 0$

$-15.71 + (20.0 \times 4.0) - (18.04 \times 6.0) + (6.0 \times 6.0 \times 3.0) - (V_E \times 6.0) = 0$

$\therefore V_E = +10.68$ kN ↑

For the complete frame:

+ve ↑ $\Sigma F_z = 0$

$+22.03 - 16.0 - 20.0 + 10.68 + V_D = 0$

$\therefore V_D = +3.29$ kN ↑

B   15.71 kNm   20.0 kN
                                    C
                4.0 m   2.0 m

                                    6.0 m

                6.0 kN/m

                18.04 kN → E
                                    V_E

$M = [(+0.5 \times 56.12) + 16.0)]$
$= 44.06$ kNm

15.71 kNm

71.83 kNm   0.27 kNm

A ⟩⟩⟩⟩⟩ B   C

56.12 kNm

$M = (-5.05 + 26.67)$
$= 21.62$ kNm

Maximum bending moment: *
$M \approx [(0.5 \times 0.27) + (6.0 \times 6.0^2)/8]$
$= 27.13$ kNm

D

*The exact value occurs slightly off mid-height but the error in this case is negligible.

Bending Moment Diagram

E

16 kN   20 kN

A ⟩⟩   B   C
  22.03 kN

17.95 kN → D

3.29 kN   6 kN/m

                18.04 kN → E
                        10.68 kN

Support reactions

A   B   C

                                point of contraflexure

tension
A ⟩⟩
  tension   B   C
                 tension

tension   D

tension

                D

                E

Deflected Shape

E

6.   Buckling Instability

6.1   *Introduction*

Structural elements which are subjected to tensile forces are inherently stable and will generally fail when the stress in the cross-section exceeds the ultimate strength of the material. In the case of elements subjected to compressive forces, secondary bending effects, (e.g. example caused by imperfections within materials and/or fabrication processes, inaccurate positioning of loads or asymmetry of the cross-section), can induce premature failure either in a part of the cross-section (local buckling), such as the web/outstand flange of an **I** section, or of the member as a whole (flexural buckling). There are numerous modes of buckling which can occur e.g.

- local buckling,
- distorsional buckling,
- flexural buckling,
- lateral torsional buckling,
- torsional buckling,
- torsional-flexural buckling,
- web buckling and
- shear buckling of plates,

as shown in Figure 6.2.

The design of most compressive members is governed by their flexural buckling resistance, i.e. the maximum compressive load which can be carried before failure occurs by excessive deflection in the plane of greatest slenderness.

Typically this occurs in columns in building frames and in trussed frameworks as shown in Figure 6.1.

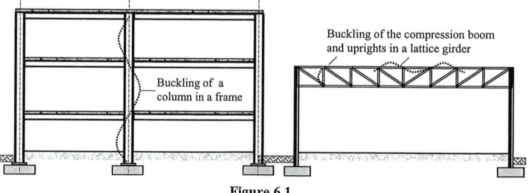

Buckling of the compression boom and uprights in a lattice girder

Buckling of a column in a frame

Figure 6.1

Only **local and flexural buckling** are considered in this text.

6.1.1   *Local Buckling*

Local buckling is characterised by localized deformation of slender cross-section elements, involving only rotation (no translation) at the nodes of a cross-section, e.g. flanges, webs etc. It is dependent on various parameters such as the size, shape, slenderness, type of stress and material properties.

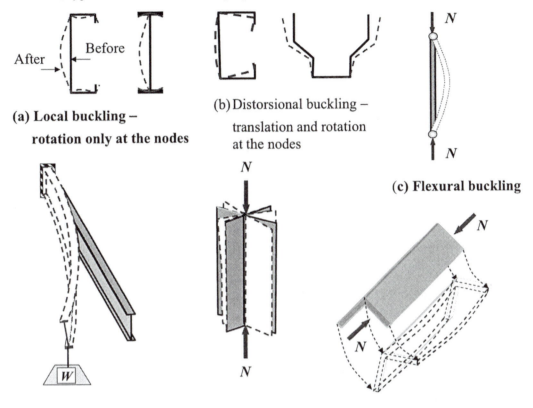

(a) Local buckling –

rotation only at the nodes

(b) Distorsional buckling –

translation and rotation
at the nodes

(c) Flexural buckling

(d) Lateral-torsional buckling   (e) Torsional buckling   (f) Torsional-flexural buckling

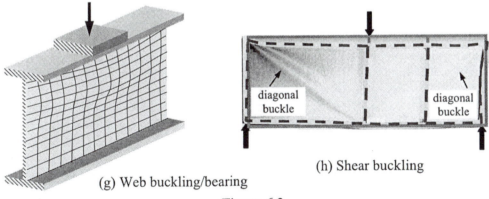

(g) Web buckling/bearing

(h) Shear buckling

Figure 6.2

The effect of local buckling on global behaviour at the ultimate limit state is such that the yield stress cannot develop in all of the fibres of the cross-section. A consequence of this is that the limiting elastic moment of resistance cannot develop. The reduction in strength is due to premature buckling of the slender elements of the cross-section which are in compression. A direct consequence of this is a reduction of stiffness in these elements and a redistribution of the stresses to the stiffer edges as shown in Figure 6.3.

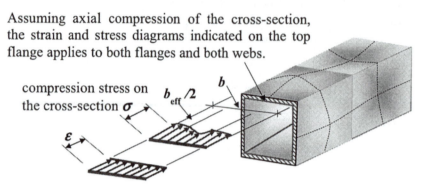

Assuming axial compression of the cross-section, the strain and stress diagrams indicated on the top flange applies to both flanges and both webs.

compression stress on the cross-section σ

$b_{eff}/2$ b

ε

Figure 6.3

Where local buckling must be taken into consideration, the formation of 'non-effective zones' in the **compression parts** of elements should be determined. The extent and position of the non-effective zones is calculated in accordance with EN 1993-1-1-5: Table 4.1/Table 4.2 to determine the effective width 'b_{eff}' and location of any slender part of the cross-section. The reduced effective plate widths and the effective area of a Class 4 element are given by the Winter formula i.e.

$b_{eff} = (\rho \times b)$ **and** $A_{c,eff} = (\rho \times A_c)$ where A_c is the gross cross-sectional area.

When any of the compression elements of a cross-section do not satisfy the requirements for a Class 3 section, local buckling must be taken into account, e.g. by using effective cross-sectional properties.

In EN 1993-1-1: Clause 5.5.2(1)/Table 5.2, four classes of cross-section in relation to local buckling are specified for internal and external elements as shown in Figure 6.5.

6.1.1.1 *Class 1 Sections*

The failure of a structure such that plastic collapse occurs is dependent on a sufficient number of plastic hinges developing within the cross-sections of the members to produce a mechanism, (i.e. the value of the internal bending moment reaching M_p at sufficient locations). For full collapse this requires one more than the number of redundancies in the structure, as illustrated in the rigid-jointed rectangular portal frame in Figure 6.4.

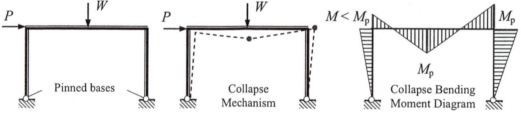

Figure 6.4

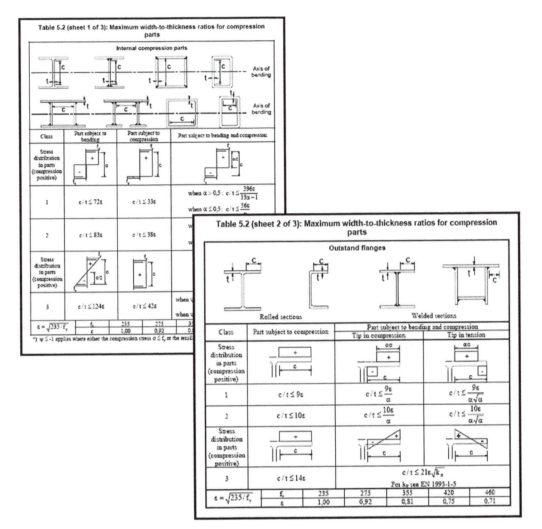

Figure 6.5 – Extract from EN 1993-1-1:2005

The required number of hinges will only develop if there is sufficient rotational capacity in the cross-section to permit the necessary redistribution of the moments within the structure. When this occurs, the stress diagram at the location of the hinge is as shown in Chapter 2: Figure 2.45(c) and the slenderness of the elements of the cross-section are low enough to prevent local buckling from occurring. Such cross-sections are defined as plastic sections and classified as **Class 1**. *Full plastic analysis and design can only be carried out using Class 1 sections.* (See Chapter 8 for plastic analysis of beams and frames.)

6.1.1.2 *Class 2 Sections*

When cross-sections can still develop the full plastic moment as in Figure 2.45(c) but are prevented by the possibility of local buckling from undergoing enough rotation to permit redistribution of the moments, the section is classified as a **Class 2 section**. *These sections can be used without restricting their capacity, except at plastic hinge positions.*

6.1.1.3 *Class 3 Sections*

Class 3 sections may be prevented from reaching their full plastic moment capacity by local buckling of one or more of the elements of the cross-section. The slenderness of the elements of the cross-section may be such that only the extreme fibre stress can attain the design strength before local buckling occurs. *Such sections are classified as* **Class 3** *and their capacity is therefore based on the limiting elastic moment* as indicated in Figure 2.45(b).

6.1.1.4 *Class 4 Sections*

When the slenderness of the element(s) of a cross-section is relatively high, then local buckling may prevent a part of the cross-section from reaching the design strength as indicated in Figure 2.45(a). *Such sections are classified as* **Class 4** *sections; their reduced capacity is based on effective cross-section properties* as specified in EN 1993-1-5: Clause 4.3.

6.1.1.5 *Section Classification*

The section classification is dependent on the aspect ratio for each of the compression plate elements in the cross-section. These elements include all component plates which are either totally or partially in compression due to the applied action effects, e.g. axial forces, bending moments etc. The plate elements are either:

♦ **internal compression parts;** considered to be simply supported along two edges parallel to the direction of the compression stress or
♦ **outstand parts;** considered to be simply supported along one edge and free on the other edge, parallel to the direction of the compression stress.

In EN 1993-1-1: Table 5.2 the limiting values are given for the aspect ratios of compression elements based on the web or flange plate slenderness for different loading conditions, i.e. bending, compression, and combined bending and compression.
These values ensure that in non-slender elements, yielding occurs before the plate critical stress 'σ_{cr}' is reached and buckling can occur.
The value of c, the flat portion of the web/flange plate, defined in EN 1993-1-1:Table 5.2 excludes the root radii for rolled sections and the weld leg length for welded sections. This enables one set of tables to be used for both rolled and welded cross-sections. For hollow sections where the internal corner radius is not known, the value of the flat portion can be taken as: $c = (b - 3t)$ or $c = (h - 3t)$. In addition, a base stress of $f_y = 235$ MPa has been adopted in the code for the limiting values given. In order to cover for all grades of steel a reduction factor is also given, i.e.

$$\varepsilon = \sqrt{235/f_y} \quad \text{EN 1993-1-1: Table 5.2}$$

The classification of a cross-section is based on the highest class of its component parts or alternatively may be defined by quoting both the flange and the web classifications as indicated in EN 1993-1-1: Clause 5.5.2.
Local buckling and effective section properties are not considered further in this text.

6.1.2 *Flexural Buckling*

Flexural buckling is characterised by out-of-plane movement of the cross-section at the critical load and is the predominant buckling mode in typical building structures using hot-rolled sections. In 1757 the Swiss engineer/mathematician Leonhard Euler developed a theoretical analysis of premature failure due to buckling.

The theory is based on the differential equation of the elastic bending of a pin-ended column which relates the applied bending moment to the curvature along the length of the column. The resulting equation for the fundamental critical load for a pin-ended column is known as the **Euler equation**, i.e.

$$\text{Euler critical buckling load} \quad N_{cr} = \frac{\pi^2 EI}{L_{cr}^2}$$

where:
L_{cr} is the critical buckling length,
I is the second moment of area about the axis of buckling,
E is Young's Modulus of elasticity.

Compression elements for this mode of buckling can be considered to be sub-divided into three groups: short elements, slender elements and intermediate elements. Each group is described separately, in Sections 6.1.2.1, 6.1.2.2 and 6.1.2.3 respectively.

6.1.2.1 *Short Elements*

Provided that the *slenderness* of an element is **low**, e.g. the length is not greater than ($10 \times$ the least cross-section dimension), the element will fail by crushing of the material induced
by predominantly axial compressive stresses as indicated in Figure 6.6(a). Failure occurs when the stress over the cross-section reaches a yield or crushing value for the material.
The failure of such a column can be represented on a stress/slenderness curve as shown in Figure 6.6(b).

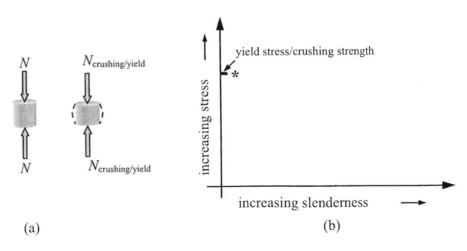

(a)   (b)

Figure 6.6

6.1.2.2 *Slender Elements*

When the ***slenderness*** of an element is **high,** the element fails by excessive lateral deflection (i.e. flexural buckling) at a value of stress considerably less than the yield or crushing values as shown in Figures 6.7(a) and (b).

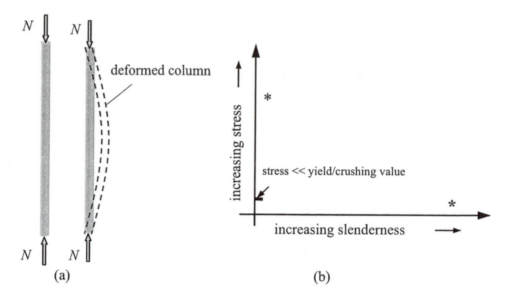

Figure 6.7

6.1.2.3 *Intermediate Elements*

The failure of an element which is neither short nor slender occurs by a combination of buckling and yielding/crushing as shown in Figures 6.8(a) and (b).

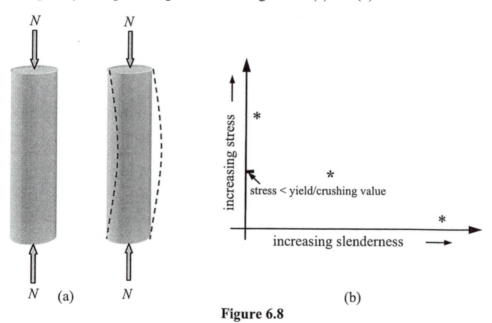

Figure 6.8

6.2   *Secondary Stresses*

As mentioned in Section 6.1, buckling is due to small imperfections within materials, application of load etc. which induce secondary bending stresses which may or may not be significant depending on the type of compression element. Consider a typical column as shown in Figure 6.9 in which there is an actual centre-line, reflecting the variations within the element, and an *assumed* centre-line along which acts an applied compressive load, *assumed* to be concentric.

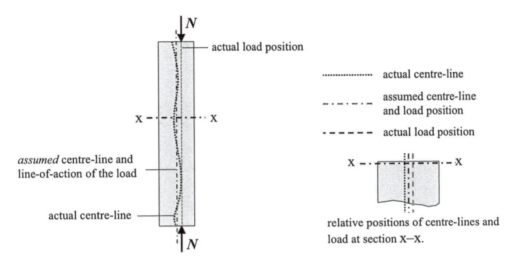

Figure 6.9

At any given cross-section the point of application of the load N will be eccentric to the actual centre-line of the cross-section at that point, as shown in Figure 6.10.

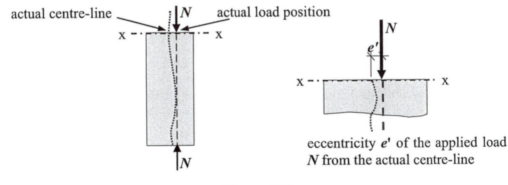

Figure 6.10

The resultant eccentric load produces a secondary bending moment in the cross-section. The cross-section is therefore subject to a combination of an axial stress due to N and a bending stress due to (Ne) where e is the eccentricity from the **assumed** centre-line as indicated in Figure 6.11

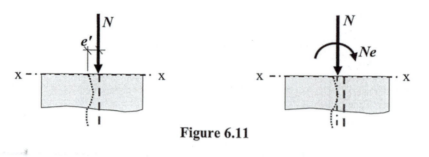

Figure 6.11

The combined axial and bending stress is given by: $\sigma = \left(\dfrac{N}{A} \pm \dfrac{Ne}{W} \right)$

where:

σ is the combined stress,
N is the applied load,
e is the eccentricity from the assumed centre-line,
A is the cross-sectional area of the section, and
W is the elastic section modulus about the axis of bending.

This equation, which includes the effect of secondary bending, can be considered in terms of each of the types of element.

6.2.1 Effect on Short Elements

In short elements the value of the *bending* stress in the equation is insignificant when compared to the axial stress i.e. $\left(\dfrac{N}{A} \right) \gg \left(\dfrac{Ne}{W} \right)$ and consequently the lateral movement and buckling effects can be ignored.

6.2.2 Effect on Slender Elements

In slender elements the value of the *axial* stress in the equation is insignificant when compared to the bending stress i.e. $\left(\dfrac{N}{A} \right) \ll \left(\dfrac{Ne}{W} \right)$ particularly since the eccentricity during buckling is increased considerably due to the lateral deflection; consequently the lateral movement and bending effects determine the structural behaviour.

6.2.3 Effect on Intermediate Elements

Most practical columns are considered to be in the *intermediate* group and consequently *both* the axial and bending effects are significant in the column behaviour, i.e. both terms in the equation $\sigma = \left(\dfrac{N}{A} \pm \dfrac{Ne}{W} \right)$ are important.

6.3   Critical Stress (σ_{cr})

In each case described in Sections 6.2.1 to 6.2.3 the critical load N_{cr} (i.e. critical stress $\times$ cross-sectional area) must be estimated for design purposes. Since the critical stress depends on the slenderness it is convenient to quantify slenderness in mathematical terms as:

slenderness $\lambda = \dfrac{L_E}{i}$

where:

L_E is the effective buckling length,

i is the radius of gyration $= \sqrt{\dfrac{I}{A}}$ and

I and A are the second moment of area about the axis of bending and the cross-sectional area of the section as before.

6.3.1 Critical Stress for Short Columns

Short columns fail by yielding/crushing of the material and $\sigma_{cr} = f_y$, the yield stress of the material. If, as stated before, columns can be assumed short when the length is not greater than (10 × the least cross-section dimension) then for a typical rectangular column of cross-section $(b \times h)$ and length $L \approx 10b$, a limit of slenderness can be determined as follows:

radius of gyration $\qquad i = \sqrt{\dfrac{I}{A}} = \sqrt{\dfrac{hb^3}{12 \times (b \times h)}} = \dfrac{b}{2\sqrt{3}}$

slenderness $\qquad \lambda = \dfrac{L}{i} \approx \dfrac{10b}{b/2\sqrt{3}} = 30 \sim 35$

From this we can consider that short columns correspond with a value of slenderness less than or equal to approximately 30 to 35.

6.3.2 Critical Stress for Slender Columns

Slender columns fail by buckling and the applied compressive stress $\sigma_{cr} \ll f_y$.
The critical load in this case is governed by the bending effects induced by the lateral deformation.

6.3.3 Euler Equation

The Euler theory of premature failure due to flexural buckling is based on the differential equation of the elastic bending of a pin-ended column which relates the applied bending moment to the curvature along the length of the column, i.e.

Bending Moment $\quad M = -EI\left(\dfrac{d^2z}{dx^2}\right)$

where $\left(\dfrac{d^2z}{dx^2}\right)$ approximates to the curvature of the deformed column.

Since this expression for bending moment only applies to linearly elastic materials, it is only valid for stress levels equal to and below the elastic limit of proportionality. This therefore defines an *upper* limit of stress for which the Euler analysis is applicable. Consider the deformed shape of the assumed centre-line of a column in equilibrium under the action of its critical load N_{cr} as shown in Figure 6.12.

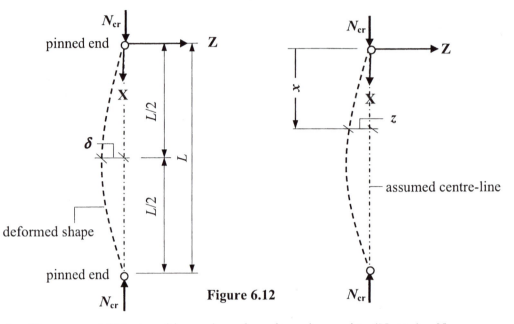

Figure 6.12

The bending moment (M) at position x along the column is equal to $(N_{cr} \times z) = N_{cr} z$

and hence $M = -EI\left(\dfrac{d^2 z}{dx^2}\right) = N_{cr} z$ $\therefore EI\left(\dfrac{d^2 z}{dx^2}\right) + N_{cr} z = 0$

This is a 2nd order differential equation of the form: $a\dfrac{d^2 z}{dx^2} + bz = 0$

The solution of this equation can be shown to be: $N_{cr} = n^2 \dfrac{\pi^2 EI}{L^2}$

where:
n is 0,1,2,3 … etc.
E, I and L are as before.

This expression for N_{cr} defines the Euler critical load for a pin-ended column in flexural buckling. The value of $n = 0$ is meaningless since it corresponds to a value of $N_{cr} = 0$. All other values of n correspond to the 1st, 2nd, 3rd …etc. harmonics (i.e. buckling mode shapes) for the sinusoidal curve represented by the differential equation. The first three harmonics are indicated in Figure 6.13.

The higher level harmonics are only possible if columns are restrained at the appropriate levels, e.g. mid-height point in the case of the 2nd harmonic and the third-height points in the case of the 3rd harmonic.

The *fundamental* critical load (i.e. $n = 1$) for a *pin-ended* column is therefore given by:

Euler Critical Load $N_E = \dfrac{\pi^2 EI}{L^2}$

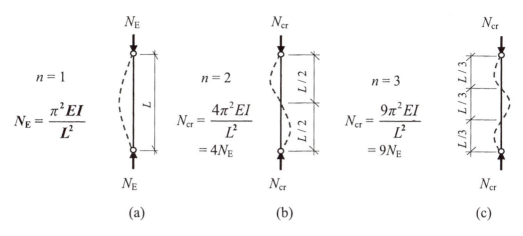

Figure 6.13 – Buckling mode-shapes for pin-ended columns

The fundamental case can be modified to determine the critical load for a column with different end-support conditions by defining an ***effective buckling length*** equivalent to that of a pin-ended column by identifying points of contraflexure in the column.

6.3.4 Effective Buckling Length (L_E)

The Euler critical load for the fundamental buckling mode is dependent on the *buckling length* between pins and/or points of contraflexure as indicated in Figure 6.13. In the case of columns which are **not** pin-ended, a modification to the boundary conditions when solving the differential equation of bending given previously yields different mode shapes and critical loads as shown in Figure 6.14.

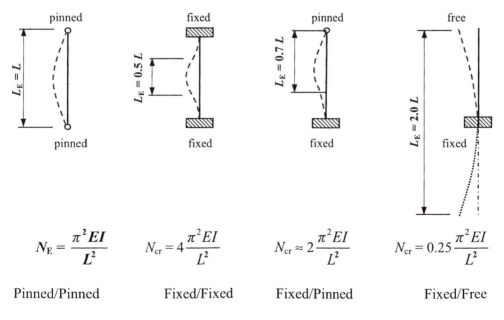

Figure 6.14 – Effective buckling lengths for different end conditions

The Euler critical stress (σ_E) corresponding to the Euler buckling load for a pin-ended column is given by:

$$\sigma_E = \frac{N_E}{\text{Area}(A)} = \frac{\pi^2 EI}{L^2 A} \quad \text{and} \quad I = Ai^2 \quad \therefore \sigma_E = \frac{\pi^2 E}{(L/i)^2}$$

where (L/i) is the slenderness λ as before.

Note: In practical design it is very difficult to achieve full fixity as assumed for the end conditions. This is allowed for by modifying the effective length coefficients e.g. increasing the value of $0.5L$ to $0.7L$ and $0.7L$ to $0.85L$.

A lower limit to the slenderness for which the Euler Equation is applicable can be found by substituting the stress at the proportional limit σ_e for σ_{cr} as shown in the following example with a steel column.
Assume that $\sigma_e = 200$ MPa (N/mm^2) and that $E = 210$ kN/mm^2

$$\therefore 200 = \frac{\pi^2 \times 210 \times 10^3}{(L/i)^2} \quad \therefore (L/i) = \sqrt{\frac{\pi^2 \times 210 \times 10^3}{200}} \approx 100$$

In this case the Euler load is only applicable for values of slenderness $\geq \approx 100$ and can be represented on a stress/slenderness curve in addition to that determined in Section 6.3.1 for short columns as shown in Figure 6.15.

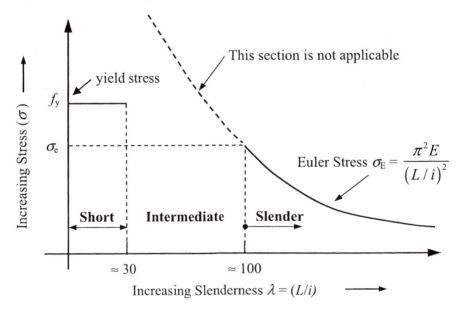

Figure 6.15

The Euler buckling load has very limited direct application in terms of practical design because of the following assumptions and limiting conditions:

- the column is subjected to a perfectly concentric axial load only,
- the column is pin-jointed at each end and restrained against lateral loading,
- the material is perfectly elastic,
- the maximum stress does not exceed the elastic limit of the material,
- there is no initial curvature and the column is of uniform cross-section along its length,
- lateral deflections of the column are small when compared to the overall length,
- there are no residual stresses in the column,
- there is no strain hardening of the material in the case of steel columns,
- the material is assumed to be homogeneous.

Practical columns do not satisfy these criteria, and in addition in most cases are considered to be intermediate in terms of slenderness.

6.3.5 *Critical Stress for Intermediate Columns*

Since the Euler curve is unsuitable for values of stress greater than the elastic limit it is necessary to develop an analysis which overcomes the limitations outlined above and which can be applied between the previously established slenderness limits (see Figure 6.15) as shown in Figure 6.16.

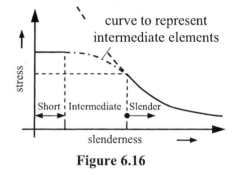

Figure 6.16

6.3.6 *Tangent Modulus Theorem*

Early attempts to develop a relationship for intermediate columns included the Tangent Modulus Theorem. Using this method a modified version of the Euler equation is adopted to determine the stress/slenderness relationship in which the value of the modulus of elasticity at any given level of stress is obtained from the stress/strain curve for the material and used to evaluate the corresponding slenderness. Consider a column manufactured from a material which has a stress/strain curve as shown in Figure 6.17(a).

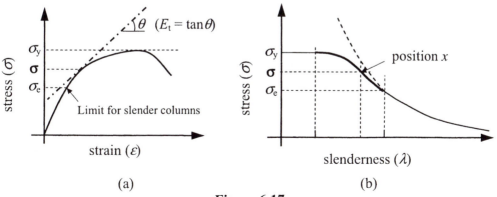

(a)   (b)

Figure 6.17

The slope of the tangent to the stress/strain curve at a value of stress equal to σ is equal to the value of the tangent modulus of elasticity E_t (**Note:** this is different from the value of E at the elastic limit). The value of E_t can be used in the Euler Equation to obtain a modified slenderness corresponding to the value of stress σ as shown at position 'x' in Figure 6.17 (b):

$$\sigma = \frac{\pi^2 E_t}{\left(L/i\right)^2} \qquad \therefore \text{ Slenderness } \lambda \text{ at position } x = (L/i) = \sqrt{\frac{\pi^2 E_t}{\sigma}}$$

If successive values of λ for values of stress between σ_e and σ_y are calculated and plotted as shown, then a curve representing the intermediate elements can be developed. This solution still has many of the deficiencies of the original Euler equation.

6.4   *Perry-Robertson Formula*

The Perry-Robertson Formula was developed to take into account the deficiencies of the Euler equation and other techniques such as the Tangent Modulus Method. This formula evolved from the assumption that all practical imperfections could be represented by a hypothetical initial curvature of the column.

As with the Euler analysis a 2nd order differential equation is established and solved using known boundary conditions, and the extreme fibre stress in the cross-section at mid-height (the assumed critical location) is evaluated. The extreme fibre stress, which includes both axial and bending effects, is then equated to the yield value. Clearly the final result is dependent on the initial hypothetical curvature.

Consider the deformed shape of the assumed centre-line of a column in equilibrium under the action of its critical load N_{cr} and an assumed initial curvature as shown in Figure 6.18.

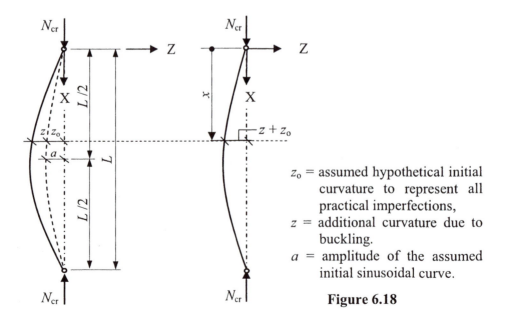

z_o = assumed hypothetical initial curvature to represent all practical imperfections,
z = additional curvature due to buckling.
a = amplitude of the assumed initial sinusoidal curve.

Figure 6.18

The bending moment at position x along the column is equal to $= N_{cr}(z + z_o)$

and hence the bending moment $M = -EI\left(\dfrac{d^2z}{dx^2}\right) = N_{cr}(z + z_o)$

$$\therefore \left(\frac{d^2z}{dx^2}\right) + \left(\frac{N_{cr}}{EI}\right)z = -\left(\frac{N_{cr}}{EI}\right)z_o$$

If the initial curvature is assumed to be sinusoidal, then $z_o = a\sin\left(\dfrac{\pi x}{L}\right)$ where a is the amplitude of the initial displacement and the equation becomes:

$$\therefore \left(\frac{d^2z}{dx^2}\right) + \left(\frac{N_{cr}}{EI}\right)z = -\left(\frac{N_{cr}}{EI}\right)a\sin\frac{\pi x}{L}$$

The solution to this differential equation is:

$$z = A\cos\left(\frac{N_{cr}}{EI}x\right) + B\sin\left(\frac{N_{cr}}{EI}x\right) + \frac{\dfrac{N_{cr}}{EI}a}{\left(\dfrac{\pi^2}{L^2} - \dfrac{N_{cr}}{EI}\right)}\sin\left(\frac{\pi x}{L}\right)$$

The constants A and B are determined by considering the boundary values at the pinned ends, i.e. when $x = 0$ $z = 0$ and when $x = L$ $z = 0$.

Substitution of the boundary conditions in the equation gives:

$x = 0$ $z = 0$ $\therefore A = 0$

$x = L$ $z = 0$ $\therefore B\sin\left(\dfrac{N_{cr}}{EI}L\right) = 0$ For $\left(\dfrac{N_{cr}}{EI}\right)$ not equal to zero, then $B = 0$

$z = \dfrac{\dfrac{N_{cr}}{EI}a}{\left(\dfrac{\pi^2}{L^2} - \dfrac{N_{cr}}{EI}\right)}\sin\left(\dfrac{\pi x}{L}\right)$ If the equation is divided throughout by $\left(\dfrac{N_{cr}}{EI}\right)$ then

$z = \dfrac{a\sin\left(\dfrac{\pi x}{L}\right)}{\left(\dfrac{\pi^2 EI}{N_{cr}L^2} - 1.0\right)}$ The Euler load $N_E = \dfrac{\pi^2 EI}{L^2}$ $\therefore z = \dfrac{a\sin\left(\dfrac{\pi x}{L}\right)}{\left(\dfrac{N_E}{N_{cr}} - 1.0\right)}$

The value of the stress at mid-height is the critical value since the maximum eccentricity of the load (and hence maximum bending moment) occurs at this position:

when $x = L / 2$, $\sin\left(\dfrac{\pi x}{L}\right) = 1.0$ and $z_{\text{mid-height}} = \dfrac{a\sin\left(\dfrac{\pi x}{L}\right)}{\left(\dfrac{N_E}{N_{\text{cr}}} - 1.0\right)} = \dfrac{a}{\left(\dfrac{N_E}{N_{\text{cr}}} - 1.0\right)}$

(**Note:** z_0 at mid-height is equal to the amplitude a of the assumed initial curvature).

The maximum bending moment $M = N_{\text{cr}}(a + z_{\text{mid-height}}) = N_{\text{cr}}\,a\left[1 + \dfrac{1}{\left(\dfrac{N_E}{N_{\text{cr}}} - 1.0\right)}\right]$

The maximum combined stress at this point is given by:

$$\sigma_{\text{maximum}} = \left(\frac{\text{axial load}}{A} + \frac{\text{bending moment} \times c}{I}\right) = \left(\frac{N_{\text{cr}}}{A} + \frac{M \times c}{Ai^2}\right)$$

where c is the distance from the neutral axis of the cross-section to the extreme fibres.
The *maximum stress* is equal to the yield value, i.e. $\sigma_{\text{maximum}} = \sigma_y$

$$\therefore \ \sigma_y = \left(\frac{N_{\text{cr}}}{A} + \frac{M \times c}{Ai^2}\right) = \frac{N_{\text{cr}}}{A} + N_{\text{cr}}\,a\left[1 + \frac{1}{\left(\dfrac{N_E}{N_{\text{cr}}} - 1.0\right)}\right] \times \frac{c}{Ai^2}$$

The *average stress* over the cross-section is the load divided by the area, i.e. (N_{cr}/A)

$$\sigma_y = \sigma_{\text{average}} + \sigma_{\text{average}}\left(\frac{N_E}{N_E - N_{\text{cr}}}\right) \times \frac{ac}{i^2} = \sigma_{\text{average}}\left[1 + \left(\frac{N_E}{N_E - N_{\text{cr}}}\right) \times \frac{ac}{i^2}\right]$$

$\sigma_{\text{average}} = (N_{\text{cr}}/A)$ and $\sigma_E = (N_E/A)$

$$\sigma_y = \sigma_{\text{average}}\left[1 + \left(\frac{\sigma_E}{\sigma_E - \sigma_{\text{average}}}\right) \times \frac{ac}{i^2}\right]$$

The (ac/i^2) term is dependent upon the assumed initial curvature and is normally given the symbol η.

$$\sigma_y = \sigma_{\text{average}}\left[1 + \left(\frac{\eta\sigma_E}{\sigma_E - \sigma_{\text{average}}}\right)\right]$$

This equation can be rewritten as a quadratic equation in terms of the average stress:

$$\sigma_y(\sigma_E - \sigma_{\text{average}}) = \sigma_{\text{average}}[(1 + \eta)\sigma_E - \sigma_{\text{average}}]$$

$$\sigma^2_{\text{average}} - \sigma_{\text{average}}[\sigma_y + (1 + \eta)\sigma_E] + \sigma_y\sigma_E = 0$$

The solution of this equation in terms of σ_{average} is:

$$\sigma_{\text{average}} = \frac{[\sigma_y + (1 + \eta)\sigma_E] - \sqrt{[\sigma_y + (1 + \eta)\sigma_E]^2 - 4\sigma_y\sigma_E}}{2.0}$$

This equation represents the *average* value of stress in the cross-section which will induce the yield stress at mid-height of the column for any given value of η. Experimental evidence obtained by Perry and Robertson indicated that the hypothetical initial curvature of the column could be represented by;

$$\eta = 0.3(L_E/100i^2)$$

which was combined with a load factor of 1.7 and used for many years in design codes to determine the critical value of average compressive stress below which overall buckling would not occur. The curve of stress/slenderness for this equation is indicated in Figure 6.19 for comparison with the Euler and Tangent Modulus solutions.

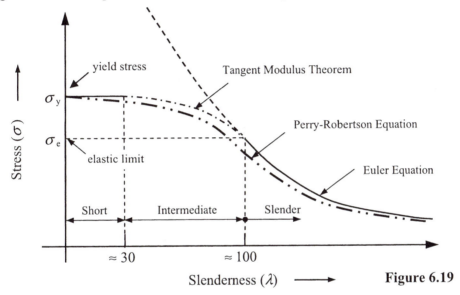

Figure 6.19

6.5 *European Column Curves*

Whilst the Perry-Robertson formula does take into account many of the deficiencies of the Euler and Tangent Modulus approaches, it does not consider all of the factors which influence the failure of columns subjected to compressive stress. In the case of steel columns for example, the effects of residual stresses induced during fabrication, the type of section being considered (i.e. the cross-section shape), the material thickness, the axis of buckling, the method of fabrication (i.e. rolled or welded), etc. are not allowed for.
A more realistic formula of the critical load capacity of columns has been established following extensive full-scale testing both in the UK and in other European countries. The Perry-Robertson formula has in effect been modified and adopted by the Eurocodes.
The Euler critical buckling stress and the 'slenderness' can be written as:

$$\sigma_{cr} = \frac{N_{cr}}{A} = \frac{\pi^2 EAi^2}{L_{cr}^2} \times \frac{1}{A} = \frac{\pi^2 Ei^2}{L_{cr}^2} = \frac{\pi^2 E}{L_{cr}^2/i^2} = \frac{\pi^2 E}{\lambda^2}$$

where λ is the slenderness and L_{cr} is the critical buckling length (L_E).

This can be re-written such that the slenderness $\lambda = \pi \sqrt{\dfrac{E}{\sigma_{cr}}}$. A graph of critical stress versus slenderness, i.e. the Euler curve is shown in Figure 6.20.

The critical stress on the Euler curve is limited by the yield stress f_y of the material. The slenderness corresponding with this value is known as the Euler slenderness (λ_1).

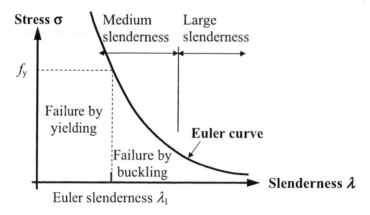

Figure 6.20

6.9.9 Non-dimensional Slenderness

EN 1993-1-1: Clause 6.3.1.2 defines a non-dimensional slenderness $\bar{\lambda}$ including material properties (E, f_y) which affect the theoretical buckling load, (Note: $\lambda = L_{cr}/i$).

The use of a non-dimensional slenderness allows a more direct comparison of susceptibility to flexural buckling for columns with different material strengths and requires only one set of curves. A typical non-dimensional buckling curve is shown in Figure 6.21. In the EN 1993-1-1, non-dimensional slenderness is defined in terms of forces rather than stresses as above, e.g. for flexural buckling $\bar{\lambda} = \sqrt{\dfrac{A f_y}{N_{cr}}}$ and χ is a reduction factor.

$\chi = \sigma / f_y$ — Plateau indicating the limiting non-dimensional slenderness below which flexural buckling need not be considered; i.e. $\bar{\lambda} \leq 0{,}2$

1,0 — Failure predicted by Euler

Actual test results

Safe lower-bound design curve in EC3

Not to scale

0,2   1,0

Non-dimensional slenderness $\bar{\lambda}$ **Figure 6.21**

The buckling curves given in EN 1993-1-1 are the result of more than 1000 tests on various types of cross-sections with values of slenderness ranging from 55 to 160. The curves include the effects of imperfections such as initial out-of-straightness, residual stresses, eccentricity of applied axial load and strain-hardening.

In total five curves, (a), (b), (c), (d) and (a)$_0$ are given, the first four relating to the following steel grades: S 235, S 275, S 355 and S 420. The latter curve (a)$_0$ relates to the higher grade steel S 460.

- **Curve (a):** represents quasi-p erfect shapes e.g. some hot-rolled **I**-sections with buckling perpendicular to the major axis and hot-rolled hollow sections.
- **Curve (b):** represents shapes with medium imperfections e.g. some hot-rolled **I**-sections with medium flange thickness, welded **I**-sections with buckling perpendicular to the minor axis, most welded box-sections and angle sections.
- **Curve (c):** represents shapes with many imperfections e.g. some hot-rolled **I**-sections, welded **I**-sections with buckling perpendicular to the minor axis, thick welded box-sections, **U, T** and solid sections.
- **Curve (d):** represents shapes with maximum imperfections e.g. hot-rolled **I**-sections with very thick flanges and thick welded I-section buckling about the minor axis.

The selection of a particular curve is given in EN 1993-1-1: 2005: Table 6.2 for various cross-sections as indicated in Figure 6.22.

Table 6.2: Selection of buckling curve for a cross-section

Cross section		Limits	Buckling about axis	Buckling curve S 235 S 275 S 355 S 420	S 460
Rolled sections	$h/b > 1.2$	$t_f \le 40$ mm	y–y z–z	a b	a$_0$ a$_0$
		40 mm $< t_f \le 100$	y–y z–z	b c	a a
	$h/b \le 1.2$	$t_f \le 100$ mm	y–y z–z	b c	a a
		$t_f > 100$ mm	y–y z–z	d d	c c
Welded I-sections		$t_f \le 40$ mm	y–y z–z	b c	b c
		$t_f > 40$ mm	y–y z–z	c d	c d
Hollow sections		hot finished	any	a	a$_0$
		cold formed	any	c	c
Welded box sections		generally (except as below)	any	b	b
		thick welds: a > 0.5t$_f$ b/t$_f$ < 30 h/t$_w$ < 30	any	c	c
U, T- and solid sections			any	c	c
L-sections			any	b	b

Figure 6.22

The non-dimensional slenderness for flexural buckling is presented in two forms in EN 1993-1-1: Clause 6.3.1.3 of the code:

$$\bar{\lambda} = \sqrt{\frac{Af_y}{N_{cr}}} = \frac{L_{cr}}{i\lambda_1} \qquad \text{EN 1993-1-1: Equation (6.50)}$$

$$\left(\text{i.e. } \bar{\lambda} = \sqrt{\frac{Af_y L_{cr}^2}{\pi^2 EI}} = \sqrt{\frac{f_y L_{cr}^2}{\pi^2 Ei^2}} = \frac{L_{cr}}{i\lambda_1} \right)$$

$$\lambda_1 = \sqrt{\frac{\pi^2 E}{f_y}} \quad \text{and } \varepsilon = \sqrt{\frac{235}{f_y}} \quad \therefore \sqrt{f}_y = \frac{\sqrt{235}}{\varepsilon} \quad \text{and}$$

The Euler slenderness $\lambda_1 = \varepsilon\sqrt{\dfrac{\pi^2 \times 210000}{235}} = 93{,}9\varepsilon$

There is very little guidance given in EN 1993-1-1 in relation to L_{cr}, the critical buckling length, other than in Annex BB for some triangulated and lattice structures. Engineers can use the recommendations given in BS 5950: Part 1 as indicated in Figure 6.14 for no-sway columns and in Figure 6.23 for columns in which sway can occur.

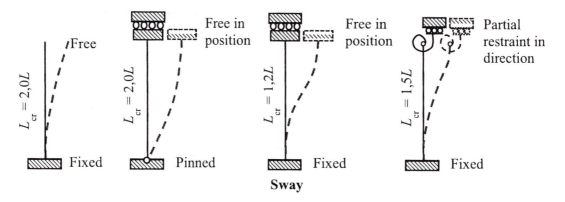

Figure 6.23

The buckling curves give values of the reduction factor 'χ' to be applied to the compressive resistance of an element cross-section as a function of the non-dimensional slenderness and is defined by EN 1993-1-1: Equation (6.49) as follows:

$$\chi = \frac{1}{\Phi + \sqrt{\Phi^2 - \bar{\lambda}^2}} \quad \text{but } \chi \le 1{,}0 \qquad \text{EN 1993-1-1: Equation (6.49)}$$

where

$$\Phi = 0{,}5\left[1 + \alpha\left(\bar{\lambda} - 0{,}2\right) + \bar{\lambda}^2\right]$$

α is an imperfection factor which is dependent on the shape of the element cross-section, the axis of buckling and the fabrication process, (hot-rolled, welded or cold-formed) and increases with increasing degree of imperfection as indicated in Table 6.1 below.

Imperfection factors for buckling curves					
Buckling curve	a_0	a	b	c	d
Imperfection factor α	0.31	0.21	0.34	0.49	0.76

Table 6.1

The formulation of EN 1993-1-1: Equation (6.49) is as follows:

A pin-ended column including an assumed initial deformation of magnitude z_0 is shown in Figure 6.24.

Assuming this to have a sinusoidal wave form, it can be written as:

$$z_0 = a\sin\frac{\pi x}{L}$$

where a is the amplitude of the sine wave. The differential equation representing the final deformation is:

$$\frac{d^2z}{d^2x} + \frac{N(z+z_0)}{EI} = 0$$

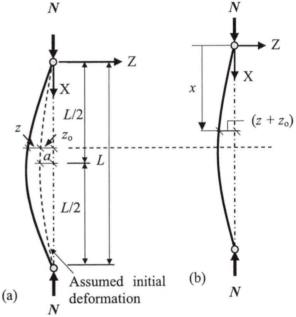

(a)   Assumed initial deformation   (b)

Figure 6.24

Substituting for the initial imperfection z_0 and applying the boundary conditions, (the member is pinned at the ends), the solution of this 2nd order differential equation is:

$$z = \frac{a}{(N_{cr}/N - 1)}\sin\frac{\pi x}{L}$$

The maximum total deflection 'e' of the column occurs when $x = L/2$ and is given by:

$$e_{\text{mid-height}} = z_0 + z_{\text{mid-height}} \quad \therefore e_{\text{mid-height}} = a + \frac{a}{(N_{cr}/N - 1)}\sin\frac{\pi}{2} = \frac{a}{(1 - N/N_{cr})}$$

The eccentricity of the applied axial load due to the deformation induces a secondary bending moment equal to $(N \times e)$. The maximum stress at the mid-height position (f_y) is then given by:

$$f_y = \frac{N}{A} + \frac{N \times e}{W} = \frac{N}{A} + \left(\frac{N}{A} \times \frac{A \times e}{W} \right) = \sigma_b + \sigma_b \frac{Ae}{W}$$

Substituting for $e = e_{\text{mid-height}} = \dfrac{a}{(1 - N_b/N_{cr})}$ and $N/N_{cr} = \sigma_b/\sigma_{cr}$

$$f_y = \sigma_b + \frac{\sigma_b a}{1 - (\sigma_b/\sigma_{cr})} \times \frac{A}{W} \quad \text{where } \sigma_{crit} \text{ is the Euler critical stress} = \pi^2 E/\lambda^2$$

$$= (\sigma_{cr} - \sigma_b)(f_y - \sigma_b) = \sigma_b \sigma_{cr} a \frac{A}{W}$$

This equation can be further modified to include other effects such as accidental eccentricity of the applied load, and residual stresses. The classical form of this equation, which is known as the Ayrton-Perry formula is:

$$(\sigma_{cr} - \sigma_b)(f_y - \sigma_b) = \eta \sigma_b \sigma_{cr}$$

where η is a factor to allow for the imperfections and σ_b is the buckling stress. An alternative representation of this equation is given by dividing by $(f_y)^2$

$$\left(\frac{\sigma_{cr}}{f_y} - \frac{\sigma_b}{f_y} \right) \left(1 - \frac{\sigma_b}{f_y} \right) = \eta \frac{\sigma_b}{f_y} \frac{\sigma_{cr}}{f_y}$$

The reduction factor 'χ' $= \sigma_b/f_y$ and $\overline{\lambda} = \sqrt{\dfrac{f_y}{\sigma_{cr}}}$ $\therefore \overline{\lambda}^2 = \dfrac{f_y}{\sigma_{cr}}$

Multiplying the equation above by f_y/σ_{cr} gives the following quadratic equation in χ:

$$(1 - \chi \overline{\lambda}^2)(1 - \chi) = \eta \chi \quad \therefore \overline{\lambda}^2 \chi^2 - (\overline{\lambda}^2 + \eta + 1)\chi + 1 = 0$$

Assuming $\eta = \alpha(\overline{\lambda} - \overline{\lambda}_0)$, the smallest solution for this quadratic equation is:

$$\chi = \frac{1 + \alpha(\overline{\lambda} - \overline{\lambda}_0) + \overline{\lambda}^2 - \sqrt{\left\{ \left[1 + \alpha(\overline{\lambda} - \overline{\lambda}_0) + \overline{\lambda}^2 \right]^2 - 4\overline{\lambda}^2 \right\}}}{2\overline{\lambda}^2}$$

Let $\overline{\lambda}_0 = 0,2$ and $\Phi = 0,5\left[1 + \alpha(\overline{\lambda} - 0,2) + \overline{\lambda}^2 \right]$

Multiplying by the conjugate, the equation for χ can be re-written as:

$$\chi = \frac{2\Phi - \sqrt{\left\{[2\Phi]^2 - 4\overline{\lambda}^2\right\}}}{2\overline{\lambda}^2} \times \frac{2\Phi + \sqrt{\left\{[2\Phi]^2 - 4\overline{\lambda}^2\right\}}}{2\Phi + \sqrt{\left\{[2\Phi]^2 - 4\overline{\lambda}^2\right\}}} = \frac{4\Phi^2 - 4\Phi^2 + 4\overline{\lambda}^2}{2\overline{\lambda}^2 \times 2\left[\Phi + \sqrt{\left(\Phi^2 - \overline{\lambda}^2\right)}\right]}$$

$$= \frac{1}{\left[\Phi + \sqrt{\left(\Phi^2 - \overline{\lambda}^2\right)}\right]} \leq 1,0$$

This equation is given in EN 1993-1-1: Clause 6.3.1.2 with the value of $\overline{\lambda}_0 = 0.2$ (i.e. the end of the horizontal plateau) and factor α for both geometrical and mechanical imperfections as given in EN1993-1-1: Table 6.1, i.e.

$$\chi = \frac{1}{\Phi + \sqrt{\Phi^2 - \overline{\lambda}^2}} \quad \textbf{but} \quad \chi \leq 1.0 \qquad \text{EN 1993-1-1: Equation (6.49)}$$

where $\quad \Phi = 0.5\left[1 + \alpha\left(\overline{\lambda} - 0.2\right) + \overline{\lambda}^2\right]$

$$\overline{\lambda} = \sqrt{\frac{Af_y}{N_{cr}}} \quad \text{for Class1, 2 and 3 cross-sections and}$$

$$\overline{\lambda} = \sqrt{\frac{A_{eff}f_y}{N_{cr}}} \quad \text{for Class 4 cross-sections}$$

The value of the reduction factor χ can be determined using EN 1993-1-1: Equation (6.49), in EN 1993-1-1: Figure 6.4 (see Figure 6.25 in this text), or alternatively from Table 6.2 given below. (Note: this table is not given in EN 1993-1-1.)

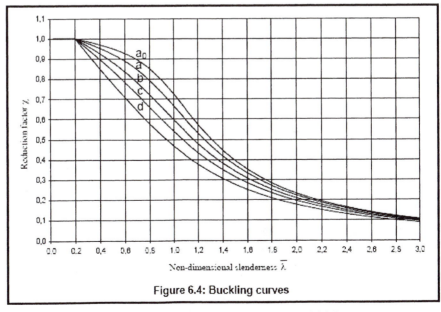

Figure 6.4: Buckling curves

Figure 6.25 – Extract from EN 1993-1-1:2005

	Reduction Factor χ					Reduction Factor χ			
	Curve (a)	Curve (b)	Curve (c)	Curve (d)		Curve (a)	Curve (b)	Curve (c)	Curve (d)
0,1	1,00	1,00	1,00	1,00	**1,6**	0,33	0,30	0,28	0,25
0,2	1,00	1,00	1,00	1,00	**1,7**	0,30	0,27	0,25	0,22
0,3	0,97	0,96	0,95	0,92	**1,8**	0,27	0,25	0,23	0,21
0,4	0,95	0,92	0,89	0,85	**1,9**	0,24	0,23	0,21	0,19
0,5	0,92	0,88	0,84	0,78	**2,0**	0,22	0,21	0,19	0,17
0,6	0,89	0,83	0,78	0,71	**2,1**	0,20	0,19	0,18	0,16
0,7	0,84	0,78	0,72	0,64	**2,2**	0,18	0,17	0,16	0,15
0,8	0,79	0,72	0,66	0,58	**2,3**	0,17	0,16	0,15	0,14
0,9	0,73	0,66	0,60	0,52	**2,4**	0,15	0,15	0,14	0,13
1,0	0,66	0,59	0,54	0,46	**2,5**	0,14	0,14	0,13	0,12
1,1	0,59	0,53	0,48	0,41	**2,6**	0,13	0,13	0,12	0,11
1,2	0,53	0,47	0,43	0,37	**2,7**	0,12	0,12	0,11	0,10
1,3	0,47	0,42	0,38	0,33	**2,8**	0,11	0,11	0,10	0,10
1,4	0,41	0,38	0,35	0,30	**2,9**	0,11	0,10	0,10	0,09
1,5	0,37	0,34	0,31	0,27	**3,0**	0,10	0,10	0,09	0,08

Table 6.2

The design of the majority of concrete and timber column members is usually based on square, rectangular or circular cross-sections, similarly with masonry columns square or rectangular sections are normally used. In the case of structural steelwork there is a wide variety of cross-sections which are adopted, the most common of which are shown in Figure 6.26.

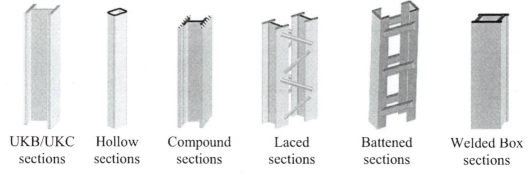

| UKB/UKC sections | Hollow sections | Compound sections | Laced sections | Battened sections | Welded Box sections |

Figure 6.26

In all cases, irrespective of the material or member cross-section, an assessment of end and intermediate restraint conditions must be made in order to estimate critical buckling lengths (L_{cr}) and hence the non-dimensional slenderness $\bar{\lambda}$.

It is important to recognise that the critical buckling length is not necessarily the same about all axes. Typically, it is required to determine two L_{cr} and $\bar{\lambda}$ values (e.g. $L_{cr,y}$, $\bar{\lambda}_y$ and $L_{cr,z}$, $\bar{\lambda}_z$), and subsequently determine the critical compressive stress relating to each one; the lower value being used to calculate the compressive resistance of a member. In the case of angle sections other axes are also considered. The application of the Ayrton-Perry formula to various steel columns is illustrated in Examples 6.1 to 6.4 and Problems 6.1 to 6.5.

6.6 Example 6.1 Slenderness

The hollow square column section shown in Figure 6.27 is pinned about both the y-y, and z-z axes at the top and fixed about both axes at the bottom. An additional restraint is to be provided to both axes at a height of L_1 above the base. Determine the required value of L_1 to optimize the compression resistance of the section.

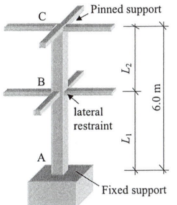

For optimum compression resistance the maximum non-dimensional slenderness for lengths AB and BC must be the same. i.e. $\bar{\lambda}_{AB} = \bar{\lambda}_{BC}$

Note: $\bar{\lambda} = \sqrt{\dfrac{Af_y}{N_{cr}}} = \dfrac{L_{cr}}{i\lambda_1}$ i.e. $\bar{\lambda} \; \alpha \; \dfrac{L_{cr}}{i}$ and hence:

$L_{cr,AB}/i_{yy} = L_{cr,BC}/i_{zz}$

Since the section is square $i_{yy} = i_{zz}$ and
$L_{cr,AB} = L_{cr,BC}$

Figure 6.27

Consider the critical lengths of AB and BC

$L_{cr,AB} = 0.85L_1$ and $L_{cr,BC} = 1.0L_2$ $\therefore 0.85L_1 = 1.0L_2$

The total height of the column $(L_1 + L_2) = 6.0$ m

$\therefore L_1 + 0.85L_1 = 6.0$ and hence $L_1 = 3.24$ m
$L_2 = 2.76$ m
The required value of $L_1 = 3.24$ m

6.7 Example 6.2 Rolled Universal Column Section

A column which is subjected to a concentric axial load 'N' is shown in Figure 6.28. Restraint against lateral movement but not rotation, is provided about both axes at the top and the bottom of the column. Additional lateral restraint is also provided about the z-z axis at mid-height as shown.

Considering flexural buckling only and using the data provided, determine the compression resistance of the column using the EN 1993-1-1 flexural buckling formulae.

Data:

Yield Stress f_y = 355 MPa (N/mm²) E = 210 kN/mm²

Section Property	Section 203 × 203 × 60 UKC Universal Column
Cross-sectional area (A)	76.40 cm²
Radius of gyration (i_{zz})	5.20 cm
Radius of gyration (i_{yy})	8.96 cm
Depth of the section (h)	209.60 mm
Width of the section (b)	205.80 mm
Flange thickness (t_f)	14.20 mm
Web thickness (t_w)	9.40 mm
Root radius (r)	10.20 mm

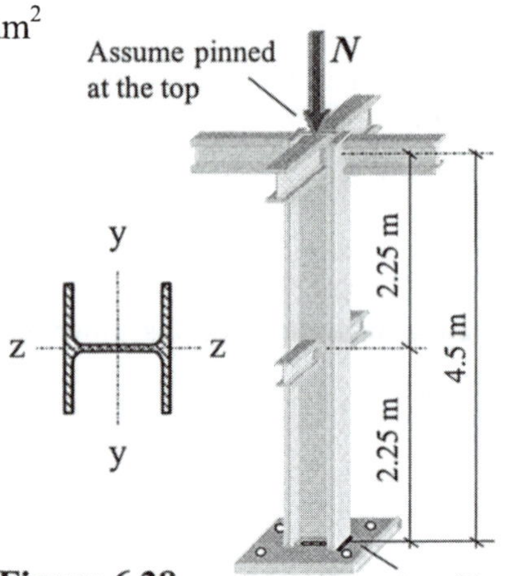

Figure 6.28

Solution:

EN 10025-2:2004

Table 7   For S355 steel thickness t_f = 14.2 mm   ($\leq$ 16 mm)
                        f_y = 355 MPa
                        (Note: for thicknesses of steel $\leq$ 16 mm, f_y = steel grade)
                        Section classification for a 203 x 203 x 60 UKB S355 (compression)

EN 1993-1-1:2005

$$\varepsilon = \sqrt{235/f_y} = \sqrt{235/355} = 0,81$$

Table 5.2(1)   Web:
                        $c = [h - 2(t_f + r)] = [209.6 - 2 \times (14.2 + 10,2)] = 160.8$ mm
                        $c/t = (160.8/9.4) = 17.1$
                        $33\varepsilon = (33 \times 0,81) = 26.73$ (See Figure 6.5 in this text)
                        $c/t < 33\varepsilon$ $\therefore$ The web is Class 1

Table 5.2(2)   Flange:
                        $c = [b - (t_w + 2r)]/2 = [205.8 - (9.4 + 2 \times 10,2)]/2 = 88.0$ mm
                        $c/t = (88.0/14.2) = 6.20$
                        $9\varepsilon = (9 \times 0,81) = 7.29$ (See Figure 6.5 in this text)
                        $c/t < 9\varepsilon$ $\therefore$ The flanges are Class 1
                                                                            Section is Class 1

Flexural buckling resistance

Clause 6.3.1.3(1) $\bar{\lambda} = \sqrt{\dfrac{Af_y}{N_{cr}}} = \dfrac{L_{cr}}{i} \times \dfrac{1}{\lambda_1}$ where $\lambda_1 = 93.9\varepsilon = 93.9 \times 0.81 = 76.06$

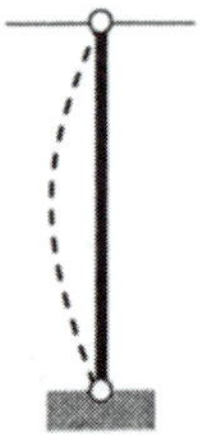

Consider the y-y axis: assume $L_{cr,y}$ = (1.0 × 4.50) = 4.50 m

Equation (6.50) $\bar{\lambda}_y = \dfrac{L_{cr}}{i_y} \times \dfrac{1}{\lambda_1} = \dfrac{4500}{89.6} \times \dfrac{1}{76.06} = 0.66$

Table 6.2 $h/b = 209.60/205.80 = 1.02 < 1,2$ and $t_f < 100$ mm

For buckling about the y-y axis use curve b
(Figure 6.22 in this text)

Equation (6.49) $\chi = \dfrac{1}{\Phi + \sqrt{\Phi^2 - \bar{\lambda}^2}}$ but $\chi \leq 1.0$

where $\Phi = 0.5\left[1 + \alpha\left(\bar{\lambda} - 0.2\right) + \bar{\lambda}^2\right]$

Table 6.1   Imperfection factor for curve b: $\alpha = 0.34$
(see Figure 6.24 in this text)

$$\Phi = 0.5\left[1 + 0.34\left(0.66 - 0.2\right) + 0.66^2\right] = 0.796$$

$$\chi = \frac{1}{0.796 + \sqrt{0.796^2 - 0.66^2}} = 0.81 \text{ but } \chi \leq 1,0$$

Alternatively use the curves given in EN 1993-1-1: Figure 6.4
(see Figure 6.26 in this text)

Consider the z-z axis: assume $L_{cr,z} = (1.0 \times 2.25) = 2.25$ m

Equation (6.50) $\bar{\lambda}_z = \dfrac{L_{cr}}{i_z} \times \dfrac{1}{\lambda_1} = \dfrac{2250}{52.0} \times \dfrac{1}{76.06} = 0.57$

Table 6.2 $h/b = 209.60/205.80 = 1.02 < 1,2$ and $t_f < 100$ mm

For buckling about the z-z axis use curve c
(Figure 6.22 in this text)

Table 6.1   Imperfection factor for curve b: $\alpha = 0.49$
(see Figure 6.24 in this text)

Equation (6.49) $\Phi = 0.5\left[1 + 0.49\left(0.57 - 0.2\right) + 0.57^2\right] = 0.753$

$$\chi = \frac{1}{0.753 + \sqrt{0.753^2 - 0.57^2}} = 0.80 \text{ but } \chi \leq 1,0$$

Alternatively use the curves given in EN 1993-1-1: Figure 6.4
(see Figure 6.26 in this text)

Critical value $\chi_z \approx 0,80$

Equation (6.47) $N_{b,z,Rd} = \dfrac{\chi A f_y}{\gamma_{M1}} = \dfrac{0,80 \times 7640 \times 355}{1,0 \times 10^3} = 2169.8$ kN

The maximum design axial load with respect to flexural buckling = 2169.8 kN

Note: EN 1993-1-1: Clause 6.3.1.4(1) *'For members with open cross-sections account should be taken of the possibility that the resistance of the member to either torsional or torsional-flexural buckling could be less than its resistance to flexural buckling.'*

6.8   *Example 6.3 Compound Column Section*

A column ABCE is shown in Figure 6.29. The column is 15.0 m long and supports a roof beam DEF at E. The beam carries a load of w kN/m length along its full length DEF. The column is fabricated from a 152 × 152 × 23 UKC with plates welded continuously to the flanges as shown. Using the data given and considering only flexural buckling, determine:

(i)   the compression resistance of the column,   and
(ii)   the maximum value of w which can be supported.

Data:

Section Property	Section 152 × 152 × 23 UKC Universal Column
Cross-sectional area (A)	29.2 cm²
Radius of gyration (i_{zz})	3.70 cm
Radius of gyration (i_{yy})	6.54 cm
Depth of the section (h)	152.40 mm
Width of the section (b)	152.20 mm
Flange thickness (t_f)	6.80 mm
Web thickness (t_w)	5.80 mm
Root radius (r)	7.60 mm
2ⁿᵈ Moment of area I_{yy}	1250.0 cm⁴
2ⁿᵈ Moment of area I_{zz}	400.0 cm⁴

Section classification: Class 1

Yield Stress $f_y = 275$ MPa

Young's Modulus $E = 210$ kN/mm²

Buckling curve:
Assume a welded box section where all the longitudinal welds are near the corners of the cross-section.

For both the y–y axis and the z–z axis use curve b.

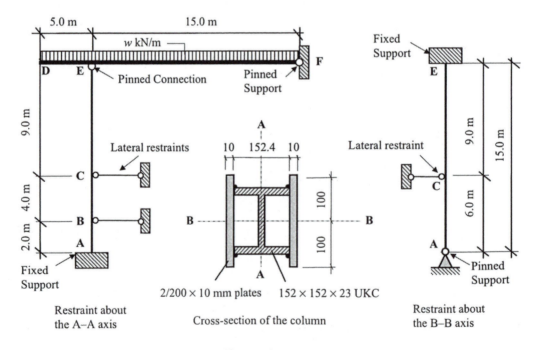

Restraint about the A–A axis

Cross-section of the column

Restraint about the B–B axis

Figure 6.29

Solution:
(i)
$A = [2920 + 2 \times (10 \times 200)] = 6.92 \times 10^3 \text{ mm}^2$

$I_{A-A} = \left[2 \times \left(\dfrac{200 \times 10^3}{12} + \left(10 \times 200 \times 81.2^2\right)\right) + 400 \times 10^4\right] = 30.41 \times 10^6 \text{ mm}^4$

$I_{B-B} = 1250 \times 10^4 + \left(2 \times \dfrac{10 \times 200^3}{12}\right) = 25.83 \times 10^6 \text{ mm}^4$

$i_{A-A} = \sqrt{\dfrac{30.41 \times 10^6}{6.92 \times 10^3}} = 66.29 \text{ mm}; \qquad i_{B-B} = \sqrt{\dfrac{25.83 \times 10^6}{6.92 \times 10^3}} = 61.10 \text{ mm}$

Flexural buckling resistance
Table 5.2 $\qquad\qquad\qquad \varepsilon = (235/275)^{0.5} = 0.92$

Clause 6.3.1.3(1) $\quad \bar{\lambda} = \sqrt{\dfrac{Af_y}{N_{cr}}} = \dfrac{L_{cr}}{i} \times \dfrac{1}{\lambda_1}$ where $\lambda_1 = 93.9\varepsilon = 93.9 \times 0.92 = 86.39$

Consider the A-A axis:
$L_{cr,A} \geq (0.85 \times 2.0) = 1.7 \text{ m}$
$\qquad\quad \geq (1.0 \times 4.0) = 4.0 \text{ m}$
$\qquad\quad \geq (1.0 \times 9.0) = 9.0 \text{ m}$
The critical buckling length $L_{cr,A} = 9.0 \text{ m}$

Equation (6.50) $\quad \bar{\lambda}_A = \dfrac{L_{cr}}{i_A} \times \dfrac{1}{\lambda_1} = \dfrac{9000}{66.29} \times \dfrac{1}{86.39} = 1.57$

Consider the B-B axis:
$L_{cr,B} \geq (1.0 \times 6.0) = 6.0 \text{ m}$
$\qquad\quad \geq (0.85 \times 9.0) = 7.65 \text{ m}$
The critical buckling length $L_{cr,B} = 7.65 \text{ m}$

Equation (6.50) $\quad \bar{\lambda}_B = \dfrac{L_{cr}}{i_B} \times \dfrac{1}{\lambda_1} = \dfrac{7650}{61.10} \times \dfrac{1}{86.39} = 1.45$

Since the same buckling curve is used for both axes the largest value of $\bar{\lambda}$ is used to determine the reduction factor, i.e. $\bar{\lambda} = \mathbf{1.57}$

Equation (6.49) $\quad \chi = \dfrac{1}{\Phi + \sqrt{\Phi^2 - \bar{\lambda}^2}}$ but $\chi \leq 1.0$

where $\quad \Phi = 0.5\left[1 + \alpha\left(\bar{\lambda} - 0.2\right) + \bar{\lambda}^2\right]$

Table 6.1 $\qquad\qquad$ Imperfection factor for curve b: $\alpha = 0.34$
$\qquad\qquad\qquad\quad$ (see Figure 6.24 in this text)
$\qquad\qquad\qquad\quad \Phi = 0.5\left[1 + 0.34\left(1.57 - 0.2\right) + 1.57^2\right] = 1.97$

$$\chi = \frac{1}{1.97 + \sqrt{1.97^2 - 1.57^2}} = 0.32 \text{ but } \chi \le 1,0$$

Alternatively use the curves given in EN 1993-1-1: Figure 6.4
(see Figure 6.26 or Table 6.1 in this text)

Critical value $\chi_z \approx 0{,}32$

Equation (6.47) $N_{b,z,Rd} = \dfrac{\chi A f_y}{\gamma_{M1}} = \dfrac{0{,}32 \times 6920 \times 275}{1{,}0 \times 10^3} = 608.96$ kN

The maximum design axial load with respect to flexural buckling = 608.96 kN

(ii)

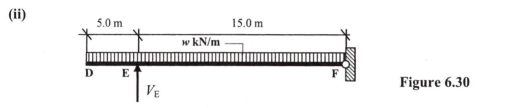

Figure 6.30

The maximum value of the vertical reaction at E = 608.96 kN

+ve $\curvearrowright$ $\Sigma M_F = 0$ $(15.0 \times V_E) - (w \times 20^2/2) = 0$

$\therefore w_{maximum} = (15.0 \times 608.96) / 200 = 45.67$ kN/m

6.9   *Built-up Compression Members*

The advantage of using built-up columns from various different elements e.g. as shown in Figure 6.31, is that they produce relatively light members with relatively high radii of gyration. The buckling of each individual element must be verified in addition to the overall column section. Built-up columns are more flexible than solid columns with the same 2nd moment of area and in addition, the shear stiffness is much smaller.

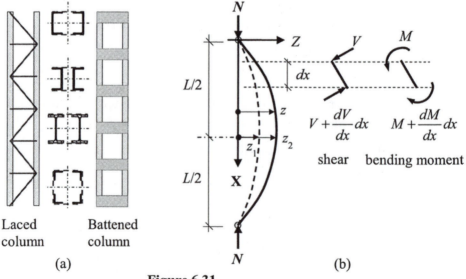

Laced column

Battened column

(a)

(b)

Figure 6.31

Consider the effect of shear deformation on the elastic critical buckling load of a column.

The moment at a point along the column $M = Nz$

The shear at a point along the beam $V = \dfrac{dM}{dx} = N\dfrac{dz}{dx}$

The deflection at mid-height is given by: $z = z_1 + z_2$ where z_1 is due to the bending moment and z_2 is due to the shear force.

The curvature due to bending is given by $\dfrac{d^2 z_1}{dx^2} = \dfrac{1}{R} = -\dfrac{M}{EI} = -\dfrac{Nz}{EI}$

The slope due to the shear force V is given by: $\dfrac{dz_2}{dx} = \beta\dfrac{V}{GA}$

where

β is a shape factor related to the cross-section (= 1,2 for a rectangle),
A is the cross-sectional area and G is the shear modulus.

The curvature due to the shear force V is given by:

$$\dfrac{d^2 z_2}{dx^2} = \dfrac{\beta}{GA}\dfrac{dV}{dx} = \dfrac{\beta N}{GA}\dfrac{d^2 z}{dx^2} \qquad \left(\text{Note: } V = N\dfrac{dz}{dx}; \ \ \dfrac{dV}{dx} = N\dfrac{d^2 z}{dx^2}\right)$$

The total curvature due to the bending moment and the shear force is given by:

$$\dfrac{d^2 z}{dx^2} = \dfrac{d^2 z_1}{dx^2} + \dfrac{d^2 z_2}{dx^2} = -\dfrac{Nz}{EI} + \dfrac{\beta N}{GA}\dfrac{d^2 z}{dx^2}$$

This 2nd order differential equation can be re-written as:

$$\dfrac{d^2 z}{dx^2} = \dfrac{N}{(1 - \beta N/GA)EI}\,z = 0$$

This can be solved to give the critical value for N:

$$\dfrac{N}{(1 - \beta N/GA)EI} = \dfrac{\pi^2}{L^2}$$

Solving for the 2nd order differential above equation results in an expression for the elastic critical load including both bending and shear deformations, i.e.

$$N_{cr,M,V} = \dfrac{1}{\dfrac{1}{N_{cr}} + \dfrac{1}{S_V}} = N_{cr}\dfrac{1}{1 + \dfrac{N_{cr}}{S_V}}$$

where:

N_{cr} is the Euler critical buckling load $= \dfrac{\pi^2 EI}{L^2}$

S_V is the shear stiffness of the column $= \dfrac{GA}{\beta}$

Clearly $N_{cr,M,V} < N_{cr}$. For solid rolled cross-sections, S_V is much greater than N_{cr} and can be neglected in design. In the case of built-up columns, S_V is much smaller than N_{cr} and can be very significant.

6.9.1 Shear Stiffness for Laced Columns

Consider a built-up column with N-shaped lacing as shown in Figure 6.32. The shear stiffness is determined by considering the elongation of one diagonal and one horizontal member as follows:

$$\frac{1}{S_v} = \gamma = \frac{\delta}{a}$$

where the flexibility of the system is represented by $1/S_v$.

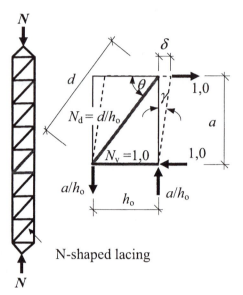

The value of the horizontal displacement δ caused by an applied unit load can be determined using the theory of virtual work.

Figure 6.32

$$\delta = \delta_{\text{horizontal member}} + \delta_{\text{diagonal member}} = \sum \frac{u^2 L}{AE}$$

$$\delta = \sum \frac{u^2 L}{AE} = \frac{N_v^2 L_v}{A_v E} + \frac{N_d^2 L_d}{A_d E} \quad \therefore \quad \delta = \frac{1,0^2 \times h_o}{A_v E} + \frac{\left(d/h_o\right)^2 d}{A_d E} = \frac{h_o}{A_v E} + \frac{d^3}{h_o^2 A_d E}$$

where:
$N_v = 1,0;$ $L_v = h_o;$ $N_d = d/h_o;$ $L_d = d;$ A_v and A_d are the shear areas of the laces.

Flexibility $\dfrac{1}{S_v} = \gamma = \dfrac{\delta}{a}$ $\therefore$ The shear stiffness $S_v = \dfrac{a}{\delta}$

δ can be re-written as: $\delta = \dfrac{d^3}{h_o^2 A_d E}\left(\dfrac{A_d h_o^3}{A_v d^3} + 1\right)$

and the shear stiffness as: $S_v = \dfrac{a}{\delta} = \dfrac{ah^2 A_d E}{d^3\left(\dfrac{A_d h_o^3}{A_v d^3} + 1\right)}$ for each plane of lacings.

This is indicated in EN 1993-1-1: Clause 6.4.2.1/Figure 6.9. Similar values are also given for alternative lacing systems.

Determination of the design load in a built-up column is based on the assumption that it can be represented by a simple elastic column with an equivalent initial imperfection and shear flexibility as shown in Figure 6.33.

The initial curvature is assumed to be sinusoidal, i.e.

$$z_o = e_o \sin(\pi x/l)$$

where e_o is the bow imperfection equal to $l/500$ as indicated in EN 1993-1-1: Clause 6.4.1(1).

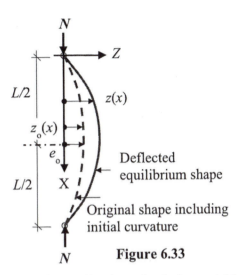

Figure 6.33

The initial geometric imperfections are amplified by the application of a design axial load N such that:

$$z(x) = z_o(x)\dfrac{1}{1-\dfrac{N}{N_{cr,M,V}}} = \dfrac{e_o}{1-\dfrac{N}{N_{cr,M,V}}}\sin(\pi x/l)$$

At the mid-span point of the column, the axial load is N and the bending moment is given by:

$$M = \left(N \times z_{(0,5L)}\right) = \dfrac{e_o N}{1-\dfrac{N}{N_{cr,M,V}}}$$

The maximum axial load in the most loaded chord member is given by:

$$N = \dfrac{N}{2} + \dfrac{M}{h} = N\left(\dfrac{1}{2} + \dfrac{e_o/h}{1-\dfrac{N}{N_{cr,M,V}}}\right) \qquad \text{where } N_{cr,M,V} = N_{cr}\dfrac{1}{1+\dfrac{N_{cr}}{S_V}}$$

$$\left(1-\dfrac{N}{N_{cr,M,V}}\right) = 1 - \dfrac{N}{N_{cr}\dfrac{1}{1+\dfrac{N_{cr}}{S_V}}} = 1 - \dfrac{N\left(1+\dfrac{N_{cr}}{S_V}\right)}{N_{cr}} = 1 - \dfrac{N}{N_{cr}} - \dfrac{N}{S_V}$$

$$N = \dfrac{N}{2} + \dfrac{M}{h} = N\left(\dfrac{1}{2} + \dfrac{e_o/h}{1-\dfrac{N}{N_{cr}}-\dfrac{N}{S_v}}\right)$$

This is similar to the formulation given in EN 1993-1-1: Equation (6.69).

In the case of laced columns the effective length is taken as equal to the system length between the lacing connections; in battened columns (disregarding any possible end restraint) the effective length of the chords is taken as equal to the distance between the centre lines of the battens.

The forces in the lacing members and in the chords adjacent to the ends are derived from the shearing force V and the axial load N.

The shear force at the ends of a built-up column is given by:

$$V = N\left(\frac{dz}{dx}\right)_{x=0} = Ne_0 \frac{\pi}{l} \times \frac{1}{1-N/N_{cr,M.V}} = \frac{\pi}{l}M$$

This is similar to the formulation given in EN 1993-1-1: Equation (6.70).

The axial force in the diagonal lacings of a built-up column is given by:

$$N = \frac{Vd}{nh_0}$$

In the case of battened columns the chords, battens and their connections to the chords, are designed to resist the bending moments, shear forces and axial loads indicated in Figure 6.34

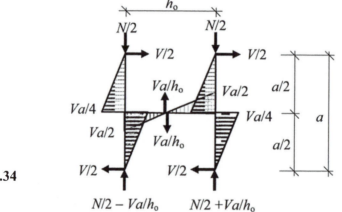

Figure 6.34

The laced built-up column shown in Figure 6.35 is required to support a design axial load of 4000 kN. The ends of the column are assumed to be pinned and the laces occur on both faces. Using the design data given:

(i)   determine the section classification for chord members,

(ii)   verify the suitability, or otherwise, of the compression resistance of the chord members,

(iii) verify the suitability, or otherwise, of the 457 x 191 x 74 UKB S275 for the flexural buckling resistance chord section,

(iv) verify the suitability, or otherwise, of the 50 mm x 10 mm thick S275 flat plates for the flexural buckling resistance of the laces.

Design data:
Chords   457 x 191 x 74 UKB section
Lacings   60 mm wide x 12 mm thick flat plates welded to flanges at 850 mm centres
Centre-to-centre distance of UKB sections (h_o)   550 mm

Section properties of each 457 x 191 x 74 UKB S275 chord member:
$h = 457.0$ mm $b = 190.4$ mm $t_w = 9.0$ mm $t_f = 14.5$ mm $A_g = 94.60$ cm^2
$i_y = 18.80$ cm $i_z = 4.20$ cm $r = 10.2$ mm $I_y = 33300$ cm^4 $I_z = 1670$ cm^4

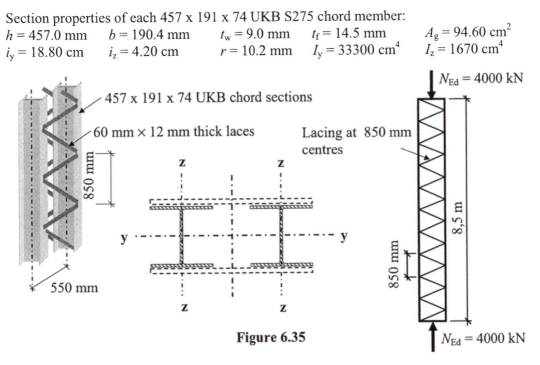

Figure 6.35

Solution:

Length of the lacings $d = (550^2 + 425^2)^{0.5} = 695.07$ mm

EN 10025-2:2004
Table 7   For S275 steel thickness $t_f = 14.5$ mm
                          $f_y = 275$ MPa and $f_u = 410$ MPa

EN 1993-1-1:2005
Clause 6.1(1) $\gamma_{M0} = 1,0$
Clause 3.2.6 $E = 210000$ MPa, $G = 81000$ MPa

(i)   Section classification for a 457 x 191 x 74 UKB S275 (compression)
                          $\varepsilon = \sqrt{235/f_y} = \sqrt{235/275} = 0.92$

Table 5.2(1) Web:
$c = [h - 2(t_f + r)] = [457.0 - 2 \times (14.5 + 10.2)] = 407.6$ mm
$c/t = (407.6/9,0) = 45.29$
$42\varepsilon = (42 \times 0.92) = 38.64$
$c/t > 42\varepsilon$ $\therefore$ The web is Class 4

Table 5.2(2) Flange:
$c = [b - (t_w + 2r)]/2 = [190.4 - (9.0 + 2 \times 10.2)]/2 = 80.50$ mm
$c/t = (80.50/14.5) = 5.55$
$9\varepsilon = (9 \times 0.92) = 8.28$
$c/t < 9\varepsilon$ $\therefore$ The flanges are Class 1
                                                    Section is Class 4

Determine the effective area of the web in accordance with EN 1993-1-5: Clause 4.4.
EN 1993-1-5:2006

Clause 4.4(1)   'The effective area, A_{eff}, is determined assuming that the cross-section is subject only to stresses due to uniform axial compression.'
$A_{c,eff} = \rho A_c$
Consider a single chord member:

The plate slenderness is given by: $\bar{\lambda}_p = \dfrac{\bar{b}/t}{28.4\varepsilon\sqrt{k_\sigma}}$

Table 4.1   For internal compression elements with uniform compression

$\psi = \sigma_2/\sigma_1 = 1.0$ and $k_\sigma = 4.0$ $\therefore \bar{\lambda}_p = \dfrac{45.29}{28.4 \times 0.92 \times \sqrt{4.0}} = 0.867$

Clause 4.4(2)   Reduction factor for plate widths (ρ)
$\rho = 1.0$ for $\bar{\lambda}_p \le 0.5 + \sqrt{0.085 - 0.055\psi} = 0.673$

$\rho = \dfrac{\bar{\lambda}_p - 0,055(3 + \psi)}{\bar{\lambda}_p^2} \le 1.0$ for $\bar{\lambda}_p > 0.5 + \sqrt{0.085 - 0.055\psi}$

For the web:
$\bar{\lambda}_p = 0.867$ and $\rho = \dfrac{0.867 - 0.055(3 + 1.0)}{0.867^2} = 0.861 \le 1.0$
$b_{eff} = (0.861 \times 407.60) = 350.94$ mm
Length of the non-effective zone = 56.66 mm

Reduction in the area = $(407.60 - 350.94) \times 9.0 = 509.94$ mm^2
$A_{c,eff,} = (A_g - 5.10) = (94.60 - 5.10) = 89.50$ cm^2

Reduction in second moment of area I_y
$= \dfrac{9.0 \times 56.66^3}{12} = 13.642 \times 10^4$ mm^4
$I_{eff,y} = (33300 - 13.642) \times 10^4 = 33286.36 \times 10^4$ mm^4

Reduction in second moment of area I_z

$$= \frac{56.66 \times 9.0^3}{12} = 0.344 \times 10^4 \text{ mm}^4$$

$$I_{\text{eff,z}} = (1670.0 - 0.344) \times 10^4 = 1669.66 \times 10^4 \text{ mm}^4$$

(ii)   Verification for compression resistance of the chord members
EN 1993-1-1:2005

Equation (6.11) $N_{\text{c,Rd}} = \dfrac{A_{\text{eff}} f_y}{\gamma_{\text{M0}}}$ for Class 4 cross-sections

Cross-sectional area $A = (2 \times 8950) = 17.90 \times 10^3 \text{ mm}^2$

Verification: $N_{\text{c,Rd}} = \dfrac{17.9 \times 10^3 \times 275}{1.0 \times 10^3} = 4922.5 \text{ kN} > 4000 \text{ kN}$

The chord section is satisfactory with respect to the compression resistance

(iii)   Verification for flexural buckling resistance of the chord members
EN 1993-1-1:2005

Second moment of area of built-up section:

Clause 6.4.2.1(4) $I_{\text{eff}} = 0.5 h_0^2 A_{\text{ch}} = 0.5 \times 550^2 \times 8950 = 1353.68 \times 10^6 \text{ mm}^4$

Shear stiffness of lacings:

Figure 6.9 $S_v = \dfrac{n E A_d a h_0^2}{2 d^3}$ where $A_d = (50 \times 10) = 500 \text{ mm}^2$ and $n = 2$

$$S_v = \frac{2 \times 210 \times 500 \times 850 \times 550^2}{2 \times 695.07^3} = 80398.5 \text{ kN/mm}$$

Chord design force at mid-height:

Clause 6.4.1(6) $N_{\text{ch,Ed}} = 0.5 N_{\text{Ed}} + \dfrac{M_{\text{Ed}} h_0 A_{\text{ch}}}{2 I_{\text{eff}}}$

Equation (6.69)   where $M_{\text{Ed}} = \dfrac{N_{\text{Ed}} e_0 + M_{\text{Ed}}^1}{1 - \dfrac{N_{\text{Ed}}}{N_{\text{cr}}} - \dfrac{N_{\text{Ed}}}{S_v}}$ and $N_{\text{cr}} = \dfrac{\pi^2 E I_{\text{eff}}}{L^2}$

$M_{\text{Ed}}^1 = 0$ and $\dfrac{M_{\text{Ed}} h_0 A_{\text{ch}}}{2 I_{\text{eff}}} = \dfrac{M_{\text{Ed}}}{h_0}$ when $I_{\text{eff}} = 0.5 h^2 A_{\text{ch}}$

Clause 6.4.1(1)   Bow imperfection $e_0 = L/500 = 8500/500 = 17 \text{ mm}$

$$N_{\text{cr}} = \frac{\pi^2 E I_{\text{eff}}}{L^2} = \frac{\pi^2 \times 210 \times 1353.68 \times 10^6}{8500^2} = 38832.6 \text{ kN}$$

$$M_{\text{Ed}} = \frac{N_{\text{Ed}} e_0 + M_{\text{Ed}}^1}{1 - \dfrac{N_{\text{Ed}}}{N_{\text{cr}}} - \dfrac{N_{\text{Ed}}}{S_v}} = \frac{4000 \times 17.0 + 0}{1 - \dfrac{4000}{38832.6} - \dfrac{4000}{80398.5}} = 80260.5 \text{ kNmm}$$

$$N_{ch,Ed} = \left(0.5N_{Ed} + \frac{M_{Ed}h_0 A_{ch}}{2I_{eff}}\right) = \left(0.5N_{Ed} + \frac{M_{Ed}}{h_0}\right)$$

$$= (0.5 \times 4000) + \frac{80260.5}{550} = 2145.93 \text{ kN}$$

Design axial load $N_{ch,Ed}$ = 2145.93 kN

Clause 6.3.1.3(1) $\bar{\lambda} = \sqrt{\dfrac{A_{eff} f_y}{N_{cr}}} = \dfrac{L_{cr}}{i_y} \times \dfrac{\sqrt{\dfrac{A_{eff\,y}}{A}}}{\lambda_1}$ where $\lambda_1 = 93.9\varepsilon = 93.9 \times \sqrt{\dfrac{235}{275}} = 86.80$

Consider the y-y axis: assume $L_{cr,y}$ = 8.50 m
Use the radius of gyration based on the gross cross-section
i_y = 18.80 cm

Equation (6.50) $\bar{\lambda}_y = \dfrac{L_{cr}}{i_y} \times \dfrac{1}{\lambda_1} = \dfrac{8500}{188.0} \times \dfrac{1}{86.8} = 0.52$

Table 6.2 h/b = 457.0/190.4 = 2.40 ≥ 1.2   and t_f < 40 mm
For buckling about the y-y axis use curve a

Figure 6.4 $\chi_y \approx 0.92$

Consider the z-z axis: assume $L_{cr,z}$ = 850 mm
Use the radius of gyration based on the gross cross-section
i_z = 4.20 cm

Equation (6.50) $\bar{\lambda}_z = \dfrac{L_{cr}}{i_z} \times \dfrac{1}{\lambda_1} = \dfrac{850}{42.0} \times \dfrac{1}{86.8} = 0.23$

Table 6.2 h/b = 457.0/190.4 = 2.40 ≥ 1.2   and t_f < 40 mm
For buckling about the z-z axis use curve b

Figure 6.4 $\chi_z \approx 0.98$

Critical value $\chi_y \approx 0.92$

Equation (6.47) $N_{y,b,Rd} = \dfrac{\chi_y A_{eff} f_y}{\gamma_{M1}} = \dfrac{0.92 \times 8950 \times 275}{1.0 \times 10^3} = 2264.35 \text{ kN}$

Equation (6.46)   **Verification:** $\dfrac{N_{Ed}}{N_{b,Rd}} = \dfrac{2145.93}{2264.35} = 0.95 < 1,0$

The chord section is satisfactory with respect to the flexural buckling resistance
(Note: the torsional and torsional–flexural buckling resistance should also be checked,
e.g. when using channel sections.)

(iv)   Verification for flexural buckling resistance of the laces

EN 1993-1-1:2005

The axial force in the laces N_{Ed} is given by:

$$N_{Ed} = \frac{V_{Ed}d}{nh_0} \quad \text{(see Figure 6.33)}$$

Equation (6.70) $V_{Ed} = \pi \dfrac{M_{Ed}}{L} = \pi \dfrac{80260.5}{8500} = 29.66 \text{ kN}$

$$\therefore \; N_{Ed} = \frac{29.66 \times 695.07}{2 \times 550} = 18.74 \text{ kN}$$

Consider the z-z axis: assume $L_{cr,z} = d = 695.07$ mm

$$i_y = i_{d,z} = \sqrt{\frac{I_{d,z}}{A_d}} = h_d/\sqrt{12} = 12.0/\sqrt{12} = 3.46 \text{ mm}$$

Equation (6.50) $\overline{\lambda}_{d,z} = \dfrac{L_{cr}}{i_z} \times \dfrac{1}{\lambda_1} = \dfrac{695.07}{3.46} \times \dfrac{1}{86.8} = 2.31$

Table 6.2   For solid sections use buckling curve c for any axis

Figure 6.4 $\chi_{dz} \approx 0.14$

Clause 6.3.1.1(3) $N_{dz,b,Rd} = \dfrac{\chi_{d,z} A f_y}{\gamma_{M1}} = \dfrac{0.14 \times 60 \times 12 \times 275}{1.0 \times 10^3} = 27.72 \text{ kN}$

Equation (6.46)   **Verification:** $\dfrac{N_{Ed}}{N_{b,Rd}} = \dfrac{18.74}{27.72} = 0.67 < 1,0$

The lacing section is satisfactory with respect to the flexural buckling resistance

6.11 *Problems: Buckling Instability*

A selection of column cross-sections is indicated in Problems 6.1 to 6.4 in addition to the position of the restraints about the y–y and z–z axes.

  (a) Considering flexural buckling only and using the data provided, determine the compression resistance of the columns using the EN 1993-1-1 flexural buckling formulae.

  (b) Verify the suitability of the chords for the laced column shown in Problem 6.5.

Data:

Problem No.	f_y (N/mm^2)	E (kN/mm^2)	Buckling curve	
			y–y axis	z–z axis
6.1	275	210	(b)	(c)
6.2	255	210	(b)	(b)
6.3	275	210	(b)	(b)
6.4	275	210	(c)	(c)
6.5	355*	210	(a)	(a)
* Hot rolled hollow sections are not available in S275 steel				

Table 6.3: Material Property and Buckling Curve Data

Section Property	Section			
	533 × 210 × 82 UKB	457 × 152 × 52 UKB	200 × 90 × 30 UKPFC (Channel)	150 × 100 × 10 RHS (Hollow Section)
Cross-sectional area (A)	105.0 cm²	66.6 cm²	37.9 cm²	44.9 cm²
Radius of gyration (i_{zz})	4.38 cm	3.11 cm	2.88 cm	3.85 cm
Radius of gyration (i_{yy})	21.30 cm	17.90 cm	8.16 cm	5.34 cm
Depth of the section (h)	528.30 mm	449.8 mm	200.0 mm	150.0 mm
Width of the section (b)	208.80 mm	152.4 mm	90.0 mm	100.0 mm
Flange thickness (t_f)	13.20 mm	10.90 mm	14.0 mm	10.0 mm
Web thickness (t_w)	9.6 mm	7.60 mm	7.0 mm	10.0 mm
Root radius (r)	12.70 mm	10.20 mm	12.0 mm	-
2nd Moment of area I_{yy}	47500.0 cm⁴	21400.0 cm⁴	2520.0 cm⁴	1280.0 cm⁴
2nd Moment of area I_{zz}	2010.0 cm⁴	645.0 cm⁴	314.0 cm⁴	665.0 cm⁴

Table 6.4- Section Property Data

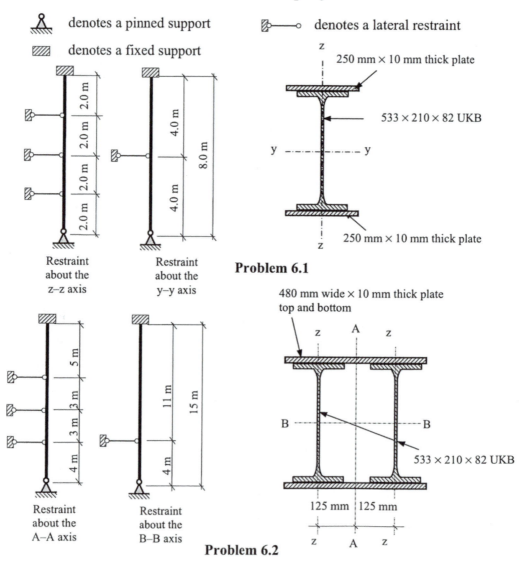

Problem 6.1

Problem 6.2

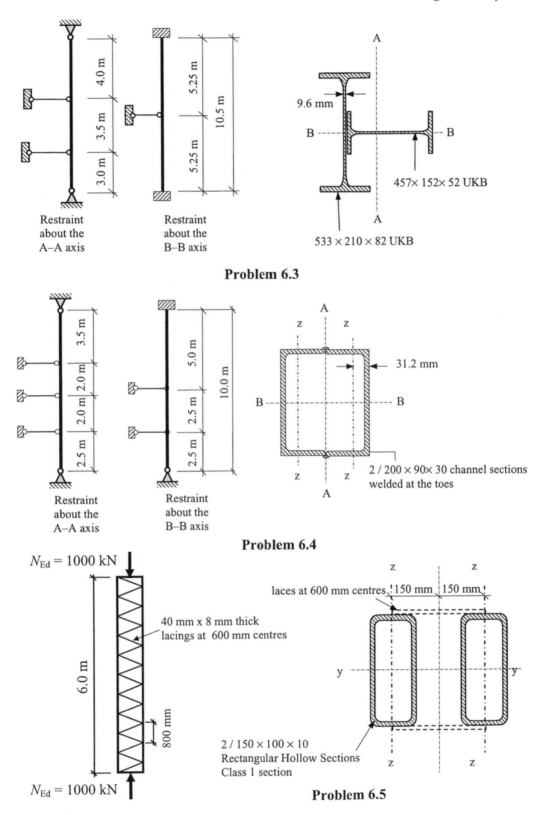

Problem 6.3

Problem 6.4

Problem 6.5

6.12   *Solutions: Buckling Instability*

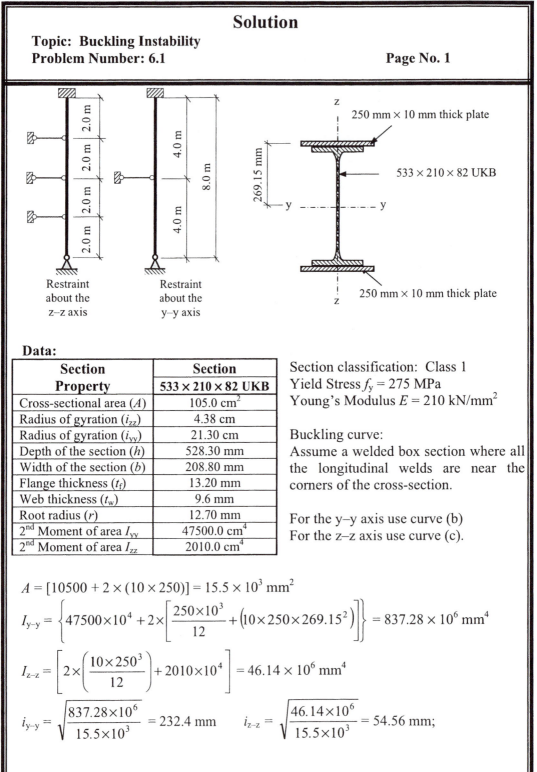

Solution

Topic: Buckling Instability
Problem Number: 6.1   **Page No. 1**

Restraint about the z–z axis

Restraint about the y–y axis

250 mm × 10 mm thick plate

533 × 210 × 82 UKB

269.15 mm

250 mm × 10 mm thick plate

Data:

Section Property	Section 533 × 210 × 82 UKB
Cross-sectional area (A)	105.0 cm^2
Radius of gyration (i_{zz})	4.38 cm
Radius of gyration (i_{yy})	21.30 cm
Depth of the section (h)	528.30 mm
Width of the section (b)	208.80 mm
Flange thickness (t_f)	13.20 mm
Web thickness (t_w)	9.6 mm
Root radius (r)	12.70 mm
2nd Moment of area I_{yy}	47500.0 cm^4
2nd Moment of area I_{zz}	2010.0 cm^4

Section classification: Class 1
Yield Stress f_y = 275 MPa
Young's Modulus E = 210 kN/mm^2

Buckling curve:
Assume a welded box section where all the longitudinal welds are near the corners of the cross-section.

For the y–y axis use curve (b)
For the z–z axis use curve (c).

$$A = [10500 + 2 \times (10 \times 250)] = 15.5 \times 10^3 \text{ mm}^2$$

$$I_{y-y} = \left\{ 47500 \times 10^4 + 2 \times \left[\frac{250 \times 10^3}{12} + \left(10 \times 250 \times 269.15^2\right) \right] \right\} = 837.28 \times 10^6 \text{ mm}^4$$

$$I_{z-z} = \left[2 \times \left(\frac{10 \times 250^3}{12} \right) + 2010 \times 10^4 \right] = 46.14 \times 10^6 \text{ mm}^4$$

$$i_{y-y} = \sqrt{\frac{837.28 \times 10^6}{15.5 \times 10^3}} = 232.4 \text{ mm} \qquad i_{z-z} = \sqrt{\frac{46.14 \times 10^6}{15.5 \times 10^3}} = 54.56 \text{ mm};$$

Solution

Topic: Buckling Instability
Problem Number: 6.1
Page No. 2

Flexural buckling resistance

EN 1993-1-1:2005

Table 5.2 $\varepsilon = (235/275)^{0.5} = 0.92$

Clause 6.3.1.3(1) $\bar{\lambda} = \sqrt{\dfrac{Af_y}{N_{cr}}} = \dfrac{L_{cr}}{i} \times \dfrac{1}{\lambda_1}$ where $\lambda_1 = 93.9\varepsilon = 93.9 \times 0.92 = 86.39$

Consider the y-y axis:

$L_{cr,y} \geq (0.85 \times 4.0) = 3.4$ m

$\qquad \geq (1.0 \times 4.0) = 4.0$ m   The critical buckling length $L_{cr,y} = 4.0$ m

Equation (6.50) $\bar{\lambda}_y = \dfrac{L_{cr}}{i_y} \times \dfrac{1}{\lambda_1} = \dfrac{3400}{232.4} \times \dfrac{1}{86.39} = 0.17$

Figure 6.4   Since $\bar{\lambda}_y \leq 0.2$ the reduction factor $\chi = 1.0$

     (see Figure 6.26/Table 6.1 in this text)

Consider the z-z axis:

$L_{cr,z} \geq (0.85 \times 2.0) = 1.7$ m

$\qquad \geq (1.0 \times 2.0) = 2.0$ m   The critical buckling length $L_{cr,z} = 2.0$ m

Equation (6.50) $\bar{\lambda}_z = \dfrac{L_{cr}}{i_z} \times \dfrac{1}{\lambda_1} = \dfrac{2000}{54.56} \times \dfrac{1}{86.39} = 0.42$

Equation (6.49) $\chi = \dfrac{1}{\Phi + \sqrt{\Phi^2 - \bar{\lambda}^2}}$ but $\chi \leq 1.0$

     where $\Phi = 0.5\left[1 + \alpha\left(\bar{\lambda} - 0.2\right) + \bar{\lambda}^2\right]$

Table 6.1   Imperfection factor for curve c: $\alpha = 0.49$
     (see Figure 6.24 in this text)

     $\Phi = 0.5\left[1 + 0.49\left(0.42 - 0.2\right) + 0.42^2\right] = 0.64$

     $\chi = \dfrac{1}{0.64 + \sqrt{0.64^2 - 0.42^2}} = 0.89$ but $\chi \leq 1,0$

     Alternatively use the curves given in EN 1993-1-1: Figure 6.4
     (see Figure 6.26 or Table 6.1 in this text)

                               Critical value $\chi_z \approx 0{,}89$

Equation (6.47) $N_{b,z,Rd} = \dfrac{\chi A f_y}{\gamma_{M1}} = \dfrac{0{,}89 \times 15.5 \times 10^3 \times 275}{1{,}0 \times 10^3} = 3793.6$ kN

The maximum design axial load with respect to flexural buckling = 3793.6 kN

Solution

Topic: Buckling Instability
Problem Number: 6.2   **Page No. 1**

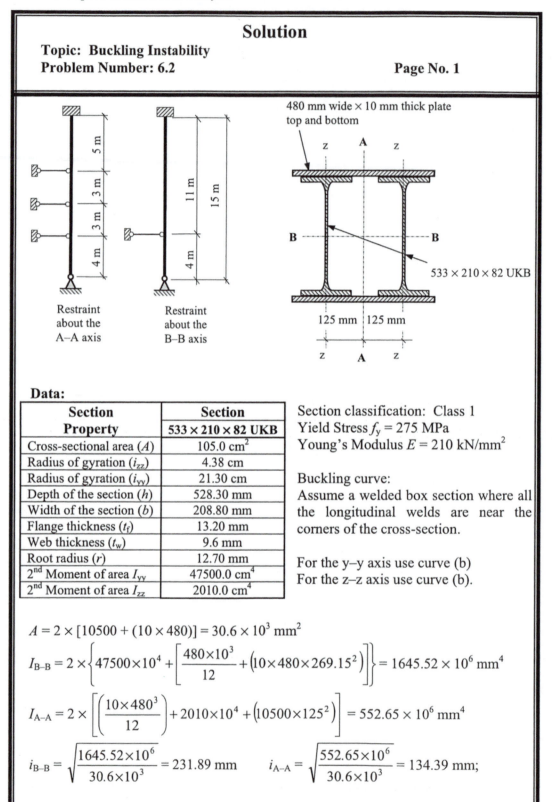

Restraint about the A–A axis

Restraint about the B–B axis

480 mm wide × 10 mm thick plate top and bottom

533 × 210 × 82 UKB

125 mm   125 mm

Data:

Section Property	Section 533 × 210 × 82 UKB
Cross-sectional area (A)	105.0 cm^2
Radius of gyration (i_{zz})	4.38 cm
Radius of gyration (i_{yy})	21.30 cm
Depth of the section (h)	528.30 mm
Width of the section (b)	208.80 mm
Flange thickness (t_f)	13.20 mm
Web thickness (t_w)	9.6 mm
Root radius (r)	12.70 mm
2nd Moment of area I_{yy}	47500.0 cm^4
2nd Moment of area I_{zz}	2010.0 cm^4

Section classification: Class 1
Yield Stress f_y = 275 MPa
Young's Modulus E = 210 kN/mm^2

Buckling curve:
Assume a welded box section where all the longitudinal welds are near the corners of the cross-section.

For the y–y axis use curve (b)
For the z–z axis use curve (b).

$$A = 2 \times [10500 + (10 \times 480)] = 30.6 \times 10^3 \text{ mm}^2$$

$$I_{B-B} = 2 \times \left\{ 47500 \times 10^4 + \left[\frac{480 \times 10^3}{12} + \left(10 \times 480 \times 269.15^2 \right) \right] \right\} = 1645.52 \times 10^6 \text{ mm}^4$$

$$I_{A-A} = 2 \times \left[\left(\frac{10 \times 480^3}{12} \right) + 2010 \times 10^4 + \left(10500 \times 125^2 \right) \right] = 552.65 \times 10^6 \text{ mm}^4$$

$$i_{B-B} = \sqrt{\frac{1645.52 \times 10^6}{30.6 \times 10^3}} = 231.89 \text{ mm} \qquad i_{A-A} = \sqrt{\frac{552.65 \times 10^6}{30.6 \times 10^3}} = 134.39 \text{ mm};$$

Solution

Topic: **Buckling Instability**
Problem Number: 6.2   **Page No. 2**

Flexural buckling resistance

EN 1993-1-1:2005

Table 5.2 $\varepsilon = (235/275)^{0.5} = 0.92$

Clause 6.3.1.3(1) $\bar{\lambda} = \sqrt{\dfrac{Af_y}{N_{cr}}} = \dfrac{L_{cr}}{i} \times \dfrac{1}{\lambda_1}$ where $\lambda_1 = 93.9\varepsilon = 93.9 \times 0.92 = 86.39$

Consider the B-B axis:

$L_{cr,B} \geq (0.85 \times 11.0) = 9.35$ m
$\qquad \geq (1.0 \times 4.0) = 4.0$ m   The critical buckling length $L_{cr,B} = 9.35$ m

Equation (6.50) $\bar{\lambda}_B = \dfrac{L_{cr}}{i_B} \times \dfrac{1}{\lambda_1} = \dfrac{9350}{231.89} \times \dfrac{1}{86.39} = 0.47$

Consider the A-A axis:

$L_{cr,A} \geq (0.85 \times 5.0) = 4.25$ m
$\qquad \geq (1.0 \times 3.0) = 3.0$ m
$\qquad \geq (1.0 \times 4.0) = 4.0$ m   The critical buckling length $L_{cr,A} = 4.25$ m

Equation (6.50) $\bar{\lambda}_A = \dfrac{L_{cr}}{i_A} \times \dfrac{1}{\lambda_1} = \dfrac{4250}{134.39} \times \dfrac{1}{86.39} = 0.37$

Since the same curve is used for both axes the critical value of $\bar{\lambda}_{cr} = \bar{\lambda}_B = 0.47$

Equation (6.49) $\chi = \dfrac{1}{\Phi + \sqrt{\Phi^2 - \bar{\lambda}^2}}$ but $\chi \leq 1.0$

$\qquad$ where $\Phi = 0.5\left[1 + \alpha\left(\bar{\lambda} - 0.2\right) + \bar{\lambda}^2\right]$

Table 6.1   Imperfection factor for curve b: $\alpha = 0.34$
$\qquad$ (see Figure 6.24 in this text)

$\qquad \Phi = 0.5\left[1 + 0.34\left(0.47 - 0.2\right) + 0.47^2\right] = 0.66$

$\qquad \chi = \dfrac{1}{0.66 + \sqrt{0.66^2 - 0.47^2}} = 0.89$ but $\chi \leq 1,0$

$\qquad$ Alternatively use the curves given in EN 1993-1-1: Figure 6.4
$\qquad$ (see Figure 6.26 or Table 6.1 in this text)

Critical value $\chi \approx 0{,}89$

Equation (6.47) $N_{b,z,Rd} = \dfrac{\chi Af_y}{\gamma_{M1}} = \dfrac{0{,}89 \times 30.6 \times 10^3 \times 275}{1{,}0 \times 10^3} = 7489.4$ kN

The maximum design axial load with respect to flexural buckling = 7489.4 kN

Solution

Topic: Buckling Instability
Problem Number: 6.3   **Page No. 1**

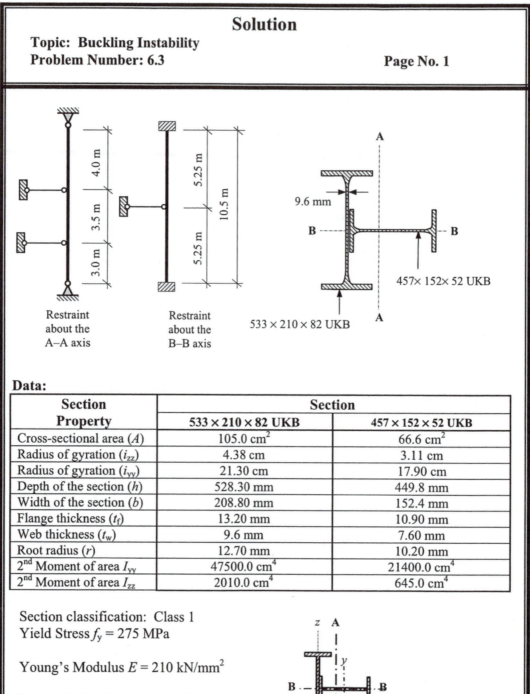

Restraint
about the
A–A axis

Restraint
about the
B–B axis

$533 \times 210 \times 82$ UKB

9.6 mm

$457 \times 152 \times 52$ UKB

Data:

Section Property	Section	
	533 × 210 × 82 UKB	**457 × 152 × 52 UKB**
Cross-sectional area (A)	105.0 cm^2	66.6 cm^2
Radius of gyration (i_{zz})	4.38 cm	3.11 cm
Radius of gyration (i_{yy})	21.30 cm	17.90 cm
Depth of the section (h)	528.30 mm	449.8 mm
Width of the section (b)	208.80 mm	152.4 mm
Flange thickness (t_f)	13.20 mm	10.90 mm
Web thickness (t_w)	9.6 mm	7.60 mm
Root radius (r)	12.70 mm	10.20 mm
2nd Moment of area I_{yy}	47500.0 cm^4	21400.0 cm^4
2nd Moment of area I_{zz}	2010.0 cm^4	645.0 cm^4

Section classification: Class 1
Yield Stress $f_y = 275$ MPa

Young's Modulus $E = 210$ kN/mm^2

For the B–B axis use curve (b)
For the A–A axis use curve (b).

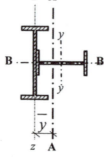

Solution

Topic: **Buckling Instability**
Problem Number: **6.3**   Page No. 2

$A = (10500 + 6660) = 17.16 \times 10^3 \text{ mm}^2$

$\bar{y} = \dfrac{\left[6660 \times (4.8 + 224.9)\right]}{17.16 \times 10^3} = 89.15 \text{ mm}$

$I_{B-B} = \left[\left(47500 \times 10^4\right) + \left(645 \times 10^4\right)\right] = 481.45 \times 10^6 \text{ mm}^4$

$I_{A-A} = \left[2010 \times 10^4 + \left(10500 \times 89.15^2\right)\right] + \left[21400 \times 10^4 + \left(6660 \times 140.55^2\right)\right]$

$\qquad = 449.11 \times 10^6 \text{ mm}^4$

$i_{B-B} = \sqrt{\dfrac{481.85 \times 10^6}{17.16 \times 10^3}} = 167.51 \text{ mm}$ $\qquad$ $i_{A-A} = \sqrt{\dfrac{449.11 \times 10^6}{17.16 \times 10^3}} = 161.78 \text{ mm};$

Flexural buckling resistance
EN 1993-1-1:2005
Table 5.2 $\qquad$ $\varepsilon = (235/275)^{0.5} = 0.92$

Clause 6.3.1.3(1) $\quad \bar{\lambda} = \sqrt{\dfrac{A f_y}{N_{cr}}} = \dfrac{L_{cr}}{i} \times \dfrac{1}{\lambda_1}$ where $\lambda_1 = 93.9\varepsilon = 93.9 \times 0.92 = 86.39$

Consider the B-B axis:
$L_{cr,B} \geq (0.85 \times 5.25) = 4.463 \text{ m}$
$\qquad\qquad\qquad\qquad$ The critical buckling length $L_{cr,B} = 4.463 \text{ m}$

Equation (6.50) $\quad \bar{\lambda}_B = \dfrac{L_{cr}}{i_B} \times \dfrac{1}{\lambda_1} = \dfrac{4463}{167.51} \times \dfrac{1}{86.39} = 0.31$

Consider the A-A axis:
$L_{cr,A} \geq (1.0 \times 3.0) = 3.0 \text{ m}$
$\qquad\quad \geq (1.0 \times 3.5) = 3.5 \text{ m}$
$\qquad\quad \geq (1.0 \times 4.0) = 4.0 \text{ m}$ $\quad$ The critical buckling length $L_{cr,A} = 4.0 \text{ m}$

Equation (6.50) $\quad \bar{\lambda}_A = \dfrac{L_{cr}}{i_A} \times \dfrac{1}{\lambda_1} = \dfrac{4000}{161.78} \times \dfrac{1}{86.39} = 0.28$

Since the same curve is used for both axes the critical value of $\bar{\lambda}_{cr} = \bar{\lambda}_B = 0.31$

Equation (6.49) $\quad \chi = \dfrac{1}{\Phi + \sqrt{\Phi^2 - \bar{\lambda}^2}}$ but $\chi \leq 1.0$

$\qquad\quad$ where $\quad \Phi = 0.5\left[1 + \alpha\left(\bar{\lambda} - 0.2\right) + \bar{\lambda}^2\right]$

Table 6.1 $\qquad$ Imperfection factor for curve b: $\alpha = 0.34$
$\qquad\qquad$ (see Figure 6.24 in this text)

Solution

Topic: Buckling Instability
Problem Number: 6.3   **Page No. 3**

$$\Phi = 0.5\left[1+0.34\left(0.31-0.2\right)+0.31^2\right]=0.57$$

$$\chi = \frac{1}{0.57+\sqrt{0.57^2-0.31^2}} = 0.95 \quad \text{but} \quad \chi \leq 1,0$$

Alternatively use the curves given in EN 1993-1-1: Figure 6.4
(see Figure 6.26 or Table 6.1 in this text)

Critical value $\chi \approx 0,95$

Equation (6.47) $N_{b,z,Rd} = \dfrac{\chi A f_y}{\gamma_{M1}} = \dfrac{0,95\times17.6\times10^3\times275}{1,0\times10^3} = 4598.0 \text{ kN}$

The maximum design axial load with respect to flexural buckling = 4598.0 kN

Solution

Topic: Buckling Instability
Problem Number: 6.4   **Page No. 1**

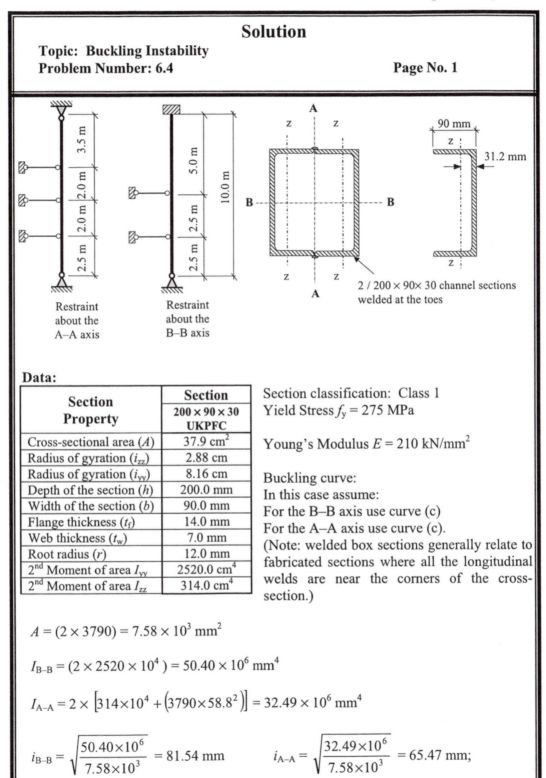

Restraint
about the
A–A axis

Restraint
about the
B–B axis

2 / 200 × 90× 30 channel sections
welded at the toes

Data:

Section Property	Section 200 × 90 × 30 UKPFC
Cross-sectional area (A)	37.9 cm²
Radius of gyration (i_{zz})	2.88 cm
Radius of gyration (i_{yy})	8.16 cm
Depth of the section (h)	200.0 mm
Width of the section (b)	90.0 mm
Flange thickness (t_f)	14.0 mm
Web thickness (t_w)	7.0 mm
Root radius (r)	12.0 mm
2ⁿᵈ Moment of area I_{yy}	2520.0 cm⁴
2ⁿᵈ Moment of area I_{zz}	314.0 cm⁴

Section classification: Class 1
Yield Stress f_y = 275 MPa

Young's Modulus E = 210 kN/mm²

Buckling curve:
In this case assume:
For the B–B axis use curve (c)
For the A–A axis use curve (c).
(Note: welded box sections generally relate to fabricated sections where all the longitudinal welds are near the corners of the cross-section.)

$$A = (2 \times 3790) = 7.58 \times 10^3 \text{ mm}^2$$

$$I_{\text{B–B}} = (2 \times 2520 \times 10^4) = 50.40 \times 10^6 \text{ mm}^4$$

$$I_{\text{A–A}} = 2 \times \left[314 \times 10^4 + \left(3790 \times 58.8^2\right)\right] = 32.49 \times 10^6 \text{ mm}^4$$

$$i_{\text{B–B}} = \sqrt{\frac{50.40 \times 10^6}{7.58 \times 10^3}} = 81.54 \text{ mm} \qquad i_{\text{A–A}} = \sqrt{\frac{32.49 \times 10^6}{7.58 \times 10^3}} = 65.47 \text{ mm};$$

Solution

Topic: Buckling Instability
Problem Number: 6.4   **Page No. 2**

Flexural buckling resistance

EN 1993-1-1:2005

Table 5.2 $\varepsilon = (235/275)^{0.5} = 0.92$

Clause 6.3.1.3(1) $\bar{\lambda} = \sqrt{\dfrac{Af_y}{N_{cr}}} = \dfrac{L_{cr}}{i} \times \dfrac{1}{\lambda_1}$ where $\lambda_1 = 93.9\varepsilon = 93.9 \times 0.92 = 86.39$

Consider the B-B axis:

$L_{cr,B} \geq (0.85 \times 5.0) = 4.25$ m

$\geq (1.0 \times 2.5) = 2.5$ m   The critical buckling length $L_{cr,B} = 4.25$ m

Equation (6.50) $\bar{\lambda}_B = \dfrac{L_{cr}}{i_B} \times \dfrac{1}{\lambda_1} = \dfrac{4250}{81.54} \times \dfrac{1}{86.39} = 0.60$

Consider the A-A axis:

$L_{cr,A} \geq (1.0 \times 2.5) = 2.5$ m

$\geq (1.0 \times 2.0) = 2.0$ m

$\geq (1.0 \times 3.5) = 3.5$ m The critical buckling length $L_{cr,A} = 3.5$ m

Equation (6.50) $\bar{\lambda}_A = \dfrac{L_{cr}}{i_A} \times \dfrac{1}{\lambda_1} = \dfrac{3500}{65.47} \times \dfrac{1}{86.39} = 0.62$

Since the same curve is used for both axes the critical value of $\bar{\lambda}_{cr} = \bar{\lambda}_A = 0.62$

Equation (6.49) $\chi = \dfrac{1}{\Phi + \sqrt{\Phi^2 - \bar{\lambda}^2}}$ but $\chi \leq 1.0$

where $\Phi = 0.5\left[1 + \alpha\left(\bar{\lambda} - 0.2\right) + \bar{\lambda}^2\right]$

Table 6.1   Imperfection factor for curve c: $\alpha = 0.49$
(see Figure 6.24 in this text)

$\Phi = 0.5\left[1 + 0.49\left(0.62 - 0.2\right) + 0.62^2\right] = 0.80$

$\chi = \dfrac{1}{0.80 + \sqrt{0.80^2 - 0.62^2}} = 0.77$ but $\chi \leq 1.0$

Alternatively use the curves given in EN 1993-1-1: Figure 6.4
(see Figure 6.26 or Table 6.1 in this text)

Critical value $\chi \approx 0{,}77$

Equation (6.47) $N_{b,z,Rd} = \dfrac{\chi A f_y}{\gamma_{M1}} = \dfrac{0{,}77 \times 7.58 \times 10^3 \times 275}{1{,}0 \times 10^3} = 1605.1$ kN

The maximum design axial load with respect to flexural buckling = 1605.1 kN

Solution

Topic: **Buckling Instability**
Problem Number: **6.5**   **Page No. 1**

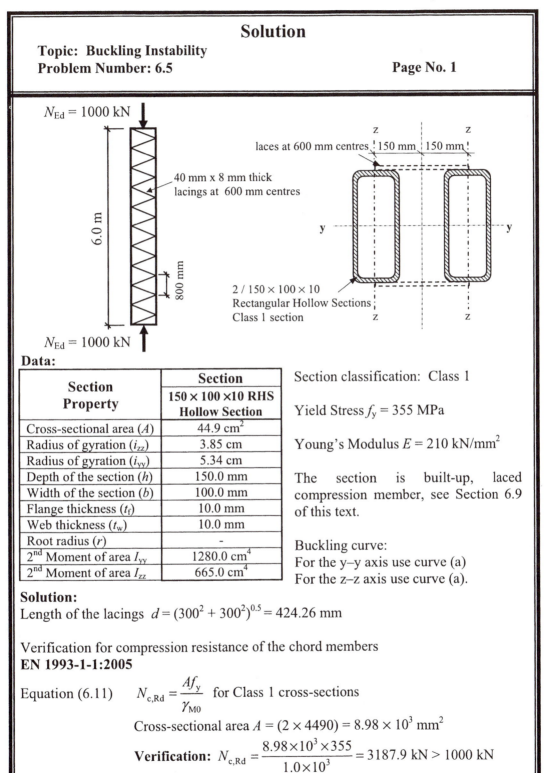

N_{Ed} = 1000 kN

40 mm x 8 mm thick
lacings at 600 mm centres

6.0 m

800 mm

N_{Ed} = 1000 kN

laces at 600 mm centres

z   z
150 mm   150 mm

y ---------------------------------- y

2 / 150 × 100 × 10
Rectangular Hollow Sections
Class 1 section

z   z

Data:

Section Property	Section 150 × 100 ×10 RHS Hollow Section
Cross-sectional area (A)	44.9 cm^2
Radius of gyration (i_{zz})	3.85 cm
Radius of gyration (i_{yy})	5.34 cm
Depth of the section (h)	150.0 mm
Width of the section (b)	100.0 mm
Flange thickness (t_f)	10.0 mm
Web thickness (t_w)	10.0 mm
Root radius (r)	-
2nd Moment of area I_{yy}	1280.0 cm^4
2nd Moment of area I_{zz}	665.0 cm^4

Section classification: Class 1

Yield Stress f_y = 355 MPa

Young's Modulus E = 210 kN/mm^2

The section is built-up, laced compression member, see Section 6.9 of this text.

Buckling curve:
For the y–y axis use curve (a)
For the z–z axis use curve (a).

Solution:
Length of the lacings $d = (300^2 + 300^2)^{0.5} = 424.26$ mm

Verification for compression resistance of the chord members
EN 1993-1-1:2005

Equation (6.11) $N_{c,Rd} = \dfrac{Af_y}{\gamma_{M0}}$ for Class 1 cross-sections

Cross-sectional area $A = (2 \times 4490) = 8.98 \times 10^3$ mm^2

Verification: $N_{c,Rd} = \dfrac{8.98 \times 10^3 \times 355}{1.0 \times 10^3} = 3187.9$ kN > 1000 kN

The chord section is satisfactory with respect to the compression resistance

---

Solution

Topic: Buckling Instability
Problem Number: 6.5   **Page No. 2**

Verification for flexural buckling resistance of the chord members

EN 1993-1-1:2005

Second moment of area of built-up section:

Clause 6.4.2.1(4) $I_{\text{eff}} = 0.5h_0^2 A_{\text{ch}} = 0.5 \times 300^2 \times 4490 = 202.05 \times 10^6 \text{ mm}^4$

Shear stiffness of lacings:

Figure 6.9 $S_v = \dfrac{nEA_d a h_0^2}{2d^3}$ where $A_d = (40 \times 8) = 320 \text{ mm}^2$ and $n = 2$

$S_v = \dfrac{2 \times 210 \times 320 \times 600 \times 300^2}{2 \times 424.26^3} = 47518.94 \text{ kN/mm}$

Chord design force at mid-height:

Clause 6.4.1(6) $N_{\text{ch,Ed}} = 0.5N_{\text{Ed}} + \dfrac{M_{\text{Ed}} h_0 A_{\text{ch}}}{2I_{\text{eff}}}$

Equation (6.69)   where $M_{\text{Ed}} = \dfrac{N_{\text{Ed}} e_o + M_{\text{Ed}}^1}{1 - \dfrac{N_{\text{Ed}}}{N_{\text{cr}}} - \dfrac{N_{\text{Ed}}}{S_v}}$ and $N_{\text{cr}} = \dfrac{\pi^2 EI_{\text{eff}}}{L^2}$

$M_{\text{Ed}}^1 = 0$ and $\dfrac{M_{\text{Ed}} h_0 A_{\text{ch}}}{2I_{\text{eff}}} = \dfrac{M_{\text{Ed}}}{h_0}$ when $I_{\text{eff}} = 0.5h^2 A_{\text{ch}}$

Clause 6.4.1(1)   Bow imperfection $e_0 = L/500 = 6000/500 = 12$ mm

$N_{\text{cr}} = \dfrac{\pi^2 EI_{\text{eff}}}{L^2} = \dfrac{\pi^2 \times 210 \times 202.05 \times 10^6}{6000^2} = 11632.56 \text{ kN}$

$M_{\text{Ed}} = \dfrac{N_{\text{Ed}} e_o + M_{\text{Ed}}^1}{1 - \dfrac{N_{\text{Ed}}}{N_{\text{cr}}} - \dfrac{N_{\text{Ed}}}{S_v}} = \dfrac{1000 \times 12.0 + 0}{1 - \dfrac{1000}{11632.56} - \dfrac{1000}{47518.94}} = 13438.0 \text{ kNmm}$

$N_{\text{ch,Ed}} = \left(0.5N_{\text{Ed}} + \dfrac{M_{\text{Ed}} h_0 A_{\text{ch}}}{2I_{\text{eff}}} \right) = \left(0.5N_{\text{Ed}} + \dfrac{M_{\text{Ed}}}{h_0} \right)$

$= (0.5 \times 1000) + \dfrac{13438.0}{300} = 544.79 \text{ kN}$

Design axial load $N_{\text{ch,Ed}} = 544.79$ kN

Solution

Topic: Buckling Instability
Problem Number: 6.5   **Page No. 3**

Clause 6.3.1.3(1) $\bar{\lambda} = \sqrt{\dfrac{Af_y}{N_{cr}}} = \dfrac{L_{cr}}{i_y} \times \dfrac{\sqrt{\dfrac{Af_y}{A}}}{\lambda_1}$ where

$\lambda_1 = 93.9\varepsilon = 93.9 \times \sqrt{\dfrac{235}{355}} = 76.40$

Consider the y-y axis: assume $L_{cr,y}$ = 6.0 m
Use the radius of gyration based on the gross cross-section
i_y = 53.4 cm

Equation (6.50) $\bar{\lambda}_y = \dfrac{L_{cr}}{i_y} \times \dfrac{1}{\lambda_1} = \dfrac{6000}{53.4} \times \dfrac{1}{76.4} = 1.47$

Table 6.2   For buckling about the y-y axis use curve a
Figure 6.4 $\chi_y \approx 0.38$

Consider the z-z axis: assume $L_{cr,z}$ = 600 mm
Use the radius of gyration based on the gross cross-section
i_z = 3.85 cm

Equation (6.50) $\bar{\lambda}_z = \dfrac{L_{cr}}{i_z} \times \dfrac{1}{\lambda_1} = \dfrac{600}{38.5} \times \dfrac{1}{76.4} = 0.20$

Table 6.2   For buckling about the z-z axis use curve a
Figure 6.4 $\chi_z \approx 1.0$
                                        Critical value $\chi_y \approx 0.38$

Equation (6.47) $N_{y,b,Rd} = \dfrac{\chi_y A f_y}{\gamma_{M1}} = \dfrac{0.38 \times 4490 \times 355}{1.0 \times 10^3} = 605.7$ kN

Equation (6.46)   **Verification:** $\dfrac{N_{Ed}}{N_{b,Rd}} = \dfrac{544.79}{605.7} = 0.90 < 1.0$

The chord section is satisfactory with respect to the flexural buckling resistance

7. Direct Stiffness Method

7.1 Direct Stiffness Method of Analysis

The '*stiffness*' method of analysis is a matrix technique on which most structural computer analysis programs are based. There are two approaches; the indirect and the direct methods. The direct method as illustrated in this chapter requires the visual recognition of the relationship between *structural* forces/displacements and the consequent *element* forces/displacements induced by the applied load system. The indirect method is primarily for use in the development of computer programs to enable the automatic correlation between these displacements.

Neither method is regarded as a hand–analysis. The direct method is included here to enable the reader to understand the concepts involved and the procedure which is undertaken during a computer analysis. The examples and problems used to illustrate these concepts have been restricted to rigid–jointed structures assuming axially–rigid elements. In addition, the structures have been limited to having no more than three degrees–of–freedom and do not have any sloping members. In both methods it is necessary to develop *element stiffness matrices*, related to a *local (element) co–ordinate system* and a *structural stiffness matrix* related to a *global co–ordinate system*. The development of these matrices and co–ordinate systems is explained in Sections 7.2 and 7.3.

7.2 Element Stiffness Matrix [k]

One of the fundamental characteristics governing the behaviour of elastic structures is the relationship between the applied loads and the displacements which these induce. This can be expressed as:

$$[F] = [k] \times [\delta]$$

where:

$[F]$ is a vector representing the forces acting on an element at its nodes i.e. the (element end forces vector),

$[k]$ is the element stiffness matrix relating to the degrees-of-freedom at the nodes relative to the local co–ordinate system,

$[\delta]$ is a vector representing the displacements (both translational and rotational) of the element at its nodes relative to the local axes co–ordinate system (element displacement vector).

Considering an element with only one degree–of–freedom, the matrix and vectors can be re-written as $k = \dfrac{F}{\delta}$, leading to a definition of stiffness as:

"The force necessary to maintain a '*unit*' displacement."

The '**axial**' stiffness of a column as shown in Figure 7.1, can be derived from the standard relationship between the elastic modulus, stress and strain as follows:

Elastic Modulus $\quad E = \dfrac{\text{stress}}{\text{strain}} = \dfrac{(F/A)}{(\delta/L)} = \dfrac{FL}{A\delta}$

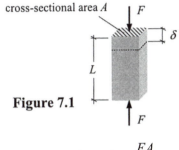

cross-sectional area A

This equation can be re-arranged to give:

$$F = \frac{EA}{L}\delta$$

Figure 7.1

hence when $\delta = 1.0$ (i.e. unit displacement) then the axial stiffness $\quad k\,(=F) = \dfrac{EA}{L}$

7.2.1 Beam Elements with Two Degrees–of–Freedom

Consider a 'beam element' of length L, Young's Modulus E and cross-sectional area A which is subject to axial forces F_1 and F_2 at the end nodes A and B as shown in Figure 7.2.

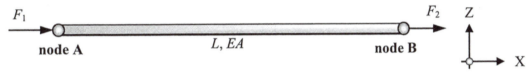

Figure 7.2

Assume that node A is displaced a distance of δ_1 in the direction of the longitudinal axis (i.e. the x-direction) and similarly node B is displaced a distance of δ_2 as shown in Figure 7.3.

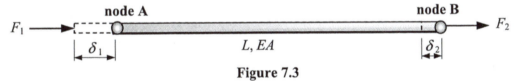

Figure 7.3

The force/displacement relationships for this element are:

$$F_1 = \left(\frac{AE}{L}\times \text{ change in length } L\right) \qquad \therefore F_1 = +\frac{AE}{L}\times(\delta_1 - \delta_2) \quad \text{(assuming } \delta_1 > \delta_2)$$

Considering equilibrium in the x direction:

$$F_2 = -F_1 \qquad\qquad \therefore F_2 = -\frac{AE}{L}\times(\delta_1 - \delta_2)$$

These two equations can be expanded and written in the form:

$$F_1 = +\frac{AE}{L}\delta_1 - \frac{AE}{L}\delta_2 \qquad\qquad \text{Equation (1)}$$

$$F_2 = -\frac{AE}{L}\delta_1 + \frac{AE}{L}\delta_2 \qquad\qquad \text{Equation (2)}$$

in matrix form this gives:
$$
\begin{bmatrix} F_1 \\ F_2 \end{bmatrix} = \begin{bmatrix} +\dfrac{AE}{L} & -\dfrac{AE}{L} \\ -\dfrac{AE}{L} & +\dfrac{AE}{L} \end{bmatrix} \times \begin{bmatrix} \delta_1 \\ \delta_2 \end{bmatrix}
$$

i.e. $[F] = [k] \times [\delta]$

where $[k]$ is the *element stiffness matrix*.

This element stiffness matrix $[k]$ representing two-degrees-of-freedom is adequate for pin-jointed structures in which it is assumed that elements are subject to purely axial loading.

7.2.2 Beam Elements with Four Degrees–of–Freedom

In the case of rigid-jointed, plane-frame structures the loading generally consists of axial, shear and bending forces, the effects of which must be determined by the axial, shear and bending effects on the elements. Consider a beam element with the following properties:

Length $= L$
Second Moment of area about the axis of bending $= I$
Modulus of Elasticity (Young's Modulus) $= E$

which is assumed to be axially rigid, (i.e. neglect axial deformations), and has four-degrees-of-freedom as indicated in Figure 7.4.

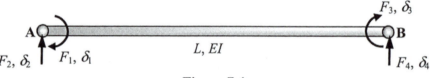

Figure 7.4

When this element is displaced within a structure each node will displace in a vertical direction and rotate as indicated in Figure 7.5, where δ_1 to δ_4 are the **nodal displacements**.

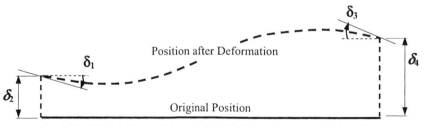

Axially-Rigid Beam Element with four degrees-of-freedom

Figure 7.5

The forces induced in this element by the loaded structure, and which maintain its' displaced form can be represented by the element end forces F_1 to F_4 as shown in Figure 7.6.

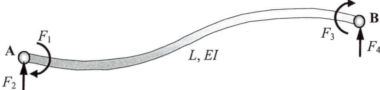

Axially-Rigid Beam Element with four element end-forces

Figure 7.6

The element end-forces can be related to the element end-displacements as in the previous case giving;

$$[F] = [k] \times [\delta]$$

$$
\begin{bmatrix} F_1 \\ F_2 \\ F_3 \\ F_4 \end{bmatrix} =
\begin{bmatrix}
k_{1,1} & k_{1,2} & k_{1,3} & k_{1,4} \\
k_{2,1} & k_{2,2} & k_{2,3} & k_{2,4} \\
k_{3,1} & k_{3,2} & k_{3,3} & k_{3,4} \\
k_{4,1} & k_{4,2} & k_{4,3} & k_{4,4}
\end{bmatrix} \times
\begin{bmatrix} \delta_1 \\ \delta_2 \\ \delta_3 \\ \delta_4 \end{bmatrix}
$$

where k_{11}, k_{12}, k_{13} etc. are the stiffness coefficients for the element.

The displacement configuration in Figure 7.5 can be considered as consisting of the superposition of four independent displacements each having only one degree-of-freedom as shown in Figure 7.8.

Similarly the element end-forces can be represented as the superposition of four sets of forces, each of which is required to maintain a displaced form as indicated in Figure 7.9

The values of $k_{1,1}$, $k_{2,1}$, $k_{3,1}$ and $k_{4,1}$ (which represent the forces necessary to maintain a unit displacement) can be evaluated using an elastic method of analysis such as *McCaulay's Method*, (see Chapter 4, Section 4.2).

Consider the case in which a unit displacement is applied in direction δ_1, (i.e. the slope at A $= -1.0$) as shown in Figure 7.7.

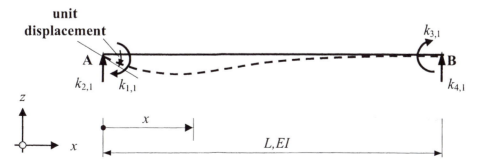

Figure 7.7

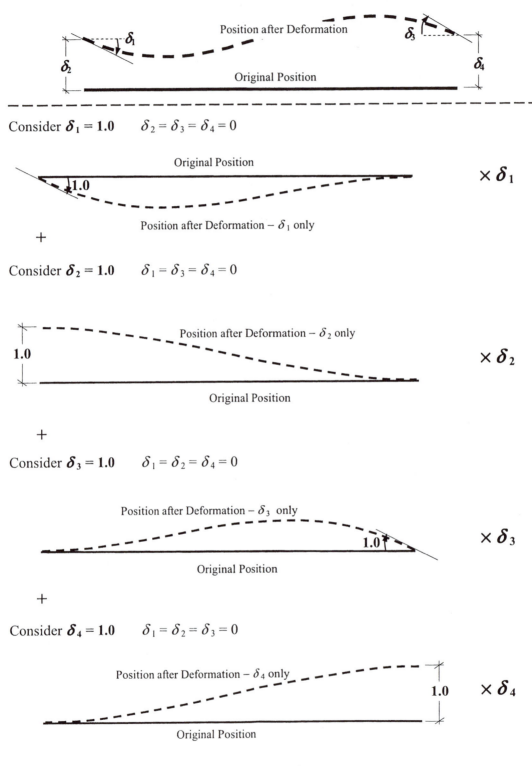

Figure 7.8

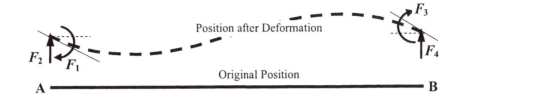

Position after Deformation

Original Position

Consider $\delta_1 = 1.0$ $\delta_2 = \delta_3 = \delta_4 = 0$

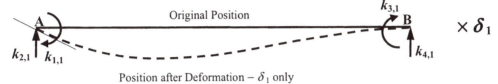

$\times \delta_1$

+

Consider $\delta_2 = 1.0$ $\delta_1 = \delta_3 = \delta_4 = 0$

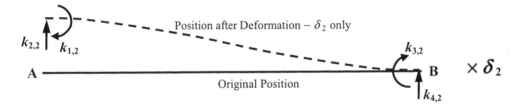

$\times \delta_2$

Consider $\delta_3 = 1.0$ $\delta_1 = \delta_2 = \delta_4 = 0$

+

$\times \delta_3$

Consider $\delta_4 = 1.0$ $\delta_1 = \delta_2 = \delta_3 = 0$

+

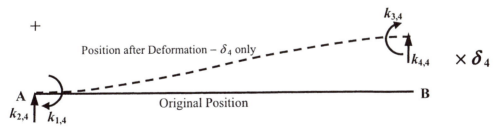

$\times \delta_4$

Figure 7.9

e.g. $F_1 = \{(k_{1,1}\,\delta_1) + (k_{1,2}\,\delta_2) + (k_{1,3}\,\delta_3) + (k_{1,4}\,\delta_4)\}$

The bending moment at any position '*x*' along the element can be expressed as:

Bending moment: $M_x = EI\dfrac{d^2z}{dx^2} = k_{1,1} + k_{2,1}x$ Equation (1)

Slope: $\left(\theta = \dfrac{dz}{dx}\right) = EI\dfrac{dz}{dx} = k_{1,1}x + \dfrac{k_{2,1}x^2}{2} + A$ Equation (2)

Deflection: $(\delta = z) = EIz = \dfrac{k_{1,1}}{2}x^2 + \dfrac{k_{2,1}}{6}x^3 + Ax + B$ Equation (3)

Boundary Conditions: when $x = 0$;   deflection $\delta = 0$ and slope $\theta = -1.0$
$\qquad\qquad\qquad\qquad x = L$; $\qquad\qquad\qquad\quad \delta = 0 \qquad\qquad\qquad \theta = 0$

Substitute for *x* and θ in equation (2): ($x = 0$, $\theta = -1.0$)

Slope: $\left(\theta = \dfrac{dz}{dx}\right) = EI\dfrac{dz}{dx} = k_{1,1}x + \dfrac{k_{2,1}x^2}{2} + A$ Equation (2a)

$EI(-1.0) = A \qquad \therefore A = -EI$

Substitute for *x* and δ in equation (3): ($x = 0$, $\delta = 0$)

Deflection: $(\delta = z) = EI\,z = \dfrac{k_{1,1}}{2}x^2 + \dfrac{k_{2,1}}{6}x^3 + Ax + B$ Equation (3a)

$EI\,(0) = B \qquad \therefore B = 0$

Re-write equations (2a) and (3a):

Slope: $\left(\theta = \dfrac{dz}{dx}\right) = EI\dfrac{dz}{dx} = k_{1,1}x + \dfrac{k_{2,1}x^2}{2} - EI$ Equation (4)

Deflection: $(\delta = z) = EI\,z = \dfrac{k_{1,1}}{2}x^2 + \dfrac{k_{2,1}}{6}x^3 - EIx$ Equation (5)

Substitute for *x* and θ in equation (4): ($x = L$, $\theta = 0$)

Slope: $\left(\theta = \dfrac{dz}{dx}\right) = 0 = k_{1,1}L + \dfrac{k_{2,1}L^2}{2} - EI$ **Equation (6)**

Substitute for *x* and δ in equation (5): ($x = L$, $\delta = 0$)

Deflection: $(\delta = z) = 0 = \dfrac{k_{1,1}}{2}L^2 + \dfrac{k_{2,1}}{6}L^3 - EIL$ **Equation (7)**

Solving equations **(6)** and **(7)** simultaneously and evaluating $\Sigma M = 0$, $\Sigma F_z = 0$ gives:

$k_{1,1} = +\dfrac{4EI}{L} \qquad k_{2,1} = -\dfrac{6EI}{L^2} \qquad k_{3,1} = +\dfrac{2EI}{L} \quad$ and $\quad k_{4,1} = +\dfrac{6EI}{L^2}$

A similar analysis considering the other three unit displacement diagrams produces the following values for the element stiffness matrix coefficients:

$$[k] = \begin{bmatrix} k_{1,1} & k_{1,2} & k_{1,3} & k_{1,4} \\ k_{2,1} & k_{2,2} & k_{2,3} & k_{2,4} \\ k_{3,1} & k_{3,2} & k_{3,3} & k_{3,4} \\ k_{4,1} & k_{4,2} & k_{4,3} & k_{4,4} \end{bmatrix} = \begin{bmatrix} \dfrac{4EI}{L} & \dfrac{-6EI}{L^2} & \dfrac{2EI}{L} & \dfrac{6EI}{L^2} \\ \dfrac{-6EI}{L^2} & \dfrac{12EI}{L^3} & \dfrac{-6EI}{L^2} & \dfrac{-12EI}{L^3} \\ \dfrac{2EI}{L} & \dfrac{-6EI}{L^2} & \dfrac{4EI}{L} & \dfrac{6EI}{L^2} \\ \dfrac{6EI}{L^2} & \dfrac{-12EI}{L^3} & \dfrac{6EI}{L^2} & \dfrac{12EI}{L^3} \end{bmatrix}$$

where:

E is Young's Modulus,

I is the Second Moment of area of the cross-section and

L is the length of the member.

This is the '*element stiffness matrix*' for a beam element with *four degrees-of-freedom* as indicated in Figure 7.10

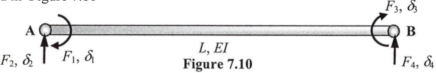

Figure 7.10

7.2.3 Local Co–ordinate System

The co–ordinate system defining the positive directions for the element end displacements and the corresponding end forces is known as the '**local co–ordinate system.**' A typical local co–ordinate system for axially rigid elements in a frame is shown in Figure 7.11.

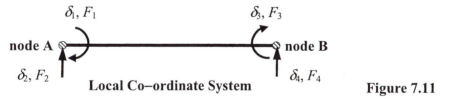

Local Co–ordinate System

Figure 7.11

7.2.4 Beam Elements with Six Degrees–of–Freedom

A typical computer analysis program for plain frame elements in rigid–jointed frames uses beam elements with six degrees–of–freedom as shown in Figure 7.12.

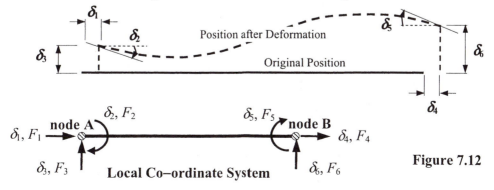

Local Co–ordinate System

Figure 7.12

The resulting stiffness matrix for such elements is:

$$[k] = \begin{bmatrix} k_{1,1} & k_{1,2} & k_{1,3} & k_{1,4} & k_{1,5} & k_{1,6} \\ k_{2,1} & k_{2,2} & k_{2,3} & k_{2,4} & k_{2,5} & k_{2,6} \\ k_{3,1} & k_{3,2} & k_{3,3} & k_{3,4} & k_{3,5} & k_{3,6} \\ k_{4,1} & k_{4,2} & k_{4,3} & k_{4,4} & k_{4,5} & k_{4,6} \\ k_{5,1} & k_{5,2} & k_{5,3} & k_{5,4} & k_{5,5} & k_{5,6} \\ k_{6,1} & k_{6,2} & k_{6,3} & k_{6,4} & k_{6,5} & k_{6,6} \end{bmatrix}$$

The values of the stiffness coefficients are as determined in Sections 7.2.1 and 7.2.2, combining the effects of both the two and four degree–of–freedom cases. The order in which the values appear in the matrix is dependent on the numerical order defined in the *local co–ordinate system,* see Figure 7.12.

$$[k] = \begin{bmatrix} +\dfrac{AE}{L} & 0 & 0 & -\dfrac{AE}{L} & 0 & 0 \\ 0 & +\dfrac{4EI}{L} & -\dfrac{6EI}{L^2} & 0 & +\dfrac{2EI}{L} & +\dfrac{6EI}{L^2} \\ 0 & -\dfrac{6EI}{L^2} & +\dfrac{12EI}{L^3} & 0 & -\dfrac{6EI}{L^2} & -\dfrac{12EI}{L^3} \\ -\dfrac{AE}{L} & 0 & 0 & +\dfrac{AE}{L} & 0 & 0 \\ 0 & +\dfrac{2EI}{L} & -\dfrac{6EI}{L^2} & 0 & +\dfrac{4EI}{L} & +\dfrac{6EI}{L^2} \\ 0 & +\dfrac{6EI}{L^2} & -\dfrac{12EI}{L^3} & 0 & +\dfrac{6EI}{L^2} & +\dfrac{12EI}{L^3} \end{bmatrix} \begin{matrix} F_1 \\ F_2 \\ F_3 \\ F_4 \\ F_5 \\ F_6 \end{matrix}$$

It is evident from the stiffness matrices developed in each case that they are symmetrical about the main diagonal, (this is a consequence of Maxwell's Reciprocal Theorem). The elements in matrices represent the force systems necessary to maintain unit displacements as indicated in Figure 7.12.

The element stiffness matrices must be modified to accommodate the orientation of any elements which are not parallel to the '*global co–ordinate system*', see Section 7.3. This is achieved by applying '*transformation matrices*' such that:

$$[k] = [T]^{\mathrm{T}}[k][T]$$

where [T] is the transformation matrix relating the rotation of the element to the global axis system. This is not considered further in this text.

7.3 *Structural Stiffness Matrix* [*K*]

The stiffness matrix for an entire structure is dependent on the number of *structural degrees-of-freedom* which corresponds with the nodal (i.e. joint) displacements, e.g. consider the structures indicated in Figure 7.13, (**Note:** assuming axial rigidity).

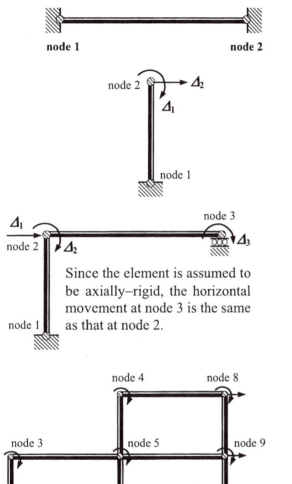

Each node is fixed with respect to translation and rotation and hence there are **NO** degrees-of-freedom.

Node 1 at the base of the cantilever is restrained in both translation and rotation.
Node 2 at the top of the cantilever is free to move in a horizontal direction and rotate.
There are **TWO** degrees-of-freedom in this structure.

Node 1 is restrained in both translation and rotation.
Node 2 is free to move in a horizontal direction and to rotate.
Node 3 is free to rotate.
There are **THREE** degrees-of-freedom in this structure.

Since the element is assumed to be axially–rigid, the horizontal movement at node 3 is the same as that at node 2.

(**Note:** since the elements are assumed to be axially rigid, the horizontal movement at node 4 is the same as that at node 8 and hence does not constitute an additional degree-of-freedom, similarly for nodes 3, 5 and 9 and nodes 2, 6, 10 and 12).

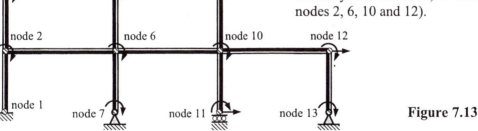

Figure 7.13

Each level of the frame can sway independently of the others and consequently there are three degrees-of-freedom due to sway (i.e. translation). In addition all of the internal joints can rotate producing nine degrees-of-freedom due to rotation.
Three of the supports can rotate whilst one i.e. the roller can also move horizontally. The total number of degrees-of-freedom when the frame is assumed to be axially rigid is equal to **SIXTEEN.**

When the axial deformations of the members is also included the number of degrees-of-freedom increases to **THIRTY ONE**.

In order to generate a structural stiffness matrix and complete the subsequent analysis it is necessary to establish a global co-ordinate system which defines the positions of the nodes and their displacements. The global co-ordinate system is also used to define the positive directions of the applied load system.

Consider a portal frame having three degrees-of-freedom as indicated in Figure 7.14.

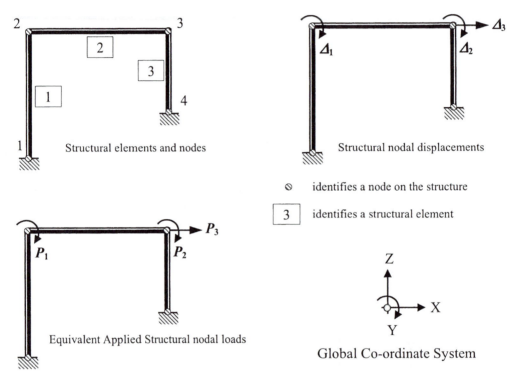

Figure 7.14

The nodal displacements in the structure can be related to the applied structural loads in the same way as those for the elements, i.e.

$$[P] = [K] \times [\Delta]$$

where:

[P]　is a vector representing the equivalent nodal loads applied to the structure (see Section 7.3) relative to the global axes – (structural load vector),

[K]　is the structural stiffness matrix relating to the degrees-of-freedom at the nodes relative to the global axes,

[Δ]　is a vector representing the displacements (both translational and rotational) of the structure at its nodes relative to the global axes, – (structural displacement vector).

The coefficients for the structural stiffness matrix (i.e. $K_{1,1}$, $K_{1,2}$, $K_{1,3}$ etc.) can be determined by evaluating the forces necessary to maintain unit displacements for each of the degrees-of-freedom in turn; in a similar manner to the element stiffness matrices.

Consider the uniform rectangular portal frame shown in Figure 7.15 which supports a number of loads as indicated.

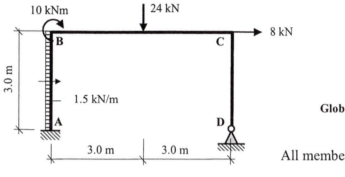

Global Co-ordinate System

All members have the same *EI* value

Figure 7.15

The structural displacements are as indicated in Figure 7.16 (assuming axially rigid members).

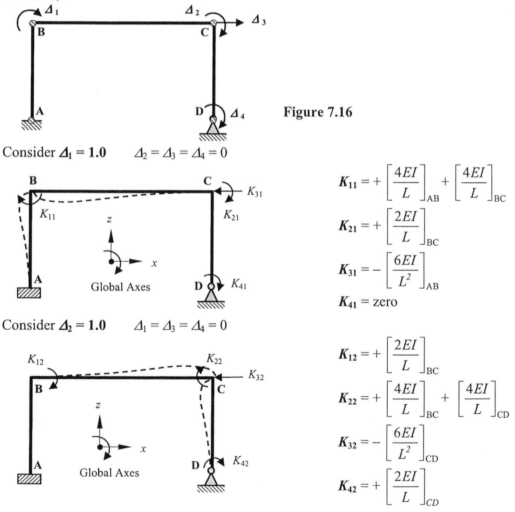

Figure 7.16

Consider $\Delta_1 = 1.0$ $\Delta_2 = \Delta_3 = \Delta_4 = 0$

$$K_{11} = + \left[\frac{4EI}{L}\right]_{AB} + \left[\frac{4EI}{L}\right]_{BC}$$

$$K_{21} = + \left[\frac{2EI}{L}\right]_{BC}$$

$$K_{31} = - \left[\frac{6EI}{L^2}\right]_{AB}$$

$$K_{41} = \text{zero}$$

Consider $\Delta_2 = 1.0$ $\Delta_1 = \Delta_3 = \Delta_4 = 0$

$$K_{12} = + \left[\frac{2EI}{L}\right]_{BC}$$

$$K_{22} = + \left[\frac{4EI}{L}\right]_{BC} + \left[\frac{4EI}{L}\right]_{CD}$$

$$K_{32} = - \left[\frac{6EI}{L^2}\right]_{CD}$$

$$K_{42} = + \left[\frac{2EI}{L}\right]_{CD}$$

Consider $\Delta_3 = 1.0$ $\Delta_1 = \Delta_2 = \Delta_4 = 0$

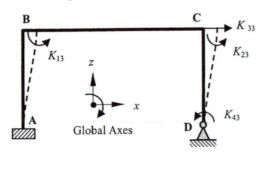

$$K_{13} = -\left[\frac{6EI}{L^2}\right]_{AB}$$

$$K_{23} = -\left[\frac{6EI}{L^2}\right]_{CD}$$

$$K_{33} = +\left[\frac{12EI}{L^3}\right]_{AB} + \left[\frac{12EI}{L^3}\right]_{CD}$$

$$K_{43} = -\left[\frac{6EI}{L^2}\right]_{CD}$$

Consider $\Delta_4 = 1.0$ $\Delta_1 = \Delta_2 = \Delta_3 = 0$

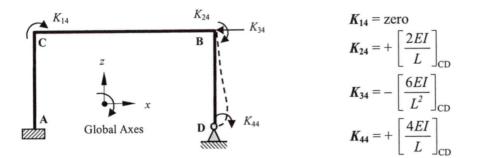

$$K_{14} = \text{zero}$$

$$K_{24} = +\left[\frac{2EI}{L}\right]_{CD}$$

$$K_{34} = -\left[\frac{6EI}{L^2}\right]_{CD}$$

$$K_{44} = +\left[\frac{4EI}{L}\right]_{CD}$$

Structural Stiffness Matrix $[K] = \begin{bmatrix} K_{1,1} & K_{1,2} & K_{1,3} & K_{1,4} \\ K_{2,1} & K_{2,2} & K_{2,3} & K_{2,4} \\ K_{3,1} & K_{3,2} & K_{3,3} & K_{3,4} \\ K_{4,1} & K_{4,2} & K_{4,3} & K_{4,4} \end{bmatrix}$

In each case the size of the structural stiffness matrix is the same as the number of degree–of–freedom.

7.4   *Structural Load Vector [P]*

In most cases the loading applied to a structure occurs within, or along the length of the elements. Since only nodal loads are used in this analysis, the applied loading must be represented as '*equivalent nodal loads*' corresponding to the degrees–of–freedom of the structure. This is easily carried out by replacing the actual load system by a set of forces **equal in magnitude and opposite in direction** to the '*fixed–end forces.*'

The 'fixed–end forces' due to the applied loads are calculated for each applied load case and only those which correspond to structural degrees–of–freedom are subsequently used to develop the structural load vector as shown in Figures 7.17. to 7.19.

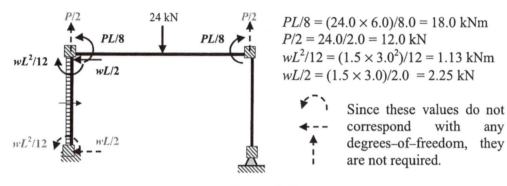

$PL/8 = (24.0 \times 6.0)/8.0 = 18.0$ kNm
$P/2 = 24.0/2.0 = 12.0$ kN
$wL^2/12 = (1.5 \times 3.0^2)/12 = 1.13$ kNm
$wL/2 = (1.5 \times 3.0)/2.0 = 2.25$ kN

Since these values do not correspond with any degrees-of-freedom, they are not required.

Figure 7.17

The structural displacements and equivalent nodal load system are as indicated in Figure 7.18, (assuming axially rigid members).

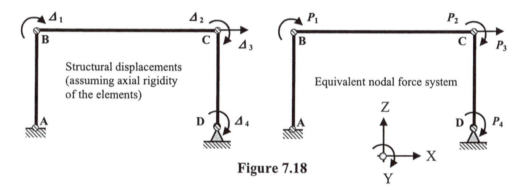

Figure 7.18

The equivalent nodal loads can be determined as follows:

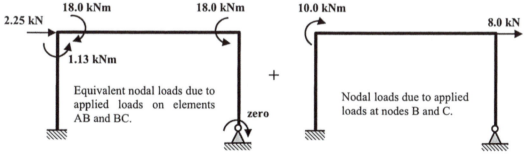

Figure 7.19

$P_1 = (-1.13 + 18.0 + 10.0) = +26.87$ kNm
$P_2 = -18.0$ kN
$P_3 = (+2.25 + 8.0) = +10.25$ kN
$P_4 =$ zero

$$\therefore [P] = \begin{bmatrix} P_1 \\ P_2 \\ P_3 \\ P_4 \end{bmatrix} = \begin{bmatrix} +26.87\text{kNm} \\ -18.0\text{kN} \\ +10.25\text{kN} \\ 0 \end{bmatrix}$$

7.5   *Structural Displacement Vector   [Δ]*

The structural displacement vector can be determined from the product of the inverse of the structural stiffness matrix and the structural load vector, i.e.

$$[\Delta] = [K]^{-1} \times [P]$$

$$
\begin{bmatrix} \Delta_1 \\ \Delta_2 \\ \Delta_3 \\ \Delta_4 \end{bmatrix} =
\begin{bmatrix}
K_{1,1} & K_{1,2} & K_{1,3} & K_{1,4} \\
K_{2,1} & K_{2,2} & K_{2,3} & K_{2,4} \\
K_{3,1} & K_{3,2} & K_{3,3} & K_{3,4} \\
K_{4,1} & K_{4,2} & K_{4,3} & K_{4,4}
\end{bmatrix}^{-1}
\begin{bmatrix} P_1 \\ P_2 \\ P_3 \\ P_4 \end{bmatrix}
$$

7.6   *Element Displacement Vector   [δ]*

An element displacement vector is required for each element and is dependent on the relationship between the structural displacements and the element nodal displacements in each case. The structural displacements in terms of the global co-ordinate system and the individual element displacements in terms of their local co-ordinate systems are shown in Figure 7.20.

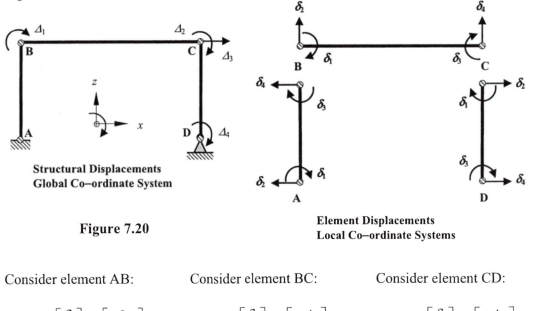

Figure 7.20

Structural Displacements
Global Co-ordinate System

Element Displacements
Local Co-ordinate Systems

Consider element AB:

$$[\delta]_{AB} = \begin{bmatrix} \delta_1 \\ \delta_2 \\ \delta_3 \\ \delta_4 \end{bmatrix} = \begin{bmatrix} 0 \\ 0 \\ +\Delta_1 \\ -\Delta_3 \end{bmatrix}$$

Consider element BC:

$$[\delta]_{BC} = \begin{bmatrix} \delta_1 \\ \delta_2 \\ \delta_3 \\ \delta_4 \end{bmatrix} = \begin{bmatrix} +\Delta_1 \\ 0 \\ +\Delta_2 \\ 0 \end{bmatrix}$$

Consider element CD:

$$[\delta]_{CD} = \begin{bmatrix} \delta_1 \\ \delta_2 \\ \delta_3 \\ \delta_4 \end{bmatrix} = \begin{bmatrix} +\Delta_2 \\ +\Delta_3 \\ +\Delta_4 \\ 0 \end{bmatrix}$$

In the direct stiffness method the correlation between the structural displacements and the element displacements is carried out visually by inspection as indicated above.

7.7   Element Force Vector   [F]_{Total}

The element end-forces due to the structural displacements can be related to the element end-displacements as indicated in Section 7.2.2.

$$[F] = [k] \times [\delta]$$

$$\begin{bmatrix} F_1 \\ F_2 \\ F_3 \\ F_4 \end{bmatrix} = \begin{bmatrix} k_{1,1} & k_{1,2} & k_{1,3} & k_{1,4} \\ k_{2,1} & k_{2,2} & k_{2,3} & k_{2,4} \\ k_{3,1} & k_{3,2} & k_{3,3} & k_{3,4} \\ k_{4,1} & k_{4,2} & k_{4,3} & k_{4,4} \end{bmatrix} \times \begin{bmatrix} \delta_1 \\ \delta_2 \\ \delta_3 \\ \delta_4 \end{bmatrix}$$

The total nodal forces developed at the nodes are given by:

$$[F]_{\text{Total}} = [F] + [\text{Fixed–End Forces}]$$

$$\begin{bmatrix} F_1 \\ F_2 \\ F_3 \\ F_4 \end{bmatrix}_{\text{Total}} = \begin{bmatrix} k_{1,1} & k_{1,2} & k_{1,3} & k_{1,4} \\ k_{2,1} & k_{2,2} & k_{2,3} & k_{2,4} \\ k_{3,1} & k_{3,2} & k_{3,3} & k_{3,4} \\ k_{4,1} & k_{4,2} & k_{4,3} & k_{4,4} \end{bmatrix} \times \begin{bmatrix} \delta_1 \\ \delta_2 \\ \delta_3 \\ \delta_4 \end{bmatrix} + \begin{bmatrix} \text{FEF}_1 \\ \text{FEF}_2 \\ \text{FEF}_3 \\ \text{FEF}_4 \end{bmatrix}$$

7.8   Example 7.1: Two–span Beam

Consider a uniform two-span beam ABC which is fully-fixed at supports A and C and simply supported at B as indicated in Figure 7.21. A uniformly distributed load of 24 kN/m is applied to span AB and a central point load of 24 kN is applied to span BC as shown.

Using the data given, the degrees-of-freedom indicated and assuming both members to be axially rigid,

(i)   generate the structural stiffness matrix [K] and the applied load vector [P],
(ii)   determine the structural displacements,
(iii)   determine the member end forces and the support reactions,
(iv)   sketch the shear force and bending moment diagrams,
(v)   sketch the deflected shape.

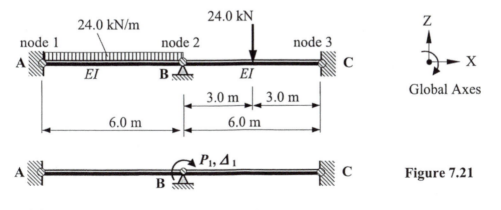

Figure 7.21

Solution:
To develop the structural stiffness matrix each degree–of–freedom is given a unit displacement in turn and the forces (corresponding to all degrees–of–freedom) necessary to maintain the displaced shape are determined. In this case there is only one degree–of–freedom and hence the stiffness matrix comprises one element.

Structural Stiffness Matrix [K]:

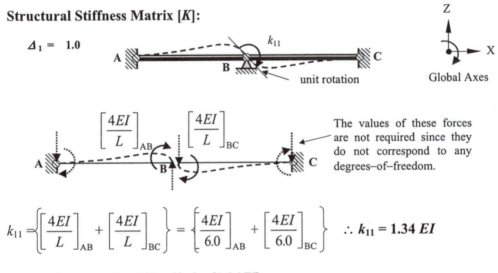

$\Delta_1 = 1.0$

unit rotation

Z

X

Global Axes

$\left[\dfrac{4EI}{L}\right]_{AB}$ $\left[\dfrac{4EI}{L}\right]_{BC}$

The values of these forces are not required since they do not correspond to any degrees–of–freedom.

$$k_{11} = \left\{\left[\frac{4EI}{L}\right]_{AB} + \left[\frac{4EI}{L}\right]_{BC}\right\} = \left\{\left[\frac{4EI}{6.0}\right]_{AB} + \left[\frac{4EI}{6.0}\right]_{BC}\right\} \quad \therefore k_{11} = 1.34\ EI$$

The stiffness matrix $[K] = [k_{11}] = [1.34EI]$

The inverse of the stiffness matrix $= [K]^{-1} = \dfrac{1}{1.34EI}$

Structural Load Vector [P]:
The structural load vector comprises coefficients equal in magnitude and opposite in direction to the fixed–end forces which correspond to the structural degrees–of–freedom. In this case, only the moment at joint B is required.

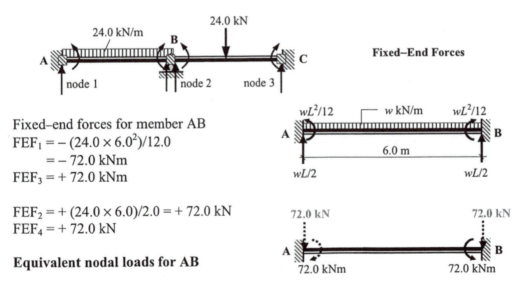

Fixed–End Forces

Fixed–end forces for member AB
$FEF_1 = - (24.0 \times 6.0^2)/12.0$
     $= - 72.0$ kNm
$FEF_3 = + 72.0$ kNm

$FEF_2 = + (24.0 \times 6.0)/2.0 = + 72.0$ kN
$FEF_4 = + 72.0$ kN

Equivalent nodal loads for AB

Fixed–end forces for member BC

$FEF_1 = -(24.0 \times 6.0)/8.0$

$\qquad = -18.0$ kNm

$FEF_3 = +18.0$ kNm

$FEF_2 = +(24.0/2.0) = +12.0$ kN

$FEF_4 = +12.0$ kN

Equivalent nodal loads for BC

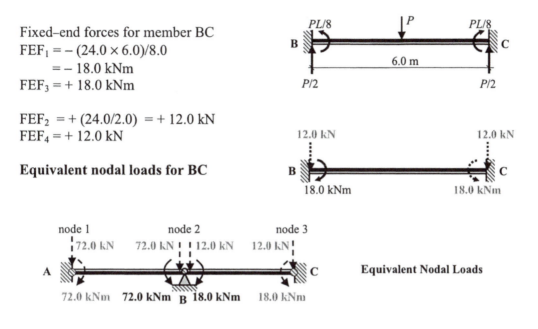

Applied load in direction of Δ_1 at joint B $= [-72.0 + 18.0] = -54.0$ kNm

Structural Load Vector $[P] = [-54.0]$

Structural Displacements [Δ]

$[\Delta_1] = [K]^{-1}[P] = \dfrac{1}{1.34EI}[-54.0]$ $\qquad \therefore \Delta_1 = -\dfrac{40.30}{EI}$ radians $\curvearrowright$

Structural Deflections

Element Stiffness Matrices [k]:

$[k] = \begin{bmatrix} +\dfrac{4EI}{L} & -\dfrac{6EI}{L^2} & +\dfrac{2EI}{L} & +\dfrac{6EI}{L^2} \\[2mm] -\dfrac{6EI}{L^2} & +\dfrac{12EI}{L^3} & -\dfrac{6EI}{L^2} & -\dfrac{12EI}{L^3} \\[2mm] +\dfrac{2EI}{L} & -\dfrac{6EI}{L^2} & +\dfrac{4EI}{L} & +\dfrac{6EI}{L^2} \\[2mm] +\dfrac{6EI}{L^2} & -\dfrac{12EI}{L^3} & +\dfrac{6EI}{L^2} & +\dfrac{12EI}{L^3} \end{bmatrix}$

Element End Forces [F]:

$$[F] = \begin{bmatrix} F_1 \\ F_2 \\ F_3 \\ F_4 \end{bmatrix} = [k][\delta] + [\mathbf{FEF}] = \begin{bmatrix} +\dfrac{4EI}{L} & -\dfrac{6EI}{L^2} & +\dfrac{2EI}{L} & +\dfrac{6EI}{L^2} \\ -\dfrac{6EI}{L^2} & +\dfrac{12EI}{L^3} & -\dfrac{6EI}{L^2} & -\dfrac{12EI}{L^3} \\ +\dfrac{2EI}{L} & -\dfrac{6EI}{L^2} & +\dfrac{4EI}{L} & +\dfrac{6EI}{L^2} \\ +\dfrac{6EI}{L^2} & -\dfrac{12EI}{L^3} & +\dfrac{6EI}{L^2} & +\dfrac{12EI}{L^3} \end{bmatrix} \begin{bmatrix} \delta_1 \\ \delta_2 \\ \delta_3 \\ \delta_4 \end{bmatrix} + \begin{bmatrix} \mathrm{FEF}_1 \\ \mathrm{FEF}_2 \\ \mathrm{FEF}_3 \\ \mathrm{FEF}_4 \end{bmatrix}$$

Consider element AB:

$$\dfrac{4EI}{L} = \dfrac{4EI}{6.0} = 0.67EI \qquad\qquad \dfrac{6EI}{L^2} = \dfrac{6EI}{6.0^2} = 0.17EI;$$

$$\dfrac{2EI}{L} = \dfrac{2EI}{6.0} = 0.34EI \qquad\qquad \dfrac{12EI}{L^3} = \dfrac{12EI}{6.0^3} = 0.06EI$$

$$[k]_{AB} = EI \begin{bmatrix} +0.67 & -0.17 & +0.34 & +0.17 \\ -0.17 & +0.06 & -0.17 & -0.06 \\ +0.34 & -0.17 & +0.67 & +0.17 \\ +0.17 & -0.06 & +0.17 & +0.06 \end{bmatrix}$$

Displacement Vector [δ]: **Fixed-End Forces Vector [FEF]:**

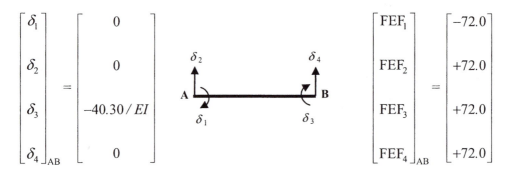

$$\begin{bmatrix} \delta_1 \\ \delta_2 \\ \delta_3 \\ \delta_4 \end{bmatrix}_{AB} = \begin{bmatrix} 0 \\ 0 \\ -40.30/EI \\ 0 \end{bmatrix} \qquad\qquad \begin{bmatrix} \mathrm{FEF}_1 \\ \mathrm{FEF}_2 \\ \mathrm{FEF}_3 \\ \mathrm{FEF}_4 \end{bmatrix}_{AB} = \begin{bmatrix} -72.0 \\ +72.0 \\ +72.0 \\ +72.0 \end{bmatrix}$$

Element End Forces [F]$_{AB}$:

$$[F]_{\text{Total}} = \begin{bmatrix} F_1 \\ F_2 \\ F_3 \\ F_4 \end{bmatrix} + \begin{bmatrix} \mathrm{FEF}_1 \\ \mathrm{FEF}_2 \\ \mathrm{FEF}_3 \\ \mathrm{FEF}_4 \end{bmatrix} = [k][\delta] + [\mathbf{FEF}]$$

$$[k][\delta] + [\text{FEF}] = EI \begin{bmatrix} +0.67 & -0.17 & +0.34 & +0.17 \\ -0.17 & +0.06 & -0.17 & -0.06 \\ +0.34 & -0.17 & +0.67 & +0.17 \\ +0.17 & -0.06 & +0.17 & +0.06 \end{bmatrix} \begin{bmatrix} 0 \\ 0 \\ -40.30/EI \\ 0 \end{bmatrix} + \begin{bmatrix} -72.0 \\ +72.0 \\ +72.0 \\ +72.0 \end{bmatrix}$$

$F_1 = -(0.34 \times 40.30) - [72.0] = -85.70 \text{ kNm}$

$F_2 = +(0.17 \times 40.30) + [72.0] = +78.85 \text{ kN}$

$F_3 = -(0.67 \times 40.30) + [72.0] = +45.0 \text{ kNm}$

$F_4 = -(0.17 \times 40.30) + [72.0] = +65.15 \text{ kN}$

Consider element BC:

$$\frac{4EI}{L} = \frac{4EI}{6.0} = 0.67EI \qquad\qquad \frac{6EI}{L^2} = \frac{6EI}{6.0^2} = 0.17EI;$$

$$\frac{2EI}{L} = \frac{2EI}{6.0} = 0.34EI \qquad\qquad \frac{12EI}{L^3} = \frac{12EI}{6.0^3} = 0.06EI$$

$$[k]_{\text{BC}} = EI \begin{bmatrix} +0.67 & -0.17 & +0.34 & +0.17 \\ -0.17 & +0.06 & -0.17 & -0.06 \\ +0.34 & -0.17 & +0.67 & +0.17 \\ +0.17 & -0.06 & +0.17 & +0.06 \end{bmatrix}$$

Displacement Vector [δ]:   **Fixed-End Forces Vector [FEF]:**

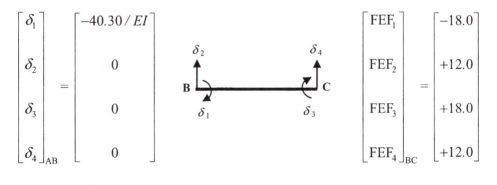

$$\begin{bmatrix} \delta_1 \\ \delta_2 \\ \delta_3 \\ \delta_4 \end{bmatrix}_{\text{AB}} = \begin{bmatrix} -40.30/EI \\ 0 \\ 0 \\ 0 \end{bmatrix} \qquad\qquad \begin{bmatrix} \text{FEF}_1 \\ \text{FEF}_2 \\ \text{FEF}_3 \\ \text{FEF}_4 \end{bmatrix}_{\text{BC}} = \begin{bmatrix} -18.0 \\ +12.0 \\ +18.0 \\ +12.0 \end{bmatrix}$$

Element End Forces [F]_BC:

$$[F]_{\text{Total}} = \begin{bmatrix} F_1 \\ F_2 \\ F_3 \\ F_4 \end{bmatrix} + \begin{bmatrix} \text{FEF}_1 \\ \text{FEF}_2 \\ \text{FEF}_3 \\ \text{FEF}_4 \end{bmatrix} = [k][\delta] + [\text{FEF}]$$

$$[k][\delta] + [FEF] = EI \begin{bmatrix} +0.67 & -0.17 & +0.34 & +0.17 \\ -0.17 & +0.06 & -0.17 & -0.06 \\ +0.34 & -0.17 & +0.67 & +0.17 \\ +0.17 & -0.06 & +0.17 & +0.06 \end{bmatrix} \begin{bmatrix} -40.30/EI \\ 0 \\ 0 \\ 0 \end{bmatrix} + \begin{bmatrix} -18.0 \\ +12.0 \\ +18.0 \\ +12.0 \end{bmatrix}$$

$F_1 = -(0.67 \times 40.30) - [18.0] = -45.0$ kNm

$F_2 = +(0.17 \times 40.30) + [12.0] = +18.85$ kN ↑

$F_3 = -(0.34 \times 40.30) + [18.0] = +4.30$ kNm

$F_4 = -(0.17 \times 40.30) + [12.0] = +5.15$ kN ↑

Reactions:

Support A:

$V_A = (F_2)_{AB} = +78.85$ kN ↑

$M_A = (F_1)_{AB} = -85.70$ kNm

Support B:

$V_B = (F_4)_{AB} + (F_2)_{BC} = +65.15 + 18.85 = 84.0$ kN ↑

$M_B = (F_3)_{AB} = (F_1)_{BC} = 45.0$ kNm

Support C:

$V_C = (F_4)_{BC} = +5.15$ kN ↑

$M_C = (F_3)_{BC} = +4.30$ kNm

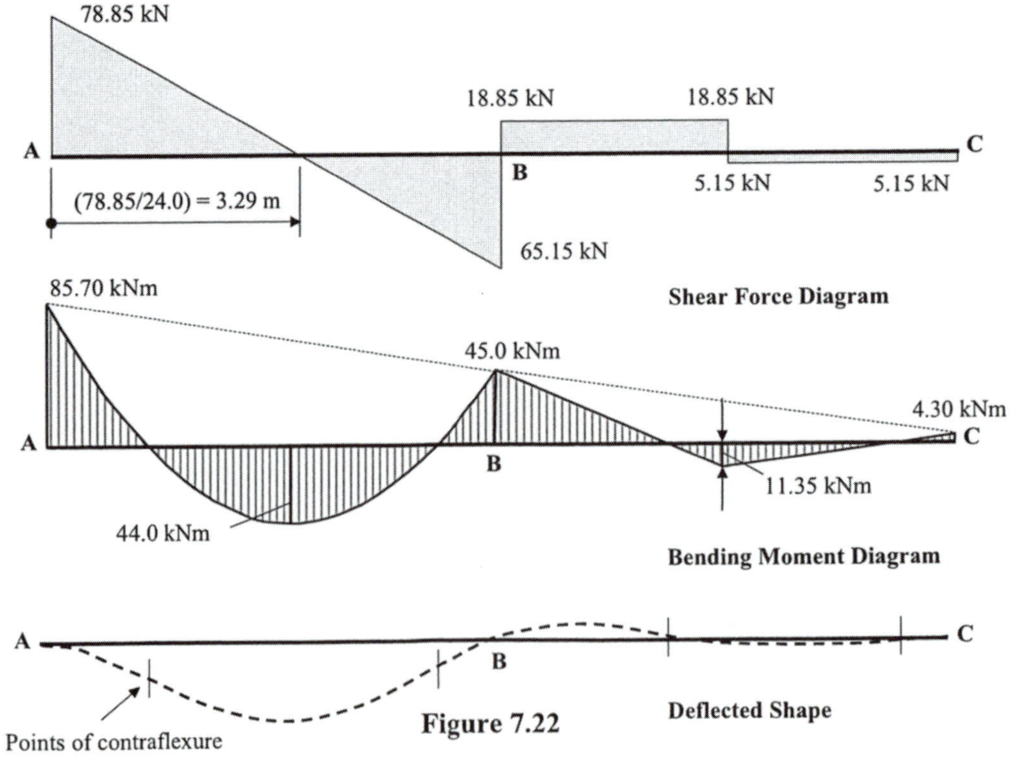

Shear Force Diagram

78.85 kN

18.85 kN   18.85 kN

A   B   C

(78.85/24.0) = 3.29 m

5.15 kN   5.15 kN

65.15 kN

Bending Moment Diagram

85.70 kNm

45.0 kNm

4.30 kNm

A   B   C

11.35 kNm

44.0 kNm

Deflected Shape

A   B   C

Points of contraflexure

Figure 7.22

7.9   *Example 7.2: Rigid–Jointed Frame*

A non–uniform, rigid–jointed frame ABCD is fully-fixed at supports A and D as indicated in Figure 7.23. A uniformly distributed load of 3 kN/m is applied to element BC a central point load of 5 kN is applied to element AB and a point load at node C as shown.

Using the data given, the degrees-of-freedom indicated and assuming all members to be axially rigid,

- (i)   generate the structural stiffness matrix [*K*] and the applied load vector [*P*],
- (ii)   determine the structural displacements,
- (iii) determine the member end forces and the support reactions,
- (iv)   sketch the shear force and bending moment diagrams,
- (vi)   sketch the deflected shape.

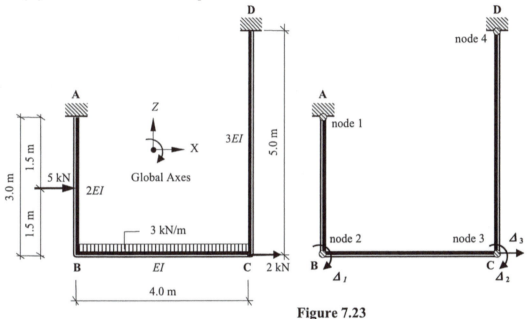

Figure 7.23

Solution:

Each degree–of–freedom is given a unit displacement in turn and the forces necessary to maintain the displacements is calculated in each case.

$\Delta_1 = 1.0$ $\Delta_2 = \Delta_3 = 0$

$$K_{11} = \left[\frac{4EI}{L}\right]_{AB} + \left[\frac{4EI}{L}\right]_{BC}$$

$$= \left[\frac{4(2.0EI)}{3.0}\right] + \left[\frac{4EI}{4.0}\right] = +3.67EI$$

$$K_{21} = \left[\frac{2EI}{L}\right]_{BC} = \left[\frac{2EI}{4.0}\right] = +0.50EI$$

$$K_{31} = \left[\frac{6EI}{L^2}\right]_{AB} = \left[\frac{6(2.0EI)}{3.0^2}\right] = +1.33EI$$

$\Delta_2 = 1.0$ $\Delta_1 = \Delta_3 = 0$

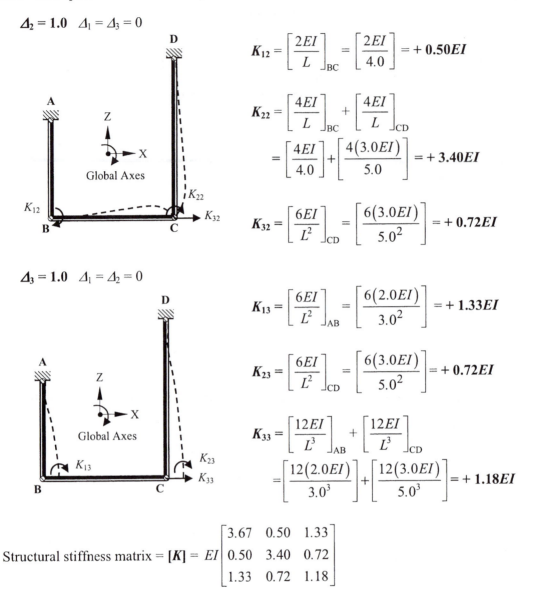

$$K_{12} = \left[\frac{2EI}{L}\right]_{BC} = \left[\frac{2EI}{4.0}\right] = +\,0.50EI$$

$$K_{22} = \left[\frac{4EI}{L}\right]_{BC} + \left[\frac{4EI}{L}\right]_{CD}$$

$$= \left[\frac{4EI}{4.0}\right] + \left[\frac{4(3.0EI)}{5.0}\right] = +\,3.40EI$$

$$K_{32} = \left[\frac{6EI}{L^2}\right]_{CD} = \left[\frac{6(3.0EI)}{5.0^2}\right] = +\,0.72EI$$

$\Delta_3 = 1.0$ $\Delta_1 = \Delta_2 = 0$

$$K_{13} = \left[\frac{6EI}{L^2}\right]_{AB} = \left[\frac{6(2.0EI)}{3.0^2}\right] = +\,1.33EI$$

$$K_{23} = \left[\frac{6EI}{L^2}\right]_{CD} = \left[\frac{6(3.0EI)}{5.0^2}\right] = +\,0.72EI$$

$$K_{33} = \left[\frac{12EI}{L^3}\right]_{AB} + \left[\frac{12EI}{L^3}\right]_{CD}$$

$$= \left[\frac{12(2.0EI)}{3.0^3}\right] + \left[\frac{12(3.0EI)}{5.0^3}\right] = +\,1.18EI$$

Structural stiffness matrix $= [K] = EI \begin{bmatrix} 3.67 & 0.50 & 1.33 \\ 0.50 & 3.40 & 0.72 \\ 1.33 & 0.72 & 1.18 \end{bmatrix}$

There are several methods for inverting matrices, the technique used here is given in Appendix 3.

The invert of a matrix is given by $[K]^{-1} = \dfrac{\left[K^C\right]^T}{|K|}$

where:
$[K^C]$ is the co-factor matrix for $[K]$
$|K|$ is the determinant of $[K]$ and
$[K^C]^T$ is the transpose of the co-factor matrix

$$EI \begin{bmatrix} \overset{+}{3.67} & \overset{-}{0.50} & \overset{+}{1.33} \\ \underset{-}{0.50} & \overset{+}{3.40} & \underset{-}{0.72} \\ \underset{+}{1.33} & \underset{-}{0.72} & \underset{+}{1.18} \end{bmatrix}$$

Co-factor Matrix: $[K^C]$

(**Note:** the transpose of a symmetric matrix is the same as the original matrix)

$k_{11}^c = + \{(3.40 \times 1.18) - (0.72 \times 0.72)\}EI^2 = +3.49EI^2$

$k_{12}^c = k_{21}^c = - \{(0.50 \times 1.18) - (1.33 \times 0.72)\}EI^2 = +0.37EI^2$

$k_{13}^c = k_{31}^c = + \{(0.50 \times 0.72) - (1.33 \times 3.40)\}EI^2 = -4.16EI^2$

$k_{22}^c = + \{(3.67 \times 1.18) - (1.33 \times 1.33)\}EI^2 = +2.56EI^2$

$k_{23}^c = k_{32}^c = - \{(3.67 \times 0.72) - (1.33 \times 0.50)\}EI^2 = -1.98EI^2$

$k_{33}^c = + \{(3.67 \times 3.40) - (0.50 \times 0.50)\}EI^2 = +12.23EI^2$

Determinant of $[K]$:

Det $[K] = EI^3 \{+ (3.67 \times 3.49) + (0.5 \times 0.37) - (1.33 \times 4.16\} = +7.46EI^3$

Inverted stiffness matrix = $[K]^{-1} = \dfrac{1}{EI}\begin{bmatrix} +0.468 & +0.050 & -0.558 \\ +0.050 & +0.343 & -0.265 \\ -0.558 & -0.265 & +1.639 \end{bmatrix}$

Structural Load Vector: $[P]$:

Fixed–end forces for member AB

FEF$_1$ = + (5.0 × 3.0)/8.0

= + 1.88 kNm

FEF$_3$ = – 1.88 kNm

FEF$_2$ = (5.0/2.0) = 2.5 kN

FEF$_4$ = 2.5 kN

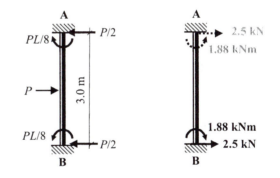

Equivalent nodal loads for AB

Fixed–end forces for member BC

FEF$_1$ = – (3.0 × 4.0^2)/12.0

= – 4.0 kNm

FEF$_3$ = + 4.0 kNm

FEF$_2$ = (3.0 × 4.0)/2.0) = 6.0 kN

FEF$_4$ = 6.0 kN

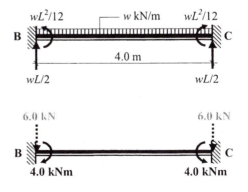

Equivalent nodal loads for BC

Applied nodal load at C = 2.0 kN →

The equivalent nodal loads required are those which correspond with the nodal degree–of–freedom as follows:

$P_1 = (+1.88 + 4.0) = +5.88$ kNm
$P_2 = -4.0 = -4.0$ kNm
$P_3 = (+2.5 + 2.0) = +4.5$ kNm

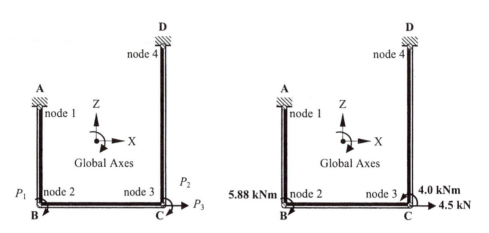

Equivalent Nodal Loads

Structural Load Vector $[P] = \begin{bmatrix} +5.88 \\ -4.0 \\ +4.5 \end{bmatrix}$

Structural Displacements $[\Delta]$:

$$[\Delta] = [K]^{-1}[P] \qquad \begin{bmatrix} \Delta_1 \\ \Delta_2 \\ \Delta_3 \end{bmatrix} = \frac{1}{EI} \begin{bmatrix} +0.468 & +0.050 & -0.558 \\ +0.050 & +0.343 & -0.265 \\ -0.558 & -0.265 & +1.639 \end{bmatrix} \begin{bmatrix} +5.88 \\ -4.0 \\ +4.50 \end{bmatrix}$$

$$\Delta_1 = \frac{1}{EI}\left[(0.467\times5.88)-(0.05\times4.0)-(0.558\times4.5)\right] = +\frac{0.03}{EI} \text{ radians } \circlearrowright$$

$$\Delta_2 = \frac{1}{EI}\left[+(0.05\times5.88)-(0.343\times4.0)-(0.265\times4.5)\right] = -\frac{2.27}{EI} \text{ radians } \circlearrowleft$$

$$\Delta_3 = \frac{1}{EI}\left[-(0.558\times5.88)+(0.265\times4.0)+(1.639\times4.5)\right] = +\frac{5.15}{EI} \text{ m } \longrightarrow$$

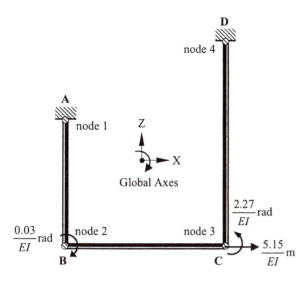

Structural Deflections

Element Stiffness Matrices [*k*]:

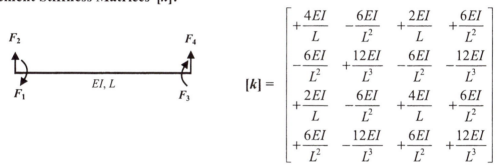

$$[k] = \begin{bmatrix} +\dfrac{4EI}{L} & -\dfrac{6EI}{L^2} & +\dfrac{2EI}{L} & +\dfrac{6EI}{L^2} \\ -\dfrac{6EI}{L^2} & +\dfrac{12EI}{L^3} & -\dfrac{6EI}{L^2} & -\dfrac{12EI}{L^3} \\ +\dfrac{2EI}{L} & -\dfrac{6EI}{L^2} & +\dfrac{4EI}{L} & +\dfrac{6EI}{L^2} \\ +\dfrac{6EI}{L^2} & -\dfrac{12EI}{L^3} & +\dfrac{6EI}{L^2} & +\dfrac{12EI}{L^3} \end{bmatrix}$$

Element End Forces [*F*]_Total:

$$[F]_{\text{Total}} = \begin{bmatrix} F_1 \\ F_2 \\ F_3 \\ F_4 \end{bmatrix} + \begin{bmatrix} \text{FEF}_1 \\ \text{FEF}_2 \\ \text{FEF}_3 \\ \text{FEF}_4 \end{bmatrix} = [k][\delta] + [\text{FEF}]$$

Consider element AB:

$$\frac{4EI}{L} = \frac{4 \times (2.0EI)}{3.0} = 2.67EI \qquad \frac{6EI}{L^2} = \frac{6 \times (2.0EI)}{3.0^2} = 1.33EI$$

$$\frac{2EI}{L} = \frac{2 \times (2.0EI)}{3.0} = 1.33EI \qquad \frac{12EI}{L^3} = \frac{12 \times (2.0EI)}{3.0^3} = 0.89EI$$

$$[k]_{\text{AB}} = EI \begin{bmatrix} +2.67 & -1.33 & +1.33 & +1.33 \\ -1.33 & +0.89 & -1.33 & -0.89 \\ +1.33 & -1.33 & +2.67 & +1.33 \\ +1.33 & -0.89 & +1.33 & +0.89 \end{bmatrix}$$

Displacement Vector [δ]:   **Fixed-End Forces Vector [FEF]:**

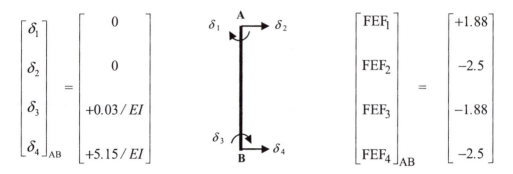

$$\begin{bmatrix} \delta_1 \\ \delta_2 \\ \delta_3 \\ \delta_4 \end{bmatrix}_{AB} = \begin{bmatrix} 0 \\ 0 \\ +0.03/EI \\ +5.15/EI \end{bmatrix} \qquad \begin{bmatrix} FEF_1 \\ FEF_2 \\ FEF_3 \\ FEF_4 \end{bmatrix}_{AB} = \begin{bmatrix} +1.88 \\ -2.5 \\ -1.88 \\ -2.5 \end{bmatrix}$$

Element End Forces [F]$_{AB}$:

$$[F]_{Total} = \begin{bmatrix} F_1 \\ F_2 \\ F_3 \\ F_4 \end{bmatrix} + \begin{bmatrix} FEF_1 \\ FEF_2 \\ FEF_3 \\ FEF_4 \end{bmatrix} = [k][\delta] + [FEF]$$

$$= EI \begin{bmatrix} +2.67 & -1.33 & +1.33 & +1.33 \\ -1.33 & +0.89 & -1.33 & -0.89 \\ +1.33 & -1.33 & +2.67 & +1.33 \\ +1.33 & -0.89 & +1.33 & +0.89 \end{bmatrix} \begin{bmatrix} 0 \\ 0 \\ +0.03/EI \\ +5.15/EI \end{bmatrix} + \begin{bmatrix} +1.88 \\ -2.5 \\ -1.88 \\ -2.5 \end{bmatrix}$$

$F_1 = [+ (1.33 \times 0.03) + (1.33 \times 5.15)] + [1.88] = + 8.77 \text{ kNm}$

$F_2 = [- (1.33 \times 0.03) - (0.89 \times 5.15)] - [2.5] = - 7.12 \text{ kN}$

$F_3 = [+ (2.67 \times 0.03) + (1.33 \times 5.15)] - [1.88] = + 5.05 \text{ kNm}$

$F_4 = [+ (1.33 \times 0.03) + (0.89 \times 5.15)] - [2.5] = + 2.12 \text{ kN}$

Consider element BC:

$$\frac{4EI}{L} = \frac{4 \times EI}{4.0} = 1.0EI \qquad \frac{6EI}{L^2} = \frac{6 \times EI}{4.0^2} = 0.38EI$$

$$\frac{2EI}{L} = \frac{2 \times EI}{4.0} = 0.5EI \qquad \frac{12EI}{L^3} = \frac{12 \times EI}{4.0^3} = 0.19EI$$

$$[k]_{BC} = EI \begin{bmatrix} +1.0 & -0.38 & +0.50 & +0.38 \\ -0.38 & +0.19 & -0.38 & -0.19 \\ +0.50 & -0.38 & +1.0 & +0.38 \\ +0.38 & -0.19 & +0.38 & +0.19 \end{bmatrix}$$

Displacement Vector [δ]:

$$\begin{bmatrix} \delta_1 \\ \delta_2 \\ \delta_3 \\ \delta_4 \end{bmatrix}_{BC} = \begin{bmatrix} +0.03 / EI \\ 0 \\ -2.27 / EI \\ 0 \end{bmatrix}$$

Fixed-End Forces Vector [FEF]:

$$\begin{bmatrix} FEF_1 \\ FEF_2 \\ FEF_3 \\ FEF_4 \end{bmatrix}_{BC} = \begin{bmatrix} -4.0 \\ +6.0 \\ +4.0 \\ +6.0 \end{bmatrix}$$

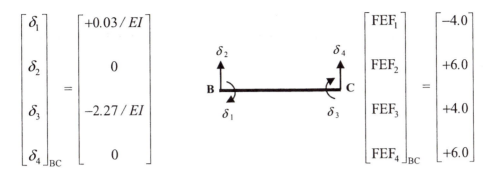

Element End Forces [F]$_{BC}$:

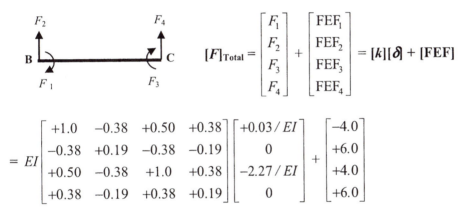

$$[F]_{Total} = \begin{bmatrix} F_1 \\ F_2 \\ F_3 \\ F_4 \end{bmatrix} + \begin{bmatrix} FEF_1 \\ FEF_2 \\ FEF_3 \\ FEF_4 \end{bmatrix} = [k][\delta] + [FEF]$$

$$= EI \begin{bmatrix} +1.0 & -0.38 & +0.50 & +0.38 \\ -0.38 & +0.19 & -0.38 & -0.19 \\ +0.50 & -0.38 & +1.0 & +0.38 \\ +0.38 & -0.19 & +0.38 & +0.19 \end{bmatrix} \begin{bmatrix} +0.03 / EI \\ 0 \\ -2.27 / EI \\ 0 \end{bmatrix} + \begin{bmatrix} -4.0 \\ +6.0 \\ +4.0 \\ +6.0 \end{bmatrix}$$

$F_1 = [+ (1.0 \times 0.03) - (0.5 \times 2.27)] - [4.0] = - 5.11$ kNm

$F_2 = [- (0.38 \times 0.03) + (0.38 \times 2.27)] + [6.0] = + 6.85$ kN

$F_3 = [+ (0.50 \times 0.03) - (1.0 \times 2.27)] + [4.0] = + 1.75$ kNm

$F_4 = [+ (0.38 \times 0.03) - (0.38 \times 2.27)] + [6.0] = + 5.15$ kN

Consider element DC:

$$\frac{4EI}{L} = \frac{4 \times 3.0 EI}{5.0} = 2.4 EI \qquad \frac{6EI}{L^2} = \frac{6 \times 3.0 EI}{5.0^2} = 0.72 EI$$

$$\frac{2EI}{L} = \frac{2 \times 3.0 EI}{5.0} = 1.2 EI \qquad \frac{12EI}{L^3} = \frac{12 \times 3.0 EI}{5.0^3} = 0.29 EI$$

$$[k]_{DC} = EI \begin{bmatrix} +2.40 & -0.72 & +1.20 & +0.72 \\ -0.72 & +0.29 & -0.72 & -0.29 \\ +1.20 & -0.72 & +2.40 & +0.72 \\ +0.72 & -0.29 & +0.72 & +0.29 \end{bmatrix}$$

Displacement Vector [δ]:   **Fixed-End Forces Vector [FEF]:**

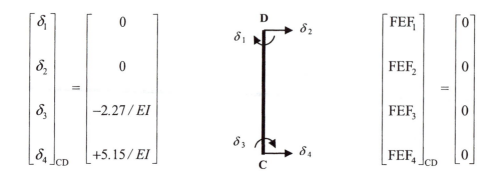

$$\begin{bmatrix} \delta_1 \\ \delta_2 \\ \delta_3 \\ \delta_4 \end{bmatrix}_{CD} = \begin{bmatrix} 0 \\ 0 \\ -2.27/EI \\ +5.15/EI \end{bmatrix}$$

$$\begin{bmatrix} FEF_1 \\ FEF_2 \\ FEF_3 \\ FEF_4 \end{bmatrix}_{CD} = \begin{bmatrix} 0 \\ 0 \\ 0 \\ 0 \end{bmatrix}$$

Element End Forces [F]_{DC}:

$$[F]_{Total} = \begin{bmatrix} F_1 \\ F_2 \\ F_3 \\ F_4 \end{bmatrix} + \begin{bmatrix} FEF_1 \\ FEF_2 \\ FEF_3 \\ FEF_4 \end{bmatrix} = [k][\delta] + [FEF]$$

$$= EI \begin{bmatrix} +2.40 & -0.72 & +1.20 & +0.72 \\ -0.72 & +0.29 & -0.72 & -0.29 \\ +1.20 & -0.72 & +2.40 & +0.72 \\ +0.72 & -0.29 & +0.72 & +0.29 \end{bmatrix} \begin{bmatrix} 0 \\ 0 \\ -2.27/EI \\ +5.15/EI \end{bmatrix} + \begin{bmatrix} 0 \\ 0 \\ 0 \\ 0 \end{bmatrix}$$

$F_1 = [-(1.20 \times 2.27) + (0.72 \times 5.15)] + [0] = +0.98$ kNm

$F_2 = [+(0.72 \times 2.27) - (0.29 \times 5.15)] + [0] = +0.14$ kN

$F_3 = [-(2.40 \times 2.27) + (0.72 \times 5.15)] + [0] = -1.74$ kNm

$F_4 = [-(0.72 \times 2.27) + (0.29 \times 5.15)] + [0] = -0.14$ kN

Reactions:

Support A:

$V_A = (F_2)_{BC} = 6.85$ kN ↑ $H_A = (F_2)_{AB} = 7.12$ kN ←

$M_A = (F_1)_{AB} = +8.77$ kNm

Support D:

$V_D = (F_4)_{BC} = 5.15$ kN ↑ $H_D = (F_2)_{DC} = 0.14$ kN →

$M_D = (F_1)_{DC} = +0.98$ kNm

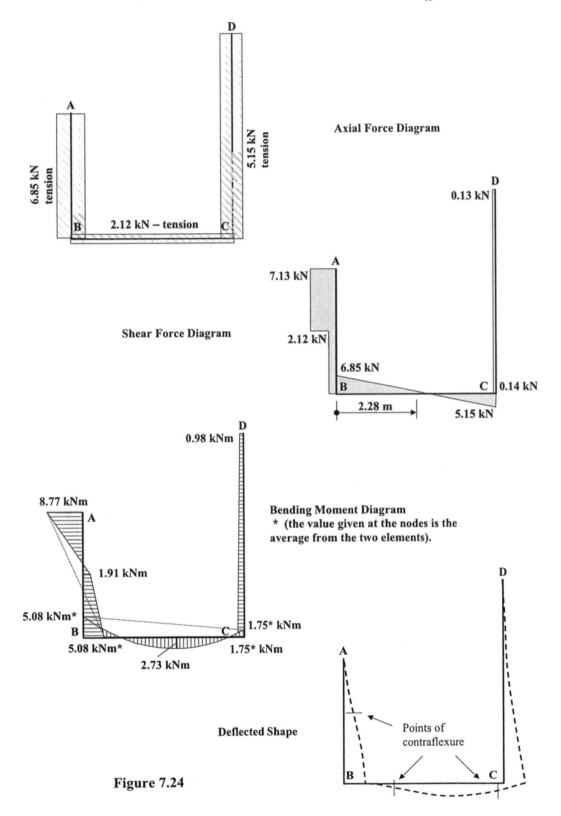

Axial Force Diagram

Shear Force Diagram

Bending Moment Diagram
 * (the value given at the nodes is the
average from the two elements).

Deflected Shape

Figure 7.24

7.10 *Problems: Direct Stiffness Method*

A series of indeterminate structures are indicated in Problems 7.1 to 7.6 in which the assumed degrees-of-freedom at the nodes and the relative *EI* values for the members are given. In each case for the data indicated:

(i)   generate the structural stiffness matrix [*K*] and the applied load vector [*P*],
(ii) determine the structural displacements [*Δ*],
(iii) determine the member end forces [*F*],
(iv) determine the support reactions,
(v)   sketch the axial load, shear force, and bending moment diagrams and the deflected shape for each structure.

Assume all members to be axially rigid.

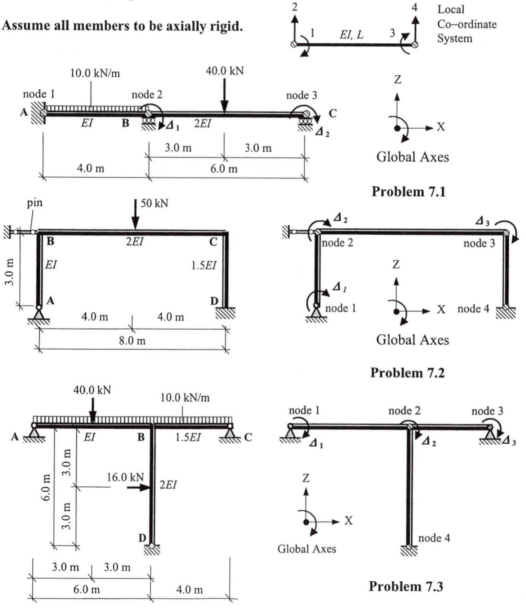

Problem 7.1

Problem 7.2

Problem 7.3

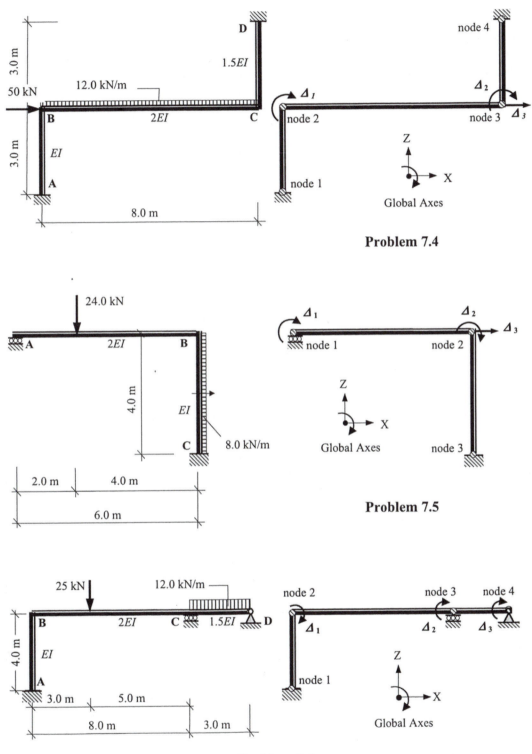

Problem 7.4

Problem 7.5

Problem 7.6

Solution

Topic: **Direct Stiffness Method**
Problem Number: **7.1**   **Page No. 1**

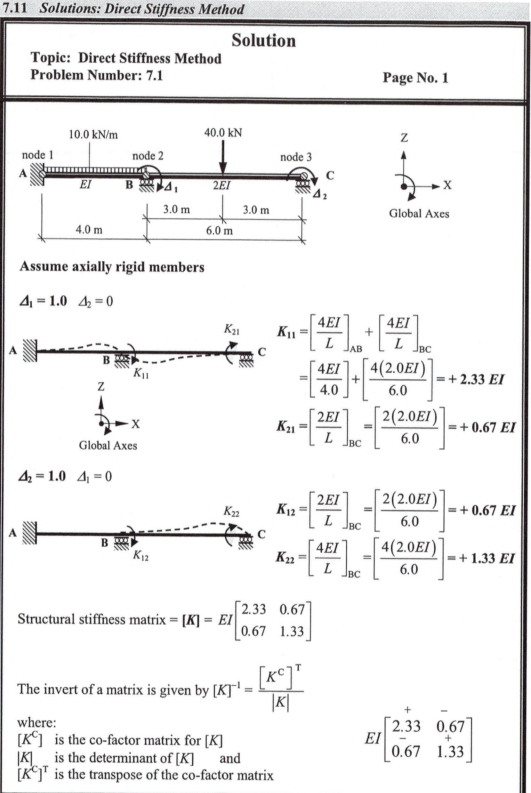

Assume axially rigid members

$\Delta_1 = 1.0$ $\Delta_2 = 0$

$$K_{11} = \left[\frac{4EI}{L}\right]_{AB} + \left[\frac{4EI}{L}\right]_{BC}$$

$$= \left[\frac{4EI}{4.0}\right] + \left[\frac{4(2.0EI)}{6.0}\right] = +\,2.33\ EI$$

$$K_{21} = \left[\frac{2EI}{L}\right]_{BC} = \left[\frac{2(2.0EI)}{6.0}\right] = +\,0.67\ EI$$

$\Delta_2 = 1.0$ $\Delta_1 = 0$

$$K_{12} = \left[\frac{2EI}{L}\right]_{BC} = \left[\frac{2(2.0EI)}{6.0}\right] = +\,0.67\ EI$$

$$K_{22} = \left[\frac{4EI}{L}\right]_{BC} = \left[\frac{4(2.0EI)}{6.0}\right] = +\,1.33\ EI$$

Structural stiffness matrix = $[K] = EI\begin{bmatrix} 2.33 & 0.67 \\ 0.67 & 1.33 \end{bmatrix}$

The invert of a matrix is given by $[K]^{-1} = \dfrac{\left[K^{C}\right]^{T}}{|K|}$

where:
$[K^{C}]$ is the co-factor matrix for $[K]$
$|K|$ is the determinant of $[K]$ and
$[K^{C}]^{T}$ is the transpose of the co-factor matrix

$$EI\begin{bmatrix} \overset{+}{2.33} & \overset{-}{0.67} \\ \underset{0.67}{-} & \underset{1.33}{+} \end{bmatrix}$$

Solution

Topic: **Direct Stiffness Method**
Problem Number: **7.1**   **Page No. 2**

Co-factor Matrix: $[K^C]$
(**Note:** the transpose of a symmetric matrix is the same as the original matrix)

$k_{11}^c = + 1.33EI$

$k_{12}^c = k_{21}^c = - 0.67EI$

$k_{22}^c = + 2.33EI$

Determinant of $[K]$:

Det $[K] = EI^2 \{+ (2.33 \times 1.33) - (0.67 \times 0.67)\} = + 2.65 \, EI^2$

Inverted stiffness matrix = $[K]^{-1} = \dfrac{1}{EI} \begin{bmatrix} 0.502 & -0.253 \\ -0.253 & 0.879 \end{bmatrix}$

Structural Load Vector: $[P]$:

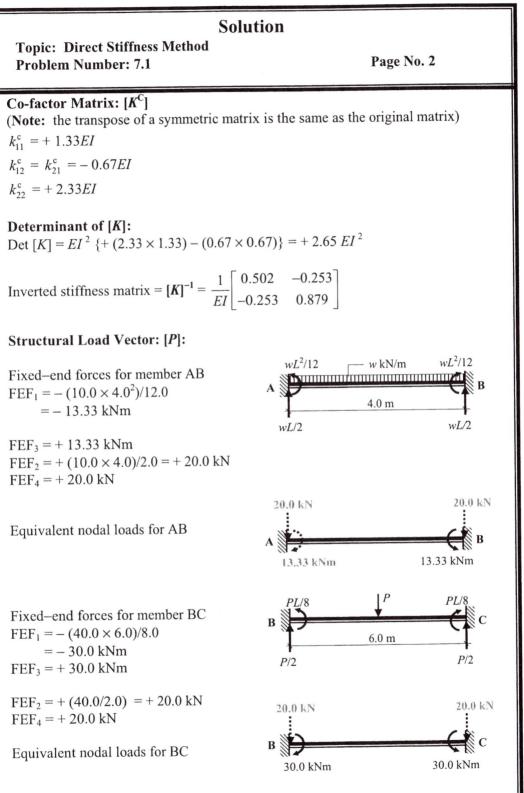

Fixed–end forces for member AB
$FEF_1 = - (10.0 \times 4.0^2)/12.0$
$\quad = - 13.33$ kNm

$FEF_3 = + 13.33$ kNm
$FEF_2 = + (10.0 \times 4.0)/2.0 = + 20.0$ kN
$FEF_4 = + 20.0$ kN

Equivalent nodal loads for AB

Fixed–end forces for member BC
$FEF_1 = - (40.0 \times 6.0)/8.0$
$\quad = - 30.0$ kNm
$FEF_3 = + 30.0$ kNm

$FEF_2 = + (40.0/2.0) = + 20.0$ kN
$FEF_4 = + 20.0$ kN

Equivalent nodal loads for BC

Solution

Topic: Direct Stiffness Method
Problem Number: 7.1   Page No. 3

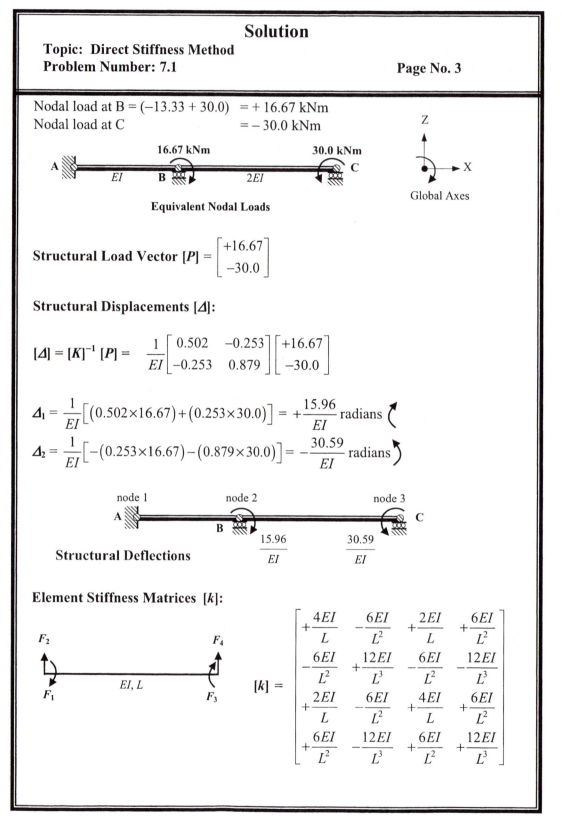

Nodal load at B = (−13.33 + 30.0) = + 16.67 kNm
Nodal load at C   = − 30.0 kNm

16.67 kNm   **30.0 kNm**

A ▨━━━━━━━━━━━━━━━━━━━━━━━━━━━━━━━━ C
       EI   B   2EI

Z
│
│
●────► X

Global Axes

Equivalent Nodal Loads

Structural Load Vector $[P] = \begin{bmatrix} +16.67 \\ -30.0 \end{bmatrix}$

Structural Displacements $[\Delta]$:

$$[\Delta] = [K]^{-1}[P] = \frac{1}{EI}\begin{bmatrix} 0.502 & -0.253 \\ -0.253 & 0.879 \end{bmatrix}\begin{bmatrix} +16.67 \\ -30.0 \end{bmatrix}$$

$$\Delta_1 = \frac{1}{EI}\left[(0.502 \times 16.67) + (0.253 \times 30.0)\right] = +\frac{15.96}{EI} \text{ radians } \circlearrowleft$$

$$\Delta_2 = \frac{1}{EI}\left[-(0.253 \times 16.67) - (0.879 \times 30.0)\right] = -\frac{30.59}{EI} \text{ radians } \circlearrowright$$

node 1   node 2   node 3

A ▨━━━━━━━━━━━━━━━━━━━━━━━━━━━━━━━━ C
            B
                   15.96   30.59
Structural Deflections   ─────   ─────
                     EI   EI

Element Stiffness Matrices $[k]$:

F_2 F_4

F_1 EI, L F_3

$$[k] = \begin{bmatrix} +\dfrac{4EI}{L} & -\dfrac{6EI}{L^2} & +\dfrac{2EI}{L} & +\dfrac{6EI}{L^2} \\ -\dfrac{6EI}{L^2} & +\dfrac{12EI}{L^3} & -\dfrac{6EI}{L^2} & -\dfrac{12EI}{L^3} \\ +\dfrac{2EI}{L} & -\dfrac{6EI}{L^2} & +\dfrac{4EI}{L} & +\dfrac{6EI}{L^2} \\ +\dfrac{6EI}{L^2} & -\dfrac{12EI}{L^3} & +\dfrac{6EI}{L^2} & +\dfrac{12EI}{L^3} \end{bmatrix}$$

Solution

Element End Forces $[F]_{Total}$:

$$[F]_{Total} = \begin{bmatrix} F_1 \\ F_2 \\ F_3 \\ F_4 \end{bmatrix} + \begin{bmatrix} FEF_1 \\ FEF_2 \\ FEF_3 \\ FEF_4 \end{bmatrix} = [k][\delta] + [FEF]$$

$$= \begin{bmatrix} +\dfrac{4EI}{L} & -\dfrac{6EI}{L^2} & +\dfrac{2EI}{L} & +\dfrac{6EI}{L^2} \\[2mm] -\dfrac{6EI}{L^2} & +\dfrac{12EI}{L^3} & -\dfrac{6EI}{L^2} & -\dfrac{12EI}{L^3} \\[2mm] +\dfrac{2EI}{L} & -\dfrac{6EI}{L^2} & +\dfrac{4EI}{L} & +\dfrac{6EI}{L^2} \\[2mm] +\dfrac{6EI}{L^2} & -\dfrac{12EI}{L^3} & +\dfrac{6EI}{L^2} & +\dfrac{12EI}{L^3} \end{bmatrix} \begin{bmatrix} \delta_1 \\ \delta_2 \\ \delta_3 \\ \delta_4 \end{bmatrix} + \begin{bmatrix} FEF_1 \\ FEF_2 \\ FEF_3 \\ FEF_4 \end{bmatrix}$$

Consider element AB:

$$\frac{4EI}{L} = \frac{4 \times EI}{4.0} = 1.0EI \qquad \frac{6EI}{L^2} = \frac{6 \times EI}{4.0^2} = 0.38EI$$

$$\frac{2EI}{L} = \frac{2 \times EI}{4.0} = 0.50EI \qquad \frac{12EI}{L^3} = \frac{12 \times EI}{4.0^3} = 0.19EI$$

$$[k]_{AB} = EI \begin{bmatrix} +1.00 & -0.38 & +0.50 & +0.38 \\ -0.38 & +0.19 & -0.38 & -0.19 \\ +0.50 & -0.38 & +1.00 & +0.38 \\ +0.38 & -0.19 & +0.38 & +0.19 \end{bmatrix}$$

Displacement Vector $[\delta]$:   **Fixed-End Forces Vector [FEF]:**

$$\begin{bmatrix} \delta_1 \\ \delta_2 \\ \delta_3 \\ \delta_4 \end{bmatrix}_{AB} = \begin{bmatrix} 0 \\ 0 \\ +15.96/EI \\ 0 \end{bmatrix} \qquad\qquad \begin{bmatrix} FEF_1 \\ FEF_2 \\ FEF_3 \\ FEF_4 \end{bmatrix}_{AB} = \begin{bmatrix} -13.33 \\ +20.0 \\ +13.33 \\ +20.0 \end{bmatrix}$$

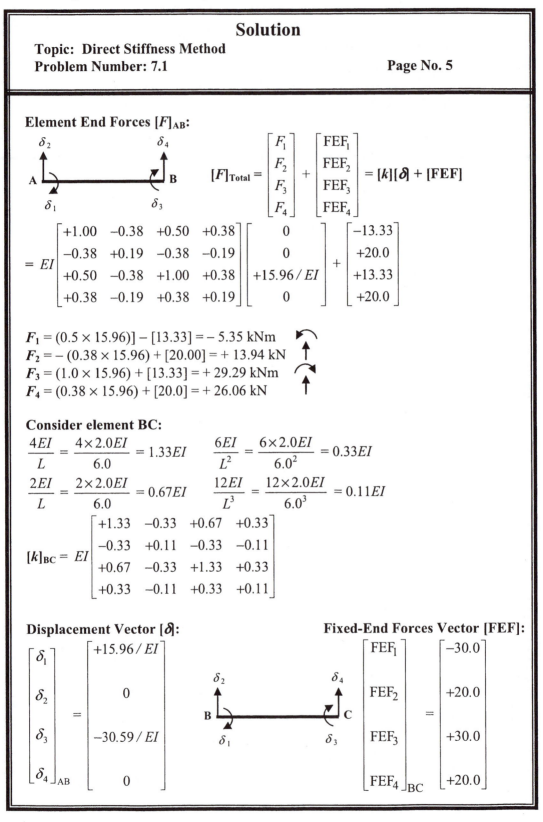

Solution

Topic: Direct Stiffness Method
Problem Number: 7.1   **Page No. 5**

Element End Forces $[F]_{AB}$:

$$[F]_{Total} = \begin{bmatrix} F_1 \\ F_2 \\ F_3 \\ F_4 \end{bmatrix} + \begin{bmatrix} FEF_1 \\ FEF_2 \\ FEF_3 \\ FEF_4 \end{bmatrix} = [k][\delta] + [FEF]$$

$$= EI \begin{bmatrix} +1.00 & -0.38 & +0.50 & +0.38 \\ -0.38 & +0.19 & -0.38 & -0.19 \\ +0.50 & -0.38 & +1.00 & +0.38 \\ +0.38 & -0.19 & +0.38 & +0.19 \end{bmatrix} \begin{bmatrix} 0 \\ 0 \\ +15.96/EI \\ 0 \end{bmatrix} + \begin{bmatrix} -13.33 \\ +20.0 \\ +13.33 \\ +20.0 \end{bmatrix}$$

$F_1 = (0.5 \times 15.96)] - [13.33] = -5.35$ kNm

$F_2 = -(0.38 \times 15.96) + [20.00] = +13.94$ kN

$F_3 = (1.0 \times 15.96) + [13.33] = +29.29$ kNm

$F_4 = (0.38 \times 15.96) + [20.0] = +26.06$ kN

Consider element BC:

$$\frac{4EI}{L} = \frac{4 \times 2.0EI}{6.0} = 1.33EI \qquad \frac{6EI}{L^2} = \frac{6 \times 2.0EI}{6.0^2} = 0.33EI$$

$$\frac{2EI}{L} = \frac{2 \times 2.0EI}{6.0} = 0.67EI \qquad \frac{12EI}{L^3} = \frac{12 \times 2.0EI}{6.0^3} = 0.11EI$$

$$[k]_{BC} = EI \begin{bmatrix} +1.33 & -0.33 & +0.67 & +0.33 \\ -0.33 & +0.11 & -0.33 & -0.11 \\ +0.67 & -0.33 & +1.33 & +0.33 \\ +0.33 & -0.11 & +0.33 & +0.11 \end{bmatrix}$$

Displacement Vector $[\delta]$:   **Fixed-End Forces Vector [FEF]:**

$$\begin{bmatrix} \delta_1 \\ \delta_2 \\ \delta_3 \\ \delta_4 \end{bmatrix}_{AB} = \begin{bmatrix} +15.96/EI \\ 0 \\ -30.59/EI \\ 0 \end{bmatrix}$$

$$\begin{bmatrix} FEF_1 \\ FEF_2 \\ FEF_3 \\ FEF_4 \end{bmatrix}_{BC} = \begin{bmatrix} -30.0 \\ +20.0 \\ +30.0 \\ +20.0 \end{bmatrix}$$

Solution

Topic: Direct Stiffness Method
Problem Number: 7.1   **Page No. 6**

Element End Forces $[F]_{BC}$:

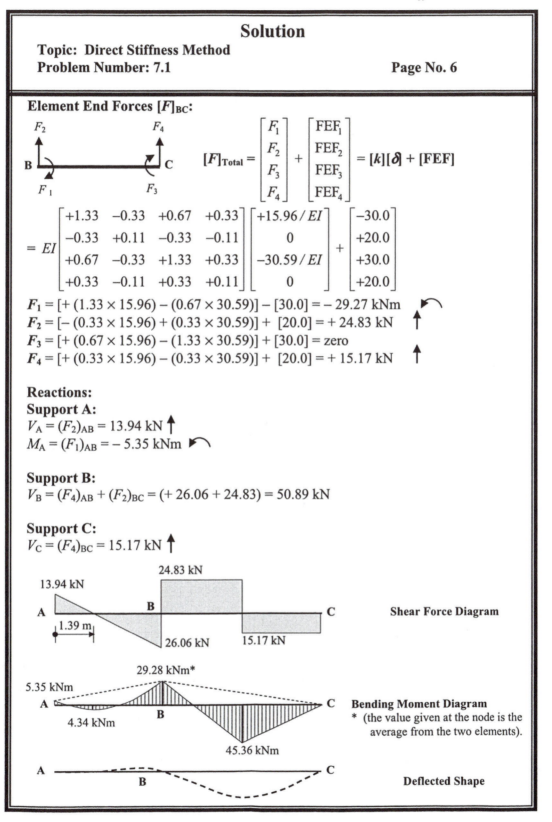

$$[F]_{Total} = \begin{bmatrix} F_1 \\ F_2 \\ F_3 \\ F_4 \end{bmatrix} + \begin{bmatrix} FEF_1 \\ FEF_2 \\ FEF_3 \\ FEF_4 \end{bmatrix} = [k][\delta] + [FEF]$$

$$= EI \begin{bmatrix} +1.33 & -0.33 & +0.67 & +0.33 \\ -0.33 & +0.11 & -0.33 & -0.11 \\ +0.67 & -0.33 & +1.33 & +0.33 \\ +0.33 & -0.11 & +0.33 & +0.11 \end{bmatrix} \begin{bmatrix} +15.96/EI \\ 0 \\ -30.59/EI \\ 0 \end{bmatrix} + \begin{bmatrix} -30.0 \\ +20.0 \\ +30.0 \\ +20.0 \end{bmatrix}$$

$F_1 = [+ (1.33 \times 15.96) - (0.67 \times 30.59)] - [30.0] = -29.27$ kNm

$F_2 = [- (0.33 \times 15.96) + (0.33 \times 30.59)] + [20.0] = +24.83$ kN

$F_3 = [+ (0.67 \times 15.96) - (1.33 \times 30.59)] + [30.0] = $ zero

$F_4 = [+ (0.33 \times 15.96) - (0.33 \times 30.59)] + [20.0] = +15.17$ kN

Reactions:

Support A:

$V_A = (F_2)_{AB} = 13.94$ kN ↑

$M_A = (F_1)_{AB} = -5.35$ kNm

Support B:

$V_B = (F_4)_{AB} + (F_2)_{BC} = (+26.06 + 24.83) = 50.89$ kN

Support C:

$V_C = (F_4)_{BC} = 15.17$ kN ↑

Shear Force Diagram

Bending Moment Diagram
* (the value given at the node is the average from the two elements).

Deflected Shape

Solution

Topic: Direct Stiffness Method
Problem Number: 7.2   **Page No. 1**

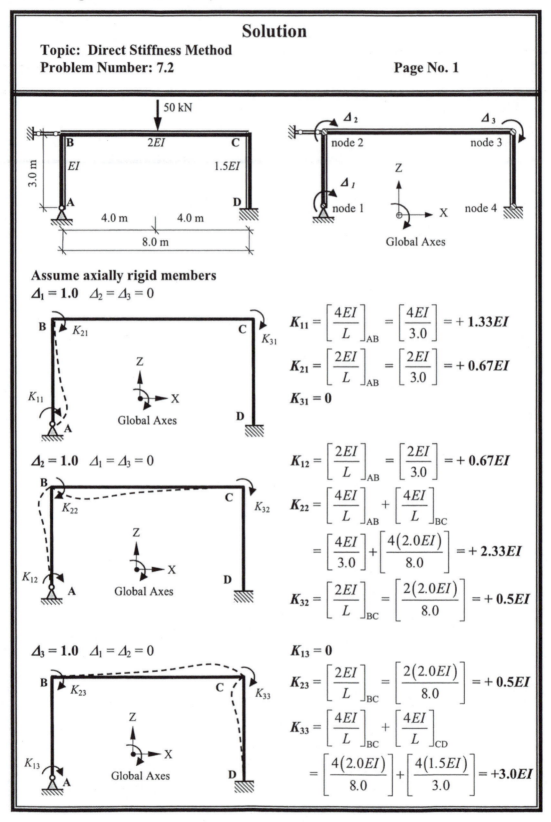

Assume axially rigid members
$\Delta_1 = 1.0$ $\Delta_2 = \Delta_3 = 0$

$$K_{11} = \left[\frac{4EI}{L}\right]_{AB} = \left[\frac{4EI}{3.0}\right] = +1.33EI$$

$$K_{21} = \left[\frac{2EI}{L}\right]_{AB} = \left[\frac{2EI}{3.0}\right] = +0.67EI$$

$$K_{31} = 0$$

$\Delta_2 = 1.0$ $\Delta_1 = \Delta_3 = 0$

$$K_{12} = \left[\frac{2EI}{L}\right]_{AB} = \left[\frac{2EI}{3.0}\right] = +0.67EI$$

$$K_{22} = \left[\frac{4EI}{L}\right]_{AB} + \left[\frac{4EI}{L}\right]_{BC}$$

$$= \left[\frac{4EI}{3.0}\right] + \left[\frac{4(2.0EI)}{8.0}\right] = +2.33EI$$

$$K_{32} = \left[\frac{2EI}{L}\right]_{BC} = \left[\frac{2(2.0EI)}{8.0}\right] = +0.5EI$$

$\Delta_3 = 1.0$ $\Delta_1 = \Delta_2 = 0$

$$K_{13} = 0$$

$$K_{23} = \left[\frac{2EI}{L}\right]_{BC} = \left[\frac{2(2.0EI)}{8.0}\right] = +0.5EI$$

$$K_{33} = \left[\frac{4EI}{L}\right]_{BC} + \left[\frac{4EI}{L}\right]_{CD}$$

$$= \left[\frac{4(2.0EI)}{8.0}\right] + \left[\frac{4(1.5EI)}{3.0}\right] = +3.0EI$$

Solution

Topic: Direct Stiffness Method
Problem Number: 7.2   **Page No. 2**

Structural stiffness matrix = $[K]$ = $EI \begin{bmatrix} 1.33 & 0.67 & 0 \\ 0.67 & 2.33 & 0.50 \\ 0 & 0.50 & 3.0 \end{bmatrix}$

The invert of a matrix is given by $[K]^{-1} = \dfrac{\left[K^C\right]^T}{|K|}$

where:
$[K^C]$ is the co-factor matrix for $[K]$
$|K|$ is the determinant of $[K]$ and
$[K^C]^T$ is the transpose of the co-factor matrix

$$EI \begin{bmatrix} \overset{+}{1.33} & \overset{-}{0.67} & \overset{+}{0} \\ \underset{}{0.67} & \overset{+}{2.33} & 0.5 \\ \underset{}{0} & \overset{-}{0.5} & \overset{+}{3.0} \end{bmatrix}$$

Co-factor Matrix: $[K^C]$
(**Note:** the transpose of a symmetric matrix is the same as the original matrix)

$k_{11}^c = + \{(2.33 \times 3.0) - (0.5 \times 0.5)\}EI^2 = + 6.74EI^2$

$k_{12}^c = k_{21}^c = - \{(0.67 \times 3.0) - (0 \times 0.5)\}EI^2 = - 2.0EI^2$

$k_{13}^c = k_{31}^c = + \{(0.67 \times 0.5) - (0 \times 2.33)\}EI^2 = + 0.34EI^2$

$k_{22}^c = + \{(1.33 \times 3.0) - 0\}EI^2 = + 4.0EI^2$

$k_{23}^c = k_{32}^c = - \{(1.33 \times 0.5) - (0 \times 0.67)\}EI^2 = - 0.67EI^2$

$k_{33}^c = + \{(1.33 \times 2.33) - (0.67 \times 0.67)\}EI^2 = + 2.65EI^2$

Determinant of $[K]$:
Det $[K] = EI^3 \{+ (1.33 \times 6.74) - (0.67 \times 2.0) + 0\} = + 7.62\ EI^3$

Inverted stiffness matrix = $[K]^{-1} = \dfrac{1}{EI} \begin{bmatrix} 0.885 & -0.264 & 0.044 \\ -0.264 & 0.524 & -0.087 \\ 0.044 & -0.087 & 0.348 \end{bmatrix}$

Structural Load Vector: $[P]$:
Fixed–end forces for member BC

$FEF_1 = - (50.0 \times 8.0)/8.0$
$\quad = - 50.0$ kNm

$FEF_3 = + 50.0$ kNm

$FEF_2 = + (50.0/2.0) = + 25.0$ kN
$FEF_4 = + 25.0$ kN

Equivalent nodal loads for BC

Solution

Topic: Direct Stiffness Method
Problem Number: 7.2   **Page No. 3**

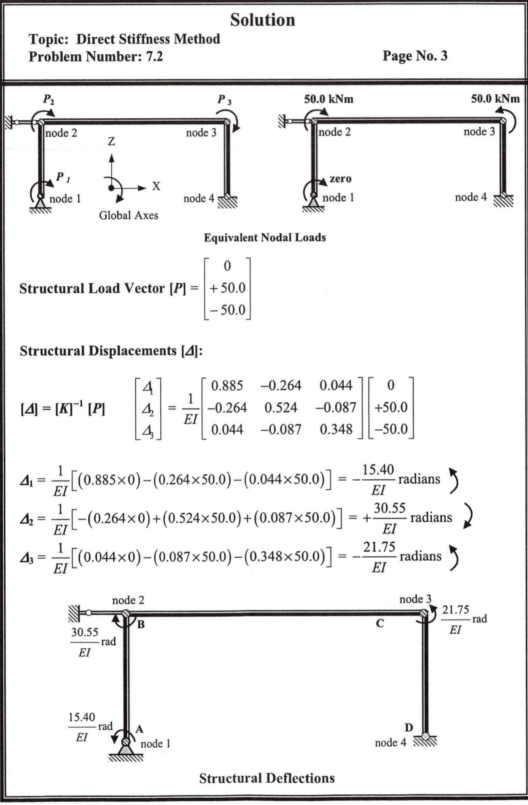

Structural Load Vector $[P] = \begin{bmatrix} 0 \\ +50.0 \\ -50.0 \end{bmatrix}$

Structural Displacements $[\Delta]$:

$$[\Delta] = [K]^{-1}[P] \qquad \begin{bmatrix} \Delta_1 \\ \Delta_2 \\ \Delta_3 \end{bmatrix} = \frac{1}{EI} \begin{bmatrix} 0.885 & -0.264 & 0.044 \\ -0.264 & 0.524 & -0.087 \\ 0.044 & -0.087 & 0.348 \end{bmatrix} \begin{bmatrix} 0 \\ +50.0 \\ -50.0 \end{bmatrix}$$

$$\Delta_1 = \frac{1}{EI}\left[(0.885 \times 0) - (0.264 \times 50.0) - (0.044 \times 50.0)\right] = -\frac{15.40}{EI}\ \text{radians} \curvearrowright$$

$$\Delta_2 = \frac{1}{EI}\left[-(0.264 \times 0) + (0.524 \times 50.0) + (0.087 \times 50.0)\right] = +\frac{30.55}{EI}\ \text{radians} \curvearrowright$$

$$\Delta_3 = \frac{1}{EI}\left[(0.044 \times 0) - (0.087 \times 50.0) - (0.348 \times 50.0)\right] = -\frac{21.75}{EI}\ \text{radians} \curvearrowright$$

Structural Deflections

Solution

Topic: Direct Stiffness Method
Problem Number: 7.2   **Page No. 4**

Element Stiffness Matrices [k]:

$$[k] = \begin{bmatrix} +\dfrac{4EI}{L} & -\dfrac{6EI}{L^2} & +\dfrac{2EI}{L} & +\dfrac{6EI}{L^2} \\[2mm] -\dfrac{6EI}{L^2} & +\dfrac{12EI}{L^3} & -\dfrac{6EI}{L^2} & -\dfrac{12EI}{L^3} \\[2mm] +\dfrac{2EI}{L} & -\dfrac{6EI}{L^2} & +\dfrac{4EI}{L} & +\dfrac{6EI}{L^2} \\[2mm] +\dfrac{6EI}{L^2} & -\dfrac{12EI}{L^3} & +\dfrac{6EI}{L^2} & +\dfrac{12EI}{L^3} \end{bmatrix}$$

Element End Forces $[F]_{\text{Total}}$:

$$[F]_{\text{Total}} = \begin{bmatrix} F_1 \\ F_2 \\ F_3 \\ F_4 \end{bmatrix} + \begin{bmatrix} \text{FEF}_1 \\ \text{FEF}_2 \\ \text{FEF}_3 \\ \text{FEF}_4 \end{bmatrix} = [k][\delta] + [\text{FEF}]$$

$$= \begin{bmatrix} +\dfrac{4EI}{L} & -\dfrac{6EI}{L^2} & +\dfrac{2EI}{L} & +\dfrac{6EI}{L^2} \\[2mm] -\dfrac{6EI}{L^2} & +\dfrac{12EI}{L^3} & -\dfrac{6EI}{L^2} & -\dfrac{12EI}{L^3} \\[2mm] +\dfrac{2EI}{L} & -\dfrac{6EI}{L^2} & +\dfrac{4EI}{L} & +\dfrac{6EI}{L^2} \\[2mm] +\dfrac{6EI}{L^2} & -\dfrac{12EI}{L^3} & +\dfrac{6EI}{L^2} & +\dfrac{12EI}{L^3} \end{bmatrix} \begin{bmatrix} \delta_1 \\ \delta_2 \\ \delta_3 \\ \delta_4 \end{bmatrix} + \begin{bmatrix} \text{FEF}_1 \\ \text{FEF}_2 \\ \text{FEF}_3 \\ \text{FEF}_4 \end{bmatrix}$$

Consider element AB:

$$\frac{4EI}{L} = \frac{4 \times EI}{3.0} = 1.33EI \qquad\qquad \frac{6EI}{L^2} = \frac{6 \times EI}{3.0^2} = 0.67EI$$

$$\frac{2EI}{L} = \frac{2 \times EI}{3.0} = 0.67EI \qquad\qquad \frac{12EI}{L^3} = \frac{12 \times EI}{3.0^3} = 0.44EI$$

$$[k]_{\text{AB}} = EI \begin{bmatrix} +1.33 & -0.67 & +0.67 & +0.67 \\ -0.67 & +0.44 & -0.67 & -0.44 \\ +0.67 & -0.67 & +1.33 & +0.67 \\ +0.67 & -0.44 & +0.67 & +0.44 \end{bmatrix}$$

Solution

Topic: Direct Stiffness Method
Problem Number: 7.2   **Page No. 5**

Displacement Vector [δ]:   **Fixed-End Forces Vector [FEF]:**

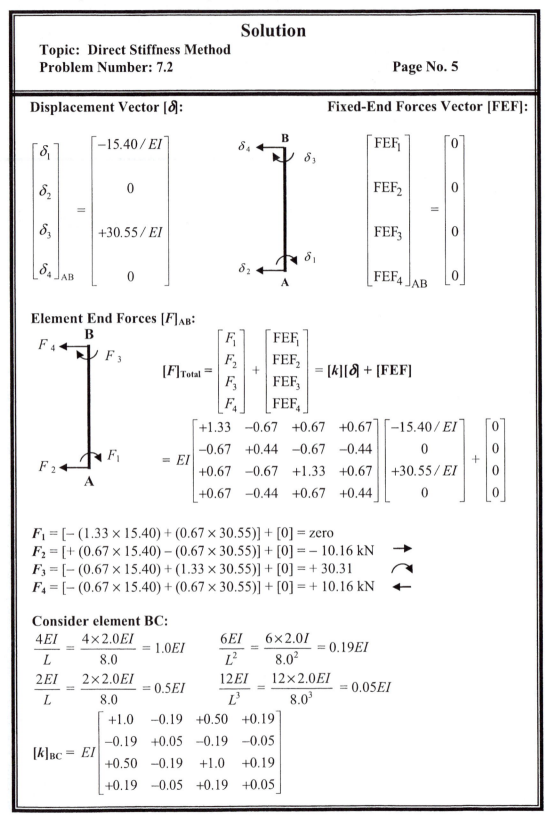

$$
\begin{bmatrix} \delta_1 \\ \delta_2 \\ \delta_3 \\ \delta_4 \end{bmatrix}_{AB} = \begin{bmatrix} -15.40/EI \\ 0 \\ +30.55/EI \\ 0 \end{bmatrix}
\qquad
\begin{bmatrix} FEF_1 \\ FEF_2 \\ FEF_3 \\ FEF_4 \end{bmatrix}_{AB} = \begin{bmatrix} 0 \\ 0 \\ 0 \\ 0 \end{bmatrix}
$$

Element End Forces [F]$_{AB}$:

$$
[F]_{Total} = \begin{bmatrix} F_1 \\ F_2 \\ F_3 \\ F_4 \end{bmatrix} + \begin{bmatrix} FEF_1 \\ FEF_2 \\ FEF_3 \\ FEF_4 \end{bmatrix} = [k][\delta] + [FEF]
$$

$$
= EI\begin{bmatrix} +1.33 & -0.67 & +0.67 & +0.67 \\ -0.67 & +0.44 & -0.67 & -0.44 \\ +0.67 & -0.67 & +1.33 & +0.67 \\ +0.67 & -0.44 & +0.67 & +0.44 \end{bmatrix}\begin{bmatrix} -15.40/EI \\ 0 \\ +30.55/EI \\ 0 \end{bmatrix} + \begin{bmatrix} 0 \\ 0 \\ 0 \\ 0 \end{bmatrix}
$$

$F_1 = [-(1.33 \times 15.40) + (0.67 \times 30.55)] + [0] = \text{zero}$

$F_2 = [+(0.67 \times 15.40) - (0.67 \times 30.55)] + [0] = -10.16 \text{ kN } \rightarrow$

$F_3 = [-(0.67 \times 15.40) + (1.33 \times 30.55)] + [0] = +30.31 \quad \curvearrowright$

$F_4 = [-(0.67 \times 15.40) + (0.67 \times 30.55)] + [0] = +10.16 \text{ kN } \leftarrow$

Consider element BC:

$$
\frac{4EI}{L} = \frac{4 \times 2.0EI}{8.0} = 1.0EI
\qquad
\frac{6EI}{L^2} = \frac{6 \times 2.0I}{8.0^2} = 0.19EI
$$

$$
\frac{2EI}{L} = \frac{2 \times 2.0EI}{8.0} = 0.5EI
\qquad
\frac{12EI}{L^3} = \frac{12 \times 2.0EI}{8.0^3} = 0.05EI
$$

$$
[k]_{BC} = EI\begin{bmatrix} +1.0 & -0.19 & +0.50 & +0.19 \\ -0.19 & +0.05 & -0.19 & -0.05 \\ +0.50 & -0.19 & +1.0 & +0.19 \\ +0.19 & -0.05 & +0.19 & +0.05 \end{bmatrix}
$$

Solution

Topic: Direct Stiffness Method
Problem Number: 7.2   **Page No. 6**

Displacement Vector [δ]:   **Fixed-End Forces Vector [FEF]:**

$$
\begin{bmatrix} \delta_1 \\ \delta_2 \\ \delta_3 \\ \delta_4 \end{bmatrix}_{BC}
=
\begin{bmatrix} +30.55/EI \\ 0 \\ -21.75/EI \\ 0 \end{bmatrix}
\qquad
\begin{bmatrix} FEF_1 \\ FEF_2 \\ FEF_3 \\ FEF_4 \end{bmatrix}_{BC}
=
\begin{bmatrix} -50.0 \\ +25.0 \\ +50.0 \\ +25.0 \end{bmatrix}
$$

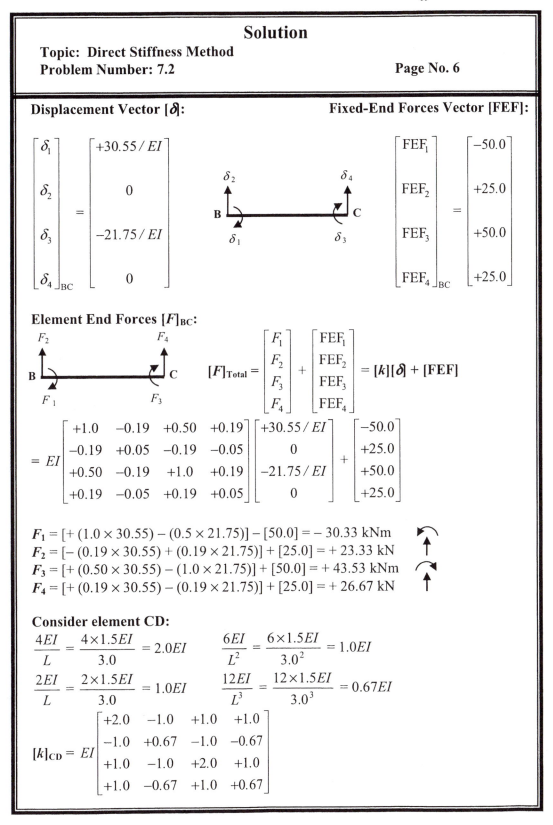

Element End Forces [F]$_{BC}$:

$$
[F]_{Total} =
\begin{bmatrix} F_1 \\ F_2 \\ F_3 \\ F_4 \end{bmatrix}
+
\begin{bmatrix} FEF_1 \\ FEF_2 \\ FEF_3 \\ FEF_4 \end{bmatrix}
= [k][\delta] + [FEF]
$$

$$
= EI
\begin{bmatrix}
+1.0 & -0.19 & +0.50 & +0.19 \\
-0.19 & +0.05 & -0.19 & -0.05 \\
+0.50 & -0.19 & +1.0 & +0.19 \\
+0.19 & -0.05 & +0.19 & +0.05
\end{bmatrix}
\begin{bmatrix} +30.55/EI \\ 0 \\ -21.75/EI \\ 0 \end{bmatrix}
+
\begin{bmatrix} -50.0 \\ +25.0 \\ +50.0 \\ +25.0 \end{bmatrix}
$$

$F_1 = [+ (1.0 \times 30.55) - (0.5 \times 21.75)] - [50.0] = -30.33 \text{ kNm}$

$F_2 = [- (0.19 \times 30.55) + (0.19 \times 21.75)] + [25.0] = +23.33 \text{ kN}$

$F_3 = [+ (0.50 \times 30.55) - (1.0 \times 21.75)] + [50.0] = +43.53 \text{ kNm}$

$F_4 = [+ (0.19 \times 30.55) - (0.19 \times 21.75)] + [25.0] = +26.67 \text{ kN}$

Consider element CD:

$$\frac{4EI}{L} = \frac{4 \times 1.5EI}{3.0} = 2.0EI \qquad \frac{6EI}{L^2} = \frac{6 \times 1.5EI}{3.0^2} = 1.0EI$$

$$\frac{2EI}{L} = \frac{2 \times 1.5EI}{3.0} = 1.0EI \qquad \frac{12EI}{L^3} = \frac{12 \times 1.5EI}{3.0^3} = 0.67EI$$

$$
[k]_{CD} = EI
\begin{bmatrix}
+2.0 & -1.0 & +1.0 & +1.0 \\
-1.0 & +0.67 & -1.0 & -0.67 \\
+1.0 & -1.0 & +2.0 & +1.0 \\
+1.0 & -0.67 & +1.0 & +0.67
\end{bmatrix}
$$

Solution

Topic: Direct Stiffness Method
Problem Number: 7.2   **Page No. 7**

Displacement Vector $[\delta]$:   **Fixed-End Forces Vector [FEF]:**

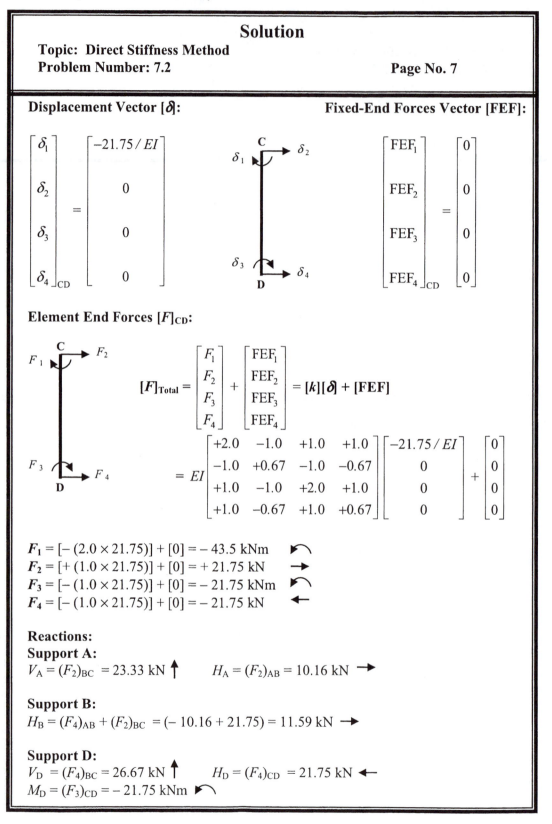

$$\begin{bmatrix} \delta_1 \\ \delta_2 \\ \delta_3 \\ \delta_4 \end{bmatrix}_{CD} = \begin{bmatrix} -21.75/EI \\ 0 \\ 0 \\ 0 \end{bmatrix}$$

$$\begin{bmatrix} FEF_1 \\ FEF_2 \\ FEF_3 \\ FEF_4 \end{bmatrix}_{CD} = \begin{bmatrix} 0 \\ 0 \\ 0 \\ 0 \end{bmatrix}$$

Element End Forces $[F]_{CD}$:

$$[F]_{Total} = \begin{bmatrix} F_1 \\ F_2 \\ F_3 \\ F_4 \end{bmatrix} + \begin{bmatrix} FEF_1 \\ FEF_2 \\ FEF_3 \\ FEF_4 \end{bmatrix} = [k][\delta] + [FEF]$$

$$= EI\begin{bmatrix} +2.0 & -1.0 & +1.0 & +1.0 \\ -1.0 & +0.67 & -1.0 & -0.67 \\ +1.0 & -1.0 & +2.0 & +1.0 \\ +1.0 & -0.67 & +1.0 & +0.67 \end{bmatrix}\begin{bmatrix} -21.75/EI \\ 0 \\ 0 \\ 0 \end{bmatrix} + \begin{bmatrix} 0 \\ 0 \\ 0 \\ 0 \end{bmatrix}$$

$F_1 = [-(2.0 \times 21.75)] + [0] = -43.5 \text{ kNm}$

$F_2 = [+(1.0 \times 21.75)] + [0] = +21.75 \text{ kN}$

$F_3 = [-(1.0 \times 21.75)] + [0] = -21.75 \text{ kNm}$

$F_4 = [-(1.0 \times 21.75)] + [0] = -21.75 \text{ kN}$

Reactions:
Support A:
$V_A = (F_2)_{BC} = 23.33 \text{ kN} \uparrow \qquad H_A = (F_2)_{AB} = 10.16 \text{ kN} \rightarrow$

Support B:
$H_B = (F_4)_{AB} + (F_2)_{BC} = (-10.16 + 21.75) = 11.59 \text{ kN} \rightarrow$

Support D:
$V_D = (F_4)_{BC} = 26.67 \text{ kN} \uparrow \qquad H_D = (F_4)_{CD} = 21.75 \text{ kN} \leftarrow$
$M_D = (F_3)_{CD} = -21.75 \text{ kNm}$

Solution

Topic: Direct Stiffness Method
Problem Number: 7.2   **Page No. 8**

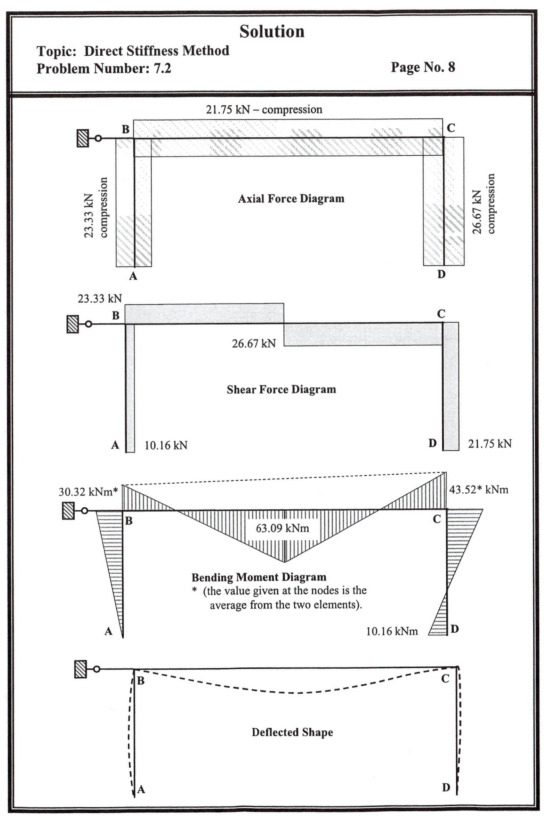

21.75 kN – compression

B   C

23.33 kN compression

Axial Force Diagram

26.67 kN compression

A   D

23.33 kN

B   C

26.67 kN

Shear Force Diagram

A   10.16 kN   D   21.75 kN

30.32 kNm*   43.52* kNm

B   C

63.09 kNm

Bending Moment Diagram
* (the value given at the nodes is the
average from the two elements).

A   10.16 kNm   D

B   C

Deflected Shape

A   D

Solution

Topic: Direct Stiffness Method
Problem Number: 7.3   **Page No. 1**

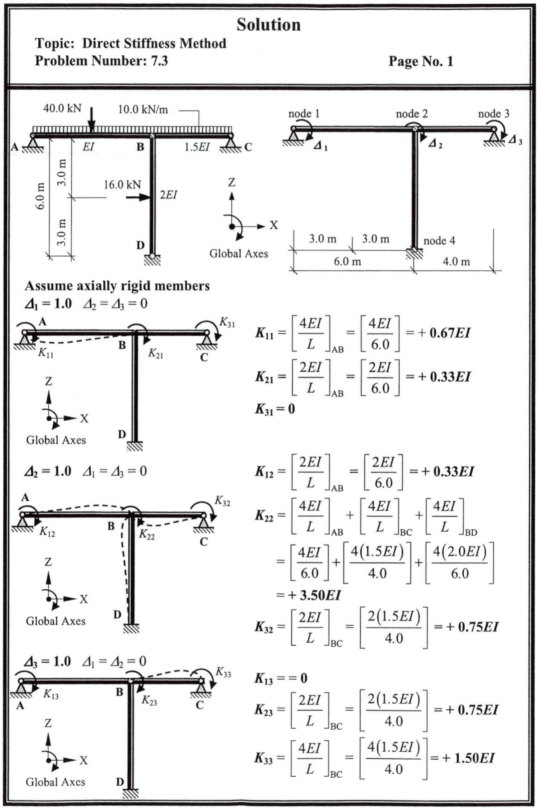

Assume axially rigid members

$\Delta_1 = 1.0$ $\Delta_2 = \Delta_3 = 0$

$$K_{11} = \left[\frac{4EI}{L}\right]_{AB} = \left[\frac{4EI}{6.0}\right] = +\mathbf{0.67}EI$$

$$K_{21} = \left[\frac{2EI}{L}\right]_{AB} = \left[\frac{2EI}{6.0}\right] = +\mathbf{0.33}EI$$

$$K_{31} = \mathbf{0}$$

$\Delta_2 = 1.0$ $\Delta_1 = \Delta_3 = 0$

$$K_{12} = \left[\frac{2EI}{L}\right]_{AB} = \left[\frac{2EI}{6.0}\right] = +\mathbf{0.33}EI$$

$$K_{22} = \left[\frac{4EI}{L}\right]_{AB} + \left[\frac{4EI}{L}\right]_{BC} + \left[\frac{4EI}{L}\right]_{BD}$$

$$= \left[\frac{4EI}{6.0}\right] + \left[\frac{4(1.5EI)}{4.0}\right] + \left[\frac{4(2.0EI)}{6.0}\right]$$

$$= +\mathbf{3.50}EI$$

$$K_{32} = \left[\frac{2EI}{L}\right]_{BC} = \left[\frac{2(1.5EI)}{4.0}\right] = +\mathbf{0.75}EI$$

$\Delta_3 = 1.0$ $\Delta_1 = \Delta_2 = 0$

$$K_{13} = = \mathbf{0}$$

$$K_{23} = \left[\frac{2EI}{L}\right]_{BC} = \left[\frac{2(1.5EI)}{4.0}\right] = +\mathbf{0.75}EI$$

$$K_{33} = \left[\frac{4EI}{L}\right]_{BC} = \left[\frac{4(1.5EI)}{4.0}\right] = +\mathbf{1.50}EI$$

Solution

Topic: Direct Stiffness Method
Problem Number: 7.3   **Page No. 2**

Structural stiffness matrix = $[K]$ = $EI \begin{bmatrix} 0.67 & 0.33 & 0 \\ 0.33 & 3.50 & 0.75 \\ 0 & 0.75 & 1.50 \end{bmatrix}$

The invert of a matrix is given by $[K]^{-1} = \dfrac{\left[K^C\right]^T}{|K|}$

where:
$[K^C]$ is the co-factor matrix for $[K]$
$|K|$ is the determinant of $[K]$ and
$[K^C]^T$ is the transpose of the co-factor matrix

$EI \begin{bmatrix} \overset{+}{0.67} & \overset{-}{0.33} & \overset{+}{0} \\ \underset{-}{0.33} & \underset{+}{3.50} & 0.75 \\ 0 & \underset{-}{0.75} & \underset{+}{1.50} \end{bmatrix}$

Co-factor Matrix: $[K^C]$
(**Note:** the transpose of a symmetric matrix is the same as the original matrix)

$k_{11}^c = + \{(3.50 \times 1.50) - (0.75 \times 0.75)\}EI^2 = + 4.69EI^2$

$k_{12}^c = k_{21}^c = - \{(0.33 \times 1.50) - (0 \times 0.75)\}EI^2 = - 0.50EI^2$

$k_{13}^c = k_{31}^c = + \{(0.33 \times 0.75) - (0 \times 3.50)\}EI^2 = + 0.25EI^2$

$k_{22}^c = + \{(0.67 \times 1.50) - 0\}EI^2 = + 1.0EI^2$

$k_{23}^c = k_{32}^c = - \{(0.67 \times 0.75) - (0 \times 0.33)\}EI^2 = - 0.50EI^2$

$k_{33}^c = + \{(0.67 \times 3.50) - (0.33 \times 0.33)\}EI^2 = + 2.24EI^2$

Determinant of $[K]$:
Det $[K] = EI^3 \{+ (0.67 \times 4.69) - (0.33 \times 0.5) + 0\} = + 2.98 \, EI^3$

Inverted stiffness matrix = $[K]^{-1} = \dfrac{1}{EI} \begin{bmatrix} 1.573 & -0.168 & 0.084 \\ -0.168 & 0.336 & -0.168 \\ 0.084 & -0.168 & 0.752 \end{bmatrix}$

Structural Load Vector: $[P]$:
Fixed–end forces for member AB

$FEF_1 = - (40.0 \times 6.0)/8.0 - (10.0 \times 6.0^2)/12 = - 60.0$ kNm
$FEF_3 = + 60.0$ kNm

$FEF_2 = + (40.0/2.0) + (10.0 \times 6.0)/2.0 = + 50.0$ kN
$FEF_4 = + 50.0$ kN

Solution

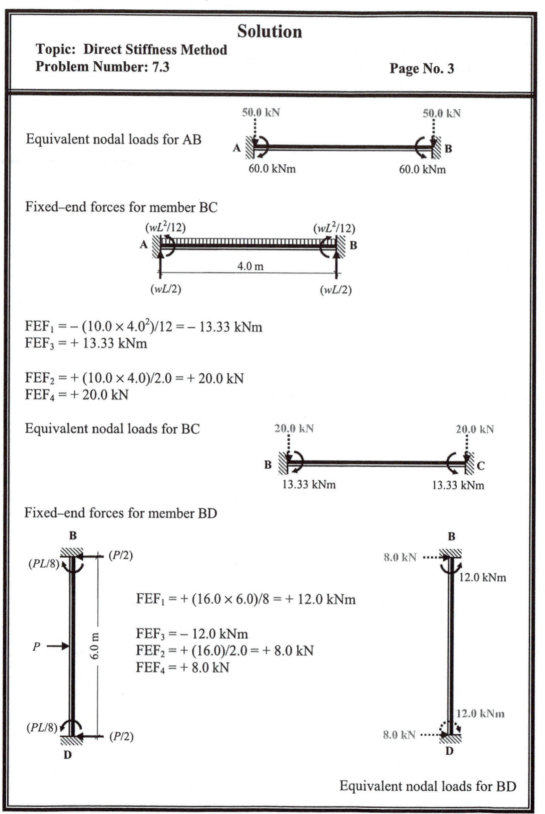

Equivalent nodal loads for AB

Fixed–end forces for member BC

$\text{FEF}_1 = -(10.0 \times 4.0^2)/12 = -13.33 \text{ kNm}$
$\text{FEF}_3 = +13.33 \text{ kNm}$

$\text{FEF}_2 = +(10.0 \times 4.0)/2.0 = +20.0 \text{ kN}$
$\text{FEF}_4 = +20.0 \text{ kN}$

Equivalent nodal loads for BC

Fixed–end forces for member BD

$\text{FEF}_1 = +(16.0 \times 6.0)/8 = +12.0 \text{ kNm}$

$\text{FEF}_3 = -12.0 \text{ kNm}$
$\text{FEF}_2 = +(16.0)/2.0 = +8.0 \text{ kN}$
$\text{FEF}_4 = +8.0 \text{ kN}$

Equivalent nodal loads for BD

Solution

Topic: Direct Stiffness Method
Problem Number: 7.3   **Page No. 4**

$P_1 = +60$ kNm, $P_2 = (-60.0 + 13.33 - 12.0) = -58.67$, $P_3 = -13.33$ kNm

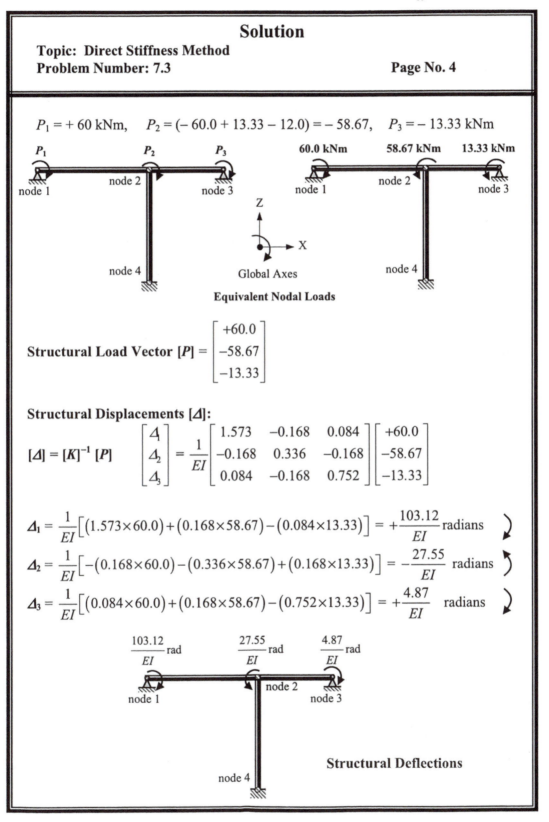

Equivalent Nodal Loads

Structural Load Vector $[P] = \begin{bmatrix} +60.0 \\ -58.67 \\ -13.33 \end{bmatrix}$

Structural Displacements $[\Delta]$:

$$[\Delta] = [K]^{-1}[P] \qquad \begin{bmatrix} \Delta_1 \\ \Delta_2 \\ \Delta_3 \end{bmatrix} = \frac{1}{EI}\begin{bmatrix} 1.573 & -0.168 & 0.084 \\ -0.168 & 0.336 & -0.168 \\ 0.084 & -0.168 & 0.752 \end{bmatrix}\begin{bmatrix} +60.0 \\ -58.67 \\ -13.33 \end{bmatrix}$$

$$\Delta_1 = \frac{1}{EI}\big[(1.573\times 60.0)+(0.168\times 58.67)-(0.084\times 13.33)\big] = +\frac{103.12}{EI}\ \text{radians}$$

$$\Delta_2 = \frac{1}{EI}\big[-(0.168\times 60.0)-(0.336\times 58.67)+(0.168\times 13.33)\big] = -\frac{27.55}{EI}\ \text{radians}$$

$$\Delta_3 = \frac{1}{EI}\big[(0.084\times 60.0)+(0.168\times 58.67)-(0.752\times 13.33)\big] = +\frac{4.87}{EI}\ \text{radians}$$

Structural Deflections

Solution

Topic: Direct Stiffness Method
Problem Number: 7.3

Element Stiffness Matrices [k]:

$$[k] = \begin{bmatrix} +\dfrac{4EI}{L} & -\dfrac{6EI}{L^2} & +\dfrac{2EI}{L} & +\dfrac{6EI}{L^2} \\[2mm] -\dfrac{6EI}{L^2} & +\dfrac{12EI}{L^3} & -\dfrac{6EI}{L^2} & -\dfrac{12EI}{L^3} \\[2mm] +\dfrac{2EI}{L} & -\dfrac{6EI}{L^2} & +\dfrac{4EI}{L} & +\dfrac{6EI}{L^2} \\[2mm] +\dfrac{6EI}{L^2} & -\dfrac{12EI}{L^3} & +\dfrac{6EI}{L^2} & +\dfrac{12EI}{L^3} \end{bmatrix}$$

Element End Forces [F]$_{\text{Total}}$:

$$[F]_{\text{Total}} = \begin{bmatrix} F_1 \\ F_2 \\ F_3 \\ F_4 \end{bmatrix} + \begin{bmatrix} \text{FEF}_1 \\ \text{FEF}_2 \\ \text{FEF}_3 \\ \text{FEF}_4 \end{bmatrix} = [k][\delta] + [\text{FEF}]$$

$$= \begin{bmatrix} +\dfrac{4EI}{L} & -\dfrac{6EI}{L^2} & +\dfrac{2EI}{L} & +\dfrac{6EI}{L^2} \\[2mm] -\dfrac{6EI}{L^2} & +\dfrac{12EI}{L^3} & -\dfrac{6EI}{L^2} & -\dfrac{12EI}{L^3} \\[2mm] +\dfrac{2EI}{L} & -\dfrac{6EI}{L^2} & +\dfrac{4EI}{L} & +\dfrac{6EI}{L^2} \\[2mm] +\dfrac{6EI}{L^2} & -\dfrac{12EI}{L^3} & +\dfrac{6EI}{L^2} & +\dfrac{12EI}{L^3} \end{bmatrix} \begin{bmatrix} \delta_1 \\ \delta_2 \\ \delta_3 \\ \delta_4 \end{bmatrix} + \begin{bmatrix} \text{FEF}_1 \\ \text{FEF}_2 \\ \text{FEF}_3 \\ \text{FEF}_4 \end{bmatrix}$$

Consider element AB:

$$\frac{4EI}{L} = \frac{4 \times EI}{6.0} = 0.67EI \qquad \frac{6EI}{L^2} = \frac{6 \times EI}{6.0^2} = 0.17EI$$

$$\frac{2EI}{L} = \frac{2 \times EI}{6.0} = 0.33EI \qquad \frac{12EI}{L^3} = \frac{12 \times EI}{6.0^3} = 0.06EI$$

$$[k]_{\text{AB}} = EI \begin{bmatrix} +0.67 & -0.17 & +0.33 & +0.17 \\ -0.17 & +0.06 & -0.17 & -0.06 \\ +0.33 & -0.17 & +0.67 & +0.17 \\ +0.17 & -0.06 & +0.17 & +0.06 \end{bmatrix}$$

Solution

Topic: Direct Stiffness Method
Problem Number: 7.3   **Page No. 6**

Displacement Vector [δ]:

$$\begin{bmatrix} \delta_1 \\ \delta_2 \\ \delta_3 \\ \delta_4 \end{bmatrix}_{AB} = \begin{bmatrix} +103.12/EI \\ 0 \\ -27.55/EI \\ 0 \end{bmatrix}$$

Fixed-End Forces Vector [FEF]:

$$\begin{bmatrix} FEF_1 \\ FEF_2 \\ FEF_3 \\ FEF_4 \end{bmatrix}_{AB} = \begin{bmatrix} -60.0 \\ +50.0 \\ +60.0 \\ +50.0 \end{bmatrix}$$

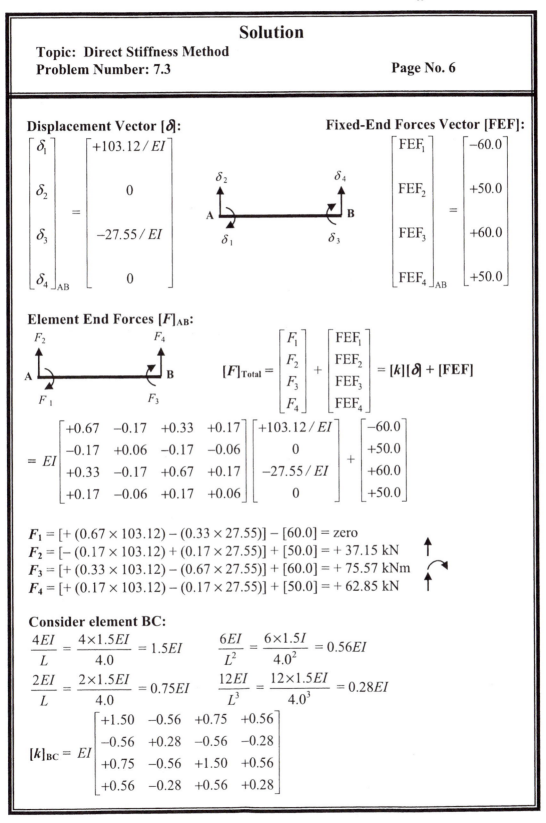

Element End Forces [F]_AB:

$$[F]_{Total} = \begin{bmatrix} F_1 \\ F_2 \\ F_3 \\ F_4 \end{bmatrix} + \begin{bmatrix} FEF_1 \\ FEF_2 \\ FEF_3 \\ FEF_4 \end{bmatrix} = [k][\delta] + [FEF]$$

$$= EI \begin{bmatrix} +0.67 & -0.17 & +0.33 & +0.17 \\ -0.17 & +0.06 & -0.17 & -0.06 \\ +0.33 & -0.17 & +0.67 & +0.17 \\ +0.17 & -0.06 & +0.17 & +0.06 \end{bmatrix} \begin{bmatrix} +103.12/EI \\ 0 \\ -27.55/EI \\ 0 \end{bmatrix} + \begin{bmatrix} -60.0 \\ +50.0 \\ +60.0 \\ +50.0 \end{bmatrix}$$

$F_1 = [+ (0.67 \times 103.12) - (0.33 \times 27.55)] - [60.0] =$ zero

$F_2 = [- (0.17 \times 103.12) + (0.17 \times 27.55)] + [50.0] = + 37.15$ kN ↑

$F_3 = [+ (0.33 \times 103.12) - (0.67 \times 27.55)] + [60.0] = + 75.57$ kNm ↻

$F_4 = [+ (0.17 \times 103.12) - (0.17 \times 27.55)] + [50.0] = + 62.85$ kN ↑

Consider element BC:

$$\frac{4EI}{L} = \frac{4 \times 1.5EI}{4.0} = 1.5EI \qquad \frac{6EI}{L^2} = \frac{6 \times 1.5I}{4.0^2} = 0.56EI$$

$$\frac{2EI}{L} = \frac{2 \times 1.5EI}{4.0} = 0.75EI \qquad \frac{12EI}{L^3} = \frac{12 \times 1.5EI}{4.0^3} = 0.28EI$$

$$[k]_{BC} = EI \begin{bmatrix} +1.50 & -0.56 & +0.75 & +0.56 \\ -0.56 & +0.28 & -0.56 & -0.28 \\ +0.75 & -0.56 & +1.50 & +0.56 \\ +0.56 & -0.28 & +0.56 & +0.28 \end{bmatrix}$$

Solution

Topic: Direct Stiffness Method
Problem Number: 7.3   **Page No. 7**

Displacement Vector [δ]:   **Fixed-End Forces Vector [FEF]:**

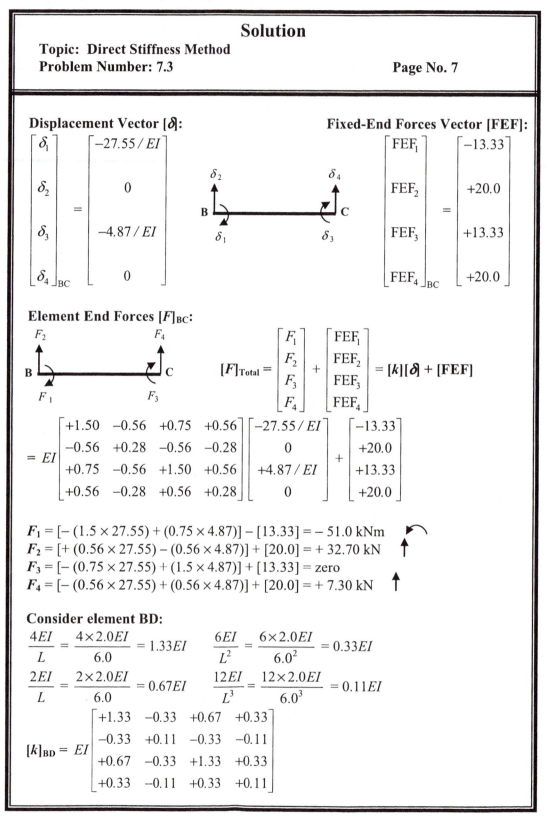

$$\begin{bmatrix} \delta_1 \\ \delta_2 \\ \delta_3 \\ \delta_4 \end{bmatrix}_{BC} = \begin{bmatrix} -27.55/EI \\ 0 \\ -4.87/EI \\ 0 \end{bmatrix} \qquad \begin{bmatrix} FEF_1 \\ FEF_2 \\ FEF_3 \\ FEF_4 \end{bmatrix}_{BC} = \begin{bmatrix} -13.33 \\ +20.0 \\ +13.33 \\ +20.0 \end{bmatrix}$$

Element End Forces [F]$_{BC}$:

$$[F]_{Total} = \begin{bmatrix} F_1 \\ F_2 \\ F_3 \\ F_4 \end{bmatrix} + \begin{bmatrix} FEF_1 \\ FEF_2 \\ FEF_3 \\ FEF_4 \end{bmatrix} = [k][\delta] + [FEF]$$

$$= EI \begin{bmatrix} +1.50 & -0.56 & +0.75 & +0.56 \\ -0.56 & +0.28 & -0.56 & -0.28 \\ +0.75 & -0.56 & +1.50 & +0.56 \\ +0.56 & -0.28 & +0.56 & +0.28 \end{bmatrix} \begin{bmatrix} -27.55/EI \\ 0 \\ +4.87/EI \\ 0 \end{bmatrix} + \begin{bmatrix} -13.33 \\ +20.0 \\ +13.33 \\ +20.0 \end{bmatrix}$$

$F_1 = [-(1.5 \times 27.55) + (0.75 \times 4.87)] - [13.33] = -51.0$ kNm

$F_2 = [+(0.56 \times 27.55) - (0.56 \times 4.87)] + [20.0] = +32.70$ kN ↑

$F_3 = [-(0.75 \times 27.55) + (1.5 \times 4.87)] + [13.33] = $ zero

$F_4 = [-(0.56 \times 27.55) + (0.56 \times 4.87)] + [20.0] = +7.30$ kN ↑

Consider element BD:

$$\frac{4EI}{L} = \frac{4 \times 2.0EI}{6.0} = 1.33EI \qquad \frac{6EI}{L^2} = \frac{6 \times 2.0EI}{6.0^2} = 0.33EI$$

$$\frac{2EI}{L} = \frac{2 \times 2.0EI}{6.0} = 0.67EI \qquad \frac{12EI}{L^3} = \frac{12 \times 2.0EI}{6.0^3} = 0.11EI$$

$$[k]_{BD} = EI \begin{bmatrix} +1.33 & -0.33 & +0.67 & +0.33 \\ -0.33 & +0.11 & -0.33 & -0.11 \\ +0.67 & -0.33 & +1.33 & +0.33 \\ +0.33 & -0.11 & +0.33 & +0.11 \end{bmatrix}$$

Solution

Topic: Direct Stiffness Method
Problem Number: 7.3   **Page No. 8**

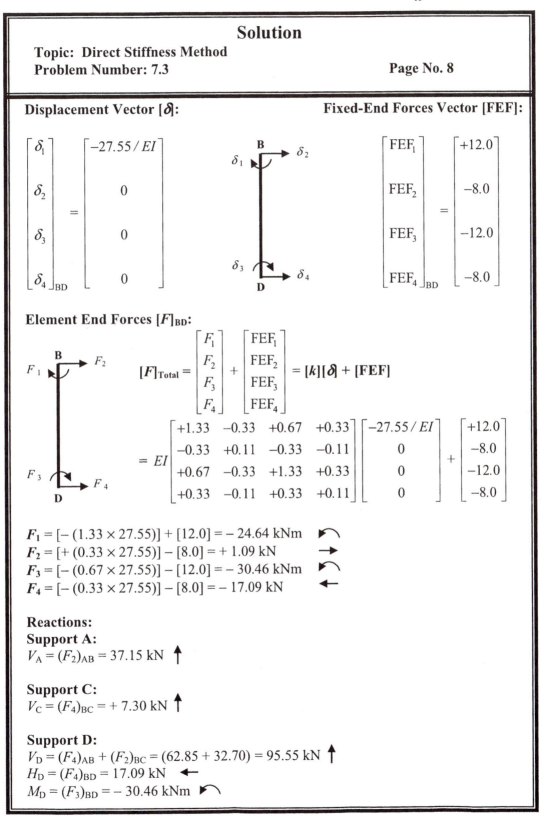

Displacement Vector [δ]:   **Fixed-End Forces Vector [FEF]:**

$$\begin{bmatrix} \delta_1 \\ \delta_2 \\ \delta_3 \\ \delta_4 \end{bmatrix}_{BD} = \begin{bmatrix} -27.55/EI \\ 0 \\ 0 \\ 0 \end{bmatrix}$$

$$\begin{bmatrix} FEF_1 \\ FEF_2 \\ FEF_3 \\ FEF_4 \end{bmatrix}_{BD} = \begin{bmatrix} +12.0 \\ -8.0 \\ -12.0 \\ -8.0 \end{bmatrix}$$

Element End Forces [F]$_{BD}$:

$$[F]_{Total} = \begin{bmatrix} F_1 \\ F_2 \\ F_3 \\ F_4 \end{bmatrix} + \begin{bmatrix} FEF_1 \\ FEF_2 \\ FEF_3 \\ FEF_4 \end{bmatrix} = [k][\delta] + [FEF]$$

$$= EI\begin{bmatrix} +1.33 & -0.33 & +0.67 & +0.33 \\ -0.33 & +0.11 & -0.33 & -0.11 \\ +0.67 & -0.33 & +1.33 & +0.33 \\ +0.33 & -0.11 & +0.33 & +0.11 \end{bmatrix}\begin{bmatrix} -27.55/EI \\ 0 \\ 0 \\ 0 \end{bmatrix} + \begin{bmatrix} +12.0 \\ -8.0 \\ -12.0 \\ -8.0 \end{bmatrix}$$

$F_1 = [-(1.33 \times 27.55)] + [12.0] = -24.64$ kNm ↰

$F_2 = [+(0.33 \times 27.55)] - [8.0] = +1.09$ kN →

$F_3 = [-(0.67 \times 27.55)] - [12.0] = -30.46$ kNm ↰

$F_4 = [-(0.33 \times 27.55)] - [8.0] = -17.09$ kN ←

Reactions:
Support A:
$V_A = (F_2)_{AB} = 37.15$ kN ↑

Support C:
$V_C = (F_4)_{BC} = +7.30$ kN ↑

Support D:
$V_D = (F_4)_{AB} + (F_2)_{BC} = (62.85 + 32.70) = 95.55$ kN ↑
$H_D = (F_4)_{BD} = 17.09$ kN ←
$M_D = (F_3)_{BD} = -30.46$ kNm ↰

Solution

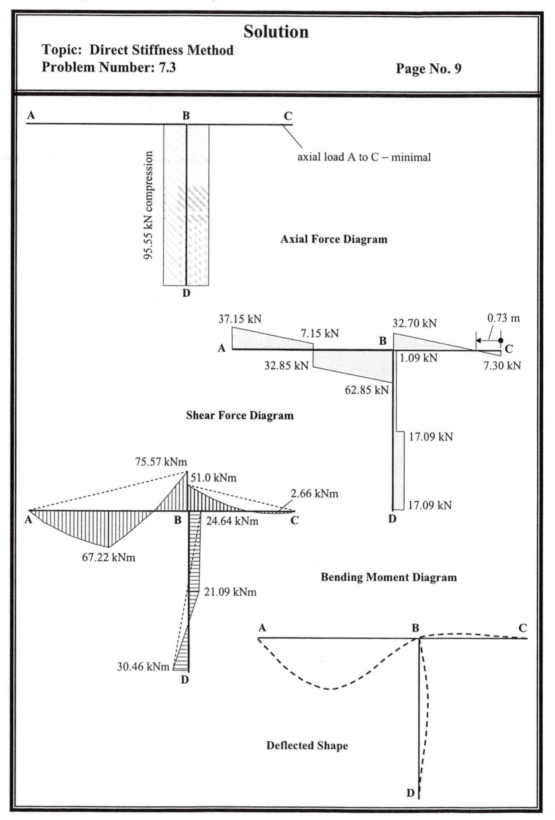

Axial Force Diagram

95.55 kN compression

axial load A to C – minimal

Shear Force Diagram

37.15 kN
7.15 kN
32.85 kN
32.70 kN
0.73 m
1.09 kN
7.30 kN
62.85 kN
17.09 kN
17.09 kN

Bending Moment Diagram

75.57 kNm
51.0 kNm
2.66 kNm
24.64 kNm
67.22 kNm
21.09 kNm
30.46 kNm

Deflected Shape

Solution

Topic: Direct Stiffness Method
Problem Number: 7.4

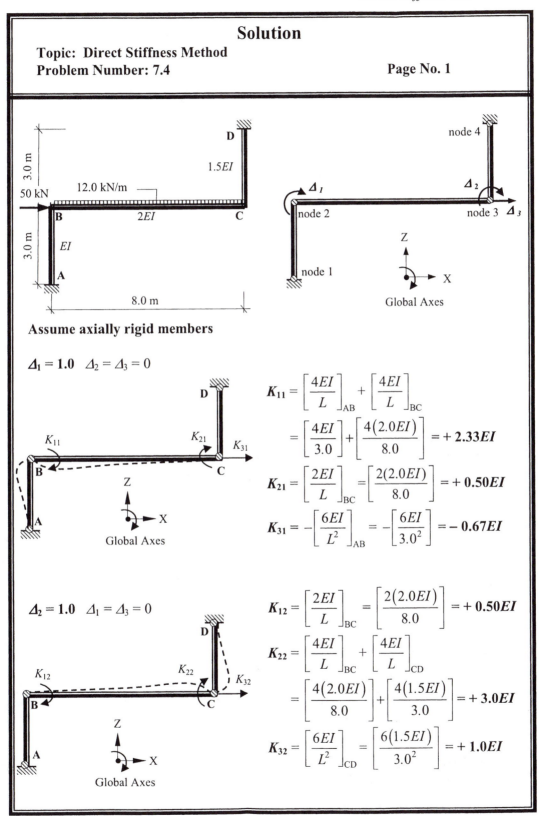

Assume axially rigid members

$\Delta_1 = 1.0$ $\Delta_2 = \Delta_3 = 0$

$$K_{11} = \left[\frac{4EI}{L}\right]_{AB} + \left[\frac{4EI}{L}\right]_{BC}$$

$$= \left[\frac{4EI}{3.0}\right] + \left[\frac{4(2.0EI)}{8.0}\right] = +\,2.33EI$$

$$K_{21} = \left[\frac{2EI}{L}\right]_{BC} = \left[\frac{2(2.0EI)}{8.0}\right] = +\,0.50EI$$

$$K_{31} = -\left[\frac{6EI}{L^2}\right]_{AB} = -\left[\frac{6EI}{3.0^2}\right] = -\,0.67EI$$

$\Delta_2 = 1.0$ $\Delta_1 = \Delta_3 = 0$

$$K_{12} = \left[\frac{2EI}{L}\right]_{BC} = \left[\frac{2(2.0EI)}{8.0}\right] = +\,0.50EI$$

$$K_{22} = \left[\frac{4EI}{L}\right]_{BC} + \left[\frac{4EI}{L}\right]_{CD}$$

$$= \left[\frac{4(2.0EI)}{8.0}\right] + \left[\frac{4(1.5EI)}{3.0}\right] = +\,3.0EI$$

$$K_{32} = \left[\frac{6EI}{L^2}\right]_{CD} = \left[\frac{6(1.5EI)}{3.0^2}\right] = +\,1.0EI$$

Solution

Topic: Direct Stiffness Method
Problem Number: 7.4

Page No. 2

$\Delta_3 = 1.0$ $\Delta_1 = \Delta_2 = 0$

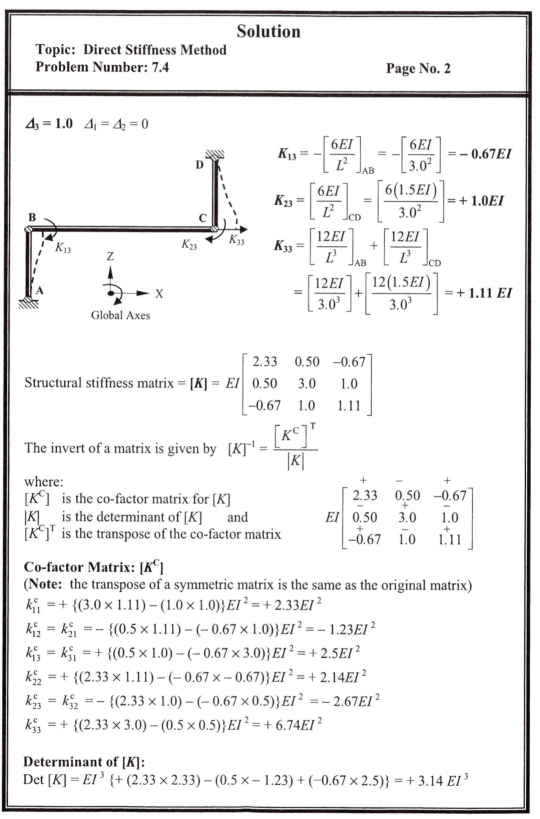

Global Axes

$$K_{13} = -\left[\frac{6EI}{L^2}\right]_{AB} = -\left[\frac{6EI}{3.0^2}\right] = -0.67EI$$

$$K_{23} = \left[\frac{6EI}{L^2}\right]_{CD} = \left[\frac{6(1.5EI)}{3.0^2}\right] = +1.0EI$$

$$K_{33} = \left[\frac{12EI}{L^3}\right]_{AB} + \left[\frac{12EI}{L^3}\right]_{CD}$$

$$= \left[\frac{12EI}{3.0^3}\right] + \left[\frac{12(1.5EI)}{3.0^3}\right] = +1.11\,EI$$

Structural stiffness matrix = $[K] = EI \begin{bmatrix} 2.33 & 0.50 & -0.67 \\ 0.50 & 3.0 & 1.0 \\ -0.67 & 1.0 & 1.11 \end{bmatrix}$

The invert of a matrix is given by $[K]^{-1} = \dfrac{\left[K^C\right]^T}{|K|}$

where:
$[K^C]$ is the co-factor matrix for $[K]$
$|K|$ is the determinant of $[K]$ and
$[K^C]^T$ is the transpose of the co-factor matrix

$$EI \begin{bmatrix} \overset{+}{2.33} & \overset{-}{0.50} & \overset{+}{-0.67} \\ \underset{}{0.50} & \overset{+}{3.0} & \underset{}{1.0} \\ \underset{}{-0.67} & \underset{}{1.0} & \overset{+}{1.11} \end{bmatrix}$$

Co-factor Matrix: $[K^C]$
(**Note:** the transpose of a symmetric matrix is the same as the original matrix)
$k_{11}^c = +\{(3.0 \times 1.11) - (1.0 \times 1.0)\}EI^2 = +2.33EI^2$
$k_{12}^c = k_{21}^c = -\{(0.5 \times 1.11) - (-0.67 \times 1.0)\}EI^2 = -1.23EI^2$
$k_{13}^c = k_{31}^c = +\{(0.5 \times 1.0) - (-0.67 \times 3.0)\}EI^2 = +2.5EI^2$
$k_{22}^c = +\{(2.33 \times 1.11) - (-0.67 \times -0.67)\}EI^2 = +2.14EI^2$
$k_{23}^c = k_{32}^c = -\{(2.33 \times 1.0) - (-0.67 \times 0.5)\}EI^2 = -2.67EI^2$
$k_{33}^c = +\{(2.33 \times 3.0) - (0.5 \times 0.5)\}EI^2 = +6.74EI^2$

Determinant of $[K]$:
Det $[K] = EI^3 \{+(2.33 \times 2.33) - (0.5 \times -1.23) + (-0.67 \times 2.5)\} = +3.14\,EI^3$

Solution

Topic: Direct Stiffness Method
Problem Number: 7.4

Inverted stiffness matrix = $[K]^{-1} = \dfrac{1}{EI}\begin{bmatrix} 0.742 & -0.392 & 0.796 \\ -0.392 & 0.682 & -0.850 \\ 0.796 & -0.850 & 2.146 \end{bmatrix}$

Structural Load Vector: [P]:
Fixed–end forces for member BC
$FEF_1 = -(12.0 \times 8.0^2)/12.0 = -64.0$ kNm
$FEF_3 = +64.0$ kNm

$FEF_2 = +(12.0 \times 8.0)/2.0 = +48.0$ kN
$FEF_4 = +48.0$ kN

Equivalent nodal loads for BC

Applied nodal load at B = 50.0 kN ⟶

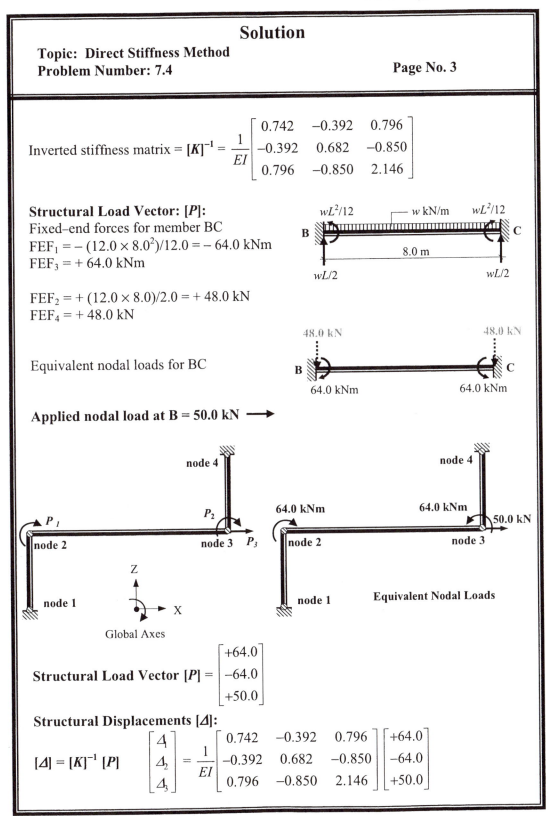

Global Axes

Structural Load Vector $[P] = \begin{bmatrix} +64.0 \\ -64.0 \\ +50.0 \end{bmatrix}$

Structural Displacements [Δ]:

$[\Delta] = [K]^{-1}[P]$ $\begin{bmatrix} \Delta_1 \\ \Delta_2 \\ \Delta_3 \end{bmatrix} = \dfrac{1}{EI}\begin{bmatrix} 0.742 & -0.392 & 0.796 \\ -0.392 & 0.682 & -0.850 \\ 0.796 & -0.850 & 2.146 \end{bmatrix}\begin{bmatrix} +64.0 \\ -64.0 \\ +50.0 \end{bmatrix}$

Solution

Topic: Direct Stiffness Method
Problem Number: 7.4

Page No. 4

$$\Delta_1 = \frac{1}{EI}\Big[(0.742\times64.0)+(0.392\times64.0)+(0.796\times50.0)\Big] = +\frac{112.38}{EI} \text{ radians}$$

$$\Delta_2 = \frac{1}{EI}\Big[-(0.392\times64.0)-(0.682\times64.0)-(0.850\times50.0)\Big] = -\frac{111.24}{EI} \text{ radians}$$

$$\Delta_3 = \frac{1}{EI}\Big[(0.796\times64.0)+(0.850\times64.0)+(2.146\times50.0)\Big] = +\frac{212.64}{EI} \text{ m}$$

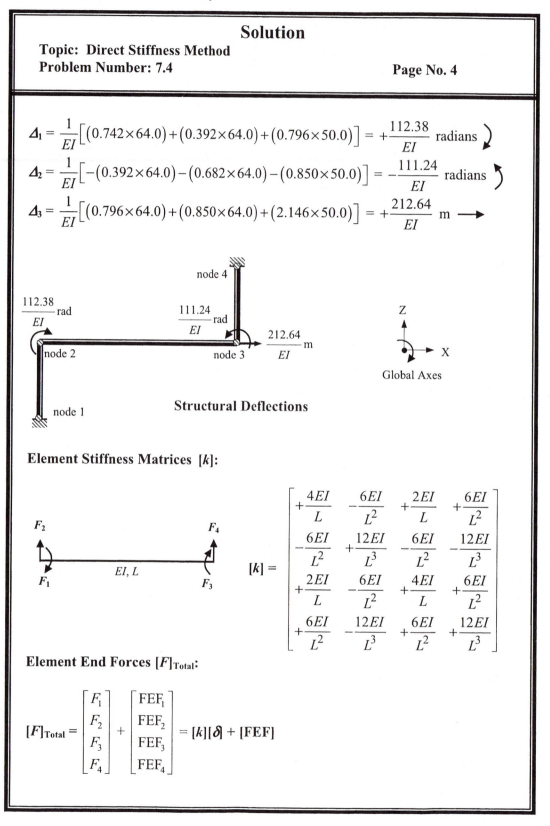

Structural Deflections

Element Stiffness Matrices [k]:

$$[k] = \begin{bmatrix} +\dfrac{4EI}{L} & -\dfrac{6EI}{L^2} & +\dfrac{2EI}{L} & +\dfrac{6EI}{L^2} \\[2mm] -\dfrac{6EI}{L^2} & +\dfrac{12EI}{L^3} & -\dfrac{6EI}{L^2} & -\dfrac{12EI}{L^3} \\[2mm] +\dfrac{2EI}{L} & -\dfrac{6EI}{L^2} & +\dfrac{4EI}{L} & +\dfrac{6EI}{L^2} \\[2mm] +\dfrac{6EI}{L^2} & -\dfrac{12EI}{L^3} & +\dfrac{6EI}{L^2} & +\dfrac{12EI}{L^3} \end{bmatrix}$$

Element End Forces [F]_Total:

$$[F]_{\text{Total}} = \begin{bmatrix} F_1 \\ F_2 \\ F_3 \\ F_4 \end{bmatrix} + \begin{bmatrix} \text{FEF}_1 \\ \text{FEF}_2 \\ \text{FEF}_3 \\ \text{FEF}_4 \end{bmatrix} = [k][\delta] + [\text{FEF}]$$

Solution

Topic: Direct Stiffness Method
Problem Number: 7.4

$$[k][\delta] + [\text{FEF}] = \begin{bmatrix} +\dfrac{4EI}{L} & -\dfrac{6EI}{L^2} & +\dfrac{2EI}{L} & +\dfrac{6EI}{L^2} \\[2mm] -\dfrac{6EI}{L^2} & +\dfrac{12EI}{L^3} & -\dfrac{6EI}{L^2} & -\dfrac{12EI}{L^3} \\[2mm] +\dfrac{2EI}{L} & -\dfrac{6EI}{L^2} & +\dfrac{4EI}{L} & +\dfrac{6EI}{L^2} \\[2mm] +\dfrac{6EI}{L^2} & -\dfrac{12EI}{L^3} & +\dfrac{6EI}{L^2} & +\dfrac{12EI}{L^3} \end{bmatrix} \begin{bmatrix} \delta_1 \\[2mm] \delta_2 \\[2mm] \delta_3 \\[2mm] \delta_4 \end{bmatrix} + \begin{bmatrix} \text{FEF}_1 \\[2mm] \text{FEF}_2 \\[2mm] \text{FEF}_3 \\[2mm] \text{FEF}_4 \end{bmatrix}$$

Consider element AB:

$$\frac{4EI}{L} = \frac{4 \times EI}{3.0} = 1.33EI \qquad \frac{6EI}{L^2} = \frac{6 \times EI}{3.0^2} = 0.67EI$$

$$\frac{2EI}{L} = \frac{2 \times EI}{3.0} = 0.67EI \qquad \frac{12EI}{L^3} = \frac{12 \times EI}{3.0^3} = 0.44EI$$

$$[k]_{AB} = EI \begin{bmatrix} +1.33 & -0.67 & +0.67 & +0.67 \\ -0.67 & +0.44 & -0.67 & -0.44 \\ +0.67 & -0.67 & +1.33 & +0.67 \\ +0.67 & -0.44 & +0.67 & +0.44 \end{bmatrix}$$

Displacement Vector [δ]:   **Fixed-End Forces Vector [FEF]:**

$$\begin{bmatrix} \delta_1 \\[2mm] \delta_2 \\[2mm] \delta_3 \\[2mm] \delta_4 \end{bmatrix}_{AB} = \begin{bmatrix} 0 \\[2mm] 0 \\[2mm] +112.38 / EI \\[2mm] -212.64 / EI \end{bmatrix}$$

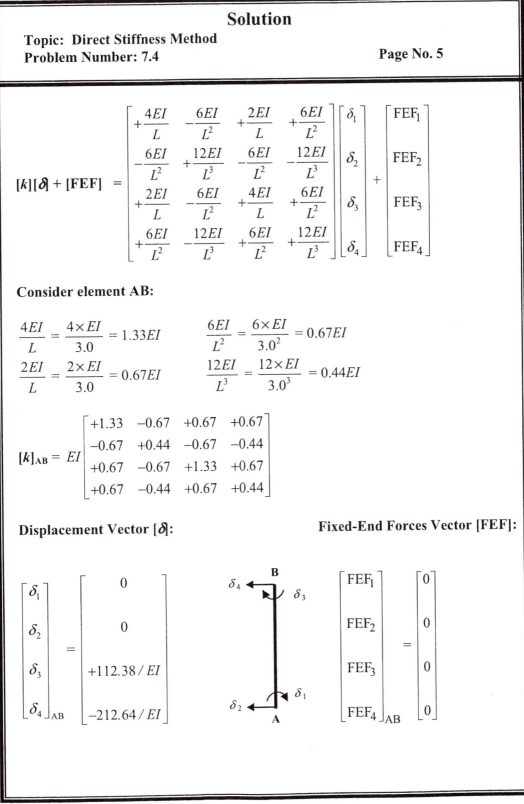

$$\begin{bmatrix} \text{FEF}_1 \\[2mm] \text{FEF}_2 \\[2mm] \text{FEF}_3 \\[2mm] \text{FEF}_4 \end{bmatrix}_{AB} = \begin{bmatrix} 0 \\[2mm] 0 \\[2mm] 0 \\[2mm] 0 \end{bmatrix}$$

Solution

Topic: Direct Stiffness Method
Problem Number: 7.4

Page No. 6

Element End Forces [F]$_{AB}$:

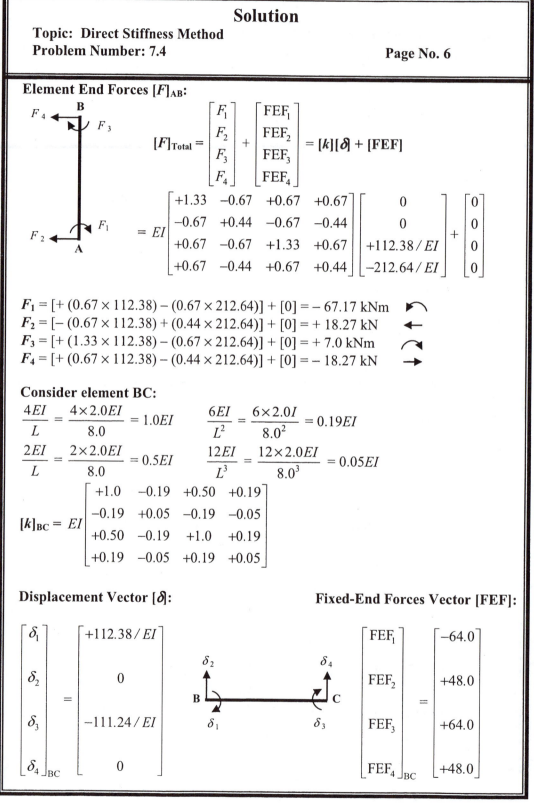

$$[F]_{\text{Total}} = \begin{bmatrix} F_1 \\ F_2 \\ F_3 \\ F_4 \end{bmatrix} + \begin{bmatrix} \text{FEF}_1 \\ \text{FEF}_2 \\ \text{FEF}_3 \\ \text{FEF}_4 \end{bmatrix} = [k][\delta] + [\text{FEF}]$$

$$= EI \begin{bmatrix} +1.33 & -0.67 & +0.67 & +0.67 \\ -0.67 & +0.44 & -0.67 & -0.44 \\ +0.67 & -0.67 & +1.33 & +0.67 \\ +0.67 & -0.44 & +0.67 & +0.44 \end{bmatrix} \begin{bmatrix} 0 \\ 0 \\ +112.38/EI \\ -212.64/EI \end{bmatrix} + \begin{bmatrix} 0 \\ 0 \\ 0 \\ 0 \end{bmatrix}$$

$F_1 = [+ (0.67 \times 112.38) - (0.67 \times 212.64)] + [0] = -67.17$ kNm

$F_2 = [- (0.67 \times 112.38) + (0.44 \times 212.64)] + [0] = +18.27$ kN

$F_3 = [+ (1.33 \times 112.38) - (0.67 \times 212.64)] + [0] = +7.0$ kNm

$F_4 = [+ (0.67 \times 112.38) - (0.44 \times 212.64)] + [0] = -18.27$ kN

Consider element BC:

$$\frac{4EI}{L} = \frac{4 \times 2.0EI}{8.0} = 1.0EI \qquad \frac{6EI}{L^2} = \frac{6 \times 2.0I}{8.0^2} = 0.19EI$$

$$\frac{2EI}{L} = \frac{2 \times 2.0EI}{8.0} = 0.5EI \qquad \frac{12EI}{L^3} = \frac{12 \times 2.0EI}{8.0^3} = 0.05EI$$

$$[k]_{BC} = EI \begin{bmatrix} +1.0 & -0.19 & +0.50 & +0.19 \\ -0.19 & +0.05 & -0.19 & -0.05 \\ +0.50 & -0.19 & +1.0 & +0.19 \\ +0.19 & -0.05 & +0.19 & +0.05 \end{bmatrix}$$

Displacement Vector [δ]:   **Fixed-End Forces Vector [FEF]:**

$$\begin{bmatrix} \delta_1 \\ \delta_2 \\ \delta_3 \\ \delta_4 \end{bmatrix}_{BC} = \begin{bmatrix} +112.38/EI \\ 0 \\ -111.24/EI \\ 0 \end{bmatrix}$$

$$\begin{bmatrix} \text{FEF}_1 \\ \text{FEF}_2 \\ \text{FEF}_3 \\ \text{FEF}_4 \end{bmatrix}_{BC} = \begin{bmatrix} -64.0 \\ +48.0 \\ +64.0 \\ +48.0 \end{bmatrix}$$

Solution

Topic: Direct Stiffness Method
Problem Number: 7.4   **Page No. 7**

Element End Forces $[F]_{BC}$:

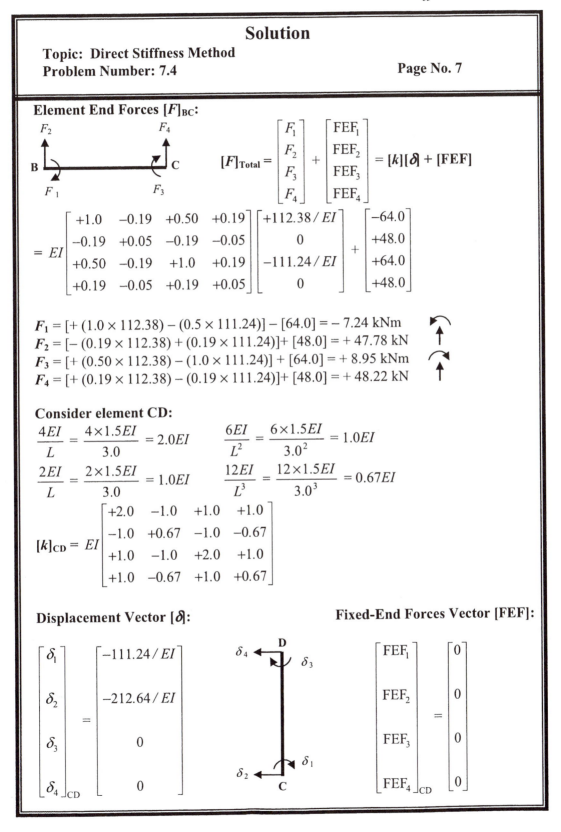

$$[F]_{Total} = \begin{bmatrix} F_1 \\ F_2 \\ F_3 \\ F_4 \end{bmatrix} + \begin{bmatrix} FEF_1 \\ FEF_2 \\ FEF_3 \\ FEF_4 \end{bmatrix} = [k][\delta] + [FEF]$$

$$= EI\begin{bmatrix} +1.0 & -0.19 & +0.50 & +0.19 \\ -0.19 & +0.05 & -0.19 & -0.05 \\ +0.50 & -0.19 & +1.0 & +0.19 \\ +0.19 & -0.05 & +0.19 & +0.05 \end{bmatrix}\begin{bmatrix} +112.38/EI \\ 0 \\ -111.24/EI \\ 0 \end{bmatrix} + \begin{bmatrix} -64.0 \\ +48.0 \\ +64.0 \\ +48.0 \end{bmatrix}$$

$F_1 = [+ (1.0 \times 112.38) - (0.5 \times 111.24)] - [64.0] = -7.24$ kNm

$F_2 = [- (0.19 \times 112.38) + (0.19 \times 111.24)] + [48.0] = +47.78$ kN

$F_3 = [+ (0.50 \times 112.38) - (1.0 \times 111.24)] + [64.0] = +8.95$ kNm

$F_4 = [+ (0.19 \times 112.38) - (0.19 \times 111.24)] + [48.0] = +48.22$ kN

Consider element CD:

$$\frac{4EI}{L} = \frac{4 \times 1.5EI}{3.0} = 2.0EI \qquad \frac{6EI}{L^2} = \frac{6 \times 1.5EI}{3.0^2} = 1.0EI$$

$$\frac{2EI}{L} = \frac{2 \times 1.5EI}{3.0} = 1.0EI \qquad \frac{12EI}{L^3} = \frac{12 \times 1.5EI}{3.0^3} = 0.67EI$$

$$[k]_{CD} = EI\begin{bmatrix} +2.0 & -1.0 & +1.0 & +1.0 \\ -1.0 & +0.67 & -1.0 & -0.67 \\ +1.0 & -1.0 & +2.0 & +1.0 \\ +1.0 & -0.67 & +1.0 & +0.67 \end{bmatrix}$$

Displacement Vector $[\delta]$:   **Fixed-End Forces Vector [FEF]:**

$$\begin{bmatrix} \delta_1 \\ \delta_2 \\ \delta_3 \\ \delta_4 \end{bmatrix}_{CD} = \begin{bmatrix} -111.24/EI \\ -212.64/EI \\ 0 \\ 0 \end{bmatrix} \qquad\qquad \begin{bmatrix} FEF_1 \\ FEF_2 \\ FEF_3 \\ FEF_4 \end{bmatrix}_{CD} = \begin{bmatrix} 0 \\ 0 \\ 0 \\ 0 \end{bmatrix}$$

Solution

Topic: Direct Stiffness Method
Problem Number: 7.4   **Page No. 8**

Element End Forces $[F]_{CD}$:

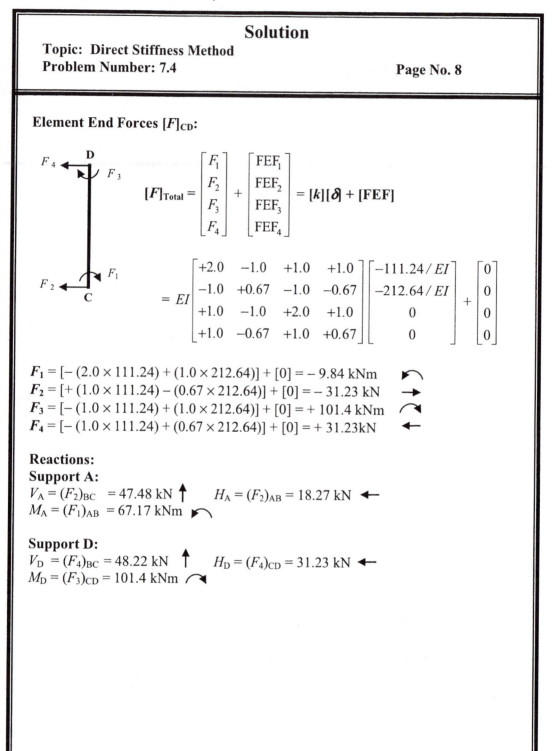

$$[F]_{Total} = \begin{bmatrix} F_1 \\ F_2 \\ F_3 \\ F_4 \end{bmatrix} + \begin{bmatrix} FEF_1 \\ FEF_2 \\ FEF_3 \\ FEF_4 \end{bmatrix} = [k][\delta] + [FEF]$$

$$= EI \begin{bmatrix} +2.0 & -1.0 & +1.0 & +1.0 \\ -1.0 & +0.67 & -1.0 & -0.67 \\ +1.0 & -1.0 & +2.0 & +1.0 \\ +1.0 & -0.67 & +1.0 & +0.67 \end{bmatrix} \begin{bmatrix} -111.24/EI \\ -212.64/EI \\ 0 \\ 0 \end{bmatrix} + \begin{bmatrix} 0 \\ 0 \\ 0 \\ 0 \end{bmatrix}$$

$F_1 = [-(2.0 \times 111.24) + (1.0 \times 212.64)] + [0] = -9.84$ kNm ↰

$F_2 = [+(1.0 \times 111.24) - (0.67 \times 212.64)] + [0] = -31.23$ kN →

$F_3 = [-(1.0 \times 111.24) + (1.0 \times 212.64)] + [0] = +101.4$ kNm ↷

$F_4 = [-(1.0 \times 111.24) + (0.67 \times 212.64)] + [0] = +31.23$ kN ←

Reactions:

Support A:
$V_A = (F_2)_{BC} = 47.48$ kN ↑ $H_A = (F_2)_{AB} = 18.27$ kN ←

$M_A = (F_1)_{AB} = 67.17$ kNm ↰

Support D:
$V_D = (F_4)_{BC} = 48.22$ kN ↑ $H_D = (F_4)_{CD} = 31.23$ kN ←

$M_D = (F_3)_{CD} = 101.4$ kNm ↷

Solution

Topic: Direct Stiffness Method
Problem Number: 7.4

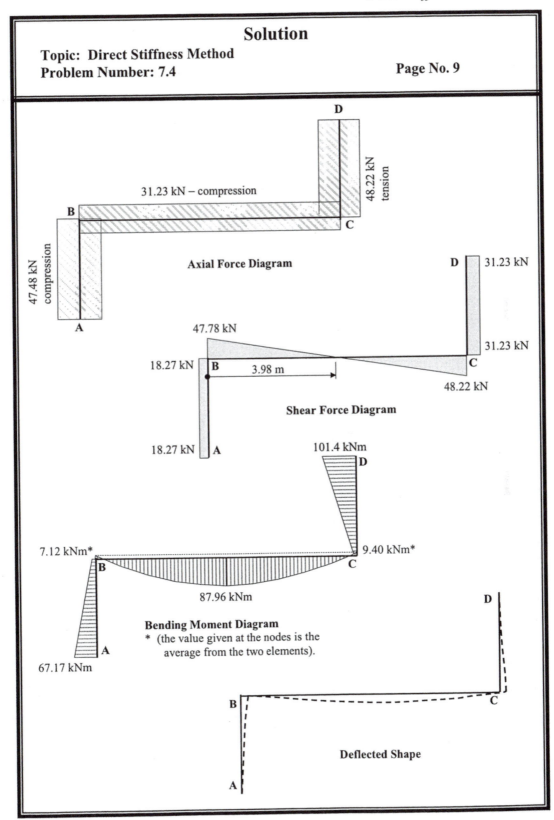

31.23 kN – compression

48.22 kN tension

47.48 kN compression

Axial Force Diagram

31.23 kN

31.23 kN

47.78 kN

18.27 kN

3.98 m

48.22 kN

18.27 kN

18.27 kN

Shear Force Diagram

101.4 kNm

7.12 kNm*

9.40 kNm*

87.96 kNm

67.17 kNm

Bending Moment Diagram
* (the value given at the nodes is the
average from the two elements).

Deflected Shape

Solution

Topic: Direct Stiffness Method
Problem Number: 7.5

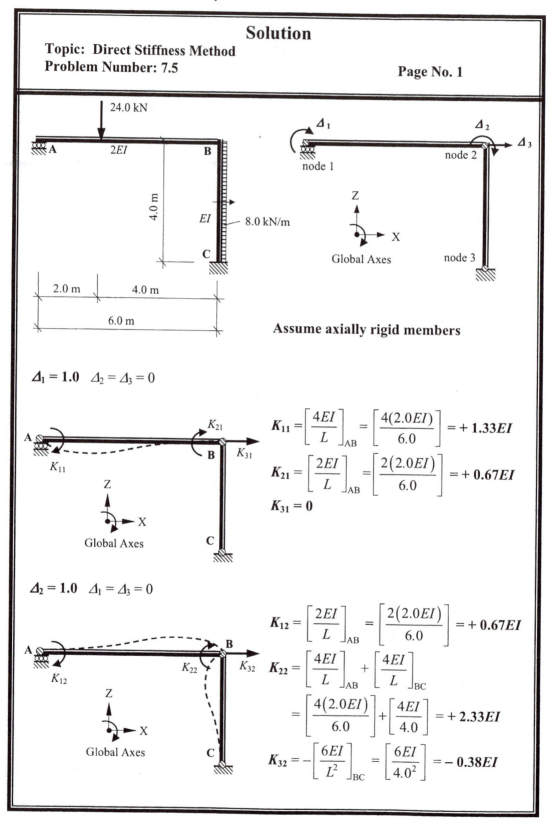

Assume axially rigid members

$\Delta_1 = 1.0 \quad \Delta_2 = \Delta_3 = 0$

$$K_{11} = \left[\frac{4EI}{L}\right]_{AB} = \left[\frac{4(2.0EI)}{6.0}\right] = +1.33EI$$

$$K_{21} = \left[\frac{2EI}{L}\right]_{AB} = \left[\frac{2(2.0EI)}{6.0}\right] = +0.67EI$$

$$K_{31} = 0$$

$\Delta_2 = 1.0 \quad \Delta_1 = \Delta_3 = 0$

$$K_{12} = \left[\frac{2EI}{L}\right]_{AB} = \left[\frac{2(2.0EI)}{6.0}\right] = +0.67EI$$

$$K_{22} = \left[\frac{4EI}{L}\right]_{AB} + \left[\frac{4EI}{L}\right]_{BC}$$

$$= \left[\frac{4(2.0EI)}{6.0}\right] + \left[\frac{4EI}{4.0}\right] = +2.33EI$$

$$K_{32} = -\left[\frac{6EI}{L^2}\right]_{BC} = \left[\frac{6EI}{4.0^2}\right] = -0.38EI$$

Solution

Topic: Direct Stiffness Method
Problem Number: 7.5

Page No. 2

$\Delta_3 = 1.0$ $\quad$ $\Delta_1 = \Delta_2 = 0$

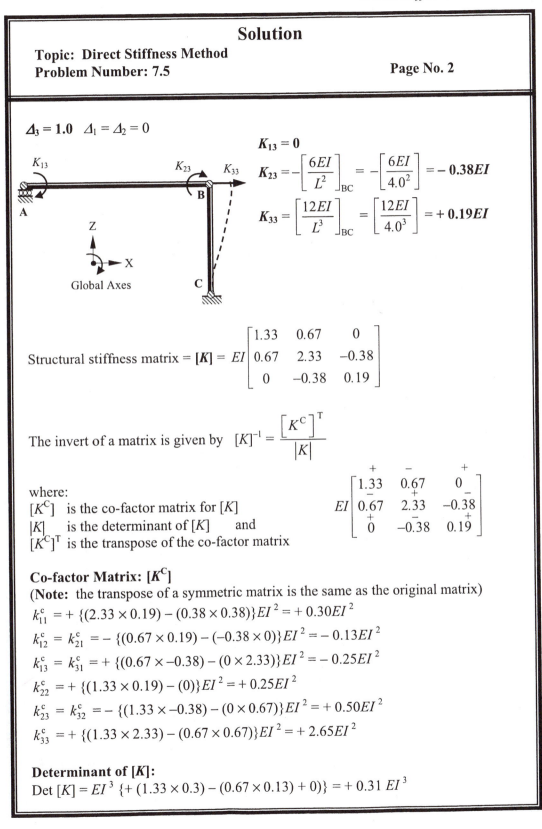

$K_{13} = 0$

$K_{23} = -\left[\dfrac{6EI}{L^2}\right]_{BC} = -\left[\dfrac{6EI}{4.0^2}\right] = -\mathbf{0.38EI}$

$K_{33} = \left[\dfrac{12EI}{L^3}\right]_{BC} = \left[\dfrac{12EI}{4.0^3}\right] = +\mathbf{0.19EI}$

Structural stiffness matrix $= [K] = EI\begin{bmatrix} 1.33 & 0.67 & 0 \\ 0.67 & 2.33 & -0.38 \\ 0 & -0.38 & 0.19 \end{bmatrix}$

The invert of a matrix is given by $\quad [K]^{-1} = \dfrac{\left[K^C\right]^T}{|K|}$

where:

$[K^C]$ is the co-factor matrix for $[K]$

$|K|$ is the determinant of $[K]$ $\quad$ and

$[K^C]^T$ is the transpose of the co-factor matrix

$$EI\begin{bmatrix} \overset{+}{1.33} & \overset{-}{0.67} & \overset{+}{0} \\ \underset{0.67}{-} & \overset{+}{2.33} & \underset{-0.38}{-} \\ \overset{+}{0} & \underset{-0.38}{-} & \overset{+}{0.19} \end{bmatrix}$$

Co-factor Matrix: $[K^C]$

(**Note:** the transpose of a symmetric matrix is the same as the original matrix)

$k_{11}^c = + \{(2.33 \times 0.19) - (0.38 \times 0.38)\}EI^2 = +0.30EI^2$

$k_{12}^c = k_{21}^c = - \{(0.67 \times 0.19) - (-0.38 \times 0)\}EI^2 = -0.13EI^2$

$k_{13}^c = k_{31}^c = + \{(0.67 \times -0.38) - (0 \times 2.33)\}EI^2 = -0.25EI^2$

$k_{22}^c = + \{(1.33 \times 0.19) - (0)\}EI^2 = +0.25EI^2$

$k_{23}^c = k_{32}^c = - \{(1.33 \times -0.38) - (0 \times 0.67)\}EI^2 = +0.50EI^2$

$k_{33}^c = + \{(1.33 \times 2.33) - (0.67 \times 0.67)\}EI^2 = +2.65EI^2$

Determinant of $[K]$:

Det $[K] = EI^3 \{+ (1.33 \times 0.3) - (0.67 \times 0.13) + 0)\} = +0.31\ EI^3$

Solution

Topic: Direct Stiffness Method
Problem Number: 7.5   **Page No. 3**

Inverted stiffness matrix = $[K]^{-1} = \dfrac{1}{EI}\begin{bmatrix} 0.968 & -0.419 & -0.806 \\ -0.419 & 0.806 & 1.613 \\ -0.806 & 1.613 & 8.548 \end{bmatrix}$

Structural Load Vector: [P]:

Fixed–end forces for member AB
$FEF_1 = -(24.0 \times 2.0 \times 4.0^2)/6.0^2$
$\quad = -21.33 \text{ kNm}$
$FEF_3 = +(24.0 \times 2.0^2 \times 4.0)/6.0^2$
$\quad = +10.67 \text{ kNm}$

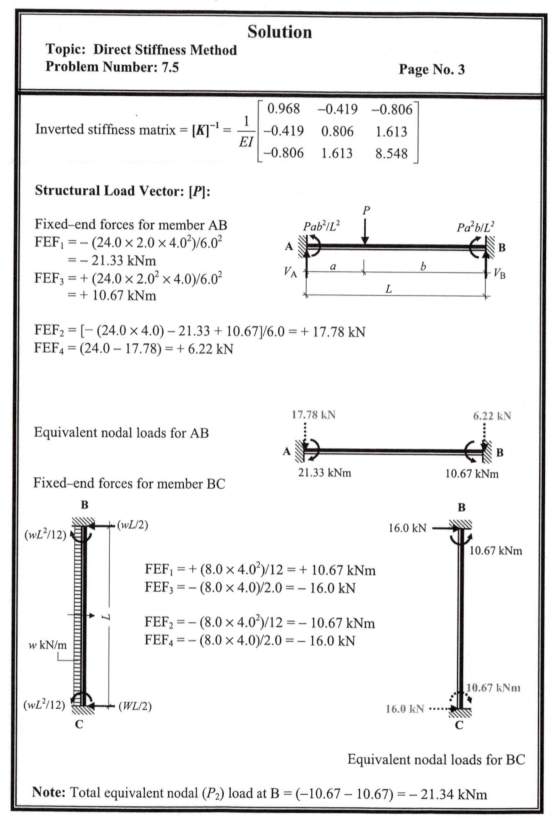

$FEF_2 = [-(24.0 \times 4.0) - 21.33 + 10.67]/6.0 = +17.78 \text{ kN}$
$FEF_4 = (24.0 - 17.78) = +6.22 \text{ kN}$

Equivalent nodal loads for AB

Fixed–end forces for member BC

$FEF_1 = +(8.0 \times 4.0^2)/12 = +10.67 \text{ kNm}$
$FEF_3 = -(8.0 \times 4.0)/2.0 = -16.0 \text{ kN}$

$FEF_2 = -(8.0 \times 4.0^2)/12 = -10.67 \text{ kNm}$
$FEF_4 = -(8.0 \times 4.0)/2.0 = -16.0 \text{ kN}$

Equivalent nodal loads for BC

Note: Total equivalent nodal (P_2) load at B $= (-10.67 - 10.67) = -21.34 \text{ kNm}$

Solution

Topic: Direct Stiffness Method
Problem Number: 7.5   **Page No. 4**

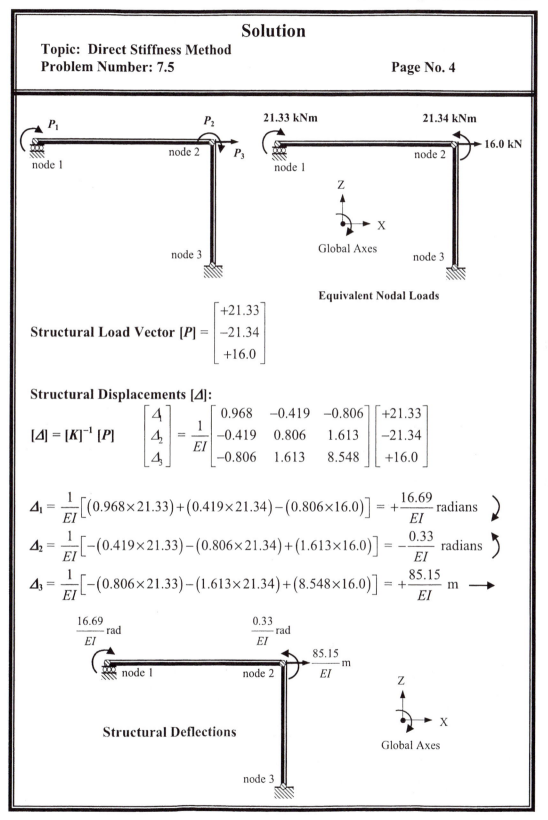

Structural Load Vector $[P] = \begin{bmatrix} +21.33 \\ -21.34 \\ +16.0 \end{bmatrix}$

Structural Displacements $[\varDelta]$:

$$[\varDelta] = [K]^{-1}[P] \qquad \begin{bmatrix} \varDelta_1 \\ \varDelta_2 \\ \varDelta_3 \end{bmatrix} = \frac{1}{EI}\begin{bmatrix} 0.968 & -0.419 & -0.806 \\ -0.419 & 0.806 & 1.613 \\ -0.806 & 1.613 & 8.548 \end{bmatrix}\begin{bmatrix} +21.33 \\ -21.34 \\ +16.0 \end{bmatrix}$$

$$\varDelta_1 = \frac{1}{EI}\left[(0.968 \times 21.33) + (0.419 \times 21.34) - (0.806 \times 16.0)\right] = +\frac{16.69}{EI} \text{ radians}$$

$$\varDelta_2 = \frac{1}{EI}\left[-(0.419 \times 21.33) - (0.806 \times 21.34) + (1.613 \times 16.0)\right] = -\frac{0.33}{EI} \text{ radians}$$

$$\varDelta_3 = \frac{1}{EI}\left[-(0.806 \times 21.33) - (1.613 \times 21.34) + (8.548 \times 16.0)\right] = +\frac{85.15}{EI} \text{ m}$$

Solution

Topic: Direct Stiffness Method
Problem Number: 7.5   **Page No. 5**

Element Stiffness Matrices [k]:

$$[k] = \begin{bmatrix} +\dfrac{4EI}{L} & -\dfrac{6EI}{L^2} & +\dfrac{2EI}{L} & +\dfrac{6EI}{L^2} \\[2mm] -\dfrac{6EI}{L^2} & +\dfrac{12EI}{L^3} & -\dfrac{6EI}{L^2} & -\dfrac{12EI}{L^3} \\[2mm] +\dfrac{2EI}{L} & -\dfrac{6EI}{L^2} & +\dfrac{4EI}{L} & +\dfrac{6EI}{L^2} \\[2mm] +\dfrac{6EI}{L^2} & -\dfrac{12EI}{L^3} & +\dfrac{6EI}{L^2} & +\dfrac{12EI}{L^3} \end{bmatrix}$$

Element End Forces [F]$_{\text{Total}}$:

$$[F]_{\text{Total}} = \begin{bmatrix} F_1 \\ F_2 \\ F_3 \\ F_4 \end{bmatrix} + \begin{bmatrix} FEF_1 \\ FEF_2 \\ FEF_3 \\ FEF_4 \end{bmatrix} = [k][\delta] + [FEF]$$

$$[k][\delta] + [FEF] = \begin{bmatrix} +\dfrac{4EI}{L} & -\dfrac{6EI}{L^2} & +\dfrac{2EI}{L} & +\dfrac{6EI}{L^2} \\[2mm] -\dfrac{6EI}{L^2} & +\dfrac{12EI}{L^3} & -\dfrac{6EI}{L^2} & -\dfrac{12EI}{L^3} \\[2mm] +\dfrac{2EI}{L} & -\dfrac{6EI}{L^2} & +\dfrac{4EI}{L} & +\dfrac{6EI}{L^2} \\[2mm] +\dfrac{6EI}{L^2} & -\dfrac{12EI}{L^3} & +\dfrac{6EI}{L^2} & +\dfrac{12EI}{L^3} \end{bmatrix} \begin{bmatrix} \delta_1 \\ \delta_2 \\ \delta_3 \\ \delta_4 \end{bmatrix} + \begin{bmatrix} FEF_1 \\ FEF_2 \\ FEF_3 \\ FEF_4 \end{bmatrix}$$

Consider element AB:

$$\frac{4EI}{L} = \frac{4 \times (2.0EI)}{6.0} = 1.33EI \qquad \frac{6EI}{L^2} = \frac{6 \times (2.0EI)}{6.0^2} = 0.33EI$$

$$\frac{2EI}{L} = \frac{2 \times (2.0EI)}{6.0} = 0.67EI \qquad \frac{12EI}{L^3} = \frac{12 \times (2.0EI)}{6.0^3} = 0.11EI$$

$$[k]_{AB} = EI \begin{bmatrix} +1.33 & -0.33 & +0.67 & +0.33 \\ -0.33 & +0.11 & -0.33 & -0.11 \\ +0.67 & -0.33 & +1.33 & +0.33 \\ +0.33 & -0.11 & +0.33 & +0.11 \end{bmatrix}$$

Displacement Vector [δ]:   **Fixed-End Forces Vector [FEF]:**

$$
\begin{bmatrix} \delta_1 \\ \delta_2 \\ \delta_3 \\ \delta_4 \end{bmatrix}_{AB} = \begin{bmatrix} +16.69/EI \\ 0 \\ -0.33/EI \\ 0 \end{bmatrix}
\qquad
\begin{bmatrix} FEF_1 \\ FEF_2 \\ FEF_3 \\ FEF_4 \end{bmatrix}_{AB} = \begin{bmatrix} -21.33 \\ +17.78 \\ +10.67 \\ +6.22 \end{bmatrix}
$$

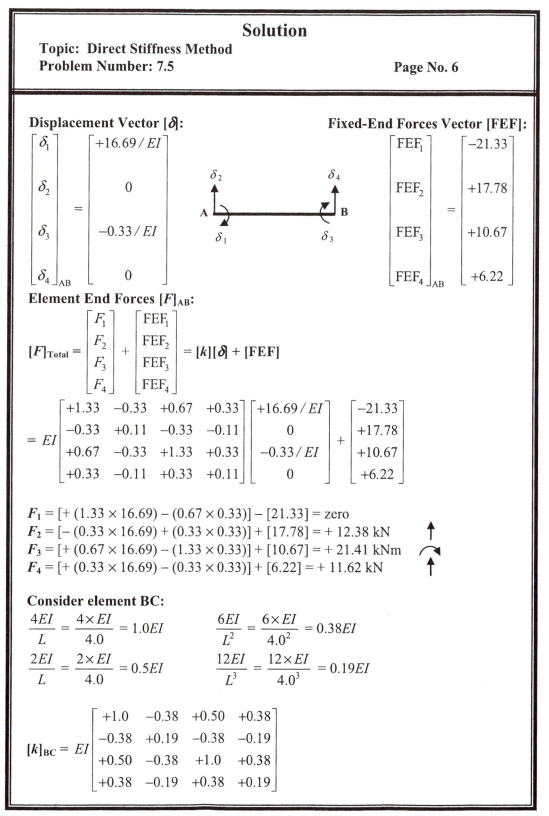

Element End Forces [F]$_{AB}$:

$$
[F]_{Total} = \begin{bmatrix} F_1 \\ F_2 \\ F_3 \\ F_4 \end{bmatrix} + \begin{bmatrix} FEF_1 \\ FEF_2 \\ FEF_3 \\ FEF_4 \end{bmatrix} = [k][\delta] + [FEF]
$$

$$
= EI \begin{bmatrix} +1.33 & -0.33 & +0.67 & +0.33 \\ -0.33 & +0.11 & -0.33 & -0.11 \\ +0.67 & -0.33 & +1.33 & +0.33 \\ +0.33 & -0.11 & +0.33 & +0.11 \end{bmatrix} \begin{bmatrix} +16.69/EI \\ 0 \\ -0.33/EI \\ 0 \end{bmatrix} + \begin{bmatrix} -21.33 \\ +17.78 \\ +10.67 \\ +6.22 \end{bmatrix}
$$

$F_1 = [+(1.33 \times 16.69) - (0.67 \times 0.33)] - [21.33] = \text{zero}$
$F_2 = [-(0.33 \times 16.69) + (0.33 \times 0.33)] + [17.78] = +12.38 \text{ kN}$
$F_3 = [+(0.67 \times 16.69) - (1.33 \times 0.33)] + [10.67] = +21.41 \text{ kNm}$
$F_4 = [+(0.33 \times 16.69) - (0.33 \times 0.33)] + [6.22] = +11.62 \text{ kN}$

Consider element BC:

$$
\frac{4EI}{L} = \frac{4 \times EI}{4.0} = 1.0EI \qquad \frac{6EI}{L^2} = \frac{6 \times EI}{4.0^2} = 0.38EI
$$

$$
\frac{2EI}{L} = \frac{2 \times EI}{4.0} = 0.5EI \qquad \frac{12EI}{L^3} = \frac{12 \times EI}{4.0^3} = 0.19EI
$$

$$
[k]_{BC} = EI \begin{bmatrix} +1.0 & -0.38 & +0.50 & +0.38 \\ -0.38 & +0.19 & -0.38 & -0.19 \\ +0.50 & -0.38 & +1.0 & +0.38 \\ +0.38 & -0.19 & +0.38 & +0.19 \end{bmatrix}
$$

Solution

Topic: Direct Stiffness Method
Problem Number: 7.5   **Page No. 7**

Displacement Vector [δ]:   **Fixed-End Forces Vector [FEF]:**

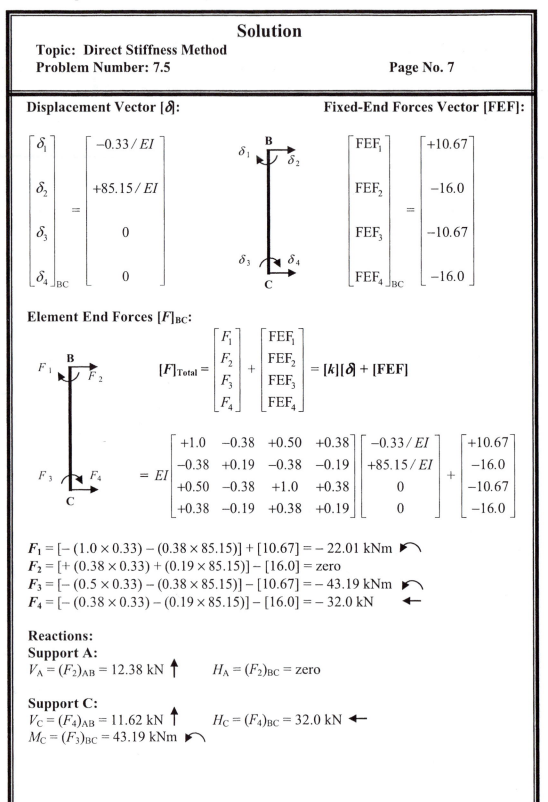

$$
\begin{bmatrix} \delta_1 \\ \delta_2 \\ \delta_3 \\ \delta_4 \end{bmatrix}_{BC} = \begin{bmatrix} -0.33/EI \\ +85.15/EI \\ 0 \\ 0 \end{bmatrix}
\qquad
\begin{bmatrix} FEF_1 \\ FEF_2 \\ FEF_3 \\ FEF_4 \end{bmatrix}_{BC} = \begin{bmatrix} +10.67 \\ -16.0 \\ -10.67 \\ -16.0 \end{bmatrix}
$$

Element End Forces [F]$_{BC}$:

$$
[F]_{Total} = \begin{bmatrix} F_1 \\ F_2 \\ F_3 \\ F_4 \end{bmatrix} + \begin{bmatrix} FEF_1 \\ FEF_2 \\ FEF_3 \\ FEF_4 \end{bmatrix} = [k][\delta] + [FEF]
$$

$$
= EI \begin{bmatrix} +1.0 & -0.38 & +0.50 & +0.38 \\ -0.38 & +0.19 & -0.38 & -0.19 \\ +0.50 & -0.38 & +1.0 & +0.38 \\ +0.38 & -0.19 & +0.38 & +0.19 \end{bmatrix} \begin{bmatrix} -0.33/EI \\ +85.15/EI \\ 0 \\ 0 \end{bmatrix} + \begin{bmatrix} +10.67 \\ -16.0 \\ -10.67 \\ -16.0 \end{bmatrix}
$$

$F_1 = [-(1.0 \times 0.33) - (0.38 \times 85.15)] + [10.67] = -22.01 \text{ kNm}$ ↰

$F_2 = [+(0.38 \times 0.33) + (0.19 \times 85.15)] - [16.0] = \text{zero}$

$F_3 = [-(0.5 \times 0.33) - (0.38 \times 85.15)] - [10.67] = -43.19 \text{ kNm}$ ↰

$F_4 = [-(0.38 \times 0.33) - (0.19 \times 85.15)] - [16.0] = -32.0 \text{ kN}$ ←

Reactions:
Support A:
$V_A = (F_2)_{AB} = 12.38 \text{ kN}$ ↑ $H_A = (F_2)_{BC} = \text{zero}$

Support C:
$V_C = (F_4)_{AB} = 11.62 \text{ kN}$ ↑ $H_C = (F_4)_{BC} = 32.0 \text{ kN}$ ←
$M_C = (F_3)_{BC} = 43.19 \text{ kNm}$ ↰

Solution

Topic: Direct Stiffness Method
Problem Number: 7.5

Page No. 8

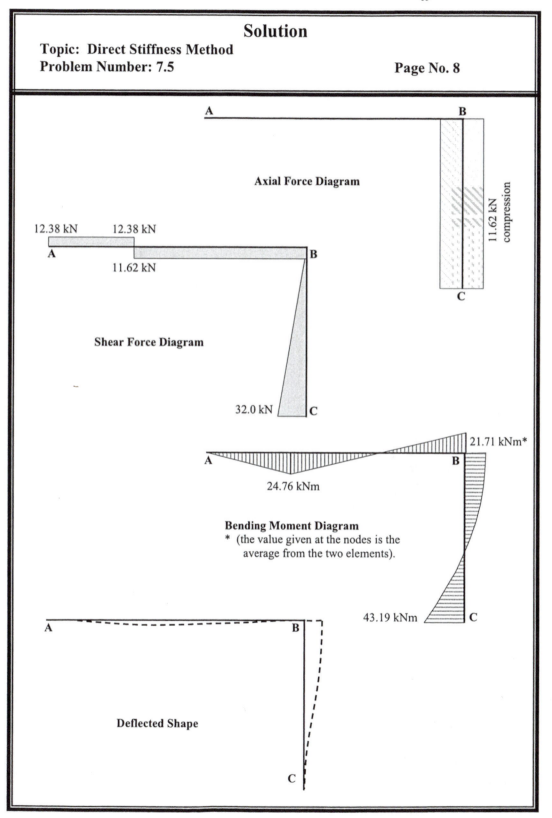

Axial Force Diagram

12.38 kN   12.38 kN

11.62 kN

11.62 kN compression

A   B

C

Shear Force Diagram

32.0 kN

Bending Moment Diagram
* (the value given at the nodes is the
average from the two elements).

24.76 kNm

21.71 kNm*

43.19 kNm

Deflected Shape

Solution

Topic: Direct Stiffness Method
Problem Number: 7.6   **Page No. 1**

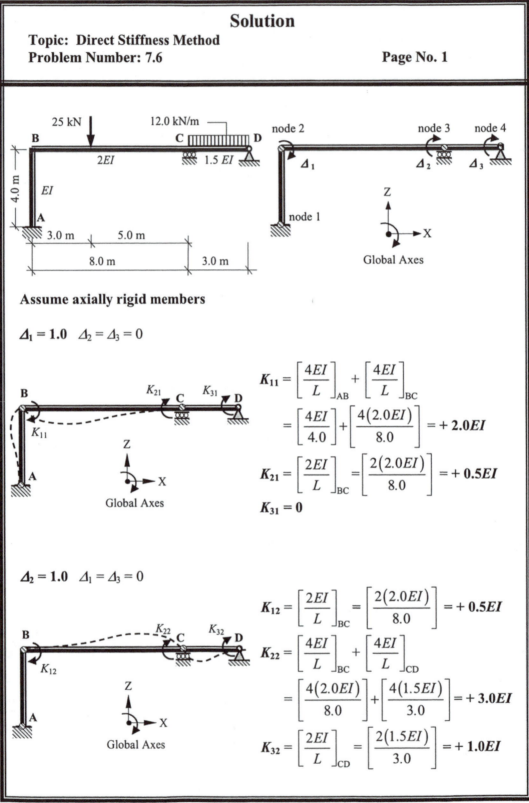

Assume axially rigid members

$\Delta_1 = 1.0$ $\Delta_2 = \Delta_3 = 0$

$$K_{11} = \left[\frac{4EI}{L}\right]_{AB} + \left[\frac{4EI}{L}\right]_{BC}$$

$$= \left[\frac{4EI}{4.0}\right] + \left[\frac{4(2.0EI)}{8.0}\right] = +2.0EI$$

$$K_{21} = \left[\frac{2EI}{L}\right]_{BC} = \left[\frac{2(2.0EI)}{8.0}\right] = +0.5EI$$

$$K_{31} = 0$$

$\Delta_2 = 1.0$ $\Delta_1 = \Delta_3 = 0$

$$K_{12} = \left[\frac{2EI}{L}\right]_{BC} = \left[\frac{2(2.0EI)}{8.0}\right] = +0.5EI$$

$$K_{22} = \left[\frac{4EI}{L}\right]_{BC} + \left[\frac{4EI}{L}\right]_{CD}$$

$$= \left[\frac{4(2.0EI)}{8.0}\right] + \left[\frac{4(1.5EI)}{3.0}\right] = +3.0EI$$

$$K_{32} = \left[\frac{2EI}{L}\right]_{CD} = \left[\frac{2(1.5EI)}{3.0}\right] = +1.0EI$$

Solution

Topic: Direct Stiffness Method
Problem Number: 7.6   **Page No. 2**

$\Delta_3 = 1.0$ $\Delta_1 = \Delta_2 = 0$

Global Axes

$K_{13} = 0$

$K_{23} = \left[\dfrac{2EI}{L} \right]_{CD} = \left[\dfrac{2(1.5EI)}{3.0} \right] = +1.0EI$

$K_{33} = \left[\dfrac{4EI}{L} \right]_{CD} = \left[\dfrac{4(1.5EI)}{3.0} \right] = +2.0EI$

Structural stiffness matrix = $[K] = EI \begin{bmatrix} 2.0 & 0.50 & 0 \\ 0.50 & 3.0 & 1.0 \\ 0 & 1.0 & 2.0 \end{bmatrix}$

The invert of a matrix is given by $[K]^{-1} = \dfrac{\left[K^C \right]^T}{|K|}$

where:
$[K^C]$ is the co-factor matrix for $[K]$
$|K|$ is the determinant of $[K]$ and
$[K^C]^T$ is the transpose of the co-factor matrix

$EI \begin{bmatrix} \overset{+}{2.0} & \overset{-}{0.50} & \overset{+}{0} \\ \underset{-}{0.50} & \overset{+}{3.0} & 1.0 \\ \overset{+}{0} & \underset{-}{1.0} & \overset{+}{2.0} \end{bmatrix}$

Co-factor Matrix: $[K^C]$
(**Note:** the transpose of a symmetric matrix is the same as the original matrix)
$k_{11}^c = + \{(3.0 \times 2.0) - (1.0 \times 1.0)\}EI^2 = +5.0EI^2$

$k_{12}^c = k_{21}^c = - \{(0.5 \times 2.0) - (0 \times 1.0)\}EI^2 = -1.0EI^2$

$k_{13}^c = k_{31}^c = + \{(0.5 \times 1.0) - (0 \times 3.0)\}EI^2 = +0.50EI^2$

$k_{22}^c = + \{(2.0 \times 2.0) - (0)\}EI^2 = +4.0EI^2$

$k_{23}^c = k_{32}^c = - \{(2.0 \times 1.0) - (0 \times 0.5)\}EI^2 = -2.0EI^2$

$k_{33}^c = + \{(2.0 \times 3.0) - (0.5 \times 0.5)\}EI^2 = +5.75EI^2$

Determinant of $[K]$:
Det $[K] = EI^3 \{+ (2.0 \times 5.0) - (0.5 \times 1.0) + 0)\} = +9.5\ EI^3$

Solution

Topic: Direct Stiffness Method
Problem Number: 7.6   **Page No. 3**

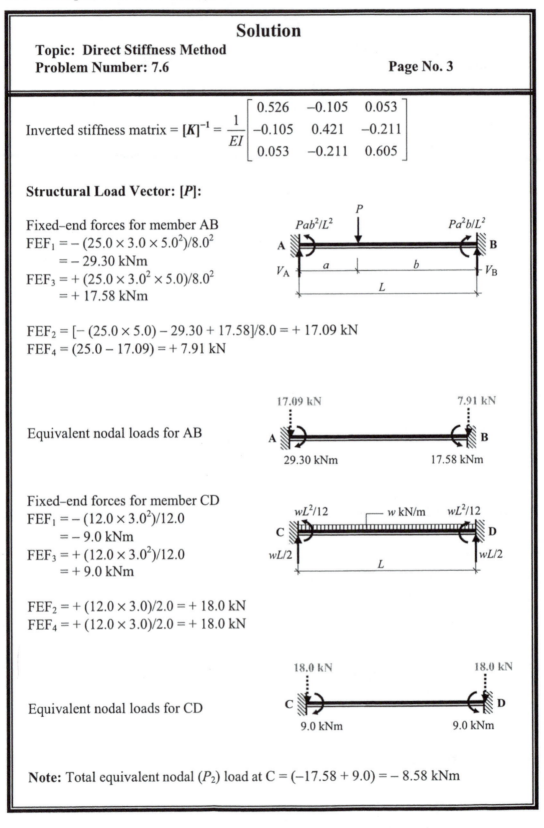

Inverted stiffness matrix $= [K]^{-1} = \dfrac{1}{EI} \begin{bmatrix} 0.526 & -0.105 & 0.053 \\ -0.105 & 0.421 & -0.211 \\ 0.053 & -0.211 & 0.605 \end{bmatrix}$

Structural Load Vector: [P]:

Fixed–end forces for member AB
$FEF_1 = -(25.0 \times 3.0 \times 5.0^2)/8.0^2$
   $= -29.30$ kNm
$FEF_3 = +(25.0 \times 3.0^2 \times 5.0)/8.0^2$
   $= +17.58$ kNm

$FEF_2 = [-(25.0 \times 5.0) - 29.30 + 17.58]/8.0 = +17.09$ kN
$FEF_4 = (25.0 - 17.09) = +7.91$ kN

Equivalent nodal loads for AB

Fixed–end forces for member CD
$FEF_1 = -(12.0 \times 3.0^2)/12.0$
   $= -9.0$ kNm
$FEF_3 = +(12.0 \times 3.0^2)/12.0$
   $= +9.0$ kNm

$FEF_2 = +(12.0 \times 3.0)/2.0 = +18.0$ kN
$FEF_4 = +(12.0 \times 3.0)/2.0 = +18.0$ kN

Equivalent nodal loads for CD

Note: Total equivalent nodal (P_2) load at C $= (-17.58 + 9.0) = -8.58$ kNm

Solution

Topic: Direct Stiffness Method
Problem Number: 7.6   **Page No. 4**

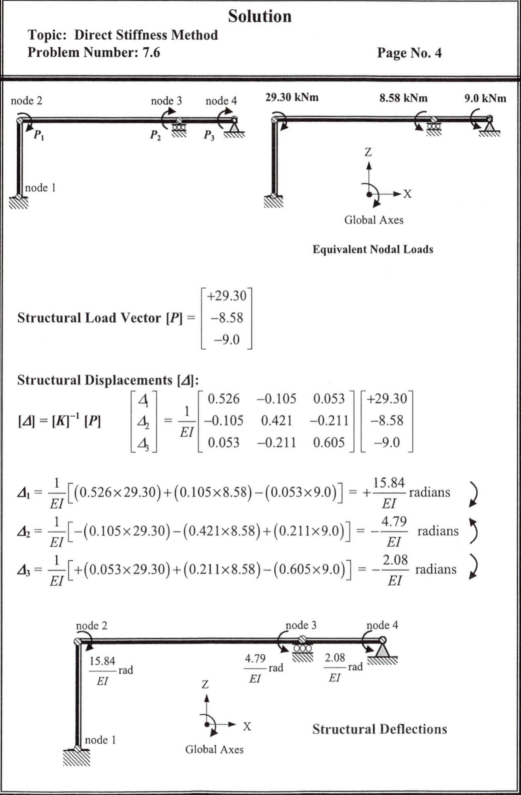

Structural Load Vector $[P] = \begin{bmatrix} +29.30 \\ -8.58 \\ -9.0 \end{bmatrix}$

Structural Displacements $[\Delta]$:

$$[\Delta] = [K]^{-1}[P] \quad \begin{bmatrix} \Delta_1 \\ \Delta_2 \\ \Delta_3 \end{bmatrix} = \frac{1}{EI}\begin{bmatrix} 0.526 & -0.105 & 0.053 \\ -0.105 & 0.421 & -0.211 \\ 0.053 & -0.211 & 0.605 \end{bmatrix}\begin{bmatrix} +29.30 \\ -8.58 \\ -9.0 \end{bmatrix}$$

$$\Delta_1 = \frac{1}{EI}\left[(0.526 \times 29.30) + (0.105 \times 8.58) - (0.053 \times 9.0)\right] = +\frac{15.84}{EI}\ \text{radians}$$

$$\Delta_2 = \frac{1}{EI}\left[-(0.105 \times 29.30) - (0.421 \times 8.58) + (0.211 \times 9.0)\right] = -\frac{4.79}{EI}\ \text{radians}$$

$$\Delta_3 = \frac{1}{EI}\left[+(0.053 \times 29.30) + (0.211 \times 8.58) - (0.605 \times 9.0)\right] = -\frac{2.08}{EI}\ \text{radians}$$

Solution

Topic: Direct Stiffness Method
Problem Number: 7.6   Page No. 5

Element Stiffness Matrices [k]:

$$[k] = \begin{bmatrix} +\dfrac{4EI}{L} & -\dfrac{6EI}{L^2} & +\dfrac{2EI}{L} & +\dfrac{6EI}{L^2} \\[2mm] -\dfrac{6EI}{L^2} & +\dfrac{12EI}{L^3} & -\dfrac{6EI}{L^2} & -\dfrac{12EI}{L^3} \\[2mm] +\dfrac{2EI}{L} & -\dfrac{6EI}{L^2} & +\dfrac{4EI}{L} & +\dfrac{6EI}{L^2} \\[2mm] +\dfrac{6EI}{L^2} & -\dfrac{12EI}{L^3} & +\dfrac{6EI}{L^2} & +\dfrac{12EI}{L^3} \end{bmatrix}$$

Element End Forces [F]$_{Total}$:

$$[F]_{Total} = \begin{bmatrix} F_1 \\ F_2 \\ F_3 \\ F_4 \end{bmatrix} + \begin{bmatrix} FEF_1 \\ FEF_2 \\ FEF_3 \\ FEF_4 \end{bmatrix} = [k][\delta] + [FEF]$$

$$[k][\delta] + [FEF] = \begin{bmatrix} +\dfrac{4EI}{L} & -\dfrac{6EI}{L^2} & +\dfrac{2EI}{L} & +\dfrac{6EI}{L^2} \\[2mm] -\dfrac{6EI}{L^2} & +\dfrac{12EI}{L^3} & -\dfrac{6EI}{L^2} & -\dfrac{12EI}{L^3} \\[2mm] +\dfrac{2EI}{L} & -\dfrac{6EI}{L^2} & +\dfrac{4EI}{L} & +\dfrac{6EI}{L^2} \\[2mm] +\dfrac{6EI}{L^2} & -\dfrac{12EI}{L^3} & +\dfrac{6EI}{L^2} & +\dfrac{12EI}{L^3} \end{bmatrix} \begin{bmatrix} \delta_1 \\ \delta_2 \\ \delta_3 \\ \delta_4 \end{bmatrix} + \begin{bmatrix} FEF_1 \\ FEF_2 \\ FEF_3 \\ FEF_4 \end{bmatrix}$$

Consider element AB:

$$\frac{4EI}{L} = \frac{4\times(EI)}{4.0} = 1.0EI \qquad\qquad \frac{6EI}{L^2} = \frac{6\times EI}{4.0^2} = 0.38EI$$

$$\frac{2EI}{L} = \frac{2\times EI}{4.0} = 0.50EI \qquad\qquad \frac{12EI}{L^3} = \frac{12\times EI}{4.0^3} = 0.19EI$$

$$[k]_{AB} = EI \begin{bmatrix} +1.0 & -0.38 & +0.50 & +0.38 \\ -0.38 & +0.19 & -0.38 & -0.19 \\ +0.50 & -0.38 & +1.0 & +0.38 \\ +0.38 & -0.19 & +0.38 & +0.19 \end{bmatrix}$$

Solution

Topic: Direct Stiffness Method
Problem Number: 7.6   **Page No. 6**

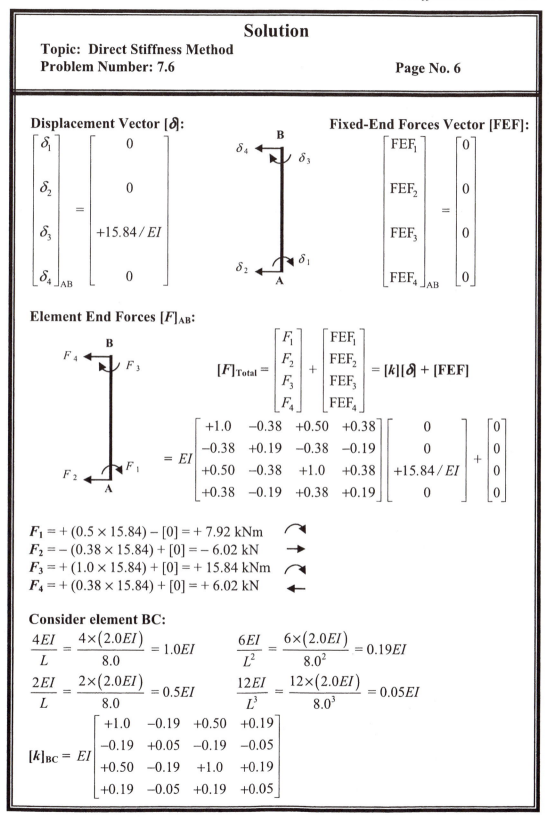

Displacement Vector [δ]:

$$\begin{bmatrix} \delta_1 \\ \delta_2 \\ \delta_3 \\ \delta_4 \end{bmatrix}_{AB} = \begin{bmatrix} 0 \\ 0 \\ +15.84/EI \\ 0 \end{bmatrix}$$

Fixed-End Forces Vector [FEF]:

$$\begin{bmatrix} FEF_1 \\ FEF_2 \\ FEF_3 \\ FEF_4 \end{bmatrix}_{AB} = \begin{bmatrix} 0 \\ 0 \\ 0 \\ 0 \end{bmatrix}$$

Element End Forces [F]$_{AB}$:

$$[F]_{Total} = \begin{bmatrix} F_1 \\ F_2 \\ F_3 \\ F_4 \end{bmatrix} + \begin{bmatrix} FEF_1 \\ FEF_2 \\ FEF_3 \\ FEF_4 \end{bmatrix} = [k][\delta] + [FEF]$$

$$= EI \begin{bmatrix} +1.0 & -0.38 & +0.50 & +0.38 \\ -0.38 & +0.19 & -0.38 & -0.19 \\ +0.50 & -0.38 & +1.0 & +0.38 \\ +0.38 & -0.19 & +0.38 & +0.19 \end{bmatrix} \begin{bmatrix} 0 \\ 0 \\ +15.84/EI \\ 0 \end{bmatrix} + \begin{bmatrix} 0 \\ 0 \\ 0 \\ 0 \end{bmatrix}$$

$F_1 = +(0.5 \times 15.84) - [0] = +7.92$ kNm

$F_2 = -(0.38 \times 15.84) + [0] = -6.02$ kN

$F_3 = +(1.0 \times 15.84) + [0] = +15.84$ kNm

$F_4 = +(0.38 \times 15.84) + [0] = +6.02$ kN

Consider element BC:

$$\frac{4EI}{L} = \frac{4 \times (2.0EI)}{8.0} = 1.0EI \qquad \frac{6EI}{L^2} = \frac{6 \times (2.0EI)}{8.0^2} = 0.19EI$$

$$\frac{2EI}{L} = \frac{2 \times (2.0EI)}{8.0} = 0.5EI \qquad \frac{12EI}{L^3} = \frac{12 \times (2.0EI)}{8.0^3} = 0.05EI$$

$$[k]_{BC} = EI \begin{bmatrix} +1.0 & -0.19 & +0.50 & +0.19 \\ -0.19 & +0.05 & -0.19 & -0.05 \\ +0.50 & -0.19 & +1.0 & +0.19 \\ +0.19 & -0.05 & +0.19 & +0.05 \end{bmatrix}$$

Solution

Topic: Direct Stiffness Method
Problem Number: 7.6   **Page No. 7**

Displacement Vector [δ]:   **Fixed-End Forces Vector [FEF]:**

$$
\begin{bmatrix} \delta_1 \\ \delta_2 \\ \delta_3 \\ \delta_4 \end{bmatrix}_{BC} = \begin{bmatrix} +15.84\,/\,EI \\ 0 \\ -4.79\,/\,EI \\ 0 \end{bmatrix}
$$

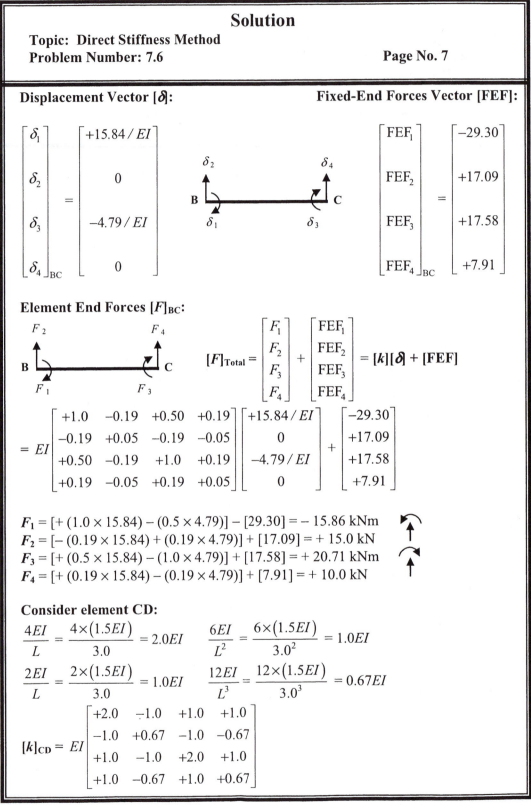

$$
\begin{bmatrix} FEF_1 \\ FEF_2 \\ FEF_3 \\ FEF_4 \end{bmatrix}_{BC} = \begin{bmatrix} -29.30 \\ +17.09 \\ +17.58 \\ +7.91 \end{bmatrix}
$$

Element End Forces [F]_BC:

$$
[F]_{Total} = \begin{bmatrix} F_1 \\ F_2 \\ F_3 \\ F_4 \end{bmatrix} + \begin{bmatrix} FEF_1 \\ FEF_2 \\ FEF_3 \\ FEF_4 \end{bmatrix} = [k][\delta] + [FEF]
$$

$$
= EI \begin{bmatrix} +1.0 & -0.19 & +0.50 & +0.19 \\ -0.19 & +0.05 & -0.19 & -0.05 \\ +0.50 & -0.19 & +1.0 & +0.19 \\ +0.19 & -0.05 & +0.19 & +0.05 \end{bmatrix} \begin{bmatrix} +15.84\,/\,EI \\ 0 \\ -4.79\,/\,EI \\ 0 \end{bmatrix} + \begin{bmatrix} -29.30 \\ +17.09 \\ +17.58 \\ +7.91 \end{bmatrix}
$$

$F_1 = [+ (1.0 \times 15.84) - (0.5 \times 4.79)] - [29.30] = -15.86 \text{ kNm}$

$F_2 = [- (0.19 \times 15.84) + (0.19 \times 4.79)] + [17.09] = +15.0 \text{ kN}$

$F_3 = [+ (0.5 \times 15.84) - (1.0 \times 4.79)] + [17.58] = +20.71 \text{ kNm}$

$F_4 = [+ (0.19 \times 15.84) - (0.19 \times 4.79)] + [7.91] = +10.0 \text{ kN}$

Consider element CD:

$$
\frac{4EI}{L} = \frac{4 \times (1.5EI)}{3.0} = 2.0EI \qquad \frac{6EI}{L^2} = \frac{6 \times (1.5EI)}{3.0^2} = 1.0EI
$$

$$
\frac{2EI}{L} = \frac{2 \times (1.5EI)}{3.0} = 1.0EI \qquad \frac{12EI}{L^3} = \frac{12 \times (1.5EI)}{3.0^3} = 0.67EI
$$

$$
[k]_{CD} = EI \begin{bmatrix} +2.0 & -1.0 & +1.0 & +1.0 \\ -1.0 & +0.67 & -1.0 & -0.67 \\ +1.0 & -1.0 & +2.0 & +1.0 \\ +1.0 & -0.67 & +1.0 & +0.67 \end{bmatrix}
$$

Solution

Topic: Direct Stiffness Method
Problem Number: 7.6   **Page No. 8**

Displacement Vector [δ]:   **Fixed-End Forces Vector [FEF]:**

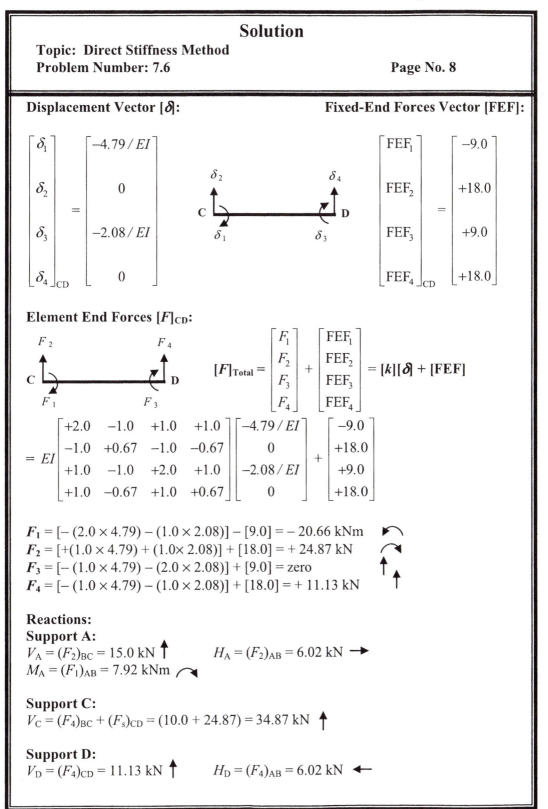

$$\begin{bmatrix} \delta_1 \\ \delta_2 \\ \delta_3 \\ \delta_4 \end{bmatrix}_{CD} = \begin{bmatrix} -4.79/EI \\ 0 \\ -2.08/EI \\ 0 \end{bmatrix} \qquad \begin{bmatrix} FEF_1 \\ FEF_2 \\ FEF_3 \\ FEF_4 \end{bmatrix}_{CD} = \begin{bmatrix} -9.0 \\ +18.0 \\ +9.0 \\ +18.0 \end{bmatrix}$$

Element End Forces [F]$_{CD}$:

$$[F]_{Total} = \begin{bmatrix} F_1 \\ F_2 \\ F_3 \\ F_4 \end{bmatrix} + \begin{bmatrix} FEF_1 \\ FEF_2 \\ FEF_3 \\ FEF_4 \end{bmatrix} = [k][\delta] + [FEF]$$

$$= EI \begin{bmatrix} +2.0 & -1.0 & +1.0 & +1.0 \\ -1.0 & +0.67 & -1.0 & -0.67 \\ +1.0 & -1.0 & +2.0 & +1.0 \\ +1.0 & -0.67 & +1.0 & +0.67 \end{bmatrix} \begin{bmatrix} -4.79/EI \\ 0 \\ -2.08/EI \\ 0 \end{bmatrix} + \begin{bmatrix} -9.0 \\ +18.0 \\ +9.0 \\ +18.0 \end{bmatrix}$$

$F_1 = [-(2.0 \times 4.79) - (1.0 \times 2.08)] - [9.0] = -20.66$ kNm

$F_2 = [+(1.0 \times 4.79) + (1.0 \times 2.08)] + [18.0] = +24.87$ kN

$F_3 = [-(1.0 \times 4.79) - (2.0 \times 2.08)] + [9.0] =$ zero

$F_4 = [-(1.0 \times 4.79) - (1.0 \times 2.08)] + [18.0] = +11.13$ kN

Reactions:
Support A:
$V_A = (F_2)_{BC} = 15.0$ kN ↑ $H_A = (F_2)_{AB} = 6.02$ kN →
$M_A = (F_1)_{AB} = 7.92$ kNm ↻

Support C:
$V_C = (F_4)_{BC} + (F_s)_{CD} = (10.0 + 24.87) = 34.87$ kN ↑

Support D:
$V_D = (F_4)_{CD} = 11.13$ kN ↑ $H_D = (F_4)_{AB} = 6.02$ kN ←

Solution

Topic: Direct Stiffness Method
Problem Number: 7.6   **Page No. 9**

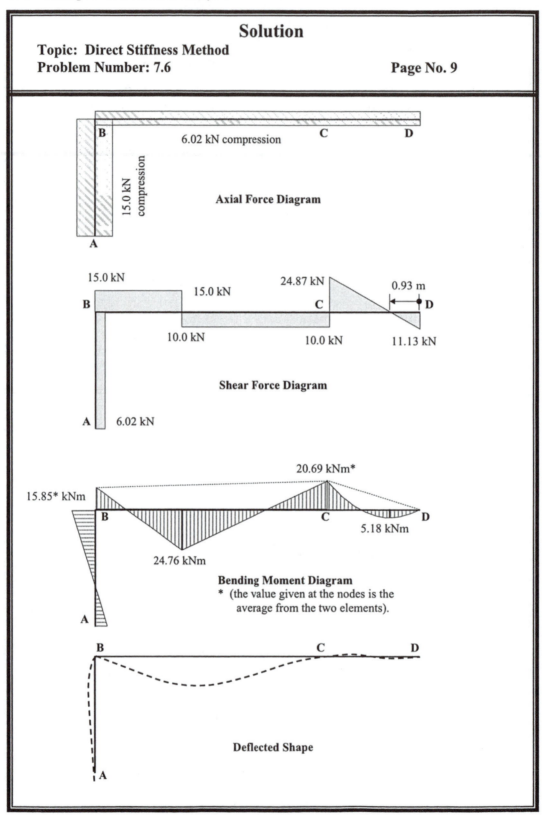

6.02 kN compression

15.0 kN compression

Axial Force Diagram

15.0 kN

15.0 kN

24.87 kN

0.93 m

10.0 kN

10.0 kN

11.13 kN

6.02 kN

Shear Force Diagram

20.69 kNm*

15.85* kNm

5.18 kNm

24.76 kNm

Bending Moment Diagram
* (the value given at the nodes is the
   average from the two elements).

Deflected Shape

8. Plastic Analysis

8.1 *Introduction*

The Plastic Moment of Resistance (M_{pl}) of individual member sections can be derived as indicated in Section 2.3 of Chapter 2. The value of M_{pl} is the maximum value of moment which can be applied to a cross–section before a plastic hinge develops. Consider structural collapse in which either individual members may fail or the entire structure may fail as a whole due to the development of plastic hinges.

According to the theory of plasticity, a structure is deemed to have reached the limit of its load carrying capacity when it forms sufficient hinges to convert it into a mechanism with consequent collapse. This is normally one hinge more than the number of degrees–of–indeterminacy (I_D) in the structure as indicated in Figure 8.1.

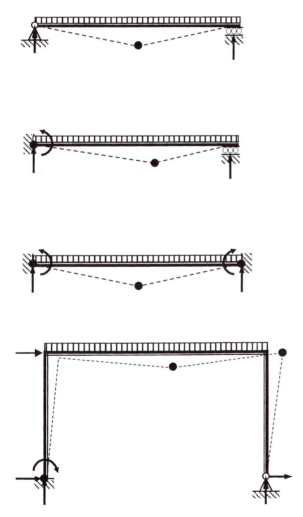

Ignoring horizontal forces:
Number of degrees-of-indeterminacy
$I_D = [(2m + r) - 2n] = 0$
Minimum number of hinges required
$(I_D + 1) = 1$

Ignoring horizontal forces:
Number of degrees-of-indeterminacy
$I_D = [(2m + r) - 2n] = 1$
Minimum number of hinges required
$(I_D + 1) = 2$

Ignoring horizontal forces:
Number of degrees-of-indeterminacy
$I_D = [(2m + r) - 2n] = 2$
Minimum number of hinges required
$(I_D + 1) = 3$

Number of degrees-of-indeterminacy
$I_D = [(3m + r) - 3n] = 2$
Minimum number of hinges required
$(I_D + 1) = 3$

Figure 8.1

8.1.1 Partial Collapse

It is possible for part of a structure to collapse whilst the rest remains stable. In this instance full collapse does not occur and the number of hinges required to cause partial collapse is less than the (I_D + 1.0). This is illustrated in the multi-span beam shown in Figure 8.2. Ignoring horizontal forces $I_D = [(2m + r) - 2n] = [(2 \times 4) + 5 - (2 \times 5)] = 3$

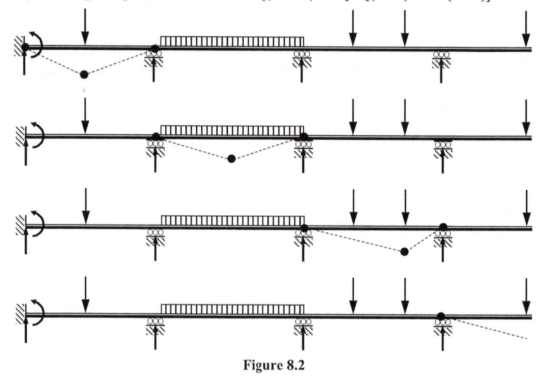

Figure 8.2

For any given design load applied to a redundant structure, more than one collapse mechanism may be possible. The correct mechanism is the one which requires the least amount of '*work done*' for its' inception.

8.1.2 Conditions for Full Collapse

There are three conditions which must be satisfied to ensure full collapse of a structure and the identification of the true collapse load, they are:

(i)   the *mechanism condition* in which there must be sufficient plastic hinges to develop a mechanism, (i.e. the number of plastic hinges $\geq [I_D + 1]$),

(ii)   the *equilibrium condition* in which the bending moments for any collapse mechanism must be in equilibrium with the applied collapse loads,

(iii)   the *yield condition* in which the magnitude of the bending moment anywhere on the structure cannot exceed the plastic moment of resistance of the member in which it occurs.

Provided that these three conditions can be satisfied then the true collapse load can be identified.

If only the mechanism and equilibrium conditions are satisfied then an **upper–bound (unsafe) solution** is obtained in which the collapse load determined is either greater than or equal to the true value.

If only the yield and equilibrium conditions are satisfied then a **lower–bound (safe) solution** is obtained in which the collapse load determined is either less than or equal to the true value.

Since the bending moment cannot exceed the M_{pl} value for a given cross–section it is evident that when hinges develop they will occur at the positions of maximum bending moment, i.e. at fixed supports, rigid–joints, under point loads and within the regions of distributed loads.

The analysis of beams and frames involves determining:

(i)   the collapse loads,
(ii)   the number of hinges required to induce collapse,
(iii)   the possible hinge positions,
(iv)   the independent collapse mechanisms and their associated M_{pl} values,
(v)   the possible combinations of independent mechanisms to obtain the highest required M_{pl} value,
(vi)   checking the validity of the calculated value with respect to mechanism, equilibrium and yield conditions.

There are two methods of analysis which are frequently used to determine the values of plastic moment of resistance for sections required for a structure to collapse at specified factored loads; they are the Static Method and the Kinematic Method. These are illustrated with respect to continuous beams in Sections 8.2 to 8.4. and with respect to frames in Sections 8.5 to 8.12.

8.2 Static Method for Continuous Beams

In the static method of analysis the '*Free Bending Moment*' diagrams for the structure are drawn and the '*Fixed Bending Moment*' diagrams are then added algebraically. The magnitude and '*sense*' +ve or −ve of the moments must be such that sufficient plastic hinges occur to cause the collapse of the whole or a part of the structure.

In addition, for collapse to occur, adjacent plastic hinges must be alternatively 'opening' and 'closing'. For uniform beams the plastic moment of resistance of each hinge will be the same i.e. M_{pl}.

8.2.1 Example 8.1: Encastré Beam

An encastré beam is 8.0 m long and supports an unfactored load of 40 kN/m as shown in Figure 8.3. Assuming that the yield stress $f_y = 460$ N/mm^2 and a load factor $\lambda = 1.7$, determine the required plastic moment of resistance and plastic section modulus.

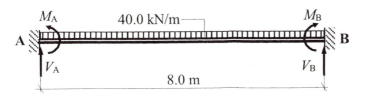

Figure 8.3

Solution:

The collapse load = $(40.0 \times 1.7) = 68.0$ kN/m

The number of hinges required to induce collapse = $(I_D + 1) = 3$ (see Figure 8.1)

The possible hinge positions are at the supports A and B and within the region of the distributed load since these are the positions where the maximum bending moments occur. Superimpose the fixed and free bending moment diagrams as shown in Figure 8.4:

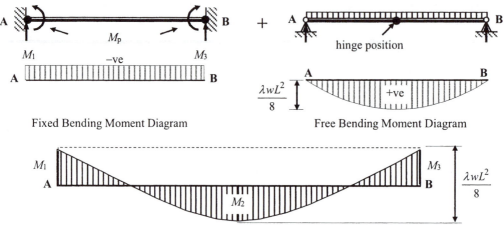

Fixed Bending Moment Diagram   Free Bending Moment Diagram

Final Bending Moment Diagram

Figure 8.4

The beam has two redundancies (ignoring horizontal components of reaction) therefore a minimum of three hinges must develop to create a mechanism. Since the beam is uniform, at failure all values of the bending moment at the hinge positions must be equal to the plastic moment of resistance and cannot be exceeded anywhere:

$$M_1 = M_2 = M_3 = M_{pl} \qquad \text{and} \qquad (M_1 + M_2) = (M_3 + M_2) = 2M_{pl} = \frac{\lambda wL^2}{8}$$

The required plastic moment of resistance $M_{pl} = \dfrac{\lambda wL^2}{16} = \dfrac{68.0 \times 8^2}{16} = 272.0$ kNm

The plastic section modulus $W_{yy} = M_{pl}/f_y = (272.0 \times 10^6)/460 = 591.3 \times 10^3$ mm³

It is evident from the above that all three conditions in Section 8.1.2 are satisfied and consequently the M_{pl} value calculated for the required collapse load is true to achieve a load factor of 1.7

8.2.2 Example 8.2: Propped Cantilever 1

A propped cantilever is 6.0 m long and supports a collapse load of 24 kN as shown in Figure 8.5. Determine the required plastic moment of resistance M_{pl}.

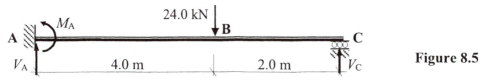

Figure 8.5

Solution:

The collapse load = 24.0 kN

The number of hinges required to induce collapse = $(I_D + 1) = 2$ (see Figure 8.1)

The possible hinge positions are at the support A and under the point load since these are the positions where the maximum bending moments occur.

The support reactions for the free bending diagram are: $V_A = 8.0$ kN and $V_C = 16.0$ kN

The maximum free bending moment at $M_{free,C} = (8.0 \times 4.0) = 32.0$ kNm

The bending moment at B due to the fixed moment $= -[M_1 \times (2.0 \times 6.0)] = -0.333 M_1$ kNm

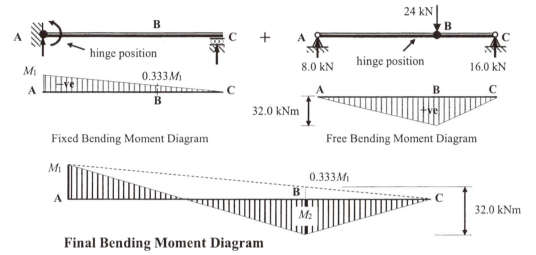

Fixed Bending Moment Diagram   Free Bending Moment Diagram

Final Bending Moment Diagram

Figure 8.6

The beam has one redundancy (ignoring horizontal components of reaction) therefore a minimum of two hinges must develop to create a mechanism. Since the beam is uniform, at failure all values of the bending moment at the hinge positions must be equal to the plastic moment of resistance and cannot be exceeded anywhere:

$M_1 = M_2 = M_{pl}$ and $(M_2 + 0.333M_1) = (M_{pl} + 0.333M_{pl}) = 1.333M_{pl} = 32.0$
The required plastic moment of resistance $M_{pl} = (32.0/1.333) = 24.0$ kNm

As in Example 8.1 all three conditions in Section 8.1.2 are satisfied and consequently the true value of M_{pl} has been calculated for the given collapse load.

8.2.3 Example 8.3: Propped Cantilever 2

A propped cantilever is L m long and supports a collapse load of w kN/m as shown in Figure 8.7. Determine the position of the plastic hinges and the required plastic moment of resistance M_{pl}.

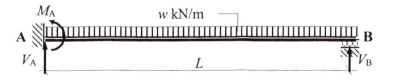

Figure 8.7

Solution:
The collapse load = w kN/m
The number of hinges required to induce collapse = $(I_D + 1) = 2$ (see Figure 8.1)

The possible hinge positions are at the support A and within the region of the distributed load since these are the positions where the maximum bending moments occur. In this case the maximum moment under the distributed load does not occur at mid–span since the bending moment diagram is not symmetrical. Consider the final bending moment diagram:

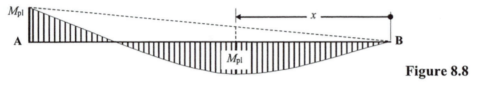

Figure 8.8

The maximum bending moment (i.e. M_{pl}) occurs at a distance 'x' from the roller support and can be determined as follows;
Since the moment is a maximum at position 'x' the shear force at 'x' is equal to zero.

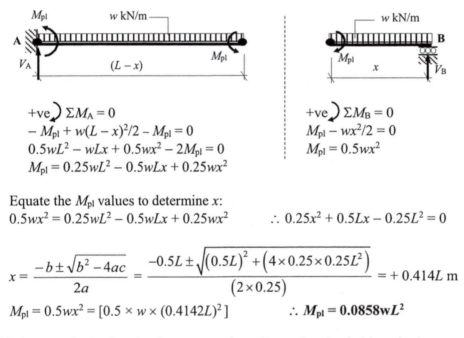

$+ve \; \circlearrowleft \; \Sigma M_A = 0$
$- M_{pl} + w(L-x)^2/2 - M_{pl} = 0$
$0.5wL^2 - wLx + 0.5wx^2 - 2M_{pl} = 0$
$M_{pl} = 0.25wL^2 - 0.5wLx + 0.25wx^2$

$+ve \; \circlearrowleft \; \Sigma M_B = 0$
$M_{pl} - wx^2/2 = 0$
$M_{pl} = 0.5wx^2$

Equate the M_{pl} values to determine x:
$0.5wx^2 = 0.25wL^2 - 0.5wLx + 0.25wx^2$ $\therefore \; 0.25x^2 + 0.5Lx - 0.25L^2 = 0$

$$x = \frac{-b \pm \sqrt{b^2 - 4ac}}{2a} = \frac{-0.5L \pm \sqrt{(0.5L)^2 + (4 \times 0.25 \times 0.25L^2)}}{(2 \times 0.25)} = +0.414L \text{ m}$$

$M_{pl} = 0.5wx^2 = [0.5 \times w \times (0.4142L)^2]$ $\therefore \; M_{pl} = 0.0858wL^2$

This is a standard value, i.e. for a propped cantilever the plastic hinge in the span occurs at a distance $x = 0.414L$ from the simply supported end and the value of the plastic moment $M_{pl} = 0.0858wL^2$

8.3   *Kinematic Method for Continuous Beams*

In this method, a displacement is imposed upon each possible collapse mechanism and an equation between external work done and internal work absorbed in forming the hinges is developed. The collapse mechanism involving the greatest plastic moment, M_{pl}, is the critical one.

Consider the previous Example 8.1 of an encastré beam with a uniformly distributed load. The hinge positions were identified as occurring at A, B and the mid–span point (since the beam and loading are symmetrical). Assuming rigid links between the hinges, the collapse mechanism of the beam when the hinges develop can be drawn as shown in Figure 8.9(c). The deformed shape is drawn grossly magnified to enable the relationship between the rotations at the hinges and the displacements of the loads to be easily identified.

A virtual work equation can be developed by equating the external work done by the applied loads to the internal work done by the formation of the hinges where:

Internal work done during the formation of a hinge = (moment × rotation)
External work done by a load during displacement = (load × displacement)
(In the case of distributed loads the average displacement is used).

The sign convention adopted is:
Tension on the **Bottom** of the beam induces a '**positive**' rotation   (i.e. +ve bending)
Tension on the **Top** of the beam induces a '**negative**' rotation   (i.e. −ve bending)

Note: the development of both −ve and +ve hinges involves +ve internal work

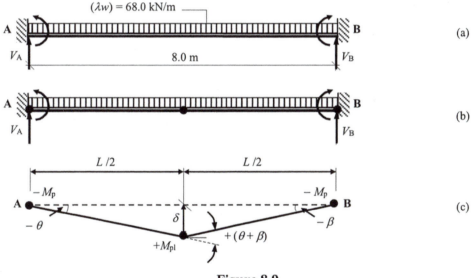

Figure 8.9

From the deformed shape in Figure 8.9:

For small values of θ and β $\delta = \dfrac{L}{2}\theta = \dfrac{L}{2}\beta$ $\therefore \beta = \theta$

The load deflects zero at the supports and δ at the centre

Average displacement of the load $= \dfrac{1}{2}\delta = \dfrac{L}{4}\theta$

The *Internal Work Done* in developing the hinges is found from the product of the moment induced (i.e. M_{pl}) and the amount of rotation (e.g. θ) for each hinge.

Internal Work Done = Moment × Rotation for each hinge position
$$= M_{pl}\theta + M_{pl}(\theta + \beta) + M_{pl}\theta = 4M_{pl}\theta$$

The *External Work Done* by the applied load system is found from the product of the load and the displacement for each load.

External Work Done = (load × ave. displacement) = $\left[(68.0 \times 8.0) \times \dfrac{8.0}{4}\theta\right]$ = 1088.0 θ

Internal Work Done = External Work Done
$4M_{pl}\theta = 1088.0\theta$
$M_{pl} = 272.0$ kNm (as before)

Consider the previous Example 8.2 of propped cantilever with a single point load. The hinge positions were identified as occurring at support A, and under the point load at B. Assuming rigid links between the hinges, the collapse mechanism of the beam when the hinges develop can be drawn as shown in Figure 8.10(c).

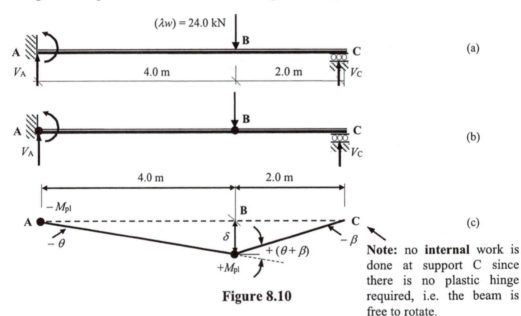

Figure 8.10

Note: no **internal** work is done at support C since there is no plastic hinge required, i.e. the beam is free to rotate.

From the deformed shape in Figure 8.10:
For small values of θ and β $\delta = 4.0\theta = 2.0\beta$ $\therefore \beta = 2.0\theta$
Displacement of the load = $\delta = 4.0\theta$

Internal Work Done = External Work Done
$M_{pl}\theta + M_{pl}(\theta + \beta) = (24.0 \times \delta)$
$4M_{pl} = 96.0\theta$
$M_{pl} = 24.0$ kNm (as before)

Consider the previous Example 8.3 of a propped cantilever with a uniformly distributed load. The hinge positions were identified as occurring at support A, and at a point load $0.4142L$ from the simple support. Assuming rigid links between the hinges, the deformed shape of the beam when the hinges develop can be drawn as shown in Figure 8.11(c).

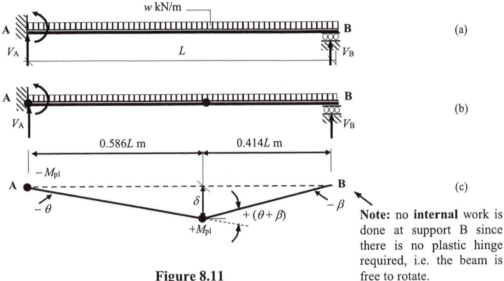

Figure 8.11

Note: no **internal** work is done at support B since there is no plastic hinge required, i.e. the beam is free to rotate.

From the deformed shape in Figure 8.11:
For small values of θ and β $\delta = 0.586L\theta = 0.414L\beta$ $\therefore \beta = 1.415\theta$
The load deflects zero at the supports and δ at a distance $0.414L$ from support B.

Average displacement of the load $= \dfrac{1}{2}\delta = \dfrac{0.586L}{2}\theta = 0.293L\theta$

Internal Work Done = External Work Done
$M_{pl}\theta + M_{pl}(\theta + \beta) = (w \times L) \times 0.293L\theta$
$3.415M_{pl}\theta = 0.293wL\theta$
$M_{pl} = 0.0858wL^2$ (as before)

8.3.1 Example 8.4: Continuous Beam

A non–uniform, three–span beam is fixed at support A, simply supported on rollers at D, F and G and carries unfactored loads as shown in Figure 8.12. Determine the minimum M_{pl} value required to ensure a minimum load factor equal to 1.7 for any span.

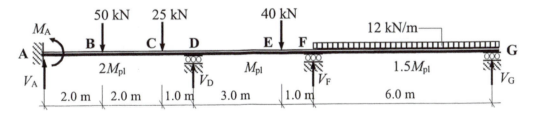

Figure 8.12

There are a number of possible elementary beam mechanisms and it is necessary to ensure all possibilities have been considered. It is convenient in multi–span beams to consider each span separately and identify the collapse mechanism involving the greatest plastic moment M_{pl}; this is the critical one and results in partial collapse.

The number of elementary independent mechanisms can be determined from evaluating (the number of possible hinge positions – the degree–of–indeterminacy).

Ignoring horizontal forces:

Number of degrees-of-indeterminacy: $I_D = [(2m + r) - 2n]$
$$= [(2 \times 3) + 5 - (2 \times 4)] = 3$$
Number of possible hinge positions $= 7$ (at A, B, C, D, E, F and between F and G)

Number of independent mechanisms $= (7 - 3) = 4$
(**Note:** In framed structures combinations of independent mechanisms must also be considered see Section 8.5).
$\lambda = 1.7$
Factored loads: $(1.7 \times 50) = 85.0$ kN $(1.7 \times 25) = 42.5$ kN $(1.7 \times 40) = 68.0$ kN
$(1.7 \times 12) = 20.4$ kN

Consider span ABCD:
In this span there are four possible hinge positions, however only three are required to induce collapse in the beam. There are two independent collapse mechanisms to consider, they are:
(i) hinges developing at A (moment $= 2M_{pl}$), B (moment $= 2M_{pl}$) and D (moment $= M_{pl}$)
(ii) hinges developing at A (moment $= 2M_{pl}$), C (moment $= 2M_{pl}$) and D (moment $= M_{pl}$)

Static Method:
The free bending moment at B $= 119.0$ kNm
The free bending moment at C $= 68.0$ kNm.

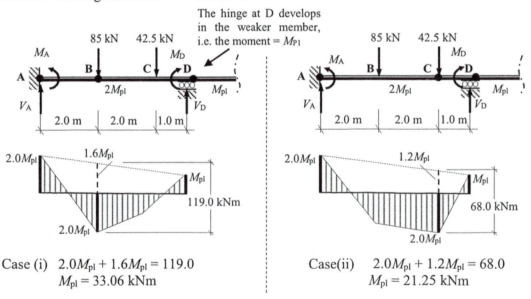

Case (i) $2.0M_{pl} + 1.6M_{pl} = 119.0$
$M_{pl} = 33.06$ kNm

Case(ii) $2.0M_{pl} + 1.2M_{pl} = 68.0$
$M_{pl} = 21.25$ kNm

In this span the critical value of M_{pl} = 33.06 kNm with hinges developing at A, B and D.
Kinematic Method:

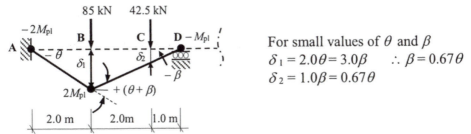

For small values of θ and β
$\delta_1 = 2.0\theta = 3.0\beta$ $\therefore \beta = 0.67\theta$
$\delta_2 = 1.0\beta = 0.67\theta$

Internal Work Done = External Work Done
$2.0M_{pl}\theta + 2.0M_{pl}(\theta + \beta) + M_{pl}\beta = (85.0 \times \delta_1) + (42.5 \times \delta_2)$
$2.0M_{pl}\theta + 3.34M_{pl}\theta + 0.67M_{pl}\theta = (85.0 \times 2.0\theta) + (42.5 \times 0.67\theta)$
$6.0M_{pl}\theta = 198.36\theta$ $\therefore M_{pl} = 33.06$ kNm (as before)

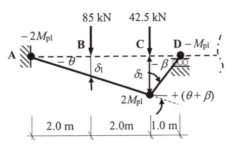

For small values of θ and β
$\delta_1 = 2.0\theta$
$\delta_2 = 4.0\theta = 1.0\beta$ $\therefore \beta = 4.0\theta$

Internal Work Done = External Work Done
$2.0M_{pl}\theta + 2.0M_{pl}(\theta + \beta) + M_{pl}\beta = (85.0 \times \delta_1) + (42.5 \times \delta_2)$
$2.0M_{pl}\theta + 10.0M_{pl}\theta + 4.0M_{pl}\theta = (85.0 \times 2.0\theta) + (42.5 \times 4.0\theta)$
$16.0M_{pl}\theta = 340.0\theta$ $\therefore M_{pl} = 21.25$ kNm (as before)
The critical value for this span is $M_{pl} = 33.06$ as before.

Consider span DEF:
In this span only three hinges are required to induce collapse in the beam.
Hinges develop at D (moment = M_{pl}), E (moment = M_{pl}) and F (moment = M_{pl})

Static Method: **Kinematic Method:**

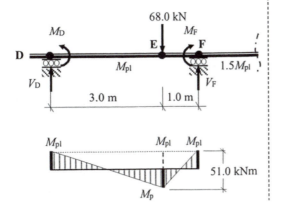

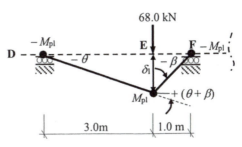

For small values of θ and β
$\delta_1 = 3.0\theta = 1.0\beta$ $\therefore \beta = 3.0\theta$

Static method:

$M_{pl} + M_{pl} = 51.0$

$M_{pl} = 25.5$ kNm

Kinematic Method:

Internal Work Done = External Work Done

$M_{pl}\theta + M_{pl}(\theta + \beta) + M_{pl}\beta = (68.0 \times \delta_1)$

$8.0M_{pl}\theta = 204\theta$

$M_{pl} = 25.5$ kNm

Consider span FG:

In this span only two hinges are required to induce collapse in the beam.

Hinges develop at F (moment = M_{pl}), and between F and G (moment = $1.5M_{pl}$)

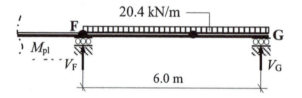

Span FG is effectively a propped cantilever and consequently the position of the hinge under the uniformly distributed load must be calculated. (**Note:** it is different from Example 8.3 since the plastic moment at each hinge position is not the same).

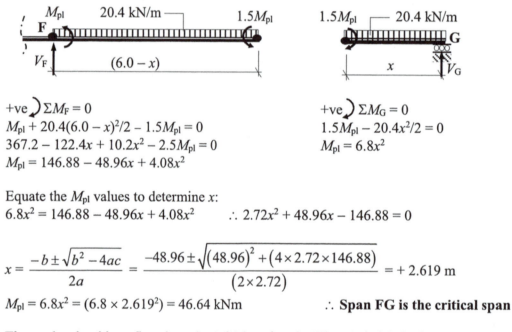

$+ve \; \curvearrowleft \; \Sigma M_F = 0$

$M_{pl} + 20.4(6.0 - x)^2/2 - 1.5M_{pl} = 0$

$367.2 - 122.4x + 10.2x^2 - 2.5M_{pl} = 0$

$M_{pl} = 146.88 - 48.96x + 4.08x^2$

$+ve \; \curvearrowleft \; \Sigma M_G = 0$

$1.5M_{pl} - 20.4x^2/2 = 0$

$M_{pl} = 6.8x^2$

Equate the M_{pl} values to determine x:

$6.8x^2 = 146.88 - 48.96x + 4.08x^2$ ∴ $2.72x^2 + 48.96x - 146.88 = 0$

$$x = \frac{-b \pm \sqrt{b^2 - 4ac}}{2a} = \frac{-48.96 \pm \sqrt{(48.96)^2 + (4 \times 2.72 \times 146.88)}}{(2 \times 2.72)} = +2.619 \text{ m}$$

$M_{pl} = 6.8x^2 = (6.8 \times 2.619^2) = 46.64$ kNm   ∴ **Span FG is the critical span**

The reader should confirm the value of M_{pl} using the Kinematic Method.

Span:	ABCD	DEF	FG
Minimum required value of M_{pl} for a load factor of 1.7	⎰ 33.06 kNm	25.5 kNm	**46.64 kNm**
Actual load factor if an M_{pl} value of 46.64 kNm is used	⎱ (1.7 × 46.64)/33.06 2.4	(1.7 × 46.64)/25.5 3.1	**1.7**
Actual M_{pl} provided	93.28 kNm	46.64 kNm	**69.96 kNm**

8.4 Problems: Plastic Analysis – Continuous Beams

A series of continuous beams are indicated in which the relative M_{pl} values and the applied **collapse** loadings are given in Problems 8.1 to 8.5. Determine the required value of M_{pl} to ensure a minimum load factor $\lambda = 1.7$.

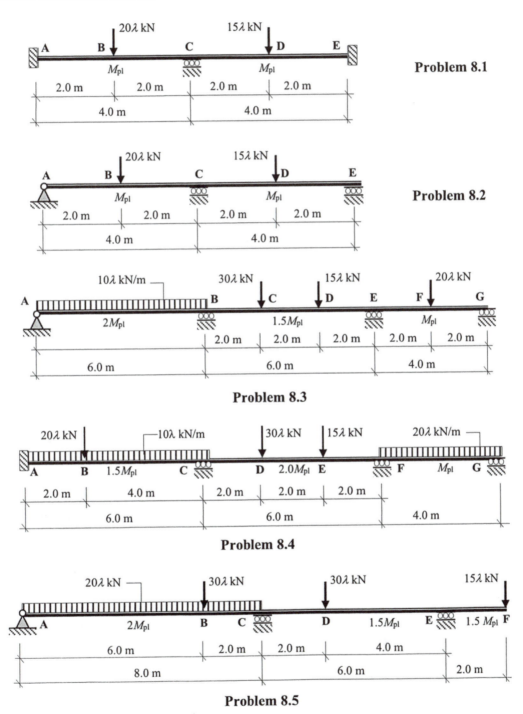

8.5   Solutions: Plastic Analysis – Continuous Beams

Solution

Topic: Plastic Analysis – Continuous Beams
Problem Number: 8.1 – Kinematic Method   **Page No. 1**

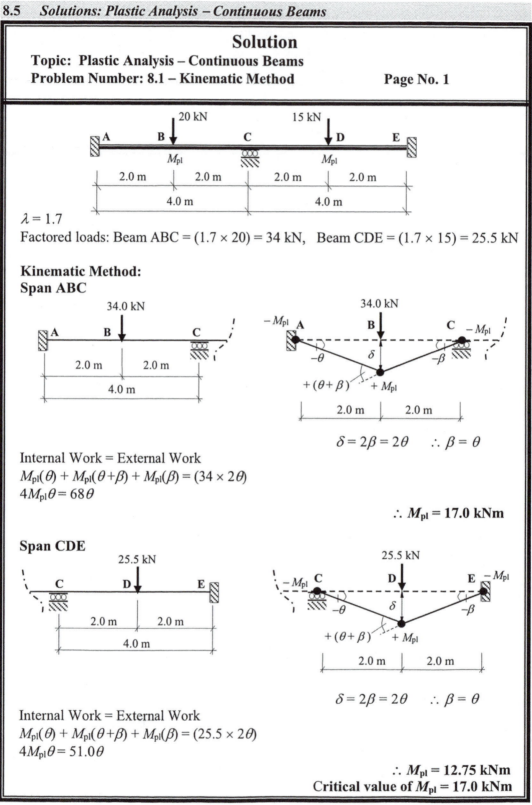

$\lambda = 1.7$
Factored loads: Beam ABC $= (1.7 \times 20) = 34$ kN,   Beam CDE $= (1.7 \times 15) = 25.5$ kN

Kinematic Method:
Span ABC

$\delta = 2\beta = 2\theta$ $\therefore \beta = \theta$

Internal Work = External Work
$M_{pl}(\theta) + M_{pl}(\theta+\beta) + M_{pl}(\beta) = (34 \times 2\theta)$
$4M_{pl}\theta = 68\theta$

$\therefore M_{pl} = 17.0$ kNm

Span CDE

$\delta = 2\beta = 2\theta$ $\therefore \beta = \theta$

Internal Work = External Work
$M_{pl}(\theta) + M_{pl}(\theta+\beta) + M_{pl}(\beta) = (25.5 \times 2\theta)$
$4M_{pl}\theta = 51.0\theta$

$\therefore M_{pl} = 12.75$ kNm
Critical value of $M_{pl} = 17.0$ kNm

Solution

Topic: Plastic Analysis – Continuous Beams
Problem Number: 8.1 – Static Method   **Page No. 2**

Static Method:

Span ABC

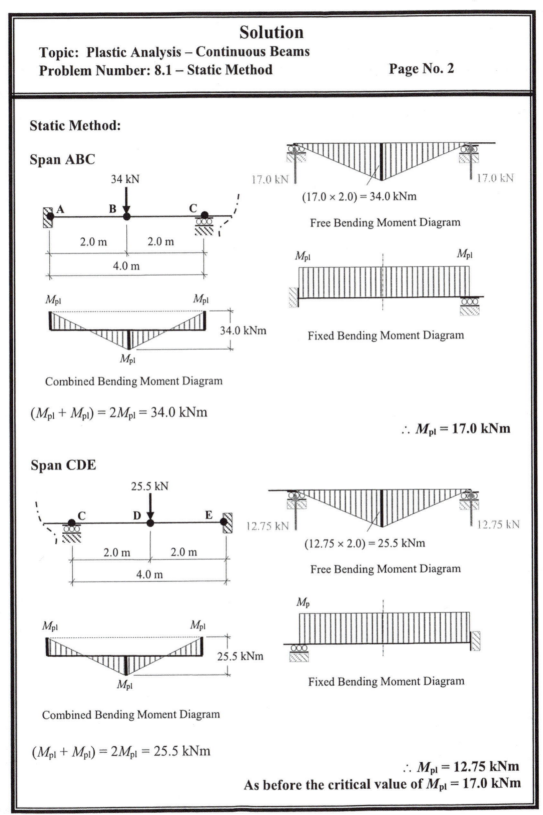

Free Bending Moment Diagram

$(17.0 \times 2.0) = 34.0$ kNm

Fixed Bending Moment Diagram

Combined Bending Moment Diagram

$(M_{pl} + M_{pl}) = 2M_{pl} = 34.0$ kNm

$\therefore M_{pl} = 17.0$ **kNm**

Span CDE

Free Bending Moment Diagram

$(12.75 \times 2.0) = 25.5$ kNm

Fixed Bending Moment Diagram

Combined Bending Moment Diagram

$(M_{pl} + M_{pl}) = 2M_{pl} = 25.5$ kNm

$\therefore M_{pl} = $ **12.75 kNm**
As before the critical value of $M_{pl} = 17.0$ kNm

Solution

Topic: Plastic Analysis – Continuous Beams
Problem Number: 8.2 – Kinematic Method   **Page No. 1**

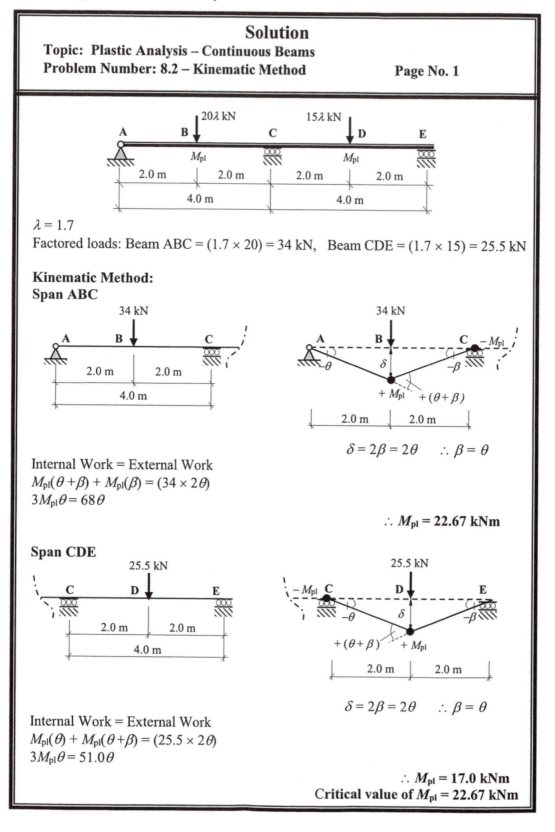

$\lambda = 1.7$

Factored loads: Beam ABC = $(1.7 \times 20) = 34$ kN,   Beam CDE = $(1.7 \times 15) = 25.5$ kN

Kinematic Method:
Span ABC

$\delta = 2\beta = 2\theta$ $\therefore \beta = \theta$

Internal Work = External Work
$M_{pl}(\theta + \beta) + M_{pl}(\beta) = (34 \times 2\theta)$
$3M_{pl}\theta = 68\theta$

$\therefore M_{pl} = 22.67$ **kNm**

Span CDE

$\delta = 2\beta = 2\theta$ $\therefore \beta = \theta$

Internal Work = External Work
$M_{pl}(\theta) + M_{pl}(\theta + \beta) = (25.5 \times 2\theta)$
$3M_{pl}\theta = 51.0\theta$

$\therefore M_{pl} = 17.0$ **kNm**
Critical value of $M_{pl} = 22.67$ kNm

Solution

Topic: Plastic Analysis – Continuous Beams
Problem Number: 8.2 – Static Method   **Page No. 2**

Static Method:

Span ABC

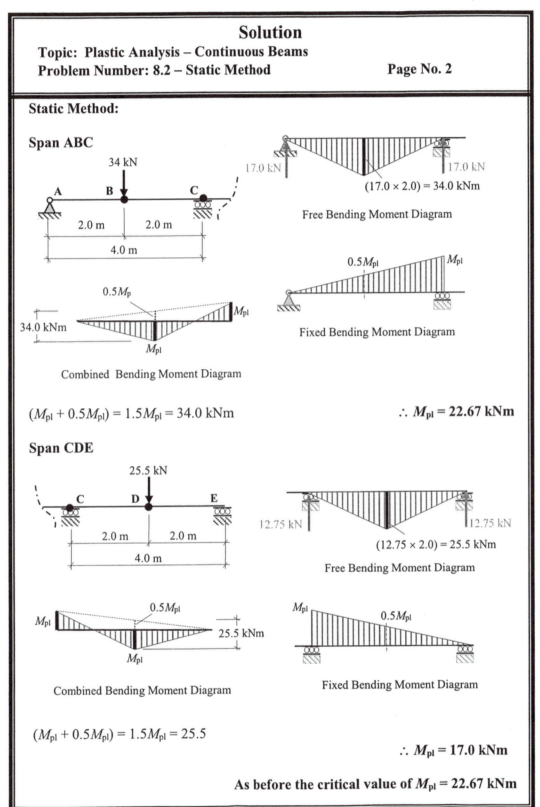

$(M_{pl} + 0.5M_{pl}) = 1.5M_{pl} = 34.0$ kNm $\therefore M_{pl} = 22.67$ kNm

Span CDE

$(M_{pl} + 0.5M_{pl}) = 1.5M_{pl} = 25.5$

$\therefore M_{pl} = 17.0$ kNm

As before the critical value of $M_{pl} = 22.67$ kNm

Solution
Topic: Plastic Analysis – Continuous Beams
Problem Number: 8.3 – Kinematic Method   **Page No. 1**

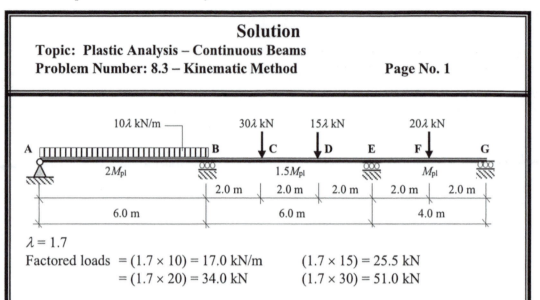

$\lambda = 1.7$

Factored loads $= (1.7 \times 10) = 17.0$ kN/m $(1.7 \times 15) = 25.5$ kN
$= (1.7 \times 20) = 34.0$ kN $(1.7 \times 30) = 51.0$ kN

Kinematic Method:
Span AB
Note: Span AB is effectively a propped cantilever and the bending moment diagram is asymmetric. The hinge between A and B does not develop at the mid-span point and should be evaluated in a manner similar to that indicated in Section 8.2.3. The reader should carry out this calculation to show that the hinge develops at a position equal to 2.582 m from the free support at A as shown below, (see page 3 of this solution).

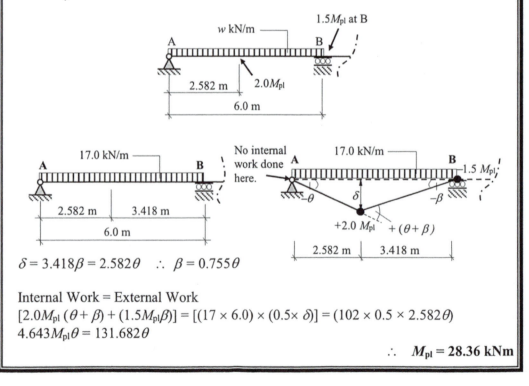

$\delta = 3.418\beta = 2.582\theta$ $\therefore$ $\beta = 0.755\theta$

Internal Work = External Work
$[2.0M_{pl}\,(\theta + \beta) + (1.5M_{pl}\beta)] = [(17 \times 6.0) \times (0.5 \times \delta)] = (102 \times 0.5 \times 2.582\theta)$
$4.643M_{pl}\theta = 131.682\theta$

$\therefore$ $M_{pl} = 28.36$ kNm

Solution
Topic: Plastic Analysis – Continuous Beams
Problem Number: 8.3 – Kinematic Method   **Page No. 2**

Span BCDE

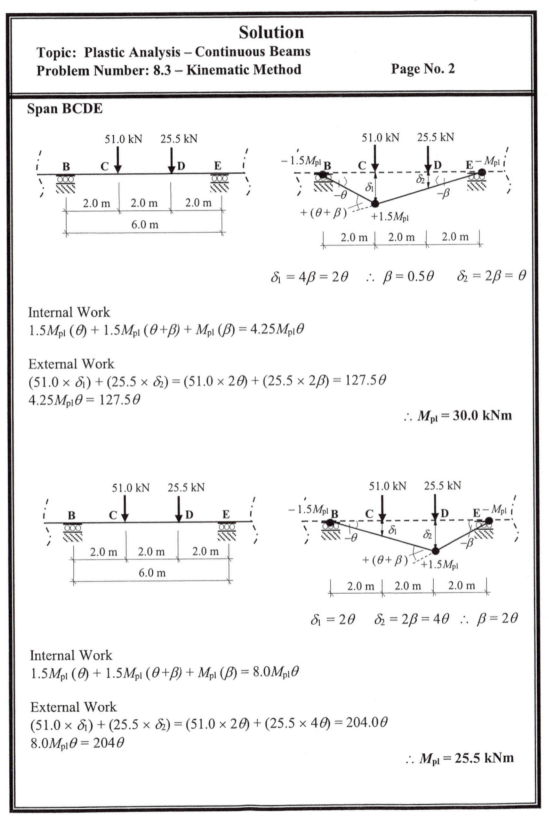

$$\delta_1 = 4\beta = 2\theta \quad \therefore \quad \beta = 0.5\theta \qquad \delta_2 = 2\beta = \theta$$

Internal Work
$$1.5M_{pl}\,(\theta) + 1.5M_{pl}\,(\theta+\beta) + M_{pl}\,(\beta) = 4.25M_{pl}\theta$$

External Work
$$(51.0 \times \delta_1) + (25.5 \times \delta_2) = (51.0 \times 2\theta) + (25.5 \times 2\beta) = 127.5\theta$$
$$4.25M_{pl}\theta = 127.5\theta$$

$$\therefore M_{pl} = 30.0 \text{ kNm}$$

$$\delta_1 = 2\theta \qquad \delta_2 = 2\beta = 4\theta \quad \therefore \quad \beta = 2\theta$$

Internal Work
$$1.5M_{pl}\,(\theta) + 1.5M_{pl}\,(\theta+\beta) + M_{pl}\,(\beta) = 8.0M_{pl}\theta$$

External Work
$$(51.0 \times \delta_1) + (25.5 \times \delta_2) = (51.0 \times 2\theta) + (25.5 \times 4\theta) = 204.0\theta$$
$$8.0M_{pl}\theta = 204\theta$$

$$\therefore M_{pl} = 25.5 \text{ kNm}$$

Solution
Topic: Plastic Analysis – Continuous Beams
Problem Number: 8.3 – Static Method   **Page No. 3**

Span EFG

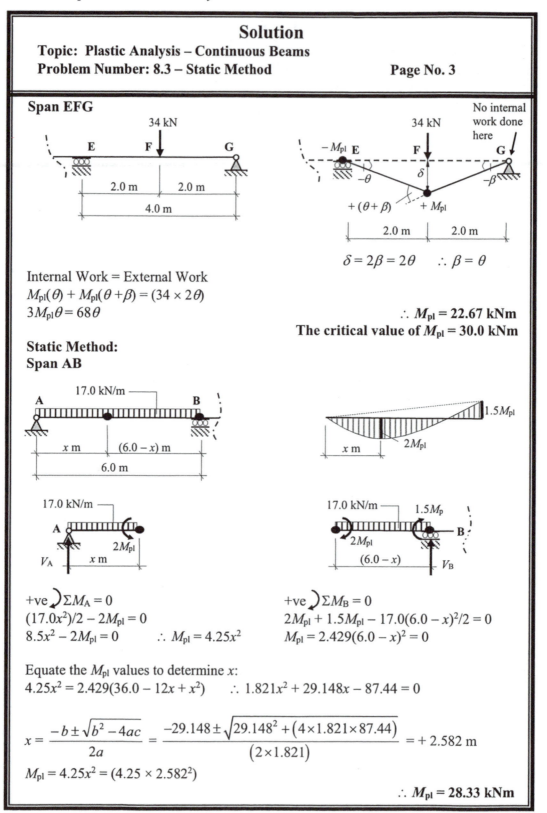

No internal work done here

$\delta = 2\beta = 2\theta \quad \therefore \beta = \theta$

Internal Work = External Work
$M_{pl}(\theta) + M_{pl}(\theta + \beta) = (34 \times 2\theta)$
$3M_{pl}\theta = 68\theta$

$\therefore M_{pl} = 22.67$ **kNm**
The critical value of $M_{pl} = 30.0$ kNm

Static Method:
Span AB

$+ve \;)\Sigma M_A = 0$
$(17.0x^2)/2 - 2M_{pl} = 0$
$8.5x^2 - 2M_{pl} = 0 \qquad \therefore M_{pl} = 4.25x^2$

$+ve \;)\Sigma M_B = 0$
$2M_{pl} + 1.5M_{pl} - 17.0(6.0 - x)^2/2 = 0$
$M_{pl} = 2.429(6.0 - x)^2 = 0$

Equate the M_{pl} values to determine x:
$4.25x^2 = 2.429(36.0 - 12x + x^2) \qquad \therefore 1.821x^2 + 29.148x - 87.44 = 0$

$$x = \frac{-b \pm \sqrt{b^2 - 4ac}}{2a} = \frac{-29.148 \pm \sqrt{29.148^2 + (4 \times 1.821 \times 87.44)}}{(2 \times 1.821)} = +2.582 \text{ m}$$

$M_{pl} = 4.25x^2 = (4.25 \times 2.582^2)$

$\therefore M_{pl} = 28.33$ **kNm**

Solution

Topic: Plastic Analysis – Continuous Beams

Problem Number: 8.3 – Static Method

Span BCDE

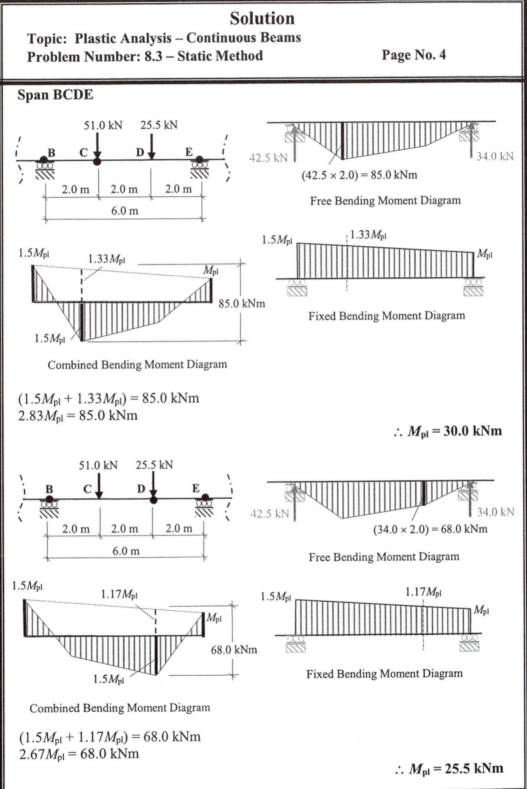

$(42.5 \times 2.0) = 85.0$ kNm

Free Bending Moment Diagram

Combined Bending Moment Diagram

Fixed Bending Moment Diagram

$(1.5M_{pl} + 1.33M_{pl}) = 85.0$ kNm
$2.83M_{pl} = 85.0$ kNm

$\therefore M_{pl} = 30.0$ kNm

$(34.0 \times 2.0) = 68.0$ kNm

Free Bending Moment Diagram

Combined Bending Moment Diagram

Fixed Bending Moment Diagram

$(1.5M_{pl} + 1.17M_{pl}) = 68.0$ kNm
$2.67M_{pl} = 68.0$ kNm

$\therefore M_{pl} = 25.5$ kNm

Solution

Topic: Plastic Analysis – Continuous Beams

Problem Number: 8.3 – Static Method   **Page No. 5**

Span EFG

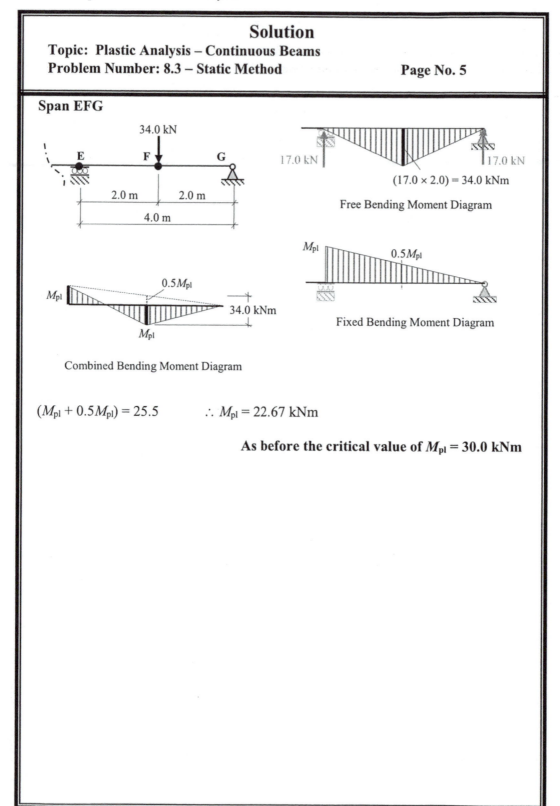

34.0 kN

E   F   G

2.0 m   2.0 m

4.0 m

17.0 kN   17.0 kN

$(17.0 \times 2.0) = 34.0$ kNm

Free Bending Moment Diagram

M_{pl} $0.5M_{pl}$

Fixed Bending Moment Diagram

M_{pl} $0.5M_{pl}$

34.0 kNm

M_{pl}

Combined Bending Moment Diagram

$(M_{pl} + 0.5M_{pl}) = 25.5$ $\therefore M_{pl} = 22.67$ kNm

As before the critical value of $M_{pl} = 30.0$ kNm

Solution

Topic: Plastic Analysis – Continuous Beams
Problem Number: 8.4 – Kinematic Method   Page No. 1

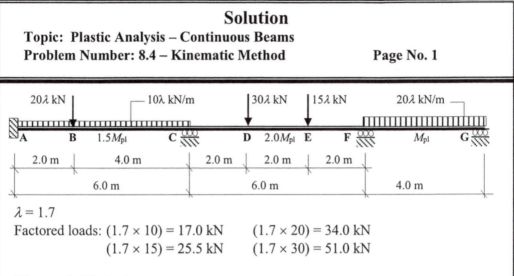

$\lambda = 1.7$

Factored loads: $(1.7 \times 10) = 17.0$ kN $(1.7 \times 20) = 34.0$ kN
                       $(1.7 \times 15) = 25.5$ kN $(1.7 \times 30) = 51.0$ kN

Kinematic Method:
Span ABC
Note: The bending moment diagram on span ABC is asymmetric and in this case the
hinge between A and C does not necessarily develop under the point load.
The position should be evaluated in a manner similar to that indicated in
Section 8.2.3. The reader should carry out this calculation to show that the hinge
develops at a position equal to 2.333 m from the support at A as shown below, (see
page 3 of this solution).

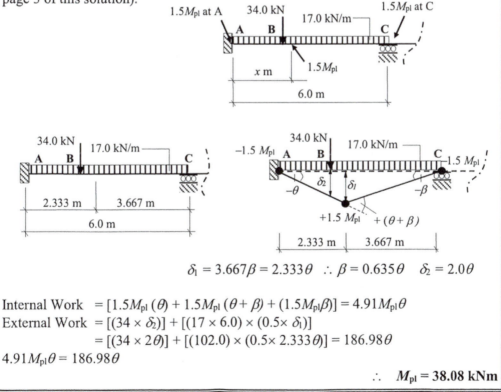

$$\delta_1 = 3.667\beta = 2.333\theta \quad \therefore \beta = 0.635\theta \quad \delta_2 = 2.0\theta$$

Internal Work $= [1.5M_{pl}\,(\theta) + 1.5M_{pl}\,(\theta + \beta) + (1.5M_{pl}\beta)] = 4.91M_{pl}\theta$
External Work $= [(34 \times \delta_2)] + [(17 \times 6.0) \times (0.5\times \delta_1)]$
$\qquad\qquad = [(34 \times 2\theta)] + [(102.0) \times (0.5\times 2.333\,\theta)] = 186.98\theta$
$4.91M_{pl}\theta = 186.98\theta$

$\qquad\qquad\qquad\qquad\qquad\qquad \therefore \quad M_{pl} = \mathbf{38.08\ kNm}$

Solution

Topic: Plastic Analysis – Continuous Beams
Problem Number: 8.4 – Kinematic Method   **Page No. 2**

Span CDEF

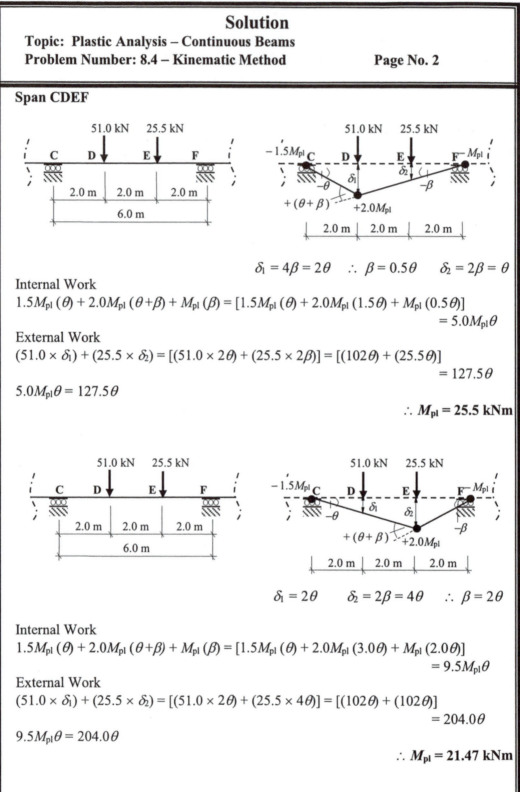

$\delta_1 = 4\beta = 2\theta$ $\therefore$ $\beta = 0.5\theta$ $\delta_2 = 2\beta = \theta$

Internal Work
$1.5M_{pl}\,(\theta) + 2.0M_{pl}\,(\theta+\beta) + M_{pl}\,(\beta) = [1.5M_{pl}\,(\theta) + 2.0M_{pl}\,(1.5\theta) + M_{pl}\,(0.5\theta)]$
$$= 5.0M_{pl}\theta$$

External Work
$(51.0 \times \delta_1) + (25.5 \times \delta_2) = [(51.0 \times 2\theta) + (25.5 \times 2\beta)] = [(102\theta) + (25.5\theta)]$
$$= 127.5\theta$$

$5.0M_{pl}\theta = 127.5\theta$

$$\therefore M_{pl} = 25.5 \text{ kNm}$$

$\delta_1 = 2\theta$ $\delta_2 = 2\beta = 4\theta$ $\therefore$ $\beta = 2\theta$

Internal Work
$1.5M_{pl}\,(\theta) + 2.0M_{pl}\,(\theta+\beta) + M_{pl}\,(\beta) = [1.5M_{pl}\,(\theta) + 2.0M_{pl}\,(3.0\theta) + M_{pl}\,(2.0\theta)]$
$$= 9.5M_{pl}\theta$$

External Work
$(51.0 \times \delta_1) + (25.5 \times \delta_2) = [(51.0 \times 2\theta) + (25.5 \times 4\theta)] = [(102\theta) + (102\theta)]$
$$= 204.0\theta$$

$9.5M_{pl}\theta = 204.0\theta$

$$\therefore M_{pl} = 21.47 \text{ kNm}$$

Solution

Topic: Plastic Analysis – Continuous Beams
Problem Number: 8.4 – Kinematic Method   **Page No. 3**

Span FG
Note: Span FG is effectively a propped cantilever and the bending moment diagram is asymmetric. The hinge between F and G develops at a position $0.4142L$ from the simply supported end as indicated in Section 8.2.3.

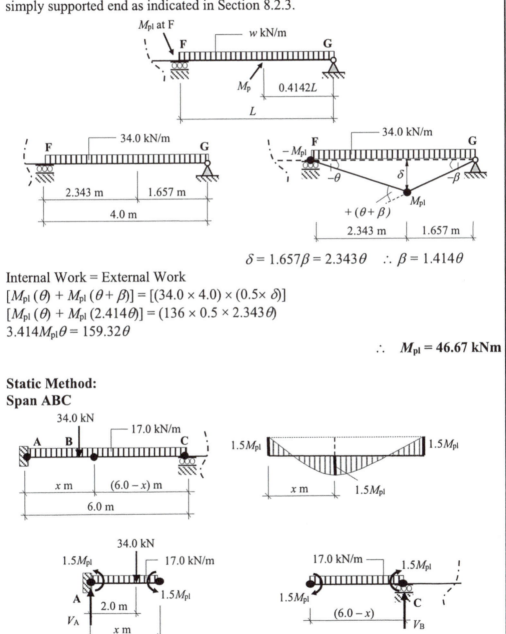

$$\delta = 1.657\beta = 2.343\theta \quad \therefore \ \beta = 1.414\theta$$

Internal Work = External Work

$[M_{pl}\,(\theta) + M_{pl}\,(\theta + \beta)] = [(34.0 \times 4.0) \times (0.5 \times \delta)]$

$[M_{pl}\,(\theta) + M_{pl}\,(2.414\theta)] = (136 \times 0.5 \times 2.343\theta)$

$3.414 M_{pl}\theta = 159.32\theta$

$$\therefore \quad M_{pl} = 46.67 \text{ kNm}$$

Static Method:
Span ABC

Solution
Topic: Plastic Analysis – Continuous Beams
Problem Number: 8.4 – Static Method   Page No. 4

+ve $\circlearrowright\Sigma M_A = 0$
$-1.5M_{pl} + (34 \times 2.0) + (17.0x^2)/2 - 1.5M_{pl} = 0$
$68.0 + 8.5x^2 - 3.0M_{pl} = 0$ $\therefore M_{pl} = 22.667 + 2.833x^2$

+ve $\circlearrowright\Sigma M_C = 0$
$1.5M_{pl} \quad - 17.0(6.0 - x)^2/2 + 1.5M_{pl} = 0$ $\therefore M_{pl} = 2.833(6.0 - x)^2$

Equate the M_{pl} values to determine x:
$22.667 + 2.833x^2 = 2.833(36.0 - 12x + x^2)$ $\therefore 33.996x - 79.321 = 0$
$x = 2.333$ m

$M_{pl} = 2.833(6.0 - x)^2 = 2.833(6.0 - 2.333)^2$

$\therefore \boldsymbol{M_{pl} = 38.09 \text{ kNm}}$

Span CDEF

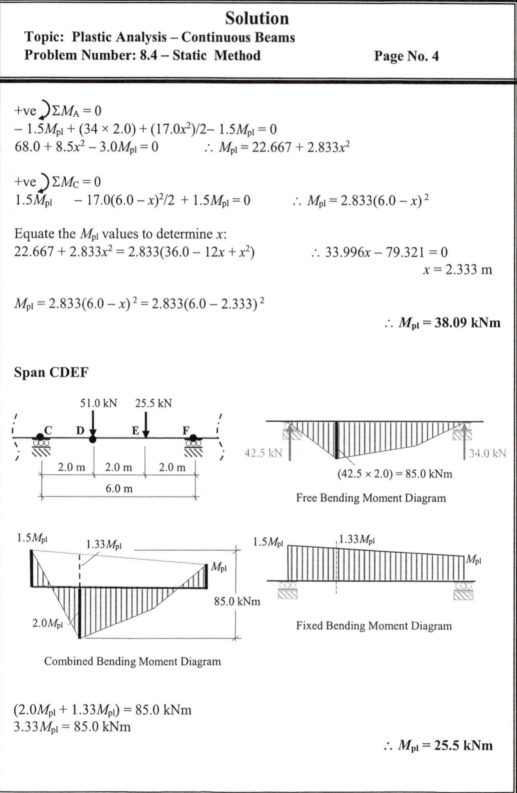

42.5 kN   34.0 kN
$(42.5 \times 2.0) = 85.0$ kNm

Free Bending Moment Diagram

Combined Bending Moment Diagram

Fixed Bending Moment Diagram

$(2.0M_{pl} + 1.33M_{pl}) = 85.0$ kNm
$3.33M_{pl} = 85.0$ kNm

$\therefore \boldsymbol{M_{pl} = 25.5 \text{ kNm}}$

Solution

Topic: Plastic Analysis – Continuous Beams
Problem Number: 8.4 – Static Method

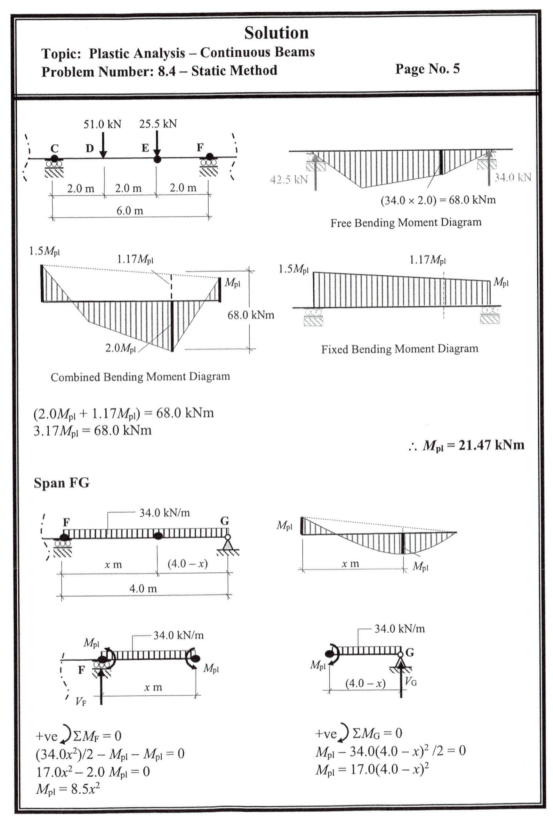

Free Bending Moment Diagram

42.5 kN 34.0 kN

$(34.0 \times 2.0) = 68.0$ kNm

Combined Bending Moment Diagram

Fixed Bending Moment Diagram

$(2.0M_{pl} + 1.17M_{pl}) = 68.0$ kNm
$3.17M_{pl} = 68.0$ kNm

$\therefore M_{pl} = \mathbf{21.47}$ **kNm**

Span FG

$+ve \;)\, \Sigma M_F = 0$
$(34.0x^2)/2 - M_{pl} - M_{pl} = 0$
$17.0x^2 - 2.0\, M_{pl} = 0$
$M_{pl} = 8.5x^2$

$+ve \;)\, \Sigma M_G = 0$
$M_{pl} - 34.0(4.0 - x)^2\,/2 = 0$
$M_{pl} = 17.0(4.0 - x)^2$

Solution

Topic: Plastic Analysis – Continuous Beams
Problem Number: 8.4 – Static Method   **Page No. 6**

Equate the M_{pl} values to determine x:
$8.5x^2 = 17.0(16.0 - 8x + x^2)$ $\therefore 8.5x^2 - 136x + 272 = 0$

$$x = \frac{-b \pm \sqrt{b^2 - 4ac}}{2a} = \frac{136 \pm \sqrt{136^2 - (4 \times 8.5 \times 272)}}{(2 \times 8.5)} = +2.343 \text{ m}$$
$M_{pl} = 8.5x^2 = (8.5 \times 2.343)^2$

$$M_{pl} = 46.67 \text{ kNm}$$

As before the critical value of M_{pl} = 46.67 kNm

Note: Span FG is the same as the standard propped cantilever in Example 8.3 in which the hinge develops at a point $0.414L$ from the simply supported end and the M_{pl} value equals $0.0858wL^2$, i.e.

Distance of hinge from support F = $[4.0 - 0.414L] = [4.0 - (0.414 \times 4.0)] = 2.344$ m
$$\therefore M_{pl} = (0.0858 \times 34.0 \times 4.0^2) = 46.67 \text{ kNm}$$

Solution

Topic: Plastic Analysis – Continuous Beams
Problem Number: 8.5 – Kinematic Method   **Page No. 1**

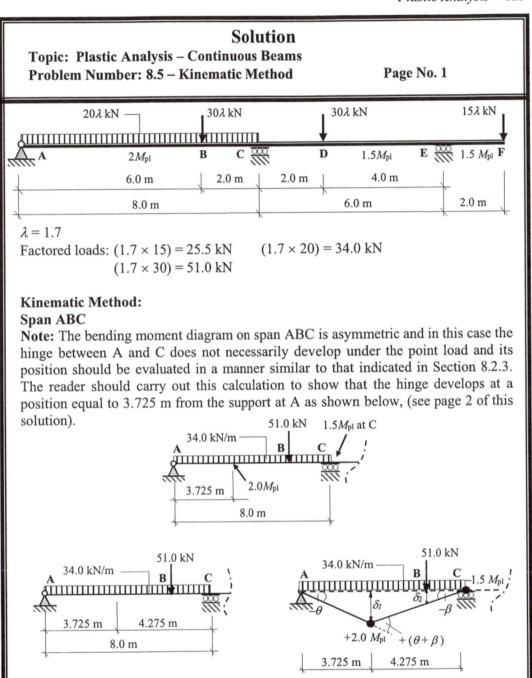

$\lambda = 1.7$

Factored loads: $(1.7 \times 15) = 25.5$ kN $(1.7 \times 20) = 34.0$ kN
$(1.7 \times 30) = 51.0$ kN

Kinematic Method:
Span ABC

Note: The bending moment diagram on span ABC is asymmetric and in this case the hinge between A and C does not necessarily develop under the point load and its position should be evaluated in a manner similar to that indicated in Section 8.2.3. The reader should carry out this calculation to show that the hinge develops at a position equal to 3.725 m from the support at A as shown below, (see page 2 of this solution).

$$\delta_1 = 4.275\beta = 3.725\theta \quad \therefore \ \beta = 0.871\theta; \qquad \delta_2 = 2.0\beta$$

Internal Work $= [\ 2.0M_{pl}\,(\theta + \beta) + (1.5M_{pl}\beta)] = 5.05M_{pl}\theta$

External Work $= [(51 \times \delta_2)] + [(34 \times 8.0) \times (0.5 \times \delta_1)]$
$\qquad\qquad = [(51 \times 1.742\theta)] + [(34.0 \times 8.0) \times (0.5 \times 3.725\theta)] = 595.44\theta$

$5.05M_{pl}\theta = 595.44\theta$

$$\therefore \quad M_{pl} = 117.91 \text{ kNm}$$

Solution
Topic: Plastic Analysis – Continuous Beams
Problem Number: 8.5 – Kinematic Method   **Page No. 2**

Span CDE

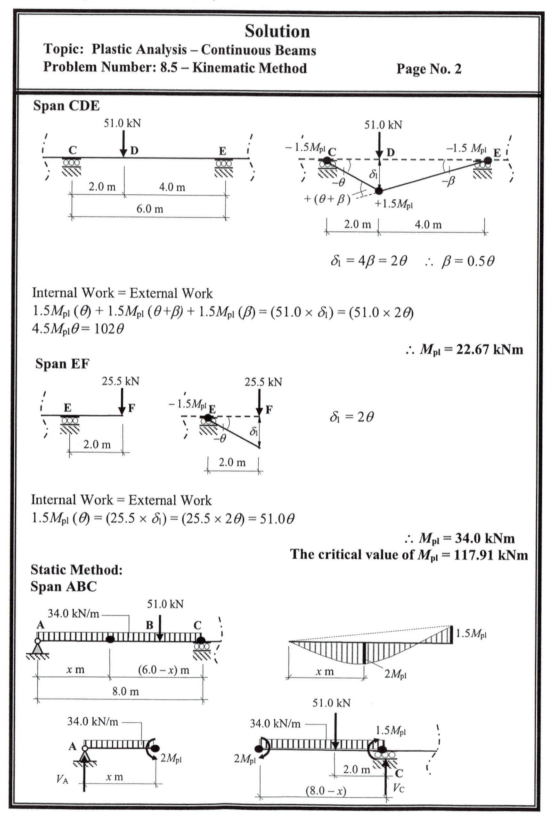

$$\delta_1 = 4\beta = 2\theta \quad \therefore \ \beta = 0.5\theta$$

Internal Work = External Work
$$1.5M_{pl}\,(\theta) + 1.5M_{pl}\,(\theta+\beta) + 1.5M_{pl}\,(\beta) = (51.0 \times \delta_1) = (51.0 \times 2\theta)$$
$$4.5M_{pl}\theta = 102\theta$$

$$\therefore \ M_{pl} = 22.67 \text{ kNm}$$

Span EF

$$\delta_1 = 2\theta$$

Internal Work = External Work
$$1.5M_{pl}\,(\theta) = (25.5 \times \delta_1) = (25.5 \times 2\theta) = 51.0\theta$$

$$\therefore \ M_{pl} = 34.0 \text{ kNm}$$
The critical value of M_{pl} = 117.91 kNm

Static Method:
Span ABC

Solution

Topic: Plastic Analysis – Continuous Beams

Problem Number: 8.5 – Static Method   **Page No. 3**

$+ve \curvearrowright \Sigma M_A = 0$ $+ve \curvearrowright \Sigma M_C = 0$

$(34.0x^2)/2 - 2M_{pl} = 0$ $2M_{pl} - 34.0(8.0 - x)^2/2 - (51.0 \times 2.0) + 1.5M_{pl} = 0$

$17.0x^2 - 2M_{pl} = 0$ $M_{pl} = 4.857(8.0 - x)^2 + 29.143$

$M_{pl} = 8.5x^2$

Equate the M_{pl} values to determine x:

$8.5x^2 = 4.857(64.0 - 16x + x^2) + 29.143$ $\therefore 3.643x^2 + 77.712x - 339.991 = 0$

$$x = \frac{-b \pm \sqrt{b^2 - 4ac}}{2a} = \frac{-77.712 \pm \sqrt{77.712^2 + (4 \times 3.643 \times 339.991)}}{(2 \times 3.643)} = +3.725 \text{ m}$$

$M_{pl} = 8.5x^2 = (8.5 \times 3.725^2)$

$$\therefore M_{pl} = 117.94 \text{ kNm}$$

Span CDE

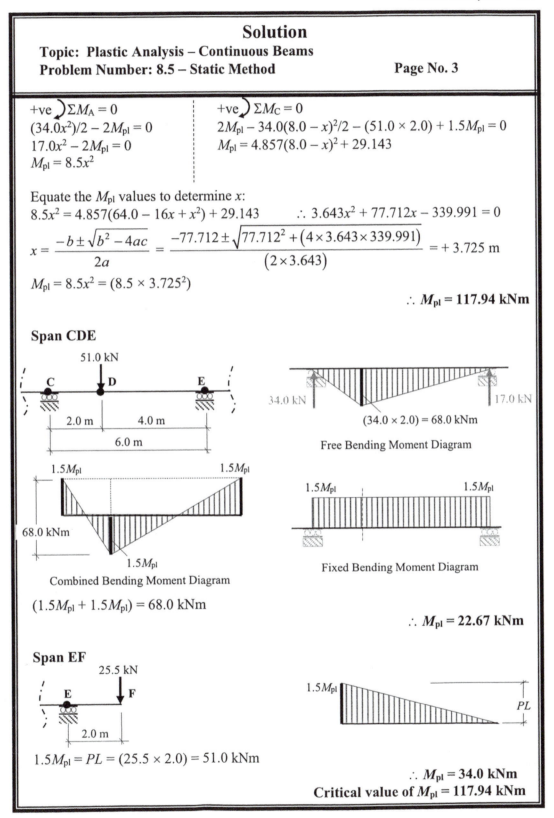

51.0 kN

C   D   E

2.0 m   4.0 m

6.0 m

34.0 kN   17.0 kN

$(34.0 \times 2.0) = 68.0$ kNm

Free Bending Moment Diagram

$1.5M_{pl}$ $1.5M_{pl}$

68.0 kNm

$1.5M_{pl}$

Combined Bending Moment Diagram

$(1.5M_{pl} + 1.5M_{pl}) = 68.0$ kNm

$1.5M_{pl}$ $1.5M_{pl}$

Fixed Bending Moment Diagram

$$\therefore M_{pl} = 22.67 \text{ kNm}$$

Span EF

25.5 kN

E   F

2.0 m

$1.5M_{pl}$

PL

$1.5M_{pl} = PL = (25.5 \times 2.0) = 51.0$ kNm

$$\therefore M_{pl} = 34.0 \text{ kNm}$$

Critical value of $M_{pl} = 117.94$ kNm

8.6   *Rigid-Jointed Frames*

In the case of beams, identification of the critical spans (i.e. in terms of M_{pl} or λ) can usually be solved quite readily by using either the static or the kinematic method and considering simple beam mechanisms. In the case of frames other types of mechanisms, such as sway, joint and gable mechanisms are also considered. Whilst both techniques can be used, the static method often proves laborious when applied to rigid frames, particularly for complex load conditions. It can be easier than the kinematic method in the case of determinate or singly redundant frames. Both methods are illustrated in this section and in the solutions to the given problems.

As mentioned previously the kinematic solution gives a lower bound to the true solution whilst the static solution gives an upper bound.

i.e. $M_{pl\ kinematic} \leq M_{pl\ true} \leq M_{pl\ static}$
$M_{pl\ kinematic} = M_{pl\ static}$ for the true solution.

Two basic types of independent mechanism are shown in Figure 8.13:

(i)   beam mechanisms

(ii)   sway mechanism

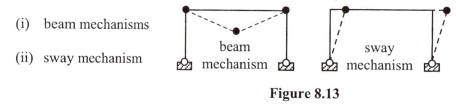

Figure 8.13

Each of these collapse mechanisms can occur independently of each other. It is also possible for a critical collapse mechanism to develop which is a combination of the independent ones such as indicated in Figure 8.14.

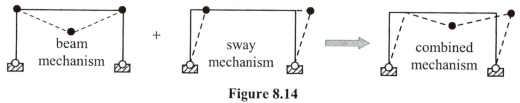

Figure 8.14

It is necessary to consider all possible combinations to identify the critical collapse mode. The M_{pl} value is determined for each independent mechanism and then combined mechanisms are evaluated to establish a maximum value of M_{pl} (i.e. minimum λ). The purpose of combining mechanisms is to eliminate sufficient hinges which exist in the independent mechanisms, leaving only the minimum number required in the resulting combination to induce collapse.

It is necessary when carrying out a kinematic solution, to draw the bending moment diagram to ensure that at no point the M_{pl} value determined, is exceeded.

8.6.1 *Example 8.5: Frame 1*

An asymmetric uniform frame is pinned at supports A and G and is subjected to a system of factored loads as shown in Figure 8.15. Assuming the $\lambda_{vertical.load} = 1.7$ and $\lambda_{horizontal\ loads} = 1.4$ determine the required plastic moment of resistance M_{pl} of the section.

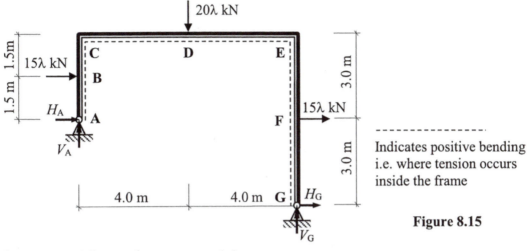

Figure 8.15

λ vertical loads $= 1.7$, λ horizontal loads $= 1.4$
Factored loads: $(1.4 \times 15) = 21.0$ kN $(1.7 \times 20) = 34.0$ kN

Number of degrees-of-indeterminacy $I_D = [(3m + r) - 3n] = [(3 \times 3) + 4) - (3 \times 4)] = 1$
Number of possible hinge positions $p = 5$ (B, C, D, E and F)
Number of independent mechanisms $= (p - I_D) = (5 - 1) = 4$
(i.e. 3 beam mechanisms and 1 sway mechanism)

Kinematic Method:
Consider each independent mechanism separately.
Mechanism (i): Beam ABC

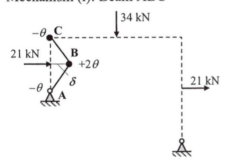

Note:
Internal work is done at **all** hinge positions.
No internal work is done at support A.
The signs of the rotations indicate tension inside or outside the frame.
$\delta = 1.5\theta$
Internal Work Done = External Work Done
$M_{pl}(2\theta + \theta) = (21.0 \times 1.5\theta)$
$3M_p\theta = 31.5\theta$
$M_{pl} = 10.5$ kNm

Mechanism (ii): Beam CDE

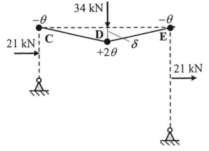

$\delta = 4.0\theta$

Internal Work Done = External Work Done
$M_{pl}(\theta + 2\theta + \theta) = (34.0 \times 4\theta)$
$4M_{pl}\theta = 136.0\theta$
$M_{pl} = 34.0$ kNm

Mechanism (iii) Beam EFG

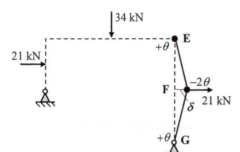

$\delta = 3.0\theta$

Note: no internal work is done at support G.

Internal Work Done = External Work Done
$M_{pl} (\theta + 2\theta) = (21.0 \times 3\theta)$
$3M_{pl}\theta = 63.0\theta$
$M_{pl} = 21.0$ kNm

Mechanism (iv): Sway

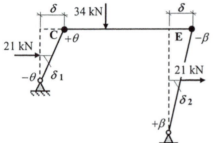

$\delta = 3.0\theta = 6.0\beta \quad \therefore \beta = 0.5\theta$
$\delta_1 = 1.5\theta$ and $\delta_2 = 3.0\beta = 1.5\theta$
Note: no internal work is done at supports A and G.

Internal Work Done = External Work Done
$M_{pl} (\theta + 0.5\theta) = (21.0 \times 1.5\theta) + (21.0 \times 1.5\theta)$
$1.5M_{pl}\theta = 63.0\theta$
$M_{pl} = 42.0$ kNm

Combinations:
Consider the independent mechanisms, their associated work equations and M_{pl} values as shown in Figure 8.16:

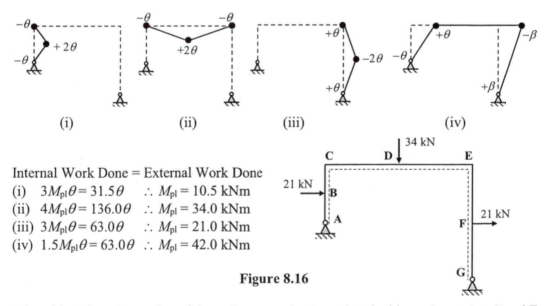

Internal Work Done = External Work Done
(i) $3M_{pl}\theta = 31.5\theta$ $\therefore M_{pl} = 10.5$ kNm
(ii) $4M_{pl}\theta = 136.0\theta$ $\therefore M_{pl} = 34.0$ kNm
(iii) $3M_{pl}\theta = 63.0\theta$ $\therefore M_{pl} = 21.0$ kNm
(iv) $1.5M_{pl}\theta = 63.0\theta$ $\therefore M_{pl} = 42.0$ kNm

Figure 8.16

It is evident from inspection of the collapse mechanisms that the hinges located at C and E can be eliminated since in some cases the rotation is negative whilst in others it is positive. The minimum number of hinges to induce total collapse is one more than the number of redundancies, i.e. $(I_D + 1) = 2$ and therefore the independent mechanisms should be combined to try and achieve this and at the same time maximize the associated M_{pl} value. It is unlikely that mechanism (i) will be included in the failure mechanism since its associated M_{pl} value is relatively small compared to the others. It is necessary to investigate several possibilities and confirm the resulting solution by checking that the bending moments do not exceed the plastic moment of resistance at any section.

Combination 1: Mechanism (v) = [(ii) + (iv)]
When combining these mechanisms the hinge at C will be eliminated and the resulting M_{pl} value can be determined by adding the work equations. It is necessary to allow for the removal of the hinge at C in the internal work done since in each equation an ($M_{pl}\theta$) term has been included, but the hinge no longer exists. A total of $2M_{pl}$ must therefore be subtracted from the resulting internal work, i.e.

Internal Work Done = External Work Done
Mechanism (ii) $4M_{pl}\theta = 136.0\theta$
Mechanism (iv) $1.5M_{pl}\theta = 63.0\theta$
less $2.0M_{pl}$ for eliminated hinge $-2.0M_{pl}\theta$
                                $3.5M_{pl}\theta = 199.0\theta$ $M_{pl} = 56.86$ kNm

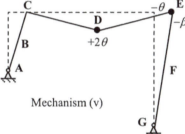

Mechanism (v)

It is possible that this is the true collapse mechanism, however this should be confirmed as indicated above by satisfying conditions (ii) and (iii) in Section 8.1.2.
An alternative solution is also possible where the hinges at C and E are eliminated, this can be a achieved if mechanism (v) is combined with mechanism (iii).

In mechanism (v) $\beta = 0.5\theta$ (see the sway calculation above) and hence the total rotation at joint E $= -(\theta + \beta) = -1.5\theta$. If this hinge is to be eliminated then the combinations of mechanisms (iii) and (v) must be in the proportions of 1.5:1.0. (**Note:** when developing mechanism (v) the proportions were 1:1).
The total value of the internal work for the eliminated hinge $= (2 \times 1.5M_{pl}) = 3.0M_{pl}$, i.e.

Internal Work Done = External Work Done
Mechanism 1.5 × (iii) $4.5M_{pl}\theta = 94.5\theta$
Mechanism (v) $3.5M_{pl}\theta = 199.0\theta$
less $3.0M_{pl}$ for eliminated hinge $-3.0M_{pl}\theta$
                                $5.0M_{pl}\theta = 293.5\theta$ $M_{pl} = 58.70$ kNm

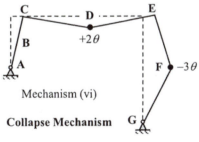

Mechanism (vi)

Collapse Mechanism

The +ve rotation indicates tension inside the frame at point D and the −ve rotation indicates tension outside the frame at point F.
This is marginally higher than the previous value and since there does not appear to be any other obvious collapse mechanism, this result should be checked as follows:

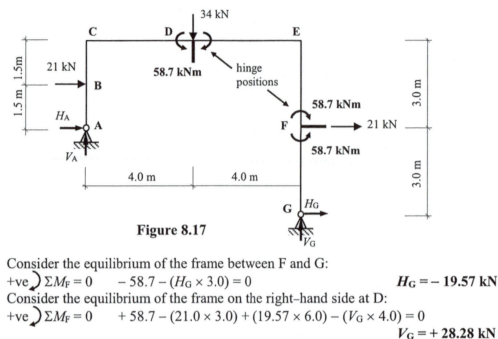

Figure 8.17

Consider the equilibrium of the frame between F and G:

$+ve \circlearrowright \Sigma M_F = 0$ $-58.7 - (H_G \times 3.0) = 0$ $\boldsymbol{H_G = -19.57 \text{ kN} \leftarrow}$

Consider the equilibrium of the frame on the right-hand side at D:

$+ve \circlearrowright \Sigma M_F = 0$ $+58.7 - (21.0 \times 3.0) + (19.57 \times 6.0) - (V_G \times 4.0) = 0$

$\boldsymbol{V_G = +28.28 \text{ kN} \uparrow}$

Consider the complete structure:

$+ve \uparrow \Sigma F_z = 0$ $V_A - 34.0 + 28.28 = 0$ $\boldsymbol{V_A = +5.72 \text{ kN} \uparrow}$

$+ve \longrightarrow \Sigma F_x = 0$ $H_A + 21.0 + 21.0 - 19.57 = 0$ $\boldsymbol{H_A = -22.43 \text{ kN} \leftarrow}$

Bending moment at B $M_B = + (22.43 \times 1.5) = +33.65 \text{ kNm} \le M_{pl}$

Bending moment at C $M_C = + (22.43 \times 3.0) - (21.0 \times 1.5) = +35.79 \text{ kNm} \le M_{pl}$

Bending moment at E $M_E = - (19.57 \times 6.0) + (21.0 \times 3.0) = -54.42 \text{ kNm} \le M_{pl}$

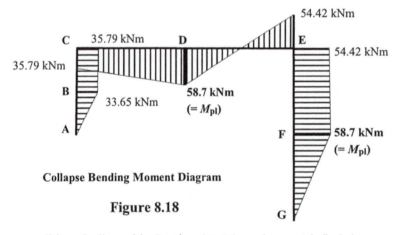

Collapse Bending Moment Diagram

Figure 8.18

The three conditions indicated in Section 8.1.2 have been satisfied: i.e.

Mechanism condition: minimum number of hinges required $= (I_D + 1) = 2$ hinges,

Equilibrium condition: the internal moments are in equilibrium with the collapse loads,

Yield condition: the bending moment does not exceed M_{pl} anywhere in the frame.

$M_{pl \text{ kinematic}} = M_{pl \text{ static}} = M_{pl \text{ true}}$

It is often convenient to carry–out the calculation of combinations using a table as shown in Table 8.1; eliminated hinges are indicated by EH in the Table.

Independent and Combined Mechanisms for Example 8.5						
Hinge Position	(i)	(ii)	(iii)	(iv)	(v) = (ii)+(iv)	(vi) = (v)+1.5(iii)
B (M_{pl})	+ 2.0θ	–	–	–	–	-
C (M_{pl})	– θ	– θ	–	+ θ	EH (2.0$M_{pl}\theta$)	EH (2.0$M_{pl}\theta$)
D (M_{pl})	–	+ 2.0θ	–	–	+ 2.0θ	+ 2.0θ
E (M_{pl})	–	– θ	+θ	– 0.5θ	–1.5θ	EH (3.0$M_{pl}\theta$)
F (M_{pl})	–	–	– 2.0θ	–	–	– 3.0θ
External Work	31.5θ	136.0θ	63.0θ	63.0θ	199.0θ	**293.5θ**
Internal Work	3.0$M_{pl}\theta$	4.0$M_{pl}\theta$	3.0$M_{pl}\theta$	1.5$M_{pl}\theta$	5.5$M_{pl}\theta$	10.0$M_{pl}\theta$
Eliminated hinges	–	–	–	–	2.0$M_{pl}\theta$	5.0$M_{pl}\theta$
Combined $M_{pl}\theta$	–	–	–	–	3.5$M_{pl}\theta$	**5.0$M_{pl}\theta$**
M_{pl} (kNm)	10.5	34.0	21.0	42.0	56.86	**58.70**

Table 8.1

Static Method:

This frame can also be analysed readily using the static method since it only has one degree–of–indeterminacy. When using this method the frame can be considered as the superposition of two frames; one statically determinate and one involving only the assumed redundant reaction as shown in Figure 8.19. Applying the three equations of equilibrium to the two force systems results in the support reactions indicated.

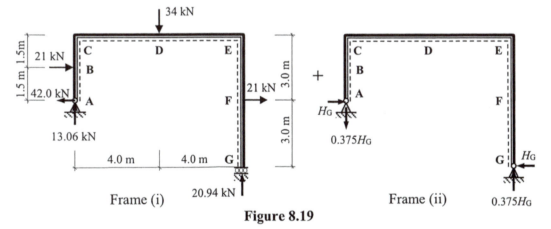

Figure 8.19

The final value of the reactions and bending moments = [Frame (i) + Frame (ii)]; e.g.

H_A = – 42.0 + H_G V_A = 13.06 – 0.375H_G V_G = 20.94 + 0.375H_G

H_G = 0 + H_G M_B = [$M_{B\ frame\ (i)}$ + $M_{B\ frame\ (ii)}$]   etc.

Equations can be developed for each of the five *possible* hinge positions in terms of the two frames as follows:

$M_B = + (42.0 \times 1.5) - (1.5H_G) = + 63.0 - 1.5H_G$ Equation (1)
$M_C = + (42.0 \times 3.0) - (21.0 \times 1.5) - (3.0H_G) = + 94.5 - 3.0H_G$ Equation (2)
$M_D = + (42.0 \times 3.0) + (13.06 \times 4.0) - (21.0 \times 1.5) - (3.0 \times H_G) - (4.0 \times 0.375H_G)$
      $= + 146.74 - 4.5H_G$ Equation (3)
$M_E = + (21.0 \times 3.0) - (6.0H_G) = + 63.0 - 6.0H_G$ Equation (4)
$M_F = - 3.0H_G$ Equation (5)

As indicated previously, only two hinges are required to induce total collapse. A collapse mechanism involving two hinge positions can be assumed and the associated equations will each have two unknown values, i.e. H_G and M_{pl} and can be solved simultaneously.
The value of the bending moment at all other hinge positions can then be checked to ensure that they do not exceed the calculated M_{pl} value. If any one does exceed the value then the assumed mechanism was incorrect and others can be checked until the true one is identified.

Assume a mechanism inducing hinges at D and E as in (v) above.

$+ 146.74 - 4.5H_G = + M_{pl}$ Equation (6)
$+ 63.0 - 6.0H_G = - M_{pl}$ Equation (7)

Add equations (6) and (7):

$+ 209.74 - 10.5H_G = 0$ $\therefore H_G = 19.98$ kN

and $M_{pl} = 56.83$ kNm

Check the value of the moments at all other possible hinge positions.

$M_B = + 63.0 - 1.5H_G = + 63.0 - (1.5 \times 19.98) = + 33.03$ kNm $\leq M_{pl}$
$M_C = + 94.5 - 3.0H_G = + 94.5 - (3.0 \times 19.98) = + 34.56$ kNm $\leq M_{pl}$
$M_F = - 3.0H_G = - (3.0 \times 19.98) = - 59.94$ **kNm** $> M_{pl}$

Since the bending moment at F is greater than M_{pl} this mechanism does not satisfy the 'yield condition' and produces an unsafe solution.

The reader should repeat the above calculation assuming hinges develop at positions D and F and confirm that the true solution is when $M_{pl} = $ **58.7 kNm** as determined previously using the kinematic method.

8.7   Problems: Plastic Analysis – Rigid-Jointed Frames 1

A series of rigid-jointed frames are indicated in Problems 8.6 to 8.9 in which the relative M_{pl} values and the applied **collapse** loads are given. In each case determine the required M_{pl} value, the value of the support reactions and sketch the bending moment diagram.

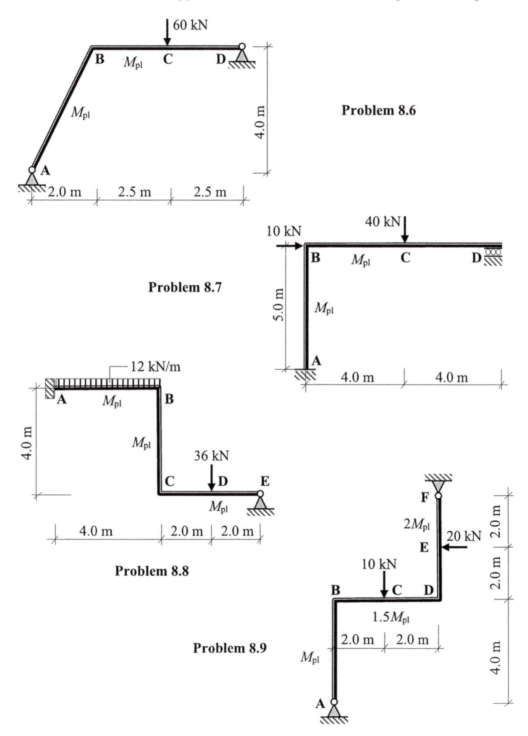

Solution

Topic: Plastic Analysis – Rigid Jointed Frames 1
Problem Number: 8.6 – Kinematic Method   Page No. 1

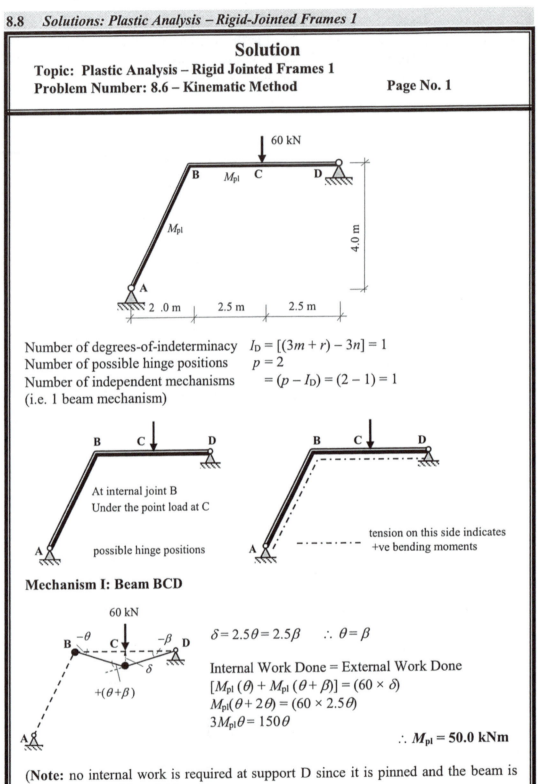

Number of degrees-of-indeterminacy $I_D = [(3m + r) - 3n] = 1$
Number of possible hinge positions $p = 2$
Number of independent mechanisms $= (p - I_D) = (2 - 1) = 1$
(i.e. 1 beam mechanism)

Mechanism I: Beam BCD

$\delta = 2.5\theta = 2.5\beta$ $\therefore \theta = \beta$

Internal Work Done = External Work Done
$[M_{pl}(\theta) + M_{pl}(\theta + \beta)] = (60 \times \delta)$
$M_{pl}(\theta + 2\theta) = (60 \times 2.5\theta)$
$3M_{pl}\theta = 150\theta$

$\therefore M_{pl} = 50.0$ kNm

(**Note:** no internal work is required at support D since it is pinned and the beam is free to rotate at this point.)

Solution

Topic: Plastic Analysis – Rigid Jointed Frames 1
Problem Number: 8.6 – Kinematic Method   Page No. 2

The value of M_{pl} obtained (50.0 kNm) should be checked by ensuring that the bending moment in the frame does not exceed this value at any location.

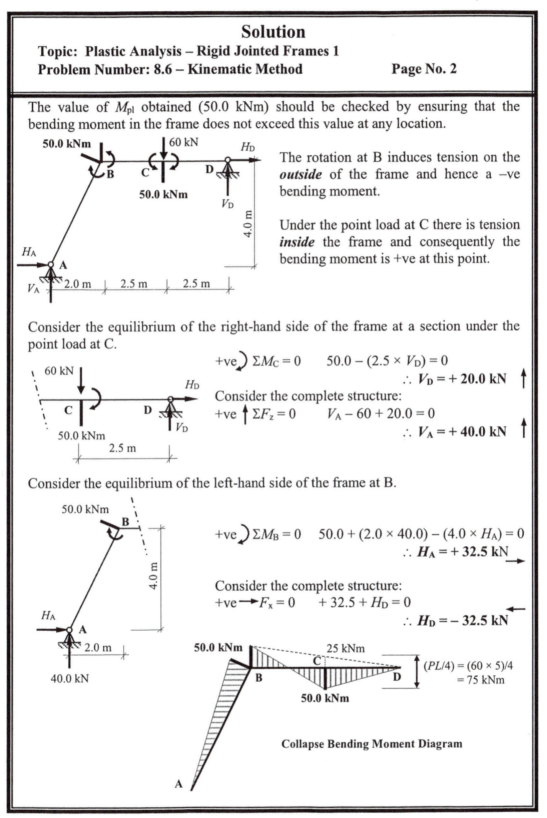

The rotation at B induces tension on the *outside* of the frame and hence a –ve bending moment.

Under the point load at C there is tension *inside* the frame and consequently the bending moment is +ve at this point.

Consider the equilibrium of the right-hand side of the frame at a section under the point load at C.

$$+ve \, \curvearrowright \, \Sigma M_C = 0 \qquad 50.0 - (2.5 \times V_D) = 0$$
$$\therefore V_D = +20.0 \text{ kN} \uparrow$$

Consider the complete structure:
$$+ve \uparrow \Sigma F_z = 0 \qquad V_A - 60 + 20.0 = 0$$
$$\therefore V_A = +40.0 \text{ kN} \uparrow$$

Consider the equilibrium of the left-hand side of the frame at B.

$$+ve \, \curvearrowright \, \Sigma M_B = 0 \qquad 50.0 + (2.0 \times 40.0) - (4.0 \times H_A) = 0$$
$$\therefore H_A = +32.5 \text{ kN} \longrightarrow$$

Consider the complete structure:
$$+ve \longrightarrow F_x = 0 \qquad +32.5 + H_D = 0$$
$$\therefore H_D = -32.5 \text{ kN} \longleftarrow$$

$$(PL/4) = (60 \times 5)/4$$
$$= 75 \text{ kNm}$$

Collapse Bending Moment Diagram

Solution

Topic: Plastic Analysis – Rigid Jointed Frames 1
Problem Number: 8.6 – Static Method   **Page No. 3**

Assume the horizontal component of reaction at support D to be the redundant reaction.

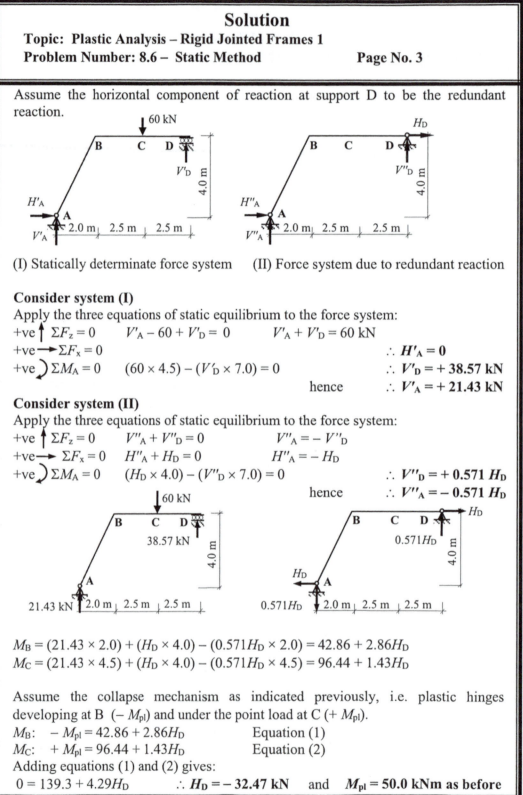

(I) Statically determinate force system   (II) Force system due to redundant reaction

Consider system (I)

Apply the three equations of static equilibrium to the force system:

$+ve \uparrow \Sigma F_z = 0$ $V'_A - 60 + V'_D = 0$ $V'_A + V'_D = 60$ kN

$+ve \rightarrow \Sigma F_x = 0$ $\therefore H'_A = 0$

$+ve \,\rangle\, \Sigma M_A = 0$ $(60 \times 4.5) - (V'_D \times 7.0) = 0$ $\therefore V'_D = + 38.57$ kN

                                    hence $\therefore V'_A = + 21.43$ kN

Consider system (II)

Apply the three equations of static equilibrium to the force system:

$+ve \uparrow \Sigma F_z = 0$ $V''_A + V''_D = 0$ $V''_A = - V''_D$

$+ve \rightarrow \Sigma F_x = 0$ $H''_A + H_D = 0$ $H''_A = - H_D$

$+ve \,\rangle\, \Sigma M_A = 0$ $(H_D \times 4.0) - (V''_D \times 7.0) = 0$ $\therefore V''_D = + 0.571 H_D$

                                    hence $\therefore V''_A = - 0.571 H_D$

$M_B = (21.43 \times 2.0) + (H_D \times 4.0) - (0.571 H_D \times 2.0) = 42.86 + 2.86 H_D$
$M_C = (21.43 \times 4.5) + (H_D \times 4.0) - (0.571 H_D \times 4.5) = 96.44 + 1.43 H_D$

Assume the collapse mechanism as indicated previously, i.e. plastic hinges developing at B $(- M_{pl})$ and under the point load at C $(+ M_{pl})$.

$M_B:$ $- M_{pl} = 42.86 + 2.86 H_D$ Equation (1)

$M_C:$ $+ M_{pl} = 96.44 + 1.43 H_D$ Equation (2)

Adding equations (1) and (2) gives:

$0 = 139.3 + 4.29 H_D$ $\therefore H_D = - 32.47$ kN   and $M_{pl} = 50.0$ **kNm as before**

Solution

Topic: Plastic Analysis – Rigid Jointed Frames 1
Problem Number: 8.7 – Kinematic Method   **Page No. 1**

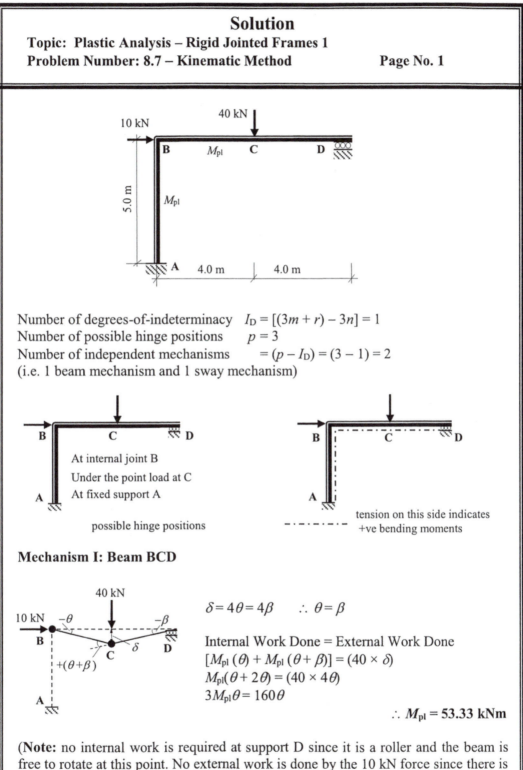

Number of degrees-of-indeterminacy $I_D = [(3m + r) - 3n] = 1$
Number of possible hinge positions $p = 3$
Number of independent mechanisms $= (p - I_D) = (3 - 1) = 2$
(i.e. 1 beam mechanism and 1 sway mechanism)

At internal joint B
Under the point load at C
At fixed support A

possible hinge positions

tension on this side indicates
+ve bending moments

Mechanism I: Beam BCD

$$\delta = 4\theta = 4\beta \qquad \therefore \ \theta = \beta$$

Internal Work Done = External Work Done
$$[M_{pl}\,(\theta) + M_{pl}\,(\theta + \beta)] = (40 \times \delta)$$
$$M_{pl}(\theta + 2\theta) = (40 \times 4\theta)$$
$$3M_{pl}\theta = 160\theta$$

$$\therefore \ M_{pl} = \textbf{53.33 kNm}$$

(**Note:** no internal work is required at support D since it is a roller and the beam is free to rotate at this point. No external work is done by the 10 kN force since there is no horizontal displacement of joint B)

Solution
Topic: **Plastic Analysis – Rigid Jointed Frames 1**
Problem Number: **8.7 – Kinematic Method**   Page No. 2

Mechanism II: Sway

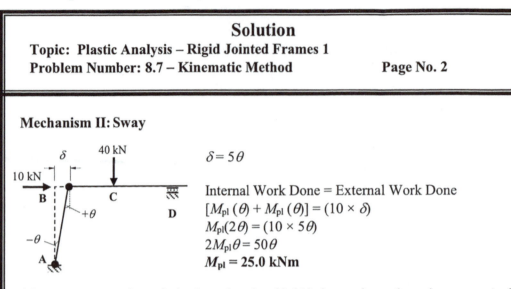

$\delta = 5\theta$

Internal Work Done = External Work Done
$[M_{pl}(\theta) + M_{pl}(\theta)] = (10 \times \delta)$
$M_{pl}(2\theta) = (10 \times 5\theta)$
$2M_{pl}\theta = 50\theta$
$M_{pl} = \textbf{25.0 kNm}$

(**Note:** no external work is done by the 40 kN force since there is no *vertical* displacement at C).

Mechanism III: Combined Beam & Sway

In this mechanism the two independent mechanisms I and II occur simultaneously to produce a collapse mechanism in which plastic hinges develop at A, and at C under the point load on beam BCD. The hinge at B is eliminated; note the −ve rotation in Mechanism I and the +ve rotation in Mechanism II at B which cancel each other out.

$\delta_1 = 5\theta$
$\delta_2 = 4\theta = 4\beta$ $\therefore \theta = \beta$

Internal Work Done = External Work Done
$[M_{pl}(\theta) + M_{pl}(\theta + \beta)] = [(10 \times \delta_1) + (40 \times \delta_2)]$
$M_{pl}(3\theta) = [(10 \times 5\theta) + (40 \times 4\theta)]$
$3M_{pl}\theta = 210\theta$
$M_{pl} = \textbf{70.0 kNm}$

The same result could have been achieved by adding, directly, the work equations for mechanisms I and II and subtracting for the internal work which no longer occurs at joint B; i.e. $M_{pl}\theta$ in each equation.

Adding equations for Mechanisms (I + II)
$3M_{pl}\theta = 160\theta$
$2M_{pl}\theta = 50\theta$
$\underline{-2M_{pl}\theta}$ (allowing for the hinge eliminated at joint B)
$3M_{pl}\theta = 210\theta$

$\therefore M_{pl} = \textbf{70.0 kNm}$

Solution
Topic: Plastic Analysis – Rigid Jointed Frames 1
Problem Number: 8.7 – Kinematic Method   **Page No. 3**

Mechanism I: Beam BCD M_{pl} = 53.33 kNm
Mechanism II: Sway M_{pl} = 25.0 kNm
Mechanism III: I & II Combined M_{pl} = 70.0 kNm

The maximum value of M_{pl} obtained (70.0 kNm) should be checked by ensuring that the bending moment in the frame does not exceed this value at any location.

The rotation at A induces tension on the **outside** of the frame and hence a –ve bending moment.

Under the point load at C there is tension **inside** the frame and consequently the bending moment is +ve at this point.

Consider the right-hand side of the frame at a section under the point load at C.

$$+ve \circlearrowleft \Sigma M_C = 0 \qquad 70.0 - (4.0 \times V_D) = 0$$
$$\therefore V_D = \textbf{17.5 kN} \uparrow$$

Consider the complete structure:
$$+ve \uparrow \Sigma F_z = 0 \qquad V_A - 40 + 17.5 = 0$$
$$\therefore V_A = \textbf{22.5 kN} \uparrow$$
$$+ve \rightarrow \Sigma F_x = 0 \qquad - H_A + 10.0 = 0$$
$$\therefore H_A = \textbf{10.0 kN} \leftarrow$$

Bending moment at B $M_B = - 70 + (5.0 \times 10.0) = - 20.0$ kNm $\leq M_{pl}$

$(PL/4) = (40 \times 8)/4$
$= 80$ kNm

Collapse Bending Moment Diagram

Solution

Topic: Plastic Analysis – Rigid Jointed Frames 1

Problem Number: 8.7 –   Static Method   **Page No. 4**

Assume the vertical component of reaction at support D to be the redundant reaction.

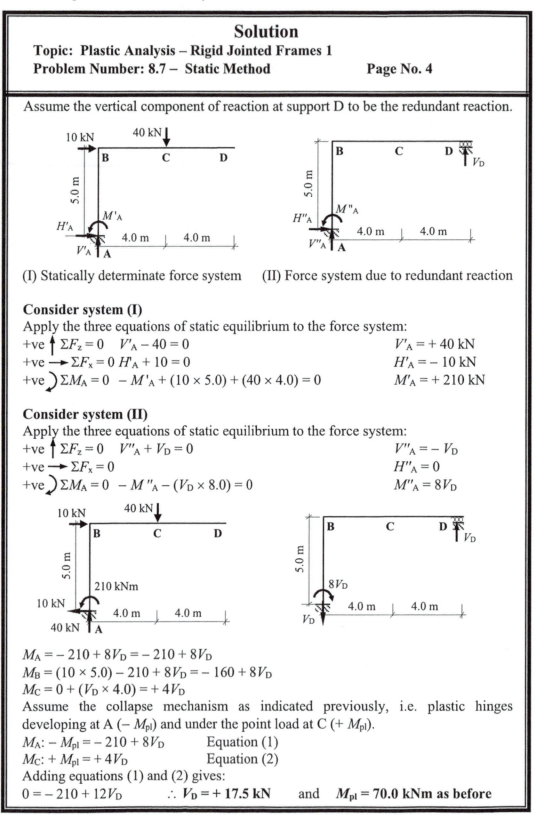

(I) Statically determinate force system   (II) Force system due to redundant reaction

Consider system (I)

Apply the three equations of static equilibrium to the force system:

+ve $\uparrow$ $\Sigma F_z = 0$ $V'_A - 40 = 0$ 　　　　　　　　　　　　$V'_A = + 40$ kN

+ve $\longrightarrow$ $\Sigma F_x = 0$ $H'_A + 10 = 0$ 　　　　　　　　　　$H'_A = - 10$ kN

+ve $\curvearrowright$ $\Sigma M_A = 0$ $- M'_A + (10 \times 5.0) + (40 \times 4.0) = 0$ 　　$M'_A = + 210$ kN

Consider system (II)

Apply the three equations of static equilibrium to the force system:

+ve $\uparrow$ $\Sigma F_z = 0$ $V''_A + V_D = 0$ 　　　　　　　　　　$V''_A = - V_D$

+ve $\longrightarrow$ $\Sigma F_x = 0$ 　　　　　　　　　　　　　　　　$H''_A = 0$

+ve $\curvearrowright$ $\Sigma M_A = 0$ $- M''_A - (V_D \times 8.0) = 0$ 　　　　$M''_A = 8V_D$

$M_A = - 210 + 8V_D = - 210 + 8V_D$

$M_B = (10 \times 5.0) - 210 + 8V_D = - 160 + 8V_D$

$M_C = 0 + (V_D \times 4.0) = + 4V_D$

Assume the collapse mechanism as indicated previously, i.e. plastic hinges developing at A $(- M_{pl})$ and under the point load at C $(+ M_{pl})$.

$M_A: - M_{pl} = - 210 + 8V_D$ 　　　　Equation (1)

$M_C: + M_{pl} = + 4V_D$ 　　　　　　Equation (2)

Adding equations (1) and (2) gives:

$0 = - 210 + 12V_D$ 　　　　$\therefore V_D = + 17.5$ kN 　　and 　**$M_{pl} = 70.0$ kNm as before**

Solution

Topic: Plastic Analysis – Rigid Jointed Frames 1
Problem Number: 8.8 – Kinematic Method   **Page No. 1**

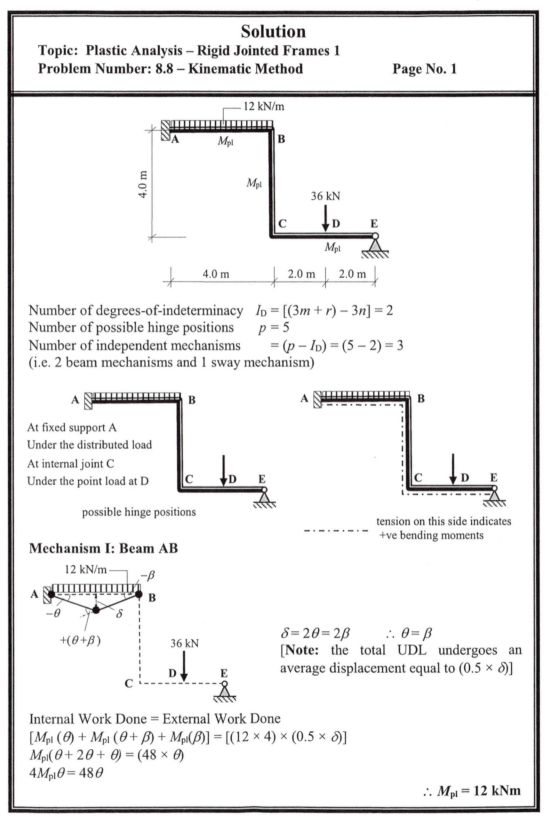

Number of degrees-of-indeterminacy $I_D = [(3m + r) - 3n] = 2$
Number of possible hinge positions $p = 5$
Number of independent mechanisms $= (p - I_D) = (5 - 2) = 3$
(i.e. 2 beam mechanisms and 1 sway mechanism)

At fixed support A
Under the distributed load
At internal joint C
Under the point load at D

possible hinge positions

tension on this side indicates
+ve bending moments

Mechanism I: Beam AB

$\delta = 2\theta = 2\beta$ $\therefore \theta = \beta$
[**Note:** the total UDL undergoes an average displacement equal to $(0.5 \times \delta)$]

Internal Work Done = External Work Done
$[M_{pl}(\theta) + M_{pl}(\theta + \beta) + M_{pl}(\beta)] = [(12 \times 4) \times (0.5 \times \delta)]$
$M_{pl}(\theta + 2\theta + \theta) = (48 \times \theta)$
$4M_{pl}\theta = 48\theta$

$\therefore M_{pl} = 12 \text{ kNm}$

Solution
Topic: Plastic Analysis – Rigid Jointed Frames 1
Problem Number: 8.8 – Kinematic Method   **Page No. 2**

Mechanism II: Beam CDE

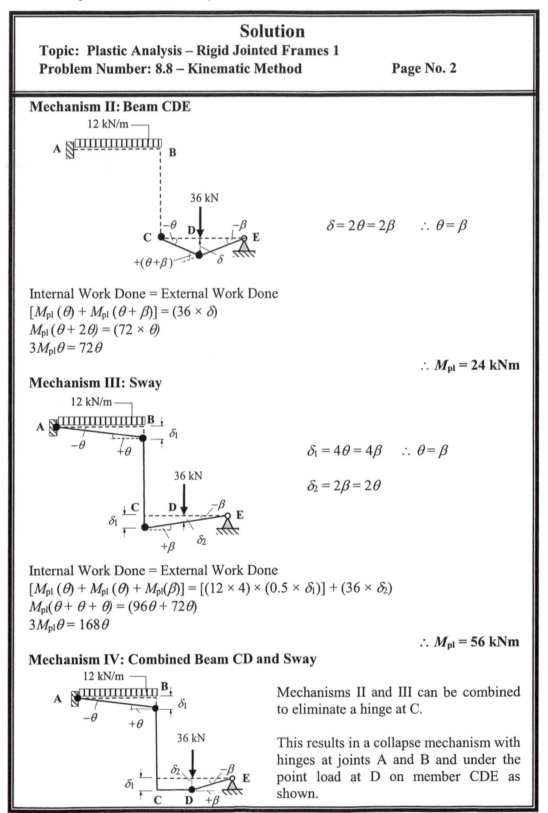

$$\delta = 2\theta = 2\beta \qquad \therefore \ \theta = \beta$$

Internal Work Done = External Work Done
$$[M_{pl}\,(\theta) + M_{pl}\,(\theta + \beta)] = (36 \times \delta)$$
$$M_{pl}\,(\theta + 2\theta) = (72 \times \theta)$$
$$3M_{pl}\theta = 72\theta$$

$$\therefore \ M_{pl} = 24 \text{ kNm}$$

Mechanism III: Sway

$$\delta_1 = 4\theta = 4\beta \quad \therefore \ \theta = \beta$$

$$\delta_2 = 2\beta = 2\theta$$

Internal Work Done = External Work Done
$$[M_{pl}\,(\theta) + M_{pl}\,(\theta) + M_{pl}(\beta)] = [(12 \times 4) \times (0.5 \times \delta_1)] + (36 \times \delta_2)$$
$$M_{pl}(\theta + \theta + \theta) = (96\theta + 72\theta)$$
$$3M_{pl}\theta = 168\theta$$

$$\therefore \ M_{pl} = 56 \text{ kNm}$$

Mechanism IV: Combined Beam CD and Sway

Mechanisms II and III can be combined to eliminate a hinge at C.

This results in a collapse mechanism with hinges at joints A and B and under the point load at D on member CDE as shown.

Solution

Topic: Plastic Analysis – Rigid Jointed Frames 1
Problem Number: 8.8 – Kinematic Method   **Page No. 3**

Adding work equations for Mechanisms (II + III)

$3M_{pl}\theta = 72\theta$

$3M_{pl}\theta = 168\theta$

$\underline{-2M_{pl}\theta}$ (allowing for the hinge eliminated at joint C)

$4M_{pl}\theta = 240\theta$

$$\therefore M_{pl} = 60.0 \text{ kNm}$$

Mechanism I: Beam AB $M_{pl} = 12.0$ kNm
Mechanism II: Beam CDE $M_{pl} = 24.0$ kNm
Mechanism III: Sway $M_{pl} = 56.0$ kNm
Mechanism IV: II & III Combined $M_{pl} = 60.0$ kNm

The maximum value of M_{pl} obtained (60.0 kNm) should be checked by ensuring that the bending moment in the frame does not exceed this value at any location.

The rotation at A induces tension on the *outside* of the frame and hence a –ve bending moment.

The rotation at B induces tension on the *inside* of the frame and hence a +ve bending moment.

Under the point load there is tension on the *underside* of beam CDE and consequently the bending moment is +ve at this point.

Consider the right-hand side of the frame at a section under the point load at D.

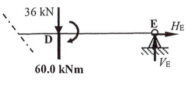

+ve $\circlearrowright$ $\Sigma M_D = 0$ $60.0 - (2.0 \times V_E) = 0$

$$\therefore V_E = 30.0 \text{ kN} \uparrow$$

Consider the complete structure:

+ve $\uparrow \Sigma F_z = 0$ $V_A - (12 \times 4) - 36 + 30.0 = 0$

$$\therefore V_A = 54.0 \text{ kN} \uparrow$$

Bending moment at C $M_C = [-(36 \times 2.0) + (30 \times 4.0)] = +48.0$ kNm $\leq M_{pl}$

Solution
Topic: Plastic Analysis – Rigid Jointed Frames 1
Problem Number: 8.8 – Kinematic Method   **Page No. 4**

Consider the right-hand side of the frame at a section at joint B.

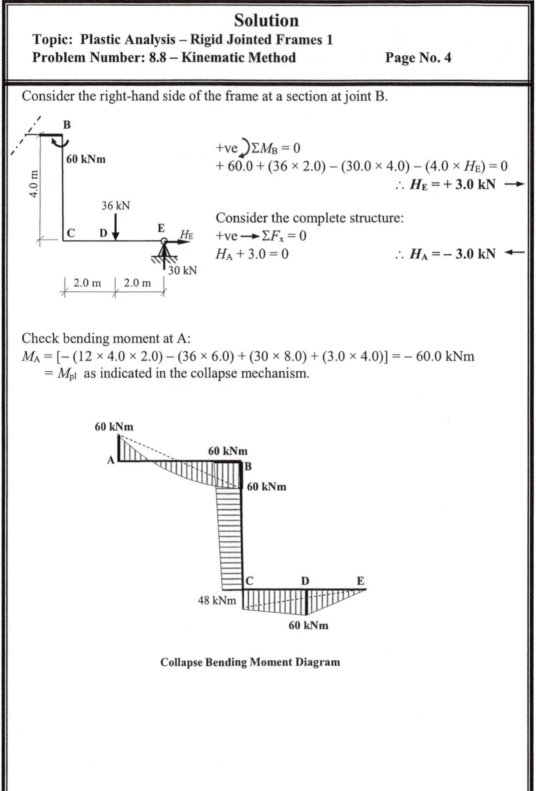

$+ve \; \curvearrowright \Sigma M_B = 0$
$+ 60.0 + (36 \times 2.0) - (30.0 \times 4.0) - (4.0 \times H_E) = 0$
$$\therefore \; H_E = + 3.0 \text{ kN} \longrightarrow$$

Consider the complete structure:
$+ve \longrightarrow \Sigma F_x = 0$
$H_A + 3.0 = 0$ $\therefore \; H_A = - 3.0 \text{ kN} \longleftarrow$

Check bending moment at A:
$M_A = [- (12 \times 4.0 \times 2.0) - (36 \times 6.0) + (30 \times 8.0) + (3.0 \times 4.0)] = - 60.0 \text{ kNm}$
  $= M_{pl}$ as indicated in the collapse mechanism.

Collapse Bending Moment Diagram

Solution

Topic: Plastic Analysis – Rigid Jointed Frames 1
Problem Number: 8.9 – Kinematic Method   **Page No. 1**

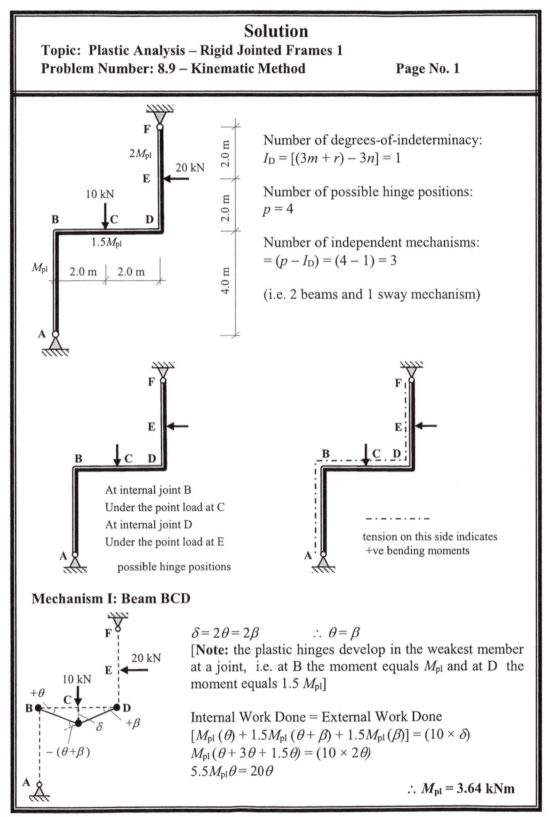

Number of degrees-of-indeterminacy:
$I_D = [(3m + r) - 3n] = 1$

Number of possible hinge positions:
$p = 4$

Number of independent mechanisms:
$= (p - I_D) = (4 - 1) = 3$

(i.e. 2 beams and 1 sway mechanism)

At internal joint B
Under the point load at C
At internal joint D
Under the point load at E

possible hinge positions

tension on this side indicates
+ve bending moments

Mechanism I: Beam BCD

$\delta = 2\theta = 2\beta$ $\therefore \theta = \beta$
[**Note:** the plastic hinges develop in the weakest member at a joint, i.e. at B the moment equals M_{pl} and at D the moment equals $1.5 M_{pl}$]

Internal Work Done = External Work Done
$[M_{pl}(\theta) + 1.5M_{pl}(\theta + \beta) + 1.5M_{pl}(\beta)] = (10 \times \delta)$
$M_{pl}(\theta + 3\theta + 1.5\theta) = (10 \times 2\theta)$
$5.5M_{pl}\theta = 20\theta$

$\therefore M_{pl} = 3.64 \text{ kNm}$

Solution
Topic: **Plastic Analysis – Rigid Jointed Frames 1**
Problem Number: **8.9 – Kinematic Method**   **Page No. 2**

Mechanism II: Beam DEF

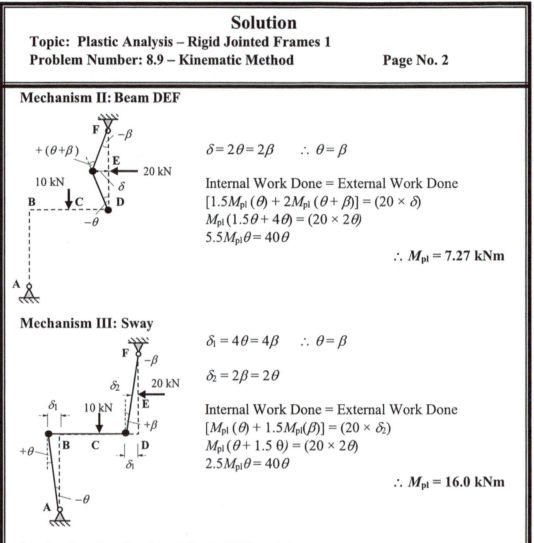

$\delta = 2\theta = 2\beta \qquad \therefore \ \theta = \beta$

Internal Work Done = External Work Done
$[1.5M_{pl}(\theta) + 2M_{pl}(\theta + \beta)] = (20 \times \delta)$
$M_{pl}(1.5\theta + 4\theta) = (20 \times 2\theta)$
$5.5M_{pl}\theta = 40\theta$

$\therefore \ M_{pl} = 7.27 \text{ kNm}$

Mechanism III: Sway

$\delta_1 = 4\theta = 4\beta \qquad \therefore \ \theta = \beta$

$\delta_2 = 2\beta = 2\theta$

Internal Work Done = External Work Done
$[M_{pl}(\theta) + 1.5M_{pl}(\beta)] = (20 \times \delta_2)$
$M_{pl}(\theta + 1.5\ \theta) = (20 \times 2\theta)$
$2.5M_{pl}\theta = 40\theta$

$\therefore \ M_{pl} = 16.0 \text{ kNm}$

Mechanism IV: Combined Beam DEF and Sway

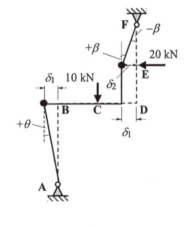

Mechanisms II and III can be combined to eliminate a hinge at D.
This results in a collapse mechanism with hinges at joint B and at E on member DEF as shown.

$\delta_1 = 4\theta = 2\beta \qquad \therefore \ \beta = 2\theta$
$\delta_2 = \delta_1 = 4\theta$

Internal Work Done = External Work Done
$[M_{pl}(\theta) + 2.0M_{pl}(\beta)] = (20 \times \delta_2)$
$M_{pl}(\theta + 4\theta) = (20 \times 4\theta)$
$5M_{pl}\theta = 80\theta$

$\therefore \ M_{pl} = 16.0 \text{ kNm}$

Solution
Topic: Plastic Analysis – Rigid Jointed Frames 1
Problem Number: 8.9 – Kinematic Method   **Page No. 3**

Adding work equations for Mechanisms (II + III)
$5.5M_{pl}\theta = 40\theta$
$2.5M_{pl}\theta = 40\theta$
$\underline{-3M_{pl}\theta}$ [allowing for the hinge eliminated at joint D i.e. $(2 \times 1.5M_{pl})$]
$5M_{pl}\theta = 80\theta$

$$\therefore M_{pl} = 16.0 \text{ kNm}$$

Mechanism I:	Beam BCD	M_{pl} = 3.64 kNm
Mechanism II:	Beam DEF	M_{pl} = 7.27 kNm
Mechanism III:	Sway	M_{pl} = 16.0 kNm
Mechanism IV:	II & III Combined	M_{pl} = 16.0 kNm

The maximum value of M_{pl} obtained (16.0 kNm) and should be checked by ensuring that the bending moment in the frame does not exceed this value at any location. Assume the combined mechanism is the failure mode.

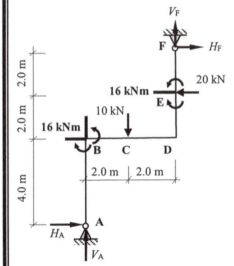

The rotation at B induces tension on the *left-hand side* of column AB and on the *top* of beam BCD and hence a +ve bending moment.

At E there is tension on the *left-hand side* of the frame and hence a +ve bending moment.

Consider the left-hand side of the frame at a section at joint B.

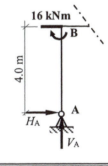

$+ve \curvearrowright \Sigma M_B = 0 + 16.0 - (4.0 \times H_A) = 0$
$$\therefore H_A = 4.0 \text{ kN} \longrightarrow$$

Consider the complete structure:
$+ve \longrightarrow \Sigma F_x = 0 \qquad H_F - 20 + 4.0 = 0$
$$\therefore H_F = 16.0 \text{ kN} \longrightarrow$$

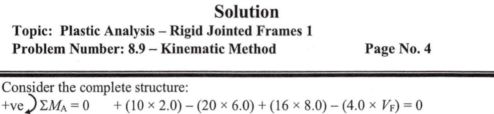

Solution
Topic: Plastic Analysis – Rigid Jointed Frames 1
Problem Number: 8.9 – Kinematic Method   **Page No. 4**

Consider the complete structure:

$+ve \curvearrowleft \Sigma M_A = 0$ $+ (10 \times 2.0) - (20 \times 6.0) + (16 \times 8.0) - (4.0 \times V_F) = 0$
$$\therefore V_F = + 7.0 \text{ kN} \uparrow$$
$+ve \uparrow \Sigma F_z = 0$ $V_A - 10.0 + 7.0 = 0$ $\therefore V_A = + 3.0 \text{ kN} \uparrow$

Consider the right-hand side of the frame at a section at joint D.

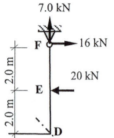

Bending moment at D $M_D = + (H_F \times 4.0) - (20 \times 2.0) = 0$
$$M_D = [(16.0 \times 4.0) - 40.0]$$
$$= + 24.0 \text{ kNm} = 1.5 M_{pl}$$

[**Note:** the bending moment at D is compared to the minimum M_{pl} value at the joint, i.e. $1.5 M_{pl}$. In this case since $M_D = 1.5 M_{pl}$ there is also a plastic hinge at joint D.]

Consider the left-hand side of the frame at a section under the point load at C on member BCD.

Bending moment at C:
$$M_C = [+ (4.0 \times 4.0) - (3.0 \times 2.0)]$$
$$= + 16.0 - 6.0$$
$$= + 10.0 \text{ kNm} \leq 1.5 M_{pl}$$

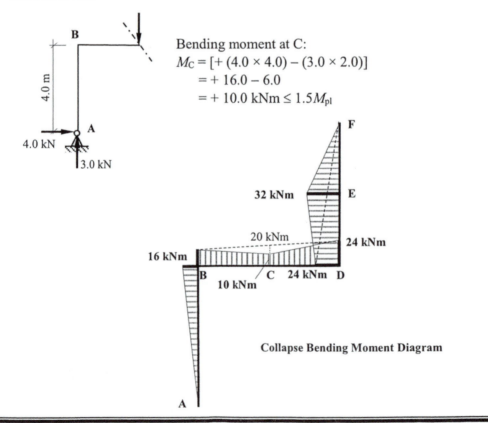

Collapse Bending Moment Diagram

Solution

Topic: Plastic Analysis – Rigid Jointed Frames 1

Problem Number: 8.9 – Static Method   **Page No. 5**

Assume the horizontal component of reaction at support F to be the redundant reaction.

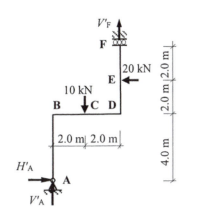

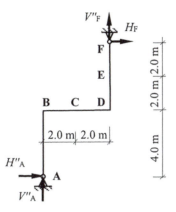

(I) Statically determinate force system   (II) Force system due to redundant reaction

Consider system (I)

Apply the three equations of static equilibrium to the force system:

+ve $\uparrow \Sigma F_z = 0$ $V'_A - 10 + V'_F = 0$

$$V'_A + V'_F = + 10 \text{ kN}$$

+ve $\longrightarrow \Sigma F_x = 0$ $H'_A - 20 = 0$

$$H'_A = + 20 \text{ kN}$$

+ve $\curvearrowright \Sigma M_A = 0$ $+ (10 \times 2.0) - (20 \times 6.0) - (V'_F \times 4.0) = 0$

$$V'_F = - 25 \text{ kN}$$
$$\therefore V'_A = + 35 \text{ kN}$$

Consider system (II)

Apply the three equations of static equilibrium to the force system:

+ve $\uparrow \Sigma F_z = 0$ $V''_A + V''_F = 0$

$$V''_A = - V''_F$$

+ve $\longrightarrow \Sigma F_x = 0$ $H''_A + H_F = 0$

$$H''_A = - H_F$$

+ve $\curvearrowright \Sigma M_A = 0$ $- (4.0 \times V''_F) + (8.0 \times H_F) = 0$

$$V''_F = + 2H_F$$
$$\therefore V''_A = - 2H_F$$

Solution

Topic: Plastic Analysis – Rigid Jointed Frames 1
Problem Number: 8.9 – Static Method   **Page No. 6**

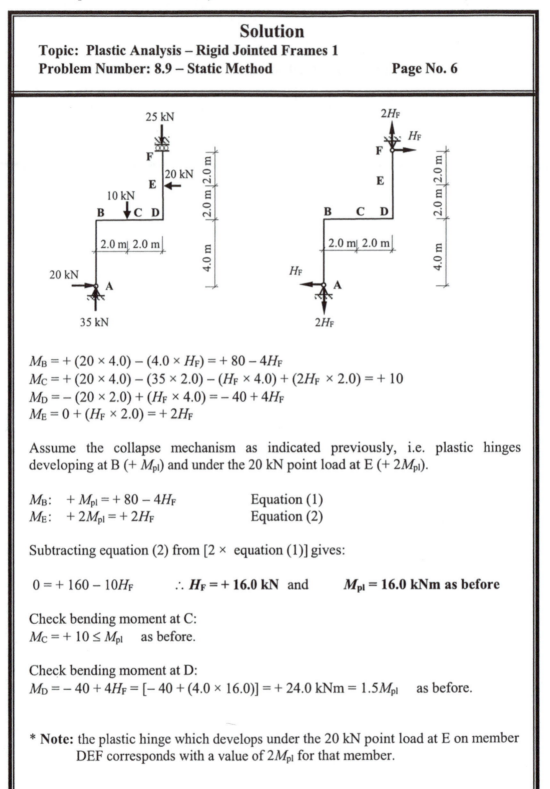

$M_B = + (20 \times 4.0) - (4.0 \times H_F) = + 80 - 4H_F$

$M_C = + (20 \times 4.0) - (35 \times 2.0) - (H_F \times 4.0) + (2H_F \times 2.0) = + 10$

$M_D = - (20 \times 2.0) + (H_F \times 4.0) = - 40 + 4H_F$

$M_E = 0 + (H_F \times 2.0) = + 2H_F$

Assume the collapse mechanism as indicated previously, i.e. plastic hinges developing at B ($+ M_{pl}$) and under the 20 kN point load at E ($+ 2M_{pl}$).

$M_B: \quad + M_{pl} = + 80 - 4H_F$ Equation (1)

$M_E: \quad + 2M_{pl} = + 2H_F$ Equation (2)

Subtracting equation (2) from [2 × equation (1)] gives:

$0 = + 160 - 10H_F$ $\therefore H_F = + 16.0$ kN   and   **$M_{pl} = 16.0$ kNm as before**

Check bending moment at C:

$M_C = + 10 \leq M_{pl}$ as before.

Check bending moment at D:

$M_D = - 40 + 4H_F = [- 40 + (4.0 \times 16.0)] = + 24.0$ kNm $= 1.5M_{pl}$ as before.

* Note: the plastic hinge which develops under the 20 kN point load at E on member DEF corresponds with a value of $2M_{pl}$ for that member.

8.9 Example 8.6: Joint Mechanism

In framed structures where there are more than two members meeting at a joint there is the possibility of a joint mechanism developing within a collapse mechanism. Consider the frame shown in Figure 8.20 with the collapse loads indicated. At joint C individual hinges can develop in members CBA, CDE and CFG giving three possible hinge positions at the joint in addition to positions B, D, F and G.

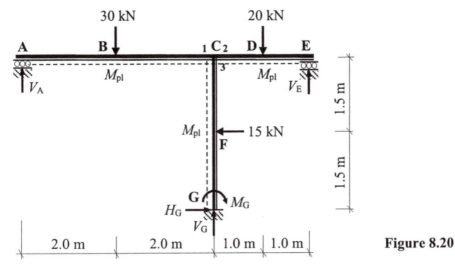

Figure 8.20

Factored loads: as given

Number of degrees-of-indeterminacy $I_D = [(3m + r) - 3n] = [(3 \times 3) + 5) - (3 \times 4)] = 2$
Number of possible hinge positions $p = 7$ (B, C_1, C_2, C_3, D, F and G)
Number of independent mechanisms $= (p - I_D) = (7 - 2) = 5$
(i.e. 3 beam mechanisms, 1 sway mechanism and 1 joint mechanism).

Kinematic Method:
Consider each independent mechanism separately.

Mechanism (i): Beam ABC

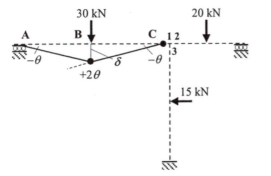

$\delta = 2.0\theta$
Note: no internal work is done at support A

Internal Work Done = External Work Done
$M_{pl}(2\theta + \theta) = (30.0 \times 2.0\theta)$
$3M_{pl}\theta = 60.0\theta$
$M_{pl} = 20.0$ kNm

The hinge at joint C is assumed to develop in member ABC at C_1.

Mechanism (ii): Beam CDE

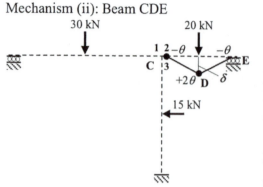

$\delta = 1.0\theta$

Note: no internal work is done at support E
Internal Work Done = External Work Done
$M_{pl}(2\theta + \theta) = (20.0 \times 1.0\theta)$
$3M_{pl}\theta = 20.0\theta$
$M_{pl} = 6.67$ kNm
The hinge at joint C is assumed to develop in member CDE at C_2.

Mechanism (iii): Beam CFG

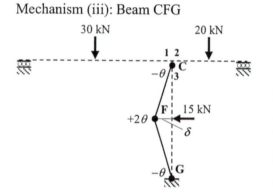

$\delta = 1.5\theta$

Internal Work Done = External Work Done
$M_{pl}(\theta + 2\theta + \theta) = (15.0 \times 1.5\theta)$
$4M_{pl}\theta = 22.5\theta$
$M_{pl} = 5.63$ kNm
The hinge at joint C is assumed to develop in member CFG at C_3.

Mechanism (iv): Sway

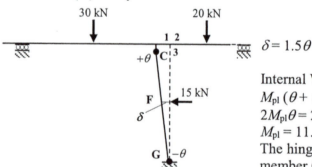

$\delta = 1.5\theta$

Internal Work Done = External Work Done
$M_{pl}(\theta + \theta) = (15.0 \times 1.5\theta)$
$2M_{pl}\theta = 22.5\theta$
$M_{pl} = 11.25$ kNm
The hinge at joint C is assumed to develop in member CFG at C_3.

Mechanism (v): Joint
The joint at C can rotate either in a clockwise direction or an anti–clockwise direction.

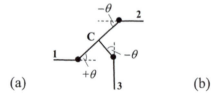

(a)

(b)

Internal Work Done = $M_{pl}(\theta + \theta + \theta) = 3M_{pl}\theta$ External Work Done = zero

The independent mechanisms can be entered into a table as before and the possible combinations investigated.

In this example $I_D = 2$ and consequently a minimum of three hinges is required to induce total collapse.

Since mechanisms (i) and (iv) have a significantly higher associated M_{pl} value these have been selected to combine with the joint mechanism to produce a possible combination:

Mechanism (vi): the addition of mechanisms (i) + (iv) + (v)(a)

Independent and Combined Mechanisms for Example 8.6							
Hinge Position	(i)	(ii)	(iii)	(iv)	(v)		(vi) = (i)+(iv)+(v)(a)
B (M_{pl})	$+2.0\theta$	$-$	$-$	$-$	(a)	(b)	$+2.0\theta$
C_1 (M_{pl})	$-\theta$	$-$	$-$	$-$	$+\theta$	$-\theta$	**EH** ($2.0M_{pl}\theta$)
C_2 (M_{pl})		$-\theta$		$-$	$-\theta$	$+\theta$	$-\theta$
C_3 (M_{pl})		$-$	$-\theta$	$+\theta$	$-\theta$	$+\theta$	**EH** ($2.0M_{pl}\theta$)
D (M_{pl})	$-$	$+2.0\theta$	$-$	$-$	$-$		$-$
F (M_{pl})	$-$	$-$	$+2.0\theta$	$-$	$-$		$-$
G (M_{pl})	$-$	$-$	$-\theta$	$-\theta$	$-$		$-\theta$
External Work	60.0θ	20.0θ	22.5θ	22.5θ	$-$		**82.5θ**
Internal Work	$3.0M_{pl}\theta$	$3.0M_{pl}\theta$	$4.0M_{pl}\theta$	$2.0M_{pl}\theta$	$3.0M_{pl}\theta$		$8.0M_{pl}\theta$
Eliminated hinges	$-$	$-$	$-$	$-$	$-$		$4.0M_{pl}\theta$
Combined $M_{pl}\theta$	$-$	$-$	$-$	$-$	$-$		**$4.0M_{pl}\theta$**
M_{pl} (kNm)	20.0	6.67	5.63	11.25	$-$		**20.63**

Table 8.2

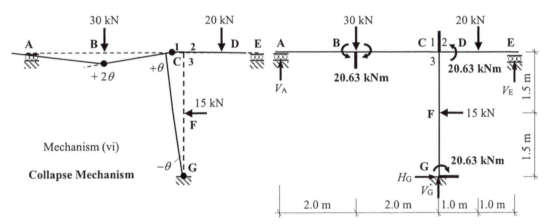

Figure 8.21

Consider the equilibrium of the frame on the left–hand side at B:

$+ve \, \circlearrowleft \, \Sigma M_B = 0 \qquad -20.63 + (V_A \times 2.0) = 0 \qquad \therefore V_A = +\,10.32\ \text{kN} \uparrow$

Consider the equilibrium of the frame on the right–hand side at C_2:

$+ve \, \circlearrowright \, \Sigma M_{C2} = 0 \qquad -20.63 + (20.0 \times 1.0) - (V_E \times 2.0) = 0 \qquad \therefore V_E = -\,0.32\ \text{kN} \downarrow$

Consider the complete structure:

$+ve \uparrow \Sigma F_z = 0 \qquad +10.32 - 30.0 - 20.0 - 0.32 + V_G = 0 \qquad \therefore V_G = +\,40.0\ \text{kN} \uparrow$

$+ve \longrightarrow \Sigma F_x = 0 \qquad H_G - 15.0 = 0 \qquad \therefore H_G = +\,15.0\ \text{kN} \longrightarrow$

Bending moment at C_1 $M_{C1} = +(10.32 \times 4.0) - (30.0 \times 2.0) = -\,18.72\ \text{kNm} \le M_{pl}$

Bending moment at C_3 $M_{C3} = +(15.0 \times 3.0) - (15.0 \times 1.5) - 20.63 = +\,1.87\ \text{kNm} \le M_{pl}$

Bending moment at D $M_D = -(0.32 \times 1.0) = -\,0.32\ \text{kNm} \le M_{pl}$

Bending moment at F $M_F = +(15.0 \times 1.5) - 20.63 = +\,1.87\ \text{kNm} \le M_{pl}$

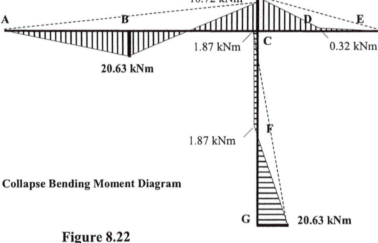

Collapse Bending Moment Diagram

Figure 8.22

The three conditions indicated in Section 8.1.2 have been satisfied: i.e.

Mechanism condition: minimum number of hinges required $= (I_D + 1) = 3$ hinges,

Equilibrium condition: the internal moments are in equilibrium with the collapse loads,

Yield condition: the bending moment does not exceed M_{pl} anywhere in the frame.

$M_{pl\ kinematic} = M_{pl\ static} = M_{pl\ true}$

8.10   *Problems: Plastic Analysis – Rigid-Jointed Frames 2*

A series of rigid-jointed frames are indicated in Problems 8.10 to 8.15 in which the relative M_{pl} values and the applied **collapse** loads are given. In each case determine the required M_{pl} value, the value of the support reactions and sketch the bending moment diagram.

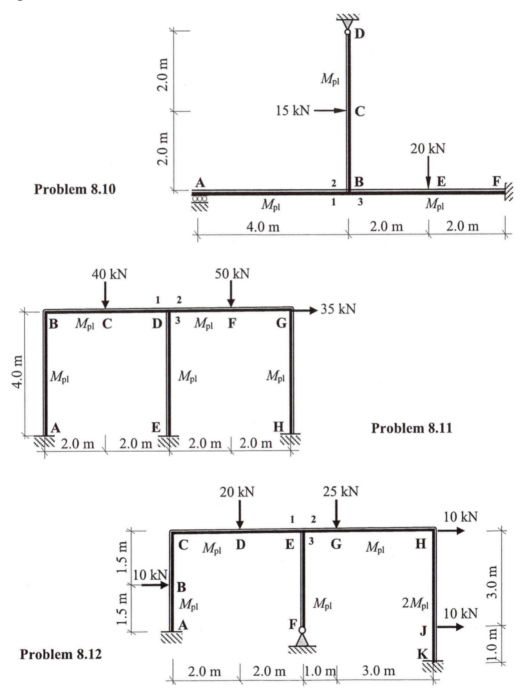

Problem 8.10

Problem 8.11

Problem 8.12

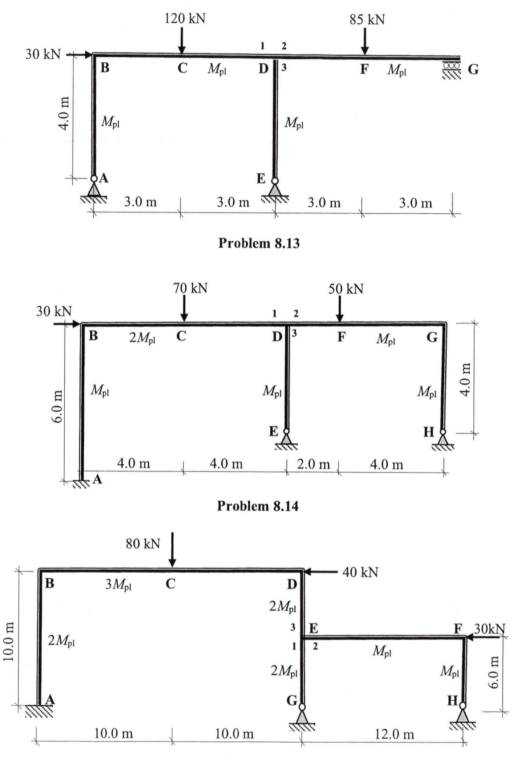

Problem 8.13

Problem 8.14

Problem 8.15

8.11   *Solutions: Plastic Analysis – Rigid-Jointed Frames 2*

Solution
Topic: Plastic Analysis – Rigid Jointed Frames 2
Problem Number: 8.10 – Kinematic Method   **Page No. 1**

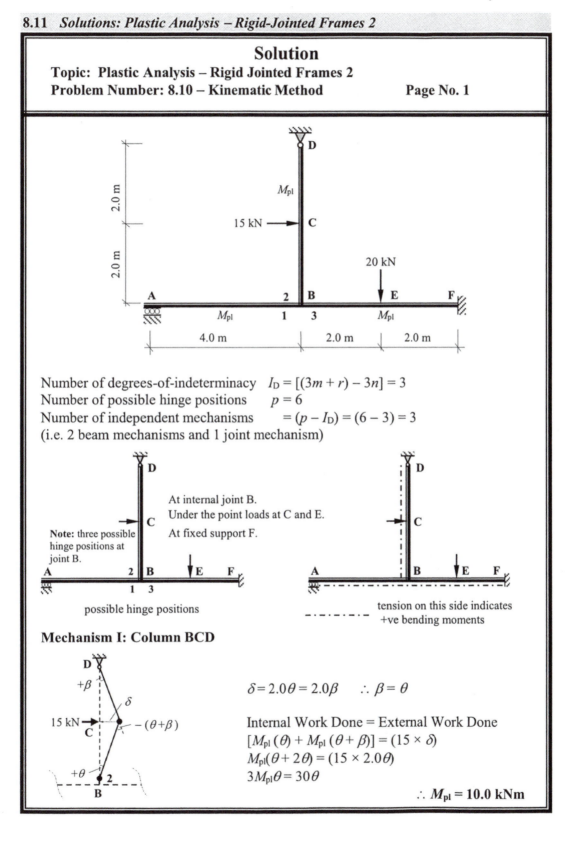

Number of degrees-of-indeterminacy $I_D = [(3m + r) - 3n] = 3$
Number of possible hinge positions $p = 6$
Number of independent mechanisms $= (p - I_D) = (6 - 3) = 3$
(i.e. 2 beam mechanisms and 1 joint mechanism)

At internal joint B.
Under the point loads at C and E.
At fixed support F.

Note: three possible hinge positions at joint B.

possible hinge positions

tension on this side indicates
+ve bending moments

Mechanism I: Column BCD

$\delta = 2.0\theta = 2.0\beta$ $\therefore \beta = \theta$

Internal Work Done = External Work Done
$[M_{pl} (\theta) + M_{pl} (\theta + \beta)] = (15 \times \delta)$
$M_{pl}(\theta + 2\theta) = (15 \times 2.0\theta)$
$3M_{pl}\theta = 30\theta$

$\therefore M_{pl} = 10.0$ **kNm**

Solution
Topic: Plastic Analysis – Rigid Jointed Frames 2
Problem Number: 8.10 – Kinematic Method   **Page No. 2**

Mechanism II: Beam BEF

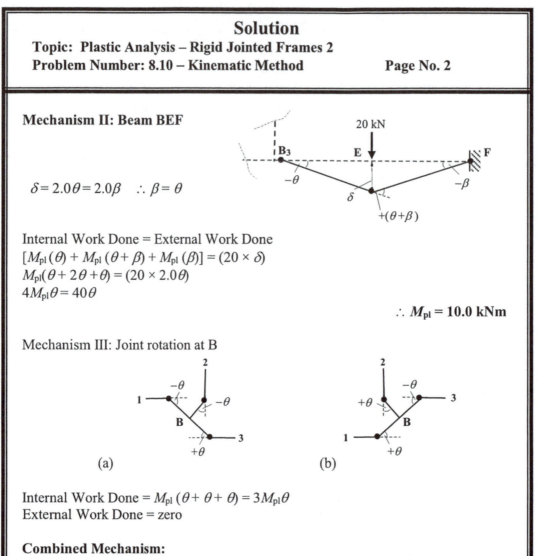

$\delta = 2.0\theta = 2.0\beta$ $\therefore \beta = \theta$

Internal Work Done = External Work Done
$[M_{pl}(\theta) + M_{pl}(\theta + \beta) + M_{pl}(\beta)] = (20 \times \delta)$
$M_{pl}(\theta + 2\theta + \theta) = (20 \times 2.0\theta)$
$4M_{pl}\theta = 40\theta$

$\therefore M_{pl} = 10.0$ kNm

Mechanism III: Joint rotation at B

(a)   (b)

Internal Work Done = $M_{pl}(\theta + \theta + \theta) = 3M_{pl}\theta$
External Work Done = zero

Combined Mechanism:
The independent mechanisms are combined to determine the maximum M_{pl} value required to induce collapse with the minimum number of hinges, (i.e. $I_D + 1$).
In this case the following combination has been evaluated
Mechanism IV = Mechanism I + Mechanism II + Mechanism III(a) eliminating hinges at B_2 and B_3, (see Table for the combinations).

Adding equations for Mechanisms [I + II + III(a)]
$3M_{pl}\theta = 30\theta$
$4M_{pl}\theta = 40\theta$
$3M_{pl}\theta = 0$
$-2M_{pl}\theta$ (allowing for the hinge eliminated at joint B_2)
$\underline{-2M_{pl}\theta}$ (allowing for the hinge eliminated at joint B_3)
$6M_{pl}\theta = 70\theta$ $\therefore M_{pl} = 11.67$ kNm

Solution
Topic: Plastic Analysis – Rigid Jointed Frames 2
Problem Number: 8.10 – Kinematic Method   **Page No. 3**

Hinge Position	Independent Mechanisms				Combined Mechanism
	I	II	III		IV
			(a)	(b)	
$B_1, (M_{pl})$			$-\theta$	$+\theta$	$-\theta$
$B_2, (M_{pl})$	$+\theta$		$-\theta$	$+\theta$	$\mathbf{EH}\ (2M_{pl}\theta)$
$B_3, (M_{pl})$		$-\theta$	$+\theta$	$-\theta$	$\mathbf{EH}\ (2M_{pl}\theta)$
$C, (M_{pl})$	-2θ				-2θ
$E, (M_{pl})$		$+2\theta$			$+2\theta$
$F, (M_{pl})$		$-\theta$			$-\theta$
$H, (M_{pl})$					
External work done	30.0θ	40.0θ			70.0θ
Internal work done	$3M_{pl}\theta$	$4M_{pl}\theta$	$3M_{pl}\theta$		$10M_{pl}\theta$
Eliminated hinges					$4M_{pl}\theta$
Combined internal work done					$6M_{pl}\theta$
M_{pl} (kNm)	10.0	10.0			11.67

Check collapse mechanism IV with hinges at B_1, C, E and F, (i.e. 4 hinges)

The value of M_{pl} obtained (11.67 kNm) should be checked by ensuring that the bending moment in the frame does not exceed the relevant M_{pl} value at any location.

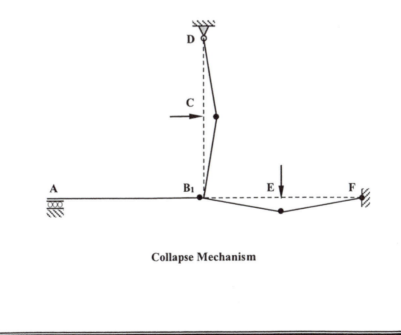

Collapse Mechanism

Solution
Topic: Plastic Analysis – Rigid Jointed Frames 2
Problem Number: 8.10 – Kinematic Method   **Page No. 4**

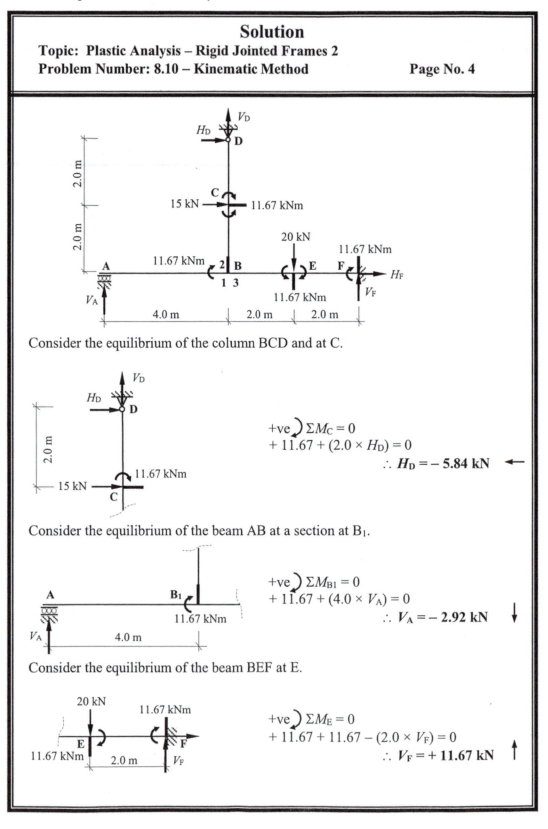

Consider the equilibrium of the column BCD and at C.

$$+ve \, \Sigma M_C = 0$$
$$+ 11.67 + (2.0 \times H_D) = 0$$
$$\therefore H_D = - 5.84 \text{ kN} \leftarrow$$

Consider the equilibrium of the beam AB at a section at B_1.

$$+ve \, \Sigma M_{B1} = 0$$
$$+ 11.67 + (4.0 \times V_A) = 0$$
$$\therefore V_A = - 2.92 \text{ kN} \downarrow$$

Consider the equilibrium of the beam BEF at E.

$$+ve \, \Sigma M_E = 0$$
$$+ 11.67 + 11.67 - (2.0 \times V_F) = 0$$
$$\therefore V_F = + 11.67 \text{ kN} \uparrow$$

Solution
Topic: Plastic Analysis – Rigid Jointed Frames 2
Problem Number: 8.10 – Kinematic Method   **Page No. 5**

Consider the vertical and horizontal equilibrium of the complete structure.

+ve ↑ $\Sigma F_z = 0$ $V_A + V_D + V_F - 20 = 0$
                  $- 2.92 + V_D + 11.67 - 20 = 0$ ∴ $V_D = + 11.25$ kN ↑

+ve → $\Sigma F_x = 0$ $H_D + H_F + 15 = 0$
                  $- 5.84 + H_F + 15 = 0$ ∴ $H_F = - 9.16$ kN ←

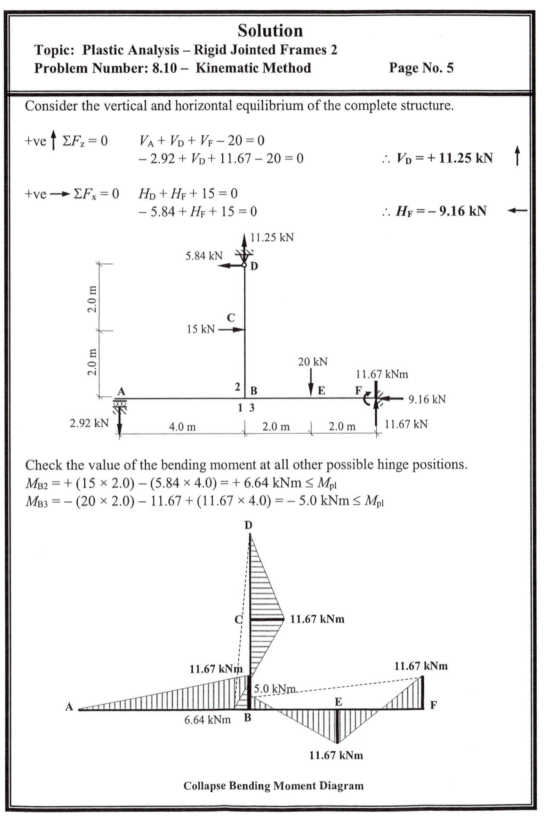

Check the value of the bending moment at all other possible hinge positions.
$M_{B2} = + (15 \times 2.0) - (5.84 \times 4.0) = + 6.64$ kNm $\leq M_{pl}$
$M_{B3} = - (20 \times 2.0) - 11.67 + (11.67 \times 4.0) = - 5.0$ kNm $\leq M_{pl}$

Collapse Bending Moment Diagram

Solution
Topic: Plastic Analysis – Rigid Jointed Frames 2
Problem Number: 8.11 – Kinematic Method   **Page No. 1**

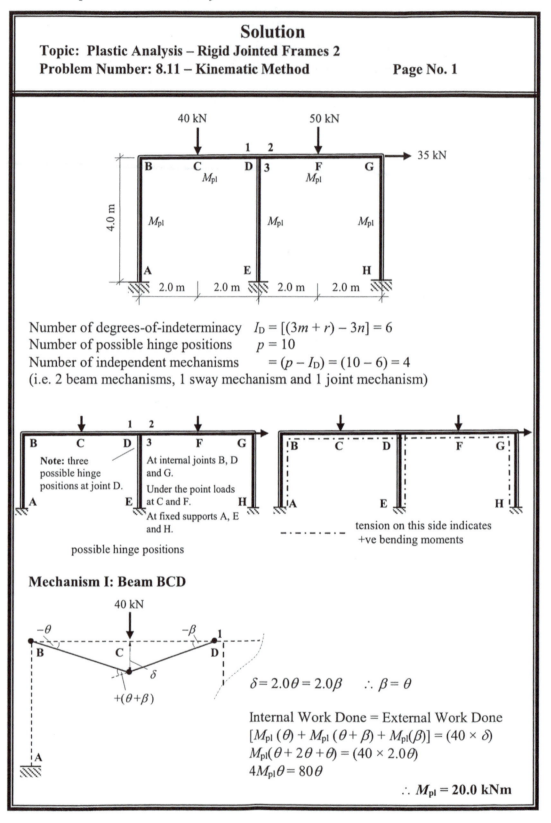

Number of degrees-of-indeterminacy $I_D = [(3m + r) - 3n] = 6$
Number of possible hinge positions $p = 10$
Number of independent mechanisms $= (p - I_D) = (10 - 6) = 4$
(i.e. 2 beam mechanisms, 1 sway mechanism and 1 joint mechanism)

Mechanism I: Beam BCD

$\delta = 2.0\theta = 2.0\beta$ $\therefore \beta = \theta$

Internal Work Done = External Work Done
$[M_{pl}\,(\theta) + M_{pl}\,(\theta + \beta) + M_{pl}(\beta)] = (40 \times \delta)$
$M_{pl}(\theta + 2\theta + \theta) = (40 \times 2.0\theta)$
$4M_{pl}\theta = 80\theta$

$\therefore M_{pl} = \textbf{20.0 kNm}$

Solution

Topic: Plastic Analysis – Rigid Jointed Frames 2
Problem Number: 8.11 – Kinematic Method   **Page No. 2**

Mechanism II: Beam DFG

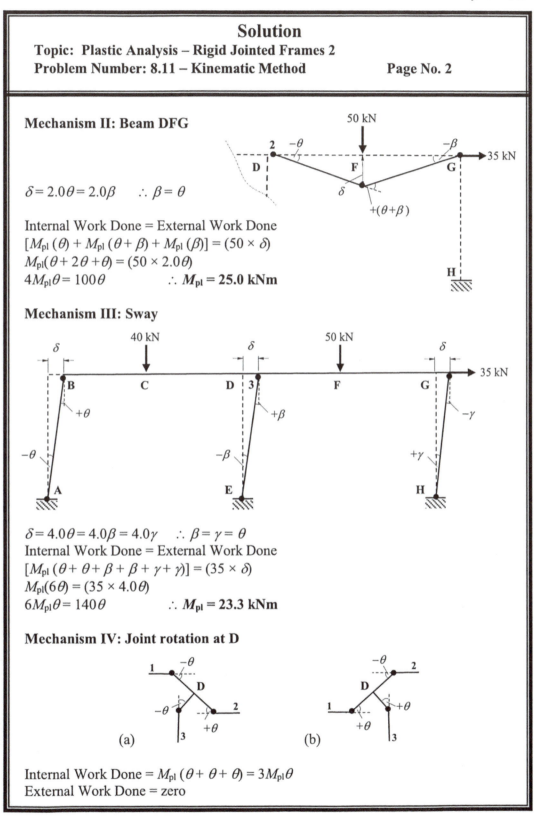

$\delta = 2.0\theta = 2.0\beta \quad \therefore \beta = \theta$

Internal Work Done = External Work Done
$[M_{pl}(\theta) + M_{pl}(\theta + \beta) + M_{pl}(\beta)] = (50 \times \delta)$
$M_{pl}(\theta + 2\theta + \theta) = (50 \times 2.0\theta)$
$4M_{pl}\theta = 100\theta \qquad \therefore M_{pl} = \textbf{25.0 kNm}$

Mechanism III: Sway

$\delta = 4.0\theta = 4.0\beta = 4.0\gamma \quad \therefore \beta = \gamma = \theta$
Internal Work Done = External Work Done
$[M_{pl}(\theta + \theta + \beta + \beta + \gamma + \gamma)] = (35 \times \delta)$
$M_{pl}(6\theta) = (35 \times 4.0\theta)$
$6M_{pl}\theta = 140\theta \qquad \therefore M_{pl} = \textbf{23.3 kNm}$

Mechanism IV: Joint rotation at D

(a)   (b)

Internal Work Done = $M_{pl}(\theta + \theta + \theta) = 3M_{pl}\theta$
External Work Done = zero

Solution

Topic: Plastic Analysis – Rigid Jointed Frames 2
Problem Number: 8.11 – Kinematic Method   **Page No. 3**

Combined Mechanisms:

The independent mechanisms are combined to determine the maximum M_{pl} value required to induce collapse with the minimum number of hinges, (i.e. $I_D + 1$).
In this case the following combinations have been evaluated:

Mechanism V = Mechanism II + Mechanism IV(a)
Mechanism VI = Mechanism V + Mechanism III
Mechanism VII = Mechanism VI + Mechanism I

Hinge Position	Independent Mechanisms			IV		Combined Mechanisms		
	I	II	III	(a)	(b)	V	VI	VII
A, (M_{pl})			$-\theta$				$-\theta$	$-\theta$
B, (M_{pl})	$-\theta$	$+\theta$					$+\theta$	EH $(2M_{pl}\theta)$
C, (M_{pl})	$+2\theta$							$+2\theta$
D$_1$, (M_{pl})	$-\theta$			$-\theta$	$+\theta$	$-\theta$	$-\theta$	-2θ
D$_2$, (M_{pl})		$-\theta$		$+\theta$	$-\theta$	EH $(2M_{pl}\theta)$	EH $(2M_{pl}\theta)$	EH $(2M_{pl}\theta)$
D$_3$, (M_{pl})			$+\theta$	$-\theta$	$+\theta$	$-\theta$	EH $(2M_{pl}\theta)$	EH $(2M_{pl}\theta)$
E, (M_{pl})		$-\theta$					$-\theta$	$-\theta$
F, (M_{pl})		$+2\theta$				$+2\theta$	$+2\theta$	$+2\theta$
G, (M_{pl})		$-\theta$	$\square\,\theta$			$-\theta$	-2θ	-2θ
H, (M_{pl})		$+\theta$					$+\theta$	$+\theta$
External work done	80θ	100θ	140θ			100θ	240θ	320θ
Internal work done	$4M_{pl}\theta$	$4M_{pl}\theta$	$6M_{pl}\theta$	$3M_{pl}\theta$		$7M_{pl}\theta$	$13M_{pl}\theta$	$17M_{pl}\theta$
Eliminated hinges						$2M_{pl}\theta$	$4M_{pl}\theta$	$6M_{pl}\theta$
Combined internal work done						$5M_{pl}\theta$	$9M_{pl}\theta$	$11M_{pl}\theta$
M_{pl} (kNm)	20.0	25.0	23.3			20.0	26.7	29.1

Check collapse mechanism VII with hinges at A, C, D$_1$, E, F, G and H (i.e. 7 hinges) The value of M_{pl} obtained (29.1 kNm) should be checked by ensuring that the bending moment in the frame does not exceed the relevant M_{pl} value at any location.

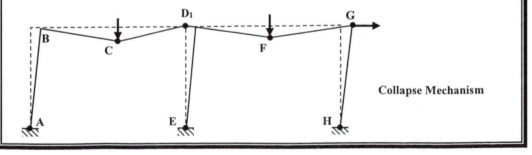

Collapse Mechanism

Solution
Topic: Plastic Analysis – Rigid Jointed Frames 2
Problem Number: 8.11 – Kinematic Method   **Page No. 4**

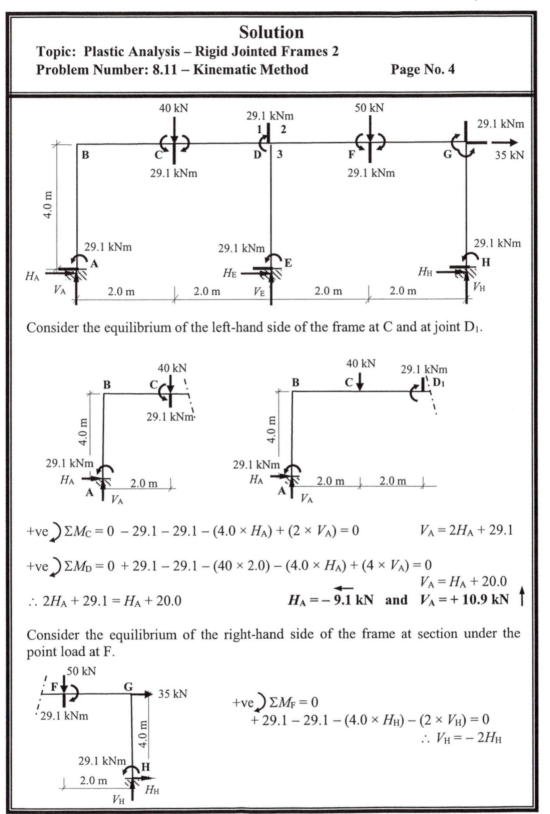

Consider the equilibrium of the left-hand side of the frame at C and at joint D_1.

$$+ve \circlearrowright \Sigma M_C = 0 \; - 29.1 - 29.1 - (4.0 \times H_A) + (2 \times V_A) = 0 \qquad V_A = 2H_A + 29.1$$

$$+ve \circlearrowright \Sigma M_D = 0 \; + 29.1 - 29.1 - (40 \times 2.0) - (4.0 \times H_A) + (4 \times V_A) = 0$$

$$V_A = H_A + 20.0$$

$$\therefore 2H_A + 29.1 = H_A + 20.0 \qquad \overleftarrow{H_A = -9.1 \text{ kN}} \quad \text{and} \quad V_A = +10.9 \text{ kN} \uparrow$$

Consider the equilibrium of the right-hand side of the frame at section under the point load at F.

$$+ve \circlearrowright \Sigma M_F = 0$$
$$+ 29.1 - 29.1 - (4.0 \times H_H) - (2 \times V_H) = 0$$
$$\therefore V_H = -2H_H$$

Solution
Topic: Plastic Analysis – Rigid Jointed Frames 2
Problem Number: 8.11 – Kinematic Method   Page No. 5

Consider the equilibrium of the right-hand side of the frame at section at joint G.

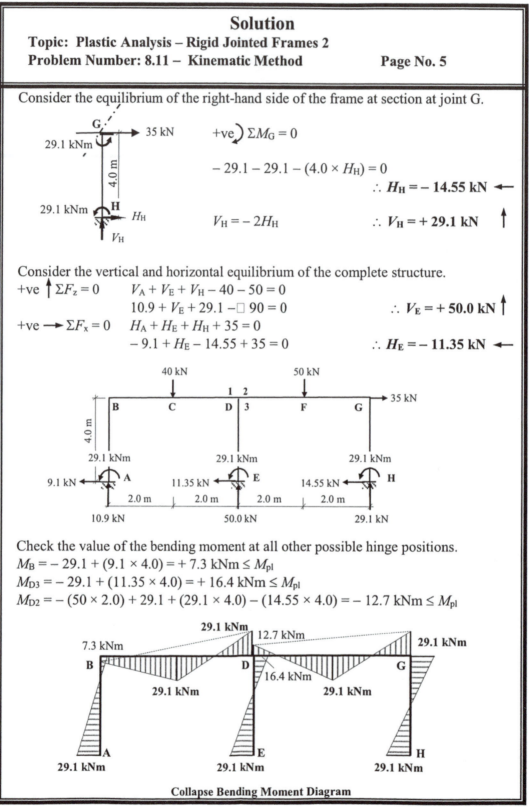

$+ve \circlearrowright \Sigma M_G = 0$

$-29.1 - 29.1 - (4.0 \times H_H) = 0$

$\therefore H_H = -14.55$ kN $\leftarrow$

$V_H = -2H_H$

$\therefore V_H = +29.1$ kN $\uparrow$

Consider the vertical and horizontal equilibrium of the complete structure.

$+ve \uparrow \Sigma F_z = 0 \qquad V_A + V_E + V_H - 40 - 50 = 0$

$10.9 + V_E + 29.1 - \square\, 90 = 0 \qquad\qquad \therefore V_E = +50.0$ kN $\uparrow$

$+ve \rightarrow \Sigma F_x = 0 \qquad H_A + H_E + H_H + 35 = 0$

$-9.1 + H_E - 14.55 + 35 = 0 \qquad\qquad \therefore H_E = -11.35$ kN $\leftarrow$

Check the value of the bending moment at all other possible hinge positions.

$M_B = -29.1 + (9.1 \times 4.0) = +7.3$ kNm $\leq M_{pl}$

$M_{D3} = -29.1 + (11.35 \times 4.0) = +16.4$ kNm $\leq M_{pl}$

$M_{D2} = -(50 \times 2.0) + 29.1 + (29.1 \times 4.0) - (14.55 \times 4.0) = -12.7$ kNm $\leq M_{pl}$

Collapse Bending Moment Diagram

Solution

Topic: Plastic Analysis – Rigid Jointed Frames 2
Problem Number: 8.12 – Kinematic Method   Page No. 1

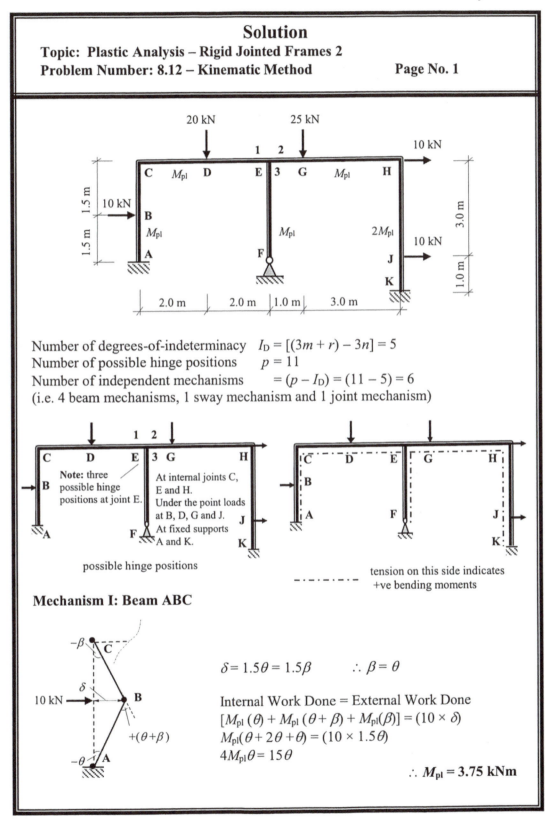

Number of degrees-of-indeterminacy $I_D = [(3m + r) - 3n] = 5$
Number of possible hinge positions $p = 11$
Number of independent mechanisms $= (p - I_D) = (11 - 5) = 6$
(i.e. 4 beam mechanisms, 1 sway mechanism and 1 joint mechanism)

Note: three possible hinge positions at joint E.

At internal joints C, E and H. Under the point loads at B, D, G and J. At fixed supports A and K.

possible hinge positions

tension on this side indicates +ve bending moments

Mechanism I: Beam ABC

$\delta = 1.5\theta = 1.5\beta$ $\therefore \beta = \theta$

Internal Work Done = External Work Done
$[M_{pl}(\theta) + M_{pl}(\theta + \beta) + M_{pl}(\beta)] = (10 \times \delta)$
$M_{pl}(\theta + 2\theta + \theta) = (10 \times 1.5\theta)$
$4M_{pl}\theta = 15\theta$

$\therefore M_{pl} = 3.75 \text{ kNm}$

Solution

Topic: Plastic Analysis – Rigid Jointed Frames 2
Problem Number: 8.12 – Kinematic Method   **Page No. 2**

Mechanism II: Beam CDE

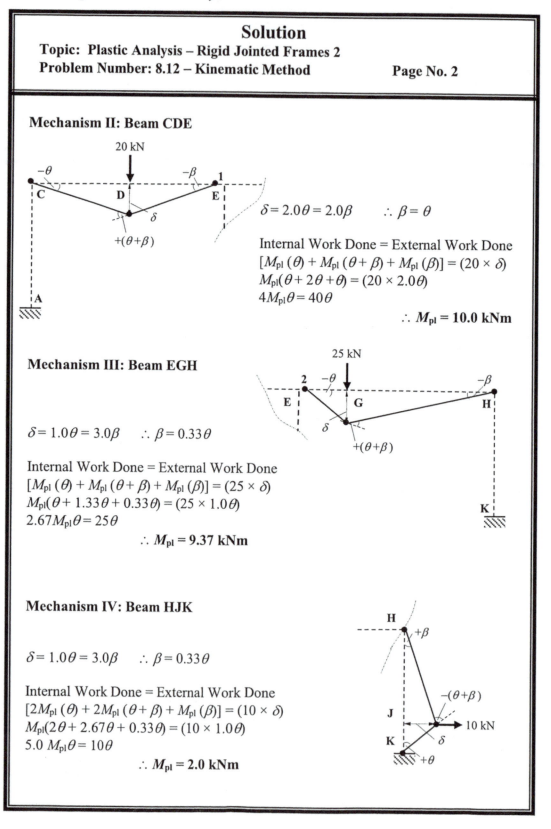

$\delta = 2.0\theta = 2.0\beta \qquad \therefore \beta = \theta$

Internal Work Done = External Work Done
$[M_{pl}\,(\theta) + M_{pl}\,(\theta + \beta) + M_{pl}\,(\beta)] = (20 \times \delta)$
$M_{pl}(\theta + 2\theta + \theta) = (20 \times 2.0\theta)$
$4M_{pl}\theta = 40\theta$

$\therefore M_{pl} = 10.0$ **kNm**

Mechanism III: Beam EGH

$\delta = 1.0\theta = 3.0\beta \qquad \therefore \beta = 0.33\theta$

Internal Work Done = External Work Done
$[M_{pl}\,(\theta) + M_{pl}\,(\theta + \beta) + M_{pl}\,(\beta)] = (25 \times \delta)$
$M_{pl}(\theta + 1.33\theta + 0.33\theta) = (25 \times 1.0\theta)$
$2.67M_{pl}\theta = 25\theta$

$\therefore M_{pl} = 9.37$ **kNm**

Mechanism IV: Beam HJK

$\delta = 1.0\theta = 3.0\beta \qquad \therefore \beta = 0.33\theta$

Internal Work Done = External Work Done
$[2M_{pl}\,(\theta) + 2M_{pl}\,(\theta + \beta) + M_{pl}\,(\beta)] = (10 \times \delta)$
$M_{pl}(2\theta + 2.67\theta + 0.33\theta) = (10 \times 1.0\theta)$
$5.0\,M_{pl}\theta = 10\theta$

$\therefore M_{pl} = 2.0$ **kNm**

Solution

Topic: **Plastic Analysis – Rigid Jointed Frames 2**
Problem Number: **8.12 – Kinematic Method**   **Page No. 3**

Mechanism V: Sway

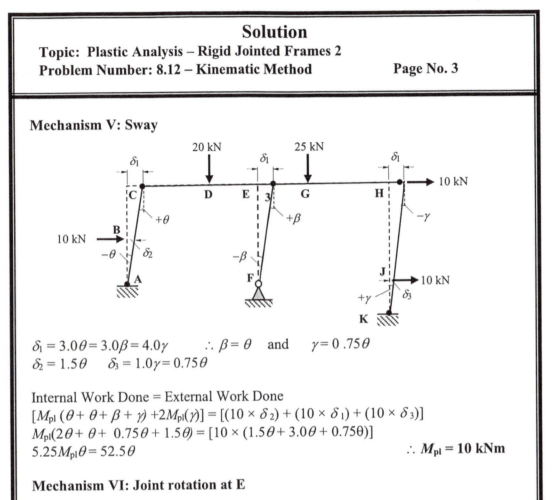

$\delta_1 = 3.0\theta = 3.0\beta = 4.0\gamma$ $\therefore \beta = \theta$ and $\gamma = 0.75\theta$
$\delta_2 = 1.5\theta$ $\delta_3 = 1.0\gamma = 0.75\theta$

Internal Work Done = External Work Done
$[M_{pl}(\theta + \theta + \beta + \gamma) + 2M_{pl}(\gamma)] = [(10 \times \delta_2) + (10 \times \delta_1) + (10 \times \delta_3)]$
$M_{pl}(2\theta + \theta + 0.75\theta + 1.5\theta) = [10 \times (1.5\theta + 3.0\theta + 0.75\theta)]$
$5.25M_{pl}\theta = 52.5\theta$ $\therefore M_{pl} = 10 \text{ kNm}$

Mechanism VI: Joint rotation at E

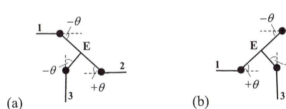

(a)   (b)

Internal Work Done = $M_{pl}(\theta + \theta + \theta) = 3M_{pl}\theta$
External Work Done = zero

Combined Mechanisms:
The independent mechanisms are combined to determine the maximum M_{pl} value required to induce collapse with the minimum number of hinges, (i.e. $I_D + 1$).
In this case the following combinations have been evaluated:
Mechanism VII = Mechanism II + Mechanism V + Mechanism VI(a)
Mechanism VIII = Mechanism VII + Mechanism III

Solution

Topic: Plastic Analysis – Rigid Jointed Frames 2
Problem Number: 8.12 – Kinematic Method   **Page No. 4**

Hinge Positions	Independent Mechanisms							Combined Mechanisms	
	I	II	III	IV	V	VI		VII	VIII
A, (M_{pl})	$-\theta$				$-\theta$			$-\theta$	$-\theta$
B, (M_{pl})	$+2\theta$								
C, (M_{pl})	$-\theta$	$-\theta$			$+\theta$			EH ($2M_{pl}\theta$)	EH($2M_{pl}\theta$)
D, (M_{pl})		$+2\theta$				(a)	(b)	$+2\theta$	$+2\theta$
E$_1$, (M_{pl})		$-\theta$				$-\theta$	$+\theta$	-2θ	-2θ
E$_2$, (M_{pl})			$-\theta$			$+\theta$	$-\theta$	$+\theta$	EH($2M_{pl}\theta$)
E$_3$, (M_{pl})					$+\theta$	$-\theta$	$+\theta$	EH ($2M_{pl}\theta$)	EH($2M_{pl}\theta$)
G, (M_{pl})			$+1.33\theta$						$+1.33\theta$
H, (M_{pl})			-0.33θ	$+0.33\theta$	-0.75θ			-0.75θ	-1.08θ
J, ($2M_{pl}$)				-1.33θ					
K, ($2M_{pl}$)				$+\theta$	$+0.75\theta$			$+0.75\theta$	$+0.75\theta$
External work done	15θ	40θ	25θ	10θ	52.5θ	0		92.5θ	**117.5θ**
Internal work done	$4M_{pl}\theta$	$4M_{pl}\theta$	$2.67M_{pl}\theta$	$5M_{pl}\theta$	$5.25M_{pl}\theta$	$3M_{pl}\theta$		$12.25M_{pl}\theta$	$14.92M_{pl}\theta$
Eliminated hinges								$4M_{pl}\theta$	$6M_{pl}\theta$
Combined internal work done								$8.25M_{pl}\theta$	**$8.92M_{pl}\theta$**
M_{pl} (kNm)	3.75	10.0	9.37	2.0	10.0			11.21	**13.17**

Check collapse mechanism VIII with hinges at A, D, E$_1$, G, H and K (i.e. 6 hinges) The value of M_{pl} obtained (13.17 kNm) should be checked by ensuring that the bending moment in the frame does not exceed the relevant M_{pl} value at any location.

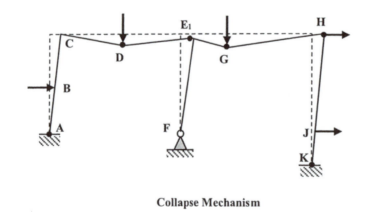

Collapse Mechanism

Solution

Topic: Plastic Analysis – Rigid Jointed Frames 2
Problem Number: 8.12 – Kinematic Method   **Page No. 5**

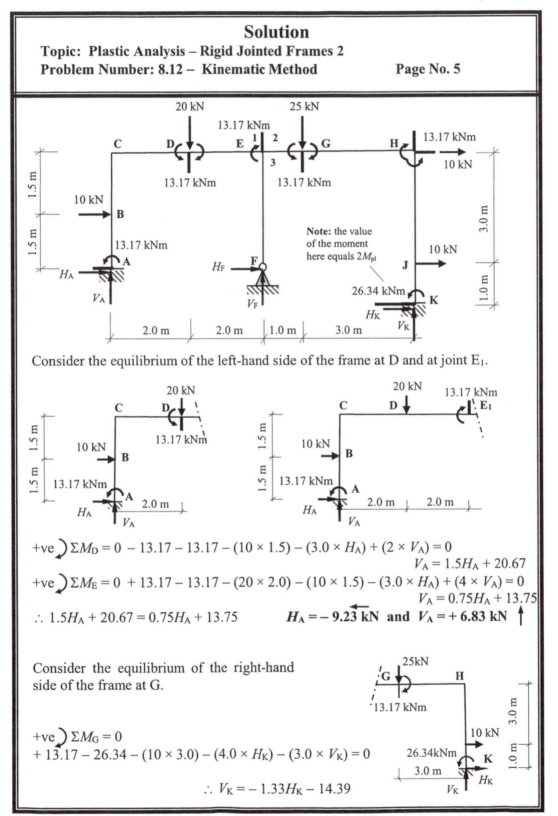

Consider the equilibrium of the left-hand side of the frame at D and at joint E_1.

$+ve \circlearrowright \Sigma M_D = 0 \; - 13.17 - 13.17 - (10 \times 1.5) - (3.0 \times H_A) + (2 \times V_A) = 0$
$$V_A = 1.5H_A + 20.67$$

$+ve \circlearrowright \Sigma M_E = 0 \; + 13.17 - 13.17 - (20 \times 2.0) - (10 \times 1.5) - (3.0 \times H_A) + (4 \times V_A) = 0$
$$V_A = 0.75H_A + 13.75$$

$\therefore \; 1.5H_A + 20.67 = 0.75H_A + 13.75$ $H_A = -\,\mathbf{9.23}$ kN and $V_A = +\,\mathbf{6.83}$ kN $\uparrow$

Consider the equilibrium of the right-hand
side of the frame at G.

$+ve \circlearrowright \Sigma M_G = 0$
$+ 13.17 - 26.34 - (10 \times 3.0) - (4.0 \times H_K) - (3.0 \times V_K) = 0$

$$\therefore \; V_K = -1.33H_K - 14.39$$

Solution

Topic: Plastic Analysis – Rigid Jointed Frames 2
Problem Number: 8.12 – Kinematic Method   **Page No. 6**

Consider the equilibrium of the right-hand side of the frame at section at joint H.

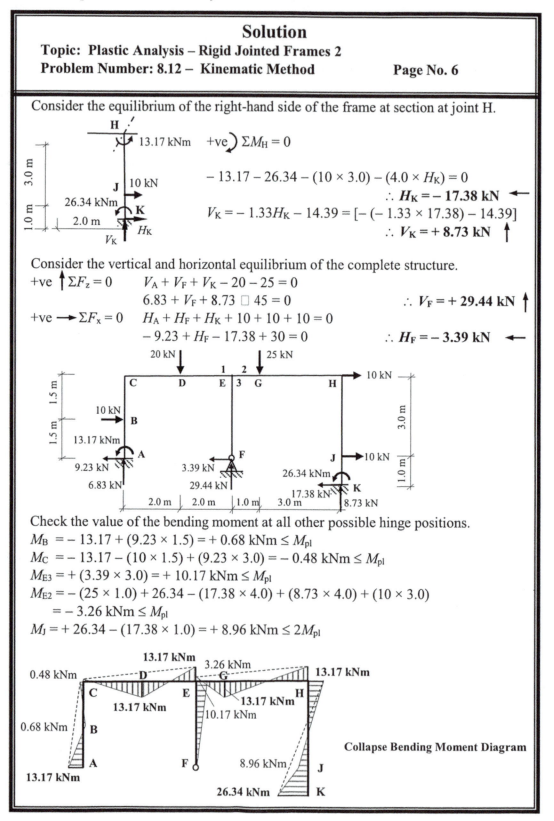

$+ve \; \curvearrowleft \; \Sigma M_H = 0$

$$-13.17 - 26.34 - (10 \times 3.0) - (4.0 \times H_K) = 0$$
$$\therefore H_K = -17.38 \text{ kN} \; \leftarrow$$
$$V_K = -1.33 H_K - 14.39 = [-(-1.33 \times 17.38) - 14.39]$$
$$\therefore V_K = +8.73 \text{ kN} \; \uparrow$$

Consider the vertical and horizontal equilibrium of the complete structure.

$+ve \; \uparrow \Sigma F_z = 0 \qquad V_A + V_F + V_K - 20 - 25 = 0$
$\qquad\qquad\qquad\quad 6.83 + V_F + 8.73 \; \square \; 45 = 0 \qquad\qquad \therefore V_F = +29.44 \text{ kN} \; \uparrow$
$+ve \; \rightarrow \Sigma F_x = 0 \qquad H_A + H_F + H_K + 10 + 10 + 10 = 0$
$\qquad\qquad\qquad\quad -9.23 + H_F - 17.38 + 30 = 0 \qquad\qquad \therefore H_F = -3.39 \text{ kN} \; \leftarrow$

Check the value of the bending moment at all other possible hinge positions.

$M_B = -13.17 + (9.23 \times 1.5) = +0.68 \text{ kNm} \le M_{pl}$
$M_C = -13.17 - (10 \times 1.5) + (9.23 \times 3.0) = -0.48 \text{ kNm} \le M_{pl}$
$M_{E3} = +(3.39 \times 3.0) = +10.17 \text{ kNm} \le M_{pl}$
$M_{E2} = -(25 \times 1.0) + 26.34 - (17.38 \times 4.0) + (8.73 \times 4.0) + (10 \times 3.0)$
$\qquad = -3.26 \text{ kNm} \le M_{pl}$
$M_J = +26.34 - (17.38 \times 1.0) = +8.96 \text{ kNm} \le 2M_{pl}$

Collapse Bending Moment Diagram

Solution

Topic: Plastic Analysis – Rigid Jointed Frames 2
Problem Number: 8.13 – Kinematic Method   **Page No. 1**

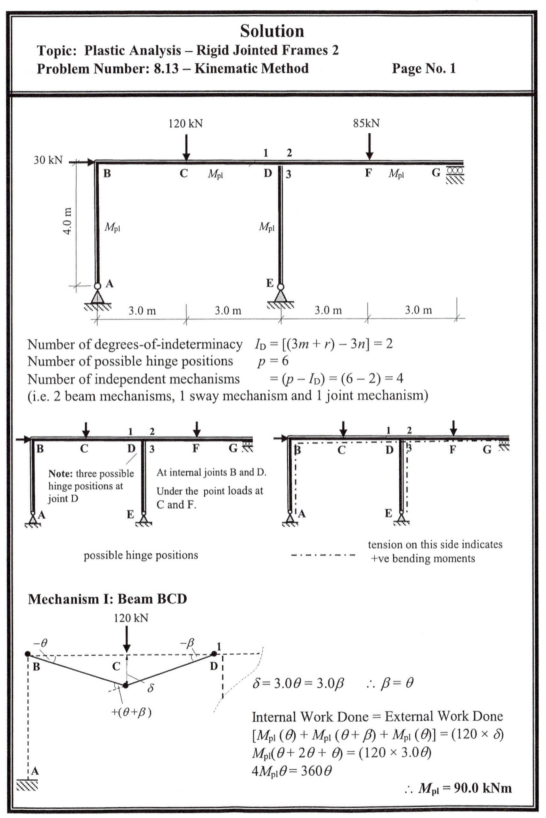

Number of degrees-of-indeterminacy $I_D = [(3m + r) - 3n] = 2$
Number of possible hinge positions $p = 6$
Number of independent mechanisms $= (p - I_D) = (6 - 2) = 4$
(i.e. 2 beam mechanisms, 1 sway mechanism and 1 joint mechanism)

possible hinge positions

tension on this side indicates
+ve bending moments

Mechanism I: Beam BCD

$\delta = 3.0\theta = 3.0\beta$ $\therefore \beta = \theta$

Internal Work Done = External Work Done
$[M_{pl}(\theta) + M_{pl}(\theta + \beta) + M_{pl}(\theta)] = (120 \times \delta)$
$M_{pl}(\theta + 2\theta + \theta) = (120 \times 3.0\theta)$
$4M_{pl}\theta = 360\theta$

$\therefore M_{pl} = 90.0$ kNm

Solution
Topic: Plastic Analysis – Rigid Jointed Frames 2
Problem Number: 8.13 – Kinematic Method   **Page No. 2**

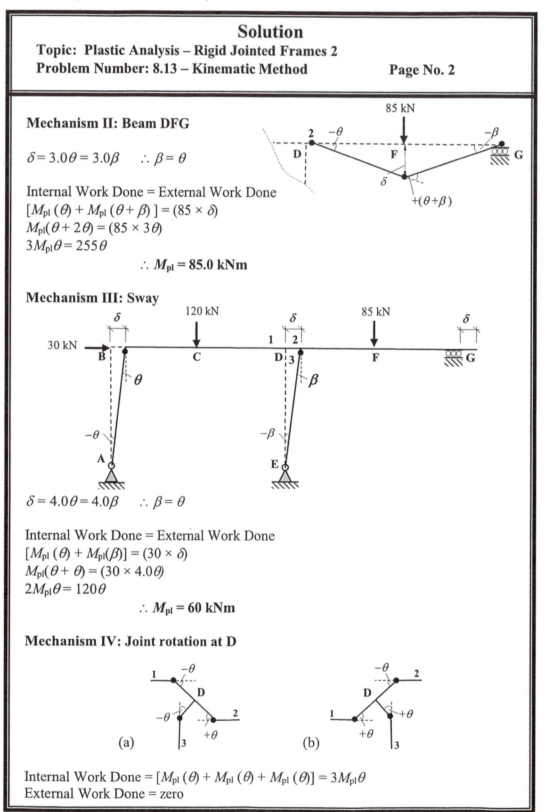

Mechanism II: Beam DFG

$\delta = 3.0\theta = 3.0\beta$ $\therefore \beta = \theta$

Internal Work Done = External Work Done
$[M_{pl}(\theta) + M_{pl}(\theta + \beta)] = (85 \times \delta)$
$M_{pl}(\theta + 2\theta) = (85 \times 3\theta)$
$3M_{pl}\theta = 255\theta$

$\qquad \therefore M_{pl} = \textbf{85.0 kNm}$

Mechanism III: Sway

$\delta = 4.0\theta = 4.0\beta$ $\therefore \beta = \theta$

Internal Work Done = External Work Done
$[M_{pl}(\theta) + M_{pl}(\beta)] = (30 \times \delta)$
$M_{pl}(\theta + \theta) = (30 \times 4.0\theta)$
$2M_{pl}\theta = 120\theta$

$\qquad \therefore M_{pl} = \textbf{60 kNm}$

Mechanism IV: Joint rotation at D

(a)   (b)

Internal Work Done = $[M_{pl}(\theta) + M_{pl}(\theta) + M_{pl}(\theta)] = 3M_{pl}\theta$
External Work Done = zero

Solution

Topic: Plastic Analysis – Rigid Jointed Frames 2
Problem Number: 8.13 – Kinematic Method   Page No. 3

Combined Mechanism:

The independent mechanisms are combined to determine the maximum M_{pl} value required to induce collapse with the minimum number of hinges, (i.e. $I_D + 1$).

In this case the following combination has been evaluated:

Mechanism V = Mechanism I + Mechanism II + Mechanism III + Mechanism IV(a)

Hinge Positions	Independent Mechanisms					Combined Mechanism
	I	II	III	IV		V
				(a)	(b)	
B, (M_{pl})	$-\theta$		$+\theta$			EH ($2M_{pl}\theta$)
C, (M_{pl})	$+2\theta$					$+2\theta$
D₁, (M_{pl})	$-\theta$			$-\theta$	$+\theta$	-2θ
D₂, (M_{pl})		$-\theta$		$+\theta$	$-\theta$	EH ($2M_{pl}\theta$)
D₃, (M_{pl})			$+\theta$	$-\theta$	$+\theta$	EH ($2M_{pl}\theta$)
F, (M_{pl})		$+2\theta$				$+2\theta$
External work done	360θ	255θ	120θ	0		735θ
Internal work done	$4M_{pl}\theta$	$3M_{pl}\theta$	$2M_{pl}\theta$	$3M_{pl}\theta$		$12M_{pl}\theta$
Eliminated hinges						$6M_{pl}\theta$
Combined internal work done						$6M_{pl}\theta$
M_{pl} (kNm)	90.0	85.0	60.0			122.50

Check collapse mechanism V with hinges at C, D₁ and F (i.e. 3 hinges)

The value of M_{pl} obtained (122.5 kNm) should be checked by ensuring that the bending moment in the frame does not exceed the relevant M_{pl} value at any location.

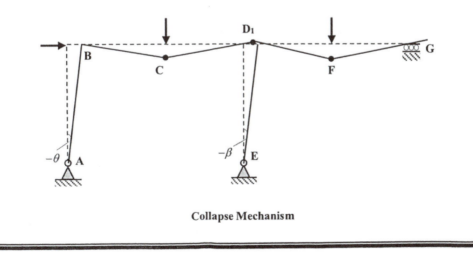

Collapse Mechanism

Solution

Topic: **Plastic Analysis – Rigid Jointed Frames 2**
Problem Number: **8.13 – Kinematic Method**   **Page No. 4**

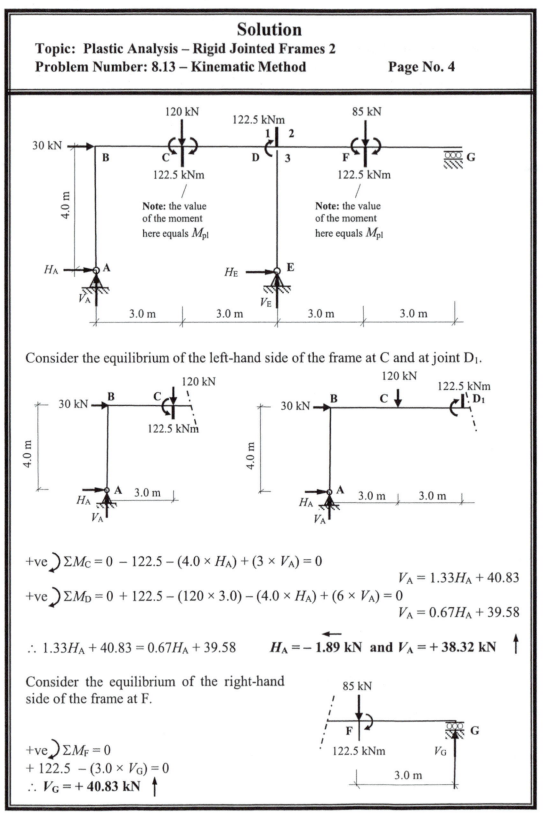

Consider the equilibrium of the left-hand side of the frame at C and at joint D_1.

$+ve \circlearrowleft \Sigma M_C = 0 - 122.5 - (4.0 \times H_A) + (3 \times V_A) = 0$

$$V_A = 1.33H_A + 40.83$$

$+ve \circlearrowleft \Sigma M_D = 0 + 122.5 - (120 \times 3.0) - (4.0 \times H_A) + (6 \times V_A) = 0$

$$V_A = 0.67H_A + 39.58$$

$\therefore 1.33H_A + 40.83 = 0.67H_A + 39.58$ $H_A = -1.89$ kN and $V_A = +38.32$ kN ↑

Consider the equilibrium of the right-hand
side of the frame at F.

$+ve \circlearrowleft \Sigma M_F = 0$
$+ 122.5 - (3.0 \times V_G) = 0$
$\therefore V_G = +40.83$ kN ↑

Solution
Topic: Plastic Analysis – Rigid Jointed Frames 2
Problem Number: 8.13 – Kinematic Method   **Page No. 5**

Consider the vertical and horizontal equilibrium of the complete structure.

+ve $\uparrow \Sigma F_z = 0$ $V_A + V_E + V_G - 120 - 85 = 0$

$38.32 + V_E + 40.83 \;\square\; 205 = 0$ $\therefore V_E = + 125.85 \text{ kN} \uparrow$

+ve $\rightarrow \Sigma F_x = 0$ $H_A + H_E + 30 = 0$

$-1.89 + H_E + 30 = 0$ $\therefore H_E = -28.11 \text{ kN} \leftarrow$

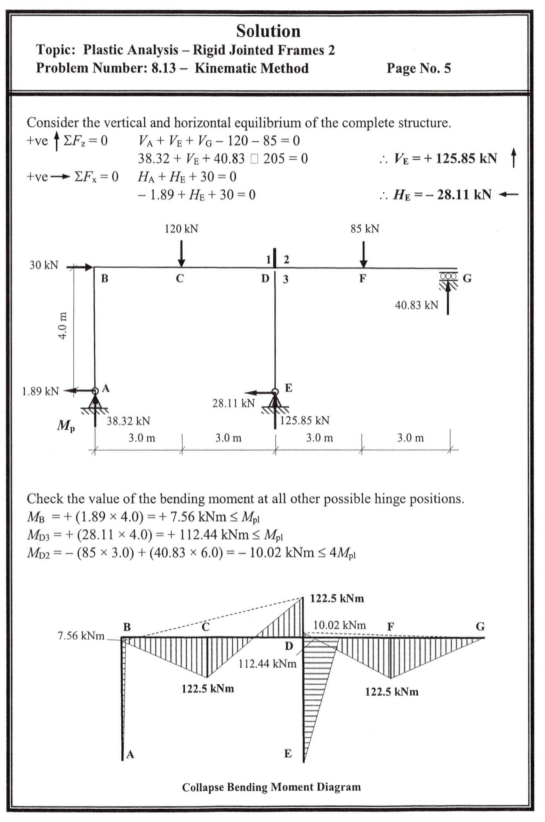

Check the value of the bending moment at all other possible hinge positions.

$M_B = + (1.89 \times 4.0) = + 7.56 \text{ kNm} \leq M_{pl}$

$M_{D3} = + (28.11 \times 4.0) = + 112.44 \text{ kNm} \leq M_{pl}$

$M_{D2} = -(85 \times 3.0) + (40.83 \times 6.0) = -10.02 \text{ kNm} \leq 4M_{pl}$

Collapse Bending Moment Diagram

Solution

Topic: Plastic Analysis – Rigid Jointed Frames 2
Problem Number: 8.14 – Kinematic Method   **Page No. 1**

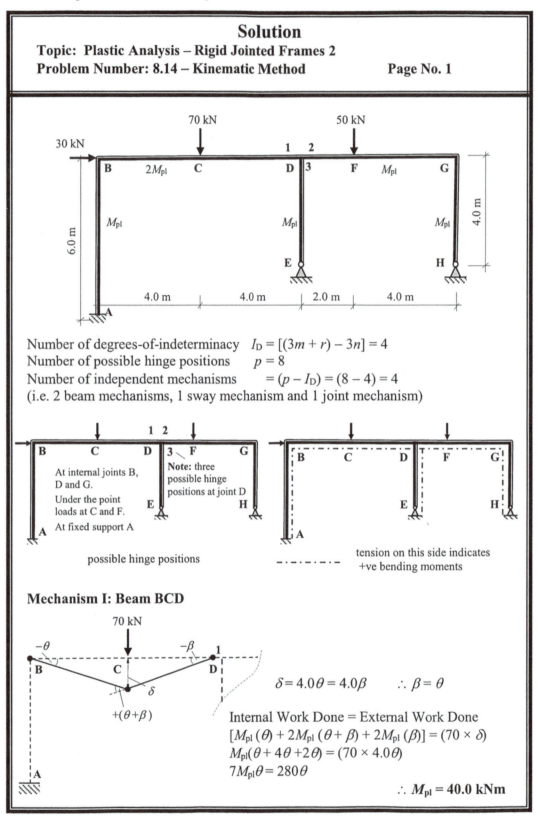

Number of degrees-of-indeterminacy $I_D = [(3m + r) - 3n] = 4$
Number of possible hinge positions $p = 8$
Number of independent mechanisms $= (p - I_D) = (8 - 4) = 4$
(i.e. 2 beam mechanisms, 1 sway mechanism and 1 joint mechanism)

At internal joints B, D and G.
Under the point loads at C and F.
At fixed support A

Note: three possible hinge positions at joint D

possible hinge positions

tension on this side indicates +ve bending moments

Mechanism I: Beam BCD

$\delta = 4.0\theta = 4.0\beta \qquad \therefore \beta = \theta$

Internal Work Done = External Work Done
$[M_{pl}\,(\theta) + 2M_{pl}\,(\theta + \beta) + 2M_{pl}\,(\beta)] = (70 \times \delta)$
$M_{pl}(\theta + 4\theta + 2\theta) = (70 \times 4.0\theta)$
$7M_{pl}\theta = 280\theta$

$\therefore M_{pl} = \mathbf{40.0\ kNm}$

Solution
Topic: Plastic Analysis – Rigid Jointed Frames 2
Problem Number: 8.14 – Kinematic Method   **Page No. 2**

Mechanism II: Beam DFG

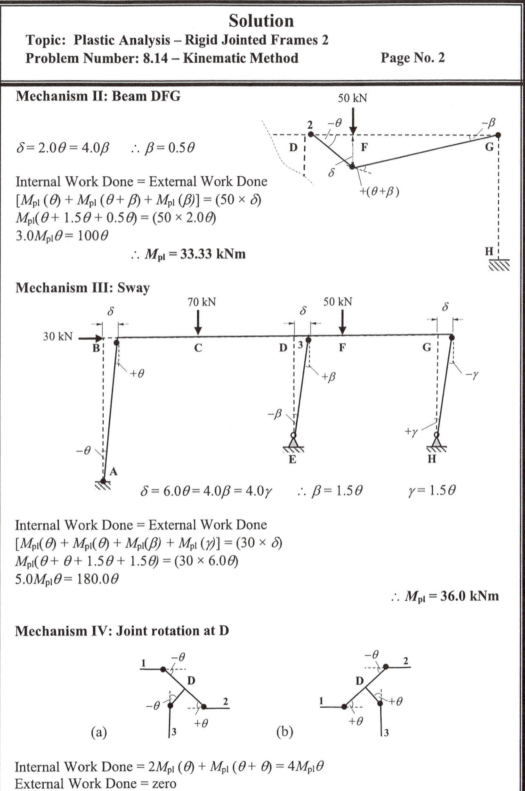

$\delta = 2.0\theta = 4.0\beta$ $\therefore \beta = 0.5\theta$

Internal Work Done = External Work Done
$[M_{pl}(\theta) + M_{pl}(\theta + \beta) + M_{pl}(\beta)] = (50 \times \delta)$
$M_{pl}(\theta + 1.5\theta + 0.5\theta) = (50 \times 2.0\theta)$
$3.0 M_{pl}\theta = 100\theta$

$\therefore M_{pl} = \textbf{33.33 kNm}$

Mechanism III: Sway

$\delta = 6.0\theta = 4.0\beta = 4.0\gamma$ $\therefore \beta = 1.5\theta$ $\gamma = 1.5\theta$

Internal Work Done = External Work Done
$[M_{pl}(\theta) + M_{pl}(\theta) + M_{pl}(\beta) + M_{pl}(\gamma)] = (30 \times \delta)$
$M_{pl}(\theta + \theta + 1.5\theta + 1.5\theta) = (30 \times 6.0\theta)$
$5.0 M_{pl}\theta = 180.0\theta$

$\therefore M_{pl} = \textbf{36.0 kNm}$

Mechanism IV: Joint rotation at D

Internal Work Done = $2M_{pl}(\theta) + M_{pl}(\theta + \theta) = 4M_{pl}\theta$
External Work Done = zero

Solution

Topic: Plastic Analysis – Rigid Jointed Frames 2
Problem Number: 8.14 – Kinematic Method　　　　　　**Page No. 3**

Combined Mechanisms:
The independent mechanisms are combined to determine the maximum M_{pl} value required to induce collapse with the minimum number of hinges, (i.e. $I_D + 1$).
In this case the following combinations have been evaluated:
Mechanism V = Mechanism I + Mechanism III + Mechanism IV(b)
Mechanism VI = Mechanism V + Mechanism II

Hinge Positions	Independent Mechanisms					Combined Mechanisms	
	I	II	III	IV		V	VI
A, (M_{pl})			$-\theta$			$-\theta$	$-\theta$
B, (M_{pl})	$-\theta$	$+\theta$				EH $(2M_{pl}\theta)$	EH $(2M_{pl}\theta)$
C, ($2M_{pl}$)	$+2\theta$			(a)	(b)	$+2\theta$	$+2\theta$
D$_1$, ($2M_{pl}$)	$-\theta$			$-\theta$	$+\theta$	EH $(4M_{pl}\theta)$	EH $(4M_{pl}\theta)$
D$_2$, (M_{pl})		$-\theta$		$+\theta$	$-\theta$	$-\theta$	-2θ
D$_3$, (M_{pl})			$+1.5\theta$	$-\theta$	$+\theta$	$+2.5\theta$	$+2.5\theta$
F, (M_{pl})		$+1.5\theta$					$+1.5\theta$
G, (M_{pl})		-0.5θ	-1.5θ			-1.5θ	-2θ
External work done	280θ	100θ	180θ	0		460θ	560θ
Internal work done	$7M_{pl}\theta$	$3M_{pl}\theta$	$5M_{pl}\theta$	$4M_{pl}\theta$		$16M_{pl}\theta$	$19M_{pl}\theta$
Eliminated hinges						$6M_{pl}\theta$	$6M_{pl}\theta$
Combined internal work done						$10M_{pl}\theta$	$13M_{pl}\theta$
M_{pl} (kNm)	40.0	33.33	36.0			46.0	43.08

Check collapse mechanism V with hinges at A, C, D$_2$, D$_3$, and G (i.e. 5 hinges).
The value of M_{pl} obtained (46.0 kNm) should be checked by ensuring that the bending moment in the frame does not exceed the relevant M_{pl} value at any location.

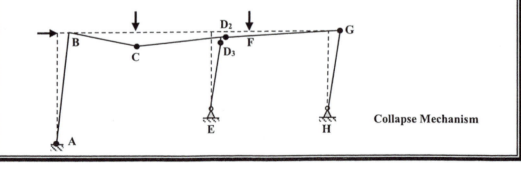

Collapse Mechanism

Solution

Topic: Plastic Analysis – Rigid Jointed Frames 2
Problem Number: 8.14 – Kinematic Method   **Page No. 5**

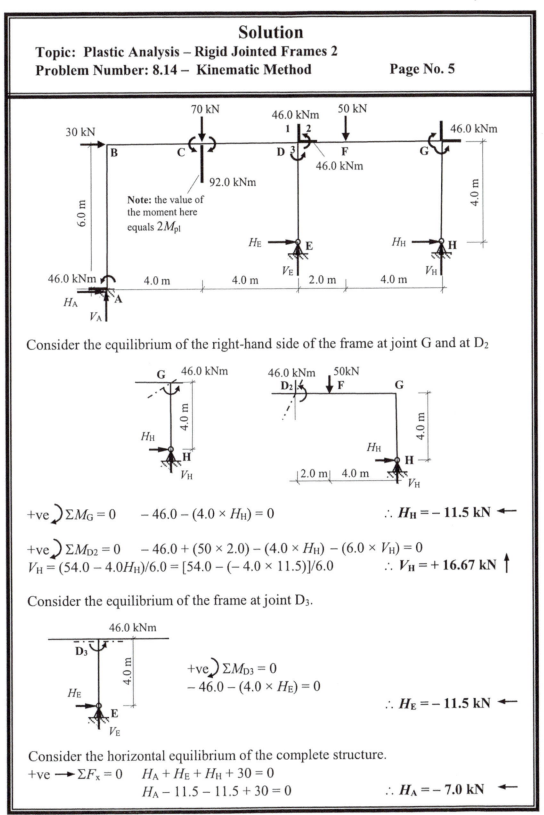

Consider the equilibrium of the right-hand side of the frame at joint G and at D_2

$+ve \,\circlearrowright \Sigma M_G = 0$ $-46.0 - (4.0 \times H_H) = 0$ $\therefore H_H = -11.5 \text{ kN} \leftarrow$

$+ve \,\circlearrowright \Sigma M_{D2} = 0$ $-46.0 + (50 \times 2.0) - (4.0 \times H_H) - (6.0 \times V_H) = 0$
$V_H = (54.0 - 4.0 H_H)/6.0 = [54.0 - (-4.0 \times 11.5)]/6.0$ $\therefore V_H = +16.67 \text{ kN} \uparrow$

Consider the equilibrium of the frame at joint D_3.

$+ve \,\circlearrowright \Sigma M_{D3} = 0$
$-46.0 - (4.0 \times H_E) = 0$

$\therefore H_E = -11.5 \text{ kN} \leftarrow$

Consider the horizontal equilibrium of the complete structure.
$+ve \longrightarrow \Sigma F_x = 0$ $H_A + H_E + H_H + 30 = 0$
$H_A - 11.5 - 11.5 + 30 = 0$ $\therefore H_A = -7.0 \text{ kN} \leftarrow$

Solution
Topic: Plastic Analysis – Rigid Jointed Frames 2
Problem Number: 8.14 – Kinematic Method   **Page No. 6**

Consider the equilibrium of the left-hand side of the frame at a section under the point load at C.

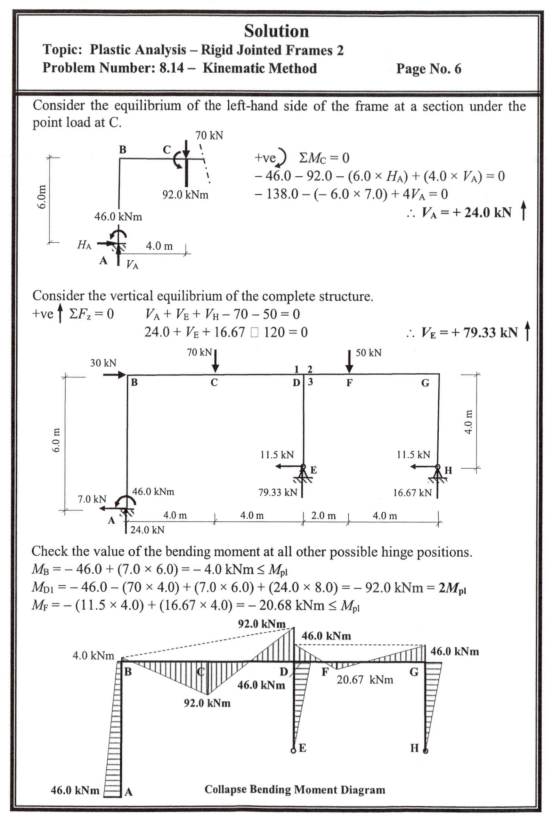

$+ve \; \curvearrowright \; \Sigma M_C = 0$
$- 46.0 - 92.0 - (6.0 \times H_A) + (4.0 \times V_A) = 0$
$- 138.0 - (- 6.0 \times 7.0) + 4V_A = 0$
$\therefore V_A = + 24.0 \text{ kN} \uparrow$

Consider the vertical equilibrium of the complete structure.
$+ve \uparrow \; \Sigma F_z = 0 \qquad V_A + V_E + V_H - 70 - 50 = 0$
$24.0 + V_E + 16.67 \; \square \; 120 = 0 \qquad\qquad \therefore V_E = + 79.33 \text{ kN} \uparrow$

Check the value of the bending moment at all other possible hinge positions.
$M_B = - 46.0 + (7.0 \times 6.0) = - 4.0 \text{ kNm} \leq M_{pl}$
$M_{D1} = - 46.0 - (70 \times 4.0) + (7.0 \times 6.0) + (24.0 \times 8.0) = - 92.0 \text{ kNm} = 2M_{pl}$
$M_F = - (11.5 \times 4.0) + (16.67 \times 4.0) = - 20.68 \text{ kNm} \leq M_{pl}$

Collapse Bending Moment Diagram

Solution

Topic: Plastic Analysis – Rigid Jointed Frames 2
Problem Number: 8.15 – Kinematic Method   **Page No. 1**

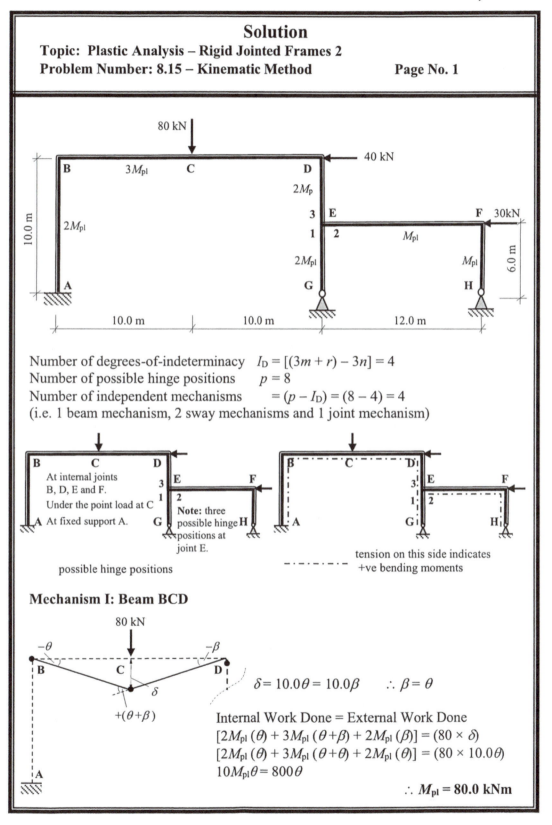

Number of degrees-of-indeterminacy $I_D = [(3m + r) - 3n] = 4$
Number of possible hinge positions $p = 8$
Number of independent mechanisms $= (p - I_D) = (8 - 4) = 4$
(i.e. 1 beam mechanism, 2 sway mechanisms and 1 joint mechanism)

At internal joints
B, D, E and F.
Under the point load at C
At fixed support A.

Note: three
possible hinge
positions at
joint E.

possible hinge positions

tension on this side indicates
+ve bending moments

Mechanism I: Beam BCD

$\delta = 10.0\theta = 10.0\beta$ $\therefore \beta = \theta$

Internal Work Done = External Work Done
$[2M_{pl}\,(\theta) + 3M_{pl}\,(\theta + \beta) + 2M_{pl}\,(\beta)] = (80 \times \delta)$
$[2M_{pl}\,(\theta) + 3M_{pl}\,(\theta + \theta) + 2M_{pl}\,(\theta)] = (80 \times 10.0\theta)$
$10M_{pl}\theta = 800\theta$

$\therefore M_{pl} = 80.0$ kNm

Solution
Topic: Plastic Analysis – Rigid Jointed Frames 2
Problem Number: 8.15 – Kinematic Method   **Page No. 2**

Mechanism II: Sway of Top Storey

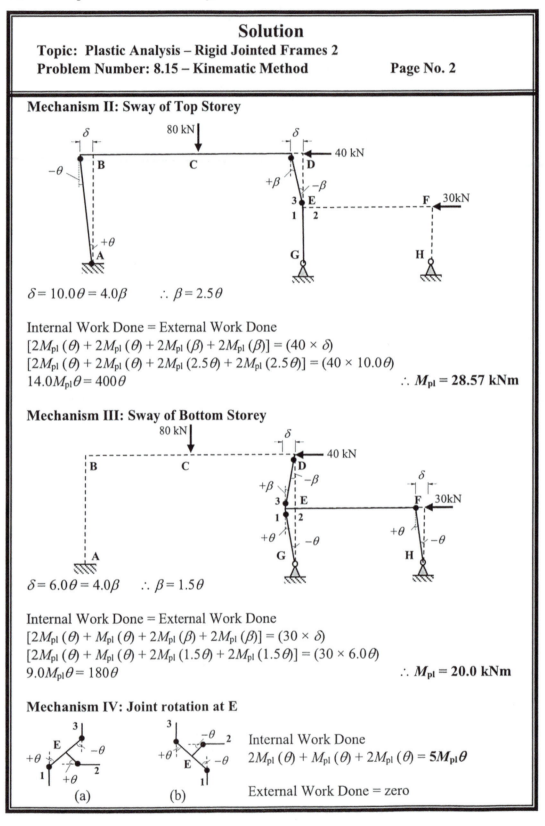

$\delta = 10.0\theta = 4.0\beta$ $\therefore \beta = 2.5\theta$

Internal Work Done = External Work Done
$[2M_{pl}\,(\theta) + 2M_{pl}\,(\theta) + 2M_{pl}\,(\beta) + 2M_{pl}\,(\beta)] = (40 \times \delta)$
$[2M_{pl}\,(\theta) + 2M_{pl}\,(\theta) + 2M_{pl}\,(2.5\theta) + 2M_{pl}\,(2.5\theta)] = (40 \times 10.0\theta)$
$14.0M_{pl}\theta = 400\theta$ $\therefore M_{pl} = \textbf{28.57 kNm}$

Mechanism III: Sway of Bottom Storey

$\delta = 6.0\theta = 4.0\beta$ $\therefore \beta = 1.5\theta$

Internal Work Done = External Work Done
$[2M_{pl}\,(\theta) + M_{pl}\,(\theta) + 2M_{pl}\,(\beta) + 2M_{pl}\,(\beta)] = (30 \times \delta)$
$[2M_{pl}\,(\theta) + M_{pl}\,(\theta) + 2M_{pl}\,(1.5\theta) + 2M_{pl}\,(1.5\theta)] = (30 \times 6.0\theta)$
$9.0M_{pl}\theta = 180\theta$ $\therefore M_{pl} = \textbf{20.0 kNm}$

Mechanism IV: Joint rotation at E

(a)   (b)

Internal Work Done
$2M_{pl}\,(\theta) + M_{pl}\,(\theta) + 2M_{pl}\,(\theta) = \textbf{5}\boldsymbol{M_{pl}\theta}$

External Work Done = zero

Solution

Topic: Plastic Analysis – Rigid Jointed Frames 2
Problem Number: 8.15 – Kinematic Method   **Page No. 3**

Combined Mechanisms:
The independent mechanisms are combined to determine the maximum M_{pl} value required to induce collapse with the minimum number of hinges, (i.e. $I_D + 1$).
In this case the following combination has been evaluated:

Mechanism V = Mechanisms [I + II + III + IV(b)]

Hinge Positions	Independent Mechanisms					Combined Mechanism
	I	II	III	IV		V
A, ($2M_{pl}$)		$+\theta$				$+\theta$
B, ($2M_{pl}$)	$-\theta$	$-\theta$				-2θ
C, ($3M_{pl}$)	$+2\theta$					$+2\theta$
D, ($2M_{pl}$)	$-\theta$	$+2.5\,\theta$	-1.5θ	(a)	(b)	**EH** ($10M_{pl}\theta$)
E$_1$, ($2M_{pl}$)			$+\theta$	$+\theta$	$-\theta$	**EH** ($4M_{pl}\theta$)
E$_2$, (M_{pl})				$+\theta$	$-\theta$	$-\theta$
E$_3$, ($2M_{pl}$)		$-2.5\,\theta$	$+1.5\theta$	$-\theta$	$+\theta$	**EH** ($10M_{pl}\theta$)
F, (M_{pl})			$+\theta$			$+\theta$
External work done	800θ	400θ	180θ	0		**1380θ**
Internal work done	$10M_{pl}\theta$	$14M_{pl}\theta$	$9M_{pl}\theta$	$5M_{pl}\theta$		$38M_{pl}\theta$
Eliminated hinges						$24M_{pl}\theta$
Combined internal work done						**$14M_{pl}\theta$**
M_{pl} (kNm)	80.0	28.57	20.0			**98.57**

Check collapse mechanism V with hinges at A, B, C, E$_2$ and F (i.e. 5 hinges).
The value of M_{pl} obtained (98.57 kNm) should be checked by ensuring that the bending moment in the frame does not exceed the relevant M_{pl} value at any location.

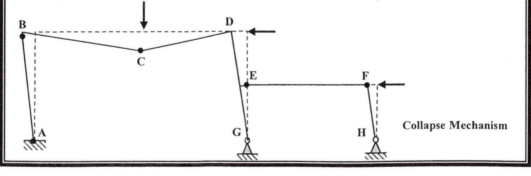

Collapse Mechanism

Solution

Topic: Plastic Analysis – Rigid Jointed Frames 2
Problem Number: 8.15 – Kinematic Method   **Page No. 4**

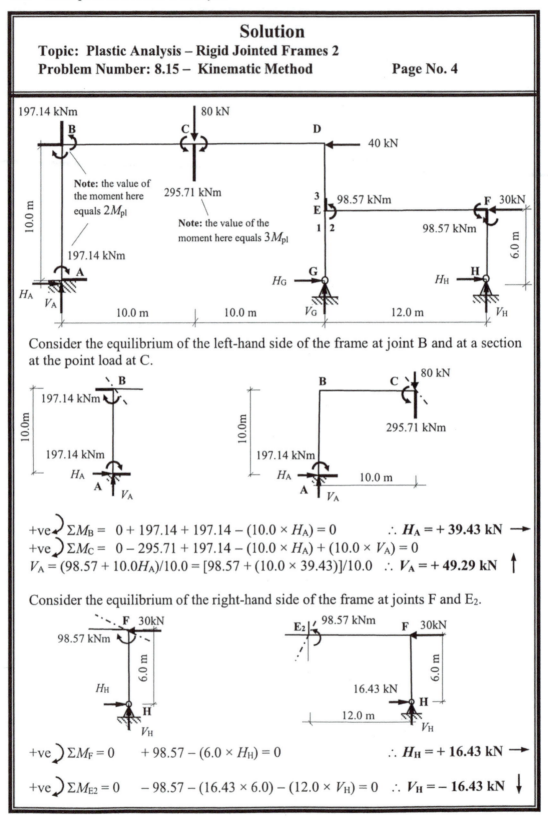

Consider the equilibrium of the left-hand side of the frame at joint B and at a section at the point load at C.

$+ve \circlearrowleft \Sigma M_B = \quad 0 + 197.14 + 197.14 - (10.0 \times H_A) = 0 \qquad \therefore H_A = +39.43 \text{ kN} \longrightarrow$

$+ve \circlearrowleft \Sigma M_C = \quad 0 - 295.71 + 197.14 - (10.0 \times H_A) + (10.0 \times V_A) = 0$

$V_A = (98.57 + 10.0 H_A)/10.0 = [98.57 + (10.0 \times 39.43)]/10.0 \quad \therefore V_A = +49.29 \text{ kN} \uparrow$

Consider the equilibrium of the right-hand side of the frame at joints F and E_2.

$+ve \circlearrowright \Sigma M_F = 0 \qquad + 98.57 - (6.0 \times H_H) = 0 \qquad \therefore H_H = +16.43 \text{ kN} \longrightarrow$

$+ve \circlearrowright \Sigma M_{E2} = 0 \qquad - 98.57 - (16.43 \times 6.0) - (12.0 \times V_H) = 0 \quad \therefore V_H = -16.43 \text{ kN} \downarrow$

Solution

Topic: Plastic Analysis – Rigid Jointed Frames 2
Problem Number: 8.15 – Kinematic Method   **Page No. 5**

Consider the horizontal equilibrium of the complete structure.

$$+v \longrightarrow \Sigma F_x = 0 \qquad H_A + H_G + H_H - 40.0 - 30.0 = 0$$
$$39.43 + H_G + 16.43 - 70.0 = 0 \qquad \therefore H_G = + \textbf{14.14 kN} \longrightarrow$$

Consider the vertical equilibrium of the complete structure.

$$+ve \uparrow \Sigma F_z = 0 \qquad V_A + V_G + V_H - 80.0 = 0$$
$$49.29 + V_G - 16.43 - 80.0 = 0 \qquad \therefore V_G = + \textbf{47.14 kN} \longrightarrow$$

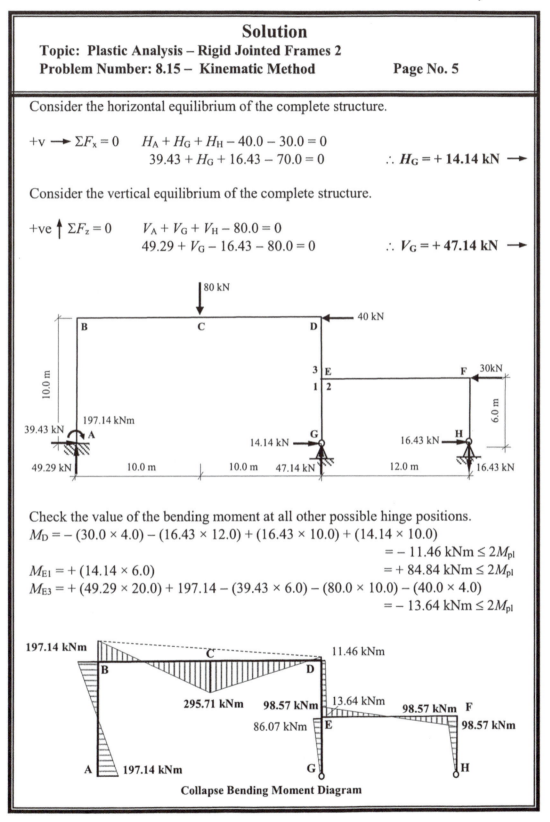

Check the value of the bending moment at all other possible hinge positions.
$$M_D = - (30.0 \times 4.0) - (16.43 \times 12.0) + (16.43 \times 10.0) + (14.14 \times 10.0)$$
$$= - 11.46 \text{ kNm} \leq 2M_{pl}$$
$$M_{E1} = + (14.14 \times 6.0) \qquad\qquad\qquad = + 84.84 \text{ kNm} \leq 2M_{pl}$$
$$M_{E3} = + (49.29 \times 20.0) + 197.14 - (39.43 \times 6.0) - (80.0 \times 10.0) - (40.0 \times 4.0)$$
$$= - 13.64 \text{ kNm} \leq 2M_{pl}$$

Collapse Bending Moment Diagram

8.12 *Gable Mechanism*

Another type of *independent* mechanism which is characteristic of pitched roof portal frames is the Gable Mechanism, as shown in Figure 8.23 with simple beam and sway mechanisms.

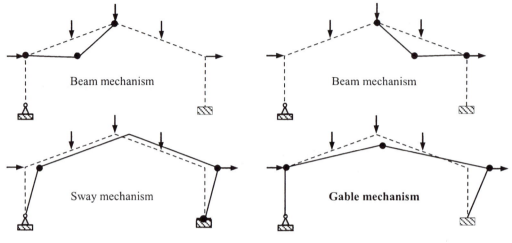

Figure 8.23

In the beam and gable mechanisms the rafter of the frame is sloping and it is necessary to evaluate the displacement in the direction of the load. i.e. not necessarily perpendicular to the member as in previous examples. Consider the typical sloping member ABC shown in Figure 8.24 (a) which is subject to a horizontal and a vertical load as indicated.

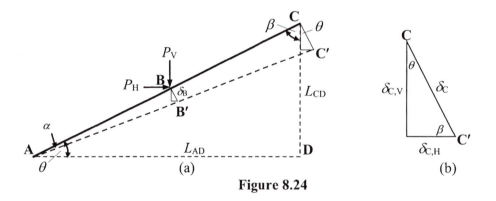

Figure 8.24

Assume that during the formation of a mechanism the centre–of–rotation of the member is point A and point C displaces in a perpendicular direction to ABC to point C'. For small rotations (α) of member ABC, $\delta_C = \text{C-C'} = L_{AC}\alpha$

The vertical and horizontal displacements of C are given by $\delta_{C,\text{vertical}} = \delta_C \cos\theta = L_{AD}\alpha$ and $\delta_{C,\text{horizontal}} = \delta_C \sin\theta = L_{CD}\alpha$ as shown in Figure 8.24(b), where θ is the angle of the member ABC to the horizontal. The vertical and horizontal displacements at point B can be determined in a similar manner.

These values can then be used in the calculation of external work for the work equation.

8.13 *Instantaneous Centre of Rotation*

In more complex frames it is convenient to use the '*instantaneous centre of rotation method*' when developing a collapse mechanism. The technique is explained below in relation to a simple rectangular portal frame and subsequently in Example 8.7.

Consider the asymmetric rectangular frame shown in Figure 8.25 in which there are two independent mechanisms, one beam and one sway. The frame requires three hinges to cause collapse. Both mechanisms can combine to produce a collapse mechanism with hinges developing at A, C and D. In this mechanism there are three rigid–links, AB′C′, C′D′ and D′E as shown.

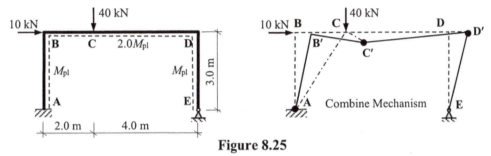

Figure 8.25

The centre–of–rotation for link AB′C′ is at A and the remote end C moves in a direction perpendicular to line AC shown. The centre–of–rotation for link D′E is at E and the remote end D moves in a direction perpendicular to line ED shown.

In the case of link C′D′, the centre–of–rotation must be determined by considering the direction of movement of each end. C′ moves in a direction perpendicular to AC and consequently the centre–of–rotation must lie on an extension of this line. Similarly, it must also lie on a line perpendicular to the movement of D, i.e. on an extension of ED. This construction is shown in Figure 8.26(a). The position of this centre–of–rotation is known as the instantaneous centre–of–rotation and occurs at the instant of collapse.

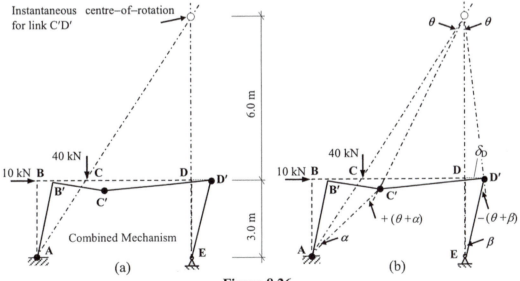

Figure 8.26

The work equations can be developed and the required M_{pl} value determined by considering the rotation of the hinges and the displacements of the loads. Consider the geometry shown in Figure 8.26 (b) and equate the displacements in terms of θ, β and α as follows:

The horizontal displacement DD' $\delta_D = 3.0\beta = 6.0\theta$ $\therefore \beta = 2.0\theta$
The rotation at the hinge at D $(\theta + \beta) = 3.0\theta$

The vertical displacement CC' $\delta_{C, \text{vertical}} = 2.0\alpha = 4.0\theta$ $\therefore \alpha = 2.0\theta$
The rotation at the hinge at C $(\theta + \alpha) = 3.0\theta$
(**Note:** equating the horizontal displacement of point C will give the same result,
                  i.e. $\delta_{C, \text{horizontal}} = 3.0\alpha = 6.0\theta$)

The rotation at the hinge at A = $\alpha = 2.0\theta$

Note: no internal work is done at support E
Internal Work Done = External Work Done
$M_{pl}(\alpha) + 2.0M_{pl}(\theta + \alpha) + M_{pl}(\theta + \beta) = (10.0 \times \delta_D) + (40.0 \times \delta_{C, \text{vertical}})$
$M_{pl}(2.0\theta) + 2.0M_{pl}(\theta + 2.0\theta) + M_{pl}(\theta + 2.0\theta) = (10.0 \times 6.0\theta) + (40.0 \times 4.0\theta)$
$11M_{pl}\theta = 220.0\theta$ $\therefore \boldsymbol{M_{pl} = 20.0 \text{ kNm}}$

The reader should confirm that this is the critical value by calculating the reactions and checking that the bending moment on the frame does not exceed the appropriate M_{pl} value for any member. (**Note:** In the case of member BCD this is equal to $2.0M_{pl} = 40$ kNm).

8.14 *Example 8.7: Pitched Roof Frame*

A non–uniform, asymmetric frame is pinned at support A, fixed at support F and is required to carry collapse loads as indicated in Figure 8.27. Determine the minimum required value of M_{pl}.

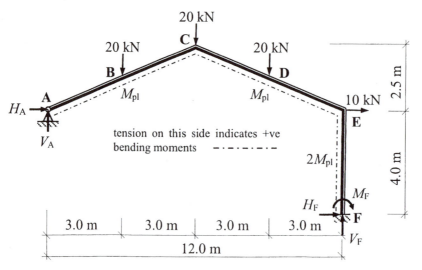

Figure 8.27

Factored loads: as given

Number of degrees-of-indeterminacy $I_D = [(3m + r) - 3n] = [(3 \times 3) + 5) - (3 \times 4)] = 2$
Number of possible hinge positions $p = 5$ (B, C, D, E and F)
Number of independent mechanisms $= (p - I_D) = (5 - 2) = 3$
(i.e. 2 beam mechanisms, 1 gable mechanism).

Kinematic Method:
Consider each independent mechanism separately.

Mechanism (i): Beam ABC

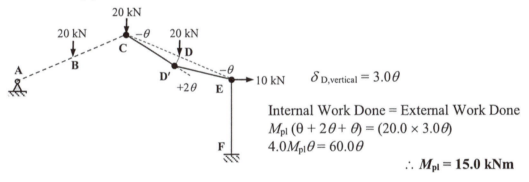

$\delta_{B,vertical} = 3.0\theta$
Note: no internal work is done at support A

Internal Work Done = External Work Done
$M_{pl}(2\theta + \theta) = (20.0 \times 3.0\theta)$
$3.0M_{pl}\theta = 60.0\theta$
$\therefore M_{pl} = \textbf{20.0 kNm}$

Mechanism (ii): Beam CDE

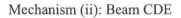

$\delta_{D,vertical} = 3.0\theta$

Internal Work Done = External Work Done
$M_{pl}(\theta + 2\theta + \theta) = (20.0 \times 3.0\theta)$
$4.0M_{pl}\theta = 60.0\theta$
$\therefore M_{pl} = \textbf{15.0 kNm}$

Mechanism (iii): Gable

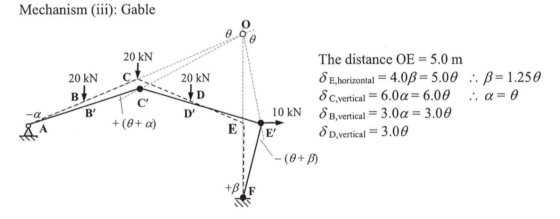

The distance OE = 5.0 m
$\delta_{E,horizontal} = 4.0\beta = 5.0\theta$ $\therefore \beta = 1.25\theta$
$\delta_{C,vertical} = 6.0\alpha = 6.0\theta$ $\therefore \alpha = \theta$
$\delta_{B,vertical} = 3.0\alpha = 3.0\theta$
$\delta_{D,vertical} = 3.0\theta$

Internal Work Done = $M_{pl}(\theta + \alpha) + M_{pl}(\theta + \beta) + 2.0M_{pl}(\beta)$
   = $M_{pl}(2.0\theta) + M_{pl}(\theta + 1.25\theta) + 2.0M_{pl}(1.25\theta) = 6.75M_{pl}\theta$

External Work Done = $(20.0 \times \delta_{B,vertical}) + (20.0 \times \delta_{C,vertical}) + (20.0 \times \delta_{D,vertical})$
   $+ (10.0 \times \delta_{E,horizontal})$
   = $(20.0 \times 3.0\theta) + (20.0 \times 6.0\theta) + (20.0 \times 3.0\theta) + (10.0 \times 5.0\theta)$
   = 290θ

Internal Work = External Work $\therefore 6.75M_{pl}\theta = 290\theta$ $\therefore M_{pl} = 42.96$ kNm

Combined Mechanism (iv): [2 × mechanism (i)] + mechanism (iii) which eliminates a hinge at C

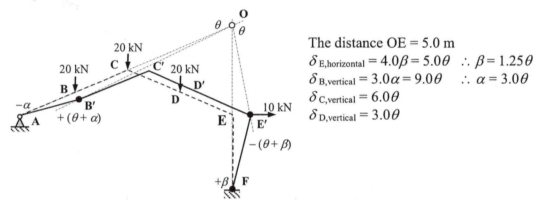

The distance OE = 5.0 m
$\delta_{E,horizontal} = 4.0\beta = 5.0\theta$ $\therefore \beta = 1.25\theta$
$\delta_{B,vertical} = 3.0\alpha = 9.0\theta$ $\therefore \alpha = 3.0\theta$
$\delta_{C,vertical} = 6.0\theta$
$\delta_{D,vertical} = 3.0\theta$

Internal Work Done = $M_{pl}(\theta + \alpha) + M_{pl}(\theta + \beta) + 2.0M_{pl}(\beta)$
   = $M_{pl}(4.0\theta) + M_{pl}(\theta + 1.25\theta) + 2.0M_{pl}(1.25\theta) = 8.75M_{pl}\theta$

External Work Done = $(20.0 \times \delta_{B,vertical}) + (20.0 \times \delta_{C,vertical}) + (20.0 \times \delta_{D,vertical})$
   $+ (10.0 \times \delta_{E,horizontal})$
   = $(20.0 \times 9.0\theta) + (20.0 \times 6.0\theta) + (20.0 \times 3.0\theta) + (10.0 \times 5.0\theta)$
   = 410θ

Internal Work = External Work $\therefore 8.75M_{pl}\theta = 410\theta$ $\therefore M_{pl} = 46.86$ kNm

The reader should confirm that this is the critical value by calculating the reactions and checking that the bending moment on the frame does not exceed the appropriate M_{pl} value for any member. (**Note:** In the case of support F this is equal to $2.0M_{pl} = 93.70$ kNm).

Alternatively, adding the virtual work equations:

Internal Work Done = External Work Done
2 × Mechanism (i) $6.0M_{pl}\theta = 120.0\theta$
Mechanism (iii) $6.75M_{pl}\theta = 290.0\theta$
less $2.0M_{pl}$ for eliminated hinge $\underline{-4.0M_{pl}\theta}$
                                 $8.75M_{pl}\theta = 410.0\theta$ $\therefore M_{pl} = 46.86$ kNm

The combined mechanism can be evaluated in a Table as shown:

Independent and Combined Mechanisms for Example 8.7				
Hinge Position	(i)	(ii)	(iii)	(v) = 2(i)+(iii)
B (M_{pl})	$+2.0\theta$	–	–	$+4.0\theta$
C (M_{pl})	$-\theta$	$-\theta$	$+2.0\theta$	EH ($4.0M_{pl}\theta$)
D (M_{pl})	–	$+2.0\theta$	–	–
E (M_{pl})	–	$-\theta$	-2.25θ	-2.25θ
F ($2M_{pl}$)	–	–	–	–
External Work	60.0θ	60.0θ	290.0θ	**410.0θ**
Internal Work	$3.0M_{pl}\theta$	$4.0M_{pl}\theta$	$6.75M_{pl}\theta$	$12.75M_{pl}\theta$
Eliminated hinges	–	–	–	$4.0M_{pl}\theta$
Combined $M_{pl}\theta$	–	–	–	**$8.75M_{pl}\theta$**
M_{pl} (kNm)	20.0	15.0	42.96	**46.86**

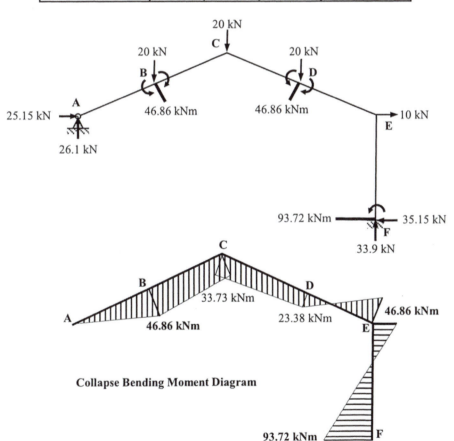

Collapse Bending Moment Diagram

Figure 8.28

8.15   *Problems: Plastic Analysis – Rigid-Jointed Frames 3*

A series of rigid-jointed frames are indicated in Problems 8.16 to 8.21 in which the relative M_{pl} values and the applied **collapse** loads are given. In each case determine the required M_{pl} value, the value of the support reactions and sketch the bending moment diagram.

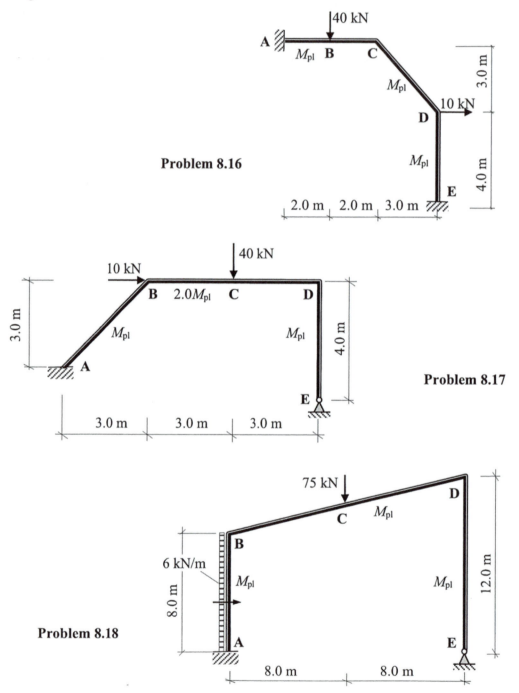

Problem 8.16

Problem 8.17

Problem 8.18

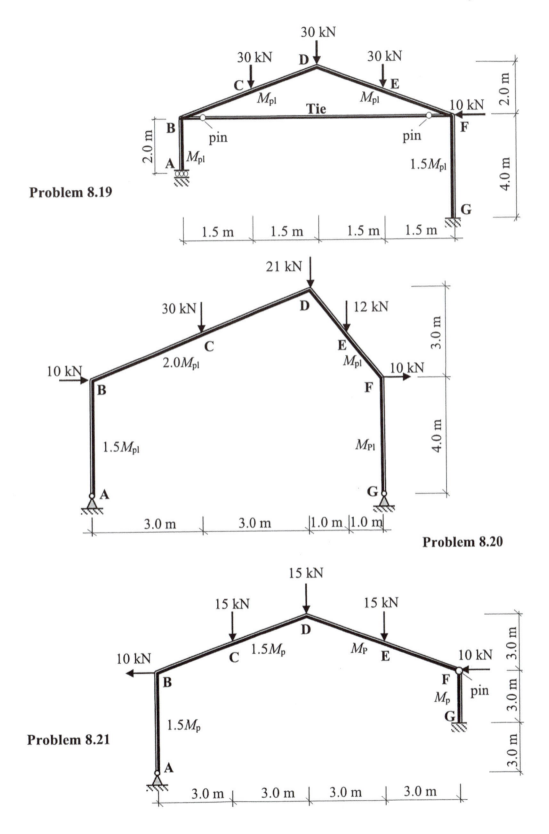

Problem 8.19

Problem 8.20

Problem 8.21

8.16   *Solutions: Plastic Analysis – Rigid-Jointed Frames 3*

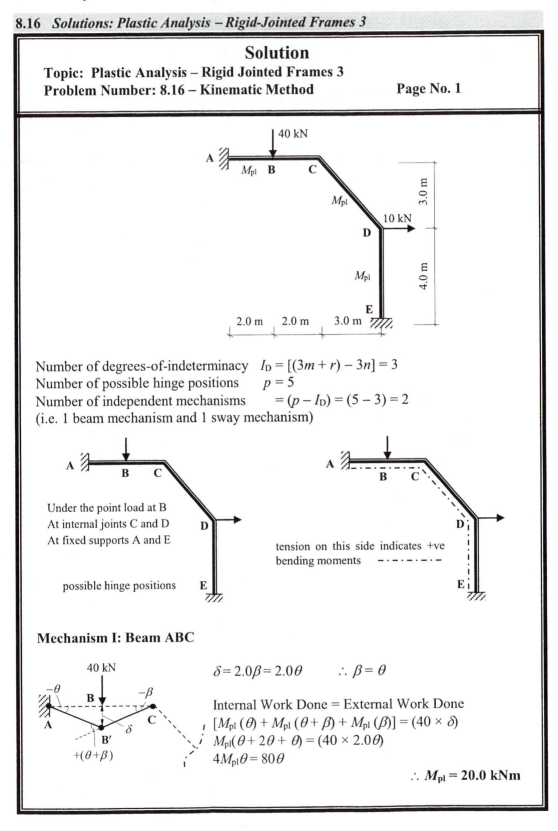

Solution

Topic: Plastic Analysis – Rigid Jointed Frames 3
Problem Number: 8.16 – Kinematic Method   **Page No. 1**

Number of degrees-of-indeterminacy $I_D = [(3m + r) - 3n] = 3$
Number of possible hinge positions $p = 5$
Number of independent mechanisms $= (p - I_D) = (5 - 3) = 2$
(i.e. 1 beam mechanism and 1 sway mechanism)

Under the point load at B
At internal joints C and D
At fixed supports A and E

possible hinge positions

tension on this side indicates +ve
bending moments $-\cdot-\cdot-\cdot-$

Mechanism I: Beam ABC

$\delta = 2.0\beta = 2.0\theta$ $\therefore \beta = \theta$

Internal Work Done = External Work Done
$[M_{pl}\,(\theta) + M_{pl}\,(\theta + \beta) + M_{pl}\,(\beta)] = (40 \times \delta)$
$M_{pl}(\theta + 2\theta + \theta) = (40 \times 2.0\theta)$
$4M_{pl}\theta = 80\theta$

$\therefore M_{pl} = 20.0 \text{ kNm}$

Solution

Topic: Plastic Analysis – Rigid Jointed Frames 3
Problem Number: 8.16 – Kinematic Method   Page No. 2

Mechanism II: Sway (Use the instantaneous centre of rotation technique)

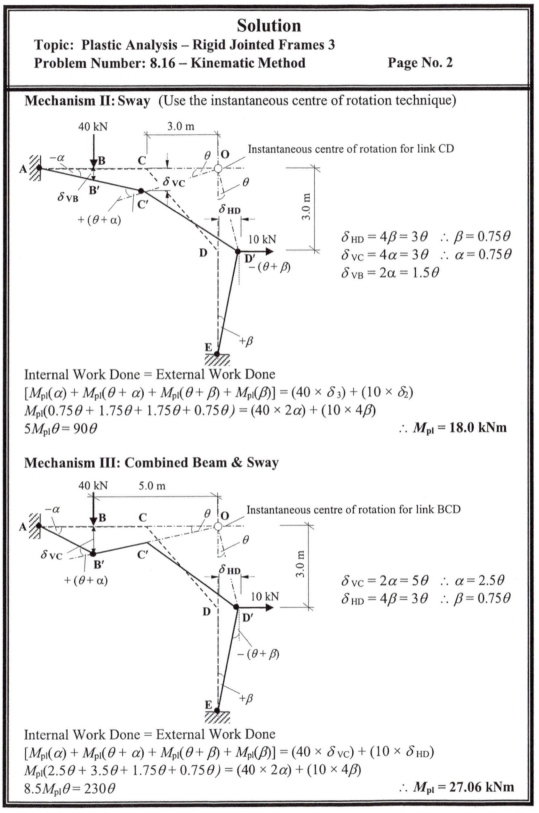

$$\delta_{HD} = 4\beta = 3\theta \quad \therefore \ \beta = 0.75\theta$$
$$\delta_{VC} = 4\alpha = 3\theta \quad \therefore \ \alpha = 0.75\theta$$
$$\delta_{VB} = 2\alpha = 1.5\theta$$

Internal Work Done = External Work Done

$[M_{pl}(\alpha) + M_{pl}(\theta + \alpha) + M_{pl}(\theta + \beta) + M_{pl}(\beta)] = (40 \times \delta_3) + (10 \times \delta_2)$
$M_{pl}(0.75\theta + 1.75\theta + 1.75\theta + 0.75\theta) = (40 \times 2\alpha) + (10 \times 4\beta)$
$5M_{pl}\theta = 90\theta$ $\therefore \ \boldsymbol{M_{pl} = 18.0 \ kNm}$

Mechanism III: Combined Beam & Sway

$$\delta_{VC} = 2\alpha = 5\theta \quad \therefore \ \alpha = 2.5\theta$$
$$\delta_{HD} = 4\beta = 3\theta \quad \therefore \ \beta = 0.75\theta$$

Internal Work Done = External Work Done

$[M_{pl}(\alpha) + M_{pl}(\theta + \alpha) + M_{pl}(\theta + \beta) + M_{pl}(\beta)] = (40 \times \delta_{VC}) + (10 \times \delta_{HD})$
$M_{pl}(2.5\theta + 3.5\theta + 1.75\theta + 0.75\theta) = (40 \times 2\alpha) + (10 \times 4\beta)$
$8.5M_{pl}\theta = 230\theta$ $\therefore \ \boldsymbol{M_{pl} = 27.06 \ kNm}$

Solution

Topic: Plastic Analysis – Rigid Jointed Frames 3
Problem Number: 8.16 – Kinematic Method   **Page No. 3**

In mechanism I the rotation at joint C $= -\beta = -\theta$
In mechanism II the rotation at joint C $= +(\theta + \alpha) = +1.75\theta$
Adding equations for Mechanisms $[(1.75 \times I) + II]$
$7M_{pl}\theta = 140\theta$
$5M_{pl}\theta = 90\theta$
$\underline{-3.5M_{pl}\theta}$ [allowing for the hinge eliminated at joint C: $(2 \times 1.75\theta)$]
$8.5M_{pl}\theta = 230\theta$ $\therefore M_{pl} = \mathbf{27.06\ kNm}$ **as before**

The value of M_{pl} obtained (27.06 kNm) should be checked by ensuring that the bending moment in the frame does not exceed this value at any location.

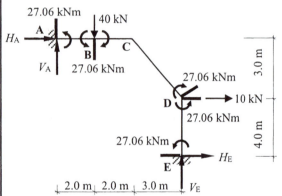

Under the point load at B and at support E there is tension *inside* the frame and consequently the bending moment is positive at these points.

The rotations at A and D induce tension on the *outside* of the frame and hence negative bending moments.

Consider the equilibrium of the right-hand side of the frame at point D and the left – hand side at B.

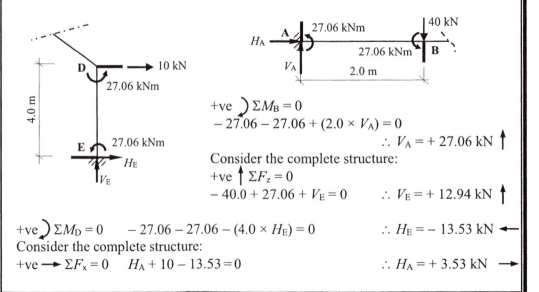

$+ve\ \circlearrowleft \Sigma M_B = 0$
$-27.06 - 27.06 + (2.0 \times V_A) = 0$
                           $\therefore V_A = +27.06\ kN\ \uparrow$
Consider the complete structure:
$+ve\ \uparrow \Sigma F_z = 0$
$-40.0 + 27.06 + V_E = 0$ $\therefore V_E = +12.94\ kN\ \uparrow$

$+ve\ \circlearrowright \Sigma M_D = 0$ $-27.06 - 27.06 - (4.0 \times H_E) = 0$ $\therefore H_E = -13.53\ kN\ \leftarrow$
Consider the complete structure:
$+ve\ \rightarrow \Sigma F_x = 0$ $H_A + 10 - 13.53 = 0$ $\therefore H_A = +3.53\ kN\ \rightarrow$

Solution

Topic: Plastic Analysis – Rigid Jointed Frames 3
Problem Number: 8.16 – Kinematic Method   **Page No. 4**

Bending moment at C (consider forces to the left–hand side) :

$$M_C = -27.06 + (27.06 \times 4.0) - (40.0 \times 2.0) = +1.18 \text{ kNm} \le M_{pl}$$

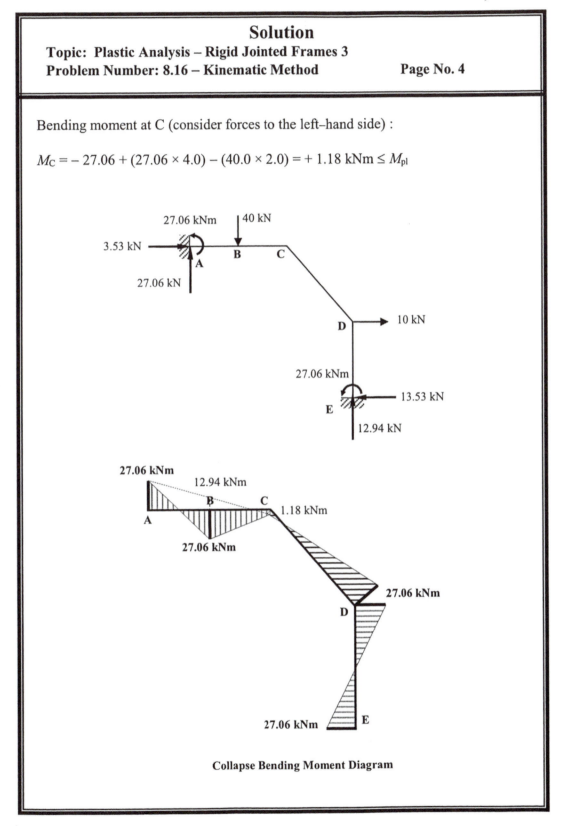

Collapse Bending Moment Diagram

Solution

Topic: Plastic Analysis – Rigid Jointed Frames 3
Problem Number: 8.17 – Kinematic Method   **Page No. 1**

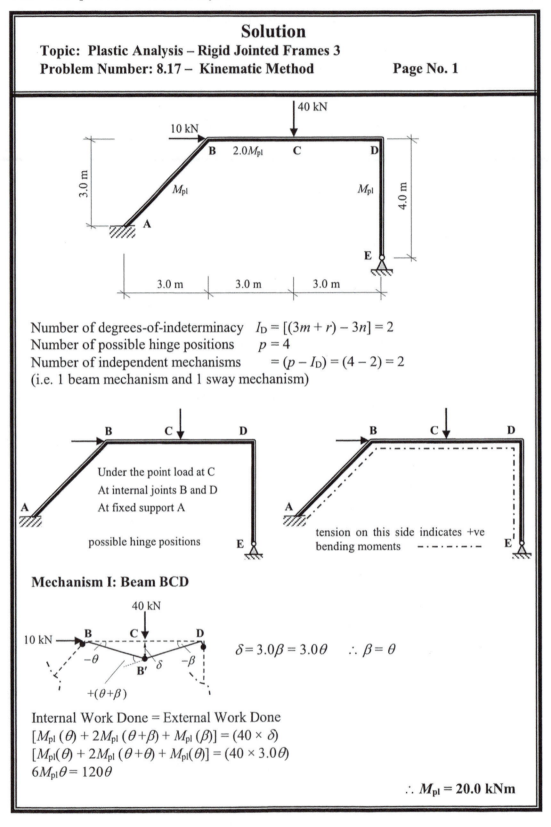

Number of degrees-of-indeterminacy $I_D = [(3m + r) - 3n] = 2$
Number of possible hinge positions $p = 4$
Number of independent mechanisms $= (p - I_D) = (4 - 2) = 2$
(i.e. 1 beam mechanism and 1 sway mechanism)

Under the point load at C
At internal joints B and D
At fixed support A

possible hinge positions

tension on this side indicates +ve
bending moments

Mechanism I: Beam BCD

$\delta = 3.0\beta = 3.0\theta$ $\therefore \beta = \theta$

Internal Work Done = External Work Done
$[M_{pl}(\theta) + 2M_{pl}(\theta + \beta) + M_{pl}(\beta)] = (40 \times \delta)$
$[M_{pl}(\theta) + 2M_{pl}(\theta + \theta) + M_{pl}(\theta)] = (40 \times 3.0\theta)$
$6M_{pl}\theta = 120\theta$

$\therefore M_{pl} = 20.0 \text{ kNm}$

Solution
Topic: Plastic Analysis – Rigid Jointed Frames 3
Problem Number: 8.17 – Kinematic Method   Page No. 2

Mechanism II: Sway (Use the instantaneous centre of rotation technique)

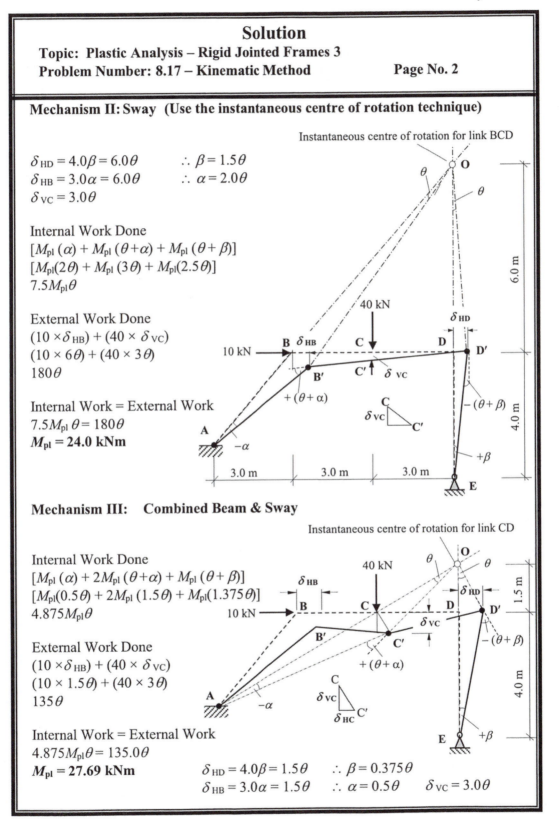

$\delta_{HD} = 4.0\beta = 6.0\theta$ $\therefore \beta = 1.5\theta$
$\delta_{HB} = 3.0\alpha = 6.0\theta$ $\therefore \alpha = 2.0\theta$
$\delta_{VC} = 3.0\theta$

Internal Work Done
$[M_{pl}(\alpha) + M_{pl}(\theta+\alpha) + M_{pl}(\theta+\beta)]$
$[M_{pl}(2\theta) + M_{pl}(3\theta) + M_{pl}(2.5\theta)]$
$7.5M_{pl}\theta$

External Work Done
$(10 \times \delta_{HB}) + (40 \times \delta_{VC})$
$(10 \times 6\theta) + (40 \times 3\theta)$
180θ

Internal Work = External Work
$7.5M_{pl}\theta = 180\theta$
$M_{pl} = \textbf{24.0 kNm}$

Mechanism III: Combined Beam & Sway

Internal Work Done
$[M_{pl}(\alpha) + 2M_{pl}(\theta+\alpha) + M_{pl}(\theta+\beta)]$
$[M_{pl}(0.5\theta) + 2M_{pl}(1.5\theta) + M_{pl}(1.375\theta)]$
$4.875M_{pl}\theta$

External Work Done
$(10 \times \delta_{HB}) + (40 \times \delta_{VC})$
$(10 \times 1.5\theta) + (40 \times 3\theta)$
135θ

Internal Work = External Work
$4.875M_{pl}\theta = 135.0\theta$
$M_{pl} = \textbf{27.69 kNm}$

$\delta_{HD} = 4.0\beta = 1.5\theta$ $\therefore \beta = 0.375\theta$
$\delta_{HB} = 3.0\alpha = 1.5\theta$ $\therefore \alpha = 0.5\theta$ $\delta_{VC} = 3.0\theta$

Solution
Topic: Plastic Analysis – Rigid Jointed Frames 3
Problem Number: 8.17 – Kinematic Method   **Page No. 3**

In mechanism I the rotation at joint $B = -\theta$
In mechanism II the rotation at joint $B = +(\alpha + \theta) = +3.0\theta$
Adding equations for Mechanisms $[(3.0 \times I) + II]$
$18.0M_{pl}\theta = 360\theta$
$7.5M_{pl}\theta = 180\theta$
$\underline{-6.0M_{pl}}$ [allowing for the hinge eliminated at joint B: $(2 \times 3\theta)$]
$19.5M_{pl}\theta = 540\theta$ $\therefore M_{pl} = $ **27.69 kNm as before**

The value of M_{pl} obtained (27.69 kNm) should be checked by ensuring that the bending moment in the frame does not exceed this value at any location.

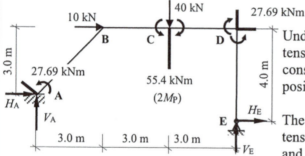

Under the point load at C there is tension *inside* the frame and consequently the bending moment is positive at this point.

The rotations at A and D induce tension on the *outside* of the frame and hence negative bending moments.

Consider the equilibrium of the right-hand side of the frame at joint D and the right–hand side at C.

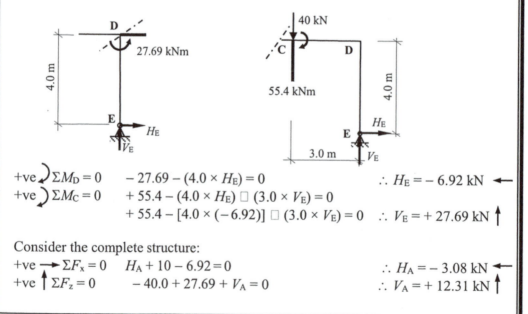

$+ve\ \circlearrowleft \Sigma M_D = 0$ $-27.69 - (4.0 \times H_E) = 0$ $\therefore H_E = -6.92 \text{ kN} \leftarrow$
$+ve\ \circlearrowright \Sigma M_C = 0$ $+55.4 - (4.0 \times H_E) \ \square\ (3.0 \times V_E) = 0$
                              $+55.4 - [4.0 \times (-6.92)] \ \square\ (3.0 \times V_E) = 0$ $\therefore V_E = +27.69 \text{ kN} \uparrow$

Consider the complete structure:
$+ve \rightarrow \Sigma F_x = 0$ $H_A + 10 - 6.92 = 0$ $\therefore H_A = -3.08 \text{ kN} \leftarrow$
$+ve \uparrow \Sigma F_z = 0$ $-40.0 + 27.69 + V_A = 0$ $\therefore V_A = +12.31 \text{ kN} \uparrow$

Solution

Topic: Plastic Analysis – Rigid Jointed Frames 3
Problem Number: 8.17 – Kinematic Method   **Page No. 4**

Consider the equilibrium of the left-hand side of the frame at joint B.

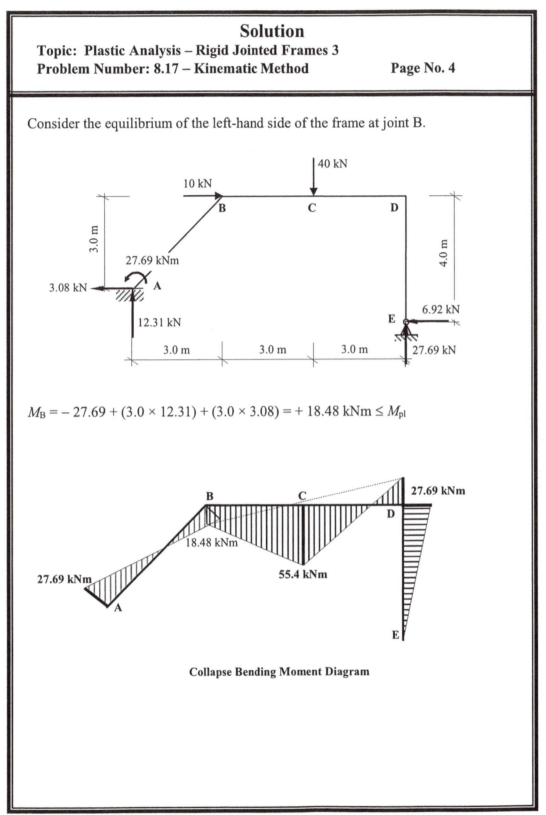

$$M_B = -27.69 + (3.0 \times 12.31) + (3.0 \times 3.08) = +18.48 \text{ kNm} \leq M_{pl}$$

Collapse Bending Moment Diagram

Solution

Topic: Plastic Analysis – Rigid Jointed Frames 3
Problem Number: 8.18– Kinematic Method   **Page No. 1**

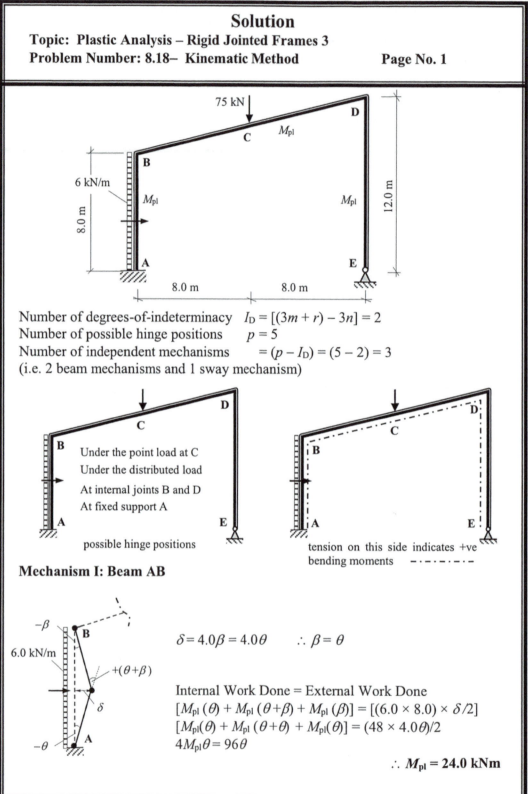

Number of degrees-of-indeterminacy $I_D = [(3m + r) - 3n] = 2$
Number of possible hinge positions $p = 5$
Number of independent mechanisms $= (p - I_D) = (5 - 2) = 3$
(i.e. 2 beam mechanisms and 1 sway mechanism)

Under the point load at C
Under the distributed load
At internal joints B and D
At fixed support A

possible hinge positions

tension on this side indicates +ve
bending moments   — · — · — · —

Mechanism I: Beam AB

$\delta = 4.0\beta = 4.0\theta$ $\therefore \beta = \theta$

Internal Work Done = External Work Done
$[M_{pl}(\theta) + M_{pl}(\theta+\beta) + M_{pl}(\beta)] = [(6.0 \times 8.0) \times \delta/2]$
$[M_{pl}(\theta) + M_{pl}(\theta+\theta) + M_{pl}(\theta)] = (48 \times 4.0\theta)/2$
$4M_{pl}\theta = 96\theta$

$\therefore M_{pl} = 24.0$ kNm

Solution
Topic: Plastic Analysis – Rigid Jointed Frames 3
Problem Number: 8.18 – Kinematic Method   **Page No. 2**

Mechanism II: Beam BCD

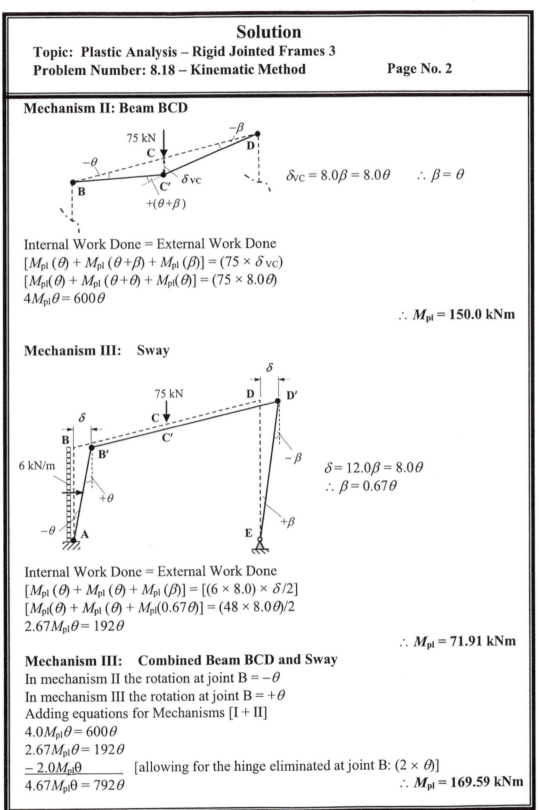

$$\delta_{VC} = 8.0\beta = 8.0\theta \qquad \therefore \beta = \theta$$

Internal Work Done = External Work Done
$$[M_{pl}(\theta) + M_{pl}(\theta+\beta) + M_{pl}(\beta)] = (75 \times \delta_{VC})$$
$$[M_{pl}(\theta) + M_{pl}(\theta+\theta) + M_{pl}(\theta)] = (75 \times 8.0\theta)$$
$$4M_{pl}\theta = 600\theta$$

$$\therefore M_{pl} = 150.0 \text{ kNm}$$

Mechanism III: Sway

$$\delta = 12.0\beta = 8.0\theta$$
$$\therefore \beta = 0.67\theta$$

Internal Work Done = External Work Done
$$[M_{pl}(\theta) + M_{pl}(\theta) + M_{pl}(\beta)] = [(6 \times 8.0) \times \delta/2]$$
$$[M_{pl}(\theta) + M_{pl}(\theta) + M_{pl}(0.67\theta)] = (48 \times 8.0\theta)/2$$
$$2.67M_{pl}\theta = 192\theta$$

$$\therefore M_{pl} = 71.91 \text{ kNm}$$

Mechanism III: Combined Beam BCD and Sway
In mechanism II the rotation at joint B = $-\theta$
In mechanism III the rotation at joint B = $+\theta$
Adding equations for Mechanisms [I + II]
$$4.0M_{pl}\theta = 600\theta$$
$$2.67M_{pl}\theta = 192\theta$$
$$\underline{-2.0M_{pl}\theta} \qquad \text{[allowing for the hinge eliminated at joint B: } (2 \times \theta)]$$
$$4.67M_{pl}\theta = 792\theta \qquad\qquad \therefore M_{pl} = 169.59 \text{ kNm}$$

Solution

Topic: Plastic Analysis – Rigid Jointed Frames 3
Problem Number: 8.18 – Kinematic Method   **Page No. 3**

Using the instantaneous centre of rotation technique.

$\delta_{HD} = 12.0\beta = 8.0\theta$ $\therefore \beta = 0.67\theta$

$\delta_{VC} = 8.0\alpha = 8.0\theta$ $\therefore \alpha = \theta$

$\delta_H = 4.0\alpha$ (average displlacement)

Internal Work Done

$[M_{pl}(\alpha) + M_{pl}(\theta + \alpha) + M_{pl}(\theta + \beta)]$
$[M_{pl}(\theta) + M_{pl}(2\theta) + M_{pl}(1.67\theta)]$
$4.67M_{pl}\theta$

External Work Done

$[(75 \times \delta_{VC}) + (6 \times 8) \times \delta_H]$
$[(75 \times 8\theta) + (48 \times 4\theta)]$
792θ

Internal Work = External Work

$4.67M_{pl}\theta = 792\theta$

$M_{pl} = 169.59$ **kNm**   **as before**

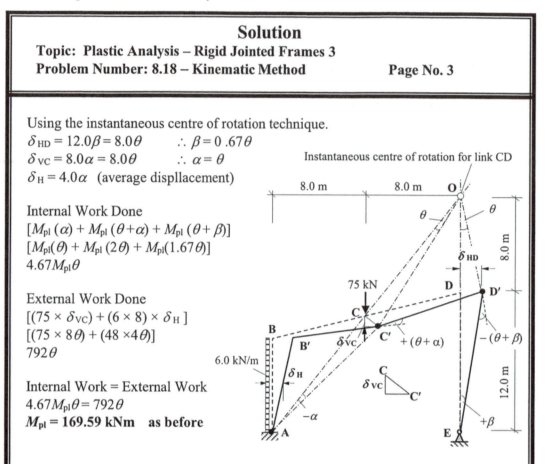

The value of M_{pl} obtained (169.59 kNm) should be checked by ensuring that the bending moment in the frame does not exceed this value at any location.

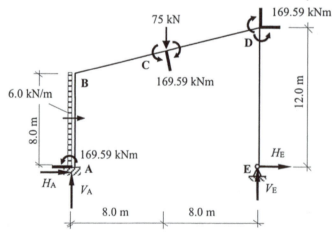

Under the point load at C there is tension *inside* the frame and consequently the bending moment is positive at this point.

The rotations at A and D induce tension on the *outside* of the frame and hence negative bending moments.

Solution

Topic: Plastic Analysis – Rigid Jointed Frames 3
Problem Number: 8.18 – Kinematic Method   **Page No. 4**

Consider the equilibrium of the right-hand side of the frame at joint D and the right–hand side at C.

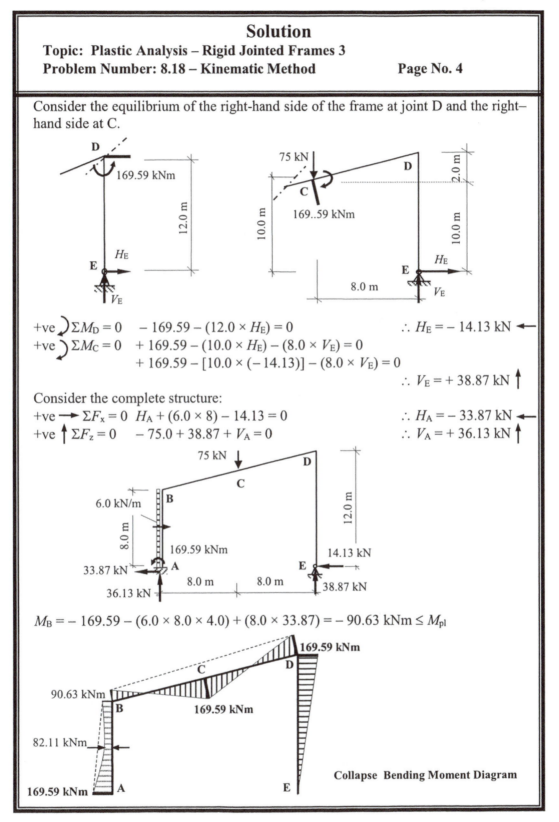

$+ve\ \curvearrowright\Sigma M_D = 0$ $-169.59 - (12.0 \times H_E) = 0$ $\therefore H_E = -14.13\ \text{kN} \leftarrow$

$+ve\ \curvearrowright\Sigma M_C = 0$ $+169.59 - (10.0 \times H_E) - (8.0 \times V_E) = 0$
               $+169.59 - [10.0 \times (-14.13)] - (8.0 \times V_E) = 0$

                                        $\therefore V_E = +38.87\ \text{kN} \uparrow$

Consider the complete structure:

$+ve \longrightarrow \Sigma F_x = 0$ $H_A + (6.0 \times 8) - 14.13 = 0$ $\therefore H_A = -33.87\ \text{kN} \leftarrow$

$+ve \uparrow \Sigma F_z = 0$ $-75.0 + 38.87 + V_A = 0$ $\therefore V_A = +36.13\ \text{kN} \uparrow$

$M_B = -169.59 - (6.0 \times 8.0 \times 4.0) + (8.0 \times 33.87) = -90.63\ \text{kNm} \le M_{pl}$

Collapse Bending Moment Diagram

Solution

Topic: Plastic Analysis – Rigid Jointed Frames 3
Problem Number: 8.19 –   Kinematic Method   **Page No. 1**

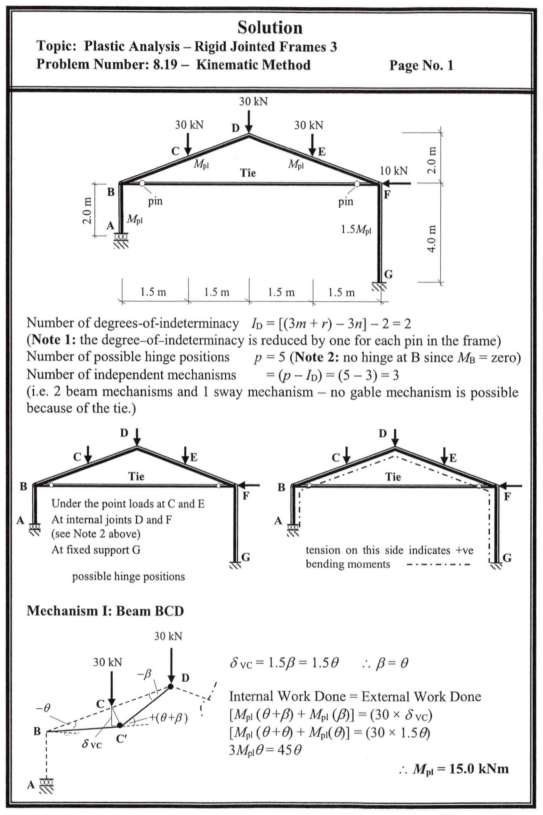

Number of degrees-of-indeterminacy $I_D = [(3m + r) - 3n] - 2 = 2$
(**Note 1:** the degree–of–indeterminacy is reduced by one for each pin in the frame)
Number of possible hinge positions $p = 5$ (**Note 2:** no hinge at B since M_B = zero)
Number of independent mechanisms $= (p - I_D) = (5 - 3) = 3$
(i.e. 2 beam mechanisms and 1 sway mechanism – no gable mechanism is possible
because of the tie.)

Under the point loads at C and E
At internal joints D and F
(see Note 2 above)
At fixed support G

possible hinge positions

tension on this side indicates +ve
bending moments   – · – · – · –

Mechanism I: Beam BCD

$\delta_{VC} = 1.5\beta = 1.5\theta$ $\therefore \beta = \theta$

Internal Work Done = External Work Done
$[M_{pl} (\theta + \beta) + M_{pl} (\beta)] = (30 \times \delta_{VC})$
$[M_{pl} (\theta + \theta) + M_{pl}(\theta)] = (30 \times 1.5\theta)$
$3M_{pl}\theta = 45\theta$

$\therefore M_{pl} = 15.0$ kNm

Solution

Topic: Plastic Analysis – Rigid Jointed Frames 3
Problem Number: 8.19 – Kinematic Method   **Page No. 2**

Mechanism II: Beam DEF

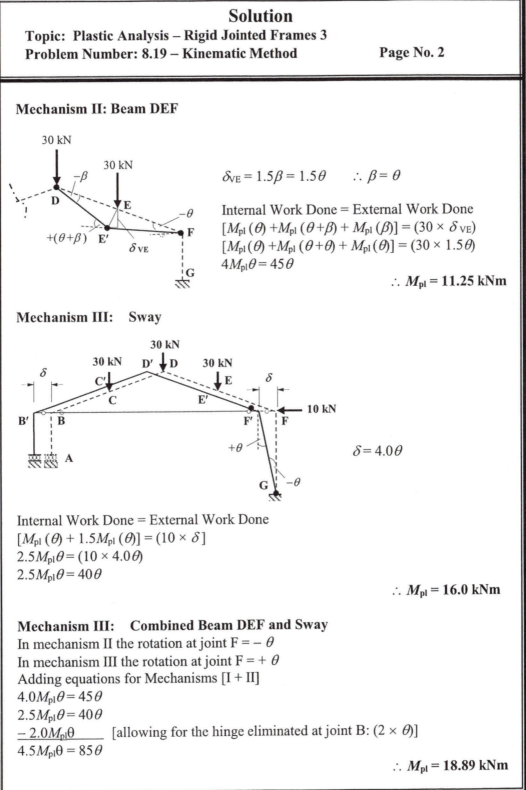

$\delta_{VE} = 1.5\beta = 1.5\theta$ $\therefore \beta = \theta$

Internal Work Done = External Work Done
$[M_{pl}(\theta) + M_{pl}(\theta+\beta) + M_{pl}(\beta)] = (30 \times \delta_{VE})$
$[M_{pl}(\theta) + M_{pl}(\theta+\theta) + M_{pl}(\theta)] = (30 \times 1.5\theta)$
$4M_{pl}\theta = 45\theta$

$\therefore M_{pl} = 11.25$ kNm

Mechanism III: Sway

$\delta = 4.0\theta$

Internal Work Done = External Work Done
$[M_{pl}(\theta) + 1.5M_{pl}(\theta)] = (10 \times \delta]$
$2.5M_{pl}\theta = (10 \times 4.0\theta)$
$2.5M_{pl}\theta = 40\theta$

$\therefore M_{pl} = 16.0$ kNm

Mechanism III: Combined Beam DEF and Sway

In mechanism II the rotation at joint F = $-\theta$
In mechanism III the rotation at joint F = $+\theta$
Adding equations for Mechanisms [I + II]
$4.0M_{pl}\theta = 45\theta$
$2.5M_{pl}\theta = 40\theta$
$\underline{-2.0M_{pl}\theta}$ [allowing for the hinge eliminated at joint B: $(2 \times \theta)$]
$4.5M_{pl}\theta = 85\theta$

$\therefore M_{pl} = 18.89$ kNm

Solution

Topic: Plastic Analysis – Rigid Jointed Frames 3

Problem Number: 8.19 – Kinematic Method   **Page No. 3**

Using the instantaneous centre of rotation technique.

$\delta_{VE} = 1.5\beta = 1.5\theta$ $\therefore \beta = \theta$
$1.5\alpha = 1.5\beta$ $\therefore \alpha = \beta$
$\delta_{HF} = 4.0\beta = 4.0\theta$

Internal Work Done
$[M_{pl}(\alpha) + M_{pl}(\theta+\beta) + 1.5M_{pl}(\beta)]$
$[M_{pl}(\theta) + M_{pl}(2\theta) + 1.5M_{pl}(\theta)]$
$4.5M_{pl}\theta$

External Work Done
$[(30 \times \delta_{VE}) + (10 \times \delta_{HF}]$
$[(30 \times 1.5\theta) + (10 \times 4\theta)]$
85θ

Internal Work = External Work
$4.5M_{pl}\theta = 85\theta$
$M_{pl} = \textbf{18.89 kNm}$ **as before**

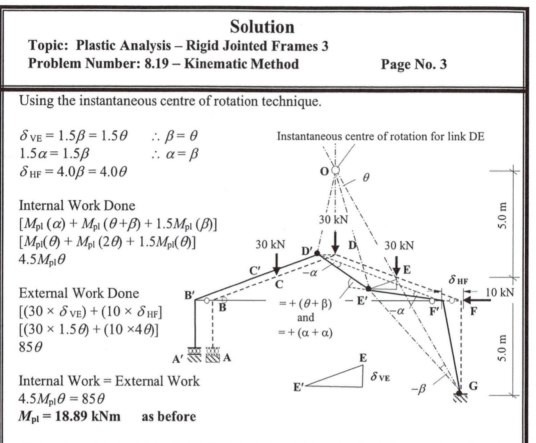

The value of M_{pl} obtained (18.89 kNm) should be checked by ensuring that the bending moment in the frame does not exceed this value at any location.

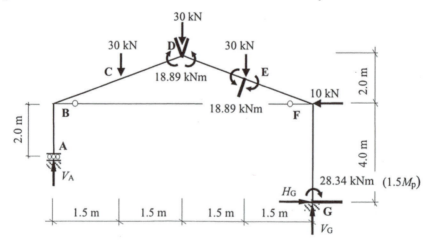

Under the point load at E there is tension *inside* the frame and consequently the bending moment is positive at this point.
The rotations at G and at joint D induce tension on the *outside* of the frame and hence negative bending moments.

Solution

Topic: Plastic Analysis – Rigid Jointed Frames 3
Problem Number: 8.19 – Kinematic Method   **Page No. 4**

Consider the complete structure:

+ve $\circlearrowright$ $\Sigma M_G = 0$

$+ 28.34 - (10.0 \times 4.0) - (30.0 \times 1.5) - (30.0 \times 3.0) - (30.0 \times 4.5) + (6.0 \times V_A) = 0$
$\therefore V_A = + 46.94$ kN $\uparrow$

+ve $\uparrow$ $\Sigma F_z = 0$ $+ 46.94 - 30.0 - 30.0 - 30.0 + V_G = 0$ $\therefore V_G = + 43.06$ kN $\uparrow$

+ve $\longrightarrow$ $\Sigma F_x = 0$ $H_G - 10.0 = 0$ $\therefore H_G = + 10.0$ kN $\longrightarrow$

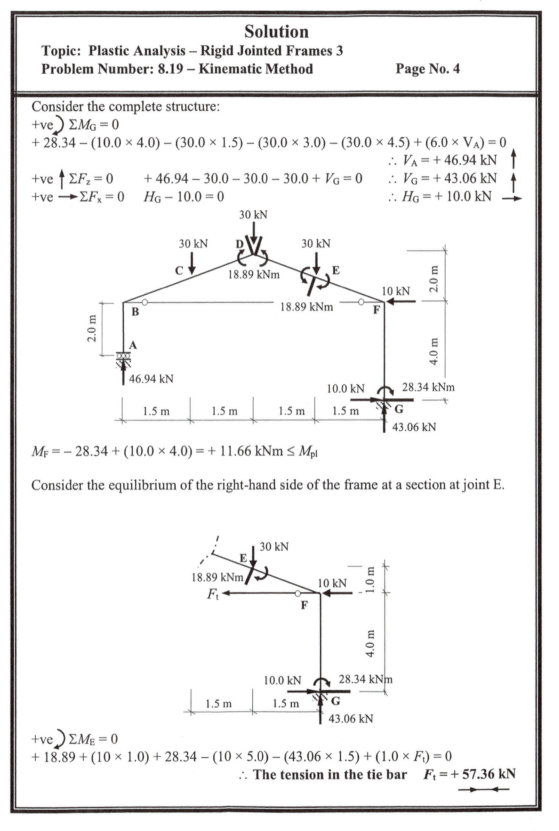

$M_F = - 28.34 + (10.0 \times 4.0) = + 11.66$ kNm $\leq M_{pl}$

Consider the equilibrium of the right-hand side of the frame at a section at joint E.

+ve $\circlearrowright$ $\Sigma M_E = 0$

$+ 18.89 + (10 \times 1.0) + 28.34 - (10 \times 5.0) - (43.06 \times 1.5) + (1.0 \times F_t) = 0$
$\therefore$ **The tension in the tie bar** $F_t = + 57.36$ **kN**

Solution
Topic: Plastic Analysis – Rigid Jointed Frames 3
Problem Number: 8.19 – Kinematic Method   **Page No. 5**

Consider the bending moment at C.

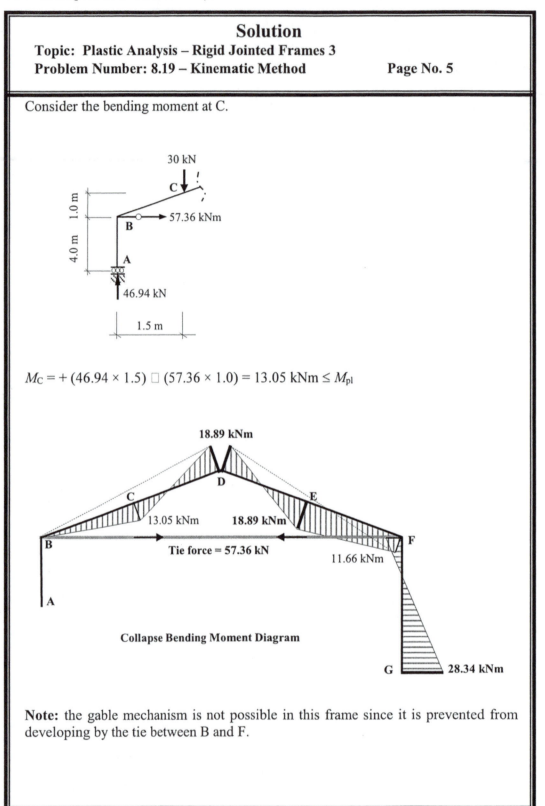

$M_C = + (46.94 \times 1.5) \; \square \; (57.36 \times 1.0) = 13.05 \text{ kNm} \le M_{pl}$

Collapse Bending Moment Diagram

Note: the gable mechanism is not possible in this frame since it is prevented from developing by the tie between B and F.

Solution

Topic: Plastic Analysis – Rigid Jointed Frames 3
Problem Number: 8.20 – Kinematic Method   **Page No. 1**

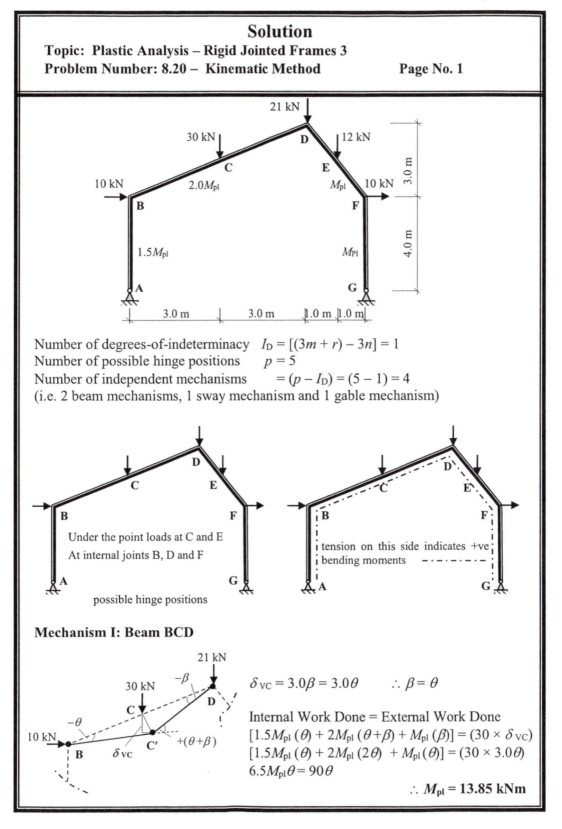

Number of degrees-of-indeterminacy $I_D = [(3m + r) - 3n] = 1$
Number of possible hinge positions $p = 5$
Number of independent mechanisms $= (p - I_D) = (5 - 1) = 4$
(i.e. 2 beam mechanisms, 1 sway mechanism and 1 gable mechanism)

Under the point loads at C and E
At internal joints B, D and F

possible hinge positions

tension on this side indicates +ve
bending moments

Mechanism I: Beam BCD

$\delta_{VC} = 3.0\beta = 3.0\theta$ $\therefore \beta = \theta$

Internal Work Done = External Work Done
$[1.5M_{pl}\,(\theta) + 2M_{pl}\,(\theta+\beta) + M_{pl}\,(\beta)] = (30 \times \delta_{VC})$
$[1.5M_{pl}\,(\theta) + 2M_{pl}\,(2\theta) + M_{pl}\,(\theta)] = (30 \times 3.0\theta)$
$6.5M_{pl}\theta = 90\theta$

$\therefore M_{pl} = 13.85$ **kNm**

Solution
Topic: Plastic Analysis – Rigid Jointed Frames 3
Problem Number: 8.20 – Kinematic Method   **Page No. 2**

Mechanism II: Beam DEF

$\delta_{VE} = 1.0\beta = 1.0\theta$ $\therefore \beta = \theta$

Internal Work Done = External Work Done
$[M_{pl}(\theta) + M_{pl}(\theta + \beta) + M_{pl}(\beta)] = (12 \times \delta_{VE})$
$[M_{pl}(\theta) + M_{pl}(\theta + \theta) + M_{pl}(\theta)] = (12 \times 1.0\theta)$
$4M_{pl}\theta = 12\theta$

$\therefore M_{pl} = \textbf{3.0 kNm}$

Mechanism III: Sway

$\delta = 4.0\theta$

Internal Work Done = External Work Done
$[1.5M_{pl}(\theta) + M_{pl}(\theta)] = [(10 \times \delta) + (10 \times \delta)]$
$2.5M_{pl}\theta = (20 \times 4.0\theta) = 80\theta$ $\therefore M_{pl} = \textbf{32.0 kNm}$

Mechanism IV: Gable

$\delta_{HF} = 4.0\beta = 4.0\theta$
$\therefore \beta = \theta$

$\delta_{VD} = 6.0\alpha = 2.0\theta$
$\therefore \alpha = 0.33\theta$

$\delta_{VC} = 3.0\alpha = \theta$

$\delta_{VE} = 1.0\theta$

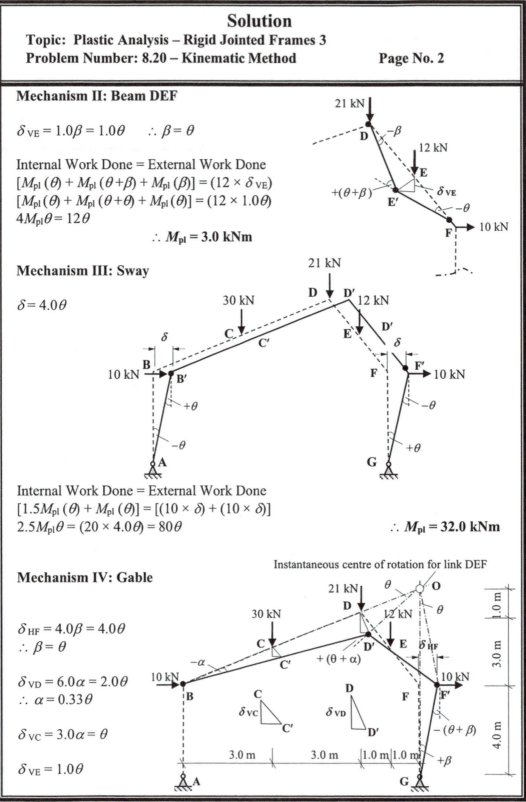

Solution

Topic: Plastic Analysis – Rigid Jointed Frames 3
Problem Number: 8.20 – Kinematic Method   **Page No. 3**

Internal Work Done $= [1.5M_{pl}\,(\alpha) + M_{pl}\,(\theta+\alpha) + M_{pl}\,(\theta+\beta)]$
$$= [(0.5M_{pl}\theta) + (1.33M_{pl}\theta) + [(2.0M_{pl}\theta)] = 3.83M_{pl}\theta$$
External Work Done $= [(30 \times \delta_{VC}) + (21 \times \delta_{VD}) + (12 \times \delta_{VE}) + (10 \times \delta_{HF})]$
$$= [(30 \times \theta) + (21 \times 2\theta) + (12 \times \theta) + (10 \times 4\theta)] = 124\theta$$

Internal Work Done = External Work Done
$3.83M_{pl}\theta = 124\theta$ ∴ $M_{pl} = $ **32.38 kNm**

Mechanism V: Combined Beam BCD, Gable and Sway

Instantaneous centre of rotation for link CDEF

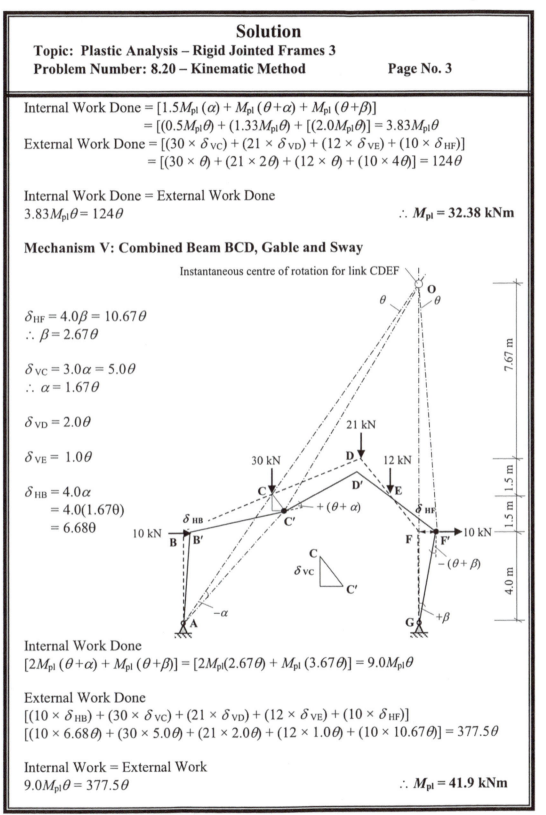

$\delta_{HF} = 4.0\beta = 10.67\theta$
∴ $\beta = 2.67\theta$

$\delta_{VC} = 3.0\alpha = 5.0\theta$
∴ $\alpha = 1.67\theta$

$\delta_{VD} = 2.0\theta$

$\delta_{VE} = 1.0\theta$

$\delta_{HB} = 4.0\alpha$
$\quad = 4.0(1.67\theta)$
$\quad = 6.68\theta$

Internal Work Done
$$[2M_{pl}\,(\theta+\alpha) + M_{pl}\,(\theta+\beta)] = [2M_{pl}(2.67\theta) + M_{pl}\,(3.67\theta)] = 9.0M_{pl}\theta$$

External Work Done
$$[(10 \times \delta_{HB}) + (30 \times \delta_{VC}) + (21 \times \delta_{VD}) + (12 \times \delta_{VE}) + (10 \times \delta_{HF})]$$
$$[(10 \times 6.68\theta) + (30 \times 5.0\theta) + (21 \times 2.0\theta) + (12 \times 1.0\theta) + (10 \times 10.67\theta)] = 377.5\theta$$

Internal Work = External Work
$9.0M_{pl}\theta = 377.5\theta$ ∴ $M_{pl} = $ **41.9 kNm**

Solution

Topic: Plastic Analysis – Rigid Jointed Frames 3
Problem Number: 8.20 – Kinematic Method   **Page No. 4**

In mechanism V the hinges at B and D have been eliminated.

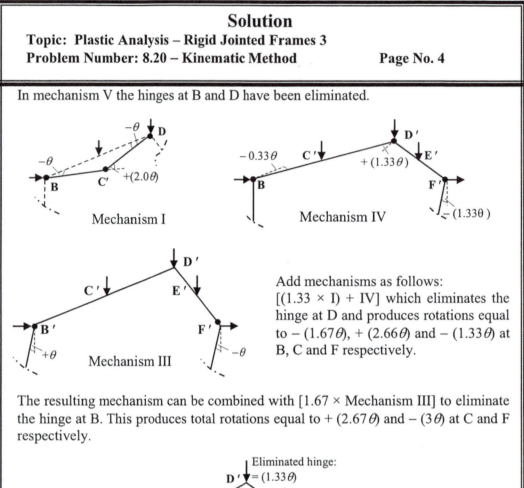

Mechanism I

Mechanism IV

Mechanism III

Add mechanisms as follows:
[(1.33 × I) + IV] which eliminates the hinge at D and produces rotations equal to $-(1.67\theta)$, $+(2.66\theta)$ and $-(1.33\theta)$ at B, C and F respectively.

The resulting mechanism can be combined with [1.67 × Mechanism III] to eliminate the hinge at B. This produces total rotations equal to $+(2.67\theta)$ and $-(3\theta)$ at C and F respectively.

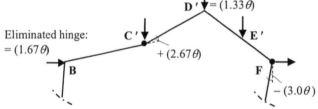

Adding equations for Mechanisms [(1.33 × I) + IV + (1.67 × III)]
$8.65M_{pl}\theta = 119.7\theta$
$3.83M_{pl}\theta = 124.0\theta$
$4.18M_{pl}\theta = 133.6\theta$
$-5.0\,M_{pl}\theta$ [allowing for the hinge eliminated at joint B: $2(1.5M_{pl} \times 1.67\theta)$]
$\underline{-2.67M_{pl}\theta}$ [allowing for the hinge eliminated at joint F: $2(M_{pl} \times 1.33\theta)$]
$9.0M_{pl}\theta = 377.3\theta$ $\therefore M_{pl} = \textbf{41.9 kNm as before}$

The value of M_{pl} obtained (41.9 kNm) should be checked by ensuring that the bending moment in the frame does not exceed this value at any location.

Solution

Topic: Plastic Analysis – Rigid Jointed Frames 3
Problem Number: 8.20 – Kinematic Method   **Page No. 5**

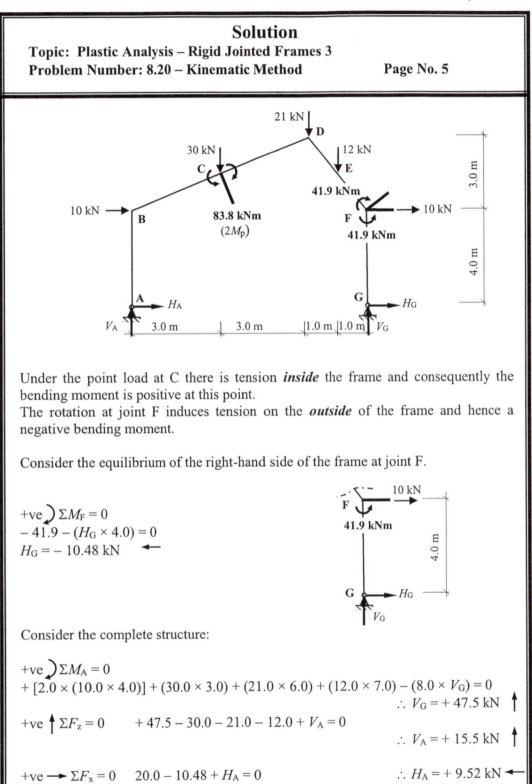

Under the point load at C there is tension *inside* the frame and consequently the bending moment is positive at this point.

The rotation at joint F induces tension on the *outside* of the frame and hence a negative bending moment.

Consider the equilibrium of the right-hand side of the frame at joint F.

$+ve \,\text{\Large)}\, \Sigma M_F = 0$
$- 41.9 - (H_G \times 4.0) = 0$
$H_G = - 10.48$ kN   ←

Consider the complete structure:

$+ve \,\text{\Large)}\, \Sigma M_A = 0$
$+ [2.0 \times (10.0 \times 4.0)] + (30.0 \times 3.0) + (21.0 \times 6.0) + (12.0 \times 7.0) - (8.0 \times V_G) = 0$
$\therefore V_G = + 47.5$ kN ↑

$+ve \uparrow \Sigma F_z = 0 \qquad + 47.5 - 30.0 - 21.0 - 12.0 + V_A = 0$
$\therefore V_A = + 15.5$ kN ↑

$+ve \longrightarrow \Sigma F_x = 0 \qquad 20.0 - 10.48 + H_A = 0$
$\therefore H_A = + 9.52$ kN ←

Solution
Topic: Plastic Analysis – Rigid Jointed Frames 3
Problem Number: 8.20 – Kinematic Method   **Page No. 6**

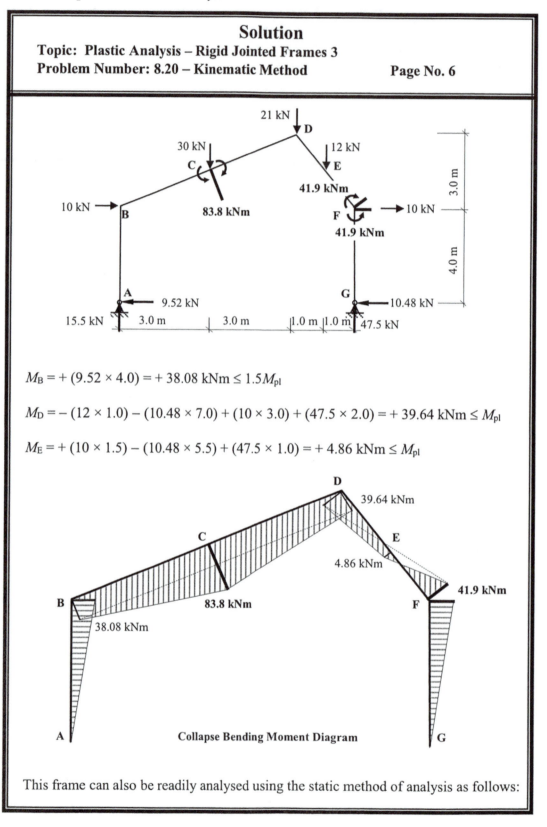

$M_B = + (9.52 \times 4.0) = + 38.08$ kNm $\leq 1.5 M_{pl}$

$M_D = - (12 \times 1.0) - (10.48 \times 7.0) + (10 \times 3.0) + (47.5 \times 2.0) = + 39.64$ kNm $\leq M_{pl}$

$M_E = + (10 \times 1.5) - (10.48 \times 5.5) + (47.5 \times 1.0) = + 4.86$ kNm $\leq M_{pl}$

Collapse Bending Moment Diagram

This frame can also be readily analysed using the static method of analysis as follows:

Solution

Topic: Plastic Analysis – Rigid Jointed Frames 3
Problem Number: 8.20– Static Method   **Page No. 7**

Assume the horizontal component of reaction at support G to be the redundant reaction.

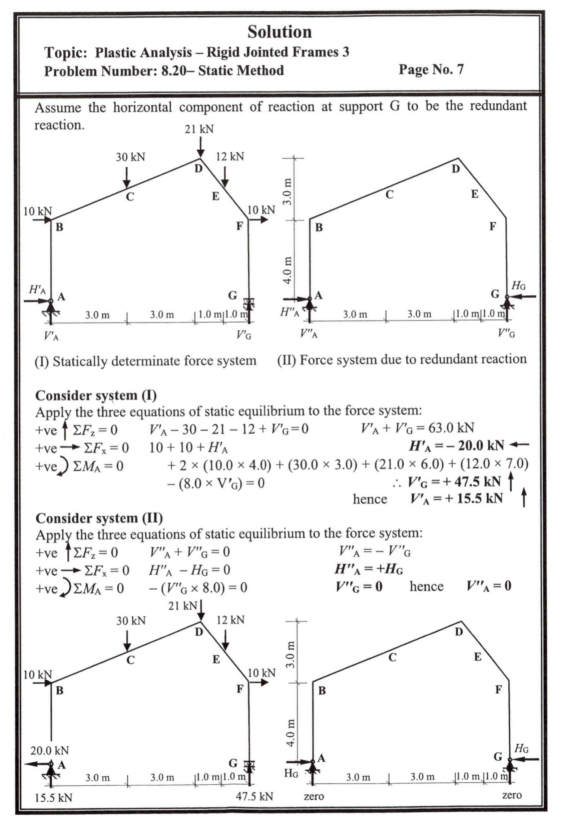

(I) Statically determinate force system   (II) Force system due to redundant reaction

Consider system (I)
Apply the three equations of static equilibrium to the force system:
+ve ↑ $\Sigma F_z = 0$ $V'_A - 30 - 21 - 12 + V'_G = 0$ $V'_A + V'_G = 63.0$ kN
+ve → $\Sigma F_x = 0$ $10 + 10 + H'_A$ **$H'_A = -20.0$ kN** ←
+ve ↻ $\Sigma M_A = 0$ $+ 2 \times (10.0 \times 4.0) + (30.0 \times 3.0) + (21.0 \times 6.0) + (12.0 \times 7.0)$
              $- (8.0 \times V'_G) = 0$ ∴ $V'_G = +47.5$ kN ↑
                                hence $V'_A = +15.5$ kN ↑

Consider system (II)
Apply the three equations of static equilibrium to the force system:
+ve ↑ $\Sigma F_z = 0$ $V''_A + V''_G = 0$ $V''_A = -V''_G$
+ve → $\Sigma F_x = 0$ $H''_A - H_G = 0$ $H''_A = +H_G$
+ve ↻ $\Sigma M_A = 0$ $- (V''_G \times 8.0) = 0$ $V''_G = 0$ hence $V''_A = 0$

Solution
Topic: Plastic Analysis – Rigid Jointed Frames 3
Problem Number: 8.20 – Static Method   **Page No. 8**

$M_B = + (20 \times 4.0) - (H_G \times 4.0) = + 80.0 - 4.0H_G$
$M_C = + (20 \times 5.5) + (15.5 \times 3.0) - (10.0 \times 1.5) - (H_G \times 5.5) = + 141.5 - 5.5H_G$
$M_D = - (12 \times 1.0) + (10 \times 3.0) + (47.5 \times 2.0) - (H_G \times 7.0) = + 113.0 - 7.0H_G$
$M_E = + (10 \times 1.5) + (47.5 \times 1.0) - (H_G \times 5.5) = + 62.5 - 5.5H_G$
$M_F = 0 - (H_G \times 4.0) = 0 - 4.0H_G$

Assume the collapse mechanism as indicated previously, i.e. plastic hinges developing under the point load at C $(+ 2.0M_{pl})$ at and joint F $(- M_{pl})$.

$M_C:$ $+ 2.0M_{pl} = + 141.5 - 5.5H_G$ Equation (1)
$M_F:$ $- M_{pl} = 0 - 4.0H_G$ Equation (2)

Adding equations (1) and [2 × (2)] gives:

$0 = 141.5 - 13.5H_G$ $\therefore H_G = + 10.48$ kN   and $\mathbf{M_{pl} = 41.9}$ **kNm as before**

Check the value of the bending moment at other possible hinge positions

$M_B = + 80.0 + 4.0H_G = + 80.0 - (4.0 \times 10.48) = 38.08$ kNm $\leq 1.5\ M_{pl}$
$M_D = + 113.0 - 7.0H_G = + 113.0 - (7.0 \times 10.48) = 39.64$ kNm $\leq M_{pl}$
$M_E = + 62.5 - 5.5H_G = + 62.5 - (5.5 \times 10.48) = 4.86$ kNm $\leq M_{pl}$

Solution

Topic: Plastic Analysis – Rigid Jointed Frames 3
Problem Number: 8.21 – Kinematic Method   **Page No. 1**

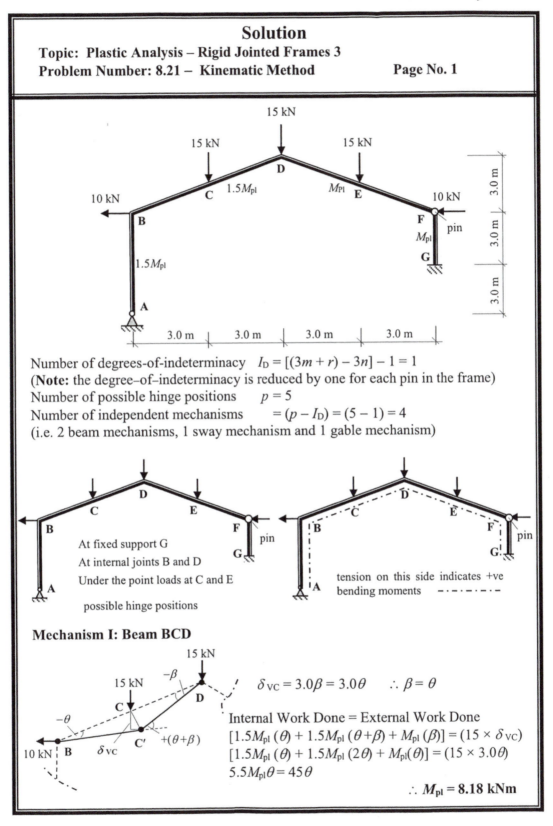

Number of degrees-of-indeterminacy $I_D = [(3m + r) - 3n] - 1 = 1$
(**Note:** the degree–of–indeterminacy is reduced by one for each pin in the frame)
Number of possible hinge positions $p = 5$
Number of independent mechanisms $= (p - I_D) = (5 - 1) = 4$
(i.e. 2 beam mechanisms, 1 sway mechanism and 1 gable mechanism)

At fixed support G
At internal joints B and D
Under the point loads at C and E

possible hinge positions

tension on this side indicates +ve
bending moments

Mechanism I: Beam BCD

$\delta_{VC} = 3.0\beta = 3.0\theta$ $\therefore \beta = \theta$

Internal Work Done = External Work Done
$[1.5M_{pl}\,(\theta) + 1.5M_{pl}\,(\theta + \beta) + M_{pl}\,(\beta)] = (15 \times \delta_{VC})$
$[1.5M_{pl}\,(\theta) + 1.5M_{pl}\,(2\theta) + M_{pl}(\theta)] = (15 \times 3.0\theta)$
$5.5M_{pl}\theta = 45\theta$

$\therefore M_{pl} = \textbf{8.18 kNm}$

Solution
Topic: Plastic Analysis – Rigid Jointed Frames 3
Problem Number: 8.21 – Kinematic Method   **Page No. 2**

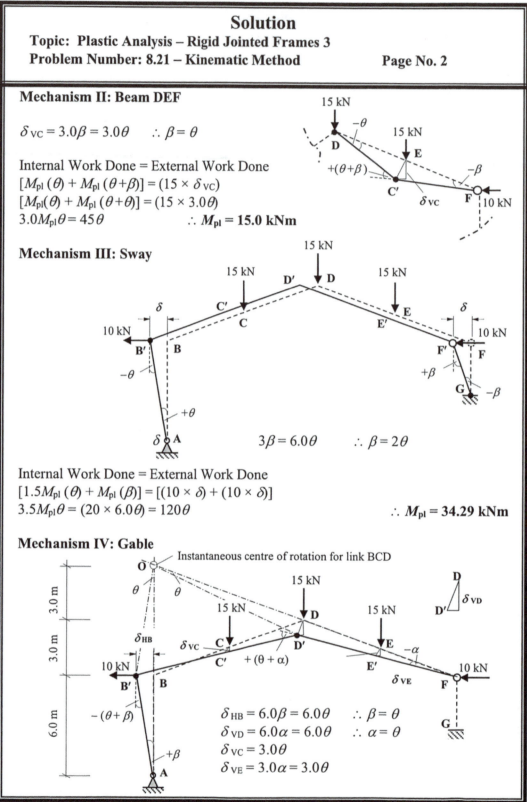

Mechanism II: Beam DEF

$\delta_{VC} = 3.0\beta = 3.0\theta$ $\therefore \beta = \theta$

Internal Work Done = External Work Done
$[M_{pl}(\theta) + M_{pl}(\theta+\beta)] = (15 \times \delta_{VC})$
$[M_{pl}(\theta) + M_{pl}(\theta+\theta)] = (15 \times 3.0\theta)$
$3.0M_{pl}\theta = 45\theta$ $\therefore M_{pl} = 15.0$ kNm

Mechanism III: Sway

$3\beta = 6.0\theta$ $\therefore \beta = 2\theta$

Internal Work Done = External Work Done
$[1.5M_{pl}(\theta) + M_{pl}(\beta)] = [(10 \times \delta) + (10 \times \delta)]$
$3.5M_{pl}\theta = (20 \times 6.0\theta) = 120\theta$ $\therefore M_{pl} = 34.29$ kNm

Mechanism IV: Gable

Instantaneous centre of rotation for link BCD

$\delta_{HB} = 6.0\beta = 6.0\theta$ $\therefore \beta = \theta$
$\delta_{VD} = 6.0\alpha = 6.0\theta$ $\therefore \alpha = \theta$
$\delta_{VC} = 3.0\theta$
$\delta_{VE} = 3.0\alpha = 3.0\theta$

Solution
Topic: Plastic Analysis – Rigid Jointed Frames 3
Problem Number: 8.21 – Kinematic Method   **Page No. 3**

Internal Work Done = $[1.5M_{pl}(\theta+\beta) + M_{pl}(\theta+\alpha)]$
$[(3.0M_{pl}\theta) + (2.0M_{pl}\theta)]$ = $5.0M_{pl}\theta$
External Work Done = $[(15 \times \delta_{VC}) + (15 \times \delta_{VD}) + (15 \times \delta_{VE}) + (10 \times \delta_{HB})]$
$[(15 \times 3\theta) + (15 \times 6\theta) + (15 \times 3\theta) + (10 \times 6\theta)] = 240\theta$

Internal Work Done = External Work Done
$5.0M_{pl}\theta = 240\theta$ $\therefore M_{pl} = 48.0$ kNm

Mechanism V: Combined Beam DEF and Gable.

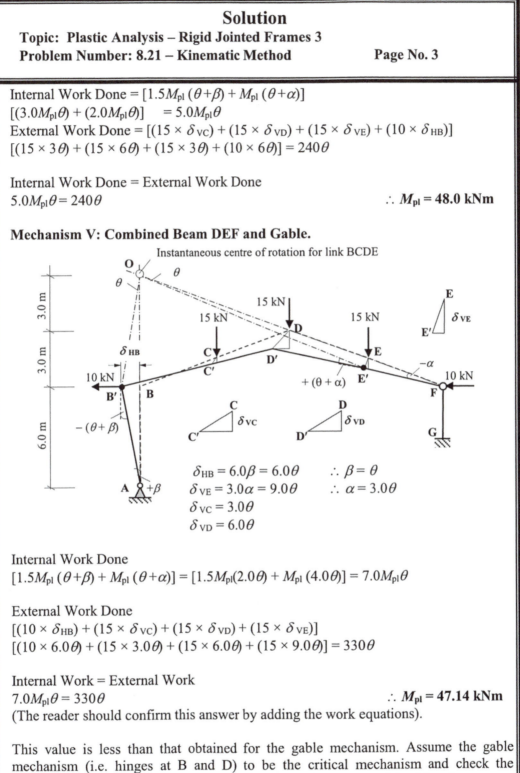

Instantaneous centre of rotation for link BCDE

$\delta_{HB} = 6.0\beta = 6.0\theta$ $\therefore \beta = \theta$
$\delta_{VE} = 3.0\alpha = 9.0\theta$ $\therefore \alpha = 3.0\theta$
$\delta_{VC} = 3.0\theta$
$\delta_{VD} = 6.0\theta$

Internal Work Done
$[1.5M_{pl}(\theta+\beta) + M_{pl}(\theta+\alpha)] = [1.5M_{pl}(2.0\theta) + M_{pl}(4.0\theta)] = 7.0M_{pl}\theta$

External Work Done
$[(10 \times \delta_{HB}) + (15 \times \delta_{VC}) + (15 \times \delta_{VD}) + (15 \times \delta_{VE})]$
$[(10 \times 6.0\theta) + (15 \times 3.0\theta) + (15 \times 6.0\theta) + (15 \times 9.0\theta)] = 330\theta$

Internal Work = External Work
$7.0M_{pl}\theta = 330\theta$ $\therefore M_{pl} = 47.14$ kNm
(The reader should confirm this answer by adding the work equations).

This value is less than that obtained for the gable mechanism. Assume the gable mechanism (i.e. hinges at B and D) to be the critical mechanism and check the bending moments at other possible hinge positions do not exceed the M_{pl} values.

Solution
Topic: Plastic Analysis – Rigid Jointed Frames 3
Problem Number: 8.21 – Kinematic Method   **Page No. 4**

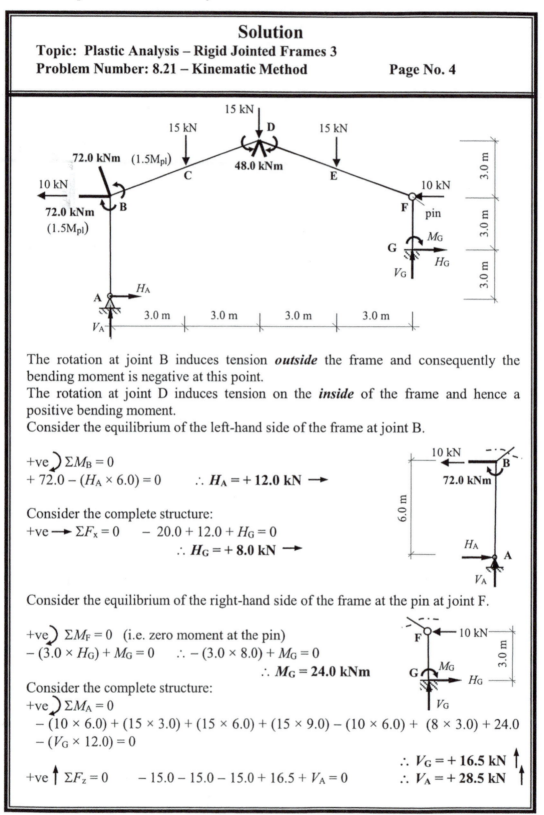

The rotation at joint B induces tension *outside* the frame and consequently the bending moment is negative at this point.

The rotation at joint D induces tension on the *inside* of the frame and hence a positive bending moment.

Consider the equilibrium of the left-hand side of the frame at joint B.

$+ve \,\curvearrowright \Sigma M_B = 0$
$+ 72.0 - (H_A \times 6.0) = 0$ $\therefore H_A = + 12.0 \text{ kN} \rightarrow$

Consider the complete structure:
$+ve \rightarrow \Sigma F_x = 0$ $- 20.0 + 12.0 + H_G = 0$
$\therefore H_G = + 8.0 \text{ kN} \rightarrow$

Consider the equilibrium of the right-hand side of the frame at the pin at joint F.

$+ve \,\curvearrowright \Sigma M_F = 0$ (i.e. zero moment at the pin)
$- (3.0 \times H_G) + M_G = 0$ $\therefore - (3.0 \times 8.0) + M_G = 0$
$\therefore M_G = 24.0 \text{ kNm}$

Consider the complete structure:
$+ve \,\curvearrowright \Sigma M_A = 0$
$- (10 \times 6.0) + (15 \times 3.0) + (15 \times 6.0) + (15 \times 9.0) - (10 \times 6.0) + (8 \times 3.0) + 24.0$
$- (V_G \times 12.0) = 0$

$\therefore V_G = + 16.5 \text{ kN} \uparrow$

$+ve \uparrow \Sigma F_z = 0$ $- 15.0 - 15.0 - 15.0 + 16.5 + V_A = 0$ $\therefore V_A = + 28.5 \text{ kN} \uparrow$

Solution

Topic: Plastic Analysis – Rigid Jointed Frames 3
Problem Number: 8.21 – Kinematic Method   **Page No. 5**

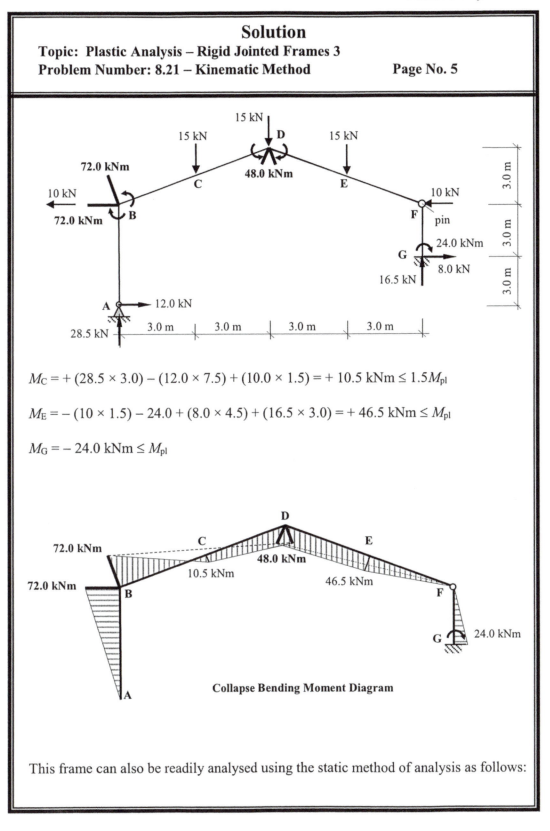

$M_C = + (28.5 \times 3.0) - (12.0 \times 7.5) + (10.0 \times 1.5) = + 10.5 \text{ kNm} \leq 1.5 M_{pl}$

$M_E = - (10 \times 1.5) - 24.0 + (8.0 \times 4.5) + (16.5 \times 3.0) = + 46.5 \text{ kNm} \leq M_{pl}$

$M_G = - 24.0 \text{ kNm} \leq M_{pl}$

Collapse Bending Moment Diagram

This frame can also be readily analysed using the static method of analysis as follows:

Solution
Topic: Plastic Analysis – Rigid Jointed Frames 3
Problem Number: 8.21 – Static Method   **Page No. 6**

Assume the horizontal component of reaction at support A to be the redundant reaction.

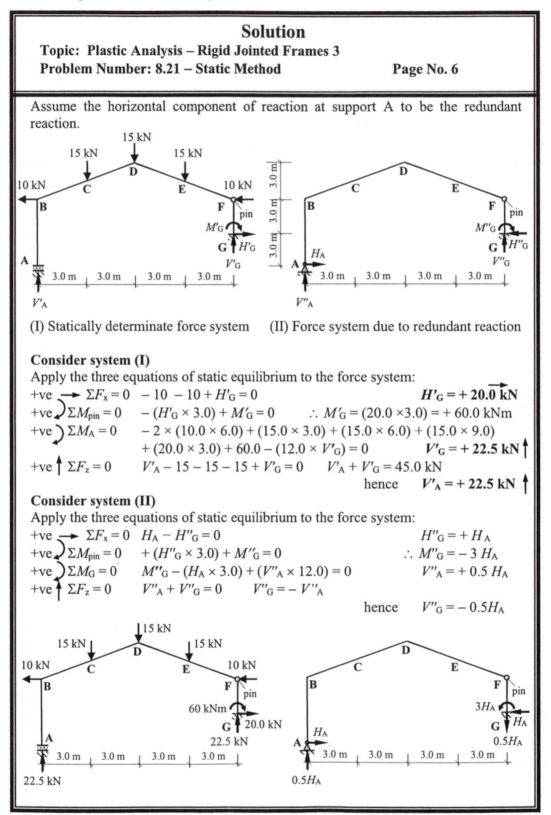

(I) Statically determinate force system   (II) Force system due to redundant reaction

Consider system (I)
Apply the three equations of static equilibrium to the force system:

$+ve \longrightarrow \Sigma F_x = 0$ $-10 - 10 + H'_G = 0$ $H'_G = + \overrightarrow{20.0}$ kN

$+ve \, \Sigma M_{pin} = 0$ $-(H'_G \times 3.0) + M'_G = 0$ $\therefore M'_G = (20.0 \times 3.0) = + 60.0$ kNm

$+ve \, \Sigma M_A = 0$ $-2 \times (10.0 \times 6.0) + (15.0 \times 3.0) + (15.0 \times 6.0) + (15.0 \times 9.0)$
$+ (20.0 \times 3.0) + 60.0 - (12.0 \times V'_G) = 0$ $V'_G = + 22.5$ kN $\uparrow$

$+ve \uparrow \Sigma F_z = 0$ $V'_A - 15 - 15 - 15 + V'_G = 0$ $V'_A + V'_G = 45.0$ kN
                              hence $V'_A = + 22.5$ kN $\uparrow$

Consider system (II)
Apply the three equations of static equilibrium to the force system:

$+ve \longrightarrow \Sigma F_x = 0$ $H_A - H''_G = 0$ $H''_G = + H_A$

$+ve \, \Sigma M_{pin} = 0$ $+ (H''_G \times 3.0) + M''_G = 0$ $\therefore M''_G = -3 H_A$

$+ve \, \Sigma M_G = 0$ $M''_G - (H_A \times 3.0) + (V''_A \times 12.0) = 0$ $V''_A = + 0.5 H_A$

$+ve \uparrow \Sigma F_z = 0$ $V''_A + V''_G = 0$ $V''_G = - V''_A$

                              hence $V''_G = - 0.5 H_A$

Solution

Topic: Plastic Analysis – Rigid Jointed Frames 3

Problem Number: 8.21 – Static Method   **Page No. 7**

$M_B = - (H_A \times 6.0) = - 6.0H_A$

$M_C = + (22.5 \times 3.0) + (10 \times 1.5) - (7.5 \times H_A) + 3(0.5H_A) = + 82.5 - 6.0H_A$

$M_D = + (22.5 \times 6.0) + (10 \times 3.0) - (15 \times 3.0) - 9H_A + (6.0 \times 0.5H_A)$
$\quad = + 120.0 - 6.0H_A$

$M_E = + (22.5 \times 9.0) + (10 \times 1.5) - (15 \times 6.0) - (15 \times 3.0) - 7.5H_A \;\; - (9.0 \times 0.5H_A)$
$\quad = + 82.5 - 3.0H_A$

$M_G = - 60.0 + (H_A \times 3.0) = - 60.0 + 3.0H_A$

Assume the collapse mechanism as indicated previously, i.e. plastic hinges developing at joint B $(- 1.5M_{pl})$ at joint D $(+ M_{pl})$ and

$M_B: - 1.5M_{pl} = 0 - 6.0H_A$ Equation (1)
$M_D: + M_{pl} = 120 - 6.0H_A$ Equation (2)

Subtracting equation (1) from equation (2) gives:

$- 2.5M_{pl} = - 120$ $\therefore M_{pl} = $ **48.0 kNm as before** and $H_A = $ **12.0 kN**

Check the value of the bending moment at other possible hinge positions

$M_C = + 82.5 - 6.0H_A = + 82.5 - (6.0 \times 12.0) = + 10.5 \text{ kNm} \leq 1.5M_{pl}$
$M_E = + 82.5 - 3.0H_A = + 82.5 - (3.0 \times 12.0) = + 46.5 \text{ kNm} \leq M_{pl}$
$M_G = - 60.0 + 3.0H_A = - 60.0 - (3.0 \times 12.0) = - 24.0 \text{ kNm} \leq M_{pl}$

9.   Influence Lines for Beams

9.1   *Introduction*

Many structures are required to support moving loads in addition to static loading, e.g. highway/railway bridges or an overhead travelling crane as shown in Figure 9.1. Whilst these moving loads are in reality dynamic in nature and their values and/or positions vary in time, the variation is slow enough for them to be considered as *'quasi-static'* loading. In such cases the behaviour of the structure at any instant in time can be determined assuming the value of the loads and load effects using the rules and principles which govern structural behaviour under static loading.

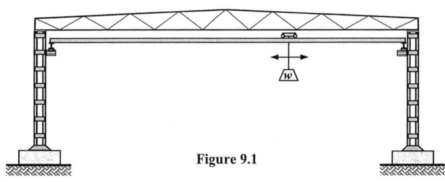

Figure 9.1

The design of such structures requires that the most critical positions of the loads are identified for various functions, e.g. the support reactions, axial loads in trusses, shear force or bending moment in beams. Consider a vehicle moving along a simply supported span as shown in Figure 9.2:

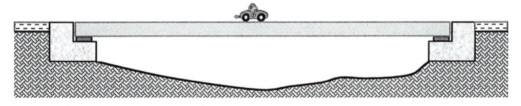

Figure 9.2

The position of the vehicle required to determine the maximum value of a function/design load effect must be identified for design purposes. This can be achieved by the use of **influence lines**. An influence line is a graph of the variation of a function e.g. the support reaction, shear force, bending moment etc. in a beam or the axial load in a pin-jointed trussed frame, at a given position in a structure as a **unit load** traverses the structure. It is important to recognise that, unlike shear and bending moment diagrams, influence lines indicate the variation of a function at a specific point in a structure.

9.2   *Example 9.1: Influence Lines for a Simply Supported Beam*

Consider the simply supported beam shown in Figure 9.3 in which a unit load traverses the structure from A to C. Influence lines (graphs) can be drawn which indicate the variation of the support reactions, the shear force and the bending moment at some general point B on the beam a distance '*a*' from the support at A. At any instant in time the load is at position '*x*' from support A.

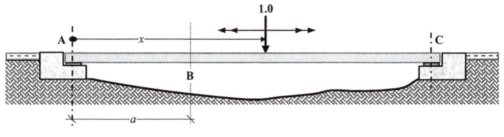

Figure 9.3

9.2.1 Influence Lines for the Support Reactions

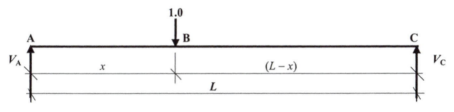

Figure 9.4

Two equations relating to the two support reactions V_A and V_C can be determined by considering the rotational and vertical equilibrium of the beam.

Consider the rotational equilibrium of the beam:

$$+ve \;\big\rangle \Sigma M_A = 0$$

$$(1.0\,x) - (V_C \times L) = 0 \qquad \therefore V_C = x/L \qquad\qquad \text{Equation (1)}$$

Consider the vertical equilibrium of the beam:

$$+ve \uparrow \Sigma F_z = 0$$

$$V_A - 1.0 + V_C = 0 \qquad \therefore V_A = (1.0 - x/L) \qquad \text{Equation (2)}$$

Equation (2) is the equation of the graph which defines the variation in V_A as the unit load traverses the beam, i.e. the equation of the influence line for V_A. Similarly, Equation (1) is the equation for the influence line for V_C. The two influence lines can be plotted by considering values of the functions V_A and V_C when: $x = 0$ and when $x = L$ respectively.

Influence Line for V_A: Equation (2) $V_A = (1.0 - x/L)$
            when $x = 0$ $V_A = 1.0$ and   when $x = L$ $V_A = 0$

Influence Line for V_C: Equation (1) $V_C = x/L$
            when $x = 0$ $V_C = 0$ and   when $x = L$ $V_C = 1.0$

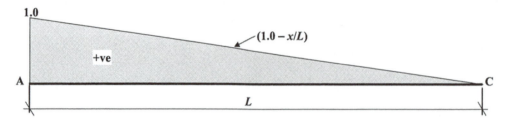

Figure 9.5: Influence line for the vertical reaction at A

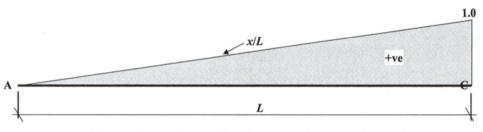

Figure 9.6: Influence line for the vertical reaction at C

9.2.2 Influence Line for the Shear Force at Point B

To develop the influence line for the shear force at a point in a beam it is convenient to consider the position of the unit point load acting in two zones as shown in Figure 9.7:

(i)   to the left of the point under consideration   and

(ii)   to the right of the point under consideration.

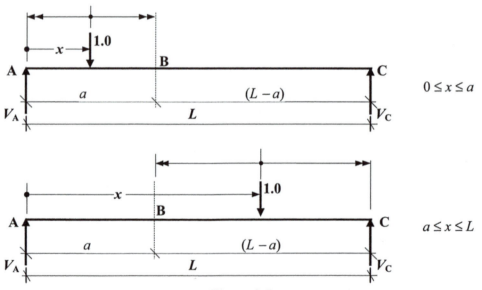

Figure 9.7

Consider $0 \le x \le a$

Shear force at B $= (V_A - 1.0)$ or   alternatively,   Shear force at B $= - V_C$

The influence line between $0 \le x \le a$ is the same as the 'inverted' (i.e. negative) influence line for V_C between these limits as shown in Figure 9.8: (Note: $V_C = x/L$).

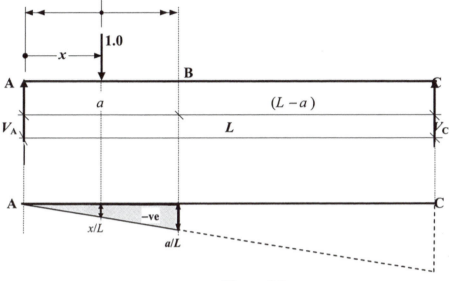

Figure 9.8

Consider $a \le x \le L$

Shear force at B $= V_A$ or   alternatively   shear force at B $= (- V_C + 1.0)$

The influence line between $a \le x \le L$ is the same as the influence line for V_A between these limits as shown in Figure 9.9:

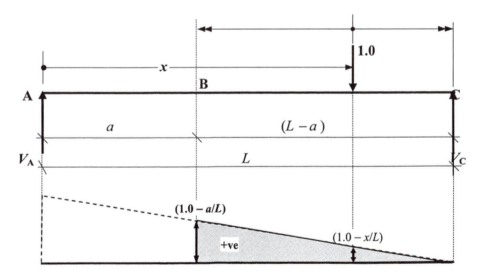

Figure 9.9

The complete influence line for the shear force at B is given by the addition of the two zones for $0 \leq x \leq a$ and $a \leq x \leq L$ as shown in Figure 9.10.

Influence Line: Shear force at B for $0 \leq x \leq L$

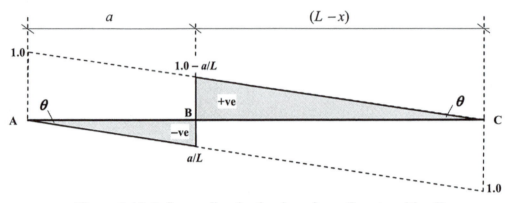

Figure 9.10: Influence line for the shear force $F_{v,B}$ at position B

9.2.3 Influence Line for the Bending Moment at Point B

The influence line for the bending moment at a point in a beam can be developed similarly to that for the shear force considering two zones, as shown in Figure 9.11:

(i)   to the left of the point under consideration   and

(ii) to the right of the point under consideration.

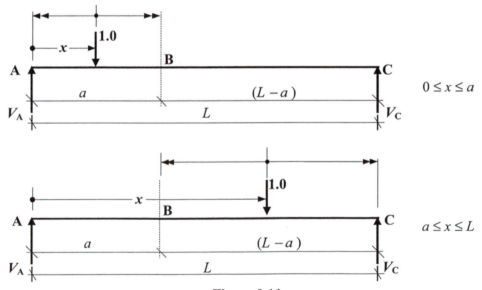

Figure 9.11

Consider $0 \le x \le a$
Bending moment at B $\quad M_B = + [aV_A - 1.0(a - x)]$ or alternatively,
Bending moment at B $\quad M_B = + V_C \times (L - a)$ $(= V_C \times \text{constant})$
The influence line for V_C between $0 \le x \le a$ is given by $V_C = x/L$.
$\therefore M_B = V_C \times (L - a) = x\,(L - a)/L$

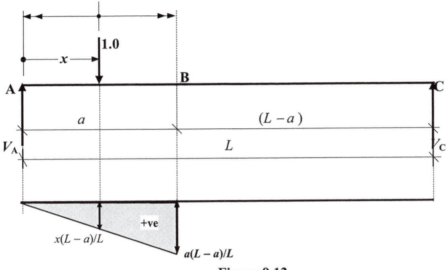

Figure 9.12

Consider $a \le x \le L$
Bending moment at B $\quad M_B = + aV_A$ $(= V_A \times \text{constant})$ or alternatively,
The influence line for V_A between $a \le x \le L$ is given by $V_A = (1.0 - x/L)$.
Bending moment at B $\quad M_B = + [V_C(L - a) - 1.0(x - a)]$
$\therefore M_B = a \times (1.0 - x/L) = a(L - x)/L$

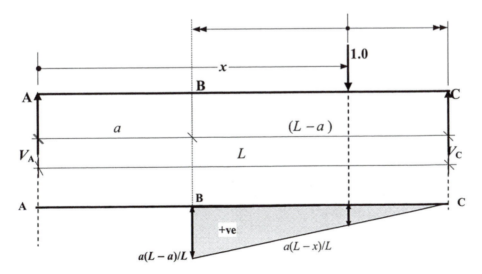

Figure 9.13

The complete influence line for the bending moment at B is given by the addition of the two zones for $0 \le x \le a$ and $a \le x \le L$ as shown in Figure 9.14.

Influence Line: Bending moment at B for $0 \le x \le L$

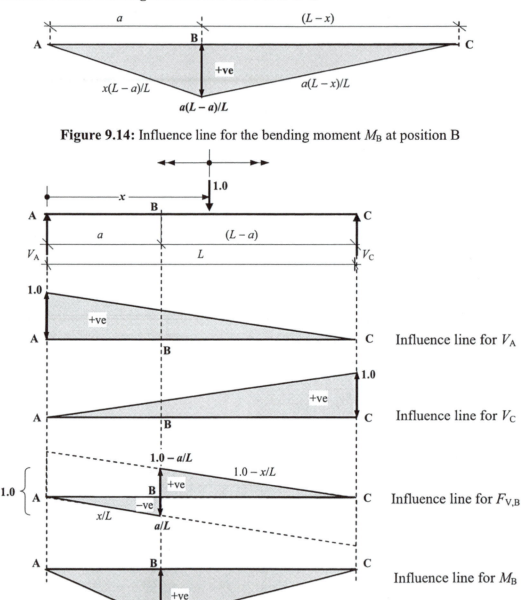

Figure 9.14: Influence line for the bending moment M_B at position B

Figure 9.15

Note: Influence lines are NOT the same as shear force and bending moment diagrams. They provide information relating to a single location on the structure.

9.3 Müller-Breslau Principle for the Influence Lines for Beams

The influence line for any response function in a beam can also be determined by the use of the Müller-Breslau principle, i.e.

"The influence line for any function is given by the deflection curve, to some scale, that results when the restraint corresponding to that function is removed and a unit displacement is induced in its place"

The principle is applicable to any type of elastic structure, i.e. both statically determinate and indeterminate structures.

9.4 Example 9.2: Influence Lines for a Statically Determinate Beam

Considering the two-span, statically determinate beam ABCDE shown in Figure 9.16, determine the influence lines for:

(i) the vertical reactions at supports A, C and E,
(ii) the shear force at point B,
(iii) the bending moment at point B and
(iv) the bending moment over support C.

The values are derived considering the articulation of the beam between the support points and any pins within the spans, for an imposed unit linear displacement (when considering support reactions or shear forces) or a unit rotational displacement (when considering a bending moment). Note: the displaced shape of statically determinate beams will be linear.

(i)
The vertical reaction at support A: Impose a unit, vertical displacement at A

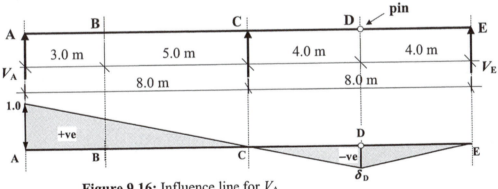

Figure 9.16: Influence line for V_A

A vertical unit displacement is imposed at A and the beam imagined to articulate between the supports at C and E and the pin at D. The value of δ_D can be determined readily by considering the geometry of the displaced shape, i.e. considering the triangles between ABC and CDE giving:

$$\frac{1.0}{8.0} = -\frac{\delta_D}{4.0} \quad \therefore \delta_D = -0.5$$

The vertical reaction at support C: Impose a unit vertical displacement at C

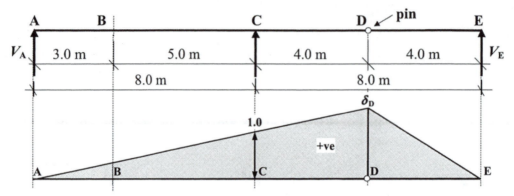

Figure 9.17: Influence line for V_C

$$\frac{1.0}{8.0} = +\frac{\delta_D}{12.0} \qquad \therefore \ \delta_D = +1.5$$

The vertical reaction at support E: Impose a unit vertical displacement at E

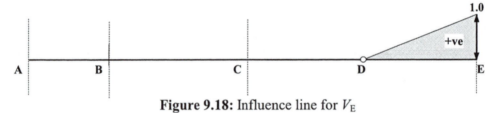

Figure 9.18: Influence line for V_E

(ii)
The shear force at point B: Impose a unit shear displacement at B

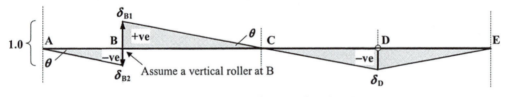

Figure 9.19: Influence line for $F_{V,B}$

Considering the triangles between AB and BC:

$$(\delta_{B,1} + \delta_{B,2}) = 1.0 \ \therefore \ \delta_{B,1} = (1.0 - \delta_{B,2}) \ \text{ and } \ \frac{\delta_{B,1}}{5.0} = -\frac{\delta_{B,2}}{3.0} \qquad \therefore \ \delta_{B,1} = -1.67\delta_{B,2}$$

$$(1.0 - \delta_{B,2}) = 1.67\delta_{B,2} \ \therefore \ \delta_{B,2} = -0.375 \ \text{ and } \ \delta_{B,1} = 0.625$$

Considering the triangles between BC and CD:

$$\frac{0.625}{5.0} = -\frac{\delta_D}{4.0} \qquad \therefore \ \delta_D = -0.5$$

(iii)
The bending moment at point B: Impose a unit rotation at B

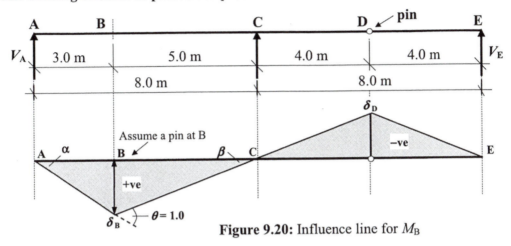

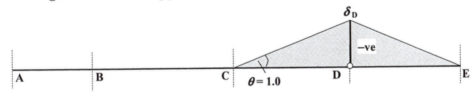

Figure 9.20: Influence line for M_B

$\theta = (\alpha + \beta) = 1.0$ $\therefore$ $\alpha = (1.0 - \beta)$ and $\delta_B = 3.0\alpha = 5.0\beta$ $\therefore$ $\alpha = 1.67\beta$
$(1.0 - \beta) = 1.67\beta$ $\therefore$ $\beta = -0.375$ and $\alpha = 0.625$ $\therefore$ $\delta_B = (3.0 \times \alpha) = 1.875$

Considering the triangles between BC and CD:
$$\frac{1.875}{5.0} = -\frac{\delta_D}{4.0} \qquad \therefore \ \delta_D = -1.5 \text{ m}$$

(iv)
The bending moment over support C: Impose a unit rotation at C

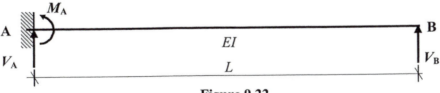

Figure 9.21: Influence line for M_C

$$\delta_D = 4.0\theta = (4.0 \times 1.0) = 4.0 \text{ m}$$

9.5   *Example 9.3: Influence Line for a Statically Indeterminate Beam*

Considering the propped cantilever AB shown in Figure 9.22, which is fixed at support A and supported on a roller at B, determine the influence line for the support reaction at B.

Figure 9.22

Consider a unit load which moves from A to B when at a general position C a distance '*x*' from B and assume an upward reaction at B as shown in Figure 9.23.

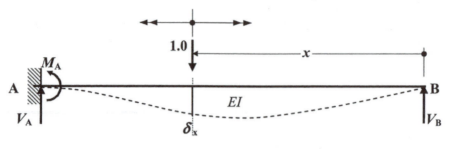

Figure 9.23

Remove the restraint at B and impose a displacement $_B\delta_B$ at B in the assumed direction of V_B using a force $F = 1.0$. This also induces a vertical deflection of $_C\delta_B$ at point C as shown in Figure 9.24.

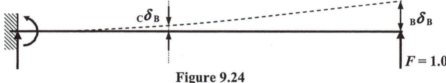

Figure 9.24

The values of $_B\delta_B$ and $_C\delta_B$ define the shape of the deflected curve for an applied unit load at B and consequently the required influence line ordinates for $V_B = \dfrac{_C\delta_B}{_B\delta_B}$, i.e. the deflected shape due to a unit load displacement applied at B. The values can be determined using standard elastic analysis e.g. MaCaulay's Method, i.e.

Bending moment at distance $x = EI\ d^2z/dx^2$
$EI\ d^2z/dx^2 = 1.0x$ Equation (1)
$EI\ dz/dx = x^2/2 + A$ Equation (2)
$\quad EIz = x^3/6 + Ax + B$ Equation (3)

Apply the boundary conditions:
when $x = 20.0$ m $dz/dx = 0$ $\therefore$ $A = -200$
when $x = 20.0$ $z = 0$ $\therefore$ $B = -20.0^3/6 + (200 \times 20.0) = 2666.7$

The deflected shape is given by: $z = (x^3/6 - 200x + 2666.7)/EI$
When $x = 0$ $_B\delta_B = [0 - (200 \times 0) + 2666.7]/EI = 2666.7/EI$
An upward displacement is regarded as a positive ordinate.

The influence line co-ordinates are given by:
$$\frac{_C\delta_B}{_B\delta_B} = \frac{\left(x^3/6 - 200x + 2666.7\right)\big/EI}{2666.7/EI} = \frac{\left(x^3/6 - 200x + 2666.7\right)}{2666.7}$$

The influence line values are determined by substituting appropriate values for x, i.e.

$x = 0$ influence ordinate $z = \dfrac{\left(0^3/6 - 200 \times 0 + 2666.7\right)}{2666.7} = \dfrac{2666.7}{2666.7} = 1.0$

$x = 4.0$ m   influence ordinate $z = \dfrac{\left(4.0^3/6 - 200 \times 4.0 + 2666.7\right)}{2666.7} = 0.704$

Similarly for $x = 8.0$ m, 12.0 m and 16.0 m as indicated in Figure 9.25.

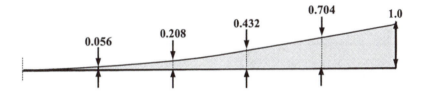

Figure 9.25: Influence line for V_B

9.6   *The use of Influence Lines*

The influence line for a function can be used to determine the critical value of that function for a variety of loading conditions, e.g. concentrated loads, distributed loads, travelling loads and trains of loads.

9.6.1 Concentrated Loads

The value of a function induced by a given concentrated load at any position on a structure, can be determined by multiplying the magnitude of the load by the ordinate at the position of the load on the influence line for that function, i.e.

Magnitude of the function = (applied load × ordinate (z) on the influence line)

The maximum positive value of a function can be determined by multiplying the magnitude of the load by the maximum positive ordinate 'z' on the influence line for that function; similarly for the maximum negative value.

9.6.2 Distributed Loads

The value of a function for any given distributed load at any position on the beam, can be determined by multiplying the magnitude of the load/metre length by the area of the influence line lying under the extent of the distributed load for that function.

Magnitude of the function = (applied load × area under the influence line)

The maximum value of the function due to a moving UDL of length smaller than the span, can be determined by positioning the moving load such that it maximises the area under the extent of the load.

9.6.3 *Example 9.4: Evaluation of Functions for a Statically Determinate Beam 1*

A 12.0 m span simply supported beam ABCD, supports a distributed load of 12.0 kN/m and a point load of 20.0 kN as shown in Figure 9.26. Draw the influence lines for:

   (i)   the vertical reactions at supports, V_A and V_D,
  (ii) the shear force $F_{V,B}$ at point B, and
 (iii) the bending moment at point C, M_C

and determine the value of each of the functions (i), (ii) and (iii) for the loading indicated.

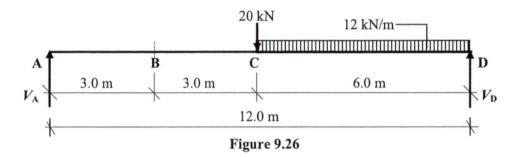

Figure 9.26

(i) The influence line for and value of V_A (see Figure 9.15).

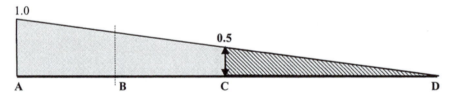

Figure 9.27: Influence line for V_A

V_A = (concentrated load × ordinate) + (distributed load × area)
    = (20.0 × 0.50) + [12.0 × (0.5 × 6.0 × 0.5)] = 28.0 kN

(ii) The influence line for and value of V_D (see Figure 9.15).

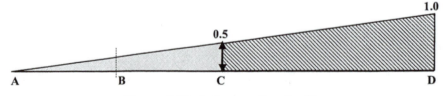

Figure 9.28: Influence line for V_D

V_A = (concentrated load × ordinate) + (distributed load × area)

    = (20.0 × 0.50) + 12.0 × [0.5 × (0.5 + 1.0) × 6.0] = 64.0 kN

(iii) The influence line for and value of $F_{v,B}$ (see Figure 9.15).

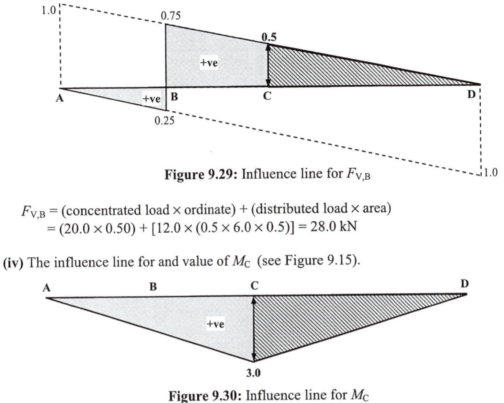

Figure 9.29: Influence line for $F_{V,B}$

$$F_{V,B} = \text{(concentrated load} \times \text{ordinate)} + \text{(distributed load} \times \text{area)}$$
$$= (20.0 \times 0.50) + [12.0 \times (0.5 \times 6.0 \times 0.5)] = 28.0 \text{ kN}$$

(iv) The influence line for and value of M_C (see Figure 9.15).

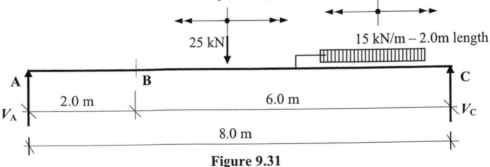

Figure 9.30: Influence line for M_C

$$M_C = \text{(concentrated load} \times \text{ordinate)} + \text{(distributed load} \times \text{area)}$$
$$= (20.0 \times 3.0) + [12.0 \times (0.5 \times 6.0 \times 3.0)] = 168.0 \text{ kNm}$$

9.6.4 *Example 9.5: Evaluation of Functions for a Statically Determinate Beam 2*

An 8.0 m span simply supported beam ABC supports a distributed load of 2.0 m length and magnitude 15.0 kN/m and a point load of 25.0 kN, both of which traverse the beam independently as shown in Figure 9.31. Using the influence lines for the shear force and the bending moment at point B, determine the maximum values of these functions when the loads can travel across the beam independently.

Figure 9.31

(i) The influence line for and value of $F_{v,B}$ (see Figure 9.15).

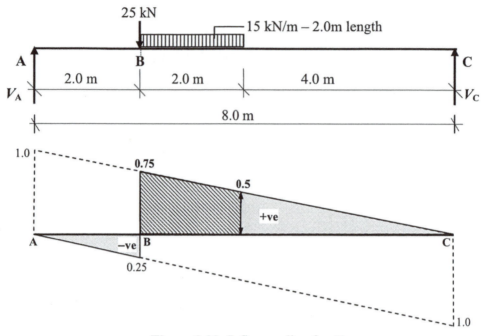

Figure 9.32: Influence line for $F_{V,B}$

$F_{V,B}$ = (concentrated load × ordinate) + (distributed load × area)
  = $(25.0 \times 0.75) + [15.0 \times 0.5 \times (0.75 + 0.5) \times 2.0)] = 37.50$ kN

(ii) The influence line for and value of M_B (see Figure 9.15).
The distributed load must be positioned as shown in Figure 9.33 such that it maximizes the value of the area under the bending moment influence line.

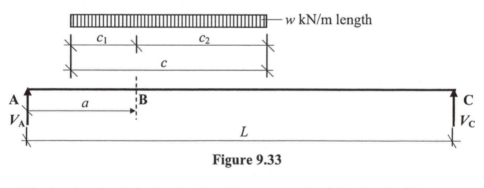

Figure 9.33

$$\frac{\text{The load to the left of point B}}{\text{The total load}} = \frac{\text{The span to the left of point B}}{\text{The total span}} = \frac{c_1}{c} = \frac{a}{L}$$

This also leads to the observation that the influence line coefficient 'z' at each end of the load has the same value.

In this example: $a = 2.0$ m; $L = 8.0$ m; $c = 2.0$ m and the load is positioned such that the value of $c_1 = 0.5$, i.e. $\dfrac{c_1}{c} = \dfrac{a}{L}$ $\therefore c_1 = \dfrac{ca}{L} = \dfrac{2.0 \times 2.0}{8.0} = 0.5$ m

$\delta_B = a(L-a)/L = 2.0 \times (8.0-2.0)/8.0 = 1.5$

$\delta_{BL} = 1.5 \times (1.50/2.0) = 1.125$ and $\delta_{BR} = 4.5 \times (1.50/6.0) = 1.125$

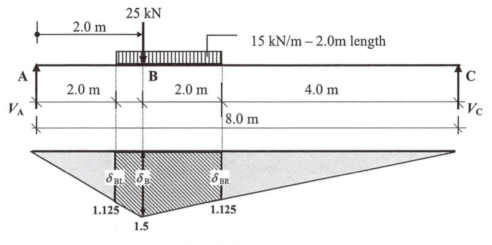

Figure 9.34: Influence line for M_B

M_B = (concentrated load × ordinate) + (distributed load × area)
$= (25.0 \times 1.5) + 15.0 \times \{(1.125 \times 2.0) + [0.5 \times 2.0 \times (1.50-1.125)]\} = 76.88$ kNm

9.7 *Example 9.6: Evaluation of Functions for a Statically Indeterminate Beam*

A two span, non-uniform beam ABC is simply supported at A, B and C as shown in Figure 9.35. Span AB carries a fixed uniformly distributed load of 40 kN/m and a concentrated load of 20 kN traverses the beam from A to C. Develop the influence line for the moment M_B at support B and using it, determine the magnitude and sense of the M_B.

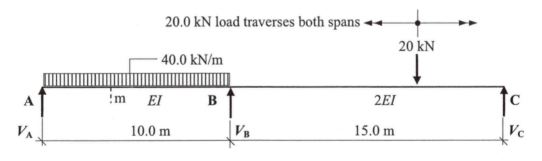

Figure 9.35

Remove the restraint which induces the moment at B by introducing a pin and impose displacements θ_{BL} and θ_{BR} at B using a moment $M = 1.0$ as shown in Figure 9.36. The unit moment induces a total displacement at B equal to the sum from both spans,

i.e. $\theta_B = (\theta_{BL} + \theta_{BR})$ and general displacements at m and n as indicated.
Positions 'm' and 'n' represent any general points along the beams AB and BC.

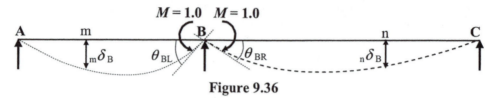

Figure 9.36

Consider each span separately and use MaCaulay's method to determine the displaced shape induced by the unit moment.

Span AB:

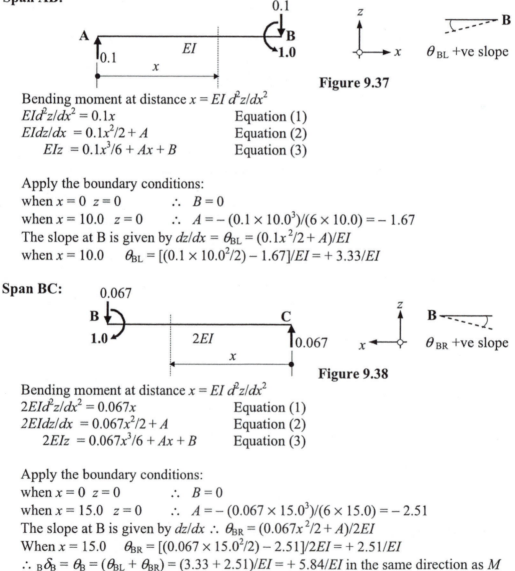

Figure 9.37

Bending moment at distance $x = EI\, d^2z/dx^2$
$EId^2z/dx^2 = 0.1x$ ⟶ Equation (1)
$EIdz/dx = 0.1x^2/2 + A$ ⟶ Equation (2)
$EIz = 0.1x^3/6 + Ax + B$ ⟶ Equation (3)

Apply the boundary conditions:
when $x = 0$ $z = 0$ ∴ $B = 0$
when $x = 10.0$ $z = 0$ ∴ $A = -(0.1 \times 10.0^3)/(6 \times 10.0) = -1.67$
The slope at B is given by $dz/dx = \theta_{BL} = (0.1x^2/2 + A)/EI$
when $x = 10.0$ $\theta_{BL} = [(0.1 \times 10.0^2/2) - 1.67]/EI = +3.33/EI$

Span BC:

Bending moment at distance $x = EI\, d^2z/dx^2$
$2EId^2z/dx^2 = 0.067x$ Equation (1)
$2EIdz/dx = 0.067x^2/2 + A$ Equation (2)
$2EIz = 0.067x^3/6 + Ax + B$ Equation (3)

Apply the boundary conditions:
when $x = 0$ $z = 0$ ∴ $B = 0$
when $x = 15.0$ $z = 0$ ∴ $A = -(0.067 \times 15.0^3)/(6 \times 15.0) = -2.51$
The slope at B is given by dz/dx ∴ $\theta_{BR} = (0.067x^2/2 + A)/2EI$
When $x = 15.0$ $\theta_{BR} = [(0.067 \times 15.0^2/2) - 2.51]/2EI = +2.51/EI$
∴ $_B\delta_B = \theta_B = (\theta_{BL} + \theta_{BR}) = (3.33 + 2.51)/EI = +5.84/EI$ in the same direction as M

For span AB the influence line co-ordinates are given by:

$$\frac{_m\delta_B}{_B\delta_B} = \frac{\left(0.0167x^3 - 1.67x\right)/EI}{5.84/EI} = 0.00286x^3 - 0.286x \text{ m}$$

The influence line values are determined by substituting appropriate values for x.

For span BCD the influence line co-ordinates are given by:

$$\frac{_n\delta_B}{_B\delta_B} = \frac{\left(0.0112x^3 - 2.51x\right)/2EI}{5.84/EI} = 0.00096x^3 - 0.215x \text{ m}$$

The influence line values are determined by substituting appropriate values for x.

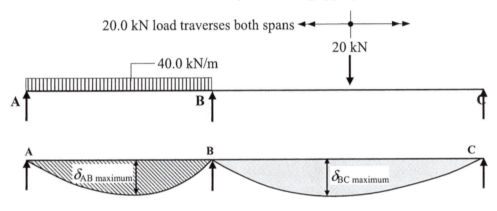

Figure 9.39: Influence line for M_B

M_B = (concentrated load × ordinate) + (distributed load × area)
The maximum ordinate in each span occurs where the slope is zero, i.e. $dz/dx = 0$

Consider span AB:
$EI\ dz/dx = 0.1x^2/2 - 1.67 = 0$ $\therefore x = 5.779$ m from A
Influence line ordinate $\delta_{AB\ maximum} = (0.00286x^3 - 0.286x) = -1.101$ m

Area under the influence line diagram is given by:
$$\int_0^{10.0}\left(0.00286x^3 - 0.286x\right)dx = \left[0.000715x^4 - 0.143x^2\right]_0^{10.0} = -7.15 \text{ m}^2$$

Consider span BC:
$EI\ dz/dx = 0.067x^2/2 - 2.51 = 0$ $\therefore x = 8.656$ m from C
Influence line ordinate $\delta_{BC\ maximum} = (0.00096x^3 - 0.215x) = -1.232$ m
The maximum value for the ordinate is in span BC and the 20 kN load should be placed at this position.

The concentrated load should be on span BC to give the worst effect, i.e.
$\delta_{BC\ maximum} = -1.232$ $\therefore M_B = -(40.0 \times 7.15) - (20.0 \times 1.232) = 310.64$ kNm

9.8   *Train of Loads*

Structures such as bridges are frequently subjected to 'train loads' i.e. a series of point loads which traverse the structure as a unit as shown in Figure 9.40:

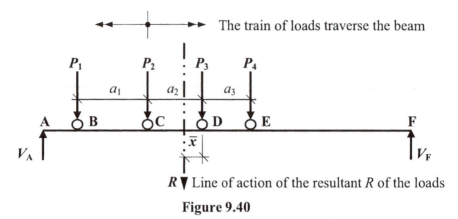

Figure 9.40

The position of the resultant of the train loads relative to e.g. P_3 as indicated in Figure 9.40 can be determined by considering rotational equilibrium of the train loads about P_3:

$$R = (P_1 + P_2 + P_3 + P_4) \quad \text{and} \quad \bar{x} = [P_1(a_1 + a_2) + P_2(a_2) - P_4(a_3)]/R$$

The maximum bending moment in the span will occur under one of the point loads adjacent to the resultant load. Consider the arrangement indicated in Figure 9.41 in which point load P_3 is assumed to be the wheel under which the maximum moment occurs and is positioned a distance 'x' to the right of the mid-span of the beam.

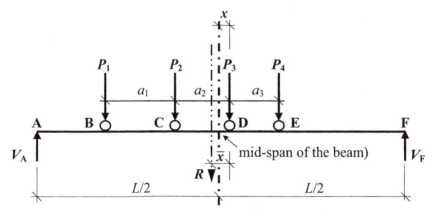

Figure 9.41

Consider the rotational equilibrium of the beam and the bending moment at D:

$$+ve \;\big)\Sigma M_F = 0 \qquad V_A \times L = R\left[L/2 + (\bar{x} - x)\right] \qquad \therefore \; V_A = \frac{R}{L}(L/2 + \bar{x} - x)$$

$$M_D = \left[V_A \times (L/2 + x)\right] - P_1(a_1 + a_2) - P_2 a_2 = \frac{R}{L}(L/2 + \bar{x} - x)(L/2 + x) - P_1(a_1 + a_2) - P_2 a_2$$

The value of x necessary to give the maximum value of the bending moment can be determined by equating $dM_D/dx = 0$, i.e.

$$\frac{dM_D}{dx} = \frac{R}{L}\left[\left(L/2+\bar{x}-x\right)(+1)+\left(L/2+x\right)(-1)\right]=0$$

$$\therefore \ \left(L/2+\bar{x}-x\right)(+1)+\left(L/2+x\right)(-1)=0$$

hence $\left(L/2+\bar{x}-x\right)=\left(L/2+x\right)$ $\ \therefore \ \bar{x}=2x$ and $x=\dfrac{\bar{x}}{2}$

The centre-line of the span must divide the distance between the resultant of all the loads in the train of loads and the load under which the maximum bending moment occurs, i.e. the load nearest the resultant.

9.8.1 Example 9.7: Evaluation of Functions for a Train of Loads

An 8.0 m span, simply supported beam AE, supports a train of point loads at B, C, and D which traverse the beam as shown in Figure 9.42. Assume that the train of loads can leave the beam. Draw the influence lines for:

   (i)   the vertical reaction V_A at support A,
   (ii) the shear force $F_{v,1/3 \text{ span}}$ at the third span point from A,
   (iii) the bending moment M_C under the point load at C to give the maximum moment

and determine the value of each of the functions (i), (ii) and (iii) for the loading indicated.

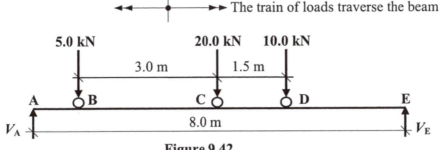

 The train of loads traverse the beam

Figure 9.42

(i) Vertical reaction at support A: V_A

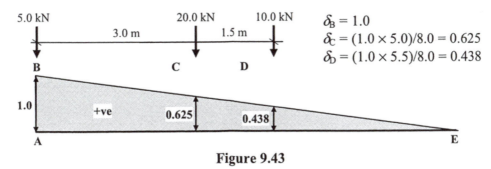

$\delta_B = 1.0$
$\delta_C = (1.0 \times 5.0)/8.0 = 0.625$
$\delta_D = (1.0 \times 5.5)/8.0 = 0.438$

Figure 9.43

$V_A = (5.0 \times 1.0) + (20.0 \times 0.625) + (10.0 \times 0.438) = 21.88$ kN

(ii) Shear force at the third span point from A: $F_{v,1/3\ span}$

The position of the loads for the maximum shear force is dependent on the load system and in most cases the maximum positive value occurs when the left-hand load is at the point being considered and the remaining loads to the right. The maximum negative value usually occurs when the right-hand load is at the point under consideration and the remaining loads to the left. It is possible e.g. in situations where the end loads are significantly less than the others, the above will not apply and trial and error will be necessary.

Case 1: consider the load at point B to be at the third-span point.

$\delta_B = (1.0 \times 5.333)/8.0 = 0.667$

$\delta_C = (1.0 \times 2.333)/8.0 = 0.292$

$\delta_D = (1.0 \times 0.833)/8.0 = 0.104$

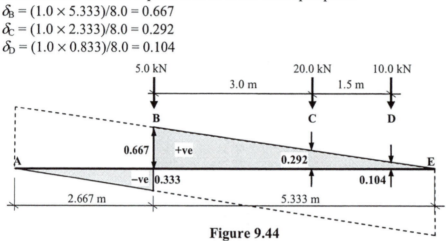

Figure 9.44

$F_{v,1/3\ span} = (5.0 \times 0.667) + (20.0 \times 0.292) + (10.0 \times 0.104) = 10.22$ kN

Case 2: consider the load at point C to be at the third-span point.

δ_B = zero (i.e. the load at B is not on the span)

$\delta_C = 0.667$

$\delta_D = (1.0 \times 3.833)/8.0 = 0.479$

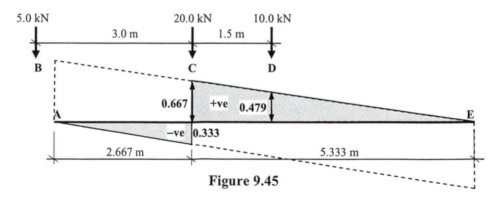

Figure 9.45

$F_{v,1/3\ span} = (20.0 \times 0.667) + (10.0 \times 0.479) = 18.13$ kN

∴ The maximum shear at the third-span point = 18.13 kN

(iii) Maximum bending moment at pont C: M_C

Assume that the maximum bending moment occurs under the point load at C and determine the position of the resultant load relative to C.

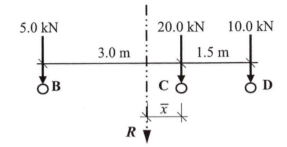

Figure 9.46

$R = (5.0 + 20.0 + 10.0) = 35.0$ kN

$\bar{x} = [P_1(a_1) - P_3(a_2)]/R = [(5.0 \times 3.0) - (10.0 \times 1.5)]/35.0 = 0.857$ m

The load at point C should be positioned such that the mid-span point of the beam bisects $\bar{x}$, i.e. $(0.857 \times 0.5) = 0.429$ m to the right of the resultant force R as shown in Figure 9.47.

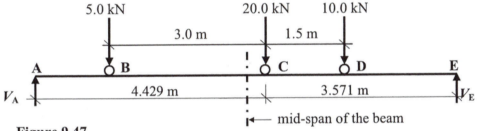

Figure 9.47

The maximum value of the influence line ordinate for the bending moment at C is given by $a(L - a)/L = 4.429 \times (8.0 - 4.429)/8.0 = 1.977$ m (see Figure 9.15).

$\delta_B = (1.977 \times 1.429)/4.429 = 0.638$ m

$\delta_C = 1.977$ m

$\delta_D = (1.977 \times 2.071)/3.571 = 1.147$ m

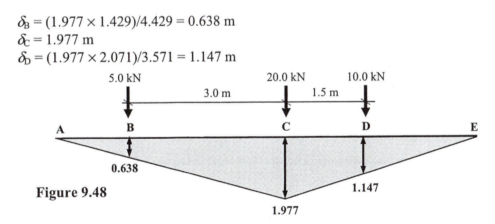

Figure 9.48

$M_C = (5.0 \times 0.638) + (20.0 \times 1.977) + (10.0 \times 1.147) = 54.20$ kNm

9.9　Problems: Influence Lines for Beams

Problem 9.1　A simply-supported beam ABCDE is supported on a roller at B and is pinned support at D as shown in Figure 9.49.

(a)　Draw the influence lines for:
　　(i)　the vertical reaction 'V_B' at support B,
　　(ii)　the shear force '$F_{v,C}$' at C,
　　(iii)　the bending moment 'M_B' at B.

(b)　Using the influence lines determine the maximum and minimum values of V_B and M_B and the maximum value of $F_{v,C}$ when a 10.0 kN/m load of length 4.0 m traverses the beam in addition to a static concentrated load as indicated. (**Note:** the 10.0 kN/m load may leave the span).

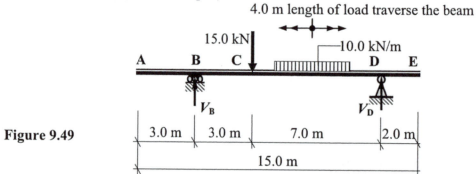

Figure 9.49

Problem 9.2　A simply-supported beam AD, is pinned at support A and supported on a roller at D as shown in Figure 9.50. A train of two 12.0 kN loads traverse the beam as indicated.

(a)　Determine the position 'x' of the loads, required to produce the maximum bending moment in the beam, assuming that it occurs under wheel C.
(b)　Draw the influence line for the support reaction at A.
(c)　Draw the influence line for the bending moment at position 'x'.
(d)　Using the influence lines developed above, determine the maximum value of the support reaction at A and the bending moment at　position 'x' for the train of loads.

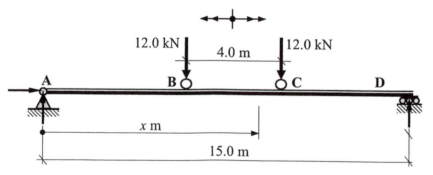

Figure 9.50

Problem 9.3 A cantilever beam ABC, is fixed at support A as shown in Figure 9.51. The beam carries two fixed uniformly distributed loads and a point load which traverses the cantilever as indicated.

(a) Draw the influence line for the vertical reaction 'V_A' at support A.
(b) Draw the influence line for the moment reaction 'M_A' at support A.
(c) Using these influence lines determine the maximum values of 'V_A' and 'M_A'.

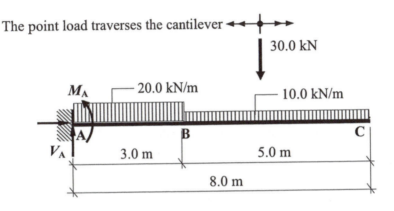

Figure 9.51

Problem 9.4 A two span, uniform beam ABCD is fixed at support A and supported on rollers at B and D as shown in Figure 9.52. Span BCD has a pin at its' mid-span point C. A fixed uniformly distributed load of 15 kN/m is supported from A to point C and a concentrated load of 20 kN traverses the beam from A to D.

(a) Draw the influence line for the moment reaction 'M_A' at support A.
(b) Using the influence line determine the maximum value and sense of M_A.
(c) Using the influence line determine the minimum value and sense of M_A.

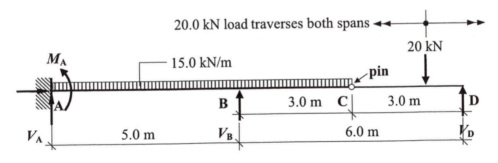

Figure 9.52

9.10   *Solutions: Influence Lines for Beams*

Solution

Topic: Influence Lines for Beams
Problem Number: 9.1   **Page No. 1**

A simply-supported beam ABCDE is supported on a roller at B and is pinned support at D as shown in Figure 9.49.

(a) Draw the influence lines for:
   (i)   the vertical reaction 'V_B' at support B,
   (ii)   the shear force '$F_{v,C}$' at C,
   (iii) the bending moment 'M_B' at B.

(b) Using the influence lines determine the maximum and minimum values of V_B and M_B and the maximum value of $F_{v,C}$ when a 10.0 kN/m load of length 4.0 m traverses the beam in addition to a static concentrated load as indicated. (**Note:** the 10.0 kN/m load may leave the span).

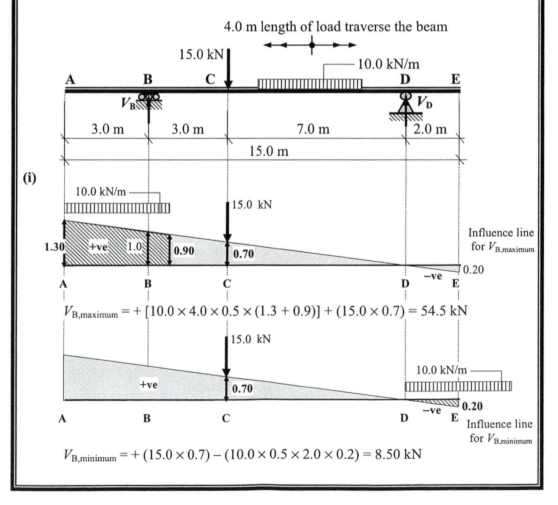

(i)

$$V_{B,maximum} = + [10.0 \times 4.0 \times 0.5 \times (1.3 + 0.9)] + (15.0 \times 0.7) = 54.5 \text{ kN}$$

$$V_{B,minimum} = + (15.0 \times 0.7) - (10.0 \times 0.5 \times 2.0 \times 0.2) = 8.50 \text{ kN}$$

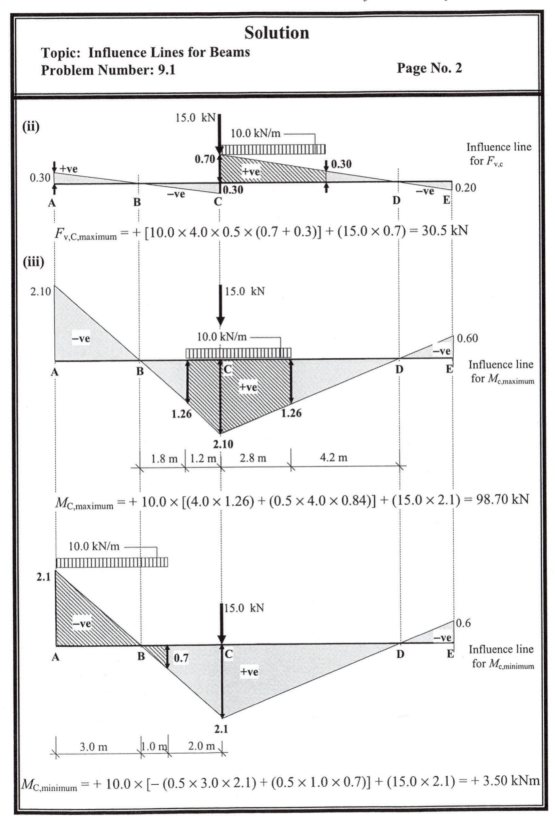

Solution

Topic: Influence Lines for Beams
Problem Number: 9.1

(ii)

15.0 kN

10.0 kN/m

Influence line for $F_{v,c}$

0.70

0.30

+ve

+ve

0.30

0.30

−ve

−ve

0.20

A   B   C   D   E

$F_{v,C,maximum} = + [10.0 \times 4.0 \times 0.5 \times (0.7 + 0.3)] + (15.0 \times 0.7) = 30.5$ kN

(iii)

2.10

15.0 kN

10.0 kN/m

0.60

−ve

−ve

+ve

Influence line for $M_{c,maximum}$

A   B   C   D   E

1.26   1.26

2.10

| 1.8 m | 1.2 m | 2.8 m | 4.2 m |

$M_{C,maximum} = + 10.0 \times [(4.0 \times 1.26) + (0.5 \times 4.0 \times 0.84)] + (15.0 \times 2.1) = 98.70$ kN

10.0 kN/m

2.1

15.0 kN

0.6

−ve

−ve

+ve

Influence line for $M_{c,minimum}$

A   B   0.7   C   D   E

2.1

| 3.0 m | 1.0 m | 2.0 m |

$M_{C,minimum} = + 10.0 \times [-(0.5 \times 3.0 \times 2.1) + (0.5 \times 1.0 \times 0.7)] + (15.0 \times 2.1) = + 3.50$ kNm

---

Solution

Topic: Influence Lines for Beams
Problem Number: 9.2   **Page No. 1**

A simply-supported beam AD, is pinned at support A and supported on a roller at D as shown in Figure 9.50. A train of two 12.0 kN loads traverse the beam as indicated.

(a) Determine the position 'x' of the loads, required to produce the maximum bending moment in the beam, assuming that it occurs under wheel C.
(b) Draw the influence line for the support reactions at A.
(c) Draw the influence line for the bending moment at position 'x'.

Using the influence lines developed above, determine the maximum value of the support reaction at A and the bending moment at position 'x' for the train of loads.

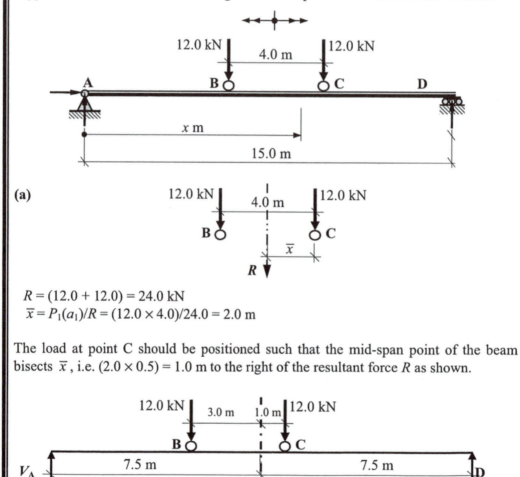

(a)

$R = (12.0 + 12.0) = 24.0$ kN
$\bar{x} = P_1(a_1)/R = (12.0 \times 4.0)/24.0 = 2.0$ m

The load at point C should be positioned such that the mid-span point of the beam bisects $\bar{x}$, i.e. $(2.0 \times 0.5) = 1.0$ m to the right of the resultant force R as shown.

Solution

Topic: Influence Lines for Beams
Problem Number: 9.2

Page No. 2

(b)

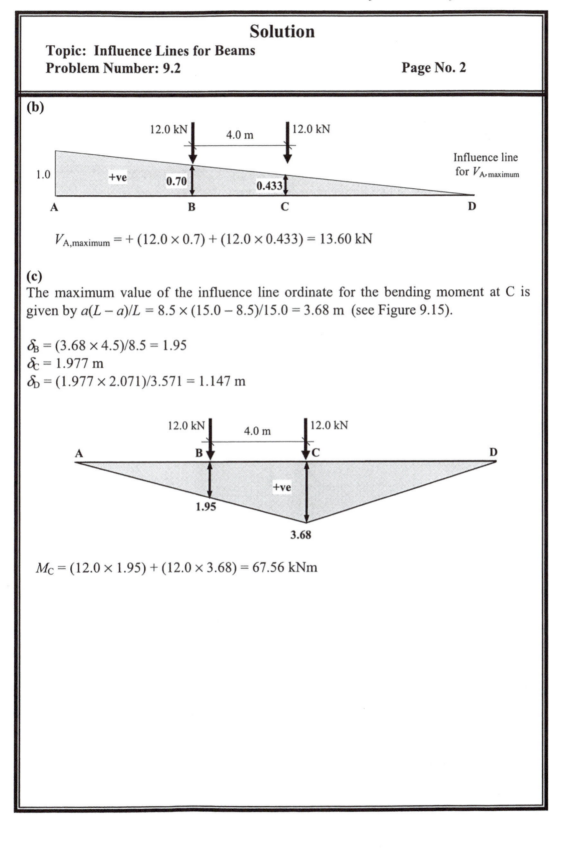

$$V_{A,maximum} = +(12.0 \times 0.7) + (12.0 \times 0.433) = 13.60 \text{ kN}$$

(c)

The maximum value of the influence line ordinate for the bending moment at C is given by $a(L - a)/L = 8.5 \times (15.0 - 8.5)/15.0 = 3.68$ m (see Figure 9.15).

$\delta_B = (3.68 \times 4.5)/8.5 = 1.95$
$\delta_C = 1.977$ m
$\delta_D = (1.977 \times 2.071)/3.571 = 1.147$ m

$$M_C = (12.0 \times 1.95) + (12.0 \times 3.68) = 67.56 \text{ kNm}$$

Solution

Topic: Influence Lines for Beams
Problem Number: 9.3   **Page No. 1**

A cantilever beam ABC, is fixed at support A as shown in Figure 9.51. The beam carries two fixed uniformly distributed loads and a point load which traverses the cantilever as indicated.

(a)   Draw the influence line for the vertical reaction 'V_A' at support A.
(b)   Draw the influence line for the moment reaction 'M_A' at support A.
(c)   Using these influence lines determine the maximum values of 'V_A' and 'M_A'.

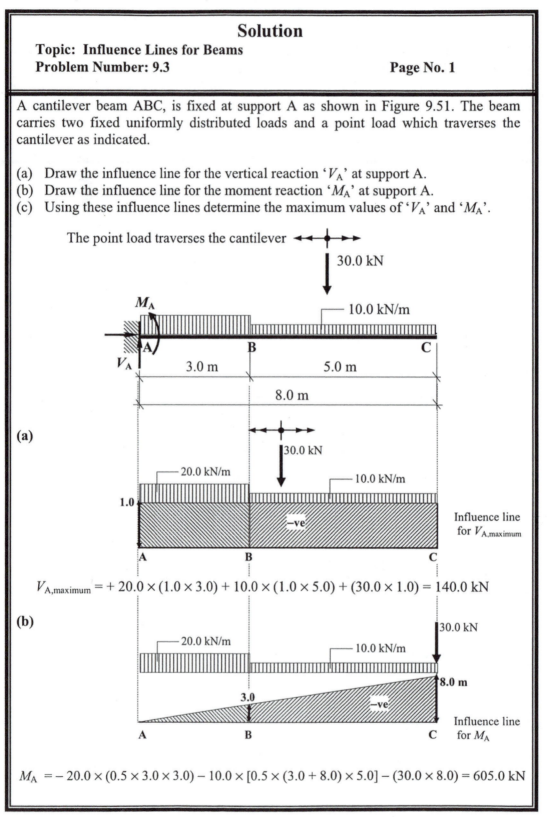

(a)

$V_{A,\text{maximum}} = + 20.0 \times (1.0 \times 3.0) + 10.0 \times (1.0 \times 5.0) + (30.0 \times 1.0) = 140.0 \text{ kN}$

(b)

$M_A = - 20.0 \times (0.5 \times 3.0 \times 3.0) - 10.0 \times [0.5 \times (3.0 + 8.0) \times 5.0] - (30.0 \times 8.0) = 605.0 \text{ kN}$

Solution

Topic: Influence Lines for Beams
Problem Number: 9.4   **Page No. 1**

A two span, uniform beam ABCD is fixed at support A and supported on rollers at B and D as shown in Figure 9.52. Span BCD has a pin at its' mid-span point C. A fixed uniformly distributed load of 15 kN/m is supported from A to point C and a concentrated load of 20 kN traverses the beam from A to D.

(a) Draw the influence line for the moment reaction 'M_A' at support A.
(b) Using the influence line determine the maximum value and sense of M_A.
(c) Using the influence line determine the minimum value and sense of M_A.

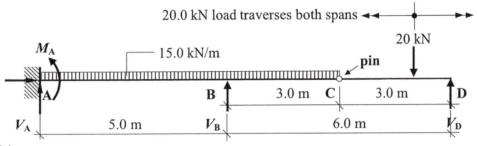

(a)
Remove the restraint which induces the moment at A by introducing a pin and impose displacement θ_A at A using a moment $M = 1.0$ The unit moment induces general displacements at m and n along the beams AB and BCD as indicated.

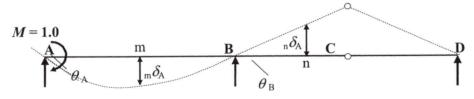

Consider span AB and use MaCaulay's method to determine the displaced shape induced by the unit moment.

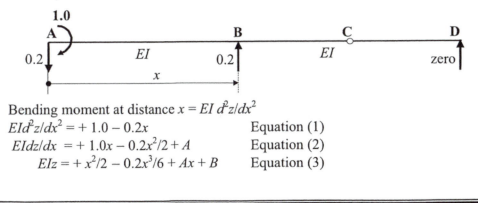

Bending moment at distance $x = EI\, d^2z/dx^2$

$EId^2z/dx^2 = + 1.0 - 0.2x$ Equation (1)
$EIdz/dx = + 1.0x - 0.2x^2/2 + A$ Equation (2)
$EIz = + x^2/2 - 0.2x^3/6 + Ax + B$ Equation (3)

Solution

Topic: Influence Lines for Beams
Problem Number: 9.4   **Page No. 2**

Apply the boundary conditions:
when $x = 0$ $z = 0$ $\therefore$ $B = 0$
when $x = 5.0$ $z = 0$ $\therefore$ $A = [-25.0/2 + (0.2 \times 5.0^3)/6]/5.0 = -1.667$

when $x = 0$ The slope at A is given by dz/dx $\therefore$ $\theta_A = {}_A\delta_A = +A/EI = -1.667/EI$
when $x = 5.0$ $\theta_B = [+(1.0 \times 5.0) - (0.2 \times 5.0^2)/2 - 1.667]/EI = +0.833/EI$

Consider the span BCD:

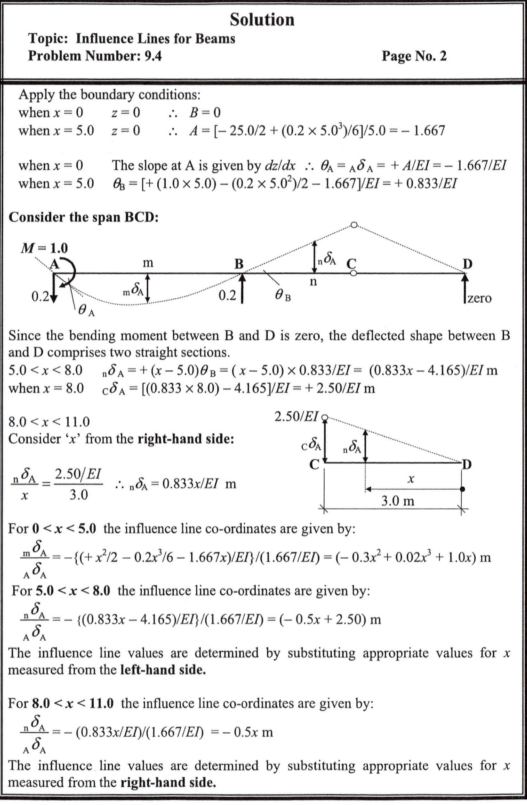

Since the bending moment between B and D is zero, the deflected shape between B and D comprises two straight sections.
$5.0 < x < 8.0$ ${}_n\delta_A = +(x - 5.0)\theta_B = (x - 5.0) \times 0.833/EI = (0.833x - 4.165)/EI$ m
when $x = 8.0$ ${}_c\delta_A = [(0.833 \times 8.0) - 4.165]/EI = +2.50/EI$ m

$8.0 < x < 11.0$
Consider 'x' from the **right-hand side:**

$$\frac{{}_n\delta_A}{x} = \frac{2.50/EI}{3.0} \quad \therefore {}_n\delta_A = 0.833x/EI \text{ m}$$

For **$0 < x < 5.0$** the influence line co-ordinates are given by:
$$\frac{{}_m\delta_A}{{}_A\delta_A} = -\{(+x^2/2 - 0.2x^3/6 - 1.667x)/EI\}/(1.667/EI) = (-0.3x^2 + 0.02x^3 + 1.0x) \text{ m}$$

For **$5.0 < x < 8.0$** the influence line co-ordinates are given by:
$$\frac{{}_n\delta_A}{{}_A\delta_A} = -\{(0.833x - 4.165)/EI\}/(1.667/EI) = (-0.5x + 2.50) \text{ m}$$

The influence line values are determined by substituting appropriate values for x measured from the **left-hand side.**

For **$8.0 < x < 11.0$** the influence line co-ordinates are given by:
$$\frac{{}_n\delta_A}{{}_A\delta_A} = -(0.833x/EI)/(1.667/EI) = -0.5x \text{ m}$$

The influence line values are determined by substituting appropriate values for x measured from the **right-hand side.**

Solution

Topic: Influence Lines for Beams
Problem Number: 9.4   **Page No. 2**

M_A = (concentrated load × ordinate) + (distributed load × area)

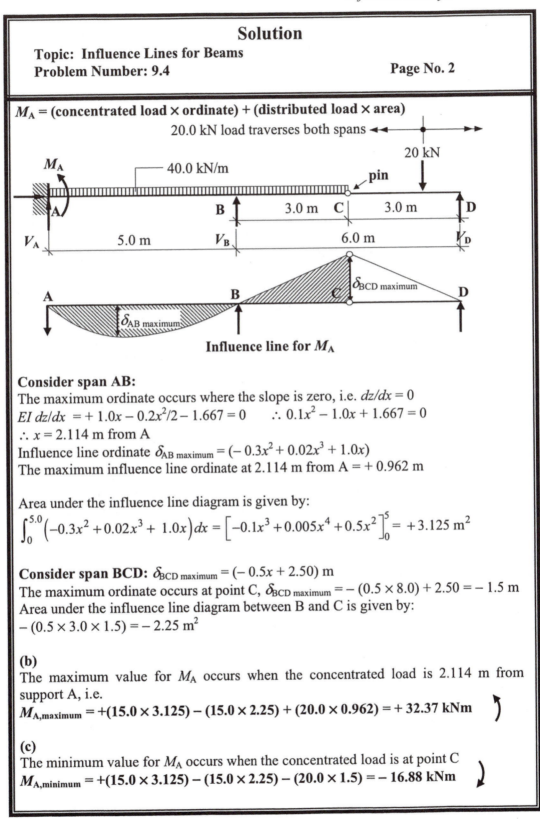

Influence line for M_A

Consider span AB:
The maximum ordinate occurs where the slope is zero, i.e. $dz/dx = 0$
$EI \, dz/dx = +1.0x - 0.2x^2/2 - 1.667 = 0$ ∴ $0.1x^2 - 1.0x + 1.667 = 0$
∴ $x = 2.114$ m from A
Influence line ordinate $\delta_{AB \, maximum} = (-0.3x^2 + 0.02x^3 + 1.0x)$
The maximum influence line ordinate at 2.114 m from A $= +0.962$ m

Area under the influence line diagram is given by:
$$\int_0^{5.0} \left(-0.3x^2 + 0.02x^3 + 1.0x\right)dx = \left[-0.1x^3 + 0.005x^4 + 0.5x^2\right]_0^5 = +3.125 \text{ m}^2$$

Consider span BCD: $\delta_{BCD \, maximum} = (-0.5x + 2.50)$ m
The maximum ordinate occurs at point C, $\delta_{BCD \, maximum} = -(0.5 \times 8.0) + 2.50 = -1.5$ m
Area under the influence line diagram between B and C is given by:
$-(0.5 \times 3.0 \times 1.5) = -2.25 \text{ m}^2$

(b)
The maximum value for M_A occurs when the concentrated load is 2.114 m from support A, i.e.
$M_{A,maximum} = +(15.0 \times 3.125) - (15.0 \times 2.25) + (20.0 \times 0.962) = +32.37$ kNm

(c)
The minimum value for M_A occurs when the concentrated load is at point C
$M_{A,minimum} = +(15.0 \times 3.125) - (15.0 \times 2.25) - (20.0 \times 1.5) = -16.88$ kNm

10. Approximate Methods of Analysis

10.1 *Introduction*

The use of computer software is invariably employed by engineers when undertaking mathematical modelling in structural analysis and design. The results from such analyses are **always** an approximation to the actual structural behaviour irrespective of the complexity and sophistication of the software. All software results are dependent on the assumptions made by the developers of that software and the limitations on use of the final product.

It is important when using such methods that design engineers can verify the accuracy of the results from the computer analysis and detect any gross errors due to e.g. incorrect input data, incorrect modelling being used or inappropriate use being made of the software. This can be achieved in many cases by the use of approximate manual calculations.

In addition to confirming the validity of computer output, it is often convenient for an engineer to obtain approximate values of design effects induced in members by the design loading when it is neither convenient nor suitable to carry-out a full, more accurate rigorous analysis, e.g. whilst on site, attending meetings with other related professionals or conducting preliminary design for initial feasibility and/or costing of a proposed project. Clearly more detailed analysis/design will be required at a later stage of the project.

Statically determinate structures are relatively straight forward to analyse requiring only the use of the three equations of static equilibrium for plane-frames to obtain axial loads, shear forces and bending moments.

Statically indeterminate structures are more complex and require knowledge of the relative member stiffness properties and the compatibility characteristics of the structure to determine accurate results. Approximate methods of analysis can be used to estimate the required member forces in a structure by consideration of the deflected form and in rigid-jointed frames, the consequent points of contraflexure, i.e. points of zero bending moment in the members. Where a point of contra-flexure can be identified and its position estimated with reasonable accuracy then this point may be regarded as a 'pin' in the structure. The existence of a pin provides an additional equation which may be used in conjunction with the three standard equations of static equilibrium. In the case of pin-jointed frames, tension only systems are sometimes used in which members are slender and designed to resist only tension forces; any member with a compression force is assumed to buckle and hence be ineffective. This arrangement is very common in the provision of cross- bracing to resist lateral wind forces.

This chapter describes various techniques which may be used to obtain approximate member forces for indeterminate, pin-jointed and rigid-jointed structures suitable for preliminary analysis or checking computer output as indicated above.

10.2 *Example 10.1 - Statically Indeterminate Pin-jointed Plane Frame 1*

The degree of indeterminacy of the pin-jointed frame shown in Figure 10.1 is given by:

$$I_D = (m + r) - 2n = (16 + 3) - (2 \times 8) = 3 \quad \text{(see Chapter 1: Section 1.5.1)}$$

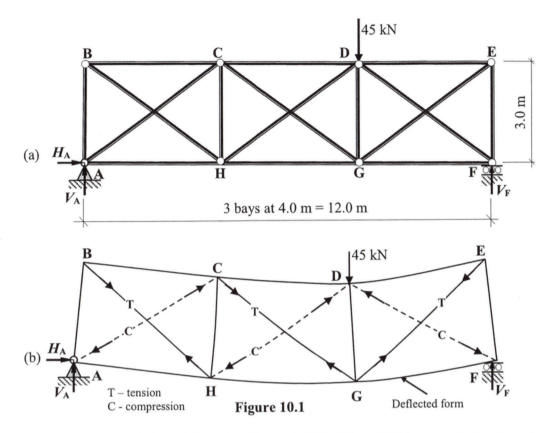

(a)

3.0 m

3 bays at 4.0 m = 12.0 m

(b)

T – tension
C - compression

Figure 10.1

Deflected form

It is evident from the deflected form that diagonals BH, CG and EG increase in length and hence are tension members whilst diagonals AC, HD and DF decrease in length and are compression members.

Since the degree of redundancy is equal to three it is necessary to make three assumptions in order to analyse the frame using the equations of equilibrium alone. There are two options as follows:

i) assume a tension bracing system in which the compression diagonals are slender and assumed to be ineffective and do not support any load. The frame is therefore reduced to a statically determinate frame and the forces determined as indicated in Chapter 3 with the forces as shown in Figure 10.2.

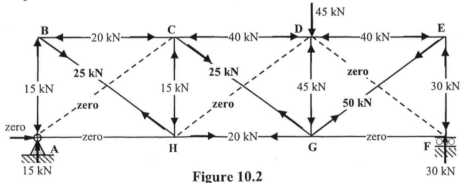

Figure 10.2

ii) assume a bracing system in which all of the diagonals are assumed to be non-slender and have the same cross-sectional area. The shear load in any given panel is assumed to be shared equally by the two diagonals in the panel, one in tension and the other in compression. Three additional equations can be obtained by considering the equilibrium of three sections 1-1, 2-2 and 3-3 indicated in Figure 10.3.

Where the cross-sectional areas of the two diagonals are not the same, distributing the shear in proportion to their axial stiffness, i.e. *EA/L*, will give a more accurate solution.

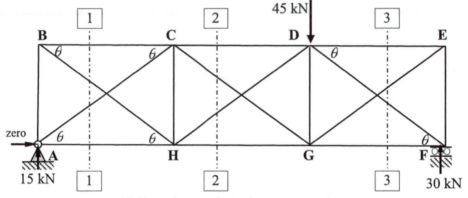

All diagonals can resist tension or compression.

Figure 10.3

$$\sin\theta = 3.0/5.0 = 0.6 \qquad \cos\theta = 4.0/5.0 = 0.8$$

Section 1-1

$$+\text{ve} \uparrow \Sigma F_z = 0$$

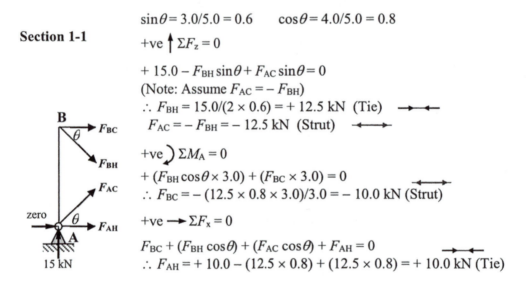

$$+ 15.0 - F_{BH}\sin\theta + F_{AC}\sin\theta = 0$$
(Note: Assume $F_{AC} = -F_{BH}$)
$$\therefore F_{BH} = 15.0/(2 \times 0.6) = +12.5 \text{ kN (Tie)}$$
$$F_{AC} = -F_{BH} = -12.5 \text{ kN (Strut)}$$

$$+\text{ve} \curvearrowright \Sigma M_A = 0$$
$$+ (F_{BH}\cos\theta \times 3.0) + (F_{BC} \times 3.0) = 0$$
$$\therefore F_{BC} = -(12.5 \times 0.8 \times 3.0)/3.0 = -10.0 \text{ kN (Strut)}$$

$$+\text{ve} \longrightarrow \Sigma F_x = 0$$
$$F_{BC} + (F_{BH}\cos\theta) + (F_{AC}\cos\theta) + F_{AH} = 0$$
$$\therefore F_{AH} = +10.0 - (12.5 \times 0.8) + (12.5 \times 0.8) = +10.0 \text{ kN (Tie)}$$

Consider the vertical equilibrium at joint A:

$$+\text{ve} \uparrow \Sigma F_z = 0$$
$$+ 15.0 + F_{AB} - F_{AC}\sin\theta = 0$$
$$\therefore F_{AB} = -15.0 + (12.5 \times 0.6) = -7.50 \text{ kN (Strut)}$$

Section 2-2

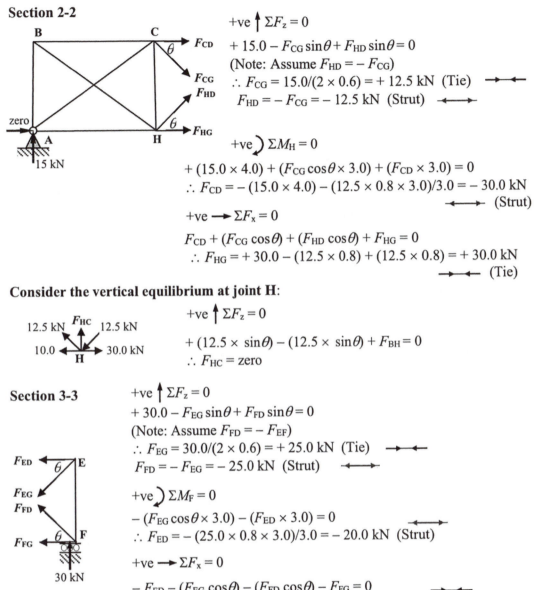

$+ve \uparrow \Sigma F_z = 0$

$+ 15.0 - F_{CG} \sin\theta + F_{HD} \sin\theta = 0$

(Note: Assume $F_{HD} = -F_{CG}$)

$\therefore F_{CG} = 15.0/(2 \times 0.6) = +12.5$ kN (Tie)

$F_{HD} = -F_{CG} = -12.5$ kN (Strut)

$+ve \,\rangle\, \Sigma M_H = 0$

$+ (15.0 \times 4.0) + (F_{CG} \cos\theta \times 3.0) + (F_{CD} \times 3.0) = 0$

$\therefore F_{CD} = -(15.0 \times 4.0) - (12.5 \times 0.8 \times 3.0)/3.0 = -30.0$ kN

(Strut)

$+ve \longrightarrow \Sigma F_x = 0$

$F_{CD} + (F_{CG} \cos\theta) + (F_{HD} \cos\theta) + F_{HG} = 0$

$\therefore F_{HG} = +30.0 - (12.5 \times 0.8) + (12.5 \times 0.8) = +30.0$ kN

(Tie)

Consider the vertical equilibrium at joint H:

12.5 kN F_{HC} 12.5 kN

10.0 ← → 30.0 kN H

$+ve \uparrow \Sigma F_z = 0$

$+ (12.5 \times \sin\theta) - (12.5 \times \sin\theta) + F_{BH} = 0$

$\therefore F_{HC} =$ zero

Section 3-3

F_{ED} E

F_{EG}

F_{FD}

F_{FG} F

30 kN

$+ve \uparrow \Sigma F_z = 0$

$+ 30.0 - F_{EG} \sin\theta + F_{FD} \sin\theta = 0$

(Note: Assume $F_{FD} = -F_{EF}$)

$\therefore F_{EG} = 30.0/(2 \times 0.6) = +25.0$ kN (Tie)

$F_{FD} = -F_{EG} = -25.0$ kN (Strut)

$+ve \,\rangle\, \Sigma M_F = 0$

$- (F_{EG} \cos\theta \times 3.0) - (F_{ED} \times 3.0) = 0$

$\therefore F_{ED} = -(25.0 \times 0.8 \times 3.0)/3.0 = -20.0$ kN (Strut)

$+ve \longrightarrow \Sigma F_x = 0$

$- F_{ED} - (F_{EG} \cos\theta) - (F_{FD} \cos\theta) - F_{FG} = 0$

$\therefore F_{FG} = +20.0 - (25.0 \times 0.8) + (25.0 \times 0.8) = +20.0$ kN (Tie)

Consider the vertical equilibrium at joints G and F:

12.5 kN F_{GD} 25.0 kN

30.0 ← → 20.0 kN G

$+ve \uparrow \Sigma F_z = 0$

$+ (12.5 \times \sin\theta) + (25.0 \times \sin\theta) + F_{GD} = 0$

$\therefore F_{HC} = -(12.5 \times 0.6) - (25.0 \times 0.6) = -22.50$ kN (Strut)

25.0 kN F_{FE}

F

20.0 kN 30 kN

$+ve \uparrow \Sigma F_z = 0$

$+ 30.0 - (25.00 \times \sin\theta) + F_{FE} = 0$

$\therefore F_{FE} = -30.0 + (25.0 \times 0.6) = -15.0$ kN (Strut)

The approximate member forces and the more accurate results using computer analysis are shown in Figure 10.4 and Figure 10.5 respectively.

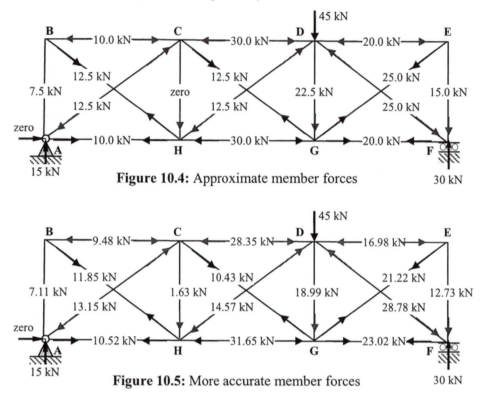

Figure 10.4: Approximate member forces

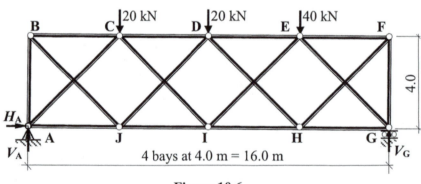

Figure 10.5: More accurate member forces

10.3   *Example 10.2 - Statically Indeterminate Pin-jointed Plane Frame 2*

In some indeterminate trusses the applied load system can be apportioned to two or more statically determinate component trusses which can be considered to make up the original truss. The number of component trusses making up the original truss is equal to $(I_D +1)$. The member forces are then determined by superposition of the two force systems.

Consider the truss shown in Figure 10.6 where $I_D = (18 + 3) - (2 \times 10) = 1$.

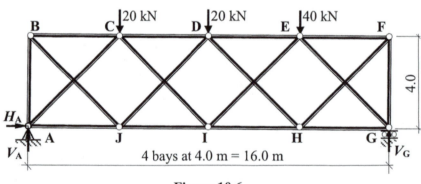

Figure 10.6

This truss can be decomposed into two statically determinate trusses (i.e. $I_D + 1$) as shown in Figure 10.7 and Figure 10.8 with their respective applied loads. The method-of-sections /joint resolution can be used to determine accurately the member forces as indicated.

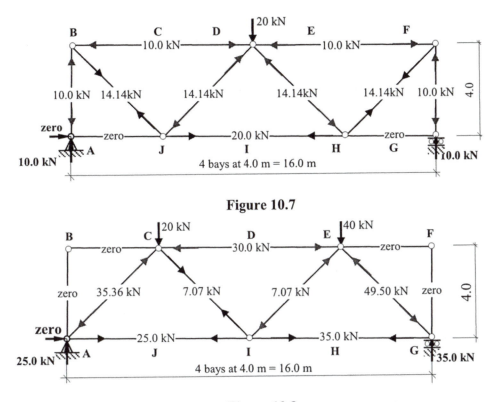

Figure 10.7

Figure 10.8

The final member forces are found by adding those determined from the individual component trusses as shown in Figure 10.9, e.g. consider members AB and JD

Total force in member AB $= -10.0 + 0 = -10.0$ kN   compression

Total force in member JI $= +20.0 + 25.0 = +45.0$ kN   tension

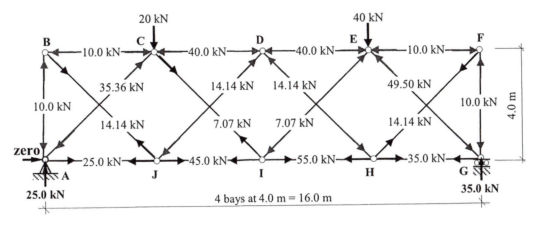

Figure 10.9

10.4   *Example 10.3 - Statically Indeterminate Single-span Beam*

Consider the statically indeterminate beam shown in Figure 10.10(a) which is fully fixed at its end supports and carries a uniformly distributed load of w kN/m length. The deflected shape and the bending moment diagram are indicated in Figure 10.10(b) and Figure 10.10(c) respectively.

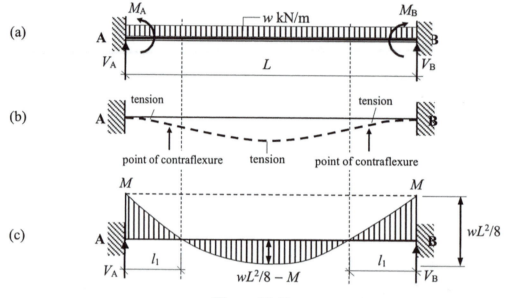

Figure 10.10

The beam has three degrees-of-indeterminacy. Generally the axial loading can be considered to be negligible and ignored and hence two assumptions are required to determine the bending moments and support reactions. It is evident from the symmetry of the beam and loading that two points of contraflexure exist; these can be regarded as pins, i.e. point of zero moment.

Since the beam in Figure 10.10 is a standard case. The value of the support moments 'M' is known to equal $\pm wL^2/12$ and each vertical reaction 'V_A and V_B' is equal to $wL/2$.

This information can be used to determine the position of the points of contraflexure for the beam, i.e.

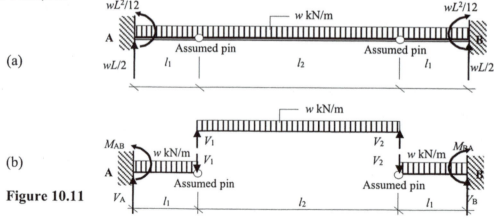

Figure 10.11

Consider the central portion of the beam with span $(L - 2l_1)$:

From Figure 10.10 : $M_{\text{maximum}} = (wL^2/8 - M) = (wL^2/8 - wL^2/12) = wL^2/24$

From Figure 10.11 : $M_{\text{maximum}} = \dfrac{w \times l_2^2}{8} \quad \therefore \quad \dfrac{w \times l_2^2}{8} = \dfrac{w \times L^2}{24}$

hence $l_2 = L/\sqrt{3} = 0.58L$ and $l_1 = (L - 0.58L)/2 = 0.21L$

Similar calculations can be carried out for other standard cases as shown in Figure 10.12. In reality it is unlikely that a support will be fully fixed. In most cases there will be some flexibility and rotation at the support points depending on the stiffnesses and loads on each span. Points of contraflexure can be assumed based on the support and loading conditions in each individual case and the values given in Figure 10.12.

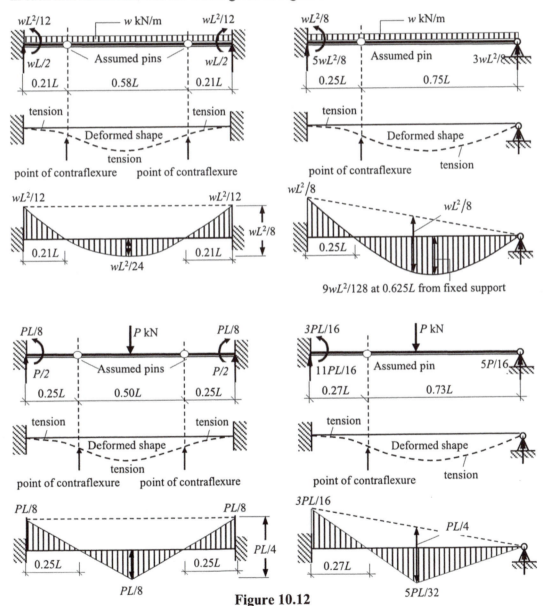

Figure 10.12

10.5   *Example 10.4 - Multi-span Beam*

A uniform, three-span continuous beam ABCD is fixed at support A and supported on rollers at B, C and D as shown in Figure 10.13(a). The deformed shape and shape of the bending moment diagram are shown in Figure 10.13(b) and Figure 10.13(c) respectively. Span BC supports a uniformly distributed at load of 10 kN/m length. Assuming suitable points of contraflexure, determine the approximate value of the support reactions.

$I_D = 3 \therefore$ 3 pins required

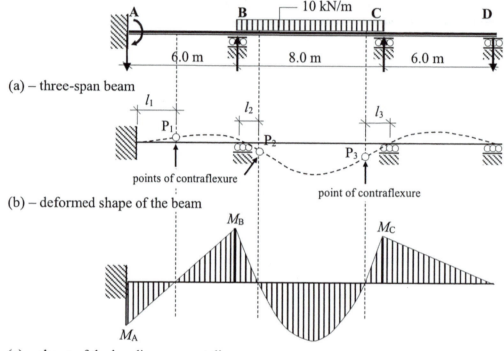

(a) – three-span beam

(b) – deformed shape of the beam

(c) – shape of the bending moment diagram

Figure 10.13

In span AB there is no loading and $M_A = M_B/2$ (i.e. the carry-over moment – see Chapter 4: 4.7.2) and consequently the point-of-contraflexure must be one third of the span from A, i.e. $l_1 = 6.0/3 = 2.0$ m. In span BC it is reasonable to assume the same values for l_1 and l_3 since both ends have a significant continuity moment and the difference in rotation at each support will be small. It is typical in such situations to assume a value of $0.15L$ to allow for the rotation of the joint, i.e. $l_2 = l_3 = (0.15 \times 8.0) = 1.2$ m. The beam can be considered in four sections as shown in Figure 10.14.

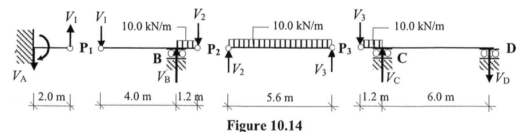

Figure 10.14

Consider section P_2 to P_3:

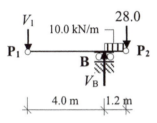

$+ve \uparrow \Sigma F_z = 0$

$- (10.0 \times 5.6) + V_2 + V_3 = 0$ and $V_2 = V_3$

$\therefore V_2 = V_3 = 56.0/2 = 28.0 \text{ kN}$

Consider section P_1 to P_2:

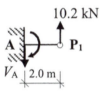

$+ve \curvearrowright \Sigma M_B = 0$

$- (V_1 \times 4.0) + (10.0 \times 1.2^2)/2 + (28.0 \times 1.2) = 0$

$\therefore V_1 = 10.2 \text{ kN}$

$+ve \uparrow \Sigma F_z = 0$

$V_B - 10.2 - (10.0 \times 1.2) - 28.0 = 0$ $\therefore V_B = 50.2 \text{ kN} \uparrow$

$M_B = - (10.2 \times 4.0) = - 40.8 \text{ kNm}$

Consider section A to P_1:

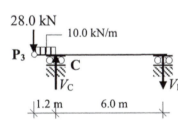

$+ve \uparrow \Sigma F_z = 0$

$- V_A + 10.2 = 0$ $\therefore V_A = 10.2 \text{ kN} \downarrow$

$M_A = + (10.2 \times 2.0) = + 20.4 \text{ kNm}$

Consider section P_3 to D:

$+ve \curvearrowright \Sigma M_C = 0$

$+ (V_D \times 6.0) - (10.0 \times 1.2^2)/2 - (28.0 \times 1.2) = 0$

$\therefore V_D = 6.8 \text{ kN} \downarrow$

$+ve \uparrow \Sigma F_z = 0$

$V_C - 28.0 - (10.0 \times 1.2) - 6.8 = 0$ $\therefore V_C = 46.8 \text{ kN} \uparrow$

$M_C = - (6.8 \times 6.0) = - 40.8 \text{ kNm}$

Check: Total applied vertical load $= (10.0 \times 8.0) = 80.0 \text{ kN} \downarrow$

Total vertical reaction $= (- 10.2 + 50.2 + 46.8 - 6.8) = 80.0 \text{ kN} \uparrow$

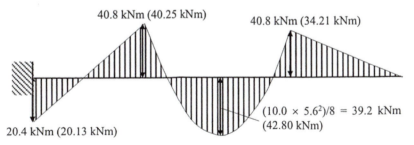

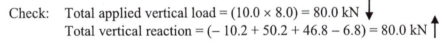

Figure10.15

The assumed points of contraflexure in continuous beams are given in some design codes, e.g. in the Eurocode for the design of structural concrete, EN 1992-1-1:2004: Figure 5.2, the value of $0.15L$ is adopted as indicated in Figure 10.16. where l_o is the distance between points of zero moment for calculating the effective flange width of T and L beams.

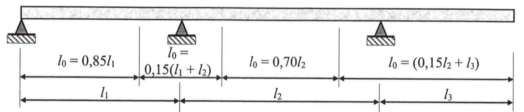

where the length of the cantilever, l_3, should be less than half the adjacent span and the ratio of adjacent spans should lie between 2/3 and 1.5.

Figure10.16: Extract from EN 1992:1-1:2004

10.6　*Rigid-jointed Frames Subjected to Vertical Loads*

In rigid-jointed frames it is unlikely that the joints will be fully rigid. Generally there will be some flexibility at the support points of the beams to the columns. Since the actual beam/column joint will be neither free (i.e. zero moment) nor fully fixed it is common practice to assume that the points of contraflexure occur between the assumed fixed-ends and the positions indicated in Figure 10.12, e.g. at an average value equal to $(0 + 0.21)L \approx 0.1L$ from the column.

It is also acceptable to ignore the axial load effects in the beams since they are negligible and have no significant effect on the deformed shape of the beams. Consider the three-storey, three-bay rigid-jointed frame indicated in Example 10.5.

10.6.1　*Example 10.5 – Multi-storey Rigid-jointed Frame 1*

A three-bay, three-storey rigid-jointed frame is shown in Figure 10.17. Using the data given determine the approximate
  (i)　determine the member forces,
  (ii)　determine the support reactions and
  (iii)　draw the approximate bending moment diagram for the frame.

Note: all members have the same *EI* value.

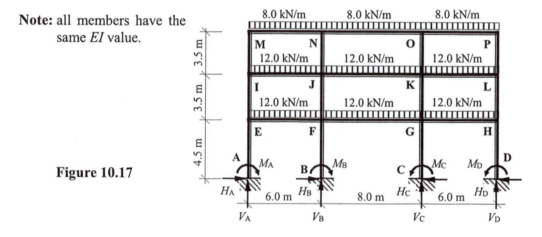

Figure 10.17

The deformed shape of the frame, neglecting axial deformation, is shown in Figure 10.18 with the assumed points of contraflexure of the beams as indicated.

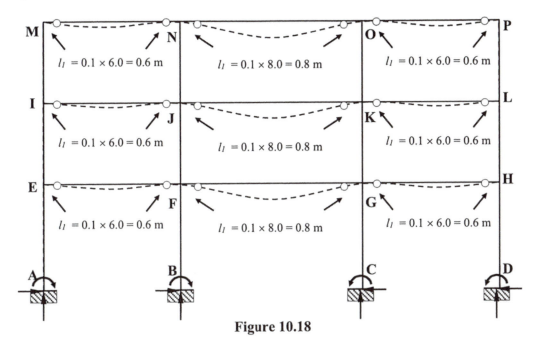

Figure 10.18

Consider the force systems indicated in Figure 10.19 in which the points of contraflexure are assumed to be pins.

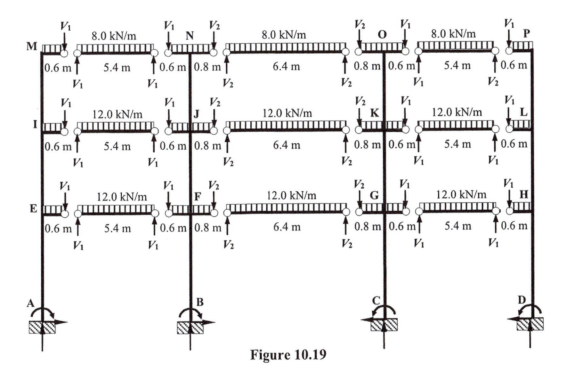

Figure 10.19

Application of the equations of equilibrium as in Example 10.4 yield the member forces determined as follows:

Consider the first floor level shown in Figure 10.20.

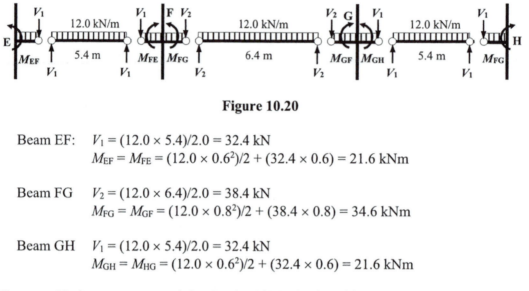

Figure 10.20

Beam EF: $V_1 = (12.0 \times 5.4)/2.0 = 32.4$ kN
$M_{EF} = M_{FE} = (12.0 \times 0.6^2)/2 + (32.4 \times 0.6) = 21.6$ kNm

Beam FG $V_2 = (12.0 \times 6.4)/2.0 = 38.4$ kN
$M_{FG} = M_{GF} = (12.0 \times 0.8^2)/2 + (38.4 \times 0.8) = 34.6$ kNm

Beam GH $V_1 = (12.0 \times 5.4)/2.0 = 32.4$ kN
$M_{GH} = M_{HG} = (12.0 \times 0.6^2)/2 + (32.4 \times 0.6) = 21.6$ kNm

The out-of-balance moment at joint E = 21.6 kNm is shared by the columns *EI* and *EA* in proportion to their flexural stiffnesses k_{EI} and k_{EA}, i.e. *EI/L* values.
$k_{EI} = I/3.5/(I/3.5 + I/4.5) = 0.56$ and $k_{EA} = I/4.5/(I/3.5 + I/4.5) = 0.44$
$M_{EI} = (0.56 \times 21.6) = 12.1$ kNm and $M_{EA} = (0.44 \times 21.6) = 9.5$ kNm

The moments at the bases of the columns are normally assumed to be the same as the values at their top ends. Assume the support moment $M_{AE} = 9.5$ kNm

The out-of-balance moment at joint F = $(34.6 - 21.6) = 13.0$ kNm is shared by the columns FJ and FB in proportion to their flexural stiffnesses k_{FJ} and k_{FB}.

$k_{FJ} = 0.56$ and $k_{FB} = 0.44$
$M_{FJ} = (0.56 \times 13.0) = 7.3$ kNm and $M_{FB} = (0.44 \times 13.0) = 5.7$ kNm
Assume the support moment $M_{BF} = 5.7$ kNm

Since the structure and loading are symmetrical joint G is the same as joint F.
$M_{GK} = 7.3$ kNm and $M_{GC} = 5.7$ kNm
Assume the support moment $M_{CG} = 5.7$ kNm

Due to the symmetry joint H is the same as joint E.
$M_{HL} = 12.1$ kNm and $M_{HD} = = 9.5$ kNm
Assume the support moment $M_{DH} = 9.5$ kNm

Consider the second floor level shown in Figure 10.21.

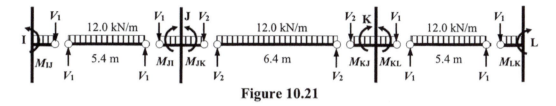

Figure 10.21

As for the first floor $V_1 = 32.4$ kN $M_{IJ} = M_{JI} = 21.6$ kNm
$V_2 = 38.4$ kN $M_{JK} = M_{KJ} = 34.6$ kNm
$V_1 = 32.4$ kN $M_{KL} = M_{LK} = 21.6$ kNm

The out-of-balance moment at joint I = 21.6 kNm is shared by the columns IM and IE in proportion to their flexural stiffnesses k_{IM} and k_{IE}.
$k_{IM} = I/3.5/(I/3.5 + I/3.5) = 0.50$ and $k_{IE} = I/3.5/(I/3.5 + I/3.5) = 0.50$
$M_{IM} = (0.50 \times 21.6) = 10.8$ kNm and $M_{IE} = (0.50 \times 21.6) = 10.8$ kNm

The out-of-balance moment at joint J = 13.0 kNm is shared by the columns JN and JF in proportion to their flexural stiffnesses k_{JN} and k_{JF}.

$k_{JN} = 0.50$ and $k_{JF} = 0.50$
$M_{JN} = (0.50 \times 13.0) = 6.5$ kNm and $M_{JF} = (0.50 \times 13.0) = 6.5$ kNm

Since the structure and loading are symmetrical joint K is the same as joint J.
$M_{KO} = 6.5$ kNm and $M_{KG} = 6.5$ kNm

Due to the symmetry joint L is the same as joint I.
$M_{LP} = 10.8$ kNm and $M_{LH} == 10.8$ kNm

Consider the roof level shown in Figure 10.22.

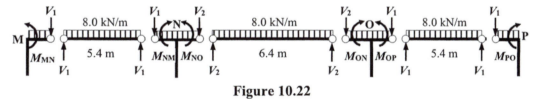

Figure 10.22

Beam MN: $V_1 = (8.0 \times 5.4)/2.0 = 21.6$ kN
$M_{MN} = M_{NM} = (8.0 \times 0.6^2)/2 + (21.6 \times 0.6) = 14.4$ kNm

Beam NO $V_2 = (8.0 \times 6.4)/2.0 = 25.6$ kN
$M_{NO} = M_{ON} = (8.0 \times 0.8^2)/2 + (25.6 \times 0.8) = 23.0$ kNm

Beam OP $V_1 = (8.0 \times 5.4)/2.0 = 21.6$ kN
$M_{OP} = M_{PO} = (8.0 \times 0.6^2)/2 + (21.6 \times 0.6) = 14.4$ kNm

The out-of-balance moment at joint M = 14.4 kNm is balanced by the column MI.
M_{MI} = 14.4 kNm

The out-of-balance moment at joint N = (23.0 − 14.4) = 8.6 kNm is balanced by the column NJ.
M_{NJ} = 8.6 kNm

Since the structure and loading are symmetrical joint O is the same as joint N.
M_{OK} = 8.6 kNm

Due to the symmetry joint P is the same as joint M.
M_{HL} = 14.4 kNm

The bending moment at the mid-span of beams EF, IJ, GH, and KL is given by:
$M \approx (wl^2/8 = (12.0 \times 5.4^2)/8 = 43.7$ kNm

The bending moment at the mid-span of beam MN is given by:
$M \approx (wl^2/8 = (8.0 \times 5.4^2)/8 = 29.1$ kNm

The bending moment at the mid-span of beams FG and JK is given by:
$M \approx (wl^2/8 = (12.0 \times 6.4^2)/8 = 61.4$ kNm

The bending moment at the mid-span of beams NO is given by:
$M \approx (wl^2/8 = (8.0 \times 6.4^2)/8 = 41.0$ kNm

The approximate bending moment diagram is indicated in Figure 10.23.

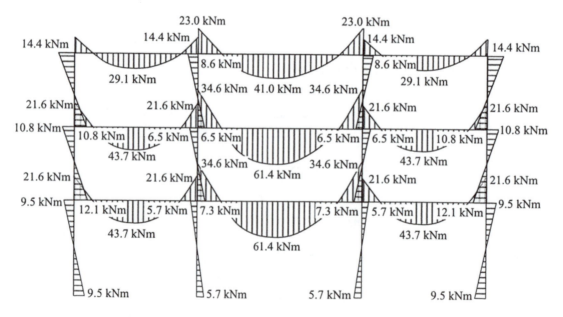

Figure 10.23: Approximate Bending Moment Diagram

The level of accuracy of the results is dependent on the assumed positions of the points of contraflexure, i.e. the assumed value of *l*. The limiting values range from very flexible joints with zero moment ($l = 0$) to fully-fixed joints with *l* equal to the values indicated in Figure 10.12 (e.g. for a distributed load $l = 0.21L$ where the joint moment is equal to $wL^2/12$). The assumption of $0.1L$ in this Example 10.5 will give results which lie between those limits and which are the correct order of magnitude.

The column axial loads and support reactions are determined readily by summation of the roof and floor loads and rotational equilibrium of the bottom level of columns as indicated in Figure 10.24.

Column Axial Loads:
$N_{MI} = N_{PL} \approx (8.0 \times 3.0) = 24.0$ kN
$N_{IE} = N_{LH} \approx 24.0 + (12.0 \times 3.0) = 60.0$ kN
$N_{EA} = N_{HD} \approx 60.0 + (12.0 \times 3.0) = 96.0$ kN
$N_{NJ} = N_{OK} \approx (8.0 \times 7.0) = 56.0$ kN
$N_{JF} = N_{KG} \approx 56.0 + (12.0 \times 7.0) = 140.0$ kN
$N_{EA} = N_{HD} \approx 140.0 + (12.0 \times 7.0) = 224.0$ kN

Support Reactions:
$V_A = V_D \approx 96.0$ kN; $V_B = V_C \approx 224.0$ kN
$H_A = H_D \approx (2 \times 9.5)/4.5 = 4.2$ kN ; $H_B = H_C \approx (2 \times 5.7)/4.5 = 2.5$ kN ;

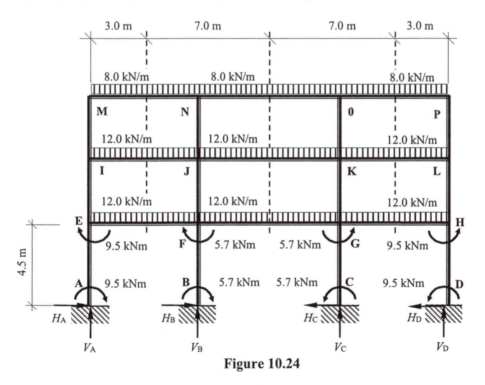

Figure 10.24

Note: the side-sway induced by gravity loads is normally very small (or zero where both the frame and loading are symmetrical) and consequently ignored in approximate analyses.

10.6.2 *Approximate Analysis of Multi-storey Rigid-jointed Frames Using Sub-frames*

Many reinforced concrete structures are cast in-situ resulting in a loadbearing frame in which the slabs, beam and columns act as a continuum to resist and transfer applied loads to the foundations. In braced concrete and steel structures, elements such as the cross-bracing, shear-walls and or shear-cores are designed to resist the lateral wind loading in transverse and longitudinal directions whilst the slabs, beams and columns are designed to resist the vertical gravity loading.

The design of rigid-frames is based on an analysis to determine maximum sagging and hogging bending moments, maximum shear forces and/or axial loads in the members. The continuity of the structure requires an analysis to be carried out for multi-span beams and/or slabs in addition to multi-storey columns.

As an alternative to the method indicated in Section 10.6.1, the structure can be considered as a series of sub-frames. Consider the multi-storey frame indicated in Figure 10.25 in which it is assumed that the lateral loading is resisted by separate elements, not indicated, such as shear-cores. The slabs, beams and columns are assumed to transfer only vertical loads by rigid-frame action. The sub-frames may be analysed using the method of moment distribution.

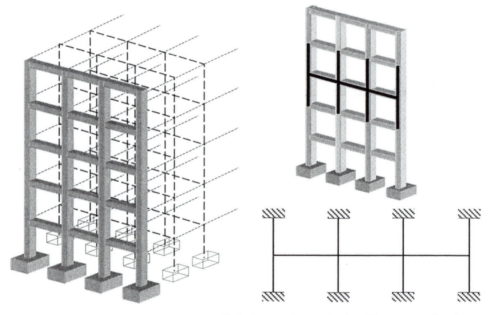

Sub-frame for analysis of beams and columns

Figure 10.25

Figure 10.26

10.6.2.1 Simplification into Sub-frames

Each sub-frame may be taken to consist of the beams at one level together with the columns above and below. The ends of the columns remote from the beams may generally be assumed to be fixed unless the assumption of a pinned end is clearly more reasonable (for example, where a foundation detail is considered unable to develop moment restraint). This is illustrated in Figure 10.26.

10.6.2.2 Alternative Simplification for Individual Beams and Associated Columns

The moments and forces in each individual beam may be found by considering a simplified sub-frame consisting only of that beam, the columns attached to the end of that beam and the beams on either side, if any.

The column and beam ends remote from the beam under consideration may generally be assumed to be fixed unless the assumption of a pinned end is clearly more reasonable. The stiffness of the beams on either side of the beam considered should be taken as half their actual values if they are taken to be fixed at their outer ends.

The moments in an individual column may also be found from this simplified sub-frame provided that the sub-frame has as its central beam the longer of the two spans framing into the column under consideration. This is illustrated in Figure 10.27.

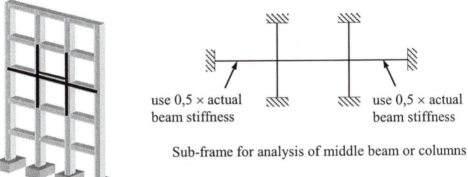

use 0,5 × actual beam stiffness   use 0,5 × actual beam stiffness

Sub-frame for analysis of middle beam or columns

Figure 10.27

10.6.2.3 'Continuous Beam' Simplification

The moments and forces in the beams at one level may also be obtained by considering the beams as a continuous beam over supports, providing no restraint to rotation. This is illustrated in Figure 10.28.

Sub-frame for analysis of beams at any one level

Figure 10.28

10.6.2.4 Asymmetrically-loaded columns where a beam has been analysed in accordance with the 'Continuous Beam' Simplification in Section 10.6.2.3 above.

The ultimate moments may be calculated by simple moment distribution procedures, on the assumption that the column and beam ends remote from the junction under consideration are fixed and that the beams possess half their actual stiffness. The arrangement of the design ultimate imposed load should be such as to cause the maximum moment in the column. This is illustrated in Figure 10.29.

A number of critical load patterns must be considered in all of the above sub-frames to determine the design values of shear and bending. In the case of columns it is necessary to include load patterns which will produce (i) the maximum axial effect combined with its coincident bending effect and (ii) the maximum bending effect combined with its coincident axial effect.

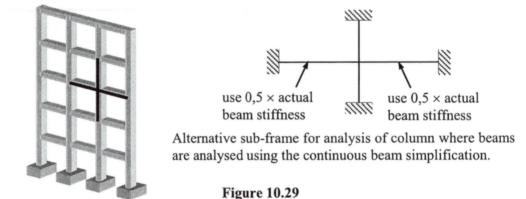

use $0{,}5 \times$ actual beam stiffness   use $0{,}5 \times$ actual beam stiffness

Alternative sub-frame for analysis of column where beams are analysed using the continuous beam simplification.

Figure 10.29

10.6.3 Simple Portal Frames with Pinned Bases Subjected to Horizontal Loads

Simple rectangular portal frames with pinned bases as shown in Figure 10.30 are singly redundant and consequently require one assumption in order to determine the member forces. The deflected shape indicates a point of contraflexure assumed to be at the mid-span of the beam. Analysis using this assumption results in the horizontal support reactions being equal to $P/2$. This is consistent with frames in which the columns are identical where the lateral load divides in proportion to the flexural stiffness of the columns producing equal horizontal reactions at the base.

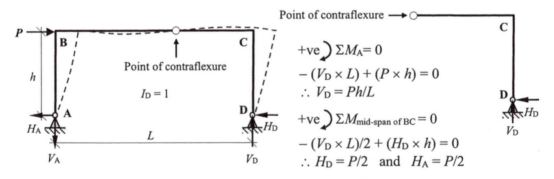

Point of contraflexure ⟶ ○

$+ve \;\curvearrowright\; \Sigma M_A = 0$

$-(V_D \times L) + (P \times h) = 0$

$\therefore\; V_D = Ph/L$

$+ve \;\curvearrowright\; \Sigma M_{\text{mid-span of BC}} = 0$

$-(V_D \times L)/2 + (H_D \times h) = 0$

$\therefore\; H_D = P/2 \quad \text{and} \quad H_A = P/2$

Figure 10.30

10.6.3.1 Example 10.6: Simple Rectangular Portal Frame – Pinned Bases

A rigid-jointed, simple portal frame with pinned bases is subjected to a horizontal load of 15.0 kN as shown in Figure 10.31. Determine the approximate values of the support reactions and sketch the approximate bending moment diagram.

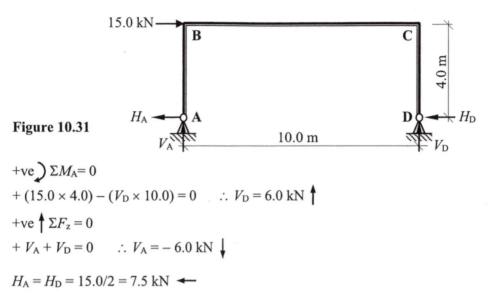

Figure 10.31

$+ve \, \circlearrowright \, \Sigma M_A = 0$

$+ (15.0 \times 4.0) - (V_D \times 10.0) = 0 \quad \therefore V_D = 6.0 \text{ kN} \uparrow$

$+ve \uparrow \Sigma F_z = 0$

$+ V_A + V_D = 0 \quad \therefore V_A = -6.0 \text{ kN} \downarrow$

$H_A = H_D = 15.0/2 = 7.5 \text{ kN} \leftarrow$

Bending moment at joints B and C:

$M_B = + (7.5 \times 4.0) = 30.0 \text{ kNm} \quad$ (tension inside the frame)

$M_C = - (7.5 \times 4.0) - 30.0 \text{ kNm} \quad$ (tension outside the frame)

(Note: in this case the results are the exact values with the symmetrical frame and points of contraflexure at the supports.)

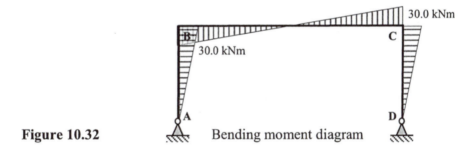

Figure 10.32   Bending moment diagram

10.6.4 *Simple Portal Frames with Fixed Bases Subjected to Horizontal Loads*

Simple rectangular portal frames with fixed bases as shown in Figure 10.33 have three degrees-of-indeterminacy and consequently require three assumptions in order to determine the member forces. The deflected shape indicates points of contraflexure assumed to be at the mid-span point of the beam and in the columns.

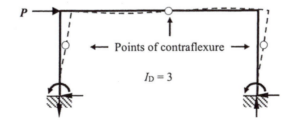

Figure 10.33

The position of the point of contraflexure in the columns is dependent on the relative flexural stiffness between the columns and the beam. This would occur at the mid-height position of the column if the beam and the fixed support were infinitely stiff. For a typical frame there will be some flexibility in both the beam and the support and the point of contraflexure will be slightly higher than mid-height. It is reasonable to assume a position equal to 0.6 of the column height from the base.

If the columns are identical then it can be assumed that the horizontal support reactions and the shear forces in the columns are equal to $P/2$. This, in combination with the assumed points of contraflexure at a height of $0.6h$, will enable the member forces to be determined. (Note: the point of contraflexure in the beam is not required in this case).

10.6.4.1 Example 10.7: Simple Rectangular Portal Frame – Fixed Bases

A rigid-jointed, simple portal frame with fixed bases is subjected to a horizontal load of 15.0 kN as shown in Figure 10.34. Determine the approximate values of the support reactions and sketch the approximate bending moment diagram.

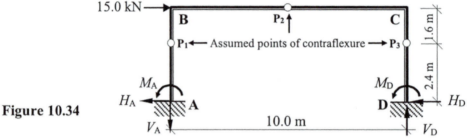

Figure 10.34

Consider the equilibrium of the frame above a horizontal section through the points of contraflexure in the columns:

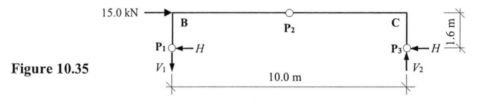

Figure 10.35

Assume that the column shear forces are equal $\therefore$ $H = 15.0/2 = 7.5$ kN ←

+ve $\curvearrowright$ $\Sigma M_{P1} = 0$

$+ (15.0 \times 1.6) - (V_2 \times 10.0) = 0$ $\therefore$ $V_2 \approx 2.4$ kN ↑

+ve ↑ $\Sigma F_z = 0$

$+ V_2 + V_1 = 0$ $\therefore$ $V_1 \approx -2.4$ kN ↓

Bending moment at joints B and C:
$M_B \approx + (7.5 \times 1.6) = + 12.0$ kNm   (tension inside the frame)
$M_C \approx - (7.5 \times 4.0) = - 12.0$ kNm   (tension outside the frame)

Consider the equilibrium of the frame below a horizontal section through the points of contraflexure in the columns:

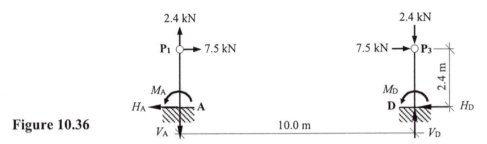

Figure 10.36

Assume that the horizontal support reactions are equal:
$$\therefore \ H_A = H_D = 15.0/2 = 7.5 \text{ kN} \ \leftarrow$$

$+\text{ve} \ \rbrace \ \Sigma M_{P1} = 0$
$+ (7.5 \times 2.4) - M_A = 0 \qquad \therefore \ M_A \approx 18.0 \text{ kNm} \ \curvearrowright$

$+\text{ve} \ \uparrow \ \Sigma F_z = 0$
$+ 2.4 - V_A = 0 \qquad \therefore \ V_A \approx - 2.4 \text{ kN} \ \downarrow$

Similarly for column CD

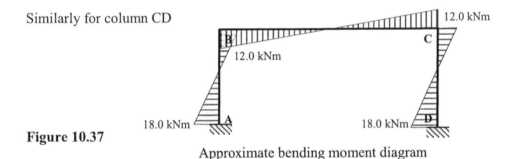

Figure 10.37

Approximate bending moment diagram

10.7 *Multi-storey, Rigid-jointed Frames Subjected to Horizontal Loads*

The behaviour of multi-storey, rigid-frames when subjected to lateral loading is different from that under vertical loading. The deformed shape of the structure indicates a single point of contraflexure in the beams in addition to the columns as shown in Figure 10.38. There are generally two methods of approximate analysis which are used, they are:

(i) the *portal method* which is more suitable for low-rise buildings, e.g. where the width of the frame is equal to or greater than the height of the frame and

(ii) the *cantilever method* which is more suitable for taller, slender buildings.

10.7.1 *Portal Method*

In the case of multi-storey, multi-bay frames an assumption in addition to the points of contraflexure must be made, i.e. the horizontal shear force is divided among all the columns on the basis that each interior column resists twice as much shear force as the exterior columns. The shear force distributed to columns is approximately in proportion to their flexural stiffness (*EI/L*). In many cases the interior columns support twice as much floor area as the exterior columns and consequently tend to be larger. In situations where this is not the case, e.g. where the exterior columns support masonry infills/cladding rather than glazing units, the distribution of shear may be modified accordingly.

In general terms for a multi-bay frame where '*n*' is the number of bays and '*P*' is the total shear above the storey being considered, the number of interior columns is equal to (*n* − 2) each of which resist *P/n* whilst the two exterior columns resist *P/2n* as shown in Figure 10.38.

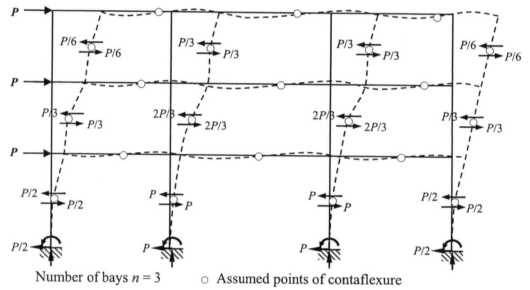

Number of bays *n* = 3   o Assumed points of contaflexure

Figure 10.38

The method involves consideration of the equilibrium of a series of horizontal sections taken through the points of contraflexure at each floor level to determine the shear, axial force and bending moments in each member as illustrated in Example 10.8.

10.7.1.1 Example 10.8: Multi-storey Rigid-jointed Frame 2

A three-bay, three-storey rigid-jointed frame is shown in Figure 10.39.
Using the portal method determine the approximate values of the member forces and sketch the bending moment diagram.

Figure 10.39

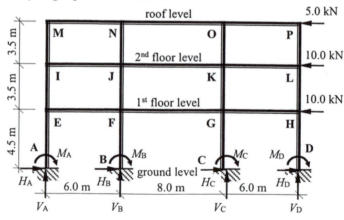

Consider the equilibrium of the frame at a series of horizontal sections through the points of contraflexure in the columns at each floor level assuming axial loading in the columns and the shear force distributed as indicated in Figure 10.40:

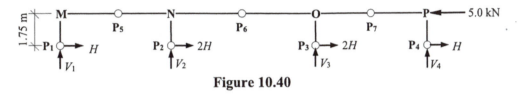

Figure 10.40

$$(H + 2H + 2H + H) = 6.0H = 5.0 \therefore H = 5.0/6.0 = 0.83 \text{ kN}$$

Consider the first bay M-N

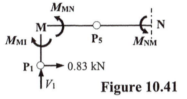

Figure 10.41

$+ve \,\rangle\, \Sigma M_{P5} = 0$

$+ (3.0 \times V_1) - (0.83 \times 1.75) = 0$

$\therefore V_1 \approx 1.45/3.0 = 0.48 \text{ kN} \uparrow (\text{compression})$

Bending moments at joint M:

$M_{MI} \approx (0.83 \times 1.75) = 1.45 \text{ kNm}$ (tension outside the frame) $\smallsmile$
$M_{MN} = M_{MI} \approx 1.45 \text{ kNm}$ $\rangle$

(Note: The shear is constant along member MN and $M_{NM} = M_{MN}$)
$M_{NM} \approx 1.45 \text{ kNm}$ (tension inside the frame) $\langle$

Consider the first and second bays M-N-O

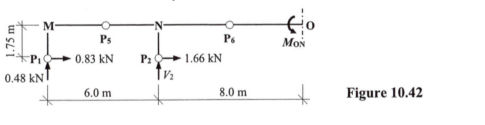

Figure 10.42

$+ve \,\rangle\, \Sigma M_{P6} = 0$

$+ (10.0V_1) + (4.0V_2) - (0.83 \times 1.75) - (1.66 \times 1.75) = 0$

$+ (10.0 \times 0.48) + (4.0 \times V_2) - 4.36 = 0 \quad \therefore V_2 \approx -0.44/4.0 = -0.11 \text{ kN} \downarrow (\text{tension})$

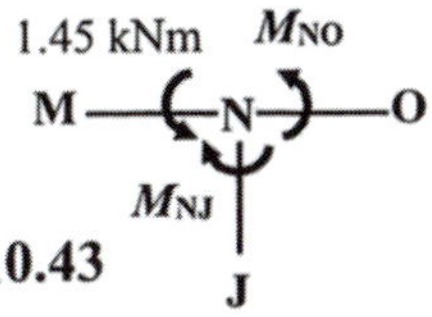

Bending moment at joint N:
$M_{NM} \approx 1.45 \text{ kNm from above}$ $\langle$(tension inside the frame)
$M_{NJ} \approx (1.66 \times 1.75) = 2.90 \text{ kNm}$ $\smallsmile$

Figure 10.43

Considering rotational equilibrium of joint N:
$M_{NO} \approx (-1.45 + 2.90) = 1.45 \text{ kNm}$ $\rangle$ (tension outside the frame)
$M_{ON} = M_{NO} \approx 1.45 \text{ kNm}$ $\langle$ (tension inside the frame)

Consider the third bay O-P

$+ve$ ⟩ $\Sigma M_{P7} = 0$

$-(3.0 \times V_4) - (0.83 \times 1.75) = 0$

$\therefore V_4 \approx -1.45/3.0 = -0.48$ kN $\downarrow$(tension)

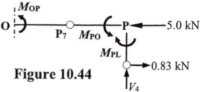

Figure 10.44

Bending moments at joint P:

$M_{PL} \approx (0.83 \times 1.75) = 1.45$ kNm   (tension inside the frame) $\smile$

$M_{PO} = M_{PL} \approx 1.45$ kNm ζ

(Note: The shear is constant along member OP and $M_{OP} = M_{PO}$)

$M_{OP} \approx 1.45$ kNm   (tension outside the frame) $\boldsymbol{)}$

Considering rotational equilibrium of joint O:

$M_{ON} \approx 1.46$ kNm

$M_{OP} \approx 1.45$ kNm

$M_{OK} \approx (1.45 + 1.45) = 2.90$ kNm

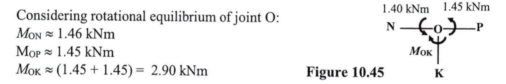

Figure 10.45

Considering the vertical equilibrium of the sub-frame shown in Figure 10.40:

$(V_1 + V_2 + V_3 + V_4 = 0)$ $\therefore V_3 = -(0.48 - 0.11 - 0.48) = +0.11$ (compression)

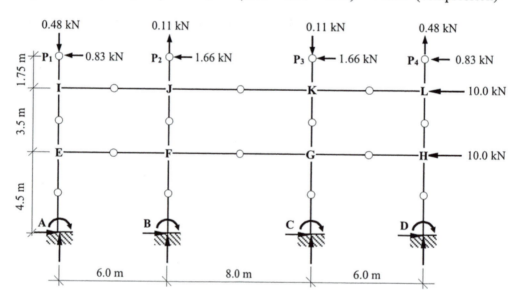

Figure 10.46

$M_{IM} = (0.83 \times 1.75) = 1.45$ kNm   ⤳

$M_{JN} = (1.663 \times 1.75) = 2.90$ kNm   ⤳

$M_{KO} = (1.66 \times 1.75) = 2.90$ kNm   ⤳

$M_{LP} = (0.83 \times 1.75) = 1.45$ kNm   ⤳

Consider the equilibrium of the sub-frame above horizontal sections through the points of contraflexure in the columns between the 1ˢᵗ/2ⁿᵈ floor and 2ⁿᵈ floor/roof levels:

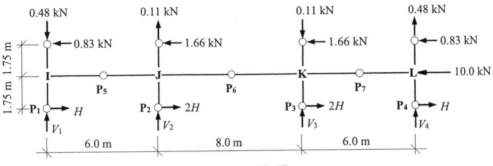

Figure 10.47

$$(H + 2H + 2H + H) = 6.0H = 15.0 \therefore H = 15.0/6.0 = 2.50 \text{ kN}$$

Consider the first bay I-J

$+ve \curvearrowright \Sigma M_{P5} = 0$

$+ (3.0 \times V_1) - (3.0 \times 0.48) - (2.50 \times 1.75) - (0.83 \times 1.75) = 0$

$\therefore V_1 \approx 7.27/3.0 = 2.42 \text{ kN} \uparrow (\text{compression})$

Bending moments at joint I:

$M_{IM} \approx 1.45 \text{ kNm} \curvearrowright$

$M_{IE} \approx (2.50 \times 1.75) = 4.38 \text{ kNm} \curvearrowleft$

$M_{IJ} \approx (1.45 + 4.38) = 5.83 \text{ kNm} \curvearrowright$

$M_{JI} = M_{IJ} \approx 5.83 \text{kNm} \curvearrowleft$

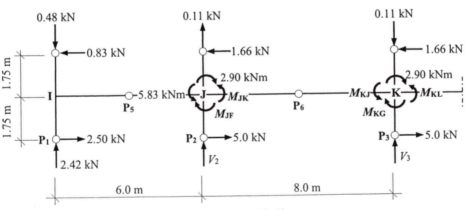

Figure 10.48

Consider the first and second bays I-J-K

Figure 10.49

$+ve\,\big)\,\Sigma M_{P6} = 0$

$+ 10.0\,(2.42 - 0.48) + 4.0\,(V_2 + 0.11) - 1.75\,(0.83 + 2.50 + 1.66 + 5.0) = 0$
$\therefore\ V_2 \approx -1.92/4.0 - 0.11 = -0.59$ kN ↓(tension)

Bending moment at joint J:
$M_{JF} \approx (5.0 \times 1.75) = 8.75$ kNm ↶
$M_{JN} \approx 2.90$ kNm ↷
$M_{JI} \approx 5.83$ kNm ↺

Considering rotational equilibrium of joint J:
$M_{JK} \approx (+2.90 + 8.75 - 5.83) = 5.82$ kNm ↻
$M_{KJ} = M_{JK} \approx 5.82$ kNm ↺

$M_{FJ} = M_{JF} \approx 8.75$ kNm ↷

Figure 10.50

Consider the third bay K-L

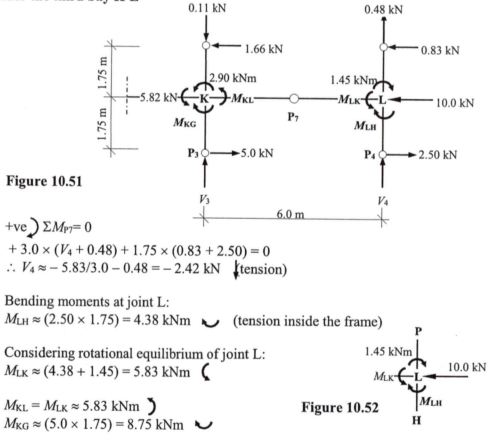

Figure 10.51

$+ve\,\big)\,\Sigma M_{P7} = 0$

$+ 3.0 \times (V_4 + 0.48) + 1.75 \times (0.83 + 2.50) = 0$
$\therefore\ V_4 \approx -5.83/3.0 - 0.48 = -2.42$ kN ↓(tension)

Bending moments at joint L:
$M_{LH} \approx (2.50 \times 1.75) = 4.38$ kNm ↶ (tension inside the frame)

Considering rotational equilibrium of joint L:
$M_{LK} \approx (4.38 + 1.45) = 5.83$ kNm ↺

$M_{KL} = M_{LK} \approx 5.83$ kNm ↻
$M_{KG} \approx (5.0 \times 1.75) = 8.75$ kNm ↶

Figure 10.52

Considering the vertical equilibrium of the sub-frame shown in Figure 10.47:
$(V_1 + V_2 + V_3 + V_4 - 0.48 + 0.11 - 0.11 + 0.48) = 0$
$\therefore\ V_3 = -(2.42 - 0.59 - 2.42) = +0.59$ ↑(compression)

Consider the equilibrium of the frame above horizontal sections through the points of contraflexure in the columns between the ground/1st floor and 1st/2nd floor levels:

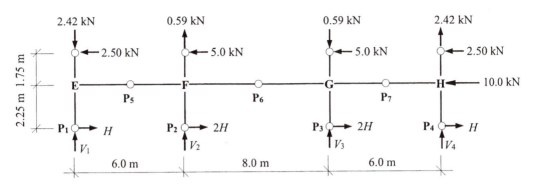

Figure 10.53

$$(H + 2H + 2H + H) = 6.0H = 25.0 \therefore H = 25.0/6.0 = 4.17 \text{ kN}$$

Consider the first bay E-F

$$+ve \curvearrowright \Sigma M_{P5} = 0$$
$$+ (3.0 \times V_1) - (3.0 \times 2.42) - (4.17 \times 2.25) - (2.50 \times 1.75) = 0$$
$$\therefore V_1 \approx 21.02/3.0 = 7.0 \text{ kN} \uparrow \text{(compression)}$$

Bending moments at joint E:
$$M_{EI} = M_{IE} \approx 4.38 \text{ kNm} \curvearrowright$$
$$M_{EA} \approx (4.17 \times 2.25) = 9.38 \text{ kNm} \curvearrowleft$$
$$M_{EF} = (M_{EA} + M_{EI}) \approx (9.38 + 4.38) = 13.76 \text{ kNm} \curvearrowright$$
$$M_{FE} = M_{EF} \approx 13.76 \text{ kNm} \curvearrowleft$$

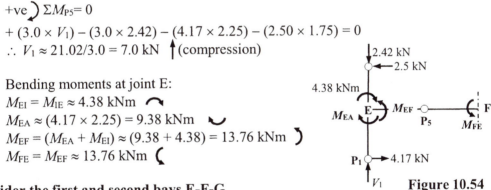

Figure 10.54

Consider the first and second bays E-F-G

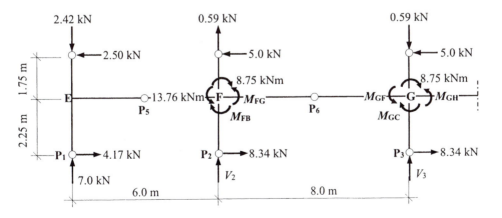

Figure 10.55

+ve $\curvearrowright$ $\Sigma M_{P6} = 0$

+ 10.0 (7.0 − 2.42) + 4.0 (V_2 + 0.59) − 1.75 (2.5 + 5.0) + 2.25 (4.17 + 8.34) = 0
∴ $V_2 \approx − 4.53/4.0 − 0.59 = − 2.69$ kN $\downarrow$(tension)

Bending moment at joint F:
$M_{FJ} \approx 8.75$ kNm $\curvearrowright$
$M_{FE} \approx 13.76$ kNm $\curvearrowleft$
$M_{FB} \approx (8.34 \times 2.25) = 18.77$ kNm $\curvearrowright$

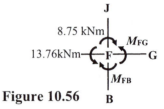

Figure 10.56

Considering rotational equilibrium of joint F:
$M_{FG} \approx (+ 8.75 + 18.77 − 13.76) = 13.76$ kNm $\curvearrowright$
$M_{GF} = M_{FG} \approx 13.76$ kNm $\curvearrowleft$

$M_{GC} \approx (8.34 \times 2.25) = 18.77$ kNm $\curvearrowright$

Consider the third bay G-H

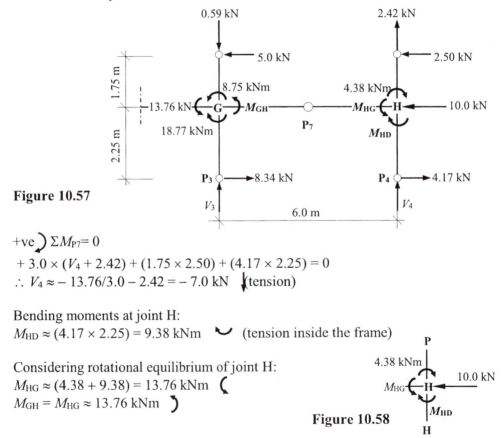

Figure 10.57

+ve $\curvearrowright$ $\Sigma M_{P7} = 0$
+ 3.0 × (V_4 + 2.42) + (1.75 × 2.50) + (4.17 × 2.25) = 0
∴ $V_4 \approx − 13.76/3.0 − 2.42 = − 7.0$ kN $\downarrow$(tension)

Bending moments at joint H:
$M_{HD} \approx (4.17 \times 2.25) = 9.38$ kNm $\curvearrowright$ (tension inside the frame)

Considering rotational equilibrium of joint H:
$M_{HG} \approx (4.38 + 9.38) = 13.76$ kNm $\curvearrowleft$
$M_{GH} = M_{HG} \approx 13.76$ kNm $\curvearrowright$

Figure 10.58

Considering the vertical equilibrium of the sub-frame shown in Figure 10.53:
(V_1 + V_2 + V_3 + V_4 − 2.42 + 0.59 − 0.59 + 2.42) = 0
∴ $V_3 = − (7.0 − 2.69 − 7.0) = + 2.69$ $\uparrow$(compression)

Consider the equilibrium of the frame below a horizontal section through the points of contraflexure in the columns between the ground and the 1st floor level:

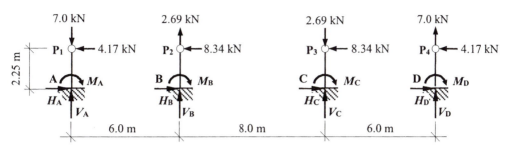

Figure 10.59

Vertical reactions:
$V_A = 7.0$ kN ↑ $V_B = -2.69$ kN ↓ $V_C = 2.69$ kN ↑ $V_D = -7.0$ kN ↓

Horizontal reactions:
$H_A = 4.17$ kN → $H_B = 8.34$ kN → $H_C = 8.34$ kN → $H_D = 4.17$ kN →

Moment reactions: (normally assumed to be the same as the top of the columns)
$M_A = 9.38$ kN ⟳ $M_B = 18.77$ kN ⟳ $M_C = 18.77$ kN ⟳ $M_D = 9.38$ kN ⟳

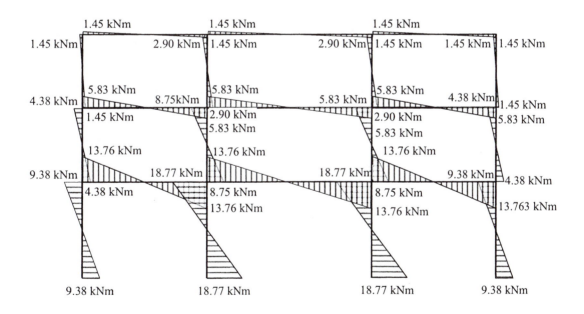

Figure 10.60: Approximate bending moment diagram

The approximate axial forces and shear forces in the members are indicated in Figure 10.61 and Figure 10.62.

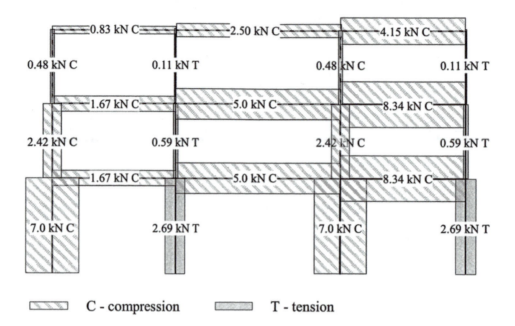

Figure 10.61: Approximate member axial force diagrams

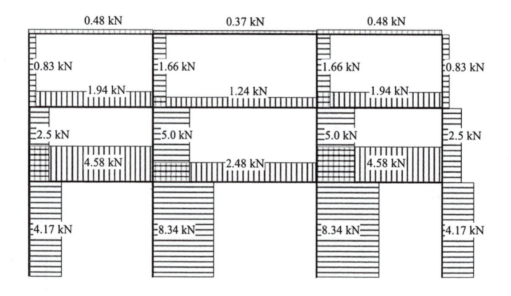

Figure 10.62: Approximate member shear force diagrams

10.7.1.2 Approximate Analysis of Vierendeel Trusses using the Portal Method

Vierendeel trusses are rigid-jointed girders in which there are no diagonal members and hence provide clear open spaces between the verticals and the chords as indicated in Figure 10.63(a). The chords are normally parallel and transmit shear and bending to the vertical members which provide the balancing moment for the sum of the chord moments.

This type of truss is usually used to support floors and/or roofs in buildings where large openings are required for services. In addition they are sometimes used for enclosed footbridges where diagonal web members are either obtrusive or undesirable for aesthetic reasons.

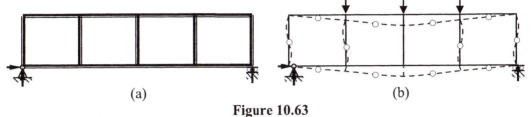

(a)   (b)

Figure 10.63

Vierendeel girders are indeterminate structures and for parallel chord girders an approximate analysis may be carried out assuming points of contraflexure at the mid-span points of the chords and the vertical of each panel. The vertical shear is assumed to be shared equally between the top and bottom chords. The chords and vertical members deform in double curvature as shown in Figure 10.63(b). In a symmetrically loaded truss with an even number of panels, the mid-span vertical member does not have any moment. The overall deflections are significantly larger than is the case for members with either solid web plates or trusses with diagonal members.

10.7.1.3 Example 10.9: Vierendeel Truss

A Vierendeel truss is pinned at support A, supported by a roller at G and carries three point loads at C, D and E as shown in Figure 10.64. Using the portal method, carry out an approximate analysis to determine the member forces in the truss.

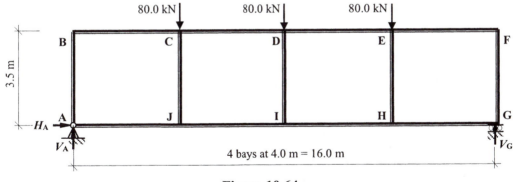

Figure 10.64

The structure is symmetrical and can be analysed considering the equilibrium of sub-frames defined by sections '1' to '4' and assuming points of contraflexure at the mid-point positions of each of the members as indicated in Figure 10.65.

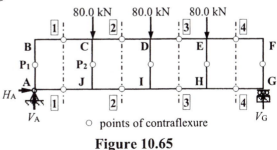

Figure 10.65

Consider the overall equilibrium of the structure:

+ve $\circlearrowleft$ $\Sigma M_A = 0$

$+ (80.0 \times 4.0) + (80.0 \times 8.0) + (80.0 \times 12.0) - 16.0V_G = 0$ $\therefore V_G = 120.0$ kN ↑

+ve ↑ $\Sigma F_z = 0$ $V_A - 80.0 - 80.0 - 80.0 + 120.0 = 0$ $\therefore V_A = 120.0$ kN ↑

+ve → $\Sigma F_x = 0$ $H_A = 0$

Consider the section 1-1: the shear force = 120.0 kN

+ve ↑ $\Sigma F_z = 0$

$120.0 - 2V = 0$ $\therefore V = 60.0$ kN ↓

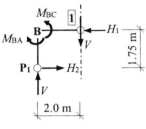

Consider the section above P1: $H_1 = H_2 = H$

+ve $\circlearrowleft$ $\Sigma M_{P1} = 0$

$- 1.75H + 2V = 0$ $\therefore H \approx (2.0 \times 60.0)/1.75 = 68.57$ kN

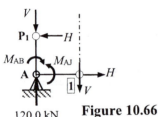

Bending moments at joint B:
$M_{BC} = M_{CB} \approx 2V = (2.0 \times 60) = 120$ kNm
$\underline{M_{BA} = M_{BA} \approx 1.75H = (1.75 \times 68.57) = 120.0 \text{ kNm}}$

Consider the section below P1
$M_{AJ} = M_{JA} \approx 120.0$ kNm

Figure 10.66

120.0 kN

Consider the section 2-2: the shear force = 40.0 kN

+ve ↑ $\Sigma F_z = 0$

$40.0 - 2V = 0$ $\therefore V = + 20.0$ kN ↓

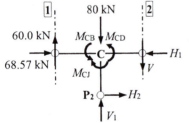

Consider the section between 1-1 and 2-2 above P2
$V_1 = (- 60.0 + 80.0 + 20.0) = + 40.0$ kN ↑

+ve $\circlearrowleft$ $\Sigma M_{P2} = 0$

$1.75 \times (68.75 - H_1) + 2.0 \times (60.0 + V) = 0$
$H_1 \approx (2.0 \times 80.0)/1.75 + 68.57 = 160.0$ kN
$H_2 = (160.0 - 68.57) = 91.43$ kN

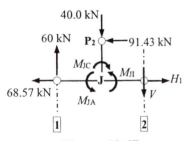

Bending moments at joint C:
$M_{CB} = M_{BC} \approx 120$ kNm
$M_{CD} = M_{DC} \approx 2V = (2.0 \times 20.0) = 40.0$ kNm
$M_{CJ} \approx 1.75H_2 = (1.75 \times 91.43) = 160.0$ kNm

Figure 10.67

Consider the section below P2
$M_{JC} = M_{CJ} \approx 160.0$ kNm
$M_{JI} = (M_{JC} - M_{JA}) = (160.0 - 120.0) \approx 40.0$ kNm

Consider the section between 2-2 and 3-3:

The truss and the loading are symmetrical about the midspan and hence:

Axial load in member DI $V_1 = (80.0 - 40.0) = 40.0$ kN
Shear force in member DI H_2 = zero
Bending moment in member DI M_{DI} = zero.

The final approximate member forces are as shown in Figures 10.69 to Figure 10.71.

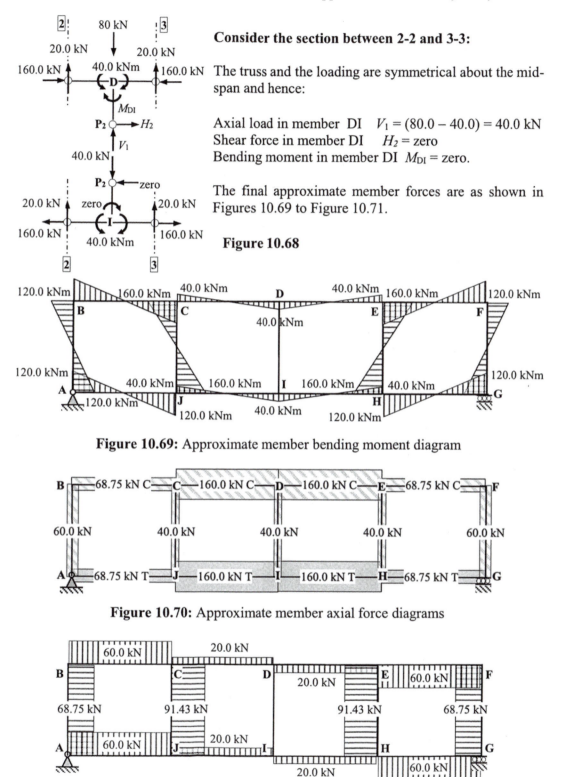

Figure 10.68

Figure 10.69: Approximate member bending moment diagram

Figure 10.70: Approximate member axial force diagrams

Figure 10.71: Approximate member shear force diagrams

10.7.2 Cantilever Method

The cantilever method, which is best suited to tall and slender buildings, is based on the same action as a vertical cantilever. The cantilever structure shown in Figure 10.72(a) is a continuum which bends in single curvature. The bending stresses can be determined using the simple elastic theory of bending assuming a linear variation over the depth of the cross-section, i.e.

Bending stress $\sigma = Mz/I_{yy}$

where:
M is the applied bending moment,
z is the distance from the centroid to the point where the stress is being determined,
I_{yy} is the second moment of area about the axis of bending for the cross-section.

Rigid-jointed frames have a different structural action in which shear deformation occurs as shown in Figure 10.72(b). Despite this different mode of deformation an approximate analysis can be carried out for such frames with fixed bases by assuming that the axial stresses in the columns have a linear variation from the centroid of the column areas and points of contraflexure occur at the mid-span of all beams and the mid-points of all columns. Use of the method is illustrated in Example 10.10.

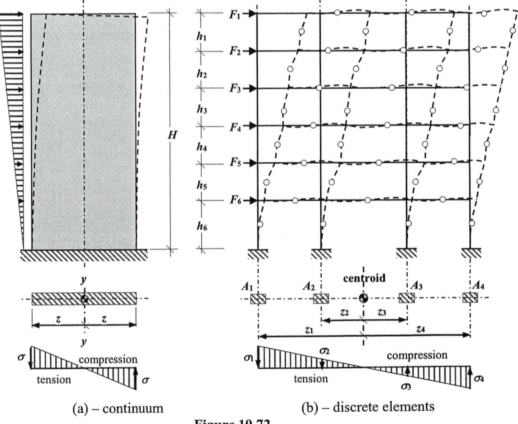

(a) – continuum   (b) – discrete elements

Figure 10.72

10.7.2.1 Example 10.10: Multi-storey Rigid-jointed Frame 3

A four-storey, two-bay asymmetric frame is shown in Figure 10.73 in which a series of horizontal loads are applied at B, C, D and E as indicated. Using the cantilever method carry out an approximate analysis to determine the member forces in the frame.

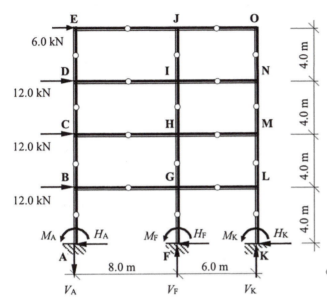

Data:
Supports A, F and K are fixed.

Relative cross-sectional areas of the columns:
AB, BC, CD and DE = 2.0A
FG, GH, HI and IJ = A
KL, LM, MN and NO = A

Assume points of contraflexure at the mid-span of all beams and the mid-points of all columns.

○ Assumed points of contraflexure.

Figure 10.73

Consider a typical horizontal cross-section through the frame and determine the position 'z' of the centroid of the column group. (Note: in this case the position is the same at each storey level).

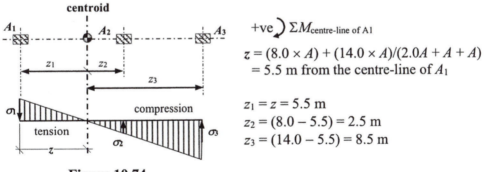

+ve $\curvearrowright$ $\Sigma M_{\text{centre-line of }A1}$

$z = (8.0 \times A) + (14.0 \times A)/(2.0A + A + A)$
$= 5.5$ m from the centre-line of A_1

$z_1 = z = 5.5$ m
$z_2 = (8.0 - 5.5) = 2.5$ m
$z_3 = (14.0 - 5.5) = 8.5$ m

Figure 10.74

Determine the approximate second moment of area of the column group about the centroid. (Neglect the $bd^3/12$ terms for each column).

$$I_{\text{centroid}} \approx \sum_{1}^{3}\left(\text{column cross-sectional area} \times \text{distance from the centroidal axis}^2\right)$$
$$= (2.0A \times 5.5^2) + (A \times 2.5^2) + (A \times 8.5^2) = 139.0A$$

Determine the column axial stresses and member forces using the simple elastic theory of bending and consideration of the equilibrium of a sub-frame defined by a horizontal section taken through the points of contraflexure for each storey.

Consider the top storey:

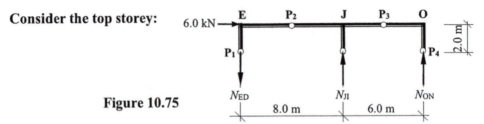

Figure 10.75

Applied moment $M = (6.0 \times 2.0) = 12.0$ kNm

Stress in column ED $\sigma_1 = Mz_1/I_{centroid} = (12.0 \times 5.5)/139.0A = 0.475/A$
Axial force in column ED $N_{ED} = (\sigma_1 \times 2.0A) = (0.475 \times 2.0) = 0.95$ kN T

Stress in column JI $\sigma_2 = Mz_2/I_{centroid} = (12.0 \times 2.5)/139.0A = 0.216/A$
Axial force in column JI $N_{ED} = (\sigma_2 \times A) = (0.216 \times 1.0) = 0.22$ kN C

Stress in column ON $\sigma_3 = Mz_3/I_{centroid} = (12.0 \times 8.5)/139.0A = 0.733/A$
Axial force in column ON $N_{ED} = (\sigma_3 \times A) = (0.733 \times 1.0) = 0.73$ kN C

(Note: the beams are assumed to have infinite axial and flexural stiffness).

Consider the equilibrium of sub-frame P_1-E-P_2

$+ve \uparrow \Sigma F_z = 0$

$-0.95 + V = 0$ $\therefore V = 0.95$ kN

$+ve \,\big)\, \Sigma M_{P1}$

$+ (6.0 \times 2.0) - (H \times 2.0) - (4.0 \times 0.95) = 0$
$H = (12.0 - 3.8)/2.0 = 4.10$ kN

$+ve \longrightarrow \Sigma F_x = 0$

$+ 6.0 - 4.10 - H_1 = 0$ $\therefore H_1 = 1.90$ kN

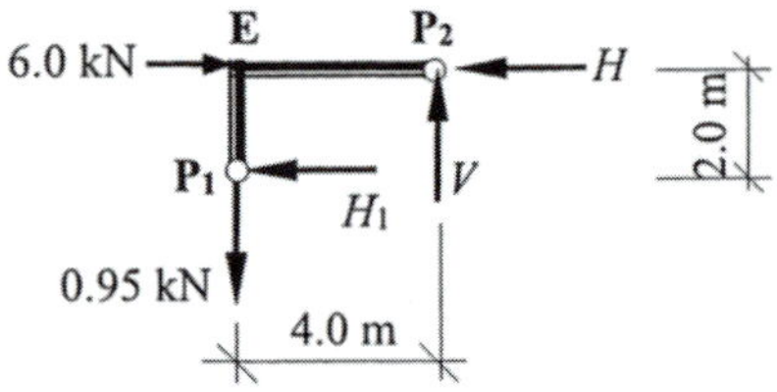

Figure 10.76

Bending moments at joint E:
$M_{ED} \approx (H_1 \times 2.0) = (1.90 \times 2.0) = 3.80$ kNm
$M_{EJ} \approx ((V \times 4.0) = (0.95 \times 4.0) = 3.80$ kNm

The reader should complete the calculations considering sub-frames P_2-J-P_3 and P_3-O-P_4 and all three sub-frames for each storey to determine the approximate axial forces, shear forces and bending moments for all members in the frame.

Appendix 1

Elastic Section Properties of Geometric Figures

A = Cross-sectional area

z_1 or z_2 = Distance to centroid

W_{yy} = Elastic section modulus about the $y-y$ axis

i_{yy} = Radius of gyration about the $y-y$ axis

I_{yy} = Second moment of area about the $y-y$ axis

Square:
$A = b^2$

$z = b/2$

$I_{yy} = \dfrac{b^4}{12}$

$W_{yy} = \dfrac{b^3}{6}$

$i_{yy} = \dfrac{b}{\sqrt{12}}$

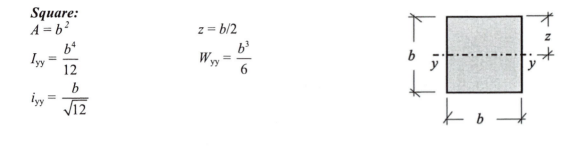

Square:
$A = b^2$

$z = b/2$

$I_{yy} = \dfrac{b^4}{3}$

$W_{yy} = \dfrac{b^3}{3}$

$i_{yy} = \dfrac{b}{\sqrt{3}}$

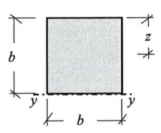

Square:
$A = b^2$

$z = \dfrac{b}{\sqrt{2}}$

$I_{yy} = \dfrac{b^4}{12}$

$W_{yy} = \dfrac{b^3}{6\sqrt{2}}$

$i_{yy} = \dfrac{b}{\sqrt{12}}$

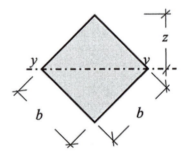

Rectangle:
$A = bh$

$z = h/2$

$I_{yy} = \dfrac{bh^3}{12}$

$W_{yy} = \dfrac{bh^2}{6}$

$i_{yy} = \dfrac{h}{\sqrt{12}}$

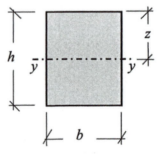

Rectangle:

$A = bh$

$I_{yy} = \dfrac{bh^3}{3}$

$i_{yy} = \dfrac{h}{\sqrt{3}}$

$z = h/2$

$W_{yy} = \dfrac{bh^2}{3}$

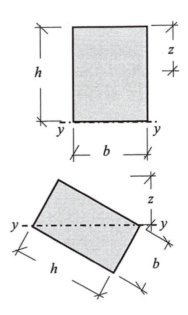

Rectangle:

$A = bh$

$I_{yy} = \dfrac{b^3 h^3}{6\left(b^2 + h^2\right)}$

$i_{yy} = \dfrac{bh}{\sqrt{6\left(b^2 + h^2\right)}}$

$z = \dfrac{bh}{\sqrt{b^2 + h^2}}$

$W_{yy} = \dfrac{b^2 h^2}{6\sqrt{b^2 + h^2}}$

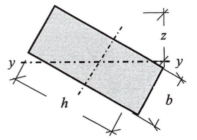

Rectangle:

$A = bh$

$I_{yy} = \dfrac{bh\left(b^2 \sin^2 \alpha + h^2 \cos^2 \alpha\right)}{12}$

$W_{yy} = \dfrac{bh\left(b^2 \sin^2 \alpha + h^2 \cos^2 \alpha\right)}{6\left(b\sin \alpha + h\cos \alpha\right)}$

$i_{yy} = \sqrt{\dfrac{b^2 \sin^2 \alpha + h^2 \cos^2 \alpha}{12}}$

$z = \dfrac{b\sin \alpha + h\cos \alpha}{2}$

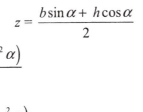

Hollow Rectangle:

$A = (bh - b_1 h_1)$

$I_{yy} = \dfrac{\left(bh^3 - b_1 h_1^3\right)}{12}$

$i_{yy} = \sqrt{\dfrac{bh^3 - b_1 h_1^3}{12A}}$

$z = h/2$

$W_{yy} = \dfrac{\left(bh^3 - b_1 h_1^3\right)}{6h}$

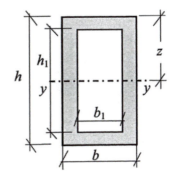

Trapezoid:

$$A = \frac{h(b+b_1)}{2}$$

$$z = \frac{h(2b+b_1)}{3(b+b_1)}$$

$$I_{yy} = \frac{h^3\left(b^2 + 4bb_1 + b_1^2\right)}{36(b+b_1)}$$

$$W_{yy} = \frac{h^2\left(b^2 + 4bb_1 + b_1^2\right)}{12(2b+b_1)}$$

$$i_{yy} = \frac{h}{6(b+b_1)}\sqrt{2\left(b^2 + 4bb_1 + b_1^2\right)}$$

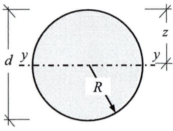

Circle:

$$A = \pi R^2$$

$$z = R = \frac{d}{2}$$

$$I_{yy} = \frac{\pi d^4}{64} = \frac{\pi R^4}{4}$$

$$W_{yy} = \frac{\pi d^3}{32} = \frac{\pi R^3}{4}$$

$$i_{yy} = \frac{d}{4} = \frac{R}{2}$$

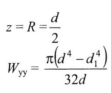

Hollow Circle:

$$A = \frac{\pi\left(d^2 - d_1^2\right)}{4}$$

$$z = R = \frac{d}{2}$$

$$I_{yy} = \frac{\pi\left(d^4 - d_1^4\right)}{64}$$

$$W_{yy} = \frac{\pi\left(d^4 - d_1^4\right)}{32d}$$

$$i_{yy} = \frac{\sqrt{d^2 + d_1^2}}{4}$$

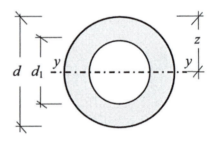

Semi-Circle:

$$A = \frac{\pi R^2}{2}$$

$$z = R\left(1 - \frac{4}{3\pi}\right)$$

$$I_{yy} = R^4\left(\frac{\pi}{8} - \frac{8}{9\pi}\right)$$

$$W_{yy} = \frac{R^3\left(9\pi^2 - 64\right)}{24(3\pi - 4)}$$

$$i_{yy} = R\frac{\sqrt{9\pi^2 - 64}}{6\pi}$$

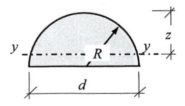

Equal Rectangles:
$$A = b(h - h_1)$$
$$z = h/2$$

$$I_{yy} = \frac{b\left(h^3 - h_1^3\right)}{12}$$

$$W_{yy} = \frac{b\left(h^3 - h_1^3\right)}{6h}$$

$$i_{yy} = \sqrt{\frac{h^3 - h_1^3}{12\left(h - h_1\right)}}$$

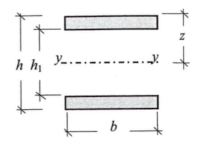

Unequal Rectangles:
$$A = bt + b_1 t_1$$
$$z = \frac{0.5bt^2 + b_1 t_1 \left(h - 0.5t_1\right)}{A}$$

$$I_{yy} = \left\{ \left(\frac{bt^3}{12} + btc^2 \right) + \left(\frac{b_1 t_1^3}{12} + b_1 t_1 c_1^2 \right) \right\}$$

$$W_{yy} = \frac{I}{z} \qquad W_{yy1} = \frac{I}{z_1}$$

$$i_{yy} = \sqrt{\frac{I_{yy}}{A}}$$

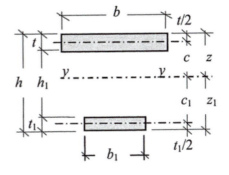

Triangle:
$$A = \frac{bh}{2}$$
$$z = \frac{2h}{3}$$

$$I_{yy} = \frac{bh^3}{36}$$
$$W_{yy} = \frac{bh^2}{24}$$

$$i_{yy} = \frac{h}{\sqrt{18}}$$

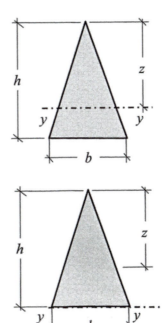

Triangle:
$$A = \frac{bh}{2}$$
$$z = \frac{2h}{3}$$

$$I_{yy} = \frac{bh^3}{12}$$
$$W_{yy} = \frac{bh^2}{12}$$

$$i_{yy} = \frac{h}{\sqrt{6}}$$

Appendix 2

Beam Reactions, Bending Moments and Deflections

Simply supported beams

Cantilever beams

Propped cantilevers

Fixed-End Beams

w = Distributed load (kN/m)   and W = Total load (kN)

Simply supported beams:

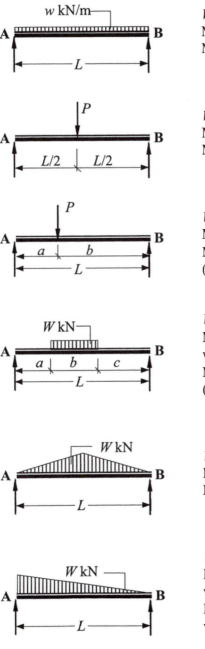

$V_A = wL/2$ $\qquad\qquad$ $V_B = wL/2$
Maximum bending moment at centre = $wL^2/8$
Maximum deflection = $(5wL^4/384EI)$

$V_A = P/2$ $\qquad\qquad$ $V_B = P/2$
Maximum bending moment at centre = $PL/4$
Maximum deflection = $(PL^3/48EI)$

$V_A = Pb/L$ $\qquad\qquad$ $V_B = Pa/L$
Maximum bending moment at centre = Pab/L
Mid-span deflection = $PL^3[(3a/L) - (4a^3/L^3)]/48EI$
(This value will be within 2.5% of the maximum)

$V_A = W(0.5b + c)/L$ $\quad$ $V_B = W(0.5b + a)/L$
Maximum bending moment at $x = W(x^2 - a^2)/2b$
where $\quad x = [a + (V_A b/W)]$ from **A**
Maximum deflection $\approx W(8L^3 - 4Lb^2 + b^3)/384EI$
(This is the value at the centre when $a = c$)

$V_A = W/2$ $\qquad\qquad$ $V_B = W/2$
Maximum bending moment at centre = $WL/6$
Maximum deflection = $WL^3/60EI$

$V_A = 2W/3$ $\qquad\qquad$ $V_B = W/3$
Maximum bending moment at $x = 0.128WL$
where $\quad x = 0.4226L$ from **A**
Maximum deflection $\approx 0.01304WL^3/384EI$
where $\quad x = 0.4807L$ from **A**

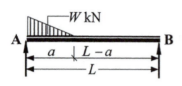

$V_A = W(3L - a)/3L$ $V_B = Wa/3L$
Maximum bending moment at x:

$$= \frac{Wa}{3}\left(1 - \frac{a}{L} + \sqrt{\frac{4a^3}{27L^3}}\right)$$

where $x = a\left(1 - \sqrt{\frac{a}{3L}}\right)$ from **A**

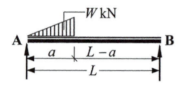

$V_A = W(3L - 2a)/3L$ $V_B = 2Wa/3L$
Maximum bending moment at x:

$$= \frac{2Wa}{3}\left(1 - \frac{2a}{3L}\right)^{\frac{3}{2}}$$

where $x = a\sqrt{1 - \frac{2a}{3L}}$ from **A**

Cantilever beams:
Anti-clockwise support moments considered negative.

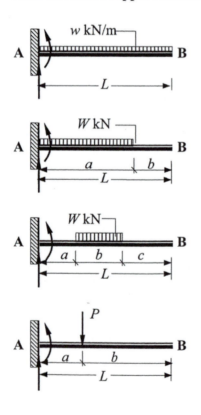

$V_A = wL$
Maximum (−ve) bending moment $M_A = -wL^2/2$
Maximum deflection $= wL^4/8EI$

$V_A = W$
Maximum (−ve) bending moment $M_A = -Wa/2$
Maximum deflection at B $= Wa^3\left(1 + \frac{4b}{3a}\right)\Big/8EI$

$V_A = W$
Maximum (−ve) bending moment $M_A = -W(a + b/2)$
Maximum deflection at B $= (W/24EI) \times k$
where $k =$
$(8a^3 + 18a^2b + 12ab^2 + 3b^3 + 12a^2c + 12abc + 4b^2c)$

$V_A = P$
Maximum (−ve) bending moment $M_A = -Pa$
Maximum deflection at B $= Pa^3\left(1 + \frac{3b}{2a}\right)\Big/3EI$

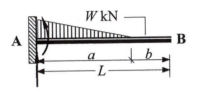

$V_A = W$

Maximum (−ve) bending moment $M_A = -\,Wa/3$

Maximum deflection at B $= Wa^3\left(1+\dfrac{5b}{4a}\right)\bigg/15EI$

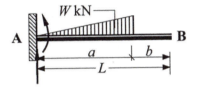

$V_A = W$

Maximum (−ve) bending moment $M_A = -\,2Wa/3$

Maximum deflection at B $= 11Wa^3\left(1+\dfrac{15b}{11a}\right)\bigg/60EI$

Propped cantilevers:
Where the support moment (M_A) is included in an expression for reactions, its value should be assumed positive.

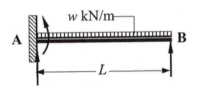

$V_A = 5wL/8 \qquad\qquad V_B = 3wL/8$

Maximum (−ve) bending moment $M_A = -\,wL^2/8$

Maximum (+ve) bending moment at $x = +\,9wL^2/128$

where $x = 0.625L$ from **A**

Maximum deflection at $y = wL^4/185EI$

where $y = 0.5785L$ from **A**

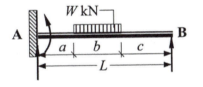

$V_A = W(0.5b + c)/L + M_A/L$

$V_B = W(0.5b + a)/L - M_A/L$

Maximum (−ve) bending moment M_A:

$= -\,Wb(b + 2c)\,[2(L^2 - c^2 - bc) - b^2)]/8L^2b$

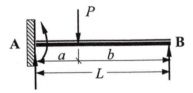

$V_A = (P - V_B)$

$V_B = Pa^2[(b + 2L)]/2L^3$

Maximum (−ve) bending moment M_A:

$= -\,Pb[(L^2 - b^2)]/2L^2$

Maximum (+ve) bending moment at point load:

$= -\,\dfrac{Pb}{2}\left(2 - \dfrac{3b}{L} + \dfrac{b^3}{L^3}\right)$

Maximum deflection at point load position:

$= \dfrac{Pa^3b^2}{12EIL^3}(4L - a)$

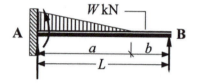

$V_A = (W - V_B)$
$V_B = Wa^2[(5L - a)]/20L^3$
Maximum (−ve) bending moment M_A:

$$= -\frac{Wa}{60L^2}\left(3a^2 - 15aL + 20L^2\right)$$

Maximum (+ve) bending moment at x:
$= [V_B x - W(x - b)^3 / 3a^2]$
where $x = b + \dfrac{a^2}{2L}\sqrt{1 - \dfrac{a}{5L}}$ from **B**

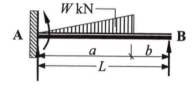

$V_A = (W - V_B)$
$V_B = Wa^2[(15L - 4a)]/20L^3$
Maximum (−ve) bending moment M_A:

$$= -Wa\left(\frac{a^2}{5L^2} - \frac{3a}{4L} + \frac{2}{3}\right)$$

Fixed-end beams:

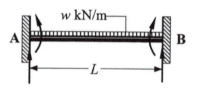

$V_A = wL/2$ $\qquad\qquad$ $V_B = wL/2$
Maximum (−ve) bending moment M_A:
$= - wL^2/12$
Maximum (+ve) bending moment at mid-span:
$= + wL^2/24$
Maximum deflection at point load:
$= wL^4/384EI$

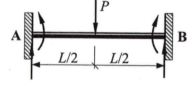

$V_A = P/2$ $\qquad\qquad$ $V_B = P/2$
Support bending moments:
$M_A = - PL/8$ $\qquad$ and $\qquad$ $M_B = + PL/8$
Maximum (+ve) bending moment at mid-span:
$= + PL/8$
Maximum deflection at mid-span $= PL^3/192EI$

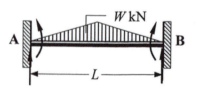

$V_A = W/2$ $\qquad\qquad$ $V_B = W/2$
Support bending moments:
$M_A = - 5WL/48$ $\qquad$ and $\qquad$ $M_B = + 5WL/48$
Maximum (+ve) bending moment at mid-span:
$= + WL/16$
Maximum deflection at mid-span $= 1.4WL^3/384EI$

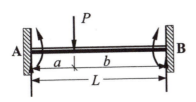

$$V_A = Pb^2 (1 + 2a/L)/L^2$$
$$V_B = Pa^2 (1 + 2b/L)/L^2$$
Support bending moments:
$$M_A = -Pab^2/L^2 \quad \text{and} \quad M_B = +Pa^2b/L^2$$
Maximum (+ve) bending moment at point load:
$$= +2Pa^2b^2/L^3$$

Maximum deflection at point $x = \dfrac{2Pa^2b^3}{3EI(3L - 2a)^2}$

where $x = \dfrac{L^2}{(3L - 2a)}$ from **A**

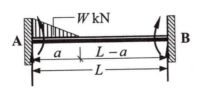

$$V_A = 0.7W \qquad V_B = 0.3W$$
Support bending moments:
$$M_A = -WL/10 \quad \text{and} \quad M_B = +WL/15$$
Maximum (+ve) bending moment at point x:
$$= +WL/23.3$$
where x is $0.45L$ from **A**
Maximum deflection at point $y = WL^3/382EI$
where y is $0.475L$ from **A**

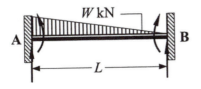

$$V_A = (W - V_B)$$
$$V_B = Wa^2[(5L - 2a)]/10L^3$$
Support bending moments:
$$M_A = -\frac{Wa}{30L^2}\left(3a^2 + 10bL\right) \quad \text{and}$$

$$M_B = +\frac{Wa^2}{30L^2}(5L - 3a)$$

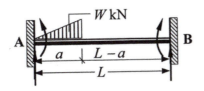

$$V_A = W[(10L^3 - 15La^2 + 8a^3)]/10L^3$$
$$V_B = (W - V_A)$$
Support bending moments:
$$M_A = -\frac{Wa}{15L^2}\left(10L^2 - 15aL + 6a^2\right) \quad \text{and}$$

$$M_B = +\frac{Wa^2}{10L^2}(5L - 4a)$$

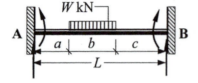

$V_A = W(0.5b + c)/L + (M_A - M_B)/L$

$V_B = W(0.5b + a)/L - (M_A - M_B)/L$

Support bending moments:

$$M_A = -\frac{W}{12L^2 b}\left\{\left[(L-a)^3 \times (L+3a)\right] - c^3(4L - 3c)\right\}$$

$$M_B = +\frac{W}{12L^2 b}\left\{\left[(L-c)^3 \times (L+3c)\right] - a^3(4L - 3a)\right\}$$

Maximum deflection at mid-span when $a = c$

$$\frac{W}{384EI}\left(L^3 + 2L^2 a + 4La^2 - 8a^3\right)$$

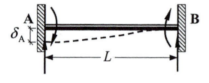

$V_A = -\dfrac{12EI}{L^3}\delta \qquad\qquad V_B = +\dfrac{12EI}{L^3}\delta$

Support bending moments:

$M_A = +\dfrac{6EI}{L^2}\delta \qquad\qquad M_B = +\dfrac{6EI}{L^2}\delta$

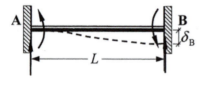

$V_A = +\dfrac{12EI}{L^3}\delta \qquad\qquad V_B = -\dfrac{12EI}{L^3}\delta$

Support bending moments:

$M_A = -\dfrac{6EI}{L^2}\delta \qquad\qquad M_B = -\dfrac{6EI}{L^2}\delta$

$V_A = -\dfrac{6EI}{L^2}\theta_A \qquad\qquad V_B = +\dfrac{6EI}{L^2}\theta_A$

Support bending moments:

$M_A = +\dfrac{4EI}{L}\theta_A \qquad\qquad M_B = +\dfrac{2EI}{L}\theta_A$

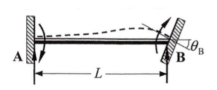

$V_A = -\dfrac{6EI}{L^2}\theta_A \qquad\qquad V_B = +\dfrac{6EI}{L^2}\theta_A$

Support bending moments:

$M_A = +\dfrac{2EI}{L}\theta_A \qquad\qquad M_B = +\dfrac{4EI}{L}\theta_A$

Appendix 3

Matrix Algebra

Product of a Matrix and a Vector:

Consider three variables a_1, a_2 and a_3 which are related to three other variables c_1, c_2 and c_3 by the three equations (1), (2) and (3) as indicated:

$$a_1 = b_{11}c_1 + b_{12}c_2 + b_{13}c_3 \qquad \text{Equation (1)}$$
$$a_2 = b_{21}c_1 + b_{22}c_2 + b_{23}c_3 \qquad \text{Equation (2)}$$
$$a_3 = b_{31}c_1 + b_{32}c_2 + b_{33}c_3 \qquad \text{Equation (3)}$$

these equations can be represented in matrix form as:

i.e.
$$[A] = [B] \times [C]$$
$$\begin{bmatrix} a_1 \\ a_2 \\ a_3 \end{bmatrix} = \begin{bmatrix} b_{1,1} & b_{1,2} & b_{1,3} \\ b_{2,1} & b_{2,2} & b_{2,3} \\ b_{3,1} & b_{3,2} & b_{3,3} \end{bmatrix} \times \begin{bmatrix} c_1 \\ c_2 \\ c_3 \end{bmatrix}$$

where b_{11}, b_{12}, b_{13} etc. are the coefficients for the square matrix $[B]$.

Clearly for known values of c_1, c_2 and c_3 the values of a_1, a_2 and a_3 can be determined directly. If however, it is required to determine the 'c' values for given 'a' values then the relationship must be re-written as:

$$[C] = [B]^{-1} \times [A]$$

and the **INVERT** of matrix $[B]$ must be obtained.

The invert of a matrix can be defined as:

$$[B]^{-1} = \frac{\operatorname{adj} B}{|B|}$$

where adj B is the **adjoint** of matrix $[B]$ and is equal to the transpose of the co-factor matrix $[B^c]$ of matrix $[B]$, i.e.

$$\operatorname{adj} [B] = [B^c]^{\mathrm{T}}$$

The co-factor matrix is given by replacing each element in the matrix by its' co-factor, i.e.

$$[B] = \begin{bmatrix} \overset{+}{b_{1,1}} & \overset{-}{b_{1,2}} & \overset{+}{b_{1,3}} \\ \overset{-}{b_{2,1}} & \overset{+}{b_{2,2}} & \overset{-}{b_{2,3}} \\ \overset{+}{b_{3,1}} & \overset{-}{b_{3,2}} & \overset{+}{b_{3,3}} \end{bmatrix} ;$$

$$[B^c] = (-1)^{i+j} \begin{bmatrix} b^c_{1,1} & b^c_{1,2} & b^c_{1,3} \\ b^c_{2,1} & b^c_{2,2} & b^c_{2,3} \\ b^c_{3,1} & b^c_{3,2} & b^c_{3,3} \end{bmatrix} \quad \text{and} \quad [B^c]^{\mathrm{T}} = \begin{bmatrix} b^c_{1,1} & b^c_{2,1} & b^c_{3,1} \\ b^c_{1,2} & b^c_{2,2} & b^c_{3,2} \\ b^c_{1,3} & b^c_{2,3} & b^c_{3,3} \end{bmatrix}$$

$$[B^c] = \begin{bmatrix} +\begin{vmatrix} b_{2,2} & b_{2,3} \\ b_{3,2} & b_{3,3} \end{vmatrix} & -\begin{vmatrix} b_{2,1} & b_{2,3} \\ b_{3,1} & b_{3,3} \end{vmatrix} & +\begin{vmatrix} b_{2,1} & b_{2,2} \\ b_{3,1} & b_{3,2} \end{vmatrix} \\ \\ -\begin{vmatrix} b_{1,2} & b_{1,3} \\ b_{3,2} & b_{3,3} \end{vmatrix} & +\begin{vmatrix} b_{1,1} & b_{1,3} \\ b_{3,1} & b_{3,3} \end{vmatrix} & -\begin{vmatrix} b_{1,1} & b_{1,2} \\ b_{3,1} & b_{3,2} \end{vmatrix} \\ \\ +\begin{vmatrix} b_{1,2} & b_{1,3} \\ b_{2,2} & b_{2,3} \end{vmatrix} & -\begin{vmatrix} b_{1,1} & b_{1,3} \\ b_{2,1} & b_{2,3} \end{vmatrix} & +\begin{vmatrix} b_{1,1} & b_{1,2} \\ b_{2,1} & b_{2,2} \end{vmatrix} \end{bmatrix}$$

where:

$|B|$ is the determinant of matrix $[B]$, which can be calculated from:

$$|B| = + \left\{ b_{1,1} \times \begin{vmatrix} b_{2,2} & b_{2,3} \\ b_{3,2} & b_{3,3} \end{vmatrix} \right\} - \left\{ b_{1,2} \times \begin{vmatrix} b_{2,1} & b_{2,3} \\ b_{3,1} & b_{3,3} \end{vmatrix} \right\} + \left\{ b_{1,3} \times \begin{vmatrix} b_{2,1} & b_{2,2} \\ b_{3,1} & b_{3,2} \end{vmatrix} \right\}$$

$$|B| = + b_{1,1}\{(b_{2,2} \times b_{3,3}) - (b_{3,2} \times b_{2,3})\} - b_{1,2}\{(b_{2,1} \times b_{3,3}) - (b_{3,1} \times b_{2,3})\}$$
$$+ b_{1,3}\{(b_{2,1} \times b_{3,2}) - (b_{3,1} \times b_{2,2})\}$$

Example A.1

Determine the values of c_1, and c_2 given that:

$$[A] = [B] \times [C] \quad \text{where:} \quad [A] = \begin{bmatrix} 40.0 \\ 45.0 \end{bmatrix} \quad \text{and} \quad [B] = \begin{bmatrix} 2.0 & 3.0 \\ 1.0 & 4.0 \end{bmatrix}$$

Solution:

Determine the co-factor matrix $[B^c] = \begin{bmatrix} +4.0 & -1.0 \\ -3.0 & +2.0 \end{bmatrix}$ $[B^c]^{\mathrm{T}} = \begin{bmatrix} +4.0 & -3.0 \\ -1.0 & +2.0 \end{bmatrix}$

$$[C] = [B]^{-1} \times [A] \quad \text{and} \quad [B]^{-1} = \frac{\mathrm{adj}\, B}{|B|} \quad \therefore \quad \begin{bmatrix} c_1 \\ c_2 \end{bmatrix} = \frac{1}{|B|} \begin{bmatrix} b^c_{1,1} & b^c_{2,1} \\ b^c_{1,2} & b^c_{2,2} \end{bmatrix} \times \begin{bmatrix} a_1 \\ a_2 \end{bmatrix}$$

The determinant of $[B]$
$$|B| = \{+(b_{1,1} \times b_{2,2}) - (b_{2,1} \times b_{1,2})\}$$
$$= \{+(2.0 \times 4.0) - (1.0 \times 3.0)\}$$
$$= + 5.0$$

$$\begin{bmatrix} c_1 \\ c_2 \end{bmatrix} = \frac{1}{|B|} \begin{bmatrix} b^c_{1,1} & b^c_{2,1} \\ b^c_{1,2} & b^c_{2,2} \end{bmatrix} \times \begin{bmatrix} a_1 \\ a_2 \end{bmatrix} = \frac{1}{5.0} \begin{bmatrix} +4.0 & -3.0 \\ -1.0 & +2.0 \end{bmatrix} \begin{bmatrix} 40.0 \\ 45.0 \end{bmatrix}$$

$$\begin{bmatrix} c_1 \\ c_2 \end{bmatrix} = \begin{bmatrix} +0.8 & -0.6 \\ -0.2 & +0.4 \end{bmatrix} \begin{bmatrix} 40.0 \\ 45.0 \end{bmatrix}$$

$$c_1 = \{+ (0.8 \times 40.0) - (0.6 \times 45.0)\} = + \mathbf{5.0}$$
$$c_2 = \{- (0.2 \times 40.0) + (0.4 \times 45.0)\} = + \mathbf{10.0}$$

Example A.2

Determine the values of c_1, c_2 and c_3 given that:

$$[A] = [B] \times [C] \quad \text{where:} \quad [A] = \begin{bmatrix} 14.0 \\ 4.0 \\ 6.0 \end{bmatrix} \text{ and } \quad [B] = \begin{bmatrix} \overset{+}{2.0} & \overset{-}{3.0} & \overset{+}{1.0} \\ \underset{}{1.0} & \overset{+}{2.0} & \underset{}{3.0} \\ \overset{+}{3.0} & \underset{}{1.0} & \overset{+}{2.0} \end{bmatrix}$$

Solution:

Determine the co-factor matrix:

$k_{11}^c = + \{(2.0 \times 2.0) - (1.0 \times 3.0)\} = + 1.0$ $\qquad$ $k_{12}^c = - \{(1.0 \times 2.0) - (3.0 \times 3.0)\} = + 7.0$

$k_{13}^c = + \{(1.0 \times 1.0) - (3.0 \times 2.0)\} = - 5.0$ $\qquad$ $k_{21}^c = - \{(3.0 \times 2.0) - (1.0 \times 1.0)\} = - 5.0$

$k_{22}^c = + \{(2.0 \times 2.0) - (3.0 \times 1.0)\} = + 1.0$ $\qquad$ $k_{23}^c = - \{(2.0 \times 1.0) - (3.0 \times 3.0)\} = + 7.0$

$k_{31}^c = + \{(3.0 \times 3.0) - (2.0 \times 1.0)\} = + 7.0$ $\qquad$ $k_{32}^c = - \{(2.0 \times 3.0) - (1.0 \times 1.0)\} = - 5.0$

$k_{33}^c = + \{(2.0 \times 2.0) - (3.0 \times 1.0)\} = + 1.0$

$$[B^c] = \begin{bmatrix} +1.0 & +7.0 & -5.0 \\ -5.0 & +1.0 & +7.0 \\ +7.0 & -5.0 & +1.0 \end{bmatrix} \qquad [B^c]^T = \begin{bmatrix} +1.0 & -5.0 & +7.0 \\ +7.0 & +1.0 & -5.0 \\ -5.0 & +7.0 & +1.0 \end{bmatrix}$$

Determinant of [B]:

$|B| = + b_{1,1} \{ (b_{2,2} \times b_{3,3}) - (b_{3,2} \times b_{2,3}) \} - b_{1,2} \{ (b_{2,1} \times b_{3,3}) - (b_{3,1} \times b_{2,3}) \} + b_{1,3} \{ (b_{2,1} \times b_{3,2}) - (b_{3,1} \times b_{2,2}) \}$

$|B| = \{+ (2.0 \times 1.0) - (3.0 \times -7.0) + (1.0 \times -5.0)\} = +18.0$

Inverted matrix $\quad [B]^{-1} = \dfrac{1}{18.0} \begin{bmatrix} +1.0 & -5.0 & +7.0 \\ +7.0 & +1.0 & -5.0 \\ -5.0 & +7.0 & +1.0 \end{bmatrix}$

$[C] = [B]^{-1} \times [A]$ and $\quad [B]^{-1} = \dfrac{\text{adj } B}{|B|} \begin{bmatrix} c_1 \\ c_2 \\ c_3 \end{bmatrix} = \dfrac{1}{18.0} \begin{bmatrix} +1.0 & -5.0 & +7.0 \\ +7.0 & +1.0 & -5.0 \\ -5.0 & +7.0 & +1.0 \end{bmatrix} \times \begin{bmatrix} 14.0 \\ 4.0 \\ 6.0 \end{bmatrix}$

$c_1 = \{+ (1.0 \times 14.0) - (5.0 \times 4.0) + (7.0 \times 6.0\}/18.0 = + \mathbf{2.0}$

$c_2 = \{+ (7.0 \times 14.0) + (1.0 \times 4.0) - (5.0 \times 6.0\}/18.0 = + \mathbf{4.0}$

$c_3 = \{- (5.0 \times 14.0) + (7.0 \times 4.0) + (1.0 \times 6.0\}/18.0 = - \mathbf{2.0}$

To check the invert determine the product $[B][B]^{-1}$ which should equal the identity matrix $[I]$ where: $[I] = \begin{bmatrix} 1.0 & 0 & 0 \\ 0 & 1.0 & 0 \\ 0 & 0 & 1.0 \end{bmatrix}$

Index